Quality Management for Services

Manfred Bruhn

Quality Management for Services

Handbook for Successful Quality Management. Principles—Concepts—Methods

Springer

Manfred Bruhn
Faculty of Business and Economics
Marketing and Corporate Management
University Basel
Basel, Switzerland

ISBN 978-3-662-67031-6   ISBN 978-3-662-67032-3 (eBook)
https://doi.org/10.1007/978-3-662-67032-3

This book is a translation of the original German „Qualitätsmanagement für Dienstleistungen" by Bruhn, Manfred, published by Springer-Verlag GmbH, DE in 2020. The translation was done with the help of an artificial intelligence machine translation tool. A subsequent human revision was done primarily in terms of content, so that the book will read stylistically differently from a conventional translation. Springer Nature works continuously to further the development of tools for the production of books and on the related technologies to support the authors.

This Springer imprint is published by the registered company Springer-Verlag GmbH, DE, part of Springer Nature.
The registered company address is: Heidelberger Platz 3, 14197 Berlin, Germany

Foreword

Ensuring high service quality continues to take a prominent position on the priority list of management of many service companies. Since the origins of the scientific discussion in the mid-1980s, the understanding of service quality as a strategic success factor of service companies has been established to a large extent in science and practice. Recent developments in the field of social media on the Internet and the various processes of digitalization—just to name two trends in the service market—present new challenges for quality management of services and underscore the still high relevance of ensuring high service quality. This includes, for example, the increasing expansion of service offerings on the Internet, to which a quality management must also take into account the further development of the Internet towards a "user-generated information and communication medium" (Web 2.0). On the one hand, customers can share their quality-relevant experiences with a service electronically (e.g. in social networks) with an almost unlimited number of people in a very short time. The negative consequences of negligence in the provision of service quality are thus amplified and can quickly develop into an unforeseeable risk. On the other hand, the consistent use of new interaction forms on the Internet opens up new potentials for a targeted quality management for providers.

Of central importance in the digitalization of service processes is the automation and personalization of services. This creates new dimensions in the interaction between service providers and service users. Automation will help to ensure that many information, search and process tasks are carried out by machines. With the help of artificial intelligence, further impulses can be expected through machine learning. In this context, the personalization of services will also play a central role in the future, because offers such as services or prices can be individually controlled by algorithms. This also paves the way for services from mass marketing to individual marketing and poses new challenges for the quality management of services.

Against this background, it is not surprising that quality initiatives often form a fixed part of the project portfolio of service companies. In doing so, a wide range of approaches to measuring service quality and various individual measures are used to increase or stabilize service quality at a high level. In corporate practice, it is possible to observe progress in terms of a coordinated use of various individual measures and methods in the sense of a holistic quality management.

However, service companies repeatedly encounter limits in their efforts to ensure a high level of service quality. The reasons for this are mainly due to three aspects. First, many quality programs fail due to their lack of strategic embedding in the company. Secondly, the failure of many quality programs is due to implementation problems. Here, the quality efforts often exhaust themselves in the recording of individual initiatives, without a company-wide anchoring of quality management being addressed. Nor is a systematic process orientation taken into account in the design and implementation of quality management. Thirdly, traditional quality management measures focus exclusively on the optimization of the quality of a company's services. In conceptual terms, however, the services provided by the provider are only one component of the service quality, as the customer perceives it. The second component is customer expectations, with which the customer compares the service received each time. Accordingly, a quality management is required which is oriented on the one hand to the optimization of the services and on the other hand to the control of the customer expectations.

In this revised edition, some updates and extensions have become necessary. The proven basic structure of the book has been retained. In addition to a conceptual, theoretical and conceptual basic part on quality management, the further chapters are based on the classic management process in the sense of an analysis, planning, implementation and control phase of quality management. This covers both strategic and implementation-oriented aspects of quality management. The updates concentrate in this revised edition primarily on the literature. The extensions relate in the tenth chapter mainly to the explanation of the re-design of the *EFQM-Model,* which is referred to as the *EFQM-Model 2020.*

This edition of "Quality Management for Services" would like to continue the tradition of previous editions and provide assistance to decision-makers in service companies. This is about a systematic and strategic conception and implementation of a holistic quality management for services, which goes beyond the isolated application of individual measures for measuring and controlling the quality of services. The book refers equally to a quality management of classic services (institutional services) and also product-accompanying services (functional services), which are not offered independently, but in combination with goods and services.

As in the past, the book provides all those who are interested in this topic with a complete overview of the "state of the art" of quality management for services in teaching, research and practice. In doing so, the book has continuously developed in the previous eleven editions in order to meet the increased demands of the field of science and practice. The book has developed into a handbook and can be used as a reference book for the different fields of quality management for services. Therefore, the individual fields of quality management or the respective chapters can be worked through independently of each other.

My thanks for this new edition go to Mrs. Karina von dem Berge, Mr. Hugo Graf from the Lucerne University of Applied Sciences and Mr. Siegfried Schmidt from the Swiss Association for Quality (SAQ) for their support in the research and suggestions for the new design of the *EFQM-Model*. A thank you as well to Verena Timmons for proofreading the translated, English edition.

The author is looking forward to further intense discussion of new directions, concepts and methods as well as experiences from science and practice with the concepts of quality management for services presented here and is grateful for any kind of suggestions.

Basel
Summer 2023

Manfred Bruhn

Contents

List of Figures

Part I

Basics of Quality Management for Services

Importance of Quality Management for Service Companies

1

Abstract
The importance of service quality is constantly increasing in science and practice. This is not only shown by the economic view of the service sector, but also by the competitive advantages that can be achieved through high service quality. The current situation of service markets places high demands on the management of service quality. The interdependencies between service quality and corporate success are shown in this chapter on the basis of the success chain of service marketing. These relationships are confirmed by numerous qualitative and quantitative research studies.

1.1 Competitive Advantages through Service Quality

The creation of high service quality has developed into a central competitive factor in recent years. Driven by an economic focus on the service sector in the 1980s, a change in marketing thinking has taken place, which is reflected in the orientation of the service offer to customer wishes (customer orientation) and in quality orientation.

The growing importance of the tertiary sector in Germany and other industrial countries is undisputed. With regard to the Federal Republic of Germany, the tertiary sector (trade, transport and hospitality, financial and insurance services, business services, real estate and housing, information and communication, as well as public and private services) already accounts for 68.2% (2018) of gross value added (1970: 45%; 1950: 39.6%) (Statista, 2019). In addition, the service sector accounted for 74.4% of employees in 2018 (Destatis, 2019).

In addition to the general increase in the importance of the service sector, various developments can be mentioned which underline the **growing importance of service quality**:

M. Bruhn, *Quality Management for Services*,
https://doi.org/10.1007/978-3-662-67032-3_1

- The entry of both young and innovative as well as internationally active and already profiled companies abroad leads to a significant **increase in competition**. Not only the number of competing providers of product-related, supporting services is growing, but also the intensity of competition in the "classic" service markets (financial services, tourism, transport, etc.) is continuously increasing. As a result, the concept of is often described in the scientific literature as a way to differentiate (Gebauer et al., 2013; Baines, 2015; Bruhn & Hadwich, 2016a).
- Due to their immateriality, services can be imitated relatively easily; therefore, it is to be expected that the **homogenization of the service offer** will increase. As a result, it becomes more difficult for many companies to achieve a differentiated positioning vis-à-vis competing companies.
- The dynamic development of information and communication technologies enables customers, for example via customer forums or blogs on the Internet, to achieve a dimension of **market transparency** that was previously unknown, making it increasingly difficult for companies to bind their customers to themselves.
- The error-free function of technical products or goods is often taken for granted. Companies that upgrade their product offering by linking it to service elements are generally expected to maintain a constant high product quality. Differentiation from competition is only possible in many cases through the quality of the additional service offerings, and it is precisely in this area that **customer expectations** are growing considerably.
- With regard to customer expectations, it is not only the functional expectations of service fulfillment, but also the expectations of the service experience. These emotional customer expectations described in connection with the so-called **Customer Experience Management** are becoming increasingly important in the face of increasing interchangeability of services (Mayer-Vorfelder, 2012).
- Consumers increasingly expect "convenience" (convenience and comfort) in connection with changes in the overall societal value system. Service companies that help consumers to achieve more quality of life through expanded service offerings, e.g. time-saving internet services such as online banking, have above-average growth. In this context, the **concept of service value** becomes increasingly relevant. This represents the value generated for the customer by the service company from the customer's perspective. The best-known approach sees the service value as a trade-off between the benefits and costs of a service for the customer (Bruhn & Hadwich, 2014).
- Their own **internationalization efforts** confront companies with different customer expectations (Meffert et al., 2010, 25 pp.), but at the center of which is always the quality of service. Accordingly, companies give even more weight to service quality in this context (Bruhn & Hadwich, 2016b).

- The growing importance of the Internet poses new challenges for the quality of service or the quality management of services. Services are increasingly provided on the Internet and can be used independently of location through the increasing mobilization of Internet access. This results in completely new areas of action and claims on the **quality of service on the Internet,** which must take into account comprehensive quality management.
- The increasing importance of the Internet is particularly reinforced by **social media** or so-called Web 2.0. This creates permanent new application areas for both service providers (e.g. in the field of market research) and service users (e.g. exchange of opinions) (Bruhn & Hadwich, 2013).

Many service markets have already reached the stage of maturity, stagnation or even shrinkage. A significant expansion of the market volume is not possible in these markets in the short and medium term without far-reaching product and service innovations. In view of these conditions, the long-term binding of existing customers in addition to the acquisition of new customers moves into the center of marketing policy considerations. Experiences have shown that only in this way can chances for the acquisition of competitive advantages be achieved. Services therefore only contribute to economic success if **competitive advantages** can actually be achieved in this way. Competitive advantages have to meet the following requirements in order to be considered success factors (Backhaus & Voeth, 2014, 22 pp.; Bruhn, 2019, p. 16):

- **Customer perception:** The performance advantages are to be designed in such a way that they are recognized by the customer (!) as a significant differentiating feature.
- **Relevance:** The advantage is to be achieved with a (core) service of the provider that is considered particularly important by the customer and is therefore relevant to the purchase.
- **Durability:** The competitive advantage must have a temporal stability and thus also the property of not being imitable in the short term.

The success of a company therefore does not depend on objectively given, but subjectively perceived positioning advantages of the customer (Simon, 1988, p. 474). The basis for success are the sustainable competitive advantages achieved from the customer's point of view, which consist in being better in the performance features important to the customer than the competition. This makes the special importance of the demand **"securing superior service quality"** and the relevance of a successful quality management for services to meet customer requirements through the service offered clear.

The question of the competitiveness of the German economy in terms of services has also been addressed by the Federal Ministry of Education and Research (BMBF) over the past two decades. With over 10 million EUR, it has supported the "Innovative

Services" program, from which numerous follow-up projects have been set up. The program was set up in the mid-1990s and completed in 2004 (BMBF, 2004). In this context, the term **Service Engineering** has emerged in Germany as a research area that primarily deals with the systematic development and design of services through the application of special models, methods and (software) tools from the field of engineering sciences (BMBF, 2004). Technical universities and Fraunhofer institutes have been involved in the research projects.

As a result of the research work, various **concepts and tools of service engineering** have been developed, such as market analysis methods, learning and knowledge management, method toolkits with guidelines, process quality assurance, etc. (BMBF, 2004). Questions of quality management are indirectly addressed, especially in the development of process innovations.

At the heart of service engineering is the transfer of engineering knowledge to services. In a **critical assessment of service engineering**, it is positive to note that engineering sciences have dealt with a interface of service research with various contributions (Hüttemann et al., 2005). Although the approach has been considerably further developed in recent years, there is a clear focus on technical services and product-related services. For this reason, the term "service products" is often used. Technical customer services such as maintenance services or specific technical operator models are in the foreground. The approaches are overall very technology-driven and consistently process-oriented. Marketing research approaches are also taken into account, such as blueprinting (see especially Chap. 5 in this book), for example in connection with service design. However, classical services and non-technically driven services (still) receive too little attention.

1.2 Success Chain of Quality Management for Services

The attention paid to the topic of service quality in science is out of proportion to the actually experienced **service quality in everyday life.** Service providers are often far from the service standards required by customers. Discussions of "service deserts" and "service oases" show in a nutshell that many companies have a need to catch up in professional quality management for services.

Poor service quality is part of everyday life in many areas. Customer orientation and service readiness are lip service in many companies, the implementation of the contents imparted to employees in seminars and training courses often does not take place, is sometimes not lived by superiors or not internalized by employees (Meffert et al., 2018).

The **consequences of poor service quality** are often left out of account, customer dissatisfaction and its consequences are still ignored in many service companies, customer acquisition is prioritized over customer retention. This is all the more surprising as the findings about the behavior of dissatisfied customers are widely known (see Reichheld & Sasser, 1990; Zeelenberg & Pieters, 2004; Dabholkar & Spaid, 2012):

- Dissatisfaction leads to the defection of current customers.
- Dissatisfied customers engage in negative word of mouth and tell a large number of other people in their immediate environment about their dissatisfaction.
- Acquiring a new customer costs the multiple of binding an existing customer.

But even neglecting these "indirect" effects, higher quality is by no means necessarily associated with higher costs. On the contrary, it is often the case that a quality-defective performance **causes follow-up costs** for the service provider, which exceed the costs of a quality-perfect performance from the outset (Crosby, 1986, p. 28; Haist & Fromm, 1991; Snieska et al., 2013). This makes it possible to offer high-quality services and to realize an "overtaking strategy" (Outpacing), in which, if necessary, improvements can be achieved in both the cost dimension (cost advantages) and the quality dimension (quality advantages).

Mobile Services of Deutsche Bahn AG

A "overtaking strategy" is often realized through technology leaps. Mobile services of transport services are an example of this. Customers of the *Deutsche Bahn* can, for example, book tickets for different means of transport with the "DB Navigator", make fast bookings and check themselves in to avoid ticket controls. You will also receive real-time information with current departure and arrival times and information on the current train formation. The app can be downloaded for free. This leads to cost advantages for the provider over the traditional counter sales and to a quality improvement for the customer through an increase in flexibility and speed when buying tickets and an extension of the service range.

(Source: Deutsche Bahn, 2019) ◀

The fact that it is much more difficult for competitors to catch up with a once achieved quality advantage than to react to other measures, such as a price reduction, should not be neglected.

In the **tension between cost, time and quality** however, only a solution that optimizes all factors will be found if it is possible for the individual company to break away from the isolated consideration of time and to make quality the subject of a permanent quality management process.

The revenue-increasing effects that can be attributed to securing superior service quality can be shown using the **Service Quality Management Success Chain** (see Fig. 1.1). Fulfilling (heterogeneous) customer expectations can lead to an increase in customer satisfaction. Customer satisfaction in turn affects customer-related behavioral effects, in particular customer loyalty. It is assumed that a satisfied customer is more likely to use the services of a company again (re-election of the provider) or even to a greater extent (e.g. in the form of cross buying). A high degree of customer loyalty—it is further assumed—has revenue and success effects at the individual customer level as a result.

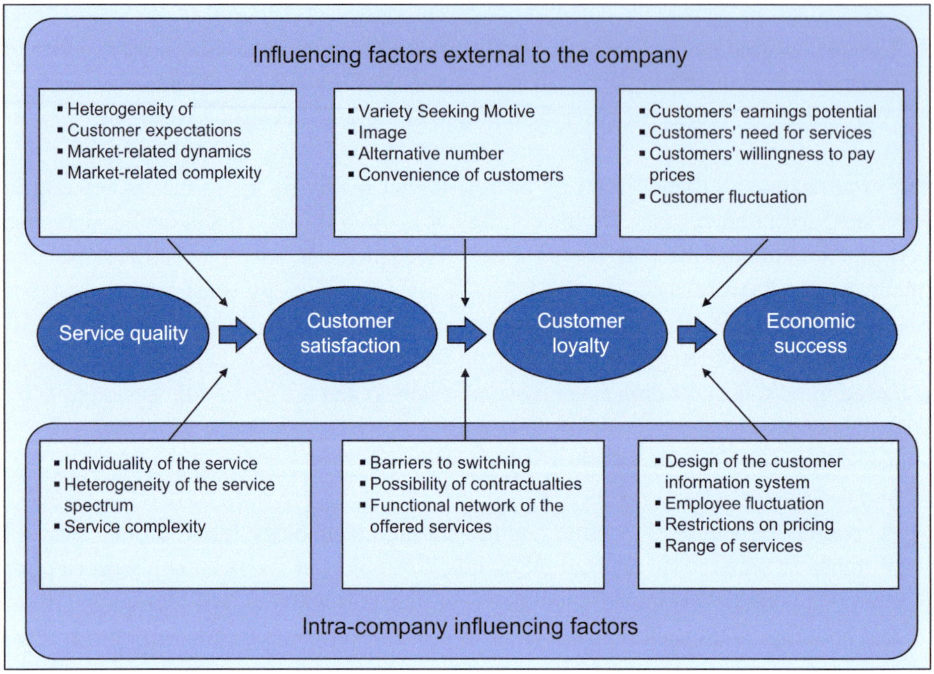

Fig. 1.1 Service quality management success chain (Source: Based on Meffert et al., 2018, p. 71)

The **quality of service** offered is a key factor for customer satisfaction. The customer has certain quality expectations that need to be met. For companies, this means on the one hand that the consideration of customer benefits and customer expectations must be regarded as a permanent task of all employees involved in the value creation process. On the other hand, due to the trust character of quality, the purchase risk must be reduced, services must be materialized and corresponding competencies must be demonstrated and communicated in the sense of a quality guarantee. A variety of trust-building and trust-securing measures must be taken in the phases of the purchase decision-making process.

In view of the mentioned relationships between customer satisfaction, customer loyalty and economic success, and in the sense of the illustrated success chain, it was recognized that—among other things from the perspective of efficiency—the assurance of service quality does not only begin at the interface between the company or its employees and the external customers ("end-of-the-pipe solution"). Rather, the **examination and optimization of the internal conditions of the company** for the creation of the service quality is also necessary for the realization of the service quality (Anosike & Eid, 2011; Nazeer et al., 2014; Hadwich & Keller, 2015).

This increases the relevance of the employees involved in the value creation process—also as a result of the consistent transfer of the marketing concept to all internal company processes—so the satisfaction and binding of "internal customers" is also given a great importance (Bruhn, 2004). An **internal customer orientation** is an important prerequisite for both the efficiency of processes within a company and the external customer orientation. Accordingly, customer orientation must be reflected in all employees—and also in services within a company (Stauss, 1999; Bruhn, 2002). Consequently, it is just as important for a company to take into account and improve the satisfaction of internal customers as it is to measure and improve the satisfaction of external customers (Gremler et al., 1994; Hadwich & Keller, 2015). Analogous to the success chain of quality management for (external) services, the economic success can also be explained by the **effect of internal service quality** (cf. Fig. 1.2).

The plausibility of this (internal and external) success chain has led to numerous companies making intensive efforts in recent years to ensure a high quality of their services. However, the effects within the success chain are not generally valid relationships. Not every expansion or intensification of quality management measures can automatically contribute to corporate success. Beyond a certain level of quality-related activity, further intensification of quality-related activities becomes unprofitable (Rust et al., 1994, p. 58). This is due to the fact that the relationships between the various links in the chain are influenced by so-called moderating variables (Homburg & Faßnacht, 2001), which can be divided into **external and internal moderating factors**:

- The existence and strength of the **relationship between quality efforts and satisfaction** is determined by external factors, such as the heterogeneity of customer expectations, and by internal factors, such as the heterogeneity of the range of services.
- The **relationship between customer satisfaction and loyalty** is influenced in its existence and intensity by external factors, such as the variety-seeking motives of service customers, and internal factors, such as existing switching barriers (Homburg & Faßnacht, 2001; Edward & Sahadev, 2011).

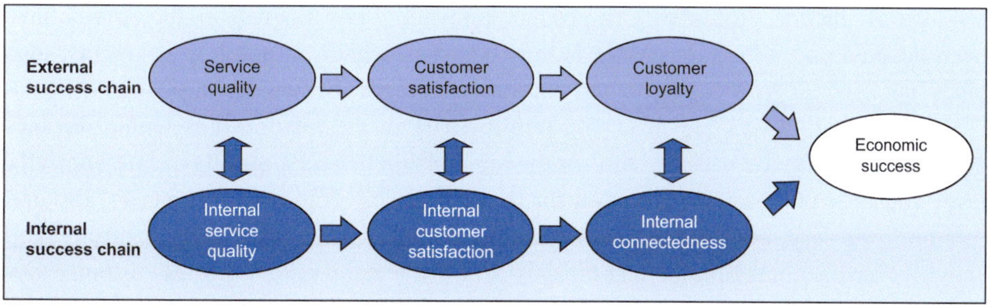

Fig. 1.2 Internal and external success chain of service quality

- The **relationship between customer loyalty and economic success** depends on external factors, such as the revenue potential of customers, and internal factors, such as employee turnover.

"Disruptive factors" of the success chain in different sectors

The variety-seeking motive of customers plays a major role especially in the tourism industry. For example, although a customer was very satisfied with a hotel, he books a different hotel for his next stay. The reason for this can be the customer's desire for a changing holiday experience, for a more varied hotel complex or a different leisure programme.

However, a customer can also continue to be bound to a company despite dissatisfaction. For example, a customer continues to book with an airline even though he is not satisfied with the service. One reason for this may be that the required flight connection is not included in the range of routes offered by other airlines.

Employee turnover as an internal disruptive factor is especially a problem for consultancies and advertising agencies. If, for example, an employee changes to a competitor company, it is possible that the clients he or she looks after will also change to that company. ◀

Due to the existence of these moderating variables, the chain of effects of quality activities on satisfaction and customer loyaltyshows. to economic success **has no general validity**. Not every quality management measure leads to an increase in customer satisfaction, not every increase in satisfaction leads to better customer loyalty and not every increase in customer loyalty leads to an increase in success. This makes it possible to understand the examples of companies that have not had any entrepreneurial success despite immense quality efforts. For example, two years after winning the prestigious *Malcolm Baldrige National Quality Award (MBNQA)*, the American Wallace Company. went bankrupt (Rust et al., 1994, p. 58). In addition, there are some examples of winners of the *MBNQA* such as. *Motorola* and *Federal Express*, which despite winning the quality award are only mediocre in terms of their financial success and winning the award often has no significant effect on company value (Jacob et al., 2004). The excessive focus of these companies on the **evaluation criteria** of the *MBNQA* led to the neglect of central aspects, such as profitability or product innovation. For this reason, the criteria have been adapted and a broader profile is now decisive, which, in addition to quality, also takes into account growth, profits and innovative strength (Brown, 2014, p. 10).

The moderating variables lead to a **nonlinearity** of the relationships within the success chain (cf. Fig. 1.3). Thus, both for the relationship between **quality-related activity level and customer satisfaction** and for the relationship between customer loyalty and economic success, an S-shaped curve can be assumed. This is because quality-related activities have a diminishing effect on satisfaction at a certain point, just as customer loyalty has a financial effect.

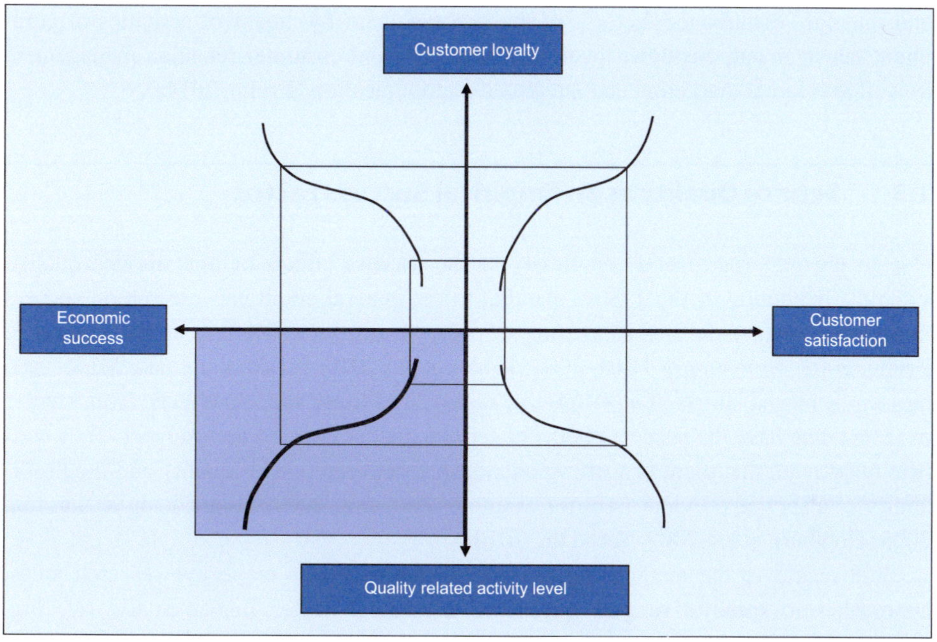

Fig. 1.3 Four-quadrant scheme for the success chain of quality management for services (Source: Bruhn, 1998, p. 9)

With regard to the **relationship between satisfaction and customer loyalty**, the assumption of a saddle-shaped curve appears to be plausible. This relationship is characterized by an indifference range at moderate customer satisfaction, while at the lower and upper ends of the curve, significant effects on customer loyalty (customer loss at the lower end or true commitment at the upper end) are to be expected (Homburg & Faßnacht, 2001).

As a result of these considerations, an **S-shaped curve** also results for the **quality-success relationship**. Beyond a certain point, further investment in quality management has an increasingly diminishing financial effect. Thus, the conclusion is that intensifying the quality-related activities of a service provider does not always have positive economic consequences. Therefore, when designing a profitability-oriented quality management, the optimization of customer satisfaction or customer loyalty is not the goal, but—taking into account the "intermediate steps" of the success chain—the optimization of the economic success of quality management.

The success of a service company depends mainly on the **management of the success chain** taking into account the moderating factors. Excellent service companies are characterized by thinking in the success chain, in the center of which is a professional quality management. In addition, however, they also succeed in controlling the external and internal "disturbance factors" of the success chain by means of activities of complaint management, customer loyalty management and customer retention management as well as internal marketing and integrated communication (Bruhn, 2014a).

1.3 Service Quality as an Empirical Success Factor

Due to the importance and complexity of the success effects of high service quality, since the beginning of the 1980s—parallel to the general empirical research on success factors in management and marketing (see for an overview Meffert, 1999; Meffert & Bruhn, 2002; Homburg & Fürst, 2005; Bruhn et al., 2010; Sivakumar et al., 2014)—can be seen increased efforts of an **empirical research of these success effects**. Even studies of recent date have the success effects of service quality as a research subject. They confirm once again that there is a strong relationship between service quality and the profitability of a service provider (see, for example, Zeithaml, 2000; Ming Chang & San Ing, 2005; Homburg et al., 2009; Yee et al., 2010).

With regard to the methodological starting points in the respective research areas, approaches to **qualitative and quantitative** research differ fundamentally (cf. Fig. 1.4). Qualitative research is more case-related and aims at the detailed and argumentative analysis of successful examples, while quantitative research is representative or at least case-aggregating and tries to make general statements at a relatively high aggregation level. Qualitative research on the success factor service quality is concretized in the effect- and task-oriented research. Within the quantitative approaches, an interfactor and an intrafactor research can be found. Within these four categories, there are approaches that provide indications of the success factor service quality.

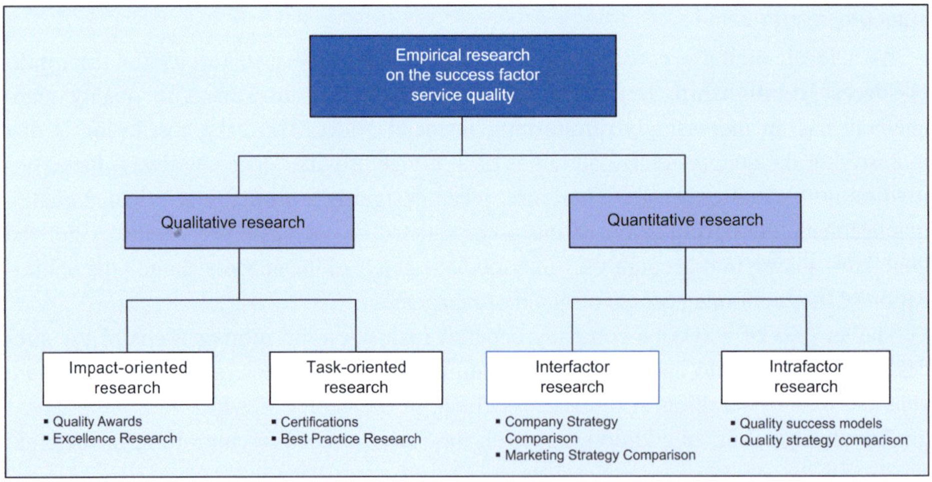

Fig. 1.4 Approaches to success factor research (Source: Based on Meffert & Bruhn, 2002, p. 8)

Within the framework of the **effect-oriented research** companies with top performances are sought, whose success can be attributed to a pronounced quality management in a verifiable manner. On the one hand, this approach is pursued in the award of quality awards (e.g. *Malcolm Baldrige National Quality Award, EFQM Excellence Award, Ludwig-Erhard-Preis* or *ESPRIX*), in which companies are assessed according to certain criteria in connection with quality and quality management. Furthermore, this approach is used in so-called excellence research (cf., for example, Simon, 1990; Meffert & Bruhn, 2002). The bestseller "In Search of Excellence" by Peters and Waterman (2004) thus conveys that excellent companies are characterized not only by "hard" (e.g. organizational structure or production technology), but also and above all by the design of "soft" success factors, such as the perceived service quality.

In contrast to performance-oriented research, **task-oriented research** analyses outstanding activities of companies in relation to service quality. In addition to certification, in which corresponding investigations are initiated by the company to be certified, the so-called best practice research makes significant contributions in this context. The study "Best Practice in Marketing" of the *Research Institute for Marketing and Trade at the University of St. Gallen* is an example. The latter uses a task-based approach and arrives at a typology of the "Best Practice in Marketing" in the so-called core tasks of customer loyalty, customer acquisition, service maintenance and service innovation (Tomczak & Reinecke, 1998). The aim of this institutionalized research program is to develop guidelines for the successful implementation of marketing strategies. It is important to find out through intensive cooperation with partner companies how it is possible to achieve competitive advantages through better marketing processes in relation to the four core tasks mentioned (University of St. Gallen, 2010).

The **interfactor research** does not consider service quality solely as a success factor, but compares it and its effect with that of other success factors. In the **comparison of business strategies**, the importance of different strategic attitudes (e.g. market orientation, cost orientation) is compared with each other (Fritz, 1993; Noble et al., 2002). Often, quality strategy is part of a broader attitude, such as market or customer orientation. Fritz's study (1993) can be cited as an example. According to the results of his study, market orientation makes a significant contribution to corporate success. However, to the same extent, production, cost and employee orientation also contribute. These findings correspond to the finding that successful companies need an outside-in perspective based on market orientation as well as an inside-out perspective based on core competencies. In the **comparison of marketing strategies**, the success effect of different marketing parameters is set against each other. The so-called PIMS project, which tried to identify the central dimensions of return on investment and cash flow of strategic business units of companies in more than 100 studies, is particularly noteworthy in this context. Of outstanding importance for marketing was the relative, perceived and superior product quality. It not only allows the implementation of relatively higher prices, but also secures higher returns through market share gains and associated cost reduction potentials (Buzzell & Gale, 1989).

The **intrafactor research** focuses on the success factor quality and examines its effects on corporate success. Within the framework of **quality success models**, the effects of high quality are analyzed in more detail. For example, one study shows that the winners of the *Malcolm Baldrige National Quality Award* perform significantly better on the stock exchange than the top 500 companies (Bruhn, 1998, pp. 16–17). In addition, there is a wealth of studies based on the so-called National Customer Barometer (Bruhn & Murmann, 1998; Bruhn, 2010), on the basis of which, for example, statements can be made about differences in the importance of individual quality features in different industries. Finally, in the **quality strategy comparison**, different orientations of quality management are placed side by side. Basically, a sales-oriented and a cost-oriented quality management can be distinguished. In a study by Rust et al. (2002), only the sales-oriented quality management has a significant impact on corporate success—as propagated by the quality management success chain.

The different studies and their central results on the success factor service quality underline the outstanding importance of quality management in the service sector and give indications for the design of a successful quality management.

1.4 The Central Paradigms of Quality Management

The following three paradigms of quality management are to be placed in the center, they emphasize the business-economic side of quality management. These are the question of the costs of quality and the achievement of an optimal quality level, the question of the costs of error correction, in particular in the time course of quality management, as well as the question of the costs of non-quality, that is, the consideration of opportunity costs.

Cost of quality and striving for an optimal quality level

It does not require special justification that the creation of quality at a certain level costs. Costs of planning, implementation, steering, control, etc. of quality. These costs arise at different points in the company (see the explanations in Chap. 15, in particular Sect. 15.1 of this book).

These costs are to be compared with the revenues of quality management. They will increase with increasing quality, but decrease to a certain level. A "maximum quality" will not pay off. So it is about an **"optimal quality",** which results in a quality level at which the difference between revenues and costs is greatest.

This generally held paradigm shows the **optimization calculus of quality management.** Costs and revenues are to be optimized economically. As with quality management, it is a cost-benefit comparison. This is shown in Fig. 1.5. Further details on the costs and benefits of quality management can be found in Chap. 15 and 16 of this book.

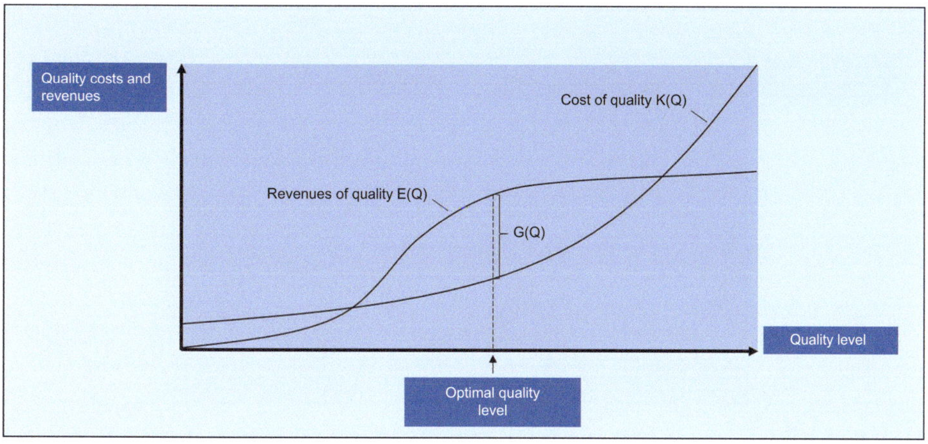

Fig. 1.5 Optimization of quality based on a cost-benefit comparison

Cost of error correction of quality management over time

The business cost consideration results in another paradigm of quality management. Ultimately, quality management—especially in the service sector—is designed to produce no errors. Corresponding concepts such as Six Sigma, zero-error principle, etc. go in this direction. If errors still occur, they are also associated with costs. So it is also a question of the **cost of error correction** in quality management. These errors need to be eliminated or avoided.

The cost of error correction—and that is what this paradigm says—is to be considered in relation to the course of quality measures. A quality project goes through different phases, such as project definition, planning, piloting, implementation, control and completion (based on Marr & Steiner, 2004). The cost of error correction increases disproportionately over the course of the project phases (concave curve). This is shown in Fig. 1.6.

Of course, it cannot be estimated in general by how much this cost increase occurs. Schmitt and Pfeifer (2015, p. 3) mention an "empirical ten-rule", but without mentioning any evidence. But maybe it's a rule of thumb that can serve as a guideline.

Cost of poor quality as opportunity cost

A wide central business paradigm is the question of which costs—more precisely, which **opportunity costs**—are incurred when poor quality is delivered. Then customers will leave and this will be reflected in lower re-purchase rates, lower up-selling, lower customer loyalty, a decline in recommendations, and other indicators that have negative economic consequences. These opportunity costs lead to a negative success of the company, that is to **costs of poor quality.** Figure 1.7 shows the relationship.

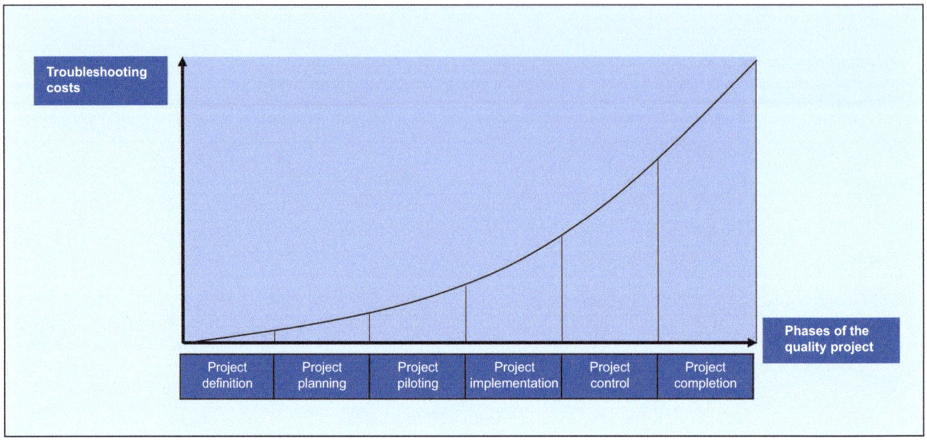

Fig. 1.6 Cost of failure correction over time

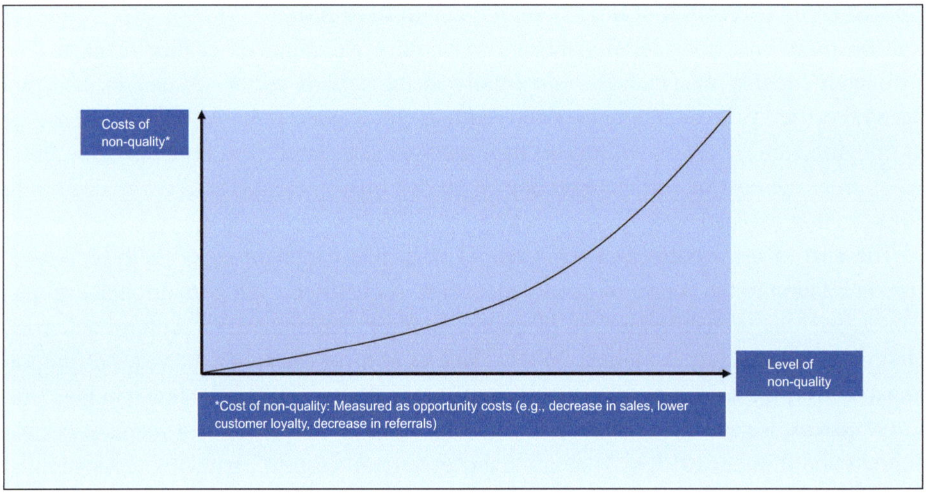

Fig. 1.7 Cost of poor quality as opportunity cost

These three central paradigms of quality management ultimately determine the discussion about quality management, also for quality management for services. Of course, there are more paradigms in detail, which will become clear in the course of the book.

References

Anosike, U. P., & Eid, R. (2011). Integrating internal customer orientation, internal service quality, and customer orientation in the banking sector: An empirical study. *The Service Industries Journal, 31*(14), 2487–2505.

Backhaus, K., & Voeth, M. (2014). *Industriegütermarketing* (10th Edn.). München: Vahlen.

Baines, T. (2015). Exploring service innovation and the servitization of the manufacturing firm. *Research-Technology Management, 58*(5), 9–11.

BMBF. (2004). Service engineering. https://www.pt-ad.pt-dlr.de/de/582.php. Accessed 15 May 2019.

Brown, M. G. (2014). *Baldrige award winning quality. How to interpret the Baldrige criteria for performance excellence*. London.

Bruhn, M. (1998). *Wirtschaftlichkeit des Qualitätsmanagements. Qualitätscontrolling für Dienstleistungen*. Berlin: Springer.

Bruhn, M. (2002). *Integrierte Kundenorientierung. Implementierung einer kundenorientierten Unternehmensführung*. Wiesbaden: Gabler.

Bruhn, M. (2004). Interne Servicebarometer als Instrument interner Kundenorientierung. Messung und Steuerung der Qualität und Zufriedenheit interner Dienstleistungen. *Marketing ZFP, 26*(4), 282–294.

Bruhn, M. (2010). Das Konzept des Internen Servicebarometers – Bestandsaufnahme, Methodik und empirische Befunde. In M. Bruhn & B. Stauss (Eds.), *Serviceorientierung im Unternehmen. Forum Dienstleistungsmanagement* (pp. 297–322). Wiesbaden: Gabler.

Bruhn, M. (2014a). *Integrierte Unternehmens- und Markenkommunikation: Strategische Planung und operative Umsetzung* (6th Edn.). Stuttgart: Schäffer-Poeschel.

Bruhn, M. (2019). *Marketing. Grundlagen für Studium und Praxis* (14th Edn.). Wiesbaden: Gabler.

Bruhn, M., & Hadwich, K. (2013). Dienstleistungsmanagement und Social Media – Eine Einführung in die theoretischen und praktischen Problemstellungen. In M. Bruhn & K. Hadwich (Eds.), *Dienstleistungsmanagement und Social Media. Forum Dienstleistungsmanagement* (pp. 3–40). Wiesbaden: Springer.

Bruhn, M., & Hadwich, K. (2014). Service Value – Eine Einführung in die theoretischen und praktischen Problemstellungen. In M. Bruhn & K. Hadwich (Eds.), *Service Value als Werttreiber. Forum Dienstleistungsmanagement* (pp. 3–31). Wiesbaden: Springer.

Bruhn, M., & Hadwich, K. (2016a). Servicetransformation – Eine Einführung in die theoretischen und praktischen Problemstellungen. In M. Bruhn & K. Hadwich (Eds.), *Servicetransformation. Forum Dienstleistungsmanagement* (pp. 3–22). Wiesbaden: Springer.

Bruhn, M., & Hadwich, K. (2016b). *Internationales Dienstleistungsmarketing: Strategien – Instrumente – Methoden*. Stuttgart: Kohlhammer.

Bruhn, M., & Murmann, B. (1998). *Nationale Kundenbarometer. Messung von Qualität und Zufriedenheit. Methodenvergleich und Entwurf eines Schweizer Kundenbarometers*. Wiesbaden: Springer.

Bruhn, M., Georgi, D., Ahlers, M., & Buess, M. (2010). Marketing-Werttreiber-Analyse – Steuerung des Marketing-Mix bei der Basler Kantonalbank. *Marketing Review St. Gallen, 27*(4), 48–53.

Buzzell, R. D., & Gale, B. T. (1989). *Das PIMS-Programm. Strategien und Unternehmenserfolg*. Wiesbaden: Springer.

Crosby, P. B. (1986). *Qualität bringt Gewinn*. Hamburg: McGraw-Hill.

Dabholkar, P. A., & Spaid, B. I. (2012). Service failure and recovery in using technology-based self-service: Effects on user attributions and satisfaction. *The Service Industries Journal, 32*(9), 1415–1432.

Destatis. (2019). https://www.destatis.de/DE/Presse/Pressemitteilungen/2019/01/PD19_001_13321.html. Accessed 02 Apr. 2019.

Deutsche Bahn. (2019). Der DB Navigator. https://www.bahn.de/p/view/service/mobile/db-navigator.shtml. Accessed 02 Apr. 2019.

Edward, M., & Sahadev, S. (2011). Role of switching costs in the service quality, perceived value, customer satisfaction and customer retention linkage. *Asia Pacific Journal of Marketing and Logistics, 23*(3), 327–345.

Fritz, W. (1993). *Marktorientierte Unternehmensführung und Unternehmenserfolg. Arbeitspapier No. 93/01*. Braunschweig.

Gebauer, H., Paiola, M., & Saccani, N. (2013). Characterizing service networks for moving from products to solutions. *Industrial Marketing Management, 42*(1), 31–46.

Gremler, D. D., Bitner, M. J., & Evans, K. R. (1994). The internal service encounter. *Logistics Information Management, 8*(4), 28–34.

Hadwich, K., & Keller, C. (2015). Interne Servicequalität in Unternehmen: eine empirische Untersuchung der Einflussfaktoren und Auswirkungen. *Zeitschrift für betriebswirtschaftliche Forschung,67*(2), 170–205.

Haist, F., & Fromm, H. (1991). *Qualität im Unternehmen. Prinzipien, Methoden, Techniken* (2nd Edn.). München: Hanser.

Homburg, C., & Faßnacht, M. (2001). Kundennähe, Kundenzufriedenheit und Kundenbindung bei Dienstleistungsunternehmen. In M. Bruhn & H. Meffert (Eds.), *Handbuch Dienstleistungsmarketing. Von der strategischen Konzeption zur praktischen Umsetzung* (2nd Edn., pp. 441–464). Wiesbaden: Springer.

Homburg, C., & Fürst, A. (2005). How organizational complaint handling drives customer loyalty: An analysis of the mechanistic and the organic approach. *Journal of Marketing, 69*(3), 95–114.

Homburg, C., Wieseke, J., & Hoyer, W. D. (2009). Social identity and the service-profit chain. *Journal of Marketing, 73*(2), 38–54.

Hüttemann, E., Rinke, S., & Ernst, G. (2005). Innovative Dienstleistungen. Bilanz im Schwerpunkt Service Engineering, PT-Paper.

Jacob, R., Madu, C. N., & Tang, C. (2004). An empirical assessment of the financial performance of Malcolm Baldrige Award winners. *International Journal of Quality Science, 21*(8), 897–914.

Marr, R., & Steiner, K. (2004). Projektmanagement. In G. Schreyögg & A. von Werder (Eds.), *Handwörterbuch Unternehmensführung und Organisation* (pp. 1196–1208). Schäffer-Poeschel.

Mayer-Vorfelder, M. (2012). *Kundenerfahrungen im Dienstleistungsprozess – eine theoretische und empirische Analyse*. Wiesbaden: Gabler.

Meffert, H. (1999). *Erfolgsfaktorenforschung im Marketing. Auf der Suche nach dem „Stein der Weisen"*. Unterlagen zum 5. Münsteraner Marketing-Symposium. (pp. 1–13). Münster: Gabler.

Meffert, H., & Bruhn, M. (2002). Wettbewerbsüberlegenheit durch exzellentes Dienstleistungsmarketing. In M. Bruhn & H. Meffert (Eds.), *Exzellenz im Dienstleistungsmarketing. Fallstudien zur Kundenorientierung* (pp. 1–26). Wiesbaden.

Meffert, H., Burmann, C., & Becker, C. (2010). *Internationales Marketing – Ein markenorientierter Managementansatz* (4th Edn.). Stuttgart: Kohlhammer.

Meffert, H., Bruhn, M., & Hadwich, K. (2018). *Dienstleistungsmarketing. Grundlagen, Konzepte, Methoden* (9th Edn.). Wiesbaden: Springer.

Ming Chang, L., & San Ing, H. (2005). Relationships among service quality, customer satisfaction and profitability in the Taiwanese banking industry. *International Journal of Management, 22*(4), 635–648.

Nazeer, S., Zahid, M. M., & Malik, F. A. (2014). Internal service quality and job performance: Does job satisfaction mediate? *Journal of Human Resources, 2*(1), 41–65.

Noble, C. H., Sinha, R. K., & Kumar, A. (2002). Market orientation and alternative strategic orientations. A longitudinal assessment of performance implications. *Journal of Marketing, 66*(4), 25–39.

Peters, T. J., & Waterman, R. H. (2004). *In search for excellence. Lessons to learn from America's best run companies*. New York.

Reichheld, F. F., & Sasser, W. E. (1990). Zero defections. Quality comes to services. *Harvard Business Review, 68*(5), 105–111.

Rust, R. T., Zahorik, A. J., & Keiningham, T. L. (1994). *Return on quality. Measuring the financial impact of your company's quest for quality*. Chicago.

Rust, R. T., Moorman, C., & Dickson, P. R. (2002). Getting return on quality. Revenue expansion, cost reduction, or both? *Journal of Marketing, 66*(4), 7–24.

Schmitt, R., & Pfeifer, T. (2015). *Qualitätsmanagement. Strategien – Methoden – Techniken* (5th Edn.). Wien: Hanser.

Simon, H. (1988). Management strategischer Wettbewerbsvorteile. *Zeitschrift für Betriebswirtschaft, 58*(4), 461–480.

Simon, H. (1990). Hidden Champions – Speerspitze der deutschen Wirtschaft. *Zeitschrift für Betriebswirtschaft, 60,* 875–890.

Sivakumar, K., Li, M., & Dong, B. (2014). Service quality: The impact of frequency, timing, proximity, and sequence of failures and delights. *Journal of Marketing, 78*(1), 41–58.

Snieska, V., Daunoriena, & Zekeviciene, A. (2013). Hidden costs in the evaluation of quality failure costs. *Engineering Economics, 24*(3), 176–186.

Statista. (2019). https://de.statista.com/statistik/daten/studie/36846/umfrage/anteil-der-wirtschaftsbereiche-am-bruttoinlandsprodukt/. Accessed 2 Apr. 2019.

Stauss, B. (1999). Internes Marketing als personalorientierte Qualitätspolitik. In M. Bruhn & B. Stauss (Eds.), *Dienstleistungsqualität. Konzepte – Methoden – Erfahrungen* (3rd Edn., pp. 204–222). Wiesbaden: Gabler.

Tomczak, T., & Reinecke, S. (1998). Best Practice in Marketing – Auf der Suche nach Marketing-Spitzenleistungen. In T. Tomczak, & S. Reinecke (Hrsg.), *Best Practice in Marketing. Erfolgsbeispiele zu den vier Kernaufgaben im Marketing* (S. 9–34). St. Gallen.

Universität St. Gallen. (2010). Best practice. http://www.ifm.unisg.ch/Praxis/Programme/Best+Practice+in+Marketing.aspx. Accessed 18 Oct. 2012.

Yee, R., Yeung, A., & Cheng, E. (2010). An empirical study of employee loyalty, service quality and firm performance in the service industry. *International Journal of Production Economics, 124*(1), 109–120.

Zeelenberg, M., & Pieters, R. (2004). Beyond valence in customer dissatisfaction. A review and new findings on behavioral responses to regret and disappointment in failed services. *Journal of Business Research, 57*(4), 445–455.

Zeithaml, V. A. (2000). Service quality, profitability, and the economic worth of customers. What we know and what we need to learn. *Journal of the Academy of Marketing Science, 28*(1), 67–85.

Conceptual Basics of Service Quality

2

Abstract

The term service and service quality can be viewed from different angles and perspectives. The chapter creates the required basic concepts. The focus is on the basics of service quality. This includes the various influencing factors of service quality. Based on empirical findings, the dimensions of service quality are shown—the potential, process and result quality. They often form the basis for the design of service quality management. On the other hand, the effects of service quality are shown; this includes psychological, behavioral and economic effects due to a high service quality.

2.1 Term and Systematization of Services

From the perspective of economics, services are usually delimited by institutional (listing of economic sectors) and occupational (classification of occupations) criteria. However, for business administration and marketing, these characteristics are only suitable for distinguishing between goods and services if an **institutional perspective** is also used (banking, trade, tourism marketing). In these companies, service marketing is to be seen as the main function. In contrast, **functional service marketing** is understood as a side function which is carried out by goods-producing companies in order to promote the sale of their goods by the services provided, so-called value added services.

When examining the term "service marketing" more closely, the first question to be examined is the **delimitation of services** from goods. The range of service offerings is extremely wide, making it difficult to delimit in individual cases. For example, both the possibility of trying on clothes in a factory outlet and the mediation of spouses can be regarded as services. While in the first case the share of the service in the total

M. Bruhn, *Quality Management for Services*,
https://doi.org/10.1007/978-3-662-67032-3_2

performance of the manufacturing company is to be seen as relatively low, the advice of a partner agency is often understood as being almost "pure" service. Figure 2.1 shows that the boundary between goods and services is asymmetrical, since while services without a goods component are conceivable, hardly any goods without a service component are conceivable. This representation is also known as the marketing mix box (Hilke, 1989, pp. 17–24).

Before transferring basic marketing principles and methods to service marketing, it is necessary to clarify which **specific characteristics** make up services, with a focus on the differences between consumer goods (Lindberg & Nordin, 2008; Maiwald et al., 2014; Nenonen et al., 2014; Corsten & Gössinger, 2015).

Services tend to have an **immaterial character,** with the actual core service being intangible. However, this does not mean that services can always be represented without physical service components. On the one hand, the result of many services is inseparably linked to physical service components (e.g. car repair using new parts), on the other hand, many services require a physical service infrastructure (e.g. telecommunications services in mobile networks), or the physical service component makes the service at least more comfortable (e.g. the equipment in a doctor's office).

However, the separation of this criterion is not as high as might be initially assumed. The intangibility of a service is largely dependent on the **underlying product understanding**—just as the understanding of the relevant market by the company. For example, a car manufacturer often sees itself as a producer of consumer goods, but from an expanded perspective, an understanding as a service provider would also be possible. According to the first approach, the company produces cars and commercial vehicles,

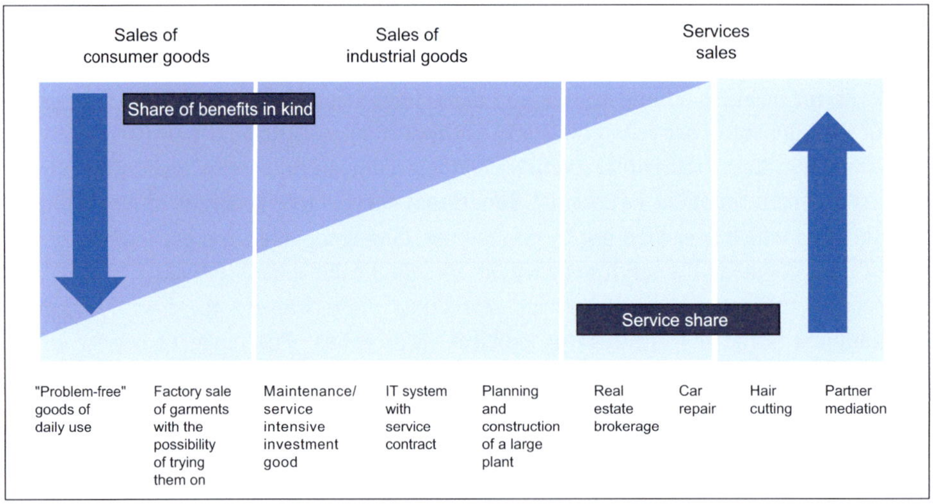

Fig. 2.1 Marketing in different sectors (Source: Based on Hilke, 1989, p. 4)

from the second perspective mobility for people and transport of goods as a service (company as "problem solver").

In close connection with the immateriality of services stands their **intangibility.** Thus, the quality of services is often only partially sensed before production and thus before purchase (e.g. restaurant visit), in some cases it is not even accessible during the service process (e.g. hair cut) or at its end (e.g. doctor's diagnosis). In addition, services are indivisible in contrast to goods, so that production and consumption take place simultaneously ("Uno-Actu-Principle"); a feature that finally also determines the criterion of transience. As a result of the simultaneity of production and exploitation, services cannot be stored (Lehmann, 1995; Mudie & Cottam, 1999, 21 pp.; Corsten & Gössinger, 2015). However, this argument is not valid for all services throughout, as results of services can be stored and partly consumed after production if the service consumption is considered between acquisition and exploitation of the service. Examples of this are the use of final reports from management consultants over the years or also vaccinations (Maleri & Frietzsche, 2008).

Basically, a **direct contact** between the provider and the customer of a service is required, but there are also forms of indirect customer relationships, for example in the financial services sector. However, it is generally spoken of the necessity of the **integration of the external factor** (Meffert et al., 2018). The external factor is either the customer himself (persons) or objects brought by the customer for the creation of the service (e.g. a car for repair). In any case, the customer's participation is required for the performance.

This peculiarity leads to the criterion of **location-boundness;** a service is not transportable (Meffert et al., 2018). It is usually created on the premises of the service provider or the customer. As with the non-storability, it should be noted here that results of services can be transported in limited cases, e.g. information. Furthermore, the location-boundness exists mainly when immobile goods are introduced as external or internal factors in the service creation process (Maleri & Frietzsche, 2008).

Finally, the **individuality or variability** as a consequence of the simultaneity of production and consumption is also a criterion for demarcation that separates services and goods. So the service is created anew for each customer so that often both under static and under dynamic aspects the scope of performance—thus also its quality—are individually different (Mudie & Cottam, 1999, 6 pp.; Corsten & Gössinger, 2015).

When considering the specific characteristics of services presented so far, four **approaches** can be distinguished at a higher level to define the **service concept** (Corsten & Gössinger, 2015, 17 pp.; Meffert et al., 2018, p. 13):

(1) Activity-oriented definition A very broad understanding of services is represented by Schüller (Schüller, 1967, 67 pp.), who classifies every **human activity** as a service. "What one does to bring his physical and mental labor into the field of human needs with or without connection to the material world of goods is a service." Due to the range of this concept, the approach is only partially convincing in some cases, as it leaves little room to derive service marketing-specific features (Meffert et al., 2018, p. 13).

(2) Potential-oriented definition The potential-oriented interpretation of services is based on the assumption that services are often seen as the **potential of a service provider** created by humans or machines to perform the service at the service consumer (Meyer & Mattmüller, 1987, 187 pp.). An example of this are the personnel and material resources of a bank that determine the service to be provided in terms of type and scope.

(3) Process-oriented definition In the process-oriented view, the service is in the focus as an activity. In particular, the **"Uno-Actu principle"** is to be taken into account here in order to do justice to the importance of the simultaneity of production and sale of the service. For the example of a banking service, this includes the advice of customers by bank employees (Meffert et al., 2018, p. 13).

(4) Result-oriented definition The result-oriented perspective finally takes into account the actually produced **intangible good** as a concretization of the service process. In the case of a bank, this is, for example, a credit or savings contract.

The three latter definitions can be subjected to a **phase-related integration** of the process-, result- and potential-oriented interpretation of the service. In this way, a combined view of the specific characteristics of services is possible (Hilke, 1984, 17 pp.). For the most complete definitional capture of the service character, it is expedient that all three phases (potential, process, result) are each represented by a separate characteristic in the service definition. In this way, the connection is taken into account that by the specific skills and the willingness of the service provider to provide a service (potential orientation) as well as the integration of the external factor by the service user as a process-initiating and -accompanying element (process orientation) finally a concrete service result results (result orientation).

Based on Meffert et al. (2018, pp. 13 f.), the following **definition** for service is used:

▶ **Services** are independent, marketable services that are associated with the provision and/or use of capabilities (potential orientation). Internal and external factors are combined in the course of the performance creation process (process orientation). The factor combination of the service provider is used with the aim of achieving benefits-creating effects on external factors—people or their objects (result orientation).

The characteristic features of services make it clear that the creation of service quality is to be understood both as a "production problem" against the background of the service creation process and as a higher-level marketing problem. This also implies that a focus on individual submix areas of marketing is too short-sighted to do justice to the multi-layered—specific—aspects of service creation and marketing. In particular, it is the **ensuring of service quality** that is to be guaranteed by service providers in order to compete in the fiercely competitive service market.

In order to better deal with the different manifestations of services with the instruments of marketing, various **differentiation approaches** have been developed in the past, with the aim of systematizing services in more or less complex form.

Corsten and Gössinger (2015, 30 pp.) have compiled numerous criteria for **one-dimensional systematization of services** that allow a—albeit not complete—classification of the forms of services that occur in practice.

Basically, personal and automated services are often distinguished in the service sector. **Personal services** are characterized by a high proportion of human performance (e.g. advisory services by doctors or lawyers, opera performances). However, this human performance component is lost when personal services are combined with a carrier medium. For example, the personal performance of an open-air concert becomes a tangible service in the form of an **"enriched service"** when an audio or video recording is made (Meyer, 1998, pp. 119 f.).

In contrast, **automated services** are services that are created by a machine. Ticket machines, telephone banking, account statement printers or fax polling are services that are increasingly being used in a wide range of service areas with the advancement of new technologies. In particular, the boom of so-called e-business that started at the turn of the millennium and continues to this day leads to more and more services being created electronically and displacing personal contact between provider and customer. Classic examples of this are internet banking, online auctions, information and advice services as well as travel bookings on the internet. The increasing use of technology is on the one hand the result of the high personnel intensity in the service sector and the high costs associated with it. On the other hand, the personal service contact is increasingly being pushed back by the use of technology. This has the positive effect that the customer's possibilities of interfering in the service creation process are reduced, which in turn can lead to a stabilization of the often fluctuating service quality.

Of particular importance for the design of the service process is the distinction between the **service object** or the subject of the service creation (external factor). So either people are recipients of a service (e.g. legal advice, public transport) or objects (e.g. cleaning services).

Finally, the distinction is often made between process-oriented and result-oriented services on the basis of the phase orientation of the service. While the service creation process is in the focus for **process-oriented services** (e.g. a city tour), the achieved effect on the service object is essential for **result-oriented services** (e.g. any repair services). However, the example of a restaurant makes it clear what kind of delimitation problems can arise. While the food intake in a fast food restaurant is primarily based on the goal of a quick and inexpensive satisfaction of the basic need "hunger" (result orientation), the food in a star restaurant usually has a different character. Of course, a basic need is also met here, but the demander is mainly interested in a holistic experience of enjoyment, which brings the process component into focus.

In summary, the individual forms of service differentiation demonstrate that the character of a service is essentially determined by **spatial, temporal** or other **situational factors**. It is this circumstance that also makes it considerably more difficult to derive generally valid, theoretically founded statements about the service sector. A summary overview of the different **forms of manifestation of services** is provided in Fig. 2.2.

Given the complexity and heterogeneity of the service sector and against the background of the desire to arrive at generally valid statements, the proposal for a general **typology** of sales objects put forward by Engelhardt et al. (1992, p. 35) (see Fig. 2.3) represents a useful approach, abandoning the distinction between goods and services.

The result of this typological delimitation of goods is four basic forms of services that can be distinguished on the basis of two main dimensions—the **degree of intangibility of the service result** and the **degree of integration of the company's service processes:**

(1) Intangible service result, highly integrated service production process Typical examples of this category are consulting services that would not be conceivable without the close involvement of the service recipient, such as medical or legal consultation or corporate consulting.

(2) Tangible service result, highly integrated service production process Again, the service is produced in close contact with the customer, but the result of the production process is predominantly tangible. Examples of such services can be found in plant and special machine construction.

(3) Tangible service result, autonomous service production process Representatives of this category of goods are classic consumer goods that are manufactured for the anonymous market without a specific customer order. The integration of customer needs takes place mainly in the design phase of the services, while the service production process has the classic characteristics of mass production.

(4) Intangible service result, autonomous service production process Examples of the fourth category of goods are telephone announcement or database services, in which the integration of the external factor is largely limited to the consumption of the services—without the involvement of the consumers.

In order to increase the clarity and decision-making relevance of this systematization, a further differentiation of the integration dimension by the sub-dimensions interaction degree and individualization degree is promising. The result is a three-dimensional service typology (Maister & Lovelock, 1988, 67 pp.; Wohlgemuth, 1989, p. 339; Corsten & Gössinger, 2015). The result of the expansion of the integration dimension is the service typology shown in Fig. 2.3 (Meffert, 1993, p. 12; Meffert et al., 2018, p. 17).

The **degree of interaction** is an expression of the basic involvement of the external factor in the process of creating value. Its manifestations range from a function mainly

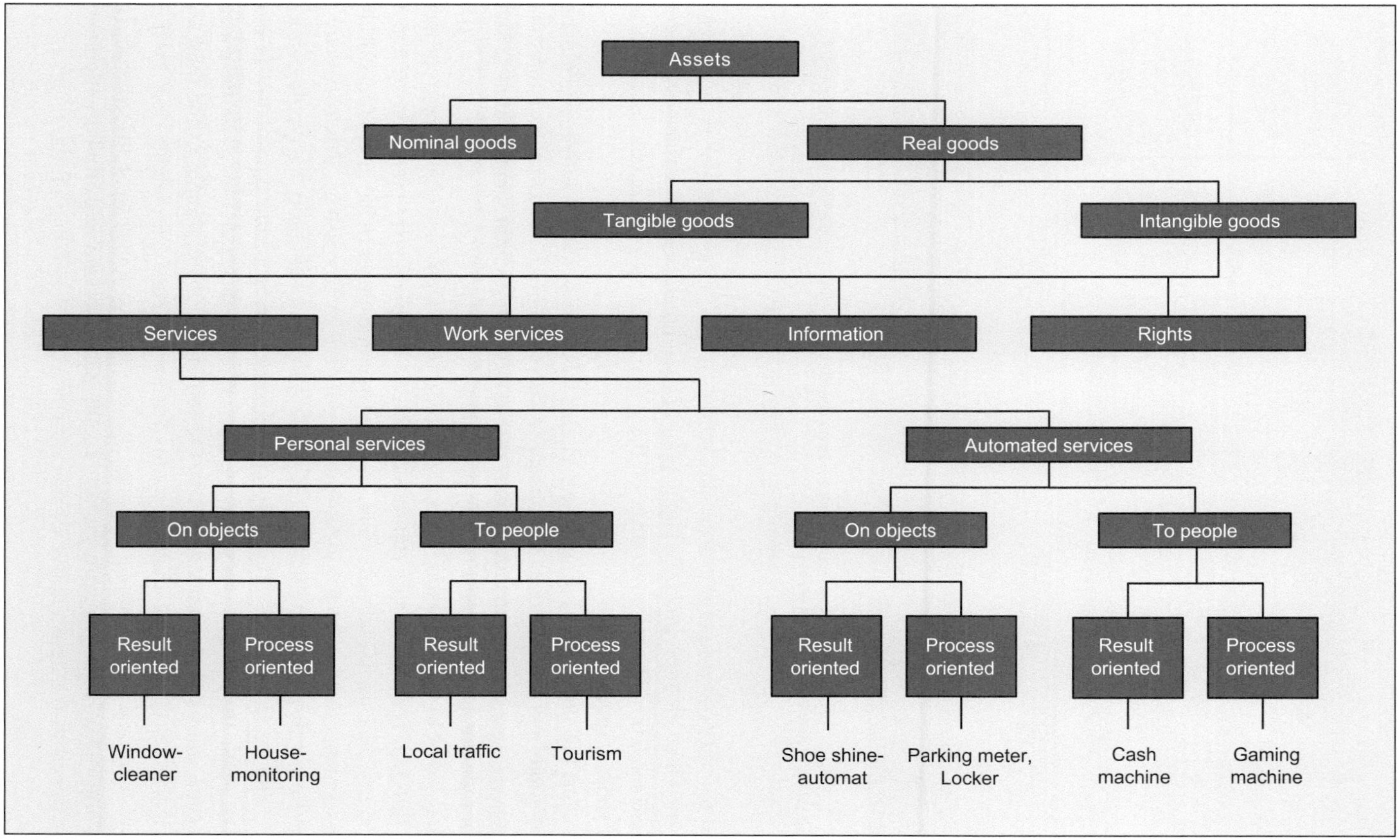

Fig. 2.2 Systematics of economic goods (with examples) (Source: Meffert & Bruhn, 2012, p. 23)

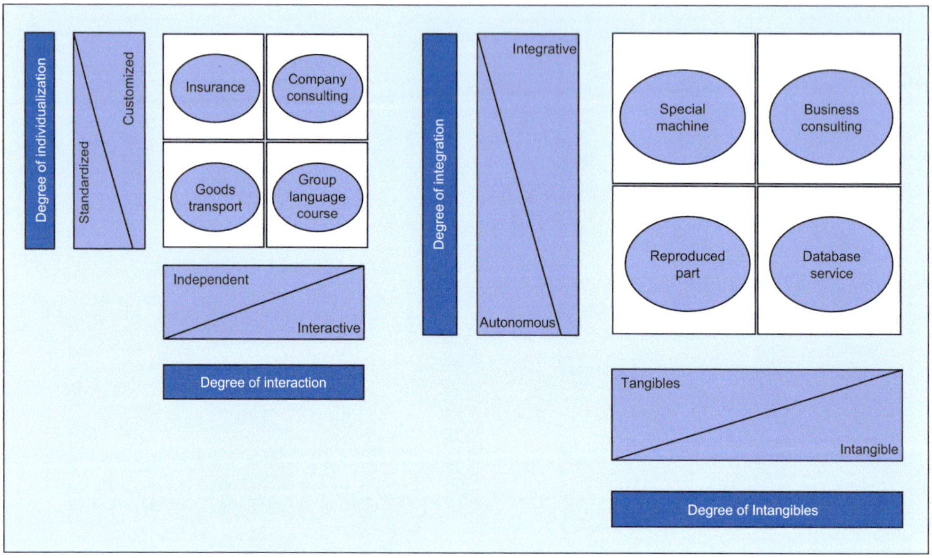

Fig. 2.3 Typology of sales objects according to Engelhardt et al. and their extension (Source: Engelhardt et al., 1992, p. 35; Meffert, 1993, p. 12)

oriented towards consumption to true interaction, in which the customer is an essential part of the value creation process. The degree of individualization describes in this context—independently of the integration of the external factor—the extent to which the created value is specialized.

However, the proposal by Engelhardt et al. (1992) to dispense with the distinction between goods and services does not appear to be sensible. The classical division rather represents the two extreme manifestations of a continuum, within which a clear assignment for individual sales objects is only possible to a limited extent (Meffert et al., 2018, 18 pp.). In order to use the marketing and quality management instruments appropriately for the different forms of manifestation of services, the previously presented systematization approaches are not yet sufficient. In order to come closer to the goal of identifying service segments that are as homogeneous as possible in themselves, it is necessary to examine further distinguishing features of services for their suitability for typologization.

For example, services are often also classified according to the type of use in **consumptive and investment services** as part of **one-dimensional systematization approaches**. For example, tourism services usually reach the consumer as the final user (consumptive service), while training events for employees of a company are increasingly characterized as investment services in the sense of a corporate investment. The type of use often leads—as with goods—to the same service (e.g. mobile telephony) being assigned to the consumptive or investment sector depending on the buyer or use situation.

The mutual assignment possibilities of similar services can also be seen in the distinction between **commercialand non-commercialservices.** For example, psychological counseling at the counseling center of a university has a non-commercial character, while a similar conversation with a practicing psychotherapist has a commercial character.

What **implications for service quality management** already result from a one-dimensional systematization can be seen using the example of consumptive or investment services. If consumptive services are characterized by a discrete time horizon (the service process is completed after a certain time), investment services often have a stronger binding character. The decision for an investment service therefore usually stands for a long-term customer relationship in which the focus is on ensuring a continuously high quality of service for the same customer. The quality experience is influenced to a much greater extent by a multitude of personal experiences than is the case when using time-limited consumptive services.

More informative for further consideration are the **two-dimensional systematization approaches.** For example, Wirtz and Lovelock distinguish between the character of services in tangible and intangible processes and between the recipient of the service in people or objects (Wirtz & Lovelock, 2016; see Fig. 2.4). According to this classification, four **types of service processes** can be distinguished:

1. Tangible processes in which the customer is the recipient of the service themselves,
2. Tangible processes in which an object of the customer is the recipient of the service,
3. Intangible processes in which the customer is the recipient of the service,
4. Intangible processes in which an object of the customer is the recipient of the service.

2.2 Approaches to the Term of Quality

The word **"quality"** has its origins in Latin ("qualis" = how procure) and describes, according to general linguistic understanding, the "nature", "grade" or "value" of an object. If this semantic analysis of the concept already indicates the general direction of the understanding, it is still not possible to derive any statement about which instance assesses the quality—the grade. And even beyond this question, the current quality discussion is characterized by misunderstandings and misinterpretations. The heterogeneous and diffuse understanding of "quality", which is often valid for both goods and services, makes it clear that it has not yet been possible to create a sustainable and generally accepted understanding of quality. The explanations for the understanding of quality range from colloquial word definitions to very abstract definitions of the quality concept for practice.

Within the international professional language of quality management, an understanding of the concept of quality has been established, which defines **quality** as "… the

What character has the service process?	Who or what is the direct recipient of the service?	
	Person	Object
Touchable (tangible)	Services directed to the human body: ▪ Health care ▪ Beauty salons ▪ Restaurants ↓ Physical presence of the customer required	Services that are applied to goods or other physical possessions are directed: ▪ Freight and transportation ▪ Repair or maintenance service ▪ Cleaning companies ▪ Waste incineration companies ↓ Physical presence of the customer not required
Untouchable (intangible)	Services directed to the intellect of man: ▪ Training ▪ Radio and TV ▪ Information services ▪ Theater ↓ Mental presence of the customer required	Services directed to untouchable assets: ▪ Banking ▪ Taxation ▪ Insurance ▪ Legal services ↓ Mental presence of the customer required only temporarily

Fig. 2.4 Character of the service process (Source: according to Wirtz & Lovelock, 2016, p. 23)

totality of characteristics (and characteristic values) of a unit with regard to its suitability to meet specified and presumed requirements" (DIN EN ISO 8402, 1995; Zollondz, 2011, 161 pp.). The concept of quality was newly formulated in the 2014 valid conceptual standard for quality management ISO 9000: 2014 and is now defined as the "… degree to which a set of inherent characteristics meets requirements" (DIN EN ISO 9000, 2014; likewise DIN EN ISO 9000, 2015). "Quality" thus means, in other words, the "realized quality of a unit with regard to the quality requirement" (cf. the discussion in Zollondz, 2011, 161 pp.).

"Quality requirement" stands for this understanding for the "totality of the individual requirements considered for the quality of a unit in the considered concretization level of the individual requirements". "Quality" describes the "totality of the characteristics and characteristic values of a unit" (Geiger, 2001, p. 802).

When considering the quality of goods, particular emphasis is placed on a clear and measurable acceptance criterion for the individual quality characteristics (Herrmann & Fritz, 2016, p. 42). This will be difficult in individual cases when considering the quality of services (e.g. characteristics such as friendliness and politeness of the contact person).

The following primarily uses the term "requirement" and focuses on the customer perspective.

The term **"Unit"** stands for products or activities and is thus often also used for services. However, this definition does not yet concretize from which institution or which group of persons (customer, employee, company) the **standards for quality assessment** are to be set (Bruhn, 1999a).

This perspective of the German Society for Quality thus combines two central **approaches to quality definition** (Bruhn, 1999a):

(1) Product-related concept of quality According to this, quality of services is often understood as the sum or level of the existing properties. This—narrow—understanding puts the consideration of objective—especially in the service sector hardly observable—criteria in the focus ("product-based").

(2) Customer-related concept of quality The quality consideration from the customer perspective ("user-based") is focused on the perception of the product properties or services by the customer. In the end, not only the objectively existing quality criteria determine the quality position of a service for a customer. The positioning takes place rather against the background of a subjective judgment about the properties he considers important (Bruhn & Hennig, 1993, pp. 216–217; Meffert et al., 2018). Problems arise for companies above all when the quality differs greatly from the objective or company-related and subjective or customer-related perspective (cf. Fig. 2.5).

Garvin expands the two basic perspectives of product and customer orientation by three **concepts** (Garvin, 1984, pp. 25–26).

This is how the **absolute quality concept** leads to an assessment that defines quality as the measure of the goodness of a performance that is often categorized by different classes (e.g. "good", "average", "bad"). This understanding is very close to the common understanding of quality.

Objective view / Subjective view	Quality objectively poor	Quality objectively good
Quality Subjectively bad	Disaster	Communication problem
Quality subjectively good	Time bomb	Ideal state

Fig. 2.5 Evaluation of a performance from an objective and subjective perspective (Source: Krämer & Maurer, 1998, p. 147)

The **manufacturing-oriented quality concept** starts from a definition of quality standards for service creation and sees these as a measure of quality control. Whether these are subjective or objective standards has not yet been determined.

From a customer perspective, a **value-oriented quality concept** can be defined that sees service quality as the result of an assessment process by the customer from the price-performance perspective. This decides whether a certain performance is "worth" its price and thus whether a specific level of service quality can be achieved in the performance creation (Garvin, 1984, 1988; Bruhn, 1999a, pp. 24–26; Meffert et al., 2018).

Despite the required emphasis on the customer perspective, the **definition of requirements for service quality** must of course not be one-dimensional or one-sided from the customer's point of view. The quality of a service is rather located in a field of tension that results from the perspective of the customers, the competitors and the own company (Bruhn, 1999a, pp. 29–31; see also Fig. 2.6).

The **requirements from the customer's perspective** must be the central criterion for determining the quality of the service for a service company. The expectations of current and potential customers are the potential (e.g. equipment of a tour bus), the process (e.g. pleasant driving style of the driver) and the result (e.g. arrival at the destination) of the service creation. Determinants of expectations are individual and situation-related demands on a problem solution as well as experiences with the service and the service provider in the past, word-of-mouth communication with the provider or other

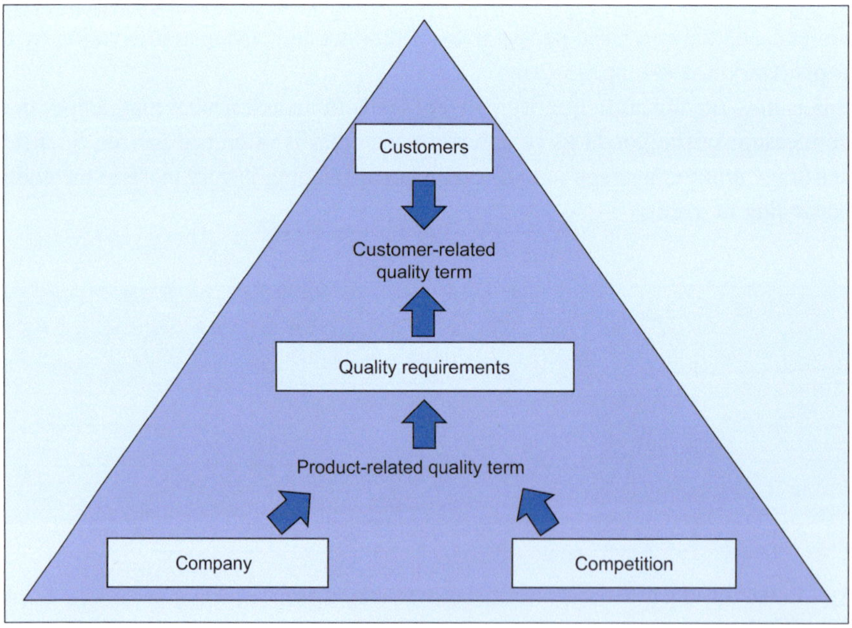

Fig. 2.6 Approaches for the definition of service quality

customers, and the external corporate communication of the provider (Parasuraman et al., 1985, pp. 44–46; Bruhn, 1999a, p. 30; Meffert et al., 2018).

In terms of **competitive requirements**, the question of profiling a service provider against (main) competitors takes center stage. The understanding of service quality in this case corresponds to the search for competitive advantages over competing providers, whose realized quality is to be fixed as the minimum quality of their own services in order to derive a comprehensive, competition-oriented quality strategy (Heskett, 1986, pp. 48–50).

From a **corporate perspective**, the requirements for service quality are finally determined by the ability and/or willingness of a service provider to ensure a certain level of service quality. The basis for these requirements are, for example, the importance of the quality factor in the company/marketing strategy, the performance and willingness of the employees, or the statements that are conveyed to the customers in the context of corporate communication about the service result or the service process.

Irrespective of these perspectives, the fundamental question of **operationalizing requirements** arises, which is necessary in order to achieve the further considerations for implementing the requirements in a targeted manner. There are proposals for formulating quality criteria for formulating requirements, i.e. "requirements for the requirements". These include uniqueness, relevance, implementability, identifiability, indivisibility, closure, verifiability, comprehensibility and traceability (Schmitt & Pfeifer, 2015, p. 148). These criteria will be easier to realize for tangible assets because they often involve technical and therefore unambiguous characteristics. With services, this will only be possible to a limited extent in this form, especially in those cases where the service has a high degree of individuality, interaction, intangibility and integration.

There are numerous methods available for **measuring requirements**. These concern both procedures that make a measurement from the customer's perspective (customer-oriented measurement approaches) and procedures that focus on a measurement from the company's and/or employees' perspective (company-oriented measurement approaches). These are presented in Chap. 5 of this book. An overview is given in Fig. 5.1.

2.3 Term of Service Quality

Due to the special features of services, a wide **interpretation of the quality term from the customer's perspective** under due consideration of the competitive and corporate perspective leads the furthest. So although services often meet the basic requirements placed on them—product-related—they are often not of good quality from the customer's perspective if they do not meet other requirements. Example: A flight that is technically perfect is often of insufficient quality—and therefore completely worthless—for a business traveller if he or she cannot keep to a scheduled appointment due to a delay.

Building on a combination of the different quality term and against the background of the characteristic or constitutive particularities of services, **service quality** is defined as follows (Bruhn, 1999a, p. 29):

▶ **Service quality** is the ability of a provider to create the quality of a primary intangible and customer-involved service in accordance with customer expectations at a certain level of requirements. It is determined by the sum of the properties or characteristics of the service, to meet certain requirements.

According to this definition, service quality is the quality of a service that represents a certain level of performance (excellent to extremely poor). The expectations of the performance level are set from the perspective of the recipient of the service—the customer. For this reason, an absolute quality concept in the service sector is often not satisfactory, it does not meet the subjective (and thus relative) claims—the "individual requirements for the quality of a unit".

For the **measurement of service quality** the combination of product and customer-oriented quality understanding is helpful. The results achieved here are the starting point for measures to ensure quality, which have their organizational background in an expanded quality management system. **Quality management** is understood as "a management system for leading and directing an organization with regard to quality" (ISO 9000:2014). In companies with a strong organizational anchoring of quality management, various stages in the quality management process are often distinguished, which can include quality planning, quality steering, quality control and quality management declaration.

2.4 Determinants of Service Quality

According to the definition of service quality, it arises in the perception of the customers by the comparison of the perceived performance and the customer expectations in relation to the performance (Parasuraman et al., 1988). Accordingly, on the definition level two central **influencing factors of the service quality** can be identified (cf. Fig. 2.7):

1. Delivered and perceived service,
2. Expectations of the service.

In the context of contacts, in which the internal factors, i.e. the performance potentials of the service provider (e.g. employees, service technologies), and the external factors, i.e. the customer himself or his disposal objects (e.g. his car during the car repair or his assets during the money investment), meet each other, the service is created. The performance of a dentist is created by the service provider, i.e. the dentist, performing

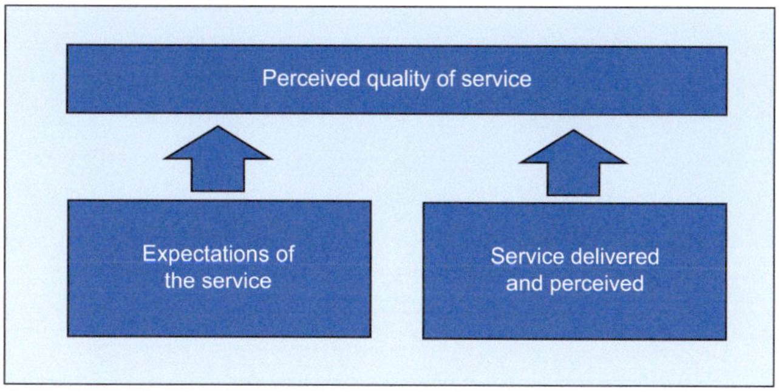

Fig. 2.7 Determinants of the perceived service quality

operations on the external factor, in this case the customer, which create benefits for the customer. This creation process, i.e. the service, is individually perceived by the customer. Positive events in the context of the service creation lead to a positive perception of the service, the customer feels that he has received a "good" service. Vice versa, if negative experiences occur during the contact with the provider in the context of the service creation. Consequently, the **delivered service** or the **perception of the service** by the customer directly affects the perceived service quality.

However, studies on service quality (see, for example, Parasuraman et al., 1988; Boulding et al., 1993; Brady & Cronin, 2001; Kettinger & Lee, 2007) show that the customer's perception process with regard to service quality—that is, not the service itself—is more complex. The same service is often perceived differently by two customers. For example, a student who usually does not stay in hotels perceives the services of a 4-star hotel differently than a CEO who usually lives in 5-star hotels. The key to understanding this complex perception process lies in the second determinant of service quality in addition to the service itself: the **customer expectations.**

In general, expectations of an individual are seen as a **psychological state** that refers to future behavioral consequences for the individual. With regard to the services of a company, customer expectations refer to a psychological state that affects the future use of the services of a service provider. When concretizing the concept of customer expectations, it is already clear from a general consideration of everyday language that the definition of customer expectations is more complex. If the supervisor demands from his employee "I expect the report by Wednesday!", a different form of expectation is addressed than with the prediction of the supervisor "I do not expect him to have the report ready by Wednesday."

Following this differentiation of the concept of expectation in general language usage, two **types of customer expectations** are generally distinguished (Liljander, 1994; Ngobo, 1997, 63 pp.; Richter, 2005, p. 17; see Fig. 2.8):

Authors[1]	Definition	Expectation type	Belonging to the basic types	Type of classification on a performance level scale		Reference object: Cross-company	Reference object: Company-related
Miller 1977; Tse and Wilton1988; Hamer1996, Sudharshanet al. 1996.	Imagination of a performance not to be surpassed	Ideal level	Normative	Given	High	X	
Parasuraman et al. 1988; Boulding et al. 1993;	Performance level that the customer wants or that the provider should deliver	Desired Level			↑		X
Cadotte et al. 1987	Perception of the quality of the services of the best provider in a category from the customer's point of view	Best Brand Level				X	
Cadotte et al. 1987	Idea of the typical or average quality of all services known to the customer in a category	Product-Type-Level				X	
Zeithaml et al. 1993	Performance level that the customer just accepts	Appropriate level			↓	X	
Miller 1977; Hamer 1996	Performance level that is just acceptable to the customer	Minimum tolerable level			Low	X	
Fisk and Young 1985; Oliver and Swan 1989.	Idea of the level required by the customer, taking into account his own effort and the given possibilities	Earned Level		Not given			X
Lin et al. 2009	Performance level that the customer expects from a particular provider	Expected Level	Predictive				X
Swan and Trawick 1980a; Oliver 1981; Prakash 1984; Spreng and Dixon 1992; Boulding et al, 1993; Ho us Zheng 2004.	Perception of the quality of the services of a particular provider	Foreseen Level					X
Miller 1977	Probability of occurrence of an event	Probable Level					X

[1] This overview lists authors who have introduced the respective expectation types in the literature or strongly advocated their use. Overviews of expectation types can also be found in e.g. Liljander and Strandvik 1993; Oliver, 1996; Ngobo 1997; Hamer 2006

Fig. 2.8 Overview of expectation types (Source: Based on Bruhn & Georgi, 2000, p. 188)

1. **Predictive expectations** have an anticipatory character in that the customer expresses them what level of performance he or she expects or considers probable before using the service (cf. Oliver, 1980, p. 460; Cadotte et al., 1987, p. 305; Tse & Wilton, 1988, p. 205; Oliver, 2010).
2. In contrast, **normative expectations** represent a customer's demand on the service provider and designate the level of performance that the customer requires from the company (cf. Parasuraman et al., 1988; Teas, 1993, p. 19; Richter, 2005, p. 17; Bruhn et al., 2006b).

Customer expectations are the central variable in assessing the quality of services (Parasuraman et al., 1988; Johnston, 1995; Spreng et al., 1996; Siddiqi, 2011). With regard to the direction, strength and directness of the influence, both different **theoretical explanation approaches** and different **empirical results** exist (cf. for example Gupta & Stewart, 1996; Bruhn & Georgi, 2000; Georgi, 2000; Richter, 2005; Shi et al., 2014). These can be summarized for the two **types of expectations** as follows (cf. Fig. 2.9):

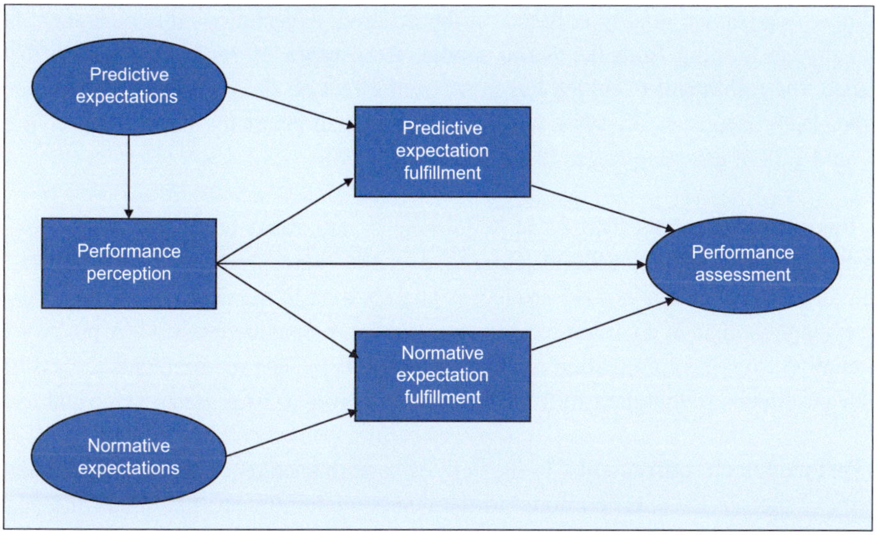

Fig. 2.9 Influence of customer expectations on performance assessment (Source: Georgi, 2000, p. 58; Richter, 2005, p. 48)

1. **Predictive expectations** have a positive effect on perceived service quality. Predictive expectations correspond to a customer's assessment of a future service to be received. If this assessment is not confirmed, the customer is wrong. However, the customer tries to avoid a mistake unconsciously. Accordingly, he adjusts the actually received service in his perception to his predictive expectations. Thus, high predictive expectations tend to lead to a more positive perception of quality.
2. **Normative expectations** are demands on the service provider. If the provider's performance does not meet these requirements, the customer "punishes" the provider in his perception with a lower quality perception. Normative expectations and perceived service quality are negatively related. The higher the customer's normative expectations are, the more difficult it is for the provider to meet them and to deliver a high quality to the customer accordingly.

Customer expectations also play a role in determining the **optimal quality level**. From an efficiency perspective, a strategy of quality optimization is to be pursued. Optimization is to be understood as the most exact alignment of the quality standards of the company to the customer expectations. If the customers can no longer perceive or assess the company's special quality position, exceeding customer expectations is counterproductive from a cost and income perspective. Thus, from a certain, individually determinable quality level, improvements in service quality lead to disproportionately high cost increases. However, a particularly high quality standard is only partially honored by the customer through increased price willingness.

Improving service quality is based on the different expectations that customers have of a service. Starting from the **Kano model**, three **types of requirements** are distinguished, the fulfillment of which has a different effect on the perceived service quality (Kano, 1984; Berger et al., 1993; see also the explanations of the Kano method in Fig. 5.27 and 5.28 of this chapter), as illustrated in Fig. 2.10:

(1) Basic requirements The basic requirements are **must-haves** of a service, the non-fulfillment of which leads to a negative quality perception by the customers. So today in many industries a good product is already a basic requirement of the customer (for example, a loan at a certain interest rate and fixed repayment terms). A positive differentiation from the competition is hardly possible here. The requirements are assumed by the customers, so that their fulfillment is not perceived as increased service quality.

(2) Performance requirements In the case of performance requirements, the perceived quality of the service is proportional to the degree of fulfillment: the higher the degree of fulfillment, the higher the perceived quality of the service. In contrast to the basic requirements, which are assumed to be self-evident by the customers and therefore not explicitly demanded, performance requirements are clearly articulated by the customers (e.g. competent advice in a bank). One also speaks of **should-have criteria** of a service.

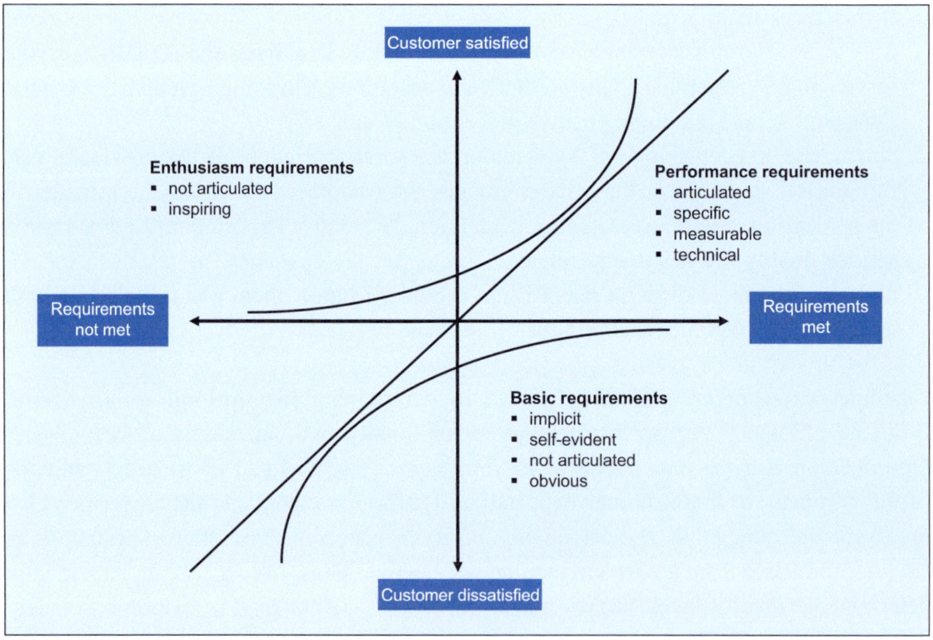

Fig. 2.10 Categorization of quality features according to the Kano model (Source: Kano, 1984, p. 41; Berger et al., 1993)

(3) Enthusiasm requirements Under enthusiasm requirements, those performance criteria are included, the fulfillment of which leads to an disproportionately high perception of the quality of the service. If these requirements are not met, this does not result in a lower quality perception. These so-called **can-have criteria** of a service are not explicitly formulated and not expected by the customers. Examples can be found here above all in an outstanding customer interaction (e.g. neutral comparison with credit offers of other banks by the consultant).

From this classification of the requirements for a service, **priorities for the design of the service offering** are derived in the sense of a quality optimization. Due to the increasing homogenization of the service offering, the competition has shifted to the level of enthusiasm requirements in recent years. Companies try to use the potential of this category of requirements in order to differentiate themselves from their competitors and to achieve a high level of satisfaction among their customers. The over-fulfillment of these requirements often leads to the perception of a high quality of service on the one hand, but on the other hand also to high costs. Another problem is that these requirements often do not have a long-term character. On the one hand, possible criteria are imitated very quickly by competitors and on the other hand the fulfillment of such enthusiasm requirements continuously leads to higher expectations among consumers, so that in the long term enthusiasm requirements become basic requirements **("expectation spiral").**

In **focusing on enthusiasm requirements**, the basic requirements are often neglected and the dissatisfaction potential if they are not met is overlooked. This often leads to customers being dissatisfied despite the fulfillment of enthusiasm requirements, as the must-have criteria remain unfulfilled. In the context of optimizing service quality, it is therefore necessary to ensure the fulfillment of basic requirements. Basic requirements form the foundation of quality perception, where the necessary quality level must be ensured. While enthusiasm requirements can be rationalized from a cost point of view, such considerations are not suitable for basic requirements.

Finally, it is important to create **awareness of the individual levels of requirements** among employees. An employee has to identify the expectations as basic, performance and/or enthusiasm requirements individually in the service creation process and to align the quality standards of the company as precisely as possible to the individual customer needs. This approach is illustrated in the following example of the "importance of quality features of the Arlberg region".

Importance of quality features of the Arlberg region in Austria

The aim of a study in St. Anton and in general in the Arlberg region was to identify differences in the importance of the service provider "Arlberg region" from the perspective of guests of the area (Matzler et al., 2001).

For this purpose, guests were interviewed during the Ski World Championship 2001 in St. Anton with regard to their

- overall satisfaction on a 5-point scale,
- satisfaction with 18 individual features on a 5-point scale,
- importance of the 18 individual features by indicating the 5 most important features by each respondent.

In addition to the direct query of importance (explicit importance), the implicit importance of the individual features has also been determined by means of a regression analysis. This was determined by measuring the influence of the individual features on overall satisfaction. This made it possible to obtain explicit and implicit importance values, with which the 18 quality features were prioritized according to their priority. The characteristics were assigned to four groups (cf. Fig. 2.11):

1. **Motivating factors** (low explicit importance, high implicit importance): Service competence of the staff (event), family and child friendliness, supporting program for the World Cup.
2. **Important performance factors** (high explicit importance, high implicit importance): Shuttle services to events, organization of the event, atmosphere at the events, service quality of the staff (event).

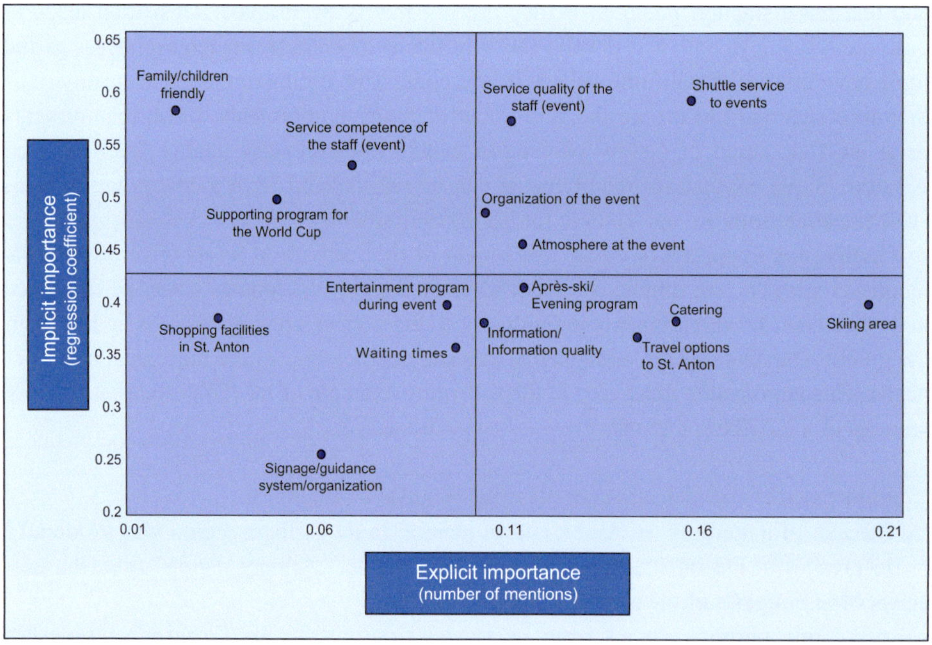

Fig. 2.11 Prioritization of quality characteristics by guests of the Arlberg region (Source: Matzler et al., 2001, p. 463)

3. **Unimportant performance factors** (low explicit importance, low implicit importance): Signage/guidance system/orientation aids, waiting times, shopping opportunities in St. Anton.
4. **Basic factors** (high explicit importance, low implicit importance): Arrival possibilities, information/information quality, catering, entertainment program during the event, skiing area, après-ski/evening program. ◀

The different conceptualizations of quality make it clear that the first step on the way to a quality management system is to determine the relevant dimensions of service quality. The different, internally and externally by company target groups perceived quality characteristics are to be understood as quality dimensions.

2.5 Dimensions of Service Quality

Both the perceptions of the customers in terms of the service and the expectations of the customers rarely relate to the entire service in general. Rather, customers perceive the service differently and form different expectations accordingly. This differentiation affects individual quality features (e.g. friendliness of the employees, speed of service). Since there are a large number of such individual features for a service and the isolated consideration of these individual features is not very productive, the features are summarized into so-called **dimensions of service quality**. This makes the quality features graspable as a whole. In the literature on service quality, there are numerous approaches to dimensioning service quality.

First, three dimensions are distinguished with regard to the **process phases of service creation** (Donabedian, 1980):

(1) Potential dimension Here, the perception of the structures and potentials of the service provider is in the focus. In the case of an insurance company, this dimension includes, in addition to size and market position, for example, the number and qualification of customer contact employees (insurance representatives).

(2) Process dimension This refers to the assessment of the processes during service creation. These include the quality of consultation, e.g. the support in the case of a car damage or also the course of administrative processes.

(3) Result dimension This perspective places the assessment of the services rendered or the result of the service process in the focus. Thus, the successful regulation of a damage is often judged on the basis of the extent of the actual service and with regard to the period of time until the regulation is complete.

When concentrating on the **type and extend of the service created**, two quality dimensions can be distinguished in a similar way (Grönroos, 1983, 1984):

(1) Technical dimension The focus of this dimension is the range of the service program (what is offered?). In addition to the scope of the insurable object areas, this also includes the possibilities for individualization with regard to a single contract (coverage, notice periods, term, etc.).

(2) Functional dimension The focus here is on the way the service is provided (how is the service offered?). For example, the professional competence of the field service employees or the politeness of the other customer contact employees are decisive for this quality dimension.

Another subdivision of quality dimensions relates to the **expectations of the customers** with regard to the service program. Two components can be distinguished here (Berry, 1986):

(1) Routine component This includes properties of services that are part of the "normal" scope of services (parking spaces in front of the insurance offices, information material, etc.). Negative deviations from this routine are often sanctioned by the customer with "penalty points".

(2) Exception component This includes additional services that the customer does not necessarily expect from the service provider (offers from special insurance, 24-hour emergency service, etc.). Customers may reward such offers with "bonus points".

Quality dimensions are often distinguished according to the **proximity of the customer to the service product** when assessing the services. Accordingly, three dimensions can be distinguished (Zeithaml, 1981):

(1) Search Elements Customers who have not yet had any experience with the service provider will look for indicators for assessment in advance (e.g. premium amount, scope of services). The potential dimension of services is of particular importance here: Due to the intangibility of services, customers base their quality assessment primarily on tangible service components that they can assess before using the service for the first time ("tangibles"), such as the appearance of a car workshop and that of the staff (see, for example, Steffen, 2006).

(2) Experience Elements If there are any experiences, assessments are often made during or at the end of the service creation process (e.g. qualification of the employees, efficiency of the administration).

(3) Credence Elements This includes all those features of a service that elude precise assessment or can only be assessed with a time delay (e.g. behaviour of the company in the event of damage, result of a medical diagnosis).

In addition, there is a classification into quality dimensions, which was not only developed conceptually, but also empirically by means of the **SERVQUAL approach** (cf. also Chap. 4 and 5). It is often interpreted as the summary of all the dimensions sketched above. Accordingly, five quality dimensions are perceived independently (Parasuraman et al., 1986, 1988; Zeithaml et al., 1992; Meinhold & Matul, 2011):

(1) Convenience of the tangible environment (Tangibles) This includes the external appearance of the service location, in particular the equipment of the rooms and the appearance of the staff (e.g. building of the insurance company, appearance of the field service employees).

(2) Reliability This dimension describes the ability of the service company to fulfil the promised services at the announced level (e.g. informative value of the consultation conversations, conclusion of a policy with an appropriate scope of protection).

(3) Responsiveness This refers to the ability of the service provider to respond to the specific needs and wishes of the customer and to be able to fulfill them. This dimension includes both the general willingness and the speed of reaction (e.g. the type and scope of the special insurance policies offered).

(4) Competence (Assurance) The competence aspect focuses on the company's basic ability to create the service under consideration, in particular the knowledge, politeness and trustworthiness of the employees (e.g. the qualifications of the customer contact employees).

(5) Empathy Empathy finally includes the willingness and ability of the company to respond to individual customer wishes or the specific situation (e.g. individual appointments, specific consulting services, behavior in case of death).

When summarizing the quality dimensions of Parasuraman/Zeithaml/Berry, three underlying dimensions can be seen (Bruhn, 1999b, p. 544):

1. **Objective quality dimensions** (e.g. punctuality, reliability, accuracy, completeness of the service).
2. **Personal quality dimensions** (e.g. openness, honesty, friendliness of the customer contact employees involved).
3. **Interpersonal quality dimensions** (e.g. courtesy, flexibility, empathy, fairness of the customer contact person in dealing with customers).

Even though the SERVQUAL dimensions found a wide echo and application for numerous industries due to their simplicity and general plausibility, they still have some **problems** (cf. Sect. 5.1.2.1). The SERVQUAL dimensions partly describe manifestations of quality dimensions or -features and not these themselves (Brady & Cronin, 2001, p. 36).

For example, reliability is not to be considered as a separate quality dimension because it affects numerous quality dimensions. Thus, for example, at a bank either the employees or the ATM can be reliable. In addition, there is agreement that service quality is a hierarchical construct that includes dimensions not only on one, but on several levels (Carman, 1990; Furrer et al., 2000).

Based on these considerations, a **hierarchical dimensioning of service quality** was developed theoretically and extensively empirically—in accordance with the procedure recommended by Churchill (1979) and, for example, applied in SERVQUAL (Brady & Cronin, 2001). Service quality is conceptualized and operationalized as a complex construct or construct of higher order (for a detailed presentation of the procedure of construct research, see Bruhn, 2016b, 74 pp.). In the understanding underlying this, it is a **three-dimensional, multi-factorial construct of third order.** A multi-factorial construct implies that the construct is not directly measurable by manifest variables, but is derived from several so-called latent variables (Homburg & Giering, 1996). This generally requires an iterative, step-by-step conceptualization of service quality from the highest level (construct level) to the lowest conceptual abstraction level (factor level).

The **construct level** is therefore the highest abstraction level at which service quality is concretized as a higher-order construct of third order. On the second level, the so-called **dimension level,** the dimensional structure of a multi-factorial construct is generally conceptualized. It is a matter of determining the content of the dimensions defining service quality. These are the three dimensions potential quality, process quality and result quality in the concrete example (see Fig. 2.12). This expands Grönroos' (1983, 1984) dimensions to the effect that functional service quality is divided into potential and process quality.

On the lowest abstraction level, the so-called **factor level**, the relevant factors (constructs of first order) of service quality are to be generally identified and designed in terms of content. In the present model, three factors are assigned to each of the three dimensions, which allow specific recommendations for quality management. These factors are then concretized again by specific items. As a criticism of this approach, it must be said that in this case the potential, process and result quality are rather considered as influencing factors than as dimensions of the service quality.

The totality of all quality dimensions finally documents the desire to gain deeper insights into the need for services. Only the precise **specification of the requirements for a service from the perspective of the consumer** finally allows the targeted use of marketing and quality instruments for the service provider (Bruhn, 1999a, p. 27). The quality dimensions have a special meaning in the context of measuring service quality. So the differentiation of service quality into different quality dimensions provides a first approach to the design of measuring and analytical instruments for the capture of service quality. For this purpose, it requires a concretization of the individual dimensions by further characteristics of the service quality (Stauss & Hentschel, 1991, p. 240). In this context, it is necessary to determine both the selection and the weighting of the individual characteristics under industry-specific aspects.

In order to ensure the quality of the service, it is essential for the management to gain knowledge of how the perception and expectation formation of customers takes place in

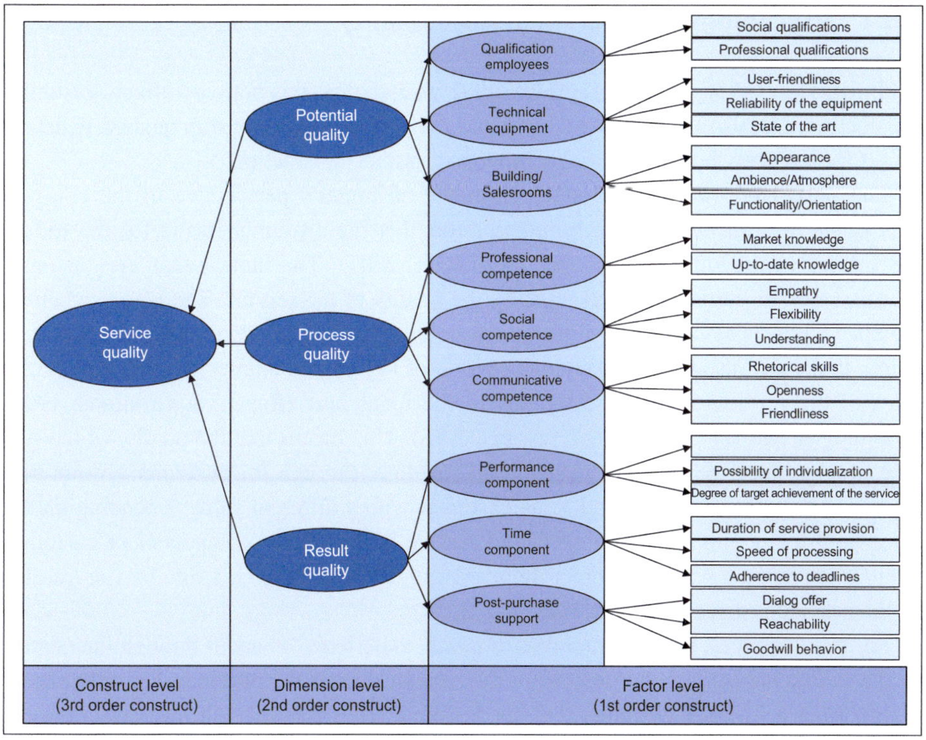

Fig. 2.12 Hierarchical dimensioning of service quality (Source: Based on Brady & Cronin, 2001, p. 37)

relation to their own service offering. Customer studies not only allow the contents and level of the expected service to be determined, but also those quality dimensions or -features to be specified, along which the quality judgment manifests itself with the customers receiving the service. A comprehensive quality management is to be set up to do so.

2.6 Effects of Service Quality

The effects of service quality can be structured—starting from the **success chain of service quality** (see Sect. 1.2). In the introductory presentation of the success chain, customer satisfaction, customer loyalty and economic success were identified as effects of service quality. In general, three **categories of effects of service quality** are distinguished:

1. Psychological effects of service quality (e.g. customer satisfaction),
2. Behavioral effects (e.g. customer loyalty),
3. Economic effects (e.g. economic success).

The different effects of service quality are explained in more detail below.

2.6.1 Psychological Effects of Service Quality

With regard to the psychological effects of service quality, this not only affects customer satisfaction, but also the perceived value and the perceived relationship quality, which in turn, together with the service quality, affects customer satisfaction.

The **perceived value** corresponds from the customer's perspective to the comparison of the benefit of a service with the effort that the customer incurs for the use of the service (Zeithaml, 1988, 13 pp.; Tuli et al., 2007). The benefit that the customer derives from the service is represented by the quality of the service. The perceived effort includes all costs associated with the purchase of a service. In addition to search, acquisition, transport, installation, maintenance, risk of failure and quality of failure costs, the price of the services used is particularly in the focus here (Ravald & Grönroos, 1996; Leischnig & Messer, 2014, p. 257; see Fig. 2.13). This means that the quality of the service is relativized by the price paid by the customer through the customer's value perception. The same service provides the customer with a different value depending on the price paid for it. Since the quality of the service is a essential component of customer value when using a service, a direct **determining of the perceived value by the quality of the service** is given (Bruhn & Hadwich, 2014).

For most services, the customer does not only experience a benefit through the quality of the individual service, but also from the relationship to the provider. For example, it is important for a bank customer that his customer advisor points out investment opportunities to him that fit the customer's portfolio. This means that the bank advisor may not only see his task in the individual transaction, i.e. the creation of a certain service, but also keep the entire customer relationship in mind in order to serve the customer in a holistic way. Against this background, the **perceived relationship quality** is defined as the cross-transactional assessment of the company's ability to shape the relationship between company and customer in the past and in the future according to the customer's requirements for the relationship (Hadwich, 2003; Beatson et al., 2008).

For a closer description of this construct, two **dimensions of relationship quality** are differentiated (Georgi, 2000; Hadwich, 2003):

1. The customer's trust in the company,
2. Familiarity between customer and company.

The **trust** of the customer represents a future-oriented component of relationship quality. Trust is defined as the customer's willingness to rely on the company with regard to its future behavior without further examination. Indicators of a trusting relationship from the customer's point of view are in particular the perceived fairness of the provider, the well-being in dealing with the provider and the perceived trustworthiness. A certain vulnerability must exist as a prerequisite for the formation of trust, i.e. decision consequences must be both uncertain and important for the trusting person (Doney & Cannon,

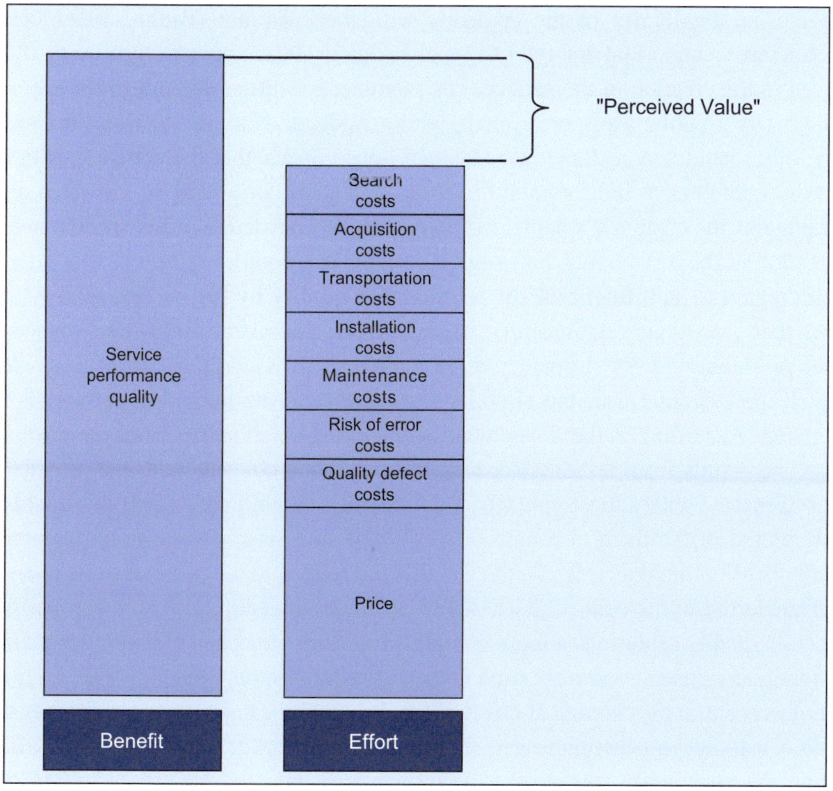

Fig. 2.13 Perceived value and service quality (Source: Based on Zeithaml, 1988, 13 pp.)

1997; Sirdeshmukh et al., 2002; Lacey, 2007; Caruana & Ramaseshan, 2015). In particular, if it is a question of so-called unmodified repurchases, but of modified repurchases, trust comes into play. This is particularly the case with individual services (e.g. project of a management consultancy for a topic not yet dealt with by a customer). The modified elements increase the degree of uncertainty and thus the perceived risk in the purchase decision.

In addition to trust, the **familiarity** of the customer with the company represents a second dimension of relationship quality. Familiarity is closely related to trust and has a past-oriented character (Luhmann, 2014). Familiarity describes the degree of familiarity with an object (e.g. situation) or subject. With regard to a company-customer relationship, familiarity denotes the degree of familiarity with the respective partner or contact person in terms of their attitudes and behavior (Georgi, 2000; Hadwich, 2003; Bruhn et al., 2006a).

Due to the **mutual dependence of customer and provider** within a relationship, the customer's familiarity not only includes his familiarity with the company, but also the perceived familiarity of the company with the customer (Bruhn, 2016b, p. 95). Therefore, it is important for the customer to know the company's processes if he is involved in the creation of the service. For example, it is often difficult to find your way around a city's public transport system, while residents of a city "master" the system. On the other hand, the customer may be very aware of whether the company is familiar with him. Examples of indicators of the company's familiarity with the customer are the knowledge of the customer's name, but above all the knowledge of his specific needs in the creation of the service (e.g. non-smoking rooms in a hotel).

With regard to an **influenceof the relationship quality by the service quality**, it can be seen that a customer relationship consists of a large number of individual transactions (Bitner & Hubbert, 1994; Liljander & Strandvik, 1995; Georgi, 2000; Meldau, 2007), in which the customer uses the provider's services (e.g. the individual restaurant visits of a guest). Accordingly, the assessment of the entire relationship from the customer's perspective results from an assessment of the individual contacts with the provider and the services used within these contacts. Theoretically, the influence of service quality can be interpreted in the form of a halo effect: When assessing more abstract phenomena, individuals use more concrete phenomena that are related to the more abstract phenomenon. Transferred to the context discussed here, the relationship quality in comparison to the service quality represents a more complex, more abstract phenomenon, the quality of which more concrete services are used in the customer's perception.

Another, central psychological effect of service quality can be seen in **customer satisfaction**. Customer satisfaction is also closely related to service quality in terms of understanding the concept, as customer expectations are also used for the definition of the latter. Customer satisfaction is also defined as the match between customer requirements (customer wishes, customer expectations) and the perception of the service delivered by the company (Oliver, 2010, pp. 6–7). The differences to the perceived service quality become apparent when considering the two variables in more detail. While customer expectations regarding individual service features play a role in the assessment of these features when it comes to service quality, customer satisfaction refers to the degree to which customer expectations are met, as experienced by the customer as a whole. With this understanding of customer satisfaction, its **influence by service quality** is obvious. If a service provider is able to create a high quality in the eyes of the customers in terms of various individual features of a service, this suggests that the customers are satisfied with the services of the provider.

The so-called **(Dis-)Confirmation Paradigm** is often used to explain the emergence of customer satisfaction (Oliver, 2010, pp. 96–99). According to this paradigm, customer satisfaction is the result of a comparison process of the actually perceived performance after the purchase of a product or the use of a service with the expectations that a customer has formed before making a purchase decision regarding a product or service (cf. Fig. 2.14). According to the paradigm, customer satisfaction exists exactly when the customer expectations are met by the delivered performance (confirmation). If the actual

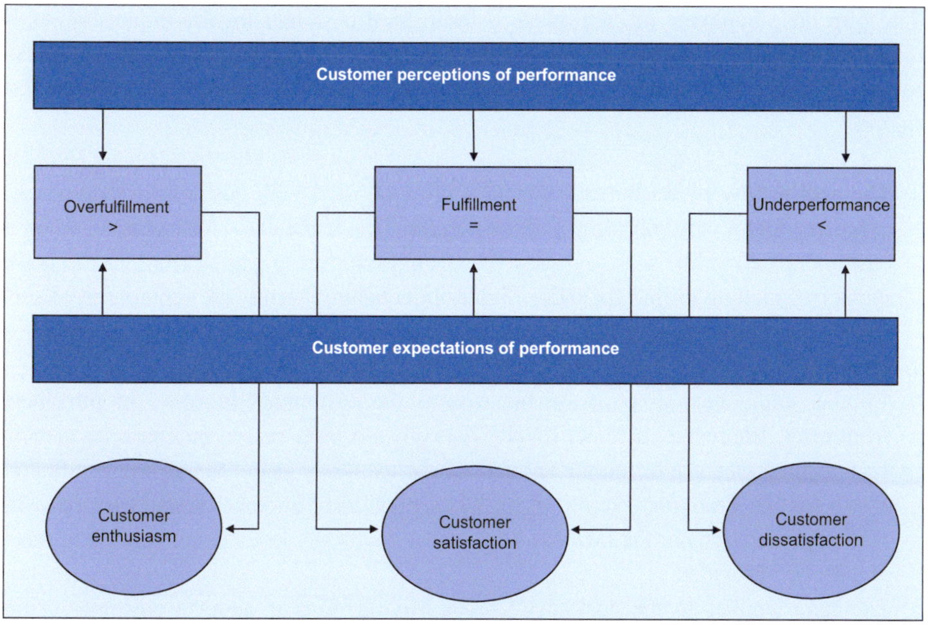

Fig. 2.14 Construct of customer satisfaction (Source: Bruhn, 2016a, p. 97)

performance deviates from the expected performance afterwards, the expectations are not confirmed (disconfirmation), which results in dissatisfaction. However, there are also approaches that do not speak of dissatisfaction in the case of an overfulfillment of customer expectations, but of enthusiasm or delight (cf. e.g. Oliver, 2010).

The (dis-)confirmation paradigm has the advantage that it is easily comprehensible and practically measurable in terms of customer satisfaction. However, it is associated with the disadvantage that it represents a static approach, which also simplifies the construct of customer satisfaction by distinguishing between satisfaction on the one hand and dissatisfaction on the other hand.

2.6.2 Behavioral Effects of Service Quality

The central consequence of service quality on the level of customer behavior is **customer loyalty.** In another version, both psychological aspects—in terms of loyalty—are subsumed under customer loyalty, while customer loyalty in the narrower sense refers to the actual behavior of the customer (Bruhn & Georgi, 1999). Two central **behaviors are counted towards customer loyalty:**

1. Purchase behavior,
2. Communication behavior.

The purchase behavior of customers is the central customer loyalty dimension, as it directly affects the economic success of the company. Three **types of purchase behavior** are distinguished, which have different effects on economic success, especially sales figures:

a) The **repurchase** by customers, which is often equated with customer loyalty itself, expresses that a customer uses a service again. This is the case, for example, when a customer of an airline who regularly flies from Frankfurt to Zurich continues to book the corresponding airline for these flights. In economic terms, the repurchase of services often contributes to the fact that a service company maintains its current sales level.
b) Another component of purchase behavior is the customer's **increase in purchase frequency**. Improved customer loyalty is expressed in increased purchase frequency, such as increased use of mobile services or the exclusive use of the airline mentioned above for the Frankfurt-Zurich route, when previously another airline was used. In contrast to repurchase, the increase in purchase frequency leads to an increase in sales of a service provider.
c) The same applies to the customer's **cross buying**. This is about a customer using additional services of a provider that he has not used before. Also in this case an increase in sales is achieved—provided that the services used so far are repurchased (Maitzen, 2016). Cross buying behavior is always relevant when a service provider offers a relatively wide range of products. An example of cross buying behavior is the credit agreement of a bank customer who has already had an account with the bank for years and has invested small amounts.
d) Finally, there is the customer's **tolerance to price increases**. This is a particularly strong form of customer loyalty. It expresses that the customer is immune to price increases to a certain extent. Customers with increased customer loyalty are more likely to accept price increases by the provider and do not react to price increases to a certain extent with a change of provider.

However, it is also necessary to take into account non-linear relationships here. An increased customer loyalty usually does not lead to the positive effects of customer loyalty described here to the same extent, since the relationships are moderated by various influencing factors, i.e. partly weakened or also strengthened (cf. Fig. 1.3).

While a customer's buying behavior directly affects sales to this customer, the customer's **communication behaviour** has an indirect effect on economic success. This is less about communication with the service provider, but rather about word-of-mouth communication with other potential customers. In particular, particularly satisfied customers report positively about a provider, while dissatisfied customers—to a greater extent—speak negatively about the provider. Positive word-of-mouth communication implies a positive attitude of the consumer, so that he is not only bound by switching barriers, but also actually connected to the company (Homburg et al., 2017). Customers who actively recommend the company to friends, acquaintances and colleagues are particularly loyal customers (Bruhn, 2016b, p. 99). Especially in the service sector, this word-of-mouth communication has a significant impact on the company's success. Since a service is usually not able to be judged in advance due to its intangibility, recommendations from third parties are often used as a quality indicator. Studies show, for example, that word-of-mouth communication has a significant impact on consumers' switching behaviour (Wangenheim & Bayón, 2004).

Success chain of an IT service provider

The usefulness of the success chain of service quality and thus the dimensions and effects of service quality can be well illustrated using the example of an IT service provider active in Switzerland and Germany. Although the IT service provider did not carry out systematic customer surveys, it did learn of individual customers' general dissatisfaction through individual customer conversations. It was therefore decided to carry out systematic satisfaction surveys and to investigate the reasons for customer dissatisfaction in a representative manner. In a preliminary study, therefore, first qualitative interviews were conducted with selected customers in order to obtain initial points of contact regarding the quality assessment and its effects. The result was a model with the following dimensions (cf. Fig. 2.15):

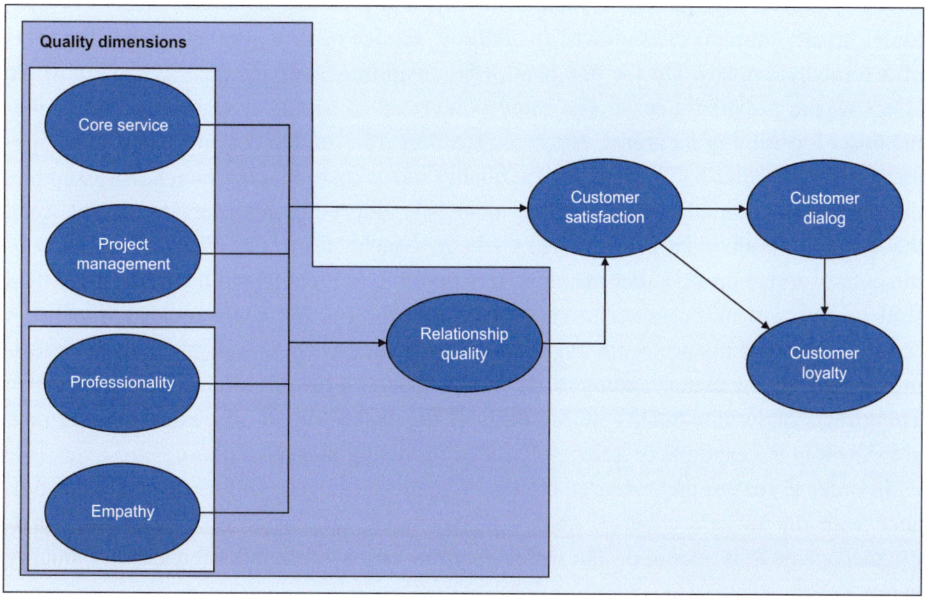

Fig. 2.15 Success chain of an IT service provider (Source: Bruhn & Frommeyer, 2004, p. 349)

- Quality dimensions: core service, project management, professionalism, empathy
- Relationship quality
- Customer satisfaction
- Customer dialogue
- Customer loyalty

In addition to measuring absolute values for the individual quality dimensions and effects, their respective importance was also determined by examining the influence of the dimensions on the subsequent variables using the statistical method of causal analysis. Interestingly, it was found that for Swiss customers, the core service was by far the most important criterion (relative importance of 66%), while in Germany it was empathy (relative importance of 47%). Based on these results, quality management measures were initiated in the two countries. For example, comprehensive empathy training was carried out in Germany, which led to a significant improvement in the assessment of this dimension in the following year. ◄

2.6.3 Economic Effects of Service Quality

As was made clear in the discussion of customer loyalty relevance, the **economic effects of service quality** play an important role in sales effects (see also Sect. 14.1.1 for economic effect sizes of service quality). In terms of the success chain, service quality affects customer satisfaction, customer loyalty, and thus the economic success of customer loyalty through sales effects. In addition, service quality also determines the costs of a service company. On the one hand, (the creation of) service quality has a positive effect on the provider's costs. The more is invested in quality management, the higher the costs for training measures, etc. (see also Sect. 15.1 on the costs of quality management). On the other hand, high service quality has a cost-reducing effect. For example, the number of complaints due to quality defects is reduced by high service quality. Also, the quality of individual performance elements contributes to the efficient operation of the entire service process. For example, it is possible to recommend the right investment strategy to the bank customer more quickly according to his needs (quality dimension "competence"), if his needs are recorded as comprehensively as possible by corresponding systems or the understanding of the bank consultant (quality dimension "empathy"). The effects of service quality on the basis of the success chain of quality management are shown in the example of a Swiss IT service provider presented above.

In order to realize these desired effects of quality management for services in accordance with the success chain of quality management, the design of a systematic quality management is required. The basis for this is a well-founded concept of quality management.

References

Beatson, A. T., Lings, I., & Gudergan, S. (2008). Employee behaviour and relationship quality: impact on customers. *The Service Industries Journal, 28*(2), 211–223.

Berger, C., Blauth, R., Boger, D., Bolster, C., Burchill, G., Du Muchel, W., Pouliout, F., Richter, R., Rubinoff, A., Shen, D., Timko, M., & Walden, D. (1993). Kano's method for understanding customer-defined quality. *Journal of the Japanese Society for Quality Control, 23*(3), 3–35.

Berry, L. L. (1986). Bii ideas in services marketing. In M. Venkatesan, D. H. Schmalensee, & C. E. Marshall (Eds.), *Creativity in services marketing, proceedings series* (pp. 6–8). Chicago: Gabler.

Bitner, M. J., & Hubbert, A. R. (1994). Encounter satisfaction versus overall satisfaction versus quality. The customer's voice. In R. T. Rust & R. L. Oliver (Eds.), *Service quality. New directions in theory and practice* (pp. 72–94). Thousand Oaks.

Boulding, W., Kalra, A., Staelin, R., & Zeithaml, V. A. (1993). A dynamic process model of service quality. From expectations to behavioral intentions. *Journal of Marketing Research, 30*(1), 7–27.

Brady, M. K., & Cronin, J. J. (2001). Some new thoughts on conceptualizing perceived service quality. A hierarchical approach. *Journal of Marketing, 65*(3), 34–49.

Bruhn, M. (1999a). Qualitätssicherung im Dienstleistungsmarketing. Eine Einführung in die theoretischen und praktischen Probleme. In M. Bruhn & B. Stauss (Eds.), *Dienstleistungsqualität. Konzepte – Methoden – Erfahrungen* (3rd Edn., pp. 21–48). Wiesbaden: Gabler.

Bruhn, M. (1999b). Verfahren zur Messung der Qualität interner Dienstleistungen. Ansätze für einen Methodentransfer aus dem (externen) Dienstleistungsmarketing. In M. Bruhn (Ed.), *Internes Marketing. Schnittstellen zwischen dem Marketing- und Personalmanagement* (2nd Edn., pp. 537–575). Wiesbaden: Gabler.

Bruhn, M. (2016a). *Kundenorientierung. Bausteine für ein exzellentes Customer Relationship Management (CRM)* (5th Edn.). München: Beck.

Bruhn, M. (2016b). *Relationship Marketing. Das Management von Kundenbeziehungen* (5th Edn.). München: Vahlen.

Bruhn, M., & Frommeyer, A. (2004). Development of relationship marketing constructs over time: Antecedents and consequences of customer satisfaction in a business-to-business environment. *Journal of Relationship Marketing, 3*(4), 61–76.

Bruhn, M., & Georgi, D. (1999). *Kosten und Nutzen des Qualitätsmanagements. Grundlagen, Methoden, Fallbeispiele.* München: Hanser.

Bruhn, M., & Georgi, D. (2000). Kundenerwartungen als Steuerungsgröße. *Marketing ZFP, 22*(3), 185–196.

Bruhn, M., & Hadwich, K. (2014). Service Value – Eine Einführung in die theoretischen und praktischen Problemstellungen. In M. Bruhn & K. Hadwich (Eds.), *Service Value als Werttreiber. Forum Dienstleistungsmanagement* (pp. 3–31). Wiesbaden: Springer.

Bruhn, M., & Hennig, K. (1993). Selektion und Strukturierung von Qualitätsmerkmalen. Auf dem Weg zu einem umfassenden Qualitätsmanagement für Kreditinstitute. *Jahrbuch der Absatz und Verbrauchsforschung, 39*(3), 214–238.

Bruhn, M., Georgi, D., & Hadwich, K. (2006a). Vertrauen und Vertrautheit als Dimensionen der Beziehungsqualität. Konzeptionalisierung, Determinanten und Wirkungen. In H. H. Bauer, M. M. Neumann, & A. Schüle (Eds.), *Konsumentenvertrauen. Konzepte und Anwendungen für ein nachhaltiges Kundenbindungsmanagement* (pp. 311–324). München: Vahlen.

Bruhn, M., Richter, M., & Georgi, D. (2006b). Dynamik von Kundenerwartungen im Dienstleistungsprozess. *Marketing ZFP, 28*(2), 116–133.

Cadotte, E. R., Woodruff, R. B., & Jenkins, R. L. (1987). Expectations and norms in models of consumer satisfaction. *Journal of Marketing Research, 24*(3), 305–314.

Carman, J. M. (1990). Consumer perceptions of service quality. An assessment of the SERVQUAL dimensions. *Journal of Retailing, 66*(1), 33–55.

Caruana, A., & Ramaseshan, B. (2015). *The effect of service quality and consumer trust on retail website loyalty.* Proceedings of the Academy of Marketing Science (AMS) Annual Conference. Denver.

Churchill, G. . (1979). Paradigm for developing better measures of marketing constructs. *Journal of Marketing Research, 16*(1), 64–73.

Corsten, H., & Gössinger, R. (2015). *Dienstleistungsmanagement* (6th Edn.). München: De Gruyter Oldenbourg.

DIN EN ISO 8402:1995. (1995). *Qualitätsmanagement und Qualitätssicherung – Begriffe*. Berlin: Beuth.

DIN EN ISO 9000:2014. (2014). *Qualitätsmanagementsysteme – Grundlagen und Begriffe*. Berlin: Beuth.

DIN EN ISO 9000:2015. (2015). *Qualitätsmanagementsysteme – Grundlagen und Begriffe*. Berlin: Beuth.

Donabedian, A. (1980). *The definition of quality and approaches to its assessment and monitoring*. Ann Arbor: Health Administration Press.

Doney, P. M., & Cannon, J. P. (1997). An examination of the nature of trust in buyer-seller relationships. *Journal of Marketing, 61*(2), 35–51.

Engelhardt, W. H., Kleinaltenkamp, M., & Reckenfelderbäumer, M. (1992). *Dienstleistungen als Absatzobjekt.* Arbeitsbericht Institut für Unternehmungsführung und Unternehmensforschung, vol. 52. Bochum: Institut für Unternehmensführung und Unternehmensforschung.

Furrer, O., Liu, B. S. C., & Sudharshan, D. (2000). The relationships between culture and service quality perceptions basis for cross-cultural market segmentation and resource allocation. *Journal of Service Research, 2*(4), 355–371.

Garvin, D. A. (1984). What does „Product Quality" really mean? *Sloan Management Review, 26*(1), 25–43.

Garvin, D. A. (1988). Die acht Dimensionen der Produktqualität. *Harvard Manager, 10*(3), 66–74.

Geiger, W. (2001). Qualität als Fachbegriff des Qualitätsmanagements. In H.-D. Zollondz (Ed.), *Lexikon Qualitätsmanagement. Handbuch des modernen Managements auf der Basis des Qualitätsmanagements* (pp. 801–810). München: Oldenbourg.

Georgi, D. (2000). *Entwicklung von Kundenbeziehungen. Theoretische und empirische Analysen unter dynamischen Aspekten*. Wiesbaden: Gabler.

Grönroos, C. (1983). *Strategic management and marketing in the service sector*. Helsinki: Marketing Science Institute.

Grönroos, C. (1984). A service quality model and its marketing implications. *European Journal of Marketing, 18*(4), 36–44.

Gupta, K., & Stewart, D. W. (1996). Customer satisfaction and customer behavior. The differential role of brand and category expectations. *Marketing Letters, 7*(3), 249–263.

Hadwich, K. (2003). *Beziehungsqualität im Relationship Marketing. Konzeption und empirische Analyse eines Wirkungsmodells*. Wiesbaden: Gabler.

Herrmann, J., & Fritz, H. (2016). *Qualitätsmanagement. Lehrbuch für Studium und Praxis.* München: Hanser.

Heskett, J. L. (1986). *Managing in the service economy* (5th Edn.). Boston: Harvard Business School Press.

Hilke, W. (1984). *Dienstleistungsmarketing aus Sicht der Wissenschaft, Diskussionsbeiträge des Betriebswirtschaftlichen Seminars der Universität Freiburg*. Freiburg.

Hilke, W. (1989). Grundprobleme und . In W. Hilke (Ed.), *Dienstleistungsmarketing* (pp. 5–44). Wiesbaden.

Homburg, C., & Giering, A. (1996). Konzeptualisierung und Operationalisierung komplexer Konstrukte. Ein Leitfaden für die Marketingforschung. *Marketing ZFP, 18*(1), 5–24.

Homburg, C., Becker, A., & Hentschel, F. (2017). Der Zusammenhang zwischen Kundenzufriedenheit und Kundenbindung. In M. Bruhn & C. Homburg (Eds.), *Handbuch Kundenbindungsmanagement. Strategien und Instrumente für ein erfolgreiches CRM* (9th Edn., pp. 99–124). Wiesbaden: Gabler.

Johnston, R. (1995). The zone of tolerance. *International Journal of Service Industry Management, 6*(2), 46–61.

Kano, N. (1984). Attractive quality and must-be quality. *Journal of the Japanese Society for Quality Control, 14*(2), 39–48.

Kettinger, W. J., & Lee, C. C. (2007). Perceived service quality and user satisfaction with the information services function. *Decision Sciences, 25*(5–6), 737–766.

Krämer, A., & Maurer, R. (1998). Mandantenbindung und Unternehmenserfolg bei Anwaltskanzleien. *Zeitschrift für die Anwaltspraxis, 9*(3), 141–152.

Lacey, R. (2007). Relationship drivers of customer commitment. *Journal of Marketing Theory and Practice, 15*(4), 315–333.

Lehmann, A. P. (1995). *Dienstleistungsmanagement. Strategien und Ansatzpunkte zur Schaffung von Servicequalität* (2nd Edn.). Stuttgart: Schäffer-Poeschel.

Leischnig, A., & Messer, U. (2014). Service Value durch Customization – Implikationen für die Gestaltung des Dienstleistungsmanagements. In M. Bruhn & K. Hadwich (Eds.), *Service Value als Werttreiber. Forum Dienstleistungsmanagement* (pp. 245–264). Wiesbaden: Gabler.

Liljander, V. (1994). Modeling perceived service quality using different comparison standards. *Journal of Consumer Satisfaction, Dissatisfaction and Complaining Behavior, 7,* 126–142.

Liljander, V., & Strandvik, T. (1995). The nature of customer relationships in services. In T. A. Swartz, D. E. Bowen, & S. W. Brown (Eds.), *Advances in services marketing and management research and practice* (4th Edn., pp. 141–167). Greenwich.

Lindberg, N., & Nordin, F. (2008). From products to services and back again: Towards a new service procurement logic. *Industrial Marketing Management, 37*(3), 292–300.

Luhmann, N. (2014). *Vertrauen. Ein Mechanismus der Reduktion sozialer Komplexität* (5th Edn.). Stuttgart: UVK.

Maister, D. H., & Lovelock, C. (1988). Managing Faciliator Services. In C. Lovelock (Ed.), *Managing services. Marketing, operations and human resources* (pp. 58–70). Englewood Cliffs.

Maitzen, P. (2016). *Attraktivität von Cross-Selling-Angeboten aus Kundensicht: Konstruktentwicklung und Überprüfung im Wirkungsmodell*. Wiesbaden: Springer.

Maiwald, K., Wieseke, J., & Everhartz, J. (2014). The dark side of providing industrial product-service systems – Perceived risk as a key challenge from a customer-centric point of view. *Procedia CIRP, 16,* 241–246.

Maleri, R., & Frietzsche, U. (2008). *Grundlagen der Dienstleistungsproduktion* (5th Edn.). Berlin: Springer.

Matzler, K., Pechlaner, H., & Siller, H. (2001). Die Ermittlung von Basis-, Leistungs- und Begeisterungsfaktoren der Gästezufriedenheit. *Tourismus Journal, 5*(4), 445–469.

Meffert, H. (1993). *Marktorientierte Führung von Dienstleistungsunternehmen. Neuere Entwicklungen in Theorie und Praxis, Arbeitspapier Nr. 78*. Münster.

Meffert, H., & Bruhn, M. (2012). *Dienstleistungsmarketing. Grundlagen, Konzepte, Methoden* (7th Edn.). Wiesbaden: Gabler.

Meffert, H., Bruhn, M., & Hadwich, K. (2018). *Dienstleistungsmarketing. Grundlagen, Konzepte, Methoden* (9th Edn.). Wiesbaden: Springer.

Meinhold, M., & Matul, C. (2011). *Qualitätsmanagement aus der Sicht von Sozialarbeit und Ökonomie*. Stuttgart: Nomos.

Meldau, S. (2007). *Qualitätsmessung in Dienstleistungscentern. Konzeptionierung und empirische Überprüfung am Beispiel eines Verkehrsflughafens*. Wiesbaden: Gabler.

Meyer, A. (1998). *Handbuch dienstleistungs-marketing* (8th Edn.). München: Schäffer-Poeschel.

Meyer, A., & Mattmüller, R. (1987). Qualität von Dienstleistungen. Entwicklung eines praxisorientierten Qualitätsmodells. *Marketing ZFP, 9*(3), 187–195.

Mudie, P., & Cottam, A. (1999). *The management and marketing of services* (2nd Edn.). Oxford: Butterworth-Heinemann.

Nenonen, S., Ahvenniemi, O., & Martinsuo, M. (2014). Image risks of servitization in collaborative service deliveries. *The Service Industries Journal, 34*(16), 1307–1329.

Ngobo, P. V. (1997). The standard issue. An accessibility diagnosticity perspective. *Journal of Consumer Satisfaction, Dissatisfaction and Complaining Behavior, 10*, 61–79.

Oliver, R. L. (1980). A cognitive model of the antecedents and consequences of satisfaction decisions. *Journal of Marketing Research, 17*(4), 460–469.

Oliver, R. L. (2010). *Satisfaction. A behavioral perspective on the consumer* (2nd Edn.). New York.

Parasuraman, A., Zeithaml, V. A., & Berry, L. L. (1985). A conceptual model of service quality and its implications for future research. *Journal of Marketing, 49*(1), 41–50.

Parasuraman, A., Zeithaml, V. A., & Berry, L. L. (1986). *SERVQUAL. A multiple-Iiem scale for measuring consumer perceptions of service quality*. Cambridge.

Parasuraman, A., Zeithaml, V. A., & Berry, L. L. (1988). SERVQUAL: A multiple-item scale for measuring consumer perceptions of service quality. *Journal of Retailing, 64*(1), 12–40.

Ravald, A., & Grönroos, C. (1996). The value concept and relationship marketing. *European Journal of Marketing, 30*(2), 19–30.

Richter, M. (2005). *Dynamik von Kundenerwartungen im Dienstleistungsprozess. Konzeptionalisierung und empirische Befunde*. Wiesbaden: Gabler.

Schmitt, R., & Pfeifer, T. (2015). *Qualitätsmanagement. Strategien – Methoden – Techniken* (5th Edn.). Wien: Hanser.

Shi, Y., Prentice, C., & He, W. (2014). Linking service quality, customer satisfaction and loyalty in Casinos, does membership matter? *International Journal of Hospitality Management, 40*, 81–91.

Siddiqi, K. O. (2011). Interrelations between service quality attributes, customer satisfaction and customer loyalty in the retail banking sector in Bangladesh. *International Journal of Business and Management, 6*(3), 12–36.

Sirdeshmukh, D., Singh, J., & Sabol, B. (2002). Consumer trust, value and loyalty in relational exchanges. *Journal of Marketing, 66*(1), 15–37.

Spreng, R. A., MacKenzie, S. B., & Olshavsky, R. W. (1996). A reexamination of the determinants of consumer satisfaction. *Journal of Marketing, 60*(3), 15–32.

Stauss, B., & Hentschel, B. (1991). Dienstleistungsqualität. *Wirtschaftswissenschaftliches Studium, 20*(5), 238–244.

Steffen, D. (2006). *Die Potenzialqualität von Dienstleistungen: Konzeptualisierung und empirische Prüfung*. Wiesbaden: Gabler.

Teas, R. K. (1993). Expectations, performance evaluation, and consumers' perceptions of quality. *Journal of Marketing, 57*(4), 18–34.

Tse, D. K., & Wilton, P. C. (1988). Models of consumer satisfaction formation. An extension. *Journal of Marketing Research, 25*(2), 204–212.

Tuli, K. R., Kohli, A. K., & Bharadwaj, S. G. (2007). Rethinking customer solutions: From product bundles to relational processes. *Journal of Marketing, 71*(3), 1–17.

v. Wangenheim, F., & Bayón, T. (2004). The effect of word of mouth on services switching: Measurement and moderating variables. *European Journal of Marketing, 38*(9/10), 1173–1185.

Wirtz, J., & Lovelock, C.H. (2016). *Services marketing. People, technology, strategy* (8th Edn.). New Jersey: World Scientific Publishing.

Wohlgemuth, A. C. (1989). Führung im Dienstleistungsbereich. Interaktionsintensität und Produktionsstandardisierung als Basis einer neuen Typologie. *Zeitschrift für Organisation, 58*(5), 339–345.

Zeithaml, V. A. (1981). How consumer evaluation processes differ between goods and services. In J. H. Donelly & W. R. George (Eds.), *Marketing of services, proceeding series* (pp. 186–190). Chicago.

Zeithaml, V. A. (1988). Consumer perceptions of price, quality, and value: A means-end model and synthesis of evidence. *Journal of Marketing, 52*(3), 2–22.

Zeithaml, V. A., Parasuraman, A., & Berry, L. L. (1992). *Qualitätsservice. Was Ihre Kunden erwarten – was Sie leisten müssen*. Frankfurt a. M.: Campus.

Zollondz, H.-D. (2011). *Grundlagen Qualitätsmanagement. Einführung in Geschichte, Begriffe, Systeme und Konzepte* (3rd Edn.). München: Oldenbourg.

Conceptual Basics of Quality Management for Services

3

Abstract

A comprehensive conceptualization approach forms the basis for professional management of service quality. For this purpose, important concepts within the framework of Total Quality Management are explained in this chapter. Based on this, the conceptual clarity is derived, what is meant by quality management for services. It turns out that the analysis of service quality, quality planning, quality implementation and quality control are important components of a successful quality management system. The chapter concludes with a discussion of new research approaches to the management of service quality. The customer journey management in the context of customer integration, the customer experience management and the management of processes of service transformation are presented.

3.1 Concepts of Total Quality Management (TQM)

In the considerations for the development and implementation of a comprehensive quality management, it is assumed from the philosophy that it is indispensable for the securing or improving the quality of products and services, that management and employees take responsibility for quality management together, on all organizational levels. This basic idea is contained in the concept of **Total Quality Management** (TQM) (Ishikawa, 1985; Oess, 1994; Döttinger & Klaiber, 1994; Sondermann, 1994; Wonigeit, 1996; George & Weimerskirch, 1998; Frehr, 1999; Zink, 2004; Gimenez-Espin et al., 2013; Schmitt & Pfeifer, 2015).

The steering committee of the German Society for Quality e. V. (1995) defines Total Quality Management as follows:

M. Bruhn, *Quality Management for Services*,
https://doi.org/10.1007/978-3-662-67032-3_3

▶ "**Total Quality Management** is a management method of an organization based on the participation of all its members. It focuses on quality and aims at long-term business success and benefits for the members of the organization as well as for society through customer satisfaction."

An extended definition is given by Oess (1993, p. 89):

▶ "Under **Total Quality Management**, all structures, processes, regulations, rules, instructions and measures are understood that serve to ensure the quality of products and services of a company in all functions (…) and all levels through the participation of all employees in a timely and cost-effective manner as well as to improve them continuously in order to enable an optimal satisfaction of the needs of consumers and society."

TQM is therefore not just part of a corporate management concept, but rather dominates all management activities. Since "quality" becomes the most important success factor in the company, with the goal of being carried by all employees, Total Quality Management is not simply a quality concept, but rather a **quality philosophy** or **quality culture** that includes the entire company (Döttinger & Klaiber, 1994, p. 258).

With regard to the **historical development of comprehensive quality management approaches**, the origin of the TQM approach can be traced back to Japan. Although after the end of the Second World War, American quality experts such as Juran (1951), Feigenbaum (1956, 1991) and Deming (1982) as well as in Germany Masing (1980) met with great receptivity for their comprehensive methods of quality assurance, the idea only gradually established itself in practice in Europe and the USA (Schildknecht, 1992, 62 pp.). While quality management in Western companies was usually a specialist department (mostly staffed by technicians), in Japan quality already developed as an internally and externally oriented management task.

With a growing number of publications on this topic, numerous concepts have arisen which can be summarized under the term TQM and the main content of which will be briefly presented below (as an overview see Zink & Schildknecht, 1989, 72 pp.; Schildknecht, 1992, 64 pp.). These concepts differ in their principles and procedures. They have in common the normative character, i.e. mostly demands are made on a quality management or on the company management (e.g. Deming: 14 Points Program; Crosby: 4 Commandments) to commit to focus on quality. Theoretically, these concepts are poorly founded; they are primarily suitable for use as "recipe knowledge" for corporate practice. So Deming, one of the most important pioneers of quality management, tried to document the goal of improving quality and productivity as part of a so-called **14 Points Program** (Deming, 1982, 16 pp.):

1. Establishment of the goal of continuous improvement of the products and services offered,
2. Adoption of the new zero-defect philosophy, which rejects the acceptance of errors,

3. Elimination of the dependence on mass testing and reliance on statistical controls,
4. Obligation of suppliers to provide statistical quality evidence,
5. Continuous improvement of production and service,
6. Ongoing training of all employees,
7. Provision of appropriate tools for the correct task performance for all employees,
8. Promotion of communication and productivity,
9. Promotion of cooperation between different departments in solving problems,
10. Elimination of messages that do not contain any specific improvements,
11. Use of statistical methods to continuously improve quality and productivity,
12. Elimination of all obstacles to providing high-quality performance,
13. Continuous offering of training measures to keep up with ongoing market changes,
14. Clear commitment of top management to quality.

Deming's focus on statistically oriented **process and procedure control** is based on the recognition that quality is not testable, but arises in the course of a process. With reference to the subjectivity of quality assessments, Deming withdraws from an exact definition and leaves quality to the individual assessment of the price-performance ratio by the customer (Deming, 1988).

A positive aspect of Deming's approach is the **process view**. This is particularly important due to the processual character of service creation. Deming's approach has been successfully applied in both the industrial and service sectors (Atul et al., 2005, p. 391). However, the approach is often not directly adopted for service companies, as it is only partially applicable due to its focus on statistical measurement methods and the processes of service creation often still elude concrete measurability and quantifiability. Here, the methods for measuring service quality need to be further refined.

Juran also has a similar quality assessment to Deming, who also proposes a customer-oriented definition of the quality of products and services with **"Fitness for use"** (Juran, 1988). However, he extends the customer concept to the dimension of the **internal customer,** so that the quality process becomes a cross-departmental task. The so-called **"Quality Trilogy"** of quality planning, -control and -improvement (Juran, 1986, p. 20) represents a comprehensive management process for continuous quality improvement, which uses mainly statistical methods and procedures. Although training programs for managers are recommended, the importance of employees is rather low in the context of this concept (Oess, 1993, p. 82).

The approaches presented already address contents of a Total Quality Management concept, but this term is only used concretely by Feigenbaum (Feigenbaum, 1991). Within the framework of a so-called **"Total Quality Control"** priorities are set in the interfunctional cooperation of the company's departments and the responsibility of all employees for product quality—understood as the fulfillment of consumer expectations (Feigenbaum, 1991). Detailed suggestions for technical quality assurance (e.g. for standards, system audits and target vs. actual comparisons) are made and quality costs are taken into account in the measurement-technical and organizational considerations (Zink & Schildknecht, 1989, pp. 82–83).

The process orientation of the approaches of Juran and Feigenbaum is more suitable for service companies, as the process quality is given a high importance in services due to the interaction of service provider and receiver (Berekoven, 1983).

The Japanese concept of **"Company Wide Quality Control (CWQC)"**, which is based on the approaches of Deming and Juran, was developed primarily by Ishikawa (Ishikawa, 1985). At the center of the concept are four quality-related aspects: the implementation of quality circles (see also Sect. 8.3.3.1) and continuous training, the "Ishikawa diagram" (Fishbone method, see also Sect. 5.2.1) and the so-called quality chain. According to Ishikawa, the participation and support of all supervisors and employees is required to establish the concept of Company Wide Quality Control as effectively as possible in the company. For this purpose, in addition to the concepts already presented, a participatory management is particularly required here—with a special emphasis on employee motivation and training. Quality for Ishikawa means meeting customer expectations, accordingly, market research and instruments for implementing quality expectations play a major role in the concept (Oess, 1993, p. 83). Analog to Feigenbaum, Deming and Juran, CWQC is also understood as a **dynamically oriented concept** whose fundamental goal is long-term, continuous quality improvement.

In contrast, Crosby does not focus on the technical area with his quality management concept, but rather understands quality management as a company-wide concept that explicitly includes administrative areas and service organizations. His general demand for a **zero-defect performance standard** according to the motto "Do it right the first time" in all areas of the company is fulfilled, in his opinion, by the establishment of a quality-oriented corporate culture, which he sees as the original task of the company management (Crosby, 1986b, pp. 82–83).

The core statements of a required cultural change are summarized in the following so-called **"four commandments"** (Crosby, 1986a, 68 pp.):

1. Quality as compliance with requirements,
2. Prevention as a basic principle,
3. Zero defects as a standard,
4. Cost of non-compliance with requirements as a measure of quality.

Crosby documents essential steps for implementing these contents in a 14-point program (Crosby, 1986b, 111 pp.), similar to Deming's mentioned approach.

The quality management approach **"Six Sigma"**, similar to Crosby's approach, which has been widely used since the mid-1990s. It is a term from statistics and means reducing variance in process flows. A error level of 0.0003% (or error-free of 99.9997%) is sought, i.e. a maximum of 3.4 errors per 1 million error possibilities. The term Six Sigma is now used as a synonym for zero-defect quality and represents a method of quality management to achieve processes with as few errors as possible (see Schmidt, 1998; Eckes, 2000; Magnusson et al., 2004; Töpfer, 2007, 2009; Töpfer & Silbermann, 2011).

The main focus of Six Sigma is on reducing **variation** within corporate processes in order to achieve strategic goals and ultimately improve business results. Six Sigma primarily focuses on an analysis of actual processes in order to identify the parameters, error possibilities and process indicators important for the processes and make them accessible to objective statistical analysis. The areas of application of Six Sigma therefore relate to process and design improvements, project management and development processes.

The **conceptual framework of Six Sigma** (cf. Fig. 3.1) comprises the four elements:

1. Commitment of the management,
2. Inclusion of the stakeholders (employees, suppliers and customers),
3. Training program and
4. Measurement system (identification and measurement of quality-critical features).

The core of the framework concept are the improvement projects. These are carried out using formalized improvement methods and are supported by improvement tools and thinking models, such as DMAIC (Define, Measure, Analyse, Improve, Control) or the DMADV method (Define, Measure, Analyse, Design, Verify) (see Magnusson et al., 2004, 19 pp.).

A modification of the Six Sigma concept is the so-called approach of the **"Lean Sigma"** (see, for example, Dahm & Heindl, 2006, 2007). It is a synthesis of Six Sigma

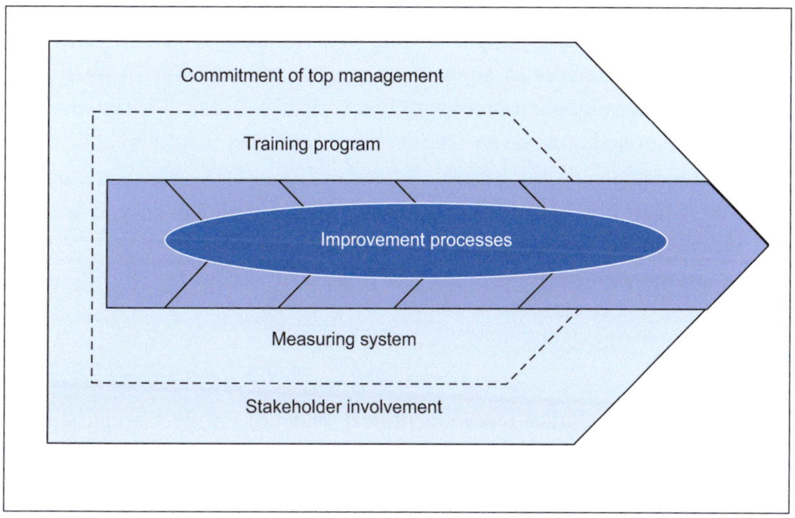

Fig. 3.1 Conceptual framework of Six Sigma (Source: Magnusson et al., 2004, p. 19)

and the management approach Lean Management. Lean Management is another process-oriented method from the 1980s, which is based on the approach of the Japanese industry, in particular *Toyota*. Its primary goal is to increase the speed of the process and to avoid waste in the process. The use of Lean Sigma aims to release the strengths of both concepts for companies: Six Sigma leads to the improvement of the quality of products and processes and thus to customer satisfaction, whereas Lean Management aims to increase the speed of the process. Due to the process orientation of both concepts, the Lean Sigma method is particularly suitable for service companies. For example, the banking institution *Credit Suisse* Private Banking was able to realize an operating profit of over 150 million Swiss francs in 2005 from its Lean Sigma initiative, which was launched in 2004 (Dahm & Heindl, 2007, p. 88).

The introduction of Lean Sigma proceeds ideally according to the following five **process steps** (Dahm & Heindl, 2007, p. 88):

1. Initiation: Announcement of the initiative in the company and to other stakeholders,
2. Selection of pilot projects and resources: Identification of pilot projects and selection of employees for initial training,
3. Implementation: Start of Lean Sigma based on pilot projects with subsequent gradual roll-out in the entire company,
4. Sustainability: Long-term prioritization of Lean Sigma,
5. Evolution in the company: Continuous development and anchoring in the corporate culture.

The Six Sigma concept is often also associated with approaches to **project management**. Here, attempts are made to apply various phases and **design for Six Sigma** for the development and implementation of projects (Chowdhury, 2006; Bergbauer et al., 2008; Toutenburg & Knöpfel, 2009; Wappis & Jung, 2013; Lunau, 2014; Schmitt & Pfeifer, 2015). This approach attempts to apply the concepts and methods of quality management early on in the development process of innovations in order to better meet customer requirements on the one hand and to save on error costs that would arise in later phases on the other. The design for Six Sigma approach differentiates between various development phases, which in turn are subdivided into sub-phases. Although it was primarily developed for the goods sector, it can also be used with modifications for the service sector. These are the following **development phases of design for Six Sigma** with the corresponding sub-phases (Schmitt & Pfeifer, 2015, 81 pp.):

1. **Define-Phase**
   with the sub-phases: Market research, project selection, project preparation, project initiation
2. **Identify-Phase**
   with the sub-phases: Collection of customer requirements, prioritization of customer requirements, technical analysis and competitive comparisons

3. **Design-Phase**
   with the sub-phases: Identification of design alternatives, evaluation of design alternatives, creation and feasibility, risk mitigation
4. **Optimize-Phase**
   with the sub-phases: Detailing and optimizing the design, determination of performance transfer functions, tolerance design, process optimization
5. **Verify-Phase**
   with the sub-phases: Project piloting, measurement and verification, performance creation preparation, project completion

In all sub-phases, methods of project development and quality management are used. This gives early consideration to the quality concept in the innovation process.

The TQM concepts presented (for a critical comparison of the approaches, see, for example, Schildknecht, 1992, 80 pp.; Oess, 1993, 86 pp.) and the associated development lines subsequently underwent various interpretations and additions, but less as part of scientific debate, and more with a view to practical considerations (for example, Garvin, 1988; Haist & Fromm, 1991; Oess, 1993; Oakland, 2004).

Notable **extensions to the TQM concepts** include

- the **Taguchi philosophy** (Taguchi, 2001), which additionally introduces a socio-political dimension of quality,
- **process management** (for example, Horatzek, 2018; Huth, 2018),
- the **Kaizen concept** (Imai, 2002) as an approach to a permanent, consistent orientation towards improvement of all processes within the company.

In particular, the **Kaizen concept** has been adopted in business practice. Kaizen means a continuous improvement of products, processes and workflows within the company in the smallest steps, initiated by the employees themselves (Zollondz, 2011, p. 291). It builds on the principle of "continuous improvement" explicitly mentioned by Deming in his 14-point program. Kaizen is a Japanese management philosophy whose goal is continuous development. The approach is not only focused on the goal of actual product improvement, but also explicitly on the customer and aims to change the thinking and behaviour of the employees involved in the organizational processes towards continuous improvement. In the German and English language areas, the Kaizen concept is also known under the term **"Continuous Improvement Process (CIP)"** and is used very effectively in companies such as *Volkswagen* (Zollondz, 2011, 289 pp.).

The CIP is based on the knowledge that the exploitation of the knowledge potential is the basis for efficiency gains to secure corporate success. The CIP is characterized by its concentrated course of action: In the form of special, usually one-week workshops, as many first improvement measures are already implemented. This includes the agreement of objectives with corresponding result variables as well as the definition of processes and their process variables in advance. The CIP does not only aim at the improvement within a certain organizational unit, but at the improvement of the complete value chain within the company. This not only requires the involvement of employees of all affected hierarchy levels in the CIP workshops, but also the works council (Kamiske & Brauer, 2011, 277 pp.). The CIP is based on the methodology of the systematic approach of planning, implementing, checking and acting for the continuous improvement of workflows and procedures (Fig. 3.2). For this purpose, the CIP approach uses the "basic principle of problem solving", which can cause many small steps to lead to lasting changes (Kosta & Kosta, 2008, p. 13).

In order to meet the cross-functional, holistic claim of a TQM in the service sector, it appears to be expedient to base both Feigenbaum's and Juran's—oriented towards the stages of the value chain—understanding of quality as well as Crosby's concept of quality-oriented corporate culture. This can be justified by the individuality of services, which makes objective quality comparisons more difficult and therefore puts quality-assuring framework conditions more in the focus of quality management. The inclusion of quality costs, as demanded by Feigenbaum, for example, proves to be problematic in connection with the immateriality of services and the integration of the external factor. Nevertheless, it is to be taken into account due to its relevance for the determination of the economic efficiency of quality measures (cf. in detail on quality costs Chap. 15).

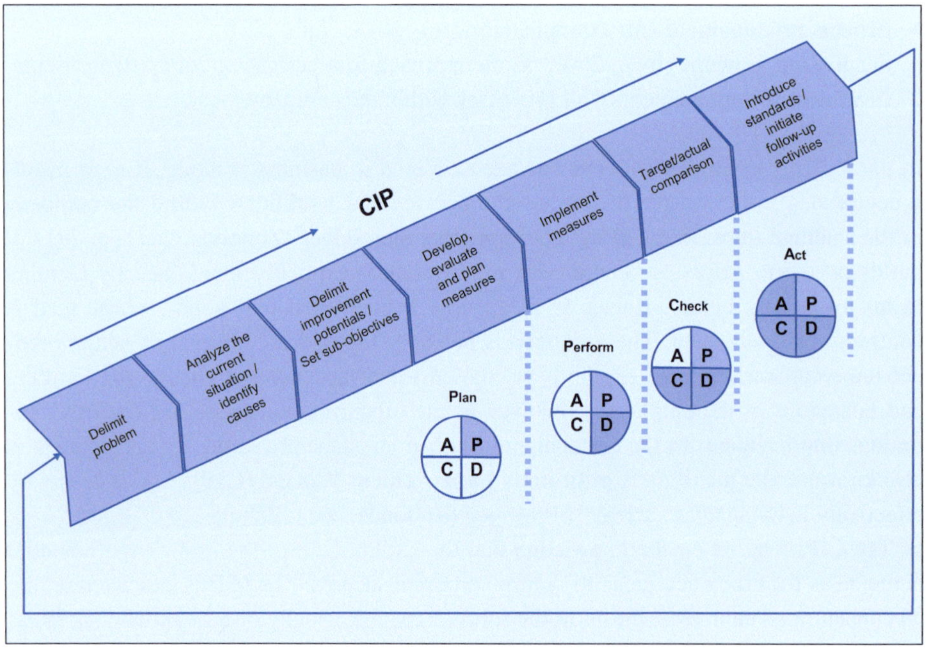

Fig. 3.2 Continuous improvement process (Source: Kosta & Kosta, 2008, p. 14)

Although already present in Deming, Juran, Feigenbaum and Crosby, it was only Ishikawa who recommended the integration of employees within the framework of a comprehensive participatory leadership concept that meets the high personnel intensity of services.

Accordingly, the following three central **building blocks of the TQM concept for services** can be derived based on the Ishikawa concept (e.g. Wonigeit, 1996, 56 pp.; Bruhn, 1999, p. 41):

1. **Total**—i.e. the inclusion of all groups of people involved in the creation of the service (employees of the company, suppliers, all customer groups) in the quality management process,
2. **Quality**—i.e. the consistent orientation of all activities of the service company to the quality requirements of the external and internal customer groups,
3. **Management**—i.e. the assumed responsibility and initiative of the top management of the service company for systematic quality assurance and improvement within the framework of a participatory-cooperative leadership style.

Total Quality Management in the Banking Sector

A study in the banking sector shows that companies that have been implementing TQM as a management concept for a long time have better performance than those that have not or have only recently implemented it. The study mainly saw greater openness to new ideas, better process optimization and the transfer of more responsibility to lower-level employees as success factors associated with TQM (Lloréns Montes & Verdú Jover, 2004). ◄

It becomes apparent that the TQM approach for services includes a comprehensive alignment of the entire company with the quality perception of internal and external service customers, with the philosophical aspect and the compliance with principles for the company clearly in the focus. Although the "founding fathers" of quality management consider the **quality philosophy of TQM** to be universal and thus also suitable for the service sector in principle, it must nevertheless be noted that the original idea of TQM was directed at the manufacturing industry. Despite the worldwide increase in the importance of the service sector, the effectiveness of the operational implementation of TQM in the service sector still lags behind that of the industrial sector (Atul et al., 2005). The reason for this is often related to the implementation of TQM in service organizations. In general, the implementation is associated with significant changes in the internal organizational structure in order to increase performance and employee-oriented quality (see also Chap. 9). More than industrial companies, service providers are constantly exposed to changing customer perceptions and preferences. In order to always meet changing customer requirements, service companies are heavily burdened with the permanent adaptation of their internal performance potentials. This ties up the resources necessary for the successful introduction of TQM in service companies and the implementation is additionally made more difficult (Huq, 2005, p. 453).

The concept of **Total Quality Service (TQS)** takes into account these and other special features of a Total Quality Management in the service sector to a comprehensive extent as a special adaptation of the TQM philosophy for service organizations (Sureshchandar et al., 2001a, b).

Based on the knowledge that the service creation process differs significantly from classical production processes in the industrial sector, the **concept of TQS** is gaining increasing importance in scientific discussion. In short, this is an attempt to apply TQM in service companies (Atul et al., 2005, p. 389). The core of TQS is to integrate a variety of service-specific aspects of quality management into a single conceptual model of TQM in service organizations (Sureshchandar et al., 2001a, b). The reason for this is the recognition that, in the context of TQM research, various service-specific aspects are examined individually and not placed in a overall context. This leads to the deficit that it is difficult to derive conclusions about a successful implementation of holistic TQM in service organizations from the research results.

Sureshchandar et al. (2001a, b) identify the following twelve **dimensions of the Total Quality Service Model,** which are critical to the success of introducing TQM in the service sector and constitute the TQS model (see Fig. 3.3):

1. Commitment of top management and visionary leadership style,
2. Human Resource Management,
3. Technical System,
4. Information and Analysis System,
5. Benchmarking,
6. Continuous Improvement,
7. Customer Orientation,
8. Employee Satisfaction,
9. Union Influence,
10. Social Responsibility,
11. Location of the service encounter between provider and customer ("Servicescapes"),
12. Service culture.

Gupta et al. (2005) finally combine Deming's TQM approach with Parasuraman et al. (1985) SERVQUAL approach and define a TQS concept from this with the **three main dimensions** of corporate culture, corporate governance and employee commitment (Gupta et al., 2005, p. 394). In this context, in particular customer needs and their expectations are taken into account within the model. The authors recommend aligning the organizational culture, corporate governance and employee commitment to the customer in order to design a successful TQS program in this way. The three dimensions are considered to be drivers for adapting business processes in order to increase the perceived service quality and thus customer satisfaction (Gupta et al., 2005).

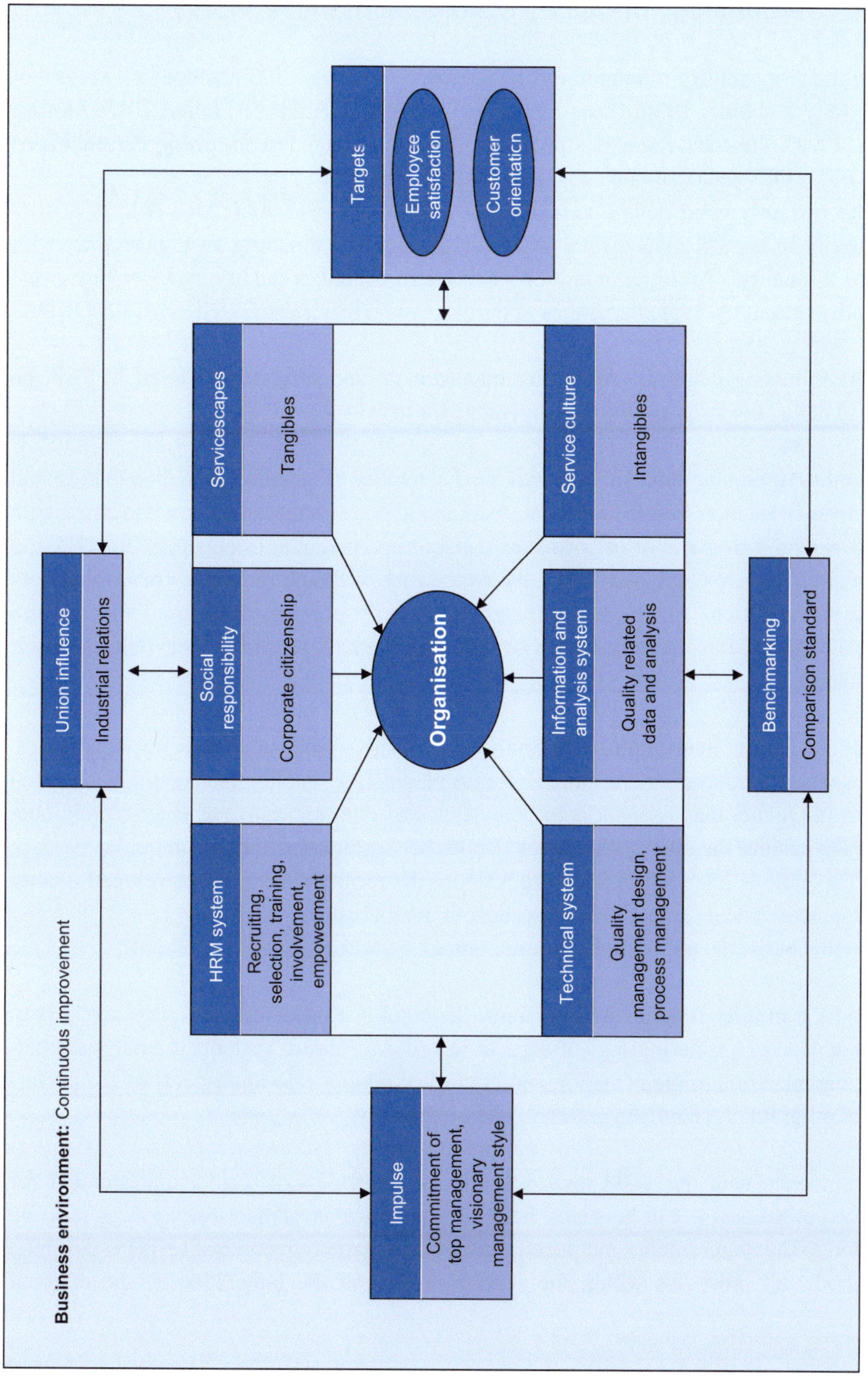

Fig. 3.3 Integrative framework concept for Total Quality Service (Source: Sureshchandar et al., 2001a, p. 356)

3.2 Terminology of Quality Management for Services

Since the term **quality management** is discussed in science and practice in a variety of ways (e.g. Stebbing, 1990; Oess, 1993; Stauss, 1994; Schmitt & Pfeifer, 2015; Meffert et al., 2018), there are numerous proposals for definitions. The following definitions of the various DIN standards are considered to be appropriate.

The currently valid definition according to DIN EN ISO 9000:2014 defines quality management as "a management system for leading and directing an organization with regard to quality". An older definition of the term describes quality management as the " totality of quality-related activities and objectives" (Deutsche Gesellschaft für Qualität e.V., 1995, p. 35).

The following definition of quality management and services (cf. Sect. 2.1) can be used to define the term **quality management for services**:

▶ **Quality management for services** is the totality of quality-related activities and objectives in relation to self-contained, marketable performances that are associated with the provision and/or use of performance capabilities (potential orientation). Internal and external factors are combined within the framework of the performance creation process (process orientation). The factor combination of the service provider is used with the aim of achieving benefits-creating effects on external factors—people or their objects (result orientation).

Under a quality management system, the organizational structure, responsibilities, processes, procedures and resources for implementing quality management are then recorded. Quality management is to be designed as comprehensive as is absolutely necessary to achieve the quality objectives (Deutsche Gesellschaft für Qualität e. V., 1995, p. 36). This note is intended to take into account the profitability of the developed quality management concept and to make cost-benefit comparisons.

On this basis, the **term quality management system** is defined as follows:

▶ Under a **quality management system**, the combination of various components is to be understood considering logical aspects in order to ensure systematic analysis, planning, organization, implementation and control of quality-relevant aspects of the performance program of a company internally and externally.

In accordance with the TQM approach and the principles of quality management for service companies, it can be stated here that quality management for services is to be adapted to the requirements and peculiarities of the market. In contrast to TQM designed for goods, all processes within the service chain and the integration of the external factor are of particular importance for the quality consideration (Bruhn, 1999, p. 44). Furthermore, the economic focus of the quality-related activities is to be ensured.

3.3 Standards and Principles of Quality Management

In Germany, Europe and worldwide, **norms** have been created to ensure cooperation for the benefit of the public through the consensual standardization of material and immaterial objects. Although these standards are voluntary in principle, they have a high degree of binding character by virtue of the agreements between the market participants. The compliance with these standards is usually agreed in contracts. Accordingly, standards for quality management have also been created and have a certain validity for quality issues.

If it is a German standard, it is called a **DIN standard** for which the German Institute for Standardization e. V. is responsible. If it is a European standard, the European Committee for Standardization (Comité Européen de Normalisation) is responsible, accordingly it is a **EN standard.** If it is an international standardization, it is an **ISO standard** of the International Organization for Standardization. If the standards are binding on all three levels mentioned, one speaks of a **DIN-EN-ISO standard.** The standard system is continuously developed, the year behind the standard and the order number indicates the year of publication. If the standards are very widespread and accepted, one speaks simply of an ISO standard.

Different interest groups are involved in the continued development, such as industry, trade, associations, testing institutes, state, etc. The standards are developed in a multi-stage process between the participants and in public circles in extensive discussions. If the standards are adopted, the market participants can have the proof of the standard application confirmed by certification (see Chap. 11 of the book).

Two standard families are of particular importance for quality management:

DIN EN ISO 9000:2015 Quality management systems—Fundamentals and vocabulary

DIN EN ISO 9001:2015 Quality management systems—Requirements

Further details on the standard system can be found in Chap. 11; Fig. 11.1. In addition to these two standard families, there are other standard systems for special topics and industries (DGQ, 2019), which should not be discussed here.

In ISO 9000, various **principles of quality management** are explained which are supposed to facilitate the application of the ISO standards 9001:2015. These are general principles which have already been discussed in Chap. 1 to 3 of this book in different models and approaches. The principles of the ISO standard are not complete and without a theoretical or conceptual basis, but a mere list without prioritization and connections. At this point, the seven principles contained in the ISO standard should be mentioned in a compact form (DIN EN ISO 9000:2015; see Fig. 3.4):

1. **Customer orientation**
   Customer orientation is seen as the central goal of quality management. This requires that a company knows the customer expectations in order to meet or perhaps exceed them. If the customer expectations are met, customer satisfaction, customer loyalty and other economic success will result.
2. **Leadership**
   As foreseen in the TQM philosophy, managers of a company have a role model function for quality issues. They have to include, train and take along employees in improvement processes of quality creation. In the end, leadership is responsible for the implementation of quality management.
3. **Engagement of people**
   Closely linked to leadership is the involvement and qualification of employees. Quality management requires qualified employees. They are to be empowered for the tasks. Accordingly, resources are to be made available for this. As instruments for increasing the engagement of people are surveys, training, improvement suggestions, internal communication programs, etc.

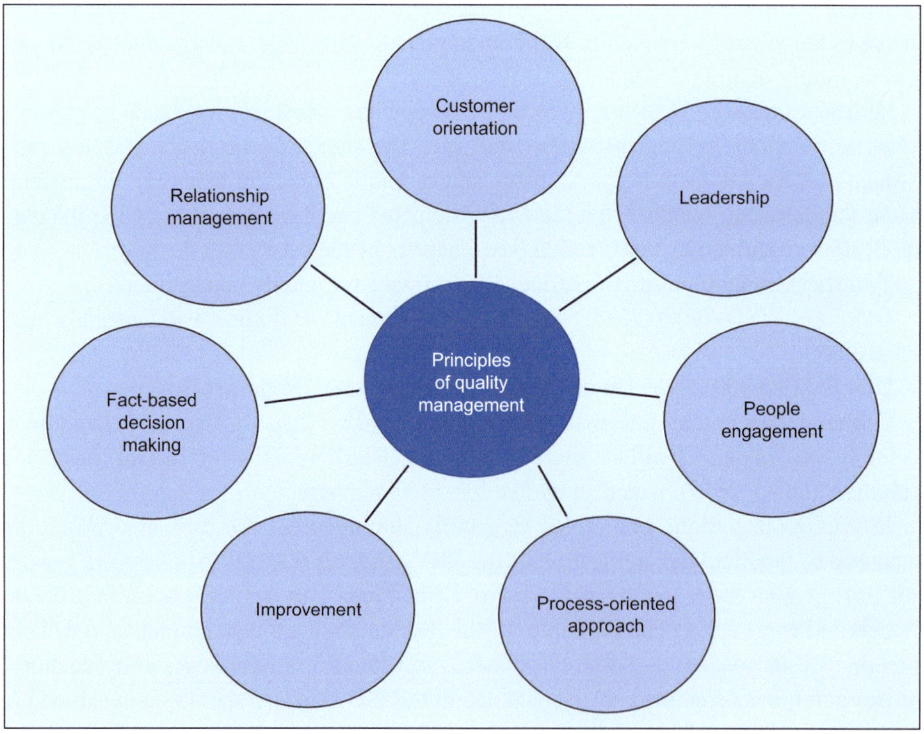

Fig. 3.4 Principles of quality management according to ISO standard 9000

4. **Process-oriented approach**
   For quality management, the company's comprehensive process flows are to be analyzed. All processes are to be coordinated and controlled with each other so that errors in different process phases can be avoided. Accordingly, careful and systematic planning of the individual processes is required.
5. **Improvement**
   At the center are the continuous improvement processes (CIP). This is intended to optimize performance, processes and products. The improvement processes are also to be approached systematically. As instruments, for example, error analyses, quality circles, improvement suggestions, feedback from market participants, surveys, etc. are available.
6. **Fact-based decision making**
   Professional quality management strives to avoid too high subjective assessments of various quality issues in terms of causes, effects and risks. Facts are necessary for decision making, which are provided by corresponding information, data, numbers and evidence. Various instruments can be used for this purpose, such as key figures, statistical surveys, experiments, etc.
7. **Relationship management**
   For a successful and long-term quality management, as many different stakeholders should be involved as possible. The aim here is not to achieve short-term advantages, but by building, maintaining and intensifying stakeholder relationships, a partnership is sought. Accordingly, the use of shared resources through cooperation with suitable market participants should also be examined.

Although these seven principles of quality management contained in the ISO 9000 family of standards are very general, they reflect some philosophies and central thinking principles of market-oriented corporate management and service management (Meffert et al., 2018) and are therefore basics of quality management of services.

3.4 Components of a Quality Management System for Services

The main task of the quality management system is the **creation and assurance of the quality capability** of the service company (Horváth & Urban, 1990, p. 14). Thus, the design of the quality management system must be targeted towards the quality capability. For this purpose, a quality management system for services must include four **components** that can be assigned to the phases of the classical management process—analysis, planning, execution and control (cf. Fig. 3.5) and which are also the basis for the further structure of the book.

The **analysis of service quality**—as an information basis for quality management for services—is characterized first and foremost by the measurement of service quality or service-specific indicators. On the one hand, it is necessary to analyze and operationalize

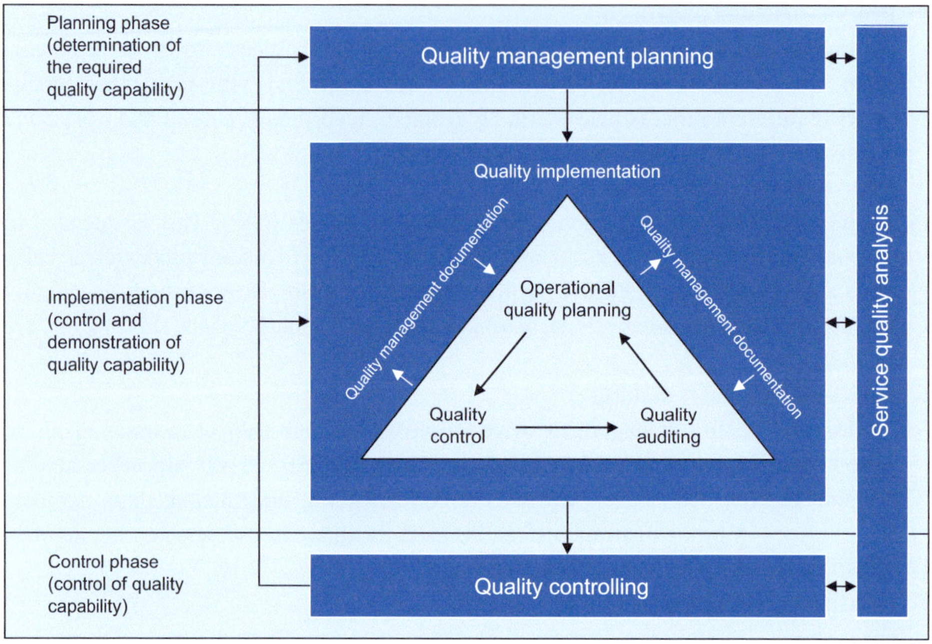

Fig. 3.5 Components of a quality management system for services (Source: Bruhn, 1998, p. 39)

the origin and effects of service quality on the basis of a suitable model (cf. Chapter 4), which makes the measurement of service quality possible in the first place. On the other hand, depending on the specific analysis goal, it is necessary to select suitable instruments for the actual measurement of service quality (cf. Chap. 5).

In the context of **planning and controlling quality management** for services, the information basis for defining the strategic orientation of quality management, which has been derived from an analysis of service quality, is used in a first step. In this context, in particular the aspects of the quality **perceived by the customer** and the **customer satisfaction** are taken into account due to their high success relevance. The formation of **customer expectations** plays a decisive role in this context, since, according to the so-called (dis-)confirmation model, a performance is judged more positively the more the customer expectations are met. For this reason, in addition to **performance-related strategies**, also **strategies of expectation management** form the basis of strategic quality management (cf. Chap. 6). In this context, market-, customer- and competition-related strategies are distinguished within the mentioned strategy areas.

In a further step it is necessary to operationalize the directions defined in the context of **strategic expectation management** (cf. Chap. 7). By defining **expectation management-related tasks and goals**, a corresponding planning process is initiated, which is implemented in the sense of the classical management process by means of analysis, planning, implementation and control.

The implementation of performance-related strategies takes place on the basis of the **quality management feedback loop**. Within the operational design of quality management (cf. Chap. 8) there is a division of the process phases according to the **quality planning, control, audting and management documentation**. Within quality planning, the required **quality capabilities** are identified and developed in accordance with the performance-related strategy. The following phase of quality control is based on the results of quality planning. In this phase, employee-, culture- and organization-related quality control instruments are used in order to meet the specified quality requirements. The degree of compliance with the specified quality requirements is collected in the phase of **quality auditing**. The achievement of the specified quality objectives is also differentiated according to internal and external company aspects and presented in the last phase of the quality management feedback loop—the **quality documentation**. Here, it is particularly important to create sufficient trust in the quality of the service company and its services, both internally and externally, by means of the interpretation of activities within the quality management system. In this sense, the aspect of **integration** of the individual phases of the quality management feedback loop into a comprehensive quality management system comes into focus.

In order to realize quality management across the company, a **systematic implementation** of quality management is required due to implementation barriers existing in the company (cf. Chap. 9). The implementation is based on the company's structures, systems and culture. In connection with a mentioned implementation, a long-term planned restructuring process is planned as a multi-stage phase concept.

In connection with the implementation process of a quality management, in particular **quality prices** represent a possible starting point (cf. Chap. 10). By using the quality prices as **quality models** internal incentives for the implementation are set in this context. Also the process of **certification** of a service company (cf. Chap. 11) provides starting points for the implementation of quality management, the success of which is documented by the actual certification.

Following the last process phase of the classical management process—the control—the perspective in the context of quality management is focused on the activities of controlling (cf. Chap. 12). **Quality control** has the task of supporting and coordinating the mentioned activities of quality management. Based on a consistent cost-benefit understanding and on the basis of a process-oriented approach, process, effectiveness and efficiency analyses are carried out within the framework of quality control (cf. Chap. 13 to 15).

From the perspective of **cost orientation**, the analysis focuses on quality-related costs and, therefore, purely monetary aspects. In contrast, the **analysis of non-monetary variables**, such as customer satisfaction or customer loyalty of a company, is opposed. For the measurement and control of these variables, **customer barometers** are available as well-developed instruments (cf. Chap. 14). The synthesis of these two perspectives is the cost-benefit analysis of quality management (cf. Chap. 15). By opposing costs as negative and the benefits as positive financial consequences, the efficiency of quality management can be represented and evaluated in the final phase of a **recurring quality management process**.

The building blocks of a quality management system for services presented here form the basis for the further composition of the book. They are dealt with in Chap. 4 to 15. Before this is done in detail, an excursion will be made to current conceptual topics of service management, which have an impact on quality management. These are developments in connection with customer contact points, customer experiences and service transformations.

3.5 Excursions: Current Conceptual Topics of Service Management

3.5.1 Management of Customer Contact Points in the Context of Customer Integration

The integration of customers into the value creation of a company has been intensively discussed for a long time as part of quality management of services (Edvardsson et al., 2012; Homburg & Kuehnl, 2014). The reason for this is that services differ from traditional products mainly in the constitutive characteristic of customer involvement (Meffert et al., 2018). As a result, the creation of most services is necessarily associated with a certain form of customer integration (Bruhn & Hadwich, 2015). Accordingly, it is also one of the central strategic and operational tasks of service quality management to determine the type and extent of customer involvement in the form of a pronounced **analysis of customer touchpoints**.

Such an analysis of the so-called **Customer Journey** seems particularly relevant against the background of an increasing saturation of markets, steadily increasing internationalization and technological developments in the context of digitalization. Here, the customer and his contact points with the company come into a completely new perspective (Narayanan, 2015). Not only do numerous new communication and distribution channels arise, but customers also increasingly act as co-designers, consultants or advertising ambassadors for companies (Bruhn & Hadwich, 2015). They contribute information that enables the provider to render services as needs-based and of high quality. Accordingly, customers actively participate in the service creation process and contribute in an intellectual (e.g. advisory services), physical (e.g. medical therapies) or emotional (e.g. music events) way (Bruhn & Hadwich, 2015).

It becomes clear that the customer's contact with the company no longer limited to the role of a passive service consumer (Ramani & Kumar, 2008; Sharma et al., 2014). Rather, it is necessary to consider the customer in his active role as a valuable resource for the company, which allows the customer to be integrated into the service creation process (Pagani et al., 2011; Singh & Sonnenburg, 2012). For the most efficient possible **design of the customer's integration** into the provider's value creation, it is therefore recommended to carry out an analysis of the Customer Journey (Tax et al., 2013). This subsumes the customer's contact with the company from the first purchase impulse to the purchase decision (Böcker, 2015; see Fig. 3.6). In the course of time, starting from the pre-consumption and development phase via the production and use phase, the customer satisfies his individual information and consumption needs before he finally evaluates the consumed service in the last phase of the post-purchase phase, especially with regard to its quality (Böcker, 2015).

In terms of the customer journey, it can be observed that an **interaction with the customer** takes place in all of its phases. So the request from a provider for a service in the pre-consumption phase is to be identified as a first interaction, while in the development phase this can already participate in another interaction in the specification of the service (co-design). In the production and use phase, the interaction is characterized by the assumption of disposition functions as well as the provision and integration of its external factors (Fliess et al., 2011). During the post-purchase phase, finally, the communication of the customers to the company and also among themselves can generate values in favor of all parties. So customers communicate their experiences with each other, recommend services further or complain if they are dissatisfied with the respective provider (Hirschman, 1970; Kozinets et al., 2010). Accordingly, the customer is finally integrated into the advertising and communication process of the provider.

The multi-layeredness of the described different characteristics of the customer's integration into the service provider's value creation is also reflected in the variety of approaches currently being discussed in the scientific literature. These include the following **concepts of customer integration:**

1. Value Co-Creation,
2. Customer Participation,
3. Co-Production,

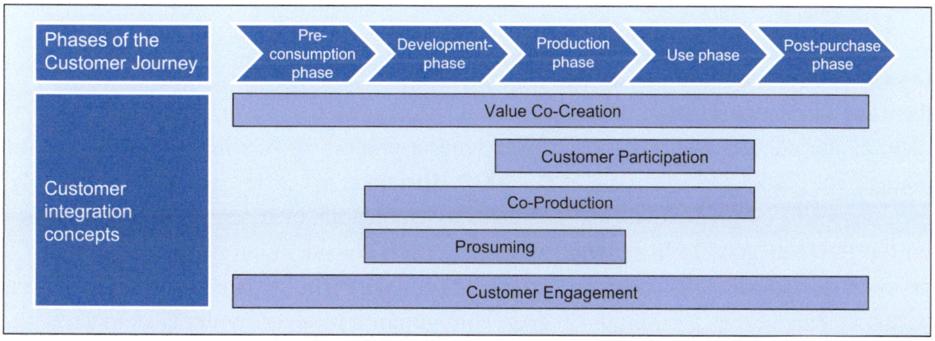

Fig. 3.6 Concepts of customer integration in the context of the Customer Journey

4. Prosuming,
5. Customer Engagement.

These approaches to customer integration for the purpose of increasing value creation and thus also improving service quality will be explained in more detail below.

(1) Value Co-Creation The term "value co-creation" includes the simultaneous consideration of the customer's and the company's participation in the company's value creation ("**customer integration**") as well as the integration of the company into the customer's processes **("supplier integration").** Accordingly, suppliers and customers cooperate on an equal footing in the context of Value Co-Creation in order to generate a high level of service quality and the highest possible benefit for both sides (Fliess et al., 2011). As a result, the customer is no longer just a supplier and bearer of external production factors, but takes over all or part of the disposal authority for the factor combination (Kleinaltenkamp, 1993; Fliess et al., 2011; Meik, 2016). With regard to this basic principle, it can be seen with regard to the phases of the customer journey and the resulting contact points of the customer to the company that customers can intervene in all phases of service provision (Fliess et al., 2011) and thus contribute to a high level of service quality.

(2) Customer Participation In the existing scientific literature, customer participation is defined as the degree to which a customer is involved in the production and delivery of (service) performance (Dabholkar, 1990, p. 484). In this context, it is possible to differentiate between physical and mental customer participation (Silpakit & Fisk, 1985) as well as joint and customer-side production (Bruhn and Hadwich 2015). Accordingly, the highest form of customer participation can be provided exclusively by the customer without the involvement of employees of a supplier (e.g. use of self-service technologies) (Bitner et al., 1997; Meuter et al., 2000). It can be seen that, with regard to the phases of the Customer Journey and the resulting contact points of the customer to the company, the concept of Customer Participation essentially relates to the production and use phase of the service.

(3) Co-Production The concept of customer co-production represents an extension of the view of customer participation presented. It includes the active, significant involvement of the customer during the service creation process by providing input factors or taking over partial services (Auh et al., 2007; Bilstein et al., 2015, p. 290). Accordingly, the concept of co-production (e.g. customer assembles his own cereal, which he orders on the Internet) goes hand in hand with an explicit consideration of the customer as a production resource (e.g. customer as service provider) (Bruhn & Hadwich, 2015). The customer thus acts as a co-producer (e.g. configuration of a car by the customer), without whose effort and time investment the production of services is not possible in most cases. Accordingly, cooperation between provider and customer is essential for successful service creation (Cowell, 1993, pp. 219–221). It becomes clear that the term customer participation refers to individual sequences of action, while the term co-production is used in a broader sense (Büttgen, 2007). Co-production can thus, for example, include several different interactions between employees and customers, while customer participation only refers to a specific interaction between provider and customer (Büttgen, 2007). Accordingly, it is also clear within the context of the co-production concept that, with regard to the phases of the customer journey and the contact points of the customer to the company that are created in the process, the concept of co-production refers to the production and use phase of the service.

(4) Prosuming The term Prosuming, which is a composition of the words Producer and Consumer, takes a similar perspective to Co-Production and Customer Participation by emphasizing the simultaneity of production and consumption (Michel, 1996). In this context, various forms of customer interaction are described, such as the partial or complete creation of services by the customer (e.g. online check-in at an airline) as well as the involvement of customers in production and development processes of the provider (e.g. the independent configuration of sneakers) (Bruhn & Hadwich, 2015). Accordingly, it becomes clear that the concept of Prosuming can be distinguished from the closely related concepts of Customer Participation and Co-Production by its exclusive focus on the development and production phase. Accordingly, the customer also takes a decisive influence on the quality of the service received at this point. One variant of Prosuming is the term **Mass Customization,** which makes use of modern information and communication systems as well as flexible production systems (Reichwald & Piller, 2009). Today, this is considered one of the essential trends for achieving competitive advantages on consumer goods markets (Piller & Müller, 2004). This is accompanied by an increase in the willingness of customers to pay and thus an improvement in the revenue base of a provider (Fliess et al., 2011). With regard to the phases of the Customer Journey and the resulting contact points of the customer with the company, the concept of Prosuming shows that this extends from the development and pre-consumption phase through the production to the use phase.

(5) Customer Engagement The concept of customer engagement in science is generally described in the existing literature as a mixture of a kind of rational loyalty and emotional attachment to a brand or company (Brodie et al., 2011, p. 266). It is necessary to distinguish here between engagement for a company or for a customer (van Doorn et al., 2010; Bruhn et al., 2015b). For example, an *IKEA* customer can engage with another customer by advising him during furniture shopping and thus avoiding his dissatisfaction (Meffert et al., 2018, p. 290). However, he engages with a company when, for example, *McDonald's* publishes a call for tenders for its own burger ideas, as part of which the best burger recipe is included in the range for a certain period of time (Meffert et al., 2018, p. 290). Furthermore, planned or spontaneous engagement can be differentiated (van Doorn et al., 2010, p. 255; Bruhn et al., 2015b). Planned engagement, for example, when a customer of *Apple* develops a new, free app for the *iPhone*. However, spontaneous engagement can be subsumed under the unexpected recommendation of a service (Meffert et al., 2018, p. 290). As the numerous examples of customer engagement show, this concept is a approach in quality management and the customer journey that can affect all customer contact points with the company.

The different terms for customer interaction in the value creation of a company summarized in Fig. 3.6 can be explained primarily by the different reference points of the concepts. While Co-Production and Prosuming have a strong production-economic focus, Customer Participation emerged from a behavioral science perspective. Value Co-Creation and Customer Engagement form comprehensive concepts of customer integration and are located in customer management (Fliess et al., 2011).

Against the background of the described theoretical concepts, the question arises both scientifically and especially from a business point of view how the Interaction with the customer can be carried out within the framework of Customer Journey can be increased and their potentials in service management can be used to improve service quality (Tax et al., 2013). Accordingly, there is a desire to better understand the customer in his different information and decision-making behavior in order to analyze the effect of individual customer contact points on the company and thus to achieve a higher level of customer understanding (Böcker, 2015). In this context, it must be taken into account that the information-seeking behavior of customers and their evaluation of the quality of the consumed service differ. As a result, it is necessary to capture the wealth of individual customer contact points and to analyze and optimize them from the customer's perspective. For this purpose, the planning tool of the **Customer Path Analysis** (Blueprinting, Service Mapping; see also Sect. 5.1.2.2) is available (Shostack, 1992). Here, the customer process is divided into sub-phases and visualized in a flowchart. This way, a sequence of customer contact points is identified from the customer's perspective with the respective interactions (Shostack, 1992; Kleinaltenkamp, 2000). In addition, it can be made clear in which areas of the service delivery system the customer can interactively participate. This creates the planning basis for a targeted coordination of the integration processes within the individual customer contact points. Figure 3.7 illustrates the Customer Path Analysis using the example of an air travel.

When creating the blueprint, it is particularly important to note that each interaction creates an experience for the customer. These **customer experiences** consist of functional features (e.g. simple, high quality, reliable, useful) and emotional impressions (e.g. feeling of appreciation, entertaining, empathetic) (Bruhn & Hadwich, 2015, p. 24). In a further step, it is therefore necessary to survey, evaluate and interpret the relevance and perception of these "experience drivers" or operational levers of individual interactions from the customer's perspective using suitable market research measures (see also Sect. 5.1.2). Accordingly, it can be concluded that the integration of the customer into

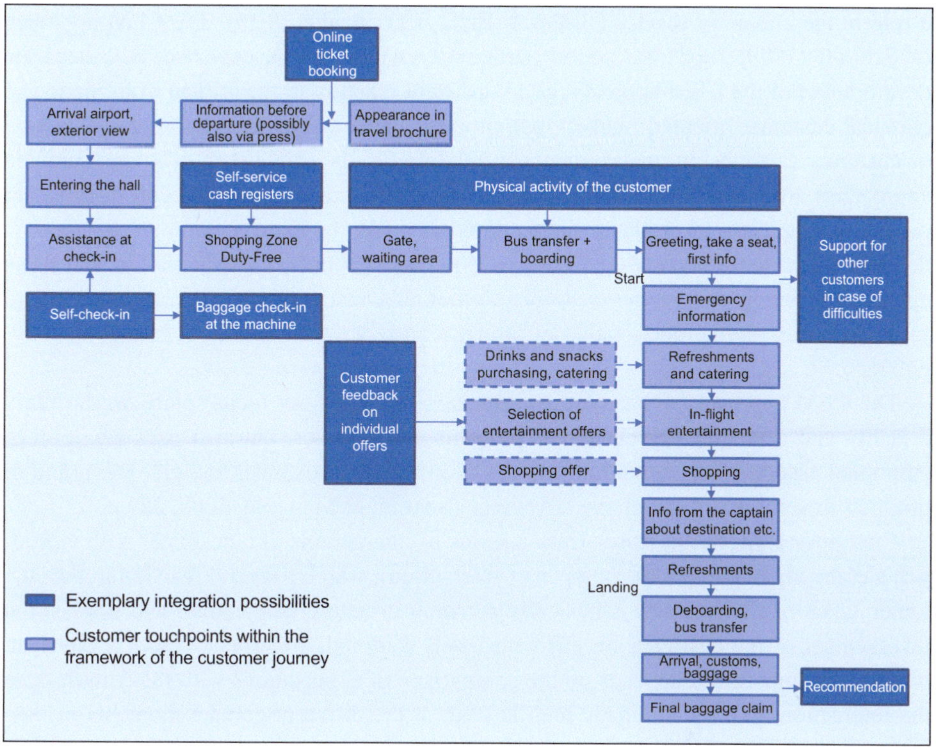

Fig. 3.7 Customer Journey using the example of an air travel (Source: Based on Meffert et al., 2018, p. 225)

the value creation processes of the service provider within the Customer Journey is an essential factor for the quality management of services and the success of the company, which must be observed and adapted precisely in the light of constantly evolving market conditions.

3.5.2 Management of Customer Experiences

In connection with the integration of the customer into the company's value creation, a customer-oriented approach called **Customer Experience Management (CEM)** is being discussed in recent marketing research (e.g. Gentile et al., 2007; Meyer & Schwager, 2007; Verhoef et al., 2009; Bruhn & Hadwich, 2012; Mayer-Vorfelder, 2012), which is gaining increasing popularity in practice and goes beyond traditional quality management. This is a customer-oriented concept in which the customer and the analysis of the customer experience form the basis for the development and implementation of a customer-centric corporate strategy. The customer is not seen as a rational decision-maker, but also symbolic, hedonistic and aesthetic consumption motives play a role in the choice of service (Bruhn & Batt, 2011; Bruhn, 2016). The CEM approach of Schmitt (1999, 2003) has found particularly widespread acceptance, who sees the development of the CEM approach as a consistent reaction to the critical examination of classical customer-oriented marketing approaches such as quality, customer satisfaction or customer relationship management. In his opinion, the established customer-oriented approaches are actually only limited customer-oriented, because in the core they rather put the management of products and technologies than the management of customers in the center (Schmitt & Mangold, 2004, p. 16). This is shown, for example, in that these approaches largely ignore the customer's emotional reactions during consumption and focus primarily on the functionality of products and services (Schmitt & Mangold, 2004, p. 20).

The CEM approach is therefore an approach that does not focus solely on the functionality and a satisfactory quality of products and services, but also pays attention to emotional aspects of consumption in order to inspire the customer and give the brand of products or services a competitive advantage (Schmitt & Mangold, 2005, 289 pp).

Customer Experience arises from a series of interactions of a customer with a product, a company or parts of the company (interaction), which trigger a reaction in the customer. Customer experience is basically personal in nature (subjective) and implies the involvement of the customer on different levels (rational, emotional, sensory, physical, mental). Its importance depends on the comparison of expectations with the stimuli from the interaction with the company and its offer at the different contact moments or customer touchpoints (Bruhn & Hadwich, 2012). In addition to generating customer enthusiasm, the main **goals of customer experience management** are the emotional customer loyalty as well as the increase in sales (Detecon International GmbH, 2010).

Customer experiences can be distinguished in terms of their experience properties and classified into five experience dimensions that are closely related to each other (Gentile et al., 2007, p. 398):

- **Sensory experiences** arise in the customer through the reception of environmental stimuli via the senses (eyes, ears, tongue, skin, nose). These stimuli trigger excitations which become conscious to the customer as sensations and cause a sensory perception (e.g. background music in the sales areas of commercial enterprises which is supposed to stimulate buying subliminally).
- Reactions of customers such as emotions and moods as well as the assessment of situations belong to the **emotional resp. affective experiences** (e.g. conveying holiday mood in posters and commercials of the tour operator *TUI*). They aim at conveying a positive feeling to the customer regarding the provider, the brand or the service.
- **Cognitive experiences** focus on the intellect of the consumer and include the reception, processing and storage of information. Their added value lies in the fact that they animate the customer to creativity and mental engagement with the brand (e.g. provocative and not unambiguous advertising messages in order to achieve intensive engagement with the advertising message).

- In the case of **behavioral experiences**, it is about conveying physical experiences, ways of using, alternative lifestyles or also interaction possibilities (e.g. privileges and advantages of a *Miles-and-More* membership with *Lufthansa*). The goal of conveying behavioral experiences is to achieve a change not only in the rational behavior but also in the lifestyle of the consumer.
- The so-called **lifestyle dimension** provides arguments for the confirmation of the values and opinions of consumers.
- **Social experiences** arise through events of interaction, contact and relationship maintenance. In addition to enabling interaction, an additional benefit is supposed to be generated for the customer through a social identity and a feeling of belonging (e.g. *Harley-Davidson* fan club).

In Fig. 3.8 the experience dimensions of the customer experience and central moderators that can influence the relationship between the dimensions and the construct customer experience are shown. The moderators are divided into those related to the situation (e.g. the current economic situation) and those related to the customer (e.g. socio-demographic characteristics such as the customer's gender and age).

A four-stage approach is proposed for targeted construction and control of customer experiences. This includes the following idealized phases of analysis, planning, implementation and control of a CEM management process: (Bruhn & Hadwich, 2012, 23 pp.):

1. Analysis of the Customer Experience Management,
2. Planning phase of the Customer Experience Management,
3. Implementation phase of the Customer Experience Management,
4. Control phase of the Customer Experience Management.

The individual **phases of theCustomer Experience Management** are explained in more detail below.

(1) Analysis of the Customer Experience Management The first stage includes the identification, analysis and measurement of individual customer experiences. Within the framework of a **causal analysis**, important experience drivers can be analyzed and subjected to an experience measurement. The service blueprint is a suitable method for determining individual experience contacts (see also Sect. 5.1.2.2).

Figure 3.9 shows an example of a structural equation model for measuring the effects of experience contacts, as they are often used in customer insight analyses. The customer insight analyses are also suitable for the control phase in order to check whether the corresponding psychological, behavioral and economic effects have been achieved by means of the measures of the customer experience management.

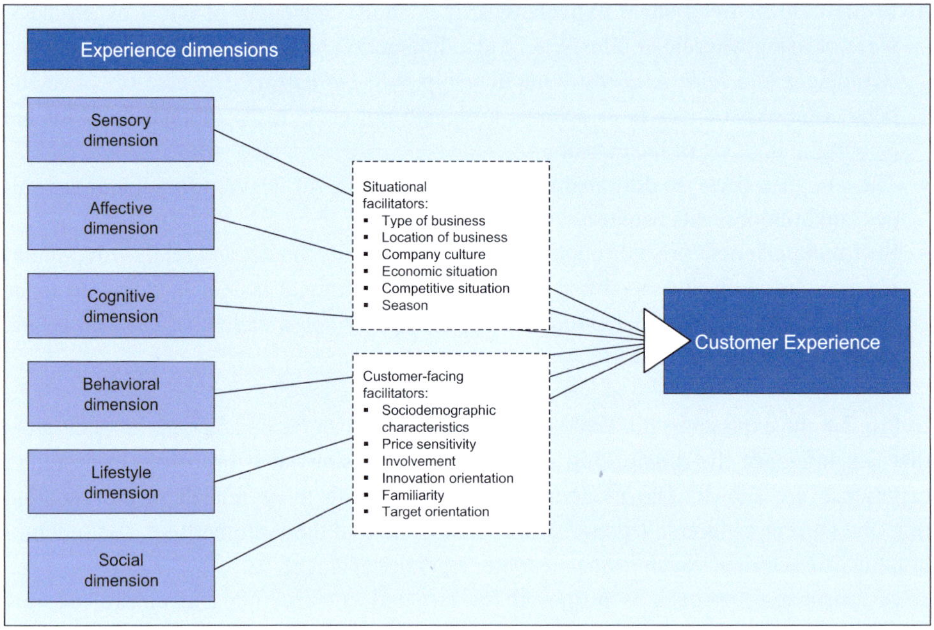

Fig. 3.8 Dimensions and moderators of the customer experience (with examples) (Source: Bruhn & Hadwich, 2012, p. 14)

(2) Planning phase of Customer Experience Management After the customer experience analysis, the planning phase follows with the definition of goals and the strategic and operational planning of customer experiences. The focus is on generating positive and unique customer experiences. In the strategic planning phase, an overarching experience motive is determined. This experience motive reflects the company's central values (Schmitt & Mangold, 2004; Mayer-Vorfelder, 2012). The operational planning phase finally defines the individual design elements to be used to generate customer experiences. It must be determined which specific measures and instruments are to be used to implement the experience motive. These include, for example, a multi-sensory approach to customers, employee-customer- interactions and the design of functional and emotional experience drivers (Zomerdijk & Voss, 2010).

Multi-sensory customer approach in retail

The transmission of a customer experience through all senses is often used in retail companies such as supermarkets, clothing stores or perfume shops. Through an attractive store and shelf design, the visual sense is stimulated, pleasant smells stimulate the sense of smell. A matching background music also contributes to the fact that the customer feels comfortable, stays longer in the store and buys more.

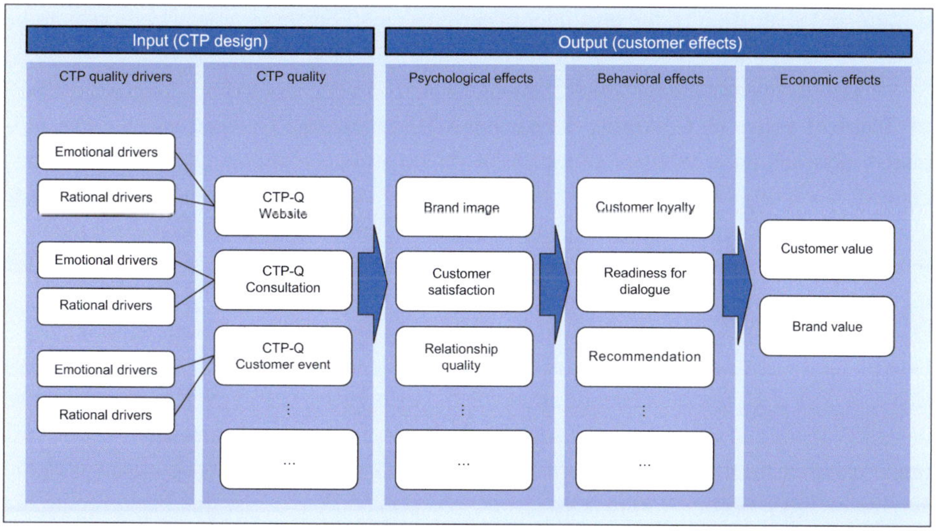

Fig. 3.9 Structural equation model for measuring the effects of customer experience contacts

For example, scientific studies confirm the effectiveness of fragrance marketing in retail. Studies in both food and shoe stores have shown that factors such as music, temperature, interior design and fragrance are important factors in the perception of atmosphere. It has been shown that in certain areas of the store, customer flows increased by 62% and spontaneous purchases from 29 to 62% (Oltmanns, 2011, p. 74).

However, it must be ensured that the individual stimuli are coordinated and convey the same message or mood. For example, a subdued atmosphere with dimmed light in a clothing boutique should not be combined with loud techno music. ◀

(3) Implementation phase of Customer Experience Management In the implementation phase, internal framework conditions are to be created that enable the emergence of positive customer experiences. In this context, an experience-oriented focus of the structures, systems and culture within the company is to be ensured. An important prerequisite is the support and acceptance at board level. The employees in customer contact play a significant role in shaping the customer experience due to the high interaction with the customer. Therefore, it is important to ensure that the implementation of a corporate culture that sees customer experiences as the foundation for achieving competitive advantages is ensured (Schmitt & Mangold, 2004, 159 pp.). In the context of the adaptation of the company systems, the introduction of analytical CEM systems for the systematic capture and evaluation of the customer experience at all customer contact points is to be ensured.

(4) Control phase of Customer Experience Management The success of CEM ultimately depends on a continuous control of the experience-enhancing measures. In this context, tracking systems based on the principle of customer barometers are used (see also Chap. 14). A regular control shows whether the measures taken to increase the customer experience pay off or which measures need to be improved (see also Fig. 3.7 for measuring the effects of customer touchpoints).

Due to the increasing importance of service brands, the construction of a brand experience world is a possible starting point for the customer experience management. The design of the brand experience for services includes the systematic management of customer experiences through the careful planning of tangible and intangible elements in the service delivery system (Pullman & Gross, 2004, p. 551). Against this background, six central **experience-centered design elements** are proposed (Zomerdijk & Voss, 2010, 68 pp.):

- The design of experience-centered services implies the creation of a **sequence of service encounters** and so-called **"key points",** which shape the context in which experiences arise. These starting points can arise at any stage of the service process and are—if perceived—combined by service customers to holistic experiences.
- Against the background that a customer experience is all the more effective the more it addresses the senses of the consumer, it is necessary within the framework of an experience-oriented service design to develop the service elements in such a way that all **senses of the consumer** are addressed.
- The direct contact between employees and customers represents a central starting point for service companies to trigger emotional reactions from the customer. Consequently, the conditions for direct **employee-customer interactions** are to be created within the framework of an experience-oriented service design.
- The provision of services is often equated with the staging of a drama due to its course, duration and underlying dramaturgy, for example, where employees can be compared to actors, customers to spectators and the tangible environment to a stage. Consequently, the **dramaturgical structure of services** must be taken into account in experience-centered design.
- The experience of customers is not only influenced by the interaction with the service provider, but also by the presence of other customers. For this reason, experience-centered service design also includes the **management of the presence of other customers.**

- In the design of service experiences, not only the employees in customer contact play an important role ("frontline employees"), but also those employees who act behind the scenes of the service provision ("back-office employees") . They perform an important function in providing the design elements of holistic customer experiences. Accordingly, an experience-centered service design requires the balanced **coordination of frontline and back-office employees.**

Examples of brand experience worlds

The origin of brand experience worlds lies in the automotive industry. All major German car manufacturers have created an experience world that allows customers to come into contact with the car and the brand *VW* up close. *VW*, for example, has created the Autostadt, an experience world in which the car and the brand *VW* are in the center. Buyers of a new car have the opportunity to pick up their vehicle directly from the factory in Wolfsburg and combine this with a visit to the Autostadt. The experience world has developed into an independent company and appears as *Autostadt GmbH* (Autostadt, 2013). Meanwhile, numerous manufacturers of consumer goods have also moved on to open so-called flagship stores, which are also brand worlds. As an example, *Beiersdorf* can be mentioned, which, among other things, have opened a flagship store for the brand *Nivea* in Hamburg. In service industries, brand experience worlds are still little known. Retailers have already started to equip their individual stores with an experience character, so that each individual store can be referred to as its own brand experience world for the respective retailer. ◀

After the theoretical and conceptual basics of quality management for services have been put into place, the next step is to carry out an analysis of service quality. For this purpose, the most important models of service quality will be presented in the next chapter.

3.5.3 Management of Processes of Service Transformation

Currently, trends of shortened product life cycles, commoditization of products, price decline and increased digitalization can be observed in many markets (Benedettini et al., 2015; Bithas et al., 2015). However, these changes are not just evolutionary, but rather structural innovations of markets and companies (Bruhn & Braun, 2016). As a result, the term "disruptive change" is used more and more often in some industries (Bruhn & Hadwich. 2016). To meet these challenges and take advantage of the resulting opportunities, such as creating new market activities, developing new business models or finding new ways of value creation (Bithas et al., 2015), the so-called **strategy of service transformation** has become increasingly common in recent years.

This is how companies change along a **process continuum from pure product manufacturer to service provider** (Oliva & Kallenberg, 2003; Brax, 2005; Nordin et al., 2011). Here, companies no longer only offer pure products on the market, but increasingly include services in their portfolio (Rada & Vandermerwe, 1988). This process finally leads to the highest level of service transformation, the benefit-generating value creation cooperation (Bruhn et al., 2015a). Accordingly, the considerations of Vargo and Lusch (2008) can be assigned, by using the extreme characteristics of Goods-Dominant and **Service-Dominant Logic** as a framework. With the term transformation, it is accordingly expressed that the process of change is in the center of the consideration (Ostrom et al., 2010). Synonymous terms for the process of service transformation form in the scientific literature servitization (Rada & Vandermerwe, 1988), service infusion (Kowalkowski et al., 2012) or service transition (Oliva & Kallenberg, 2003). Furthermore, numerous terms are used in the context of service transformation, which consider the service offering as an object of service transformation. These include hybrid bundles of services (Shankar et al., 2009), system solutions (Brax & Jonsson, 2009) and Product-Service Systems (Baines et al., 2007). While hybrid bundles of services emphasize the importance of the combination of products and services to create a higher customer benefit, system solutions focus on the customer-specific individual problem solution (Bruhn & Hadwich, 2016). It should be noted here that the variations in the conceptualization of the transformation process to the service provider contribute to a high level of research, but it is important to ensure that necessary theoretical progress and empirical studies are not restricted (Gebauer & Saul, 2014).

In the context of service transformation, it is particularly important to consider the **change processes of companies** that can be divided into four phases and value creation models (cf. Fig. 3.10). The plausibility of the four phases and their development steps was carried out on the basis of practical examples and a panel of experts from the senior management of various company departments (Bruhn et al., 2015a). The typology of the individual phases starts from the constitutive features of services and builds on the discussion of dimensions for the typology of performance by Engelhardt et al. (1992) and Meffert (1993). First, a distinction is made on a vertical level as to how far an adaptation of the service provider's performance to customer wishes is required and how high the immateriality of the service result is (2015a). The horizontal axis, on the other hand, includes the two dimensions interaction of the service provider partnership and integration level of the service offering (2015a).

In the context of this **typology** the first stage of service transformation includes the **product-oriented business model,** in which the product of the provider is the core service and the intangible service is closely related to the core service (2015a). Depending on the level of development, the provider offers additional intangible services in addition to its original product in order to promote the sale of the core service and increase productivity.

The second step of service transformation is to develop into a so-called **system solution provider** (Bruhn et al., 2015a). The aim is to offer a complete range of services from one source. Accordingly, the system solution provider expands its product range

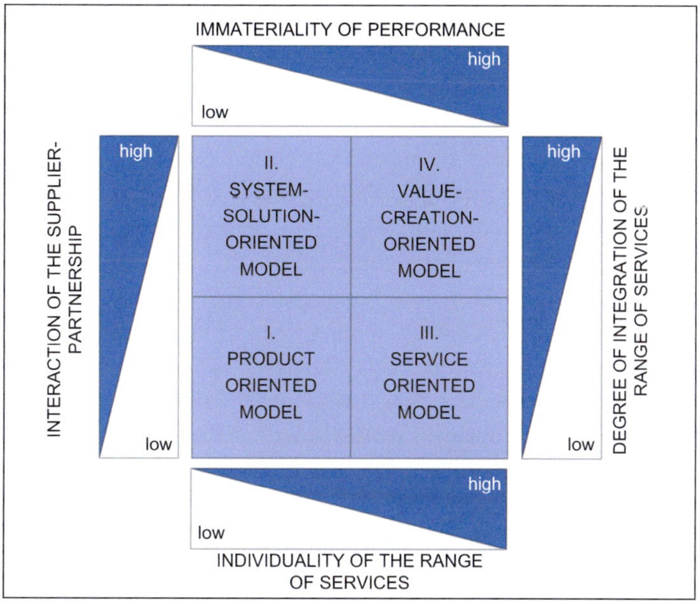

Fig. 3.10 Phase model of service transformation (Source: Bruhn et al., 2015a, p. 60)

in this transformation process step with a performance guarantee and no longer sells the product itself, but the associated service (Bruhn & Hadwich, 2016).

The third stage in the phase model of service transformation is the **service-oriented business model** (Bruhn et al., 2015a). Here the service itself is the actual core service. The task in this context is to build up new core competencies and to create independent organizational units within the company to generate a high customer and service orientation (Bruhn et al., 2015a; Bruhn & Braun, 2016).

The **value-oriented business model** finally represents the most extensive development step in service transformation. This combines the previous three stages of the product, system solution and service provider and thus offers the customer in addition to the planning, production and financing of the overall product, its operation and/or marketing (Bruhn et al., 2015a). Here, the classic customer-supplier relationship changes to the effect that now both sides act as equal partners on the market (Bruhn & Hadwich, 2016).

It can be seen that the transformation to the service provider offers companies the opportunity to differentiate themselves from existing markets with ever narrower margins and increasing competition pressure (Bruhn & Braun, 2016). The development of close customer relationships is possible and in particular economic potential can be realized (Bruhn & Hadwich, 2016). Often, however, existing business models change with the transformation and there is a **shift in value creation** from product value creation to service value creation. Difficulties in this context often arise from the only vaguely possible prediction of the success of the new business model (Parks et al., 1991), uncertainties regarding the new market and the new competition in which it is to be operated (Brinckmann et al., 2010), as well as the lack of information for decision-making (McMullen & Shepherd, 2006).

In this context, it is essential for companies to always align themselves with the needs of their customers and create corresponding capacities that allow for a high degree of customer orientation (Kindström et al., 2013). This means that, at the conceptual level, marketing activities are changing from a transaction- to a **relationship orientation.** On the operational level, however, marketing competence goes beyond customer relationship management. They now include skills that contribute to understanding customer needs, adapting solutions to customer needs, and integrating products and services into solutions (Gebauer & Saul, 2014). This is accompanied by a corresponding **adaptation of the company's strategy and business model** as well as the adaptation of existing organizational structures and processes in order to ensure a successful service transformation process (Kowalkowski et al., 2015).

In this context, a review of the various empirical studies on the development status of service transformation (Neely et al., 2011; Bielefeldt et al., 2016; Parbs & Pöppelbuß, 2016) shows that the topic of service transformation has arrived in corporate practice, but there is still considerable development potential. It also becomes clear that while the **process of service transformation** as well as the perspective of the provider with opportunities for strategy development and implementation of the transformation are already intensively discussed in the scientific literature, it is still unclear how the customer deals with the process of service transformation (Braun et al., 2016). For example, Juha and Pentti (2008) and Maiwald et al. (2014) show that the transformation to a service provider is associated with numerous perceived risks. It is possible, for example, that customers have doubts about the provider's competence to provide the service, the provider is perceived as a monopoly or there are difficulties in evaluating the new service (Lindberg & Nordin, 2008; Hogreve & Wonner, 2014; Maiwald et al., 2014).

It becomes clear that there are numerous barriers that often lead to the fact that the process of service transformation is doomed to failure. An example of this is the so-called **service paradox** (Brax, 2005; Gebauer et al., 2005). This implies that although implementing a service-oriented business model is associated with an increase in profits, the implementation of the service transformation is associated with numerous challenges that often result in companies achieving the opposite and making losses that can lead to insolvency (Brax, 2005).

The basic requirements for avoiding the service paradox are a high **market and service orientation of the company** (Bruhn & Braun, 2016). Because only this way can a comprehensive investigation of customer needs be guaranteed within the process of service transformation (Agrawal et al., 2012). In this context, Bruhn and Braun (2016) carry out a comprehensive analysis of the requirements for the service transformation process in the interface to market and service orientation. Here, on the basis of a comprehensive literature analysis and expert interviews, the authors identify important provider-side and customer-side requirements for the successful implementation of the service transformation process (cf. Fig. 3.11).

The results of this study show a total of 14 supplier-side **requirements for service transformation,** these are: changes in the thinking of the provider company, close cooperation between provider and customer, the generation of trust, a learning organization, an increase in the joint value creation between provider and customer, service orientation of the company structures, systems and culture, the transfer of knowledge, a permanent implementation of customer needs analyses, the definition of KPIs for measuring the success of the service transformation, the avoidance of the service paradox, service-/value-oriented business models, service innovations, to innovate the customer as well as the appropriate use of the increasing digitalization. With regard to the **customer-side perspective** the following ten requirements were worked out: assessability of the new services/solutions offered, customer-side acceptance of the new services/solutions offered, reduction of the perceived customer-side risk, reduction of the lack of customer-side trust and increase of the credibility of the provider, the service transformation promoting properties of the customer, the customer-side ability and willingness to enter into interaction with the provider, a high confidence of the customer, increase of the customer-side integration and involvement, reduction of perceived dependencies as well as the analysis of changes of customer-side processes, procedures and networks.

This study by Bruhn and Braun (2016) also makes it clear that the relevance of service transformation and the mastery of its challenges are undisputed not only in science, but also in corporate practice. For example, manufacturing companies supplement their products with services and invest in expanding their service business (Bruhn & Hadwich, 2016). Companies such as *ABB, Caterpillar, General Electric, IBM, Siemens,* etc. grow mainly in the service business, and the share of services in sales is currently around 50% (Bruhn & Hadwich, 2016). As a result, it is becoming apparent that all industries are currently undergoing a process of service transformation, and in many cases the existing industry participants are surprised that completely new service providers are appearing on the market (Nordin et al., 2011; Bruhn & Hadwich, 2016). It is becoming apparent that the service transformation is not only about offering services, but rather about the development of innovative service-based business models (Kastalli & van Looy, 2013). Often, the focus is on digital transformation (Akram, 2012; Lerch & Gotsch, 2014; Opresnik & Taisch, 2015). This describes corporate realignments of business models and investments in technologies in order to efficiently control customer contact points and customer experiences in the course of digitalization (Solis et al., 2014, p. 3). Examples from recent years are *Amazon* in the book trade, *Zalando* in the shoe trade, *Spotify* in the music industry, *Uber* in the taxi business, *Airbnb* in the hotel industry and *Car2Go* in the automobile industry (Bruhn & Hadwich, 2016). It is assumed that this digital transformation will extend to other traditional areas.

A closer look at the dynamics of digital transformation shows that it is characterized by five drivers (Kollmann, 2013). First, the rapid development of computer performance is an important prerequisite. Secondly, the digitalization of data increases the economic

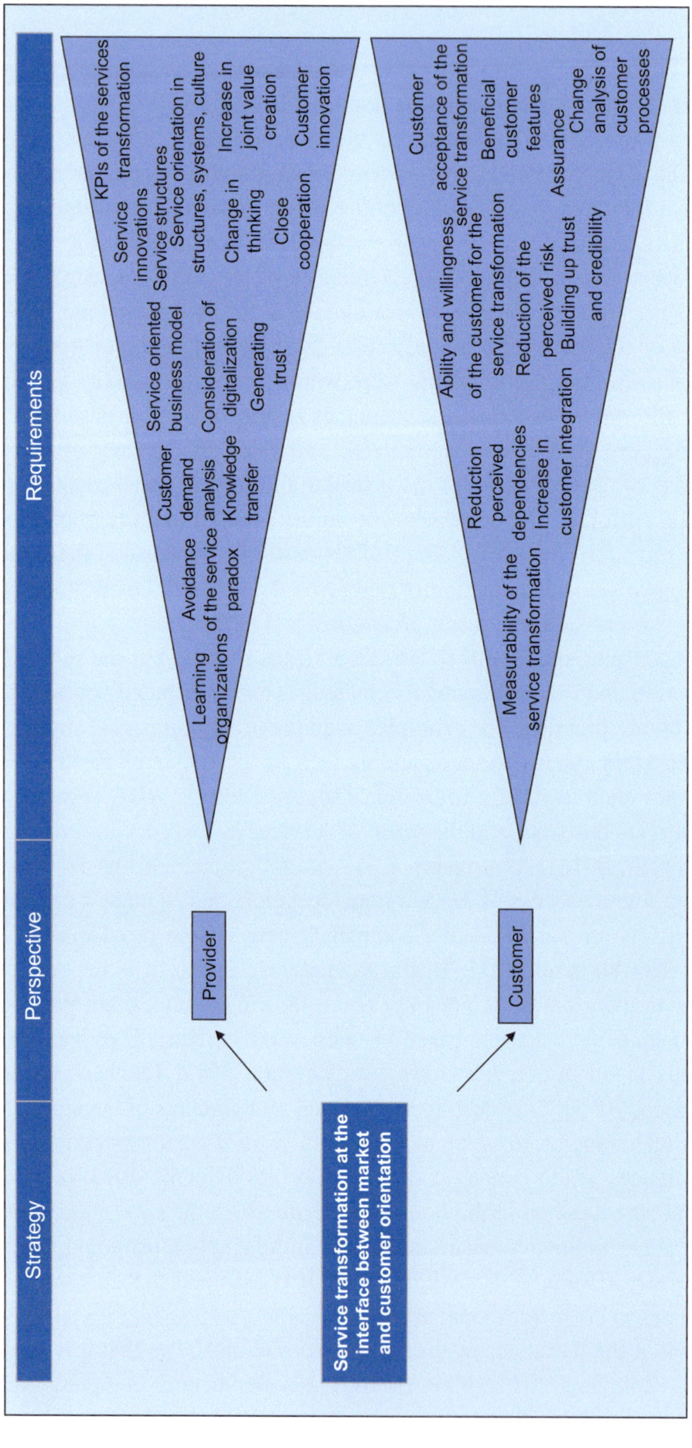

Fig. 3.11 Supplier and customer-side requirements for a successful service transformation (Source: Bruhn & Braun, 2016)

attractiveness of the use and dissemination of digital information due to cost regression effects. Third, the increasing networking is the basis for the development of new communication forms (e.g. social media and web 2.0), which enable very fast data transfer. Fourth, there is a strong growth in data volumes. Finally, the fifth driver is mobile data transfer via new terminals (Kollmann, 2013).

Accordingly, it can be seen that the strategy of service transformation not only brings about considerable changes within individual industries, but also decisively influences the quality of services. So the transformation from product manufacturer to service provider is associated with a **quality jump** that enables faster and more needs-based service creation and thus contributes decisively to increasing customer satisfaction and customer loyalty.

Against this background, the transformation from product manufacturer to service provider has developed into a prominent research topic. This is also reflected, among other things, in the continuously increasing number of scientific contributions (Braun et al., 2016; Fliess & Lexutt, 2016). However, the transformation from product to service provider also raises numerous questions that have not yet been fully answered in the scientific literature (for an overview, see, for example, Bruhn & Braun, 2016). For example, despite the large number of scientific papers, there is still no common theoretical basis or extensive empirical study of the process of service transformation available today (Baines et al., 2007; Gebauer & Saul, 2014). Accordingly, there is still an extraordinary need for research here, and it is important to follow up with further discussions on this topic.

References

Agrawal, V., Ferguson, M., Toktay, L., & Thomas, V. (2012). Is leasing greener than selling? *Management Science, 58,* 523–533.

Akram, A. (2012). *Towards servitization in the age of digital innovation: A case from vehicle industry.* The 35th Information Systems Research Seminar in Scandinavia (IRIS 2012). Sigtuna.

Atul, G., Jason, C. M., & Herath, S. K. (2005). Quality management in service firms. Sustaining structures of total quality service. *Managing Service Quality, 15*(4), 389–402.

Auh, S., Bell, S. J., McLedo, C. S., & Shihd, E. (2007). Eric Shihd, co-production and customer loyalty in financial services. *Journal of Retailing, 83*(3), 359–370.

Autostadt. (2013). Autostadt. www.autostadt.de. Accessed 22 Apr. 2013.

Baines, T. S., Lightfoot, H. W., Evans, S., Neely, A., Greenough, R., Peppard, J., Roy, R., Shehab, E., Braganza, A., Tiwari, A., Alcock, J. R., Angus, J. P., Bastl, M., Cousens, A., Irving, P., Johnson, M., Kingston, J., Lockett, H., Martinez, V., Michele, P., Tranfield, D., Walton, I. M., & Wilson, H. (2007). State-of-the-art in product-service systems. Proceedings of the institution of mechanical engineers. *Journal of Engineering Manufacture, 221*(10), 1543–1552.

Benedettini, O., Swink, M., & Neely, A. (2015). Why do servitized firms fail? A risk-based explanation. *International Journal of Operations & Production Management, 35*(6), 946–979.

Berekoven, L. (1983). *Der Dienstleistungsmarkt in der Bundesrepublik Deutschland.* Göttingen: Vandenhoeck & Ruprecht.

Bergbauer, A., Kleemann, B., & Raake, D (2008). *Six Sigma in der Praxis. Das Programm für nachhaltige Prozessverbesserungen und Ertragssteigerungen* (3rd Edn.). Renningen: Expert.

Bielefeldt, J., Thaler, V., & Herbst, H. (2016). Servicetransformation als Treiber von neuen Geschäftstätigkeiten – Eine empirische Betrachtung in unterschiedlichen Branchen. In Bruhn & Hadwich (Eds.), *Servicetransformation. Forum Dienstleistungsmanagement* (pp. 145–170). Wiesbaden: Springer.

Bilstein, N., Hogreve, J., & Bartilla, K. (2015). Auswirkungen der Co-Produktion auf die Kunden-Mitarbeitenden-Beziehung. In M. Bruhn & K. Hadwich (Eds.), *Interaktive Wertschöpfung durch Dienstleistungen. Forum Dienstleistungsmanagement* (pp. 287–308). Wiesbaden: Springer Gabler.

Bithas, G., Kutsikos, K., Sakas, D. P., & Konstantopoulos, N. (2015). Business transformation through service science: The road ahead. *Procedia-Social and Behavioral Sciences, 175,* 439–446.

Bitner, J. M., Faranda, W. T., Hubbert, A. R., & Zeithaml, V. A. (1997). Customer contributions and roles in service delivery. *International Journal of Service Industry Management, 8*(3), 193–205.

Böcker, J. (2015). Die Customer Journey – Chance für mehr Kundennähe. In Deutscher Dialogmarketing Verand e. V. (Ed.), *Dialogmarketing Perspektiven 2014/2015* (pp. 165–177). Wiesbaden: Springer.

Braun, C., Falter, M., & Hadwich, K. (2016). Understanding the transformation from products to services – A literature review and research agenda. In M. Bruhn & K. Hadwich (Eds.), *Servicetransformation. Forum Dienstleistungsmanagement* (pp. 79–108). Wiesbaden: Springer Gabler.

Brax, S. (2005). A manufacturer becoming service provider – Challenges and a paradox. *Managing Service Quality: An International Journal, 15*(2), 142–155.

Brax, S., & Jonsson, K. (2009). Developing integrated solution offerings for remote diagnostics: A comparative case study of two manufacturers. *International Journal of Operations & Production Management, 29*(5), 539–560.

Brinckmann, J., Grichnik, D., & Kapsa, D. (2010). Should entrepreneurs plan or just storm the castle? A meta-analysis on contextual factors impacting the business planning-performance relationship in small firms. *Journal of Business Venturing, 25*(1), 24–40.

Brodie, R., Hollebeek, L., Juric, B., & Ilic, A. (2011). Customer engagement: Conceptual domain, fundamental propositions, and implications for research. *Journal of Service Research, 14*(3), 252–271.

Bruhn, M. (1998). *Wirtschaftlichkeit des Qualitätsmanagements. Qualitätscontrolling für Dienstleistungen*. Berlin: Springer.

Bruhn, M. (1999). Qualitätssicherung im Dienstleistungsmarketing. Eine Einführung in die theoretischen und praktischen Probleme. In M. Bruhn & B. Stauss (Eds.), *Dienstleistungsqualität. Konzepte – Methoden – Erfahrungen* (3rd Edn., pp. 21–48). Wiesbaden: Springer.

Bruhn, M. (2016). *Kundenorientierung. Bausteine für ein exzellentes Customer Relationship Management (CRM)* (5th Edn.). München: Beck.

Bruhn, M., & Batt, V. (2011). Mitarbeitende als Erlebnisprovider – Theoretische Fundierung und empirische Untersuchung an einem Fallbeispiel. *Marketing ZFP, 33*(3), 208–220.

Bruhn, M., & Braun, C. (2016). The service transformation process: A research agenda in the customer orientation interface. *Marketing ZFP, 38*(1), 5–21.

Bruhn, M., & Hadwich, K. (2012). Customer Experience – Eine Einführung in die theoretischen und praktischen Problemstellungen. In M. Bruhn & K. Hadwich (Eds.), *Customer Experience: Forum Dienstleistungsmanagement* (pp. 3–36). Wiesbaden: Gabler.

Bruhn, M., & Hadwich, K. (2015). Interaktive Wertschöpfung durch Dienstleistungen – Eine Einführung in die theoretischen und praktischen Problemstellungen. In M. Bruhn & K. Hadwich (Eds.), *Interaktive Wertschöpfung durch Dienstleistungen. Forum Dienstleistungsmanagement* (pp. 3–29). Wiesbaden: Springer Gabler.

Bruhn, M., & Hadwich, K. (2016). Servicetransformation – Eine Einführung in die theoretischen und praktischen Problemstellungen. In M. Bruhn & K. Hadwich (Ed.), *Servicetransformation. Forum Dienstleistungsmanagement* (pp. 3–22). Wiesbaden: Springer Gabler.

Bruhn, M., Hepp, M., & Hadwich, K. (2015a). Vom Produkthersteller zum Serviceanbieter – Geschäftsmodelle der Servicetransformation. In M. Bruhn & K. Hadwich (Eds.), *Interaktive Wertschöpfung durch Dienstleistungen. Forum Dienstleistungsmanagement* (pp. 133–146), Springer.

Bruhn, M., Keller, C., & Batt, V. (2015b). Formen des Customer Engagement – Entwicklung einer Typologie. In M. Bruhn & K. Hadwich (Eds.), *Interaktive Wertschöpfung durch Dienstleistungen. Forum Dienstleistungsmanagement* (pp. 83–109). Wiesbaden: Springer Gabler.

Büttgen, M. (2007). *Kundenintegration in den Dienstleistungsprozess: eine verhaltenswissenschaftliche Untersuchung*. Wiesbaden: Springer.

Chowdhury, S. (2006). *Design for six sigma. The revolutionary process for achieving extraordinary profits*. New York: Gabler.

Cowell, M. (1993). Ecological restoration and environmental ethics. *Environmental Ethics, 15*(1), 19–32.

Crosby, P. B. (1986a). *Qualität bringt Gewinn*. Hamburg: McGraw-Hill.

Crosby, P. B. (1986b). *Qualität ist machbar*. Hamburg: McGraw-Hill.

Dabholkar, P. A. (1990). *How to improve perceived service quality by increasing customer participation*. Proceedings of the 1990 Academy of Marketing Science (AMS) Annual Conference. New Orleans: Springer.

Dahm, M., & Heindl, C. (2006). Von Six Sigma zu Lean Sigma. *ZUb Zeitschrift der Unternehmensberatung, 1*(2), 81–88.

Dahm, M., & Heindl, C. (2007). Effizienter werden mit Lean Sigma. *IO New Management, 76*(6), 86–92.

Deming, W. E. (1988). *Out of the Crisis. Quality, productivity and competitive position*. Cambridge.

Deming, W. E. (Ed.). (1982). *Quality, productivity and competitive position*. Cambridge: Massachusetts Institute of Technology.

Detecon International GmbH. (2010). Customer experience management in the telecommunications industry. http://www.detecon.com/en/publications/studies/studies_detail.html?unique_id=43778. Accessed 17 Oct. 2012.

Deutsche Gesellschaft für Qualität e. V. (1995). *Begriffe zum Qualitätsmanagement* (6th Edn.). Berlin: Beuth.

DGQ. (2019). Grundsätze des Qualitätsmanagements. https:\\www.dgq.de. Accessed 15 May 2019.

DIN EN ISO 9000:2014. (2014). *Qualitätsmanagementsysteme – Grundlagen und Begriffe*. Berlin: Beuth.

DIN EN ISO 9000:2015. (2015). *Qualitätsmanagementsysteme—Grundlagen und Begriffe*. Berlin: Beuth.

Döttinger, K., & Klaiber, E. (1994). Realisierung eines wirksamen Qualitätsmanagementsystems im Sinne des Total Quality Managements. In B. Stauss (Ed.), *Qualitätsmanagement und Zertifizierung von DIN ISO 9000 zum Total Quality Management* (pp. 255–273). Wiesbaden: Gabler.

Eckes, G. (2000). *The Six Sigma Revolution*. New York: Wiley.

Edvardsson, B., Kristensson, P., Magnusson, P., & Sundström, E. (2012). Customer integration within service development – A review of methods and an analysis of insitu and exsitu contributions. *Technovation, 32*(7), 419–429.

Engelhardt, W. H., Kleinaltenkamp, M., & Reckenfelderbäumer, M. (1992). *Dienstleistungen als Absatzobjekt*. Arbeitsbericht Institut für Unternehmungsführung und Unternehmensforschung, vol. 52. Bochum: Institut für Unternehmensführung an der Ruhr-Universität Bochum.

Feigenbaum, A. V. (1956). Total quality control. *Harvard Business Review, 34*(6), 93–101.

Feigenbaum, A. V. (1991). *Total quality control* (3rd Edn.). New York: McGraw-Hill.

Fliess, S., & Lexutt, E. (2016). Erfolgsfaktoren der Service Transition – eine systematische Literaturanalyse. In M. Bruhn & K. Hadwich (Eds.), *Servicetransformation. Forum Dienstleistungsmanagement* (pp. 49–78). Wiesbaden: Springer Gabler.

Fliess, S., Jacob, F., & Fandel, G. (2011). Von der Kundenintegration 1.0 zur Kundenintegration 2.0 – Implikationen für Praxis und Forschung. *Zeitschrift für Betriebswirtschaft, 81*(5), 5–20.

Frehr, H.-U. (1999). Total-quality-management. In W. Masing (Ed.), *Handbuch Qualitätsmanagement* (4th Edn., pp. 31–48). München: Hanser.

Garvin, D. A. (1988). *Managing quality. The strategic and competitive edge*. New York: Free Press.

Gebauer, H., & Saul, C. (2014). Eine bibliometrische Analyse des Forschungsstandes und der zukünftigen Forschungsfragen für die Transformation vom Produzenten zum Dienstleister. *Die Unternehmung, 68*(4), 229–249.

Gebauer, H., Fleisch, E., & Friedli, T. (2005). Overcoming the service paradoxon in manufacturing companies. *European Management Journal, 23*(1), 14–26.

Gentile, C., Spiller, N., & Noci, G. (2007). How to sustain the customer experience: An overview of experience components that co-create value With the customer. *European Management Journal, 25*(5), 395–410.

George, S., & Weimerskirch, A. (1998). *Total quality management. Strategies and techniques proven at today's most successful companies* (2nd Edn.). New York: Wiley .

Gimenez-Espin, J. A., Jiménez-Jiménez, D., & Martínez-Costa, M. (2013). Organizational culture for total quality management. *Total Quality Management & Business Excellence, 24*(5–6), 678–692.

Gupta, A., McDaniels, J. C., & Herath, S. K. (2005). Quality management in service firms. Sustaining structures of total quality service. *Managing Service Quality, 15*(4), 389–402.

Haist, F., & Fromm, H. (1991). *Qualität im Unternehmen. Prinzipien, Methoden, Techniken* (2nd Edn.). München: Hanser.

Hirschman, A. O. (1970). *Exit, voice, and loyalty: Responses to decline in firms, organizations, and states*. Cambridge: Harvard University Press.

Hogreve, J., & Wonner, R. (2014). Implizites Wissen als Determinante des Kaufverhaltens produktbegleitender Dienstleistungen auf Industriegütermärkten – Eine explorative Analyse. *Die Unternehmung, 68*(4), 268–281.

Homburg, C., & Kuehnl, C. (2014). Is the more always better? A comparative study of internal and external integration practices in new product and new service development. *Journal of Business Research, 67*(7), 1360–1367.

Horatzek, S. (2018). *Prozessmanagement*. Weinheim: Hanser.

Horváth, P., & Urban, G. (1990). *Qualitätscontrolling*. Stuttgart: Schäffer-Poeschel.

Huth, M. (2018). *Toolbox Prozessmanagement. Vorgehensmodell und praktische Methoden für Industrie und Dienstleistung*. München: Hanser.

Huq, Z. (2005). Managing change. A barrier to TQM implementation in service industries. *Managing Service Quality, 15*(5), 452–469.

Imai, M. (2002). *Kaizen. Der Schlüssel zum Erfolg im Wettbewerb* (2nd Edn.). München: Econ Ullstein.

Ishikawa, K. (1985). *What is quality control?* New York: Prentice Hall.

Juha, M., & Pentti, J. (2008). Managing risks in organizational purchasing through adaptation of buying centre structure and the buying process. *Journal of Purchasing and Supply Management, 14*(4), 253–262.

Juran, J. M. (1951). Basic concepts. In J. M. Juran, F. M. Gryna, & R. S Bingham (Eds.), *Quality control handbook*. New York: McGraw Hill.

Juran, J. M. (1986). The quality trilogy. *Quality Progress, 19*(8), 19–24.

Juran, J. M. (1988). *Juran on planning for quality*. New York: The Free Press.

Kamiske, G. F., & Brauer, J.-P. (2011). *Qualitätsmanagement von A bis Z. Erläuterungen moderner Begriffe des Qualitätsmanagements* (7th Edn.). München: Hanser.

Kastalli, I., & van Looy, B. (2013). Disentangling the impact of service business model innovation on manufacturing firm performance. *Journal of Operations Management, 31*(4), 169–180.

Kindström, D., Kowalkowski, C., & Sandberg, E. (2013). Enabling service innovation: A dynamic capabilities approach. *Journal of Business Research, 66*(8), 1063–1073.

Kleinaltenkamp, M. (1993). Investitionsgütermarketing als Beschaffung externer Faktoren. In E. M. Thelen & G. B. Mairamhof (Eds.), *Dienstleistungsmarketing. Tagungsband zum 2. Workshop für Dienstleistungsmarketing* (pp. 101–126). Frankfurt a. M.: Peter Lang

Kleinaltenkamp, M. (2000). Blueprinting – Grundlage des Managements von Dienstleistungsunternehmen. In H. Woratschek (Ed.), *Neue Aspekte des Dienstleistungsmarketing: Konzepte für Forschung und Praxis* (pp. 3–28). Wiesbaden: Gabler.

Kollmann, T. (2013). *Akzeptanz innovativer Nutzungsgüter und -systeme: Konsequenzen für die Einführung von Telekommunikations- und Multimediasystemen*. Wiesbaden: Gabler.

Kosta, C., & Kosta, S. (2008). *Der kontinuierliche Verbesserungsprozess: Methoden des KVP*. München: Hanser.

Kowalkowski, C., Kindström, D., Alejandro, T. B., Brege, S., & Biggemann, S. (2012). Service infusion as agile incrementalism in action. *Journal of Business Research, 65*(6), 765–772.

Kowalkowski, C., Windahl, C., Kindström, D., & Gebauer, H. (2015). What service transition? Rethinking established assumptions about manufacturers' service-ied growth strategies. *Industrial Marketing Management, 45*(2), 59–69.

Kozinets, R. V., de Valck, K., Wojnicki, A. C., & Wilner, S. J. S. (2010). Networked narratives: Understanding word-of-mouth marketing in online communities. *Journal of Marketing, 74*(2), 71–89.

Lerch, C., & Gotsch, M. (2014). Die Rolle der Digitalisierung bei der Transformation vom Produzenten zum produzierenden Dienstleister. *Die Unternehmnung, 68*(4), 250–267.

Lindberg, N., & Nordin, F. (2008). From products to services and back again: Towards a new service procurement logic. *Industrial Marketing Management, 37*(3), 292–300.

Lloréns Montes, F. J., & Verdú Jover, A. J. V. (2004). Total quality management, institutional isomorphism and performance: The case of financial services. *The Service Industries Journal, 24*(5), 103–119.

Lunau, S. (Ed.) (2014). *Six Sigma + Lean Toolset: Mindset zur erfolgreichen Umsetzung von Verbesserungsprojekten.* (5th Edn.). Berlin: Springer Gabler.

Magnusson, K., Kroslid, D., & Bergman, B. (2004). *Six Sigma umsetzen. Die neue Qualitätsstrategie für Unternehmen* (2nd Edn.). München: Hanser.

Maiwald, K., Wieseke, J., & Everhartz, J. (2014). The dark side of providing industrial product-service systems – Perceived risk as a key challenge from a customer-centric point of view. *Procedia CIRP, 16,* 241–246.

Masing, W. (Ed.) (1980). *Handbuch der Qualitätssicherung*. München: Hanser.

Mayer-Vorfelder, M. (2012). *Kundenerfahrungen im Dienstleistungsprozess – eine theoretische und empirische Analyse*. Wiesbaden: Gabler.
McMullen, J. S., & Shepherd, D. A. (2006). Entrepreneurial action and the role of uncertainty in the theory of the entrepreneur. *Academy of Management Review, 31*(1), 132–152.
Meik, J. (2016). *Kundenintegration und Kundenbeziehungen. Konzeption, Wirkungsmechanismen und Managementimplikationen.* Wiesbaden: Springer Gabler.
Meffert, H. (1993). *Marktorientierte Führung von Dienstleistungsunternehmen. Neuere Entwicklungen in Theorie und Praxis, Arbeitspapier Nr. 78.* Münster.
Meffert, H., Bruhn, M., & Hadwich, K. (2018). *Dienstleistungsmarketing. Grundlagen, Konzepte, Methoden* (9th Edn.). Wiesbaden: Springer Gabler.
Meuter, M. L., Ostrom, A. L., Roundtree, R. I., & Bitner, M. J. (2000). Self-service technologies: Understanding customer satisfaction with technology-based service encounters. *Journal of Marketing, 64*(3), 50–64.
Meyer, C., & Schwager, A. (2007). Understanding customer experience. *Harvard Business Review, 85*(2), 116–126.
Michel, S. (1996). *Prosuming-Marketing. Konzeption und Anwendung*. Bern: Haupt.
Narayanan, V. K. (2015). Customer-focused IT: A process of continuous value innovation. *Strategy & Leadership, 43*(4), 11–17.
Neely, A., Benedettini, O., & Visnjic, I. (2011). *The servitization of manufacturing: Further evidence*. Proceedings of the 18th European operations management association conference. Cambridge.
Nordin, F., Kindström, D., Kowalkowski, C., & Rehme, J. (2011). The risks of providing services: Differential risk effects of the service-development strategies of customisation, bundling, and range. *Journal of Service Management, 22*(3), 390–408.
Oakland, J. S. (2004). *Oakland on quality management* (3rd Edn.). Oxford: Butterworth-Heinemann.
Oess, A. (1993). *Total Quality Management. Die ganzheitliche Qualitätsstrategie* (3rd Edn.). Wiesbaden: Gabler.
Oess, A. (1994). Total Quality Management. Eine ganzheitliche Unternehmensphilosophie. In B. Stauss (Ed.), *Qualitätsmanagement und Zertifizierung von DIN ISO 9000 zum Total Quality Management* (pp. 199–222). Wiesbaden: Gabler.
Oliva, R., & Kallenberg, R. (2003). Managing the transition from products to services. *International Journal of Service Industry Management, 14*(2), 160–172.
Oltmanns, B. (2011). Duft weckt Emotionen. *Stores and Shops, o. V.*(5), 72–75.
Opresnik, D., & Taisch, M. (2015). The value of big data in servitization. *International Journal Production Economics, o. V.*(165), 174–184.
Ostrom, A. L., Bitner, M. J., Brown, S. W., Burkhard, K. A., Goul, M., Smith-Daniels, V., Demirkan, H., & Rabinovich, E. (2010). Moving forward and making a difference: Research priorities for the science of service. *Journal of Service Research, 13*(1), 4–36.
Pagani, M., Hofacker, C. F., & Goldsmith, R. E. (2011). The influence of personality on active and passive use of social networking sites. *Psychology & Marketing, 28*(5), 441–456.
Parasuraman, A., Zeithaml, V. A., & Berry, L. L. (1985). A conceptual model of service quality and its implications for future research. *Journal of Marketing, 49*(1), 41–50.
Parbs, H., & Pöppelbuß, J. (2016). Servicetransformation in der Windenergiebranche. In M. Bruhn & K. Hadwich (Eds.), *Servicetransformation. Forum Dienstleistungsmanagement* (pp. 773–802). Wiesbaden: Springer Gabler.
Parks, B., Olson, P. D., & Bokor, D. W. (1991). Don't mistake business plans for planning (It may be dangerous to your financial health). *Journal of Small Business Strategy, 2*(1), 15–24.

Piller, F. T., & Müller, M. (2004). A new marketing approach to mass customisation. *International Journal of Computer Integrated Manufacturing, 17*(7), 583–593.

Pullman, M. E., & Gross, M. A. (2004). Ability of experience design elements to elicit emotions and loyalty behaviors. *Decision Sciences, 35*(3), 551–578.

Rada, J., & Vandermerwe, S. (1988). Servitization of business: Adding value by adding service. *European Management Journal, 6*(4), 314–320.

Ramani, G., & Kumar, V. (2008). Interaction orientation and firm performance. *Journal of Marketing, 72*(1), 27–45.

Reichwald, R. P., & Piller, F. F. (2009). *Interaktive Wertschöpfung. Open Innovation, Individualisierung und neue Formen der Arbeitsteilung*. Wiesbaden: Gabler.

Schildknecht, R. (1992). *Total Quality Management. Konzeption und State of the Art*. New York: Campus.

Schmidt, J. (1998). A questionable bet on six sigma and the evolution of quality in product development. *Six Sigma Forum Magazine, 1*(1), 28–35.

Schmitt, B. H. (1999). *Experiential marketing – How to get customers to sense, feel, think, act and relate to your company and brands*. New York: The Free Press.

Schmitt, B. (2003). *Customer experience management: A revolutionoary approach to connecting with your customer*. New York: Wiley.

Schmitt, B. H., & Mangold, M. (2005). Customer Experience Management als zentrale Erfolgsgröße der Markenführung. In F.-R. Esch (Ed.), *Moderne Markenführung. Grundlagen, Innovative Ansätze, Praktische Umsetzungen* (4th Edn., pp. 287–303). Wiesbaden: Springer.

Schmitt, B., & Mangold, M. (2004). *Kundenerlebnis als Wettbewerbsvorteil. Mit Customer Experience Management Marken und Märkte Gewinn bringend gestalten*. Wiesbaden: Gabler.

Schmitt, R., & Pfeifer, T. (2015). *Qualitätsmanagement. Strategien, Methoden, Techniken* (5th Edn.). München: Hanser.

Shankar, V., Berry, L. L., & Dotzel, T. (2009). A practical guide to combining products and services. *Harvard Business Review, 87*(11), 94–99.

Sharma, S., Conduit, J., & Hill, S. R. (2014). Organisational capabilities for customer participation in health care service innovation. *Australasian Marketing Journal, 22*(3), 179–188.

Shostack, G. L. (1992). Understanding services through blueprinting. *Advances in Services Marketing and Management, 1*(1), 75–90.

Silpakit, P., & Fisk, R. P. (1985). Participating the service encounter: A theoretical framework. In T. M. Bloch & G. D. Upah (Eds.), *Services marketing in a changing environment* (pp. 117–121). Chicago.

Singh, S., & Sonnenburg, S. (2012). Brand performances in social media. *Journal of Interactive Marketing, 26*(4), 189–197.

Solis, B., Lieb, R., & Szymanski, J. (2014). *The 2014 state of digital transformation – How companies are investing in the digital customer experience*. San Francisco.

Sondermann, J. P. (1994). Instrumente des Total Quality Managements. Ein Überblick. In B. Stauss (Ed.), *Qualitätsmanagement und Zertifizierung von DIN ISO 9000 zum Total Quality Management* (pp. 223–253). Wiesbaden: Gabler.

Stauss, B. (1994). Qualitätsmanagement und Zertifizierung als unternehmerische Herausforderung. Eine Einführung in den Sammelband. In B. Stauss (Ed.), *Qualitätsmanagement und Zertifizierung von DIN ISO 9000 zum Total Quality Management* (pp. 11–23). Wiesbaden: Gabler.

Stebbing, L. (1990). *Quality management in the service industry*. Chichester: Ellis Horwood.

Sureshchandar, G. S., Rajendran, C., & Anantharaman, R. N. (2001a). A conceptual model for total quality management in service organizations. *Total Quality Management, 12*(3), 343–363.

Sureshchandar, G. S., Rajendran, C., & Anantharaman, R. N. (2001b). A holistic model for total quality service. *International Journal of Service Industry Management, 12*(4), 378–412.

Taguchi, G. (2001). *Introduction to quality engineering*. Tokio: Asian Productivity Organization.

Tax, S. S., McCutcheon, D., & Wilkinson, I. F. (2013). The Service Delivery Network (SDN) a customer-centric perspective of the customer journey. *Journal of Service Research, 16*(4), 454–470.

Töpfer, A. (2007). *Six Sigma. Konzeption und Erfolgsbeispiele* (4th Edn.). Berlin: Springer.

Töpfer, A. (2009). *Lean Six Sigma – Erfolgreiche Kombination von Lean Management, Six Sigma und Design for Six Sigma*. Berlin: Springer.

Töpfer, A., & Silbermann, S. (2011). *Lean Management und Six Sigma als Werkzeuge zur Steigerung der Dienstleistungsproduktivität*. Wiesbaden: Gabler.

Toutenburg, H., & Knöpfel, P. (2009). *Six Sigma* (2nd Edn.). Dordrecht: Springer.

van Doorn, J., Lemon, K. N., Mittal, V., Nass, S., Pick, D., Pirner, P., & Verhoef, P. C. (2010). Customer engagement behavior: Theoretical foundations and research directions. *Journal of Service Research, 13*(3), 253–266.

Vargo, S. L., & Lusch, R. F. (2008). From goods to service (s): Divergences and convergences of logics. *Industrial Marketing Management, 37*(3), 254–259.

Verhoef, P. C., Lemon, K. N., Parasuraman, A., Roggeveen, A., Tsiros, M., & Schlesinger, L. A. (2009). Customer experience creation: Determinants, dynamics and management strategies. *Journal of Retailing, 85*(1), 31–41.

Wappis, J., & Jung, B. (2013). *Null-Fehler-Management. Umsetzung von Six Sigma* (4th Edn.). München: Hanser.

Wonigeit, J. (1996). *Total Quality Management. Grundzüge und Effizienzanalyse* (2nd Edn.). Wiesbaden: Gabler.

Zink, K. J. (2004). *TQM als integratives Managementkonzept* (2nd Edn.). München: Hanser.

Zink, K. J., & Schildknecht, R. (1989). Total Quality Konzepte. Entwicklungslinien und Überblick. In K. J. Zink (Ed.), *Qualität als Managementaufgabe* (pp. 67–100). Landsberg am Lech: Moderne Industrie.

Zollondz, H.-D. (2011). *Grundlagen Qualitätsmanagement. Einführung in Geschichte, Begriffe, Systeme und Konzepte* (3th Edn.). München: Oldenbourg.

Zomerdijk, L. G., & Voss, C. A. (2010). Service design for experience-centric services. *Journal of Service Research, 13*(1), 67–82.

Part II

Analysis of Service Quality

Models of Service Quality

4

Abstract

Different models for the analysis of the emergence of service quality have been developed in the existing literature. Based on a comprehensive presentation and discussion of the best-known and most important models, it becomes clear in the context of this chapter that the so-called GAP model has gained international importance. As a result, the approach of the GAP model has been modified and adapted for different application areas and industries in the meantime. An explanation of these approaches is also given in the context of this chapter. In conclusion, central quality dimensions could be isolated in the scientific literature—based on the results of the GAP model—which come into effect for the customers of a service provider when assessing the experienced quality.

The models of service quality serve a more precise **analysis of the emergence of service quality.**In quality management, models represent reality in many ways. They are an image of the original, archetype, prototype or reference system of what is meant by quality. The models of service quality are an attempt to reduce the complexity of the concept of service quality in order to create the conceptual basis for dealing with quality problems (Zollondz, 2011, p. 194). In literature and practice, the following models have been particularly relevant:

1. GAP models of service quality,
2. Service quality model by Grönroos,
3. Service quality model by Meyer and Mattmüller,
4. Dynamic process model by Boulding et al.
5. Relationship quality model by Liljander and Strandvik,
6. Qualitative satisfaction model by Stauss and Neuhaus,

M. Bruhn, *Quality Management for Services*,
https://doi.org/10.1007/978-3-662-67032-3_4

7. Model of quality of e-services,
8. Model of e- health service quality.

The individual models are explained in detail in the following sections.

4.1 GAP Models of Service Quality

4.1.1 GAP Model by Parasuraman/Zeithaml/Berry

The **GAP model** is to be understood as an sector-independent model of service quality (Parasuraman et al., 1986, 131 pp.). In the mid-1980s, Parasuraman et al. characterized the status quo in three points (Parasuraman et al., 1985, p. 42; Zeithaml et al., 1992, 28 pp.):

1. Due to the often complex service offerings, it is more difficult for consumers to assess quality than for material goods.
2. In the assessment of quality, not only the result of the performance creation process, but also the process itself is of importance.
3. For the assessment of service quality, only the customer perspective is relevant.

To consolidate these findings, it was analyzed in which way the consumers rate the quality of services. **Assessments from the customer's perspective** could be collected with the help of focus group interviews about these and other related questions, whereby this stage of information gathering was deliberately broadly designed in regard to the investigated service industries. Twelve focus groups were selected, three groups each for four **industries** (Zeithaml et al., 1992, 29 pp.):

- Private customer business of banks,
- Credit card business,
- Stockbrokers,
- Repair shops.

The **results** of these interviews can be summarized as follows (Zeithaml et al., 1992, pp. 31 f.):

1. Despite the relatively large divergence with regard to the individual industries, it was observed that the quality assessment takes place according to many similar characteristics.
2. Special service quality is attributed to a service provider especially when the expectations of the customers are (over) fulfilled. Basically, it is determined by the deviation

between the expected (desired) performance and the experienced (perceived) performance level.
3. Customer expectations of service quality are mainly influenced by four impulses:

- Verbal communication of customers,
- Personal situation of customers,
- Previous experience with the provider,
- Communication of the provider.

In addition, as a central result of the surveys conducted, ten **quality dimensions** could be isolated, which come into play for customers of a service provider when assessing the quality experienced (Parasuraman et al., 1985, 46 pp.; Zeithaml et al., 1992, 34 pp.):

1. Material environment ("tangibles"),
2. Reliability,
3. Responsiveness,
4. Competence,
5. Courtesy,
6. Credibility,
7. Security,
8. Access,
9. Communication,
10. Understanding/Knowing Customers.

The basis of the GAP model is a two-part division of the interaction relationship into the levels **service provider and customer.** Figure 4.1 shows this basic model and at the same time points to the various interaction relationships between service provider and customer as well as possible conflict areas, so-called gaps. In total, the authors identify five **GAPs,** the character and influencing factors of which are presented below. In a specified form, the influencing factors are listed in Fig. 4.2.

GAP 1 The **customer expectations** towards the service provider are constituted by their individual needs, their experiences in the past and the attitudes established by word-of-mouth communication. GAP 1 can arise between the thus resulting expectations on the customer side and the **customer expectations perceived by management**if the corresponding expectations are not identical. When minimizing this discrepancy, the exact capture of the customer requirements for the service is of particular importance, since GAP 1 also determines the extent of the other gaps. GAP 1 can essentially be explained by three **influence factors** (Zeithaml et al., 1992, 66 pp.):

(1) Consideration of market research Since market research is an important means of better understanding the expectations and perceptions of customers with regard to

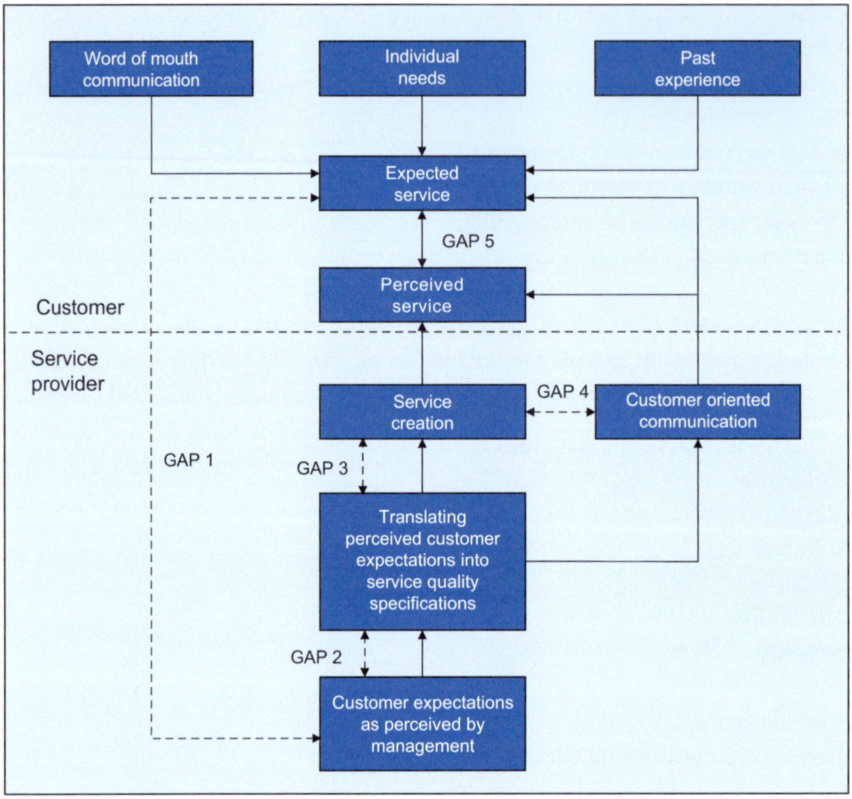

Fig. 4.1 GAP model of service quality (Source: Zeithaml et al., 1988, p. 44)

services, the extent to which market research is carried out is an important prerequisite for reducing GAP 1. Other relevant variables in connection with market research include the extent to which managers in companies take into account market research results, i.e. read, understand and use them, as well as the importance of the topic of service quality in market research studies. The consideration of market research in the company is also determined by the extent of the direct relationship between top managers and customers. Even in large service companies, top managers can spend some time "on the line", i.e. establish a direct relationship with customers and gain experience in creating services. In order to ensure its relevance, market research is to be understood as a permanent process of information gathering and processing. For service companies, this includes in particular competitive research, product research and customer analysis. In addition to the evaluation of secondary data material, the company-specific collection of primary data is also required, involving the entire spectrum of product tests, image studies, customer frequency research, customer structure analysis, customer opinion research, competitor observation to complaint and suggestion management.

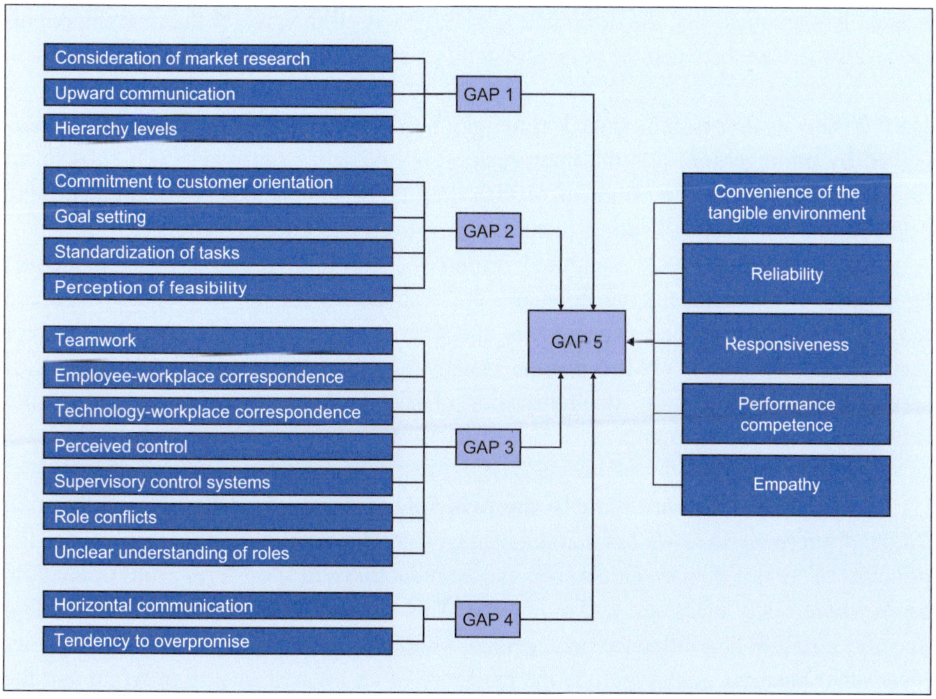

Fig. 4.2 Influencing factors of the GAP model (Source: Zeithaml et al., 1992, p. 131)

(2) Upward communication within the company Employees in direct customer contact are usually much better able to accurately predict the expectations and perceptions of customers regarding the service (Schneider & Bowen, 1999). The right customer assessment by top managers therefore depends very much on the extent and type of communication with employees who have direct customer contact as well as with employees outside the company (e.g. independent insurance agents, retailers), who represent the company and its services. Through upward communication, top managers are informed about activities and results throughout the company. Special, relevant types of communication in this context are, on the one hand, formal communication (e.g. reports on problems and expectations in the creation of services) and, on the other hand, informal communication (e.g. conversations between contact personnel and senior management). Important in connection with upward communication is its quality and effectiveness, which in turn depends on the communication medium. Personal communication is, for example, preferred if the message is difficult or unclear, or if the sender and receiver differ in their areas of experience and views.

(3) Hierarchical levels Hierarchical levels inhibit communication and understanding because they represent barriers between senders and receivers of messages. For this reason, it is assumed that the discrepancy in GAP 1 is all the greater the more hierarchical levels there are between the personnel with customer contact and the top managers.

GAP 2 This GAP 2 is concretized in the gap between the **customer expectations perceived by management** and the interpretation by the service provider with subsequent implementation in **specifications of service quality.** Managers in service companies are often confronted with difficulties in fulfilling or even exceeding customer expectations. A variety of factors—such as limited resources, short-term profit orientation, market conditions or management indifference—are responsible for the discrepancy between the perception of customer expectations and their implementation in specifications for services. The **causes** of GAP 2 are the management's commitment to the principle of service quality, goal setting, standardization of tasks and perception of their feasibility (Zeithaml et al., 1992, 86 pp.).

(1) Management's commitment to the principle of service quality One explanation for the discrepancy in GAP 2 is that the management does not feel fully committed to the principle of service quality. Goals such as cost reduction and short-term profit orientation can be more easily measured and monitored. This can lead to neglect of service quality. Specific variables that influence management's support of the principle of service quality (over other business goals) include the existence of an internal quality program and the extent to which managers believe that their efforts to improve service quality are also recognized and rewarded by the company.

(2) Goal setting The setting of goals not only improves corporate as well as personal performance, but also allows for more effective control of the entire company. Companies that have been successful in creating high service quality (e.g. *American Express, McDonald's*) are known for setting formal goals with respect to service quality. Through an analysis of customer complaints, for example, *American Express* found that punctuality, accuracy and responsibility were the main goals to be pursued. Management then identified 180 operational goals for creating service quality. Once these goals were formally established, procedures were developed to control the speed with which telephone calls were answered, complaints were received, bills were sent, and new applications were processed. The development of service goals also requires defining service quality in such a way that the service provider understands what goals the management is pursuing.

(3) Standardization of tasks The effective implementation of management perceptions into specific standards of service quality depends on the extent to which tasks to be fulfilled can be standardized. If work places and tasks are subject to routine, then specific regulations and standards can be established and effectively implemented in the company. If, however, services are oriented towards the individual customer wishes (e.g. support of investment portfolios or asset management), it is difficult to establish specific standards (such as time limits for the support of customers). Even when creating services that are highly oriented towards the individual customer wishes, there are always routine tasks that can be identified. For example, doctors can standardize non-technical aspects of their service, such as the admission of patients, invoice issuance, weighing and temperature measurement, which occur repeatedly.

(4) Perception of feasibility The discrepancy in GAP 2 also depends on the extent to which managers perceive the fulfilment of customer expectations to be feasible. For example, the interviewed managers of a repair service company were quite aware that high service quality for customers consisted in the rapid repair of equipment defects. At the same time, however, they argued that the specification for the rapid repair of equipment defects was not feasible for two reasons, firstly because the necessary time for providing a repair service could only be estimated very difficultly in advance and, secondly, because trained customer service technicians were not available to the same extent during the high season (during the summer months) as at any other time. For the perception of feasibility, therefore, the skills and systems necessary for the fulfilment of the specifications in the company as well as convincing the managers that these specifications can be economically fulfilled are of particular importance.

GAP 3 This GAP 3 represents the discrepancy between the **specifications of service quality** and the **actually created services**. GAP 3 can also be referred to as the "discrepancy of service creation", as this discrepancy reflects the extent to which the service provider does not provide the service at the level expected by management. This discrepancy occurs when employees are unable or unwilling to provide the service in the required manner. The key reasons responsible for the discrepancy in GAP 3 include teamwork, employee-workplace correspondence, technology-workplace correspondence, perceived control, supervisory control systems, role conflict, and unclear role understanding.

(1) Teamwork In groups that achieve high performance, employees work as a team and achieve their goals by enabling all group members to participate in the decision-making process and also in the success of the group. Teamwork is the focus of service quality creation programs for companies known for their outstanding customer service. The following factors are decisive for teamwork: the extent to which employees see other employees as customers; the extent to which employees believe that their superiors care for them; the extent to which employees feel that they are cooperating with other employees rather than competing; and the extent to which employees are committed and identify with the company.

(2) Employee-workplace correspondence Problems with service quality often arise because the contact personnel are not up to their tasks. Since workplaces with customer contact tend to occur in the lower part of the company structure (e.g. counter staff at car rental companies, telephone operators, repair technicians), the staff of these workplaces consists to a large extent of the less well-trained and lowest-paid employees in the company. For this reason, they may lack language, interpersonal or other skills that are required for effective customer service. In addition, there is the frequent turnover of contact personnel and the tendency of companies to fill vacancies very quickly, even at the risk of hiring employees with little experience or insufficient knowledge. Managers generally use neither enough care in the selection and hiring of staff, nor do they provide sufficient resources for this.

(3) Technology-workplace correspondence The creation of high service quality also depends on appropriate equipment and technologies that the employee uses to perform his work. Technology and equipment, such as computers and diagnostic equipment, can increase the performance of an employee in the service sector. To create high service quality, appropriate and reliable technologies must be available. Equipment failures often inhibit adequate performance of the employee.

(4) Perceived control Individuals' reactions to stress situations depend strongly on whether they can control these situations. Employees in service companies suffer less from stress if they believe they can control situations at their workplace. Less stress in turn leads to better performance. If employees in service companies feel that they can react flexibly and not mechanically to problem situations when creating services, this leads to increased control and improved performance. The perceived control can depend on the extent to which regulations, procedures and corporate culture restrict the flexibility of the contact person's decisions when dealing with customers. Service companies are usually internally organized in such a way that it is very difficult for the employees to create services quickly. If the contact person needs the approval of other departments in the company before a service can be provided, this calls the quality of the service into question. Although the contact person is actually striving to provide customer service, this task cannot be fulfilled because the creation of this service requires the approval of numerous company departments.

(5) Supervisory control systems In some companies, the performance of the contact person is measured by the respective result (e.g. the number of units produced in an hour, the number or value of sales in a week). In these cases, individual performance is recorded and monitored by so-called result control systems. Performance is based on written business processes that measure the result of the employees. However, in many service companies, the result control system may be inappropriate or insufficient for assessing the performance of employees in creating service quality. For example, most bank customers want counter employees who work carefully and quickly and are friendly. Banks that measure the performance of counter employees solely on output variables such as the daily end-of-day report overlook important aspects of work performance that are perceived by customers as service quality. In these and other service situations, performance can also be monitored by behavior control systems that consist to a large extent of observations or other reports on the employee's work and behavior instead of result measurements.

(6) Role conflict The role of the employee is defined by the expectations and requirements placed on him as well as by the pressure exerted by individuals who have a specific interest in the nature of his performance (e.g. top managers, direct superiors, customers). If the expectations of these people are contradictory or too high, the employee is forced into a role conflict, i.e. he becomes aware that he cannot meet the requirements of these individuals. Role conflicts can also arise if employees are tasked with selling additional services to customers; the employees may then feel that they are imposing services on customers and are thus caught between the expectations of the company and their desire to serve the customer. The perception of the role conflict is psychologically uncomfortable for the employee and can have negative effects on the employee's satisfaction and performance in the company. A service company that recognises conflict potentials at workstations with customer contact has good prospects of solving the problem of role conflict.

(7) Unclear role understanding Lack of information on the correct execution of work tasks leads to unclear role understanding for employees. Unclear role understanding can arise if employees are unclear about the requirements that managers and other superiors place on them as well as about the fulfilment of these requirements. Furthermore, unclear role understanding can arise if employees do not know how their performance will be assessed and rewarded. The frequency, quality and accuracy of downward communication influence the unclear role understanding of employees in the service sector. The more often managers communicate these issues clearly and unambiguously, the clearer the role understanding will be for employees.

GAP 4 GAP 4 is focused on the differences between the **actual services** and the **communication with customers** about these services. Statements about the service offering and service quality made on different channels determine the expectations of customers to a great extent. The underlying discrepancy between the generated expectation and the actual service delivery is influenced by horizontal communication and the insufficient consideration of advertising statements (Zeithaml et al., 1992, pp. 131–139).

(1) Horizontal Communication Horizontal communication refers to the exchange of information between and within departments that is necessary to achieve corporate goals. For example, the creation of advertising messages requires intensive communication between the advertising department and customer contact personnel to ensure that the image drawn by the advertising can be implemented in daily practice. The degree of customer upset at an unsatisfactory service is considerably greater if the company's communication has previously caused another—higher—expectation (Zeithaml et al., 1992, p. 135).

(2) Tendency to make exaggerated promises The need to greatly consider the advertising message is closely related to horizontal communication. Companies are often "taken at their word" by customers with regard to their advertising claims. Misleading statements that open up interpretive scope for the actual level of service quality expected create misunderstanding or upset amongst consumers in the event of non-fulfillment (Zeithaml et al., 1992, pp. 143–144).

GAP 5 This GAP is finally the central gap of the model and largely dependent on gaps one to four. The underlying **difference between the expected and the real experienced service** can be reduced by minimizing the remaining four gaps and is the key to good service when the perceived service quality meets or exceeds customer expectations (Parasuraman et al., 1985, p. 46). GAP 5 thus describes the construct of "perceived service quality" as the difference value between expected and perceived service in relation to the central dimensions of service quality presented in Sect. 2.5 (see Fig. 4.2) **Convenience of the tangible environment** ("Tangibles"), **Reliability**, **Responsiveness**, **Performance competence** ("Assurance") and **Empathy**. This is also where the so-called SERVQUAL approach ("Service Quality") comes in. Within SERVQUAL, GAP 5 is operationalized as an indicator of service quality by the five quality dimensions. The SERVQUAL approach is a widely used, industry-independent method for measuring requirements for service quality and is the methodological foundation within various measurement approaches (see Sect. 5.1.2.1).

Depending on the respective service creation, the customer will therefore form an **opinion on the quality of the experienced service**. The perceived service quality can meet, not meet or even exceed expectations. Of particular interest are the cases of exceeding or falling short of the expected service quality. While the largely corresponding expectation and perception will lead to a satisfactory experience of the service quality, a significant deviation from the performance level in the positive sense is to be expected with an attribution towards ideal quality ("Ideal Quality"), with increasing negative deviation towards unacceptable quality ("Totally Unacceptable Quality") (Parasuraman et al., 1985, 48 pp.). However, certain reservations about this attribution are appropriate, as no statements about the actual ("absolute") level of service quality are possible. It is very likely that a failure to achieve the expected service quality will be evaluated differently depending on the (subjectively) absolute quality level (Hentschel, 1999).

4.1.2 Variations of the GAP Model

The GAP model has not only experienced a wide range of applications and discussion in corporate practice; there has also been intensive scientific discussion of the model. One consequence of this discussion are various **variations of the GAP model,**which concretize the model for specific contexts. In particular, the following variations are included:

1. GAP model for direct employee-customer contact,
2. GAP model for multi-stage services,
3. GAP model for internal services,
4. Internal/external GAP model in the context of customer integration.

The various variations of the GAP model are explained below.

(1) GAP model for direct employee-customer contact The **direct employee-customer contact** is a key feature of services. This variation of the GAP model therefore does not concern a special case, but the typical case of a service. However, in Parasuraman's classical GAP model (1985), this direct contact is only implicitly taken into account by identifying various employee-related influencing factors of the gaps and thus the service quality as causes of the four internal gaps. A key function of the customer contact employee is not only in the creation of the service, but also in his role as a "market researcher" by using direct customer contact to collect information about the customer. In this context, and against the background of the GAP model and service quality, information about customer expectations is particularly in the focus.

An extension of the GAP model accordingly is shown in Fig. 4.3. Here, the perception of customer expectations by employees is in addition to the customer expectations perceived by management. This creates two additional gaps (Luk & Layton, 2002):

- GAP 6 as the gap between the expected service and the customer expectations perceived by the employees: This gap leads to the identification of further influencing factors of service quality by differentiating the market research function of the employees.
- GAP 7 as the gap between the customer expectations perceived by the employees and the customer expectations perceived by management: The influencing factors of this gap mainly concern the internal communication between employees and management.

While these two GAPS contribute to a more accurate mapping of expectations and perceptions of service quality in direct employee-customer contact, it should be noted, that the development of these GAPs took place exclusively in the hotel industry. Accordingly, it can be noted that the results of Luk and Layton (2002, p. 126) must be subjected to further empirical cross-industry studies in order to be generalized.

(2) GAP model for multi-stage services In the field of service-related quality research, one usually assumes a direct relationship structure of the interaction partners. The subject of the investigations are, for example, the relationships between customer and service company or between dealer/broker and service company (Dabholkar et al., 1996;

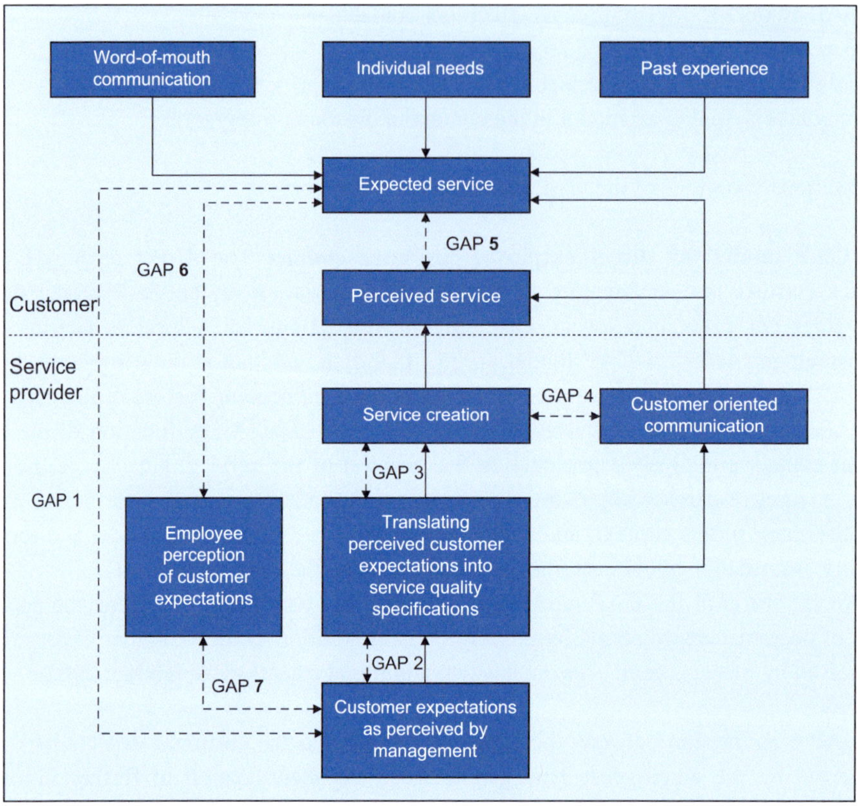

Fig. 4.3 GAP model in direct employee-customer contact (Source: Luk & Layton, 2002, p. 113)

Boshoff & Terblanche, 1997; Stanforth & Lennon, 1997; Parikh, 2006). The relationships of these interaction partners are mostly considered separately, i.e. independently of each other. This overlooks that there are both direct relationships of the service provider to the end customer and indirect customer relationships through the involvement of an intermediary level in various service sectors, such as the insurance sector.

Therefore, it must be noted that in various service sectors, such as car dealerships, insurance and travel agencies, etc., there are two types of relationships:

- **Direct relationships** of the service provider to the end customer.
- **Indirect customer relationships** through the involvement of mediators (or agents/dealers/intermediaries).

Thus, there is a parallel direct and indirect customer contact and thus a **triadic relationship structure** between end customer, intermediary and service provider (Murmann 1999).

Direct and indirect customer contact can be seen as a bundle of services of a company. On the one hand, the company has direct contact with the end customers, on the other hand, an intermediary (e.g. a broker) is involved in the relationship, so that there is also an indirect relationship to the end customer (Murmann, 1999). These triadic relationship structures exist beyond the insurance sector, for example, in the industries of building societies, health care, transport and tourism, computer technology and personnel recruitment. A special problem here is the perception of quality by end customers. A central question is whether the end customer perceives the provider and intermediary as a whole and makes a uniform quality assessment or makes separate and possibly different quality assessments. These alternatives of quality assessment can have different effects on the behavior of the end customer (Zeithaml et al., 1996).

The **determining of service quality** in a triadic relationship structure occurs in the different service interactions between customer, intermediary and service provider (Kellogg & Chase, 1995; Collier, 1996; Niranjan & Metri, 2008). Service interactions can—depending on the different market participants—be distinguished in internal and external interactions. On one hand, there are external service interactions between

- Customer and intermediary,
- Customer and service provider,
- Intermediary and service provider.

On the other hand, **internal service interactions** take place both at the level of the intermediary and at the level of the service provider (cf. also Grund, 1998, p. 78). With the model of service quality shown in Fig. 4.4, both the external and internal service interactions between customer, intermediary and service provider are comprehensively represented (Murmann, 1999). In analogy to the GAP model by Parasuraman et al. (1985, 1988), the service process is ideally represented by the variables "expected performance", "perception of expectations", "service specification", "service creation", "external service communication" and "perceived performance". Customers, intermediaries and service providers assess the quality of a performance used in the different service interactions according to the comparison of the expected performance with the perceived performance: The customer forms a quality judgment with regard to the intermediary services and another quality judgment with regard to the service provider's services to him; the intermediary evaluates the quality of the service provider's services to him, and the service provider forms a quality judgment with regard to the intermediary services to him.

Furthermore, mediators and service providers each also assess their **own services** to the different interaction partners and form an **self-image** by comparing the expectations of their own services with the perception of their own services. In addition, as a special feature of the triadic relationship structure, there are also so-called **third-party images:**The service provider evaluates the service quality of the mediator's services to the customer, and the mediator evaluates the quality of the service provider's services to the customer.

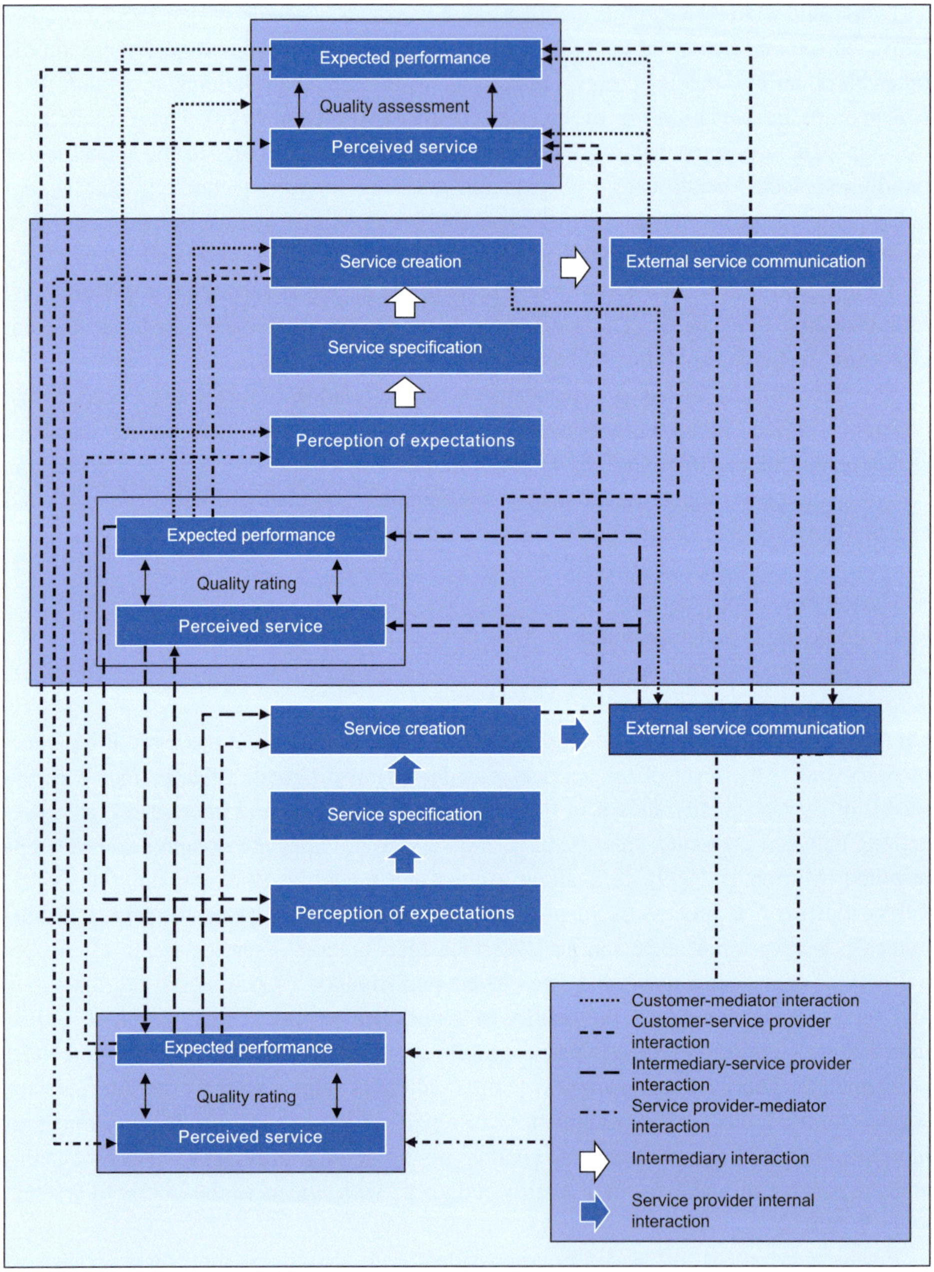

Fig. 4.4 Model of service quality in direct and indirect customer contact of the service provider (Source: Murmann, 1999, p. 77)

Within the relationship triangle, the following **interdependencies** exist:

- **Interdependence of customer perceptions:** For example, the customer's quality assessment of the mediator's services is influenced by how the customer evaluates the quality of the service provider's services. Vice versa, the customer's quality assessment of the service provider's services is influenced by how the customer evaluates the quality of the mediator's services.
- **Interdependence of customer interactions** with the interactions between mediators and service providers: The customer's quality assessment of the service provider's services is influenced by the mediator's services to the service provider—that is, the service provider's quality assessment of the mediator's services to the service provider. On the other hand, the customer's quality assessment of the mediator's services is influenced by the service provider's services to the mediator—that is, the mediator's quality assessment of the service provider's services to the mediator.
- **Interdependence of quality judgments** in the direct interactions between mediators and service providers with the third-party perceptions with regard to the indirect customer interactions: For example, the third-party perceptions of the mediator with regard to the service provider's service to the customer depends on the mediator's quality assessment of the service provider's services to him.

Empirical studies show significant GAPs between quality judgments, self-image and image of others in direct and indirect customer contact structures (see Fig. 4.4), which arise from the interdependencies shown and lead to quality defects. Ensuring the quality of services in triadic relationship structures requires an interdependent orientation of the quality management system due to the parallelism of the interactions between customer, service provider and mediator, i.e. in addition to the design of customer- and employee-related components of the quality management system, it is necessary to integrate mediator-related elements (Murmann, 1999). Accordingly, the analysis of the direct and indirect customer contact of service companies provides informative results. Nevertheless, it is recommended to check the underlying quality model in other industries in order to create a high degree of generalizability. Furthermore, it appears to be expedient to check the success of the derived quality management measures by means of long-term studies as well as to pay closer attention to the interdependence of performance relationships in the context of economic calculations (Murmann, 1999 p. 260 f.).

(3) GAP model for internal services In many service industries, both direct customer contact employees and indirectly supporting employees are involved in the creation of services. In the banking sector, the two groups of employees are often referred to as "front office" and "back office". The term "front line staff" and "support staff" is also used. In any case, there is an **internal exchange of services** between customer contact employees and supporting employees; the supporting employees can be referred to as "internal service providers", while the customer contact personnel represent "internal customers".

Based on this differentiation, the GAP model can be further differentiated by applying it to this internal exchange relationship (cf. Fig. 4.5). Building on the gap terminology from the classical gap model, three **internal gaps** are differentiated here (Frost & Kumar, 2000):

1. Internal GAP 1 as the gap between the expectations of the customer contact employees of the internal service providers and their perception in terms of these expectations,
2. Internal GAP 3 as the gap between the customer expectations translated by the internal service provider and the internal service creation,
3. Internal GAP 5 as the internal service quality and thus the discrepancy between the service expected by the customer contact personnel and the service delivered by the internal service provider.

An examination of internal quality gaps allows the identification of further differentiated **factors influencing service quality**, as this takes account of the complex (internal) processes of service creation in this way. The rectification of the internal gaps listed can therefore help to avoid a magnification of quality problems and thus prepare for high

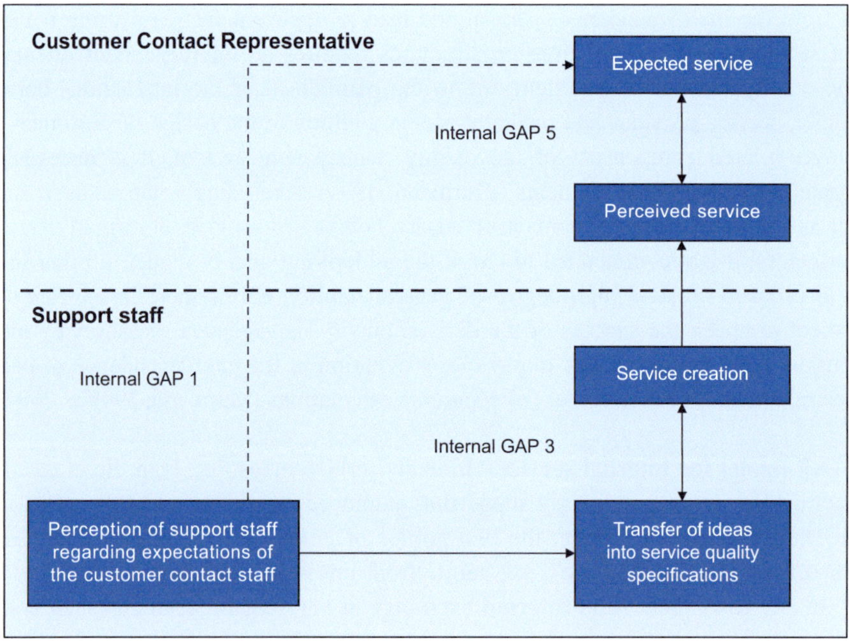

Fig. 4.5 GAP model for internal services (Source: Frost & Kumar, 2000, p. 366)

service quality internally. Nevertheless, further research is needed at this point to check whether the identified differences between the dimensions of service quality considered between internal and external customers generally exist or whether the identified differences are due to the data collected (Frost & Kumar, 2000).

(4) Internal/external GAP model in the context of customer integration In the work of Dimitriadis and Stevens (2008), an integrated GAP model is developed for the implementation and evaluation of relationship quality and customer integration. In this context, a combination of organizational success and failure factors as well as customer expectations and perceptions is made (see Fig. 4.6). A total of four **GAPs** are identified in the context of this GAP model, which decisively determine the relationship quality to the customer (Dimitriadis & Stevens, 2008, p. 504):

1. GAP 1 describes the general discrepancy between customer expectations of the relationship between supplier and customer, and the company's promises of the relationship within the framework of customer relationship management. Above all, this GAP shows the lack of implementation of customer orientation in companies. If the

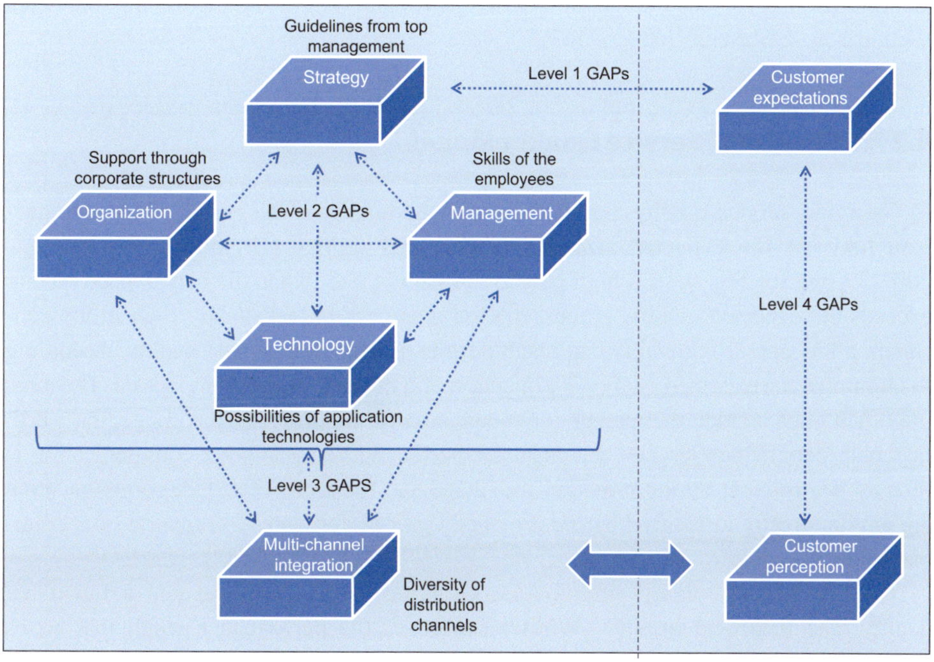

Fig. 4.6 Internal/External GAP model in the context of customer integration (Source: Dimitriadis & Stevens, 2008, p. 503)

company has no information whatsoever about customer expectations, these cannot be satisfied in the course of creating the service.

2. GAP 2, as the first company-related GAP, refers to the discrepancy between the set strategy for customer orientation and its implementation in organizational processes and systems. For example, if the information and communication systems available do not allow for a customer-oriented way of working on the part of employees, then the strategy of customer orientation cannot be implemented efficiently.
3. GAP 3, as the second company-related GAP, includes the principle that formulated strategies and available resources must be implemented in planned action. In this context, it is necessary to unify all customer touchpoints and align them consistently with the customer's needs.
4. GAP 4 finally includes, as a customer-related GAP, the discrepancy between customer expectations of the quality of the relationship with the company and the perceived relationship management efforts of the company.

The consideration of an integrated customer- and company-related GAP model for the analysis of the relationship quality allows for an effective and efficient evaluation of the customer orientation offered by the supplier and thus contributes decisively to the increase of the customer-perceived service quality. However, it is necessary to further develop the presented measurement instrument for the relationship quality and, above all, to refine and validate the identified discrepancies in GAP 2 and GAP 3 in more detail (Dimitriadis & Stevens, 2008, p. 507).

4.2 Grönroos' Service Quality Model

In Grönroos' service quality model, the perceived quality is the **result of the comparison between the expected and the experienced service** (Grönroos, 1984, 37 pp., 2007, 77 pp.; see Fig. 4.7). A high perceived quality arises when the experienced quality exceeds the expected quality. If the expected quality is unrealistically high in the consumer, a low perceived quality can result despite a good experienced quality. According to Grönroos, the expected quality is a function of a number of different factors. The most important ones include the market communication of the provider, the image, the word-of-mouth communication and the customer needs. While the market communication is directly controllable by the provider, the image and the word-of-mouth communication are only indirectly influenced by the provider's past performance— supported for example, by advertisement.

The experienced quality differentiates Grönroos into a technical and a functional quality. The **technical quality (What?)** comprises the performance result that arises through the performance creation process within the supplier-customer- interaction. This quality dimension is assigned an objective measurability. Thus, the functional car after a

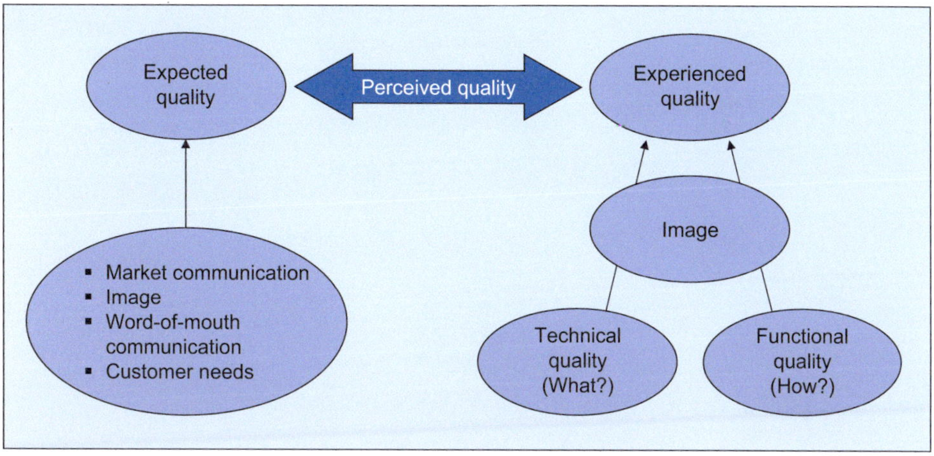

Fig. 4.7 Grönroos' service quality model (Source: Grönroos, 2007, p. 77)

repair, the diploma at the end of a study or the stock portfolio after a consultation with a bank are relatively objectively assessable service results.

In addition, for the experienced quality, it is decisive in which way this result has come about. These aspects are summarized by Grönroos under the **functional quality (How?).** For example, functional quality is dependent on the appearance and friendliness of the staff, atmospheric influences or the availability of the service provider and is subject to a subjective perception by the consumer to a greater extent than technical quality. In terms of the overall assessment of the experienced quality, Grönroos is of the opinion that functional quality is of greater importance. An exceptionally good functional quality can, under certain circumstances, even make up for deficiencies in technical quality.

In addition to the two quality dimensions, the image affects the experienced quality in different ways. On one hand, a good image can lead the consumer to "forgive" the provider for small errors. If errors occur more frequently, however, the image is damaged. On the other hand, a bad image can lead the customer to consider even small errors to be serious. The image acts in this sense as a filter that either amplifies or mitigates the technical and functional quality.

4.3 Service Quality Model by Meyer and Mattmüller

Meyer and Mattmüller take an **expansion of the quality dimensions of Donabedian** (cf. Sect. 2.5) and connect it with Grönroos's model (cf. also Meyer & Mattmüller, 1987 below). The differentiation is always related to what a service requester receives or what a service provider is able to do (cf. Fig. 4.8).

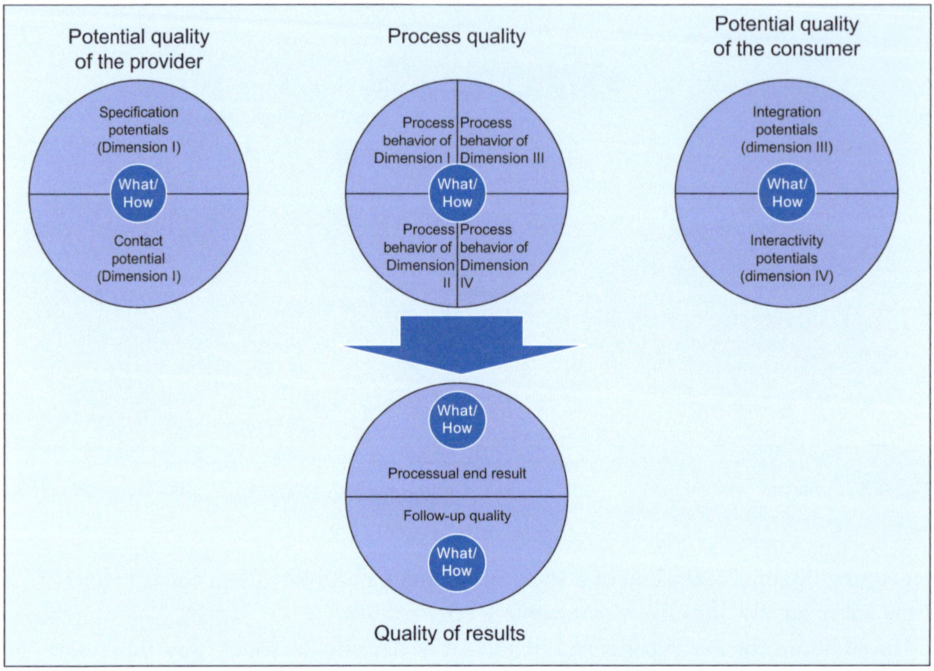

Fig. 4.8 Service quality model according to Meyer and Mattmüller (Source: Meyer & Mattmüller, 1987, p. 191)

In terms of **potential quality**, potentials are distinguished between specification and contact potentials on the part of the provider and between integration and interactivity potentials on the part of the consumer. The specification dimension expresses the potential of a provider to specify a concrete service by means of a combination of factor capabilities of the internal subjects (service providers and employees) and the supporting factors (e.g. tools, technology) available to him. Contact potential is the second determinant of supplier-related potential quality. This is particularly about the ability of a provider to always ensure a high customer orientation during customer contact situations ("Service Encounter"). With regard to the consumer, integration and interactivity potentials are distinguished. The integration dimension relates to the customer's basic physical, mental and emotional readiness and ability to get involved in the service process. The interactivity potential takes into account the possible causal relationships of the interactive contacts between different customers on the service quality.

The **process quality** takes into account the potentials contributed by the provider and consumer interactively and can be described as specific process behavior in each case.

The **quality of the result** is the processual result of this meeting, which can be observed as a time-oriented result observation, and the follow-up quality is a period-oriented consideration. The "how" of the follow-up quality is not subject to direct influence by the service provider.

The quality of the service is dependent on the **ratio of the process to the result dimension** according to Meyer/Mattmüller. A causal relationship is assumed between the partial qualities. The model makes it particularly clear that personal and interpersonal contacts between employees and customers as elements of potential quality play a special role.

It should be noted critically that the model of Meyer and Mattmüller (1987)—as well as Grönroos—is a conceptual model that has not been subjected to empirical testing.

4.4 Dynamic Process Model by Boulding et al.

The **dynamic process model** of service quality by Boulding et al. (1993) is based on the assumption that the expectations and perceptions of a customer with regard to the quality of the service can change over time and cause different patterns of behavior towards the service provider (see also Boulding et al., 1993; Boulding & Staelin, 1993 below). The perceived quality of service of a customer is influenced above all by three factors (see Fig. 4.9):

- The **will-expectations** indicate which level of performance the customer considers probable with the provider under consideration before using the service ("What will happen?").
- The **should-expectations** about the quality of service correspond to the customer's ideas about an appropriate service promised by the service provider ("What should happen?")—in contrast to the so-called ideal conceptions of the quality of service.
- The **just delivered performance** during the service process or during contact with the company.

Before the start of the service process, the customer has specific will- and should-expectations regarding every dimension of service quality. These original expectations and the then delivered service lead to a certain cumulative perception of the dimensions of service quality. The **addition of these perceived quality dimensions,** which relate to the dimensions of the GAP model, results in a global quality judgment about the service quality of the enterprise. This judgment prompts the customer to a certain behavior pattern, such as loyalty to the enterprise.

The will- and should-expectations of the customer regarding the service quality are therefore essential determinants for his quality perception. It is assumed that the customer revises or confirms his **expectations over time**. Influencing factors of the expectations are in this model the experiences from past performance transactions, relevant information about the service from external communication sources (such as communication with other service customers) and the expectations of the previous period. By

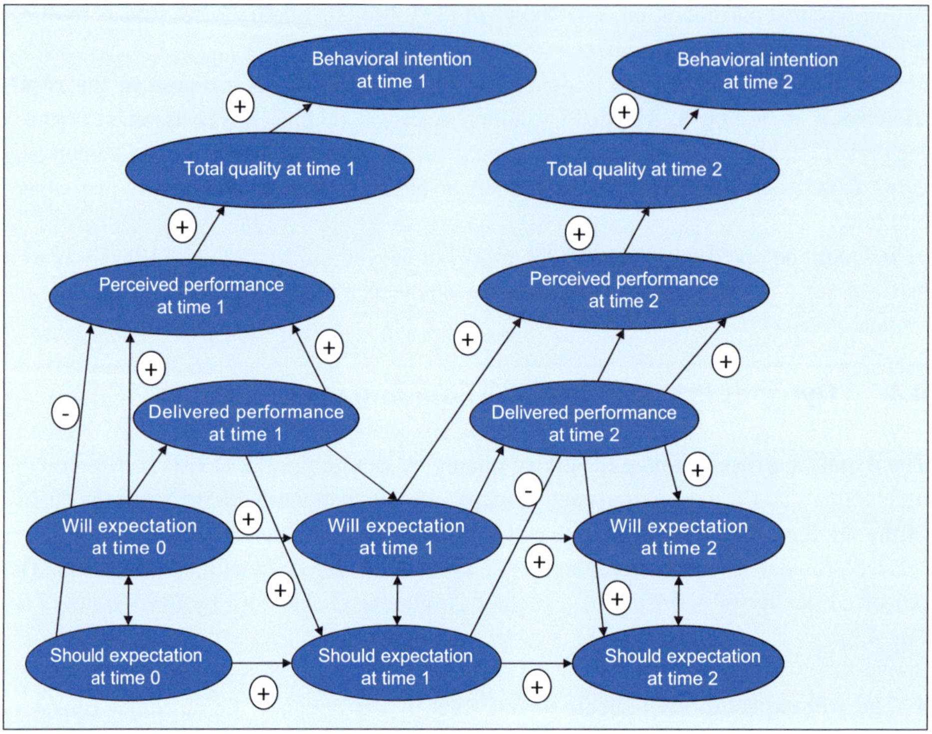

Fig. 4.9 Dynamic process model by Boulding et al (Source: Boulding et al., 1993, p. 12)

assuming that each customer has different expectations in terms of service quality, the model is able to provide a justification for different perceptions of the same service quality from the perspective of different customers.

In addition, this dynamic model includes specific **implications for quality management** of a service company. While the GAP model presented above indirectly implies that management could increase the perceived service quality by either lowering customer expectations or improving the performance considered, this model separates between will- and should-expectations and specifically presents management with the task of increasing those expectations that positively influence the perceived service quality and thus influence the customer to an action corresponding to the goals of the service company (e.g. repeated use of the service). The management is therefore confronted with the task of dealing with a specific management of customer expectations within the framework of quality management in order to exploit their possibilities of controlling customer expectations.

In **laboratory and field experiments** the following results were obtained when testing the model by Boulding et al. (1993):

- The higher the customer's will-expectations are, the better the perceived service quality is assessed by him.
- The higher the customer's should-expectations are, the weaker he assesses the perceived service quality.

The task of **management** is therefore particularly to try to increase the will-expectations and to reduce or at least keep the customer's should-expectations constant. However, the model does not provide any concrete implementation strategies for management—in contrast to the GAP model.

4.5 Relationship Quality Model by Liljander and Strandvik

The relationship quality model, which also refers to dynamic aspects of the service creation process, was developed by Liljander and Strandvik (see also Liljander & Strandvik, 1995, pp. 141–149 below). Unlike Boulding et al. (1993), Liljander and Strandvik do not try to dynamize an originally static quality model, but include numerous constructs in their consideration that are related to process orientation and the importance of the **relationship between the customer and the service provider** The starting point of their model is the assumption that a positive service quality and customer satisfaction lead to a higher customer loyalty and thus are a significant determinant of corporate success. They see service quality from an "external" and satisfaction from an "internal" perspective, i.e. that (un)satisfaction can only arise as a result of a service provider-customer contact, while the assessment of quality is also possible without the customer taking advantage of the service. For Liljander/Strandvik, satisfaction is more closely related to customer behavior than the assessment of quality, but the latter is a determinant of satisfaction.

Liljander/Strandvik distinguish between two **levels of contact** of a service provider to its customers: episodes and the relationship. An **episode** is understood as an event of the interaction between customer and service provider with clearly defined start and end points. It represents the complete act of service creation and can consist of several transactions. An episode is characterized by a financial and social exchange as well as an exchange of a service and of information.

A **relationship** generally consists of at least two episodes. Depending on the continuity and frequency of use of the service under consideration, a relationship can be designed differently. In the case of continuously used services, the first episode usually represents the beginning of a relationship, while in the case of less continuous and rarely used services, the second episode is a necessary but not sufficient condition for the beginning of a relationship. Based on these considerations, the authors assume a model of relationship quality, which is reproduced in Fig. 4.10.

Based on the **(dis-)confirmation paradigm**, it is assumed that the customer measures the respective performance on both levels of contact against a comparison standard. Within a certain tolerance zone, he thus perceives a certain quality. This is compared to

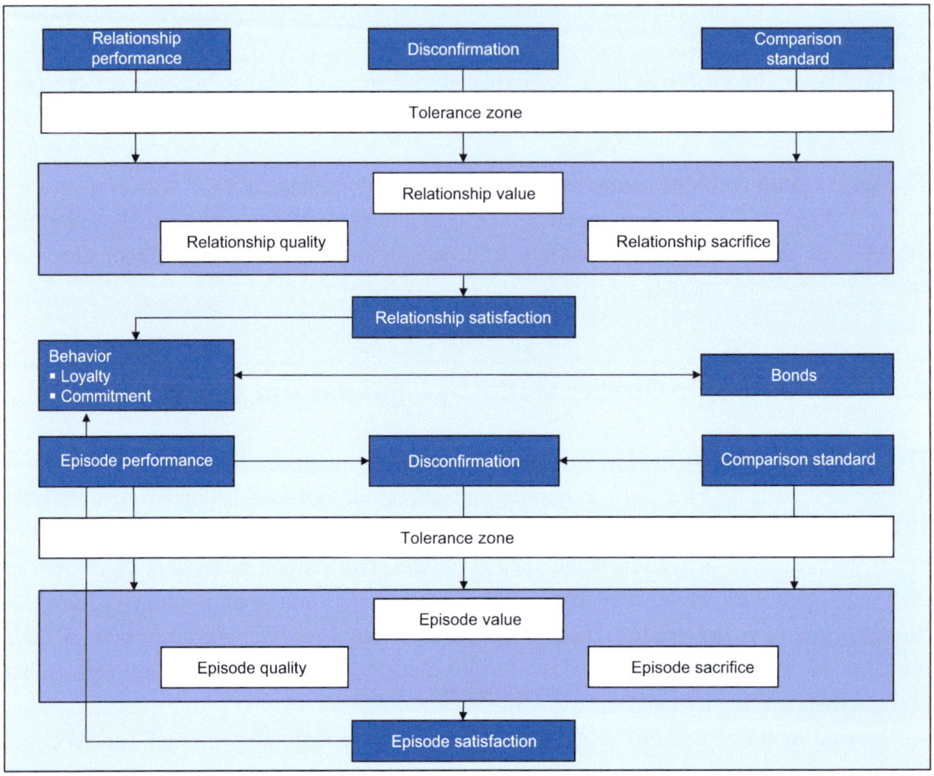

Fig. 4.10 Relationship quality model of Liljander and Strandvik (Source: Liljander and Strandvik, 1995, p. 143)

the respective "victim" of the customer, resulting in (dis-)satisfaction. The resulting **relationship quality** is an important determinant of customer behavior, which is expressed by the constructs commitment and loyalty. In close connection with the behavior are so-called bonds to the company. Such bonds can exist, which can be interpreted as exit barriers and thus perceived negatively by the customer and also hardly controllable by the service provider (e.g. legal, economic, technological, geographical or temporal bonds). To a second group belong knowledge, social, cultural, ideological and psychological bonds.

When considering the **commitment of** both the customer and the service provider, it is necessary to look at the relationship. Commitment is understood as the attitude of a person towards the interaction and the resulting intentions. If it is assumed that both the customer and the service provider can have a positive, negative or indifferent commitment to the relationship, this results in nine commitment combinations, which lead to different **relationship types** between customer and service provider from the customer's perspective and are shown in Fig. 4.11.

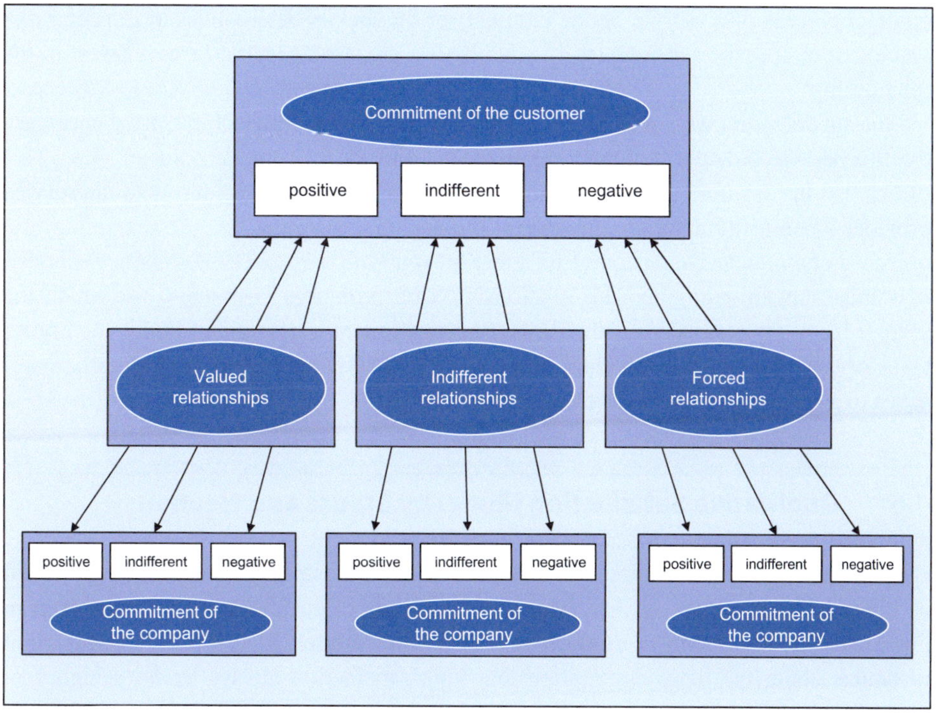

Fig. 4.11 Relationship types between customer and service provider according to Liljander and Strandvik (Source: Liljander & Strandvik, 1995, p. 155)

Valued relationships are characterized by a high commitment of the customer. Depending on the strength of the service provider's commitment, either mutual commitment or the customer is treated like all other current and potential customers, or even rejected, despite his high commitment. The latter can be the case, for example, when a bank actually does not want to serve an unprofitable customer, but is legally obliged to do so. **Indifferent relationships** are characterized in that the customer has neither a positive nor a negative commitment to the relationship with the company. This means that he has a low involvement in performance, hardly perceives any differences between the offers of different service providers and does not feel firmly bound to the company. This is often the case when the customer only uses the services of a certain company out of habit. This can be, for example, the use of a special fitness studio that is located in the same building as the workplace and is only used for reasons of time saving and practicability. The indifferent commitment of the customer can be opposed by a positive, also indifferent or negative commitment of the company to the relationship. If the customer has a negative commitment to the relationship with the service provider, one speaks of **forced relationships.** These can be based on formal bonds (e.g. minimum contract periods of a mobile phone contract) or the lack of alternatives (e.g. postal dispatch). Here, too, the service provider can have a great, no specific or no interest in the relationship.

The model provides some **implications for service practice.** First, it is necessary that the service provider defines a relationship from the customer's point of view. This means that the customer does not necessarily feel in a close relationship with the service provider when using database marketing, the use of a customer-specific marketing mix based on comprehensive customer information stored in a customer database. Secondly, it is important to strengthen the relationship with profitable customers, for which the knowledge of the reasons for the customer's commitment is essential. Finally, the importance of the relevance of individual transactions within an episode as well as certain episodes of a relationship becomes clear.

4.6 Qualitative Satisfaction Model by Stauss and Neuhaus

As postulated in the service quality success chain, customer satisfaction is the result or effect of high service quality, which in turn leads to customer loyalty. However, in the context of developing a **qualitative satisfaction model**, Stauss and Neuhaus point to doubts about the basic assumption that high satisfaction always leads to high customer loyalty. For corresponding empirical results of other researchers, he makes competing customer needs, the attractiveness of alternative services and situational factors responsible.

In the qualitative satisfaction model of Stauss/Neuhaus it is assumed that there are emotional, expectation-related and binding potentials of endangerment among satisfied customers. In analogy to the dynamic model of job satisfaction by Bruggemann (1974) Stauss and Neuhaus (1995) differentiate three **satisfaction—**and two **types of dissatisfaction** (see also Stauss & Neuhaus, 1995):

- The **"demanding satisfied"** is characterized by a high satisfaction with the service provider; However, the latter always has to strive to meet them anew because of the constantly growing customer requirements.
- Unlike the first type, the **"stable satisfied"** has a passive claim behavior.
- The **"resigned satisfied"** shows a certain indifference towards the relationship with the service provider. This can be explained above all by the lack of alternatives.
- Like the **"stable satisfied"** also the **"stable dissatisfied"** has a weak activity level, but is dissatisfied with the performance.
- The **"demanding dissatisfied"** expresses his dissatisfaction to the service provider and would not decide again for the same provider.

Based on the types of satisfaction, it is assumed that global satisfaction and type affiliation are related. In addition, the **risk potential** depends on the type affiliation. So the "resigned satisfied" will have a greater risk potential than the "demanding satisfied" and both a greater than the "stable satisfied". However, it can also be assumed that customers with a high global satisfaction belong to the "resigned satisfied". The risk potential is a significant influencing variable of the respective customer behavior, which can be expressed, for example, in customer loyalty. So "demanding satisfied" and "resigned satisfied" have probably already considered a provider change than the "stable satisfied" and are less willing to recommend the company.

The model allows some **implications for service practice.** It becomes clear that it is not enough to just ask the global satisfaction in customer satisfaction surveys. Rather, it is important to determine the risk potential of satisfied customers. Based on these analyses, marketing instruments are to be used in particular with those customers who have a high risk potential. In the context of a critical assessment of the qualitative satisfaction model, it is particularly important to point out the global measurement of customer satisfaction. Because no individual characteristics of the service are evaluated, two basic problems arise. In the context of customer satisfaction measurement, it is questionable whether the respondents evaluate a service with regard to the same criteria. In close connection with this is the problem that a service provider does not receive any specific indications as to which partial services are to be modified due to customer dissatisfaction.

The strength of the models of service quality presented here lies in their coherent representation of cause-and-effect relationships of service quality. They provide management of service companies with a helpful analytical framework for a deeper understanding of the origin and effect of service quality. They are also suitable for deriving action instructions for the management with regard to the realization of an optimal service quality level. Only if this is achieved, the service quality can develop its positive effects (cf. Sect. 2.6) to the full. In the end, they provide a sensible basis for the design of quality management and in particular for the measurement of service quality (cf. Chap. 5).

4.7 Model of E-Service Quality

The development of new information and communication technologies has brought new business fields in the service sector with it, so-called **e-services.** These are services that are not offered and provided to the customer "face-to-face", but via the Internet. In order for all phases of a service creation process to be mapped on the Internet, the classical dimensions of the potential, process and result dimension of services must also be taken into account here. Nevertheless, the context of new information and communication technologies brings new expectations and perceptions of service quality on the part of customers, which requires an adaptation of the classical model of service quality.

▶ **E-services** are independent, marketable services that aim at a utility-generating effect (result dimension) on external factors by providing the provider's electronic capabilities (potential dimension) and by integrating an external factor using electronic data exchange (process dimension) (Bruhn, 2002, p. 6).

The essential differences between classical, non-electronic services and e-services are shown in Fig. 4.12. **Main differences** lie in the ability and the need for the presence of the provider, the degree of virtualization of the services, the availability of the offer, the type of service encounter and the customer approach. Only online advertised or only partially online created classical services are to be classified in a range of uncertainty (Bruhn & Georgi, 2006, pp. 161–162).

Online counseling or information, as well as training and online customer services are typical application areas. The focus of activities in online counseling or online information is not to supply motivating information for purchase decisions, but rather the interest in responding to specific customer needs and transporting the processed information about core services to the customer. Within the marketing function, electronic customer relationship management, or E-CRM, is seen as an addition to the customer relationship

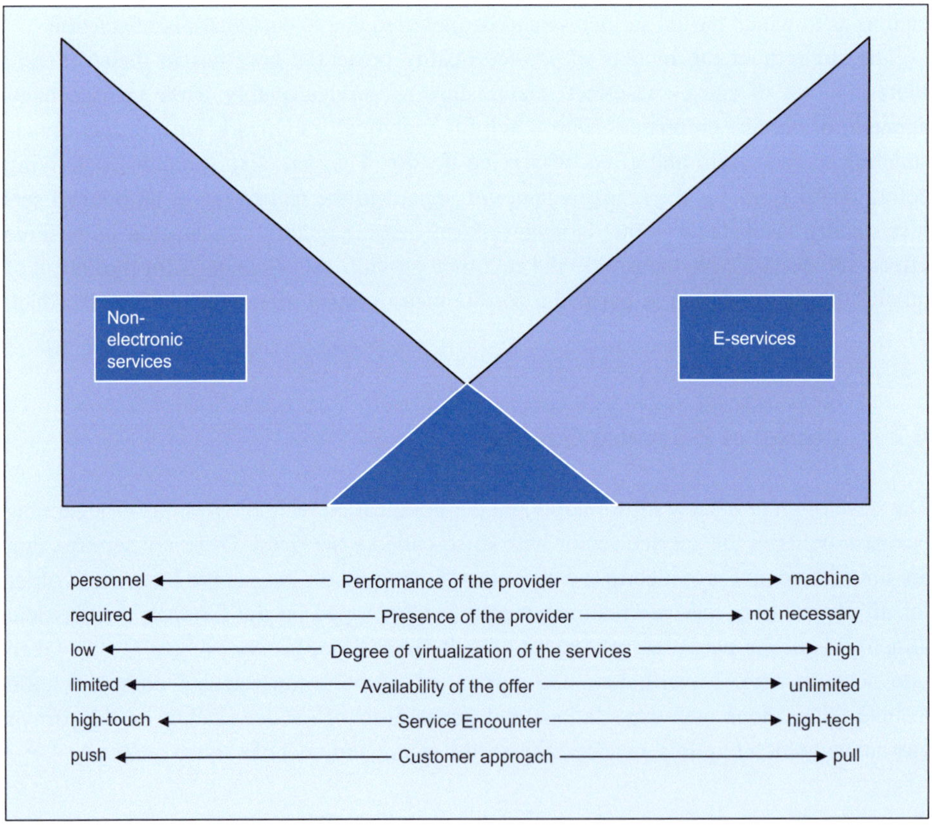

Fig. 4.12 Differentiation of e-services from traditional services (Source: Bruhn, 2002, p. 12)

management tool kit and thus links the original CRM with Internet technology (Manschwetus & Rumler, 2002). The term **"e-business"** is often used as a general term for the entire electronic business process that takes place over the Internet. "e-business" encompasses all transactions with customers and suppliers, as well as all people involved in this process.

The new possibilities of acquiring new customer segments through e-services raise the fundamental question of substitution of their traditional service area for providers of service-oriented businesses (Bitner et al., 2002). For the representation of the **substitution or extension possibilities** of traditional service offerings through e-services, it makes sense to distinguish between a existing and a new service both for traditional services as well as e-services. The resulting service strategies are shown in Fig. 4.13.

A possible combination of traditional services with e-services is the **service substitution**. Here, the existing offer of traditional service offerings is replaced by a new e-service offering, such as the online banking of a financial service provider. The new online offer opens up numerous advantages for the provider, such as the development of additional customer groups or increased efficiency and thus lower customer support costs.

The support of an existing offer of classical services by existing e-services characterizes the combination possibilities of the **service support.** An example of this is the sale of computers, which is supplemented by an online support, for example in the form of FAQs (Frequently Asked Questions), (e.g. *Dell*). Such an offer then proves to be useful if, from the customer's point of view, higher benefits are generated by the e-services.

The combination of a new offline offer with an existing e-service represents another combination possibility of the **service extension**. This is the case, for example, when a company like Amazon opens real sales outlets in addition to its online bookshop. Disadvantages of an exclusive e-service offer, such as lack of customer loyalty, are reduced by such a service extension. Furthermore, the customer benefit can be increased

		E-service offer	
		Available	New
Range of classic services	New	**Service extension** Expansion of the offering through new classic services with existing electronic support	**Service innovation** Completely new services with completely new electronic support
	Available	**Service support** Support of the existing service through existing e-services	**Service substitution** Extension/substitution of classic services by new e-services

Fig. 4.13 Matrix of the combination of classic services with e-services (Source: Bruhn, 2002, S. 26)

by the combination with existing e-services (e.g. by an online complaint option) and thus the reference and benefit basis of the offline service can be extended.

The **service innovation** is characterized by the offer of a new offline service in combination with a new e-service. An example of this is a customer club that is opened as part of the introduction of a customer loyalty management and offers both a magazine and an online information service as well as an online ordering option. Under the aim of profiling against the competition, this approach makes sense in the area of value-added services. In the area of stand-alone services, it is also recommended from the point of view of diversification.

For the successful implementation of electronic services, it is crucial to ensure a high quality of service (Parasuraman et al., 2005; Collier & Bienstock, 2006; Faßnacht & Koese, 2006). In this context, the term **e-service quality** is often used. A systematic literature review shows that authors for their measurement models of the quality of e-services mainly combine classical quality indicators (analogous to the indicators of SERVQUAL) with specific indicators for the quality of websites (van Riehl et al., 2003; Jayawardhena, 2004; Long & McMellon, 2004). Figure 4.14 provides an overview of the dimensions that are most often used in empirical research and thus also speak for a corresponding validity. The industries investigated included, amongst others, e-banking (Joseph et al., 1999; Jayawardhena, 2004), online flight services (van Riehl et al. 2003), online retailing (Yang & Jun, 2002; Zeithaml, 2002; Wolfinbarger & Gilly, 2003; Parasuraman et al., 2005; Giovanis & Athanasopoulou, 2014) and web portals (Kuo et al., 2005).

The results show that the dimensions of e-services have parallels to the three dimensions of service quality (potential, process and result dimension). Accordingly, e-service quality can be conceptualized using the dimensions and indicators shown in Fig. 4.15.

The model differentiates between the incubative and active dimension. The **incubative dimension** includes the basic structure and design of the website on which the e-services are offered. In general, these quality features can already be assessed before the actual use of the e-services (Santos, 2003, p. 238). These include user-friendliness, appearance, networking with other websites, structure and layout as well as content of the website.

The **active dimension** includes all factors that ensure a smooth service process and a satisfactory service result. This includes the factors reliability, efficiency, support, communication, security and (usage) incentive. When compared to the SERVQUAL model for services, the potential dimension can be assigned to the incubative dimension, while the result dimension corresponds to the active dimension. Factors from both the incubative (e.g. user-friendliness, networking) and the active dimension (e.g. communication, security) are part of the process dimension.

A key role for satisfaction with the service provider is the perception of e-service quality. This increases the number and frequency of visitors to the website as well as the length of stay. High scores in the active dimension increase customer loyalty and the willingness of customers to recommend (Santos, 2003, p. 238).

Quality dimensions \ Authors	Joseph et al. (1999)	Barnes und Vidgen (2001)	Cox und Dale (2001)	Jun und Cai (2001)	Barnes und Vidgen (2002)	Loiacono et al. (2002)	Madu und Madu (2002)	Yang und Jun (2002)	Zeithaml (2002)	Wolfenbarger und Gilly (2003)	Santos (2003)	Van Riel et al. (2002)	Yang et al. (2003)	Long und McMellon (2004)	Jayawardhena (2004)	Yang et al. (2004)	Parasuraman et al. (2005)	Kuo et al. (2005)	Kim et al. (2005)	Fassnacht und Koese (2006)	Lin (2007)	Li und Suomi (2007)	Li und Suomi (2009)	Li et al. (2009)	Ladhari (2010)	Finn (2011)	Zavareh et al. (2012)	Zhao et al. (2012)	Chen et al. (2013)	Thaichon et al. (2014)	Kao und Lin (2016)
Ease of use (comprehensibility, user-friendliness, logical structure)	✓					✓		✓			✓		✓			✓	✓	✓		✓	✓			✓	✓	✓			✓		
Execution (performance, accuracy, added value, compensation)							✓		✓	✓							✓		✓			✓	✓			✓	✓		✓	✓	
Efficiency	✓								✓		✓						✓		✓			✓			✓		✓				
Reliability (availability, contact)		✓		✓			✓	✓	✓	✓	✓	✓	✓	✓	✓	✓	✓			✓	✓		✓	✓	✓				✓		✓
Reliability (credibility, competence, reputation)		✓	✓				✓	✓					✓	✓	✓	✓	✓									✓					
Access (access options)	✓	✓	✓	✓				✓					✓				✓	✓	✓			✓		✓							
Responsiveness (execution speed, process management, response time)		✓		✓	✓		✓	✓	✓			✓	✓	✓	✓	✓	✓		✓		✓	✓	✓	✓	✓	✓	✓		✓		
Material environment (design, functions, usability, structure, layout, aesthetics appearance, user interface, website design, linking)		✓	✓	✓	✓	✓	✓			✓	✓	✓	✓	✓	✓		✓		✓	✓	✓		✓	✓	✓	✓	✓		✓		✓
Security / Data protection		✓					✓	✓	✓	✓	✓	✓	✓			✓	✓		✓	✓	✓		✓	✓	✓		✓			✓	
Personalization (adaptation to customer requirements, differentiation, relative customer advantage, decision attractiveness, product portfolio)	✓					✓	✓	✓				✓	✓			✓	✓						✓		✓						✓
Trust / Empathy		✓		✓	✓	✓	✓							✓	✓		✓	✓		✓	✓		✓	✓			✓	✓		✓	✓
Information quality (timeliness, relevance, completeness, comprehensibility)			✓	✓	✓	✓					✓							✓	✓	✓	✓	✓	✓					✓	✓	✓	
Communication (support, contact possibility, customer service, cooperation feedback management / complaint management	✓	✓				✓			✓	✓	✓		✓	✓								✓					✓	✓		✓	
Innovation capability (continuous improvement)							✓							✓																	

Fig. 4.14 Quality dimensions of e-services in the literature (Source: Hadwich et al., 2010, 118 pp.)

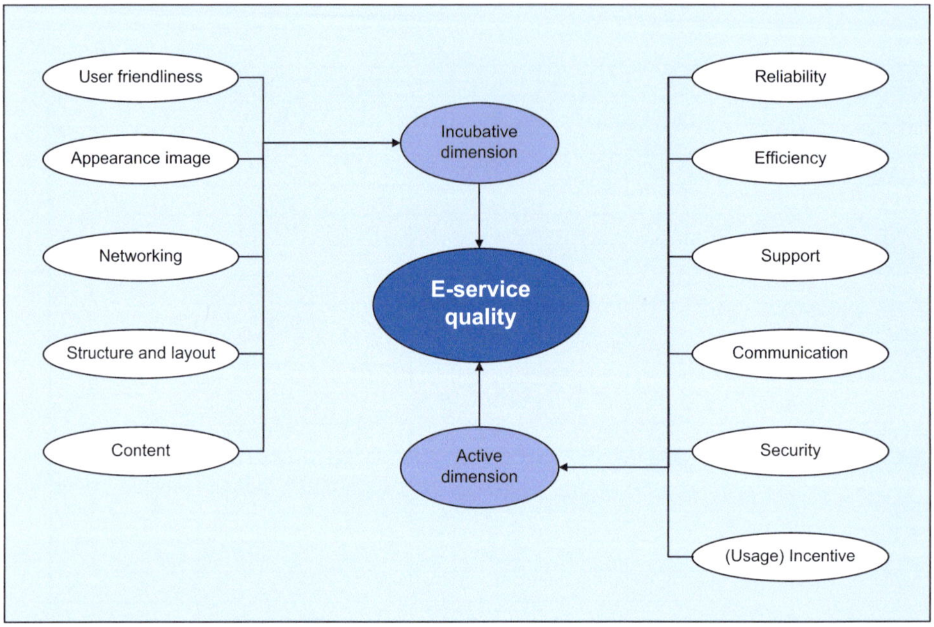

Fig. 4.15 Dimensions and indicators of e-service quality (Source: Santos, 2003, p. 239)

4.8 Model of E-Health Service Quality

The use of e- health services has been growing in popularity in recent years. Typical examples of e-health services include online pharmacies, health portals, online self-help groups, live web consultations or e-mail consultations (Kirchgeorg & Lorbeer, 2002; Bruhn & Batt, 2015). While in 2008 only 38% of households in Germany consulted the Internet as an information source for health questions, by 2011 this had already risen to more than 53%. After the doctor's visit or the doctor's recommendation, the Internet is now the most frequently used information source for health-related questions (Nielsen, 2012; Bruhn & Batt, 2015).

▶ **E-health services** can be defined as independent, marketable services in the health care sector that are provided or supported by service providers (e.g. doctors) using information and communication technologies aiming to create a beneficial effect for the customer (e.g. patients) (Meyers et al., 2002).

Despite the large number of customers who now use electronic health services, it has so far failed to establish itself on a long-term and nationwide basis (Reichwald et al., 2008; Schlötelburg & Bröckerhoff, 2010). Some doctors see the time-consuming nature of providing e-health services as the reason for this (Bruhn & Batt, 2015). The key arguments are the additional time required to answer e-mails or the time that needs to be spent calming patients who have been unsettled by misinformation on the Internet (Lerer & Rowell, 2000). Furthermore, for example data protection considerations, prevent the establishment of e-health services (Eng, 2001). The result is that more and more people are turning away from providing e-health services.

In addition to the perspective of service providers and authorities, however, the view of customers must also be taken into account with regard to the inadequate exploitation of the potential of e-health services. Here it becomes apparent that some e-health services are used while others are not used at all. This can be explained from the demand side in that not all e-health services have a satisfactory quality level (Bruhn & Batt, 2015). However, a high perceived e-health service quality (e-HSQ) is to be seen as a prerequisite for the service to be used at all (Bruhn & Batt, 2015). This is also confirmed by numerous studies from other research areas, in which the positive effect of quality perception on behavior has been demonstrated many times (e.g. Liu et al., 2011; Sousa & Voss, 2012).

Against this background, Bruhn and Batt (2015) develop a model for measuring e-health service quality. The construct of e- health service quality consists of the following six **dimensions:** availability, speed, information quality, individualization, user-friendliness and security of the e- health service (Bruhn & Batt, 2015; see Fig. 4.16). In this context, the results of the study confirm the central importance of e-health service quality for the service provider. A quality of e-health service perceived as high influences the user's intention to use the e-health service again and to recommend it, and is therefore a significant success factor (Bruhn & Batt, 2015).

The study by Bruhn and Batt (2015) represents the first engagement with the construct of e-health service quality. Due to the exploratory character of the study, the topic is consequently to be considered as not concluded. Rather, the findings provide a starting point for further scientific engagement with the construct. For example, it needs to be examined how the construct and its effects differ for specific e-health services. Thus, it is possible that when considering individual services, there are differences in the effects of e-health service quality (Bruhn & Batt, 2015).

The presented models of service quality serve both a better understanding of the dimensions and characteristics of service quality and of their relationships. In particular, it needs to be examined which model is particularly suitable for the respective object of observation. On this basis, it is then to be clarified how these dimensions and characteristics can be measured. For this purpose, there are numerous different methods available for measuring service quality.

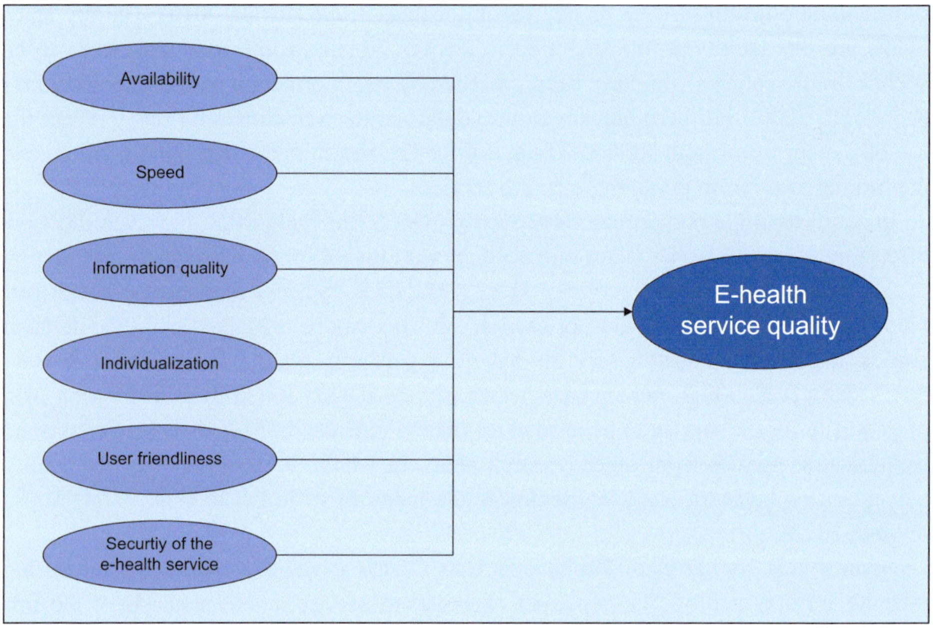

Fig. 4.16 Dimensions of the e-health service quality (Source: Bruhn & Batt, 2015)

References

Bitner, M. J., Ostrom, A. L., & Meuter, M. L. (2002). Implementing successful self-service technologies. *Academy of Management Executive, 16*(4), 96–109.

Boshoff, C., & Terblanche, N. S. (1997). Measuring retail service quality. A replication study. *South African Journal of Business Management, 28*(4), 123–129.

Boulding, W., & Staelin, R. (1993). A dynamic process model of service quality assessment. In E. E. Scheuing & W. F. Christopher (Eds.), *The service quality handbook* (pp. 177–194). New York.

Boulding, W., Kalra, A., Staelin, R., & Zeithaml, V. A. (1993). A dynamic process model of service quality. From expectations to behavioral intentions. *Journal of Marketing Research, 30*(1), 7–27.

Bruggemann, A. (1974). Zur Unterscheidung verschiedener Formen von Arbeitszufriedenheit. *Arbeit und Leistung, 28*(11), 281–284.

Bruhn, M. (2002). Electronic Services. Eine Einführung in die theoretischen und praktischen Probleme. In M. Bruhn, & B. Stauss (Eds.), *Electronic Services. Jahrbuch Dienstleistungsmanagement 2002* (pp. 3–49). Wiesbaden: Gabler.

Bruhn, M., & Batt, V. (2015). Qualität von E-Health Services: Entwicklung und empirische Überprüfung eines Messinstruments. *Marketing ZFP, 37*(1), 42–56.

Bruhn, M., & Georgi, D. (2006). *Services marketing. Managing the service value chain.* Harlow: Pearson Education Limited.

Collier, D. A. (1996). Tight and loose comprehensive customer contact (3c) plans and strategies. In T. A. Swartz, & S. W. Brown (Eds.), *Advances in services marketing and management. Research and practice* (5. Aufl., pp. 153–168). London.

Collier, J. E., & Bienstock, C. C. (2006). Measuring service quality in e-retailing. *Journal of Service Research, 8*(3), 260–275.

Dabholkar, P. A., Thorpe, D. I., & Rentz, J. O. (1996). A measure of service quality for retail stores. Scale development and validation. *Journal of the Academy of Marketing Science, 24*(1), 3 16.

Dimitriadis, S., & Stevens, E. (2008). Integrated customer relationship management for service activities: An internal/external gap model. *Managing Service Quality: An International Journal, 18*(5), 496–511.

Eng, T. R. (2001). *The e-health landscape: A terrain map of emerging information and communication technologies in health and health care.* Princeton: Robert Wood Johnson Foundation.

Faßnacht, M., & Koese, I. (2006). Quality of electronic services. *Journal of Service Research, 9*(1), 19–37.

Frost, F. A., & Kumar, M. (2000). INTSERVQUAL—An internal adaptation of the GAP model in a large service organisation. *Journal of Services Marketing, 14*(4/5), 358–377.

Giovanis, A. N., & Athanasopoulou, P. (2014). Gaining customer loyalty in the e-tailing marketplace: The role of e-service quality, e-qatisfaction and e-trust. *International Journal of Technology Marketing, 9*(3), 288–304.

Grönroos, C. (1984). A service quality model and its marketing implications. *European Journal of Marketing, 18*(4), 36–44.

Grönroos, C. (2007). *Service management and marketing. Customer management in service competition* (3. Aufl.). Chichester: Wiley.

Grund, M. A. (1998). *Interaktionsbeziehungen im Dienstleistungsmarketing. Zusammenhänge zwischen Zufriedenheit und Bindung von Kunden und Mitarbeitern.* Wiesbaden: Gabler.

Hadwich, K., Georgi, D., Tuzovic, S., Büttner, J., & Bruhn, M. (2010). Perceived quality of e-health services. A conceptual scale development of e-health service quality based on the C-OAR-SE approach. *International Journal of Pharmaceutical and Healthcare Marketing, 4*(2), 112–136.

Hentschel, B. (1999). Multiattributive Messung von Dienstleistungsqualität. In M. Bruhn, & B. Stauss (Eds.), *Dienstleistungsqualität. Konzepte – Methoden – Erfahrungen* (3rd Edn., pp. 289–320). Wiesbaden: Gabler.

Jayawardhena, C. (2004). Mesurement of service quality in internet banking. The development of an instrument. *Journal of Marketing Management, 20*(1), 185–207.

Joseph, M., McClure, C., & Joseph, B. (1999). Service quality in the banking sector: The impact of technology on service delivery. *International Journal of Bank Marketing, 20*(4/5), 182–192.

Kellogg, D. L., & Chase, R. B. (1995). Constructing an empirically derived measure for customer contact. *Management Science, 41*(11), 1734–1749.

Kirchgeorg, M., & Lorbeer, A. (2002). Kundenbindungsstrategien von E-Health Service-Anbietern. In M. Bruhn, & B. Stauss (Eds.), *Electronic Services. Jahrbuch Dienstleistungsmanagement 2002* (pp. 585–611). Wiesbaden: Gabler.

Kuo, T., Lu, I., Huang, C., & Wu, G. (2005). Measuring users' perceived portal service quality: An empirical study. *Total Quality Management, 16*(3), 309–320.

Lerer, L., & Rowell, N. (2000). *The e-health consumer, INSEAD healthcare management initiative, the healthcare 2020 Platform.* Fontainebleau.

Liljander, V., & Strandvik, T. (1995). The nature of customer relationships in services. In T. A. Swartz, D. E. Bowen, & S. W. Brown (Eds.), *Advances in services marketing and management research and practice* (4th Edn., pp. 141–167). Greenwich.

Liu, C. T., Guo, Y. M., & Lee, C. H. (2011). The effects of relationship quality and switching barriers on customer loyalty. *International Journal of Information Management, 31*(1), 71–79.

Long, M., & McMellon, C. (2004). Exploring the determinants of retail service quality on the internet. *Journal of Services Marketing, 18*(1), 78–90.
Luk, S. T., & Layton, R. (2002). Perception gaps in customer expectations: Managers versus service providers and customers. *Service Industries Journal, 22*(2), 109–128.
Manschwetus, U., & Rumler, A. (2002). *Strategisches Internetmarketing. Entwicklungen in der Net-Economy.* Wiesbaden: Springer.
Meyer, A., & Mattmüller, R. (1987). Qualität von Dienstleistungen. Entwicklung eines praxisorientierten Qualitätsmodells. *Marketing ZFP, 9*(3), 187–195.
Meyers, J., van Brunt, D., Patrick, K., & Greene, A. (2002). Personalizing medicine on the web: E-health offers hospitals several strategies for success. *Health Forum Journal, 45*(1), 22–26.
Murmann, B. (1999). *Qualität mehrstufiger Dienstleistungsinteraktionen. Besonderheiten bei Dienstleistungsunternehmen mit direktem und indirektem Kundenkontakt.* Wiesbaden: Gabler.
Nielsen (2012). Gesundheitsinformationen aus dem Internet immer wichtiger. http://nielsen.com/de/de/insights/presseseite/2012/gesundheitsinformationen-aus-dem-internet-immer-wichtiger.html. Zugegriffen: 19. Febr. 2012.
Niranjan, T. T., & Metri, B. A. (2008). Client-vendor-end user triad: A service quality model for IS/ITES outsourcing. *Journal of Services Research, 8*(1), 123–128.
Parasuraman, A., Zeithaml, V. A., & Berry, L. L. (1985). A conceptual model of service quality and its implications for future research. *Journal of Marketing, 49*(1), 41–50.
Parasuraman, A., Zeithaml, V. A., & Berry, L. L. (1986). *SERVQUAL. A multiple-item scale for measuring consumer perceptions of service quality.* Report No. 86–108. Cambridge.
Parasuraman, A., Zeithaml, V. A., & Berry, L. L. (1988). SERVQUAL: A multiple-item scale for measuring consumer perceptions of service quality. *Journal of Retailing, 64*(1), 12–40.
Parasuraman, A., Zeithaml, V. A., & Malhotra, A. (2005). ES-QUAL. A multiple-item scale for assessing electronic service quality. *Journal of Service Research, 7*(3), 213–233.
Parikh, D. (2006). Measuring retail service quality: An empirical assessment of the instrument. *Vikalpa, 31*(2), 45–55.
Reichwald, R. P., Möslein, K. M., Kölling, M., & Neyer, A. K. 2 *Service-Innovation*. Leipzig.
van Riehl, A., Semeijn, J., & Janssen, W. (2003). E-Service quality expectations. A case study. *Total Quality Management, 14*(4), 437–450.
Santos, J. (2003). E-Service quality: A model of virtual service quality dimensions. *Managing Service Quality, 13*(3), 233–246.
Schlötelburg, C., & Bröckerhoff, H.-P. (2010). *E-Health-Compendium Telemonitoring 2010/11. Themen – Anwendungen – Service*. Berlin.
Schneider, B., & Bowen, D. E. (1999). The service organization. Human resources management is critical. In J. E. G. Bateson (Eds.), *Managing services marketing. Text and readings* (3rd Edn., pp. 273–283). Texas.
Sousa, R., & Voss, C. (2012). The impacts of e-service quality on customer behaviour in multichannel e-services. *Total Quality Management & Business Excellence, 23*(7–8), 789–806.
Stanforth, N. F., & Lennon, S. J. (1997). Salesperson service and customer satisfaction. The impact of store policies. In T. A. Swartz, D. E. Bowen, & D. Iacobucci (Eds.), *Advances in services marketing and management research and practice* (6th Edn., pp. 79–99). Greenwich.
Stauss, B., & Neuhaus, P. (1995). *Das Qualitative Zufriedenheitsmodell, Diskussionsbeiträge Nr. 66*. Ingolstadt.
Wolfinbarger, M., & Gilly, M. (2003). eTailQ. Dimensionalizing, measuring and predicting eTail quality. *Journal of Retailing, 79*(3), 183–198.
Yang, Z., & Jun, M. (2002). Consumer perception of e-service quality: From internet purchaser and non-purchaser perspectives. *Journal of Business Strategies, 19*(1), 19–41.

Zeithaml, V. A. (2002). Service excellence in electronic channels. *Managing Service Quality, 12*(3), 135–138.
Zeithaml, V. A., Berry, L. L., & Parasuraman, A. (1988). Communication and control processes in the delivery of service quality. *Journal of Marketing, 52*(4), 35–48.
Zeithaml, V. A., Berry, L. L., & Parasuraman, A. (1996). The behavioral consequences of service quality. *Journal of Marketing, 60*(2), 31–46.
Zeithaml, V. A., Parasuraman, A., & Berry, L. L. (1992). *Qualitätsservice. Was Ihre Kunden erwarten – was Sie leisten müssen*. Frankfurt a. M.: Campus.
Zollondz, H.-D. (2011). *Grundlagen Qualitätsmanagement. Einführung in Geschichte, Begriffe, Systeme und Konzepte* (3rd Edn.). München: Oldenbourg.

5 Measuring Service Quality

Abstract

The key to successful quality management is primarily the measurement of service quality. As a result, this chapter shows the existing possibilities for measuring service quality and critically discusses them. It turns out that in addition to customer-oriented measurement approaches (e.g. expert observations, customer surveys, complaint analyses), also company-oriented approaches to quality measurement (e. g. benchmarking, employee surveys, suggestion schemes) are used. In the context of the consideration of practical examples and possible application of individual measurement approaches, it finally turns out that, with regard to the strengths and weaknesses shown, the SERVQUAL approach and the SERVPERF approach currently belong to the most widely used measurement approaches. These approaches are presented and evaluated in detail.

The main task of quality management is to create company performance in accordance with customer requirements. The starting point of a systematic quality management is therefore the measurement of service quality, in the context of which the customer requirements for service quality are determined. For this purpose, a variety of methods are available, which take different importance levels in companies day to day business practice. Basically, two perspectives can be distinguished, with the help of which the requirements for service quality can be measured:

1. By means of **customer-oriented measurement approaches**, a measurement is made from the customer's point of view.
2. By means of **company-oriented measurement approaches**, a measurement is made from the perspective of the company, either from the perspective of the management or the employees.

M. Bruhn, *Quality Management for Services*,
https://doi.org/10.1007/978-3-662-67032-3_5

Figure 5.1 shows a systematization of the different measurement approaches based on this. The growing need to take into account the customer perspective in service marketing is reflected in the number and degree of differentiation of customer-oriented measurement concepts. At the same time, the listed measurement concepts—to varying degrees—are instruments for the identification, direct measurement or possible prioritization or categorization of relevant dimensions of service quality. The corresponding, different use of these instruments within the control loop of quality management follows this. It goes without saying that this division is not always non-overlapping.

5.1 Customer-Oriented Measurement Approaches

Within customer-oriented approaches to measuring the requirements for service qualitydifferentiated and undifferentiated approaches can be distinguished. Within the **undifferentiated approaches** only global quality or satisfaction judgments are obtained regarding the services under consideration by means of surveys (Hentschel, 1999). Such an approach is advantageous from the perspective of simple data collection, but the gain in knowledge remains relatively low if only global statements are considered. Therefore, the **differentiated approaches** will be examined in more detail below, which allow for more detailed statements about the respective sub-qualities of services. These can be divided into objective and subjective measurement approaches according to the **degree of objectivity of the measurement**.

5.1.1 Objective Measurement Approaches

With customer-oriented objective measurement approaches, the performance quality of a company is judged from the customer's perspective, but not on the basis of the subjective assessment of individual customers. Rather, an "objectified", i.e. an intersubjectively verifiable measurement is attempted. In order to assess the service process from the customer's perspective on the basis of at least partially intersubjectively verifiable criteria, the following **methods** can be used:

1. Expert observations
2. Silent shopper approach
3. Service tests
4. Quality rankings

The different objective measurement approaches are presented below.

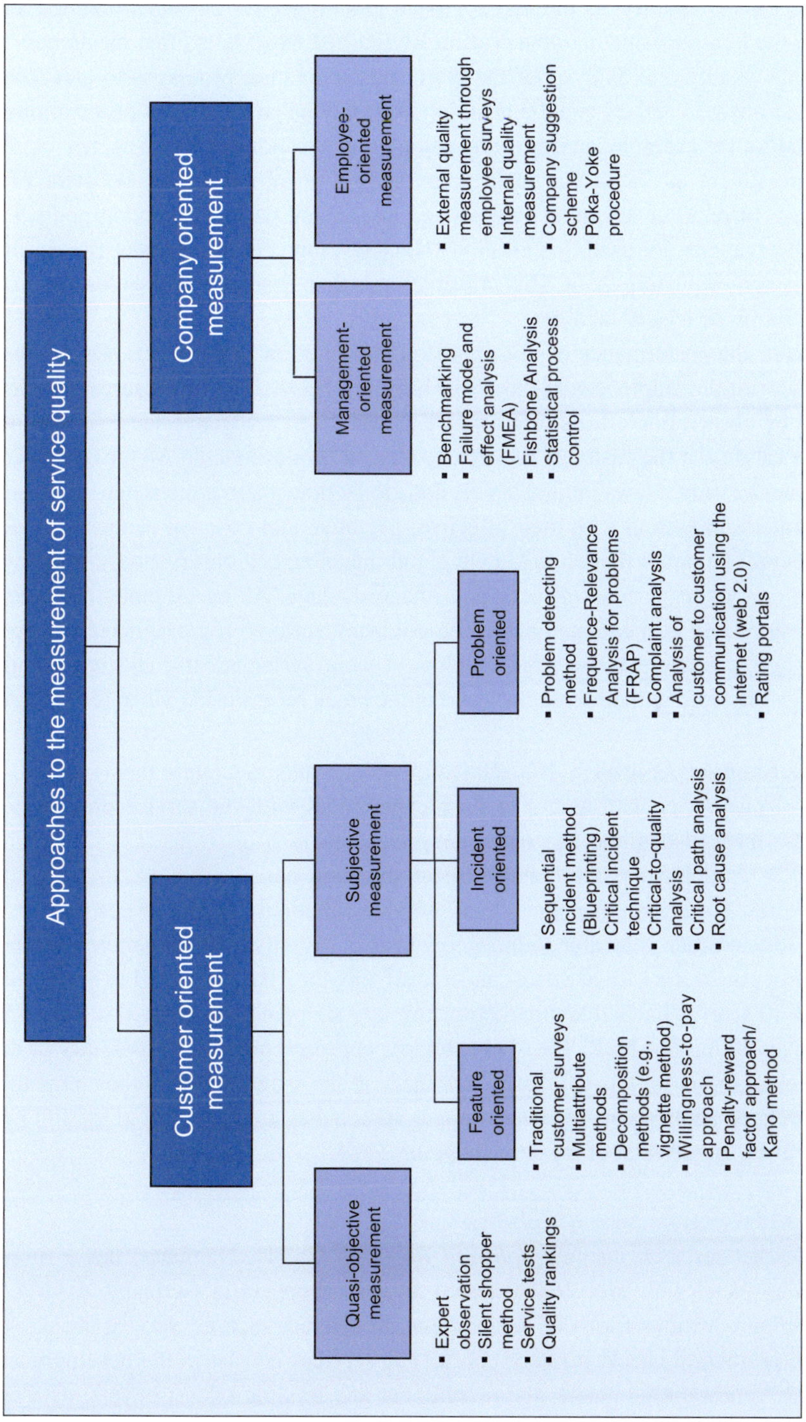

Fig. 5.1 Systematization of the approaches to measuring service quality

(1) Expert observations As the first approach to an objective customer-oriented measurement, the non-participating **observation by trained experts** is often mentioned. The aim of this capture and analysis of the performance creation process is to gain knowledge about obvious defects and the resulting customer behavior. Expert observations are often used to, for example, analyze consultation conversations of travel agency employees or investment advisers. The main focus here is on observing the behavior of the employees. In retail or leisure activities (e.g. amusement parks, swimming pools, etc.), expert observations are used, for example, by analyzing certain customer routes, dwell times, etc. and evaluating them. This information is then used as the basis for optimizing the sales rooms or leisure facilities.

However, the performance of this approach is limited in two respects. On one hand, the behavior of the employees, if they have been informed about the observation, can be distorted by observation effects in a positive as well as in a negative sense; on the other hand, the validity of the customer reactions collected is questionable. So the obvious customer behavior only allows limited conclusions to be drawn about the actual processes in the customer's in-system with their affective, cognitive and conative elements. In addition, it should be noted that an unknowing and unannounced observation of employees can have serious legal consequences and image damage. As an example, the German discount store *Lidl* is mentioned here. The company suffered a great image damage in public when it became known that employees were systematically and without their knowledge observed at their workplace and in the break rooms using video recordings.

(2) Silent Shopper Approach If a service provider wants to capture the—subjective—feelings of customers with regard to their experiences with the service process on as objective a level as possible, the **silent shopper approach** offers itself, which is often classified as participatory observation. Silent shoppers (also known as "mystery shoppers") are test buyers who simulate "real" service situations for employees that are not recognizable in order to be able to draw conclusions from them about deficiencies in the service process. In the United States, undercover shopping has been used to control sales for about 30 years. The German market for mystery shopping is estimated to have a volume of over 30 million EUR. The silent shopper approach offers the possibility of comparative observation if "test purchases" are made at the same time with the competition. Practice-related applications of silent shopper approaches are presented in the following example using an provider of mystery shopping.

The use of mystery shopping in practice

The market research institute *Splendid Research* GmbH (Hamburg) has a mystery shopping panel with over 26,000 tested mystery shoppers in Germany, Austria and Switzerland. Various forms of deployment are offered, such as: service check, store check, promotion check, mystery check. The services consist in the development of an observation questionnaire, the recruitment and training of test buyers, data analysis, etc.. For example, a large number of observations can be made during a store check, such as: Price and advertising material collection, implementation quality of advertising measures, presence of mandatory items, positioning of sales promotion materials at the POS, franchisee system conformity and -loyalty, etc.. It becomes clear that this way the retail trade is checked.

With regards to costs, test purchases in cities with about 25 questions are estimated at 85–105 EUR per check.

(Source: Splendid Research, 2019) ◀

Mystery shopping for special target groups such as millionaires

The *GfK Switzerland* has ten millionaires in use who, if necessary, are deployed as mystery shoppers or in companies within the financial services sector. For this purpose, a pool of approximately 1,000 people with freely available funds of more than one million francs is available. This ensures that the special requirements of this target group are taken into account accordingly. In a similar way, the market and opinion research institute *TNS Infratest* (Munich) has a millionaire panel of 130 millionaires who are available for mystery shopping. ◀

As part of mystery shopping, both **quantitative** (e.g. number of times the telephone rings before a call is answered (Wilson, 1998, p. 153)) and **qualitative data** (friendliness of the service staff) can be obtained. The German market research company *GfK*, for example, offers the following **criteria** to be surveyed by mystery shopping:

- Service: competence, friendliness and waiting time
- Quality of advice and product recommendations
- Compliance with service standards
- Additional sales (cross selling)
- Product presentation
- Availability of the product range
- Pricing
- Cleanliness
- Implementation of marketing activities

Data collection takes place at the interface with the customer, such as the point of sale, call center, correspondence or field service. The **use of the silent shopper approach** ideally runs according to the following pattern:

In the **first step**, it is important to define transparent goals that are to be achieved with the use of the approach. These must be taken into account when designing the checklist used by the test shoppers. To create the checklist, it is recommended to intensively examine the process of service creation and identify the critical points. In addition, it is necessary to test the location of service creation several times and at different times. This reduces the distorting effect that can occur due to special situations (e.g. sickness-related emergency staffing) (van der Wiele et al., 2005, pp. 534–535).

The **second step** deals with the information collected as part of the procedure. With regard to the data to be recorded, it is important to ensure that they relate to the general service quality dimensions (cf. Sect. 2.5) and that the key indicators defined by management are reflected in the vision and mission of the company. When selecting the test shoppers, it is important to ensure that they are independent, critical, objective and anonymous (van der Wiele et al., 2005, p. 535).

The **third step** relates to the processing of the data collected as part of the method. It is important to make the reports created quickly accessible to the responsible management team (van der Wiele et al., 2005, p. 535). A targeted communication of the test results within the organization can create additional incentives for performance improvement (van der Wiele et al., 2005, p. 537).

However, the claim of a neutral and customer-perspective-based assessment of the quality level raised for the use of test customers is often not achieved in practice (Nerdinger, 1994, pp. 209–217) due to psychological facts in the interaction of people. Empirical studies show that

- the operationalization and assessment of, for example, the "friendliness" of employees already pose requirements on the test customers which they can only meet to a limited extent (Sutton & Rafaeli, 1988),
- positive characteristics of customer contact (e.g. "friendliness") are rated negatively in different situations (e.g. queues, customer time pressure) (Sutton & Rafaeli, 1988; Böhm et al., 2006),
- the interaction between the test customer and his counterpart is by no means as genuine as it is assumed to be ideally (Richardson and Robinson 1986).

The **limitations** show that it is hardly possible to achieve an uncontaminated customer-employee-interaction. In this context, it is also necessary to take into account the intention of the test customers. They are not interested in a certain product or a certain service, but in the creation of an observable situation (Goerlich, 1994, p. 96; Nerdinger, 1994, pp. 211–213; Drees & Schiller, 2003).

Nevertheless, the silent shopping procedure is a suitable instrument for examining customer perceptions in relation to service quality. It is suitable for capturing both qualitative information (e.g. friendliness) and quantitative information (e.g. waiting times). The strength of the silent shopping approach lies above all in its combined use with other instruments; in particular with those instruments which are used after the actual measurement to improve the identified performance gaps (van der Wiele et al., 2005).

(3) Service tests More objective results are promised by the product testing investigations of services by the German *Stiftung Warentest,* which give a neutral overview of the quality of services in a competitive comparison. Since 1974 a separate department has been dedicated to the investigation of services, and also the introduction of the magazine "Finanztest" testifies to the growing importance of service investigations for the foundation (Rosenberger, 1999). Up to the year 1994 the number of tests was still around 2600 (o. V., 1994). In the year 2014 the foundation looks back on more than 5500 tests, of which around 2500 tests relate explicitly to services (Stiftung Warentest, 2016). However, the high degree of human interaction in many services makes it difficult to achieve an objective analysis. Consultation conversations with banks or building societies can be simulated in this way, but this does not yet result in a representative or objective image of the entire service offering.

Despite the increasing number of services tested, their scope is mainly limited to services with a low degree of personal interaction or to non-interaction-related characteristics (e.g. conditions of a supplementary health insurance). Interaction-specific quality characteristics (e.g. friendliness, competence of an employee) often represent a snapshot and do not provide a representative image of the entire service provider. On the basis of a randomly tested consultation conversation in a bank branch, it is not possible to make a reliable statement about the general quality level of the corresponding financial service provider.

Evaluation of the investment advice of banks by *Stiftung Warentest*

The result of an investigation of investment advice at banks carried out by *Stiftung Warentest* is given in Fig. 5.2. The corresponding banks were evaluated on the basis of criteria defined in advance. The individual criteria are weighted differently according to their relevance. The criterion "solving investment problems" is given the highest weight of 65%, while the criterion "conversation flow" is only 5% of the evaluation. ◄

(4) Quality rankings A procedure for measuring quality that is increasingly used in business practice are the so called rankings. The aim is to check the quality of services of different selected service providers according to different parameters and to rank them based on this check (Nürnberg & Schneider, 2014). Each individual parameter is assigned a maximum score to be achieved, which in sum will result in 100 parameters over all.

Since quality differences can be easily demonstrated in this way, there are numerous different rankings in practice, for example for the quality of hospitals, the quality of magazines or universities. With regard to the rankings of universities, for example, it can be seen that they can either relate to individual faculties (in terms of research, teaching or overall performance) or whole universities can be ranked in terms of their performance, quality, reputation or the like (so-called subject vs. global rankings). Furthermore, "subjective rankings" (surveys and image analyses), for example, carried out with students, professors or personnel managers, can be distinguished from "objective rankings", which have defined data as a basis (for example, publication frequency, scholarship quota, number of scientific awards, etc.), (Rosigkeit, 1997).

In the context of creating a quality ranking, it is essential to update this ranking every two to three years in order to take into account any changes in quality in the meantime (Rosigkeit, 1997). Furthermore, it is of crucial importance for the statement of rankings who draws up the lists. Only rankings that are widely accepted by the addressees lead to

Provider	Quality assessment	Solving the investment problem (65%)				Determine customer status (15%)				Product and cost information (15%)	Conversation flow	Consultation protocols not handed out (0%)
			Investment desire	Investment recommen-dation	Docu-mentation		Investment goal	Knowledge and experience	Financial and personal circumstances			
Sparkasse Hannover	Satisfactory (3.1)	+ (2.5)	O	+	⊖	+ (2.5)	++	O	O	O (3.2)	+ (1.6)	1 from 6
Hamburger Sparkasse	Satisfactory (3.3)	+ (2.2)	+	+	⊖	+ (2.1)	++	O	O	⊖ (3.6)	++ (1.3)	2 from 7
Kreissparkasse Köln	Satisfactory (3.5)	O (2.6)	O	+	⊖	+ (2.2)	++	O	O	O (2.9)	++ (1.4)	2 from 6
Mittelbrandenburgische Sparkasse	Sufficient (3.7)	+ (1.9)	+	+	–	+ (2.4)	++	⊖	+	O (3.4)	++ (1.0)	4 from 7
Sparkasse Pforzheim Calw	Sufficient (3.7)	O (2.7)	+	O	–	+ (2.4)	++	O	O	⊖ (3.6)	++ (1.1)	2 from 6
++ = very good; + = good; O = satisfactory; - = poor; -- = very poor												

Fig. 5.2 Evaluation of the investment advice of banks by *Stiftung Warentest*. (Source: Based on Stiftung Warentest, 2012)

the desired general increase in transparency. Therefore, one should try to have quality rankings always carried out by independent institutions. For example, the quality ranking of regional newspaper in Germany, such as the *ZEIT* or the *Frankfurter Allgemeine* Zeitung, is drawn up by a media economist from the *University of Hamburg* on the basis of a survey of independent media experts (Tagesspiegel, 2011). Furthermore, in order to even out the distribution of errors of individual criteria, a mix of indicators should be used. Rankings based on only one criterion may provide useful results for specialists and experts, but they are not suitable for the general public (Rosigkeit, 1997).

Despite the large number of available quality rankings, the benefit of the trial to express extremely complex issues about performance and quality in the simple and aggregated form of a figure is also critically examined. Thus, the lists are better to use, the greater the differences between the individual qualities considered. If the difference between the first and last place is very small, the question of reliability may cause difficulties (Rosigkeit, 1997). In particular, the selection of evaluation criteria should be noted. Thus, ranking studies that include several parameters are better than those that rely on only one criterion (Rosigkeit, 1997). As a result, a detailed analysis of quality rankings is always required to filter out meaningful and usable evaluations.

Example of a quality ranking by a Bavarian Railway Company

The *Bayerische Eisenbahngesellschaft (BEG)* regularly conducts surveys on the quality of companies that offer railway connections. The ranking is based on the quality measurement system of the *BEG* and includes the following criteria:

- Cleanliness of the vehicles (inside and outside)
- Passenger information in regular and emergency cases
- Functionality of the equipment
- Service orientation of the train attendants
- Customer orientation in case of complaints

Punctuality is not included in the evaluation. If a company reaches the value zero, the expectations of the BEG are just met. If a company is above, it receives a bonus. Those who receive minus points pay a fine (+100 points = maximum bonus; −100 points = maximum malus).

The following list shows selected ranking places (as of June 30, 2018).

Rank	Company	Points
1	Kissinger Stern	+ 95.80
2	Oberpfalzbahn/Waldbahn	+ 89.58
3	Agilis-Nord	+ 89.27
4	Kahlgrund	+ 77.05
5	Werdenfelsbahn	+ 71.98
6	Berchtesgadener Land Bahn	+ 60.82
7	Agilis-Mitte	+ 56.33
8	Bavarian Regiobahn	+ 53.67
9	Meridian	+ 45.67
10	Main-Saale-Express	+ 44.85
.		
.		
.		
20	Main-Spessart-Express	+ 17.65
21	Regio Nordostbayern	+ 12.63
22	Dieselnetz Allgäu Los 1	+ 11.85
23	Westfrankenbahn	+ 7.36
24	Dieselnetz Allgäu Los 2	+ 6.03
25	Bayerisches Oberland	+ 4.13
26	Regio Allgäu-Schwaben	+ 1.54
27	Dieselnetz Ulm Übergang	- 0.70
28	Alex Nord und Süd	- 4.66

(Source: BEG 2019)

5.1.2 Subjective Measurement Approaches

In contrast to objective methods, customer-oriented subjective measurement methods focus on the quality perception of a service from the perspective of individual customers. This group of methods can be further divided into:

- Feature-oriented approaches
- Incident-oriented approaches
- Problem-oriented approaches

In general, the groups of methods can be distinguished according to their purpose of use. The feature-oriented measurement approaches serve primarily to quantify quality judgments and their importance. These methods are therefore used more often when a service provider wants to control the success of measures to improve quality. A prerequisite for the use of these methods is that the provider knows the quality features of the services that are relevant to his customers. If this is not the case, incident-oriented approaches are recommended, the main purpose of which is to provide a complete picture of the quality perception of customers and to use the identified quality features as a basis for feature-oriented measurement. Incident-oriented methods are therefore often used before (initial) measurement of service quality using feature-oriented methods. The group of problem-oriented approaches focuses specifically on the analysis of critical negative events. Often, methods are used to quantify problems and their relevance, as well as to search and identify (new) problems. Depending on the goals of quality management, the measurement methods contained in the groups of methods can be used both individually and in combination.

5.1.2.1 Feature-Oriented Measurement Approaches

In feature-oriented measurement methods, the assessment of the overall performance consists of the evaluation of individual performance elements. These evaluations from the customer's perspective are usually collected by customer surveys. This group of measurement approaches includes:

1. Classical customer surveys
2. Multi-attribute methods
3. Decomposition methods
4. Willingness-to-pay approach
5. Penalty-reward-factors approach/Kano method

The feature-oriented approaches are explained in more detail below.

(1) Classical customer surveys Customer surveys are the most common form of market research. The goal of customer surveys is to determine the satisfaction with the service quality in the subjective perception of the customers. Different types of customer surveys are thinkable.

Basically, the quality can be determined on a global and on a detailed level. A global customer survey is carried out when a customer makes an overall assessment of the quality. This global assessment can be made for the entire quality of the service offered, but also for individual quality features—that is, for partial quality. Figure 5.3 shows an example of a possible form of global assessment of service quality using the example of a car dealership.

In contrast to the global customer survey, in the **detailed survey,** in addition to the perceived quality fulfillment, also the expectation of the service quality is recorded. This survey form can also be applied to the overall quality and to individual sub-qualities. In Fig. 5.4 the detailed survey of the service quality of a car dealership is shown.

Global measurement "Overall, I am satisfied with the quality of service provided by my car dealership ..."				
Very satisfied				Not satisfied at all
☐	☐	☐	☐	☐

Global survey of the overall quality of a car dealership

	Global measurement "Overall I am satisfied with the following services provided by my car dealership ..."				
Sub-qualities	Very satisfied				Not satisfied at all
Wide range of services	☐	☐	☐	☐	☐
High reliability of services	☐	☐	☐	☐	☐
Good warranty services	☐	☐	☐	☐	☐
High competence of the staff	☐	☐	☐	☐	☐
Friendliness of the staff	☐	☐	☐	☐	☐
Good value for money	☐	☐	☐	☐	☐
Fast execution/low waiting times	☐	☐	☐	☐	☐

Global survey of individual sub-qualities of a car dealership

Fig. 5.3 Global survey of service quality using the example of a car dealership. (Source: Bruhn, 2013, p. 69)

Expectation measurement "I always expect an excellent quality of service from a car dealership."					Fulfillment measurement "The service provided by my dealership consistently demonstrates excellence in service quality."				
I fully agree				I completely disagree	I fully agree				I completely disagree
☐	☐	☐	☐	☐	☐	☐	☐	☐	☐

Detailed survey of the overall quality of an auto branch

Sub-qualities	Expectation measurement "I expect the following performance from my car dealership ..."					Fulfillment measurement "My car dealership provides me with the following service ..."				
	I fully agree				I completely disagree	I fully agree				I completely disagree
Wide range of services	☐	☐	☐	☐	☐	☐	☐	☐	☐	☐
High reliability of services	☐	☐	☐	☐	☐	☐	☐	☐	☐	☐
Good warranty services	☐	☐	☐	☐	☐	☐	☐	☐	☐	☐
High competence of the staff	☐	☐	☐	☐	☐	☐	☐	☐	☐	☐
Friendliness of the staff	☐	☐	☐	☐	☐	☐	☐	☐	☐	☐
Good value for money	☐	☐	☐	☐	☐	☐	☐	☐	☐	☐
Fast execution/low waiting times	☐	☐	☐	☐	☐	☐	☐	☐	☐	☐

Detailed survey of individual sub-qualities of a car dealership

Fig. 5.4 Detailed survey of the service quality using the example of a car dealership. (Source: Bruhn, 2013, p. 70)

A detailed survey of the service quality makes it possible to derive an affective and a cognitive component of the quality. The **affective component** represents the expectations of the service quality, while the **cognitive component** is mapped by the degree of fulfillment. The affective component can also be interpreted as importance. The higher the expectations of a (partial) service, the more important it is perceived by the customers. The cognitive component, on the other hand, reflects the satisfaction with the service (performance), i.e. the degree of fulfillment. The higher the degree of fulfillment, the more satisfied the customer is.

Based on this, the individual sub-qualities can be represented in an **importance-performance matrix** (cf. Fig. 5.5). The matrix consists of four quadrants and identifies those quality areas in which there is a particularly urgent need for action.

In a next step, corresponding **standard strategies** can be derived for each of the four quadrants. The particularly critical quality features are located in **quadrant I.** These are considered by customers to be above average in importance, but the delivered quality is rated as below average in satisfaction. Therefore, these features are to be **improved** immediately, as poor quality could lead to customer attrition due to the importance of the attributes. In the example shown, "quick performance" and "value for money" are found in quadrant I. This means that, for example, internal processes need to be improved in order to reduce customer waiting times.

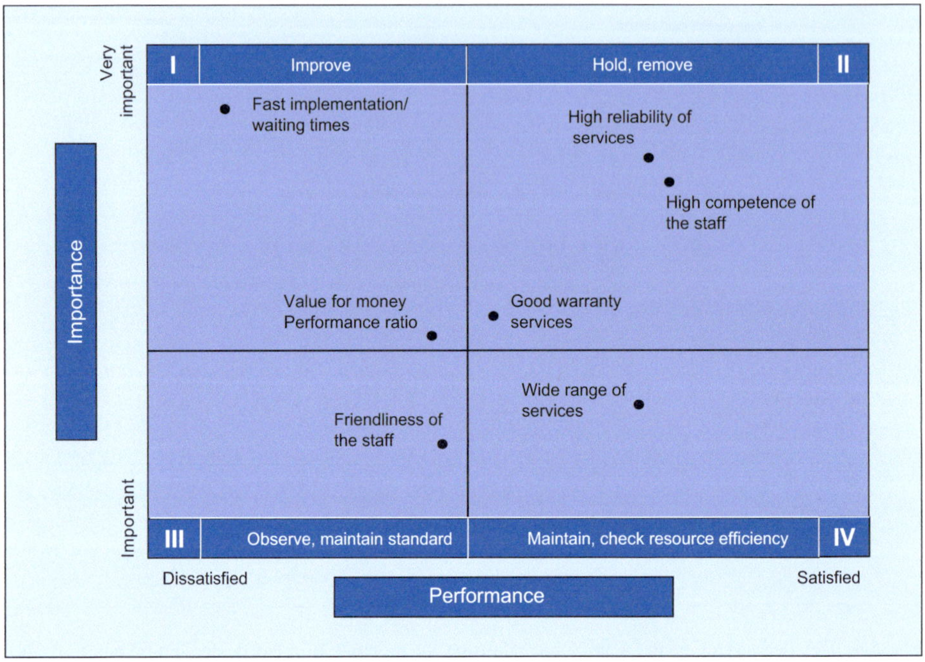

Fig. 5.5 Importance-performance matrix using the example of a car dealership

The quality attributes in **quadrant II** were rated both in terms of importance and satisfaction above average. These attributes are important to customers and are satisfactorily provided by the company. Therefore, the standard strategy here is **maintain or expand.** In the example, "reliability", "competence" and "guarantee services" can be found here.

Quadrant III describes those features that were neither particularly important nor very satisfactory to customers. In the example, this only applies to the feature "friendliness". Due to the relatively low relevance for customers, the standard strategy for these areas is **observe** and **maintain standard**. The quality features are to be observed to the extent that customer preferences are often not seen as long-term constant, but there is a possibility that customer preferences will change and high importance will be attached to these features in the near future. If this happens, the company must react immediately and adjust the quality level accordingly to the new customer requirements.

Quadrant IV is finally characterized by high satisfaction with low importance. Here the company is well rated, but the quality features play no big role in the customer's purchase decision. Therefore, the standard strategy for this quadrant is **hold** and **check resource efficiency**. In the example, this is where "wide range of service" is located. The company has a wide range of services, but this is not explicitly expected by customers. If there are bottlenecks in the company, the focus should be on this area when reducing projects, by examining which service components are required by customers and which services one could do without.

Overall, the **importance-performance analysis** provides important information for prioritizing quality features. It is recommended to repeat these analyses at regular intervals in order to find out about the change from the customer's point of view (see also Sect. 14.2.2 on the use of the importance-performance analysis in the context of the national customer barometer).

(2) Multi-attribute methods Multi-attribute measurement methods are based on the assumption that global quality assessments by service customers are the result of an individual assessment of the various quality features (Stauss & Hentschel, 1991, p. 240; Hentschel, 1999), i.e. that a global quality judgment is the sum of a multitude of evaluated quality attributes.

Multi-attribute methods follow the pattern of the following function (based on Kroeber-Riel & Gröppel-Klein, 2013, 263 pp.):

$$Q_{ij} = f(W_{ij1}, W_{ij2}, \ldots, W_{ijn})$$

with the following:

Q_{ij} = Global perception of the quality of service *j* by consumer *i*

W_{ijk} = Perception of consumer i of quality attribute *k* of service *j* ($k = 1, \ldots, n$).

The measurement of requirements for service quality can be approached from two perspectives within multi-attribute methods (cf. Fig. 5.6):

- Attitude-oriented approaches
- Satisfaction-oriented approaches

The **attitude-oriented concept** starts from the customer's predisposition that the quality assessment is to be seen as a relatively permanent, learned, positive or negative inner attitude towards an assessment object (Hentschel, 1999; Trommsdorff & Teichert, 2011, 315 pp.). Accordingly, the quality assessment from the customer's point of view results from a learning process in which both possibly own experiences with the service provider and indirect experiences (e.g. through communication with other customers) are included (Benkenstein, 1993, p. 1101).

The **satisfaction-oriented quality research** is based on the so-called "disconfirmation paradigm" (Oliver, 1980, 460 pp.; Bruhn, 1982). According to this, the quality assessment is constituted by the discrepancy between the perceived and the expected service quality, so that the satisfaction-oriented quality determination is only possible in cases where at least one transaction between the service provider and the customer has already taken place.

From a **dynamic perspective**, however, the strict separation of the attitude- and satisfaction-oriented approach appears to be questionable. It is to be expected that satisfaction with current service transactions and differentiated quality criteria will decrease over time. Thus, with increasing distance from a concrete performance process, satisfaction

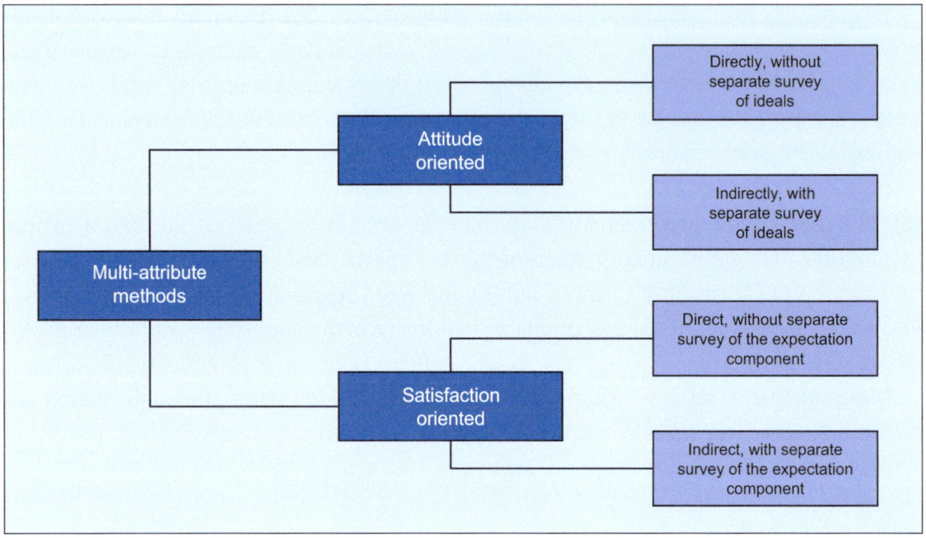

Fig. 5.6 Systematization of the multi-attribute measurement approaches of service quality. (Source: Based on Hentschel, 1999, p. 298)

will increasingly arise in a more global attitude towards the service provider (Oliver, 1980).

In addition to this distinction, another differentiation of the multi-attribute methods for measuring the requirements for service quality is possible. While in the application of **direct (single-component) approaches** only an impression of the respective quality criterion is queried, an **indirect (two-component) approach** also requires an assessment of the importance of the respective quality criterion. Figure 5.6 provides an overview of this.

In the case of direct, **attitude-oriented quality assessment**, usually a single quality component is assessed by means of an evaluation scale (e.g. degree of agreement with a statement about the property of a performance feature) by the informant. This approach goes back to the work of Fishbein (1963) in the context of attitude research. The **indirect, attitude-oriented measurement** captures, in addition to the perceived, the ideal of a criterion (e.g. degree of agreement with a statement about the ideal level of a performance feature; see also Trommsdorff, 1975). The advantage of this attitude-oriented approach is essentially the fact that the informants do not necessarily have to have experience with the services in question. However, there is a danger that, in the case of an exclusively attitude-oriented quality assessment, current deficits—in particular in the perception of actual customers—will not be taken into account to a sufficient extent.

The **direct, satisfaction-oriented determination of service quality** puts the focus on the difference between expectation and perception with regard to the performance (e.g. evaluation of a performance feature; see Brandt, 1987), while finally, by means of the **indirect, satisfaction-oriented approach**, both the expectations before and the perceptions after the service provision by the consumer are assessed. This approach is particularly advantageous if the individual service transactions can be clearly distinguished from each other (Hentschel, 1999). In addition, experienced performances are assessed, so that the results have a close content-related and temporal reference to the service provision. This recommends the use of the satisfaction-oriented capture of the requirements for service quality also from the point of view of early clarification.

Figure 5.7 shows four basic ways of measuring service quality using the fictitious example of a service provider.

When comparing these methods, reference should be made to the basic problem of indirect, satisfaction-oriented quality measurement. In many cases, a priori/a posteriori distinction with regard to the detection of quality deficiencies can be quite helpful, but it remains to be clarified in what form the **expectations of the customers** can be operationalized. Here, various options are available for interpreting expectations by the respondent (see also Sect. 2.4).

The variety of possible operationalizations of the expectation concept requires a careful design of corresponding studies. First, a **suitable expectation concept** must be used, secondly the future service customers must be known before the purchase decision and thirdly it is necessary that these customers are able to formulate their expectations accordingly (Hentschel, 1999, see also Sect. 7.2).

How much the statement power of an investigation can be negatively influenced by a diffuse definition of the expectation concept, should be illustrated by an example. Thus, Mudie and Cottam propose on a not further differentiated, global level in the sense of the disconfirmation paradigm a simple approach to determine customer satisfaction, which results from the difference between the perceived and the expected service quality (Mudie & Cottam, 1999, p. 17). However, the result of such an undifferentiated consideration is to be interpreted with caution. Customers who visit a doctor with the expectation of a three-hour waiting time would not be dissatisfied if the waiting time actually occurred. Absolutely considered, however, this duration is not justifiable, so there are other—especially more specific—approaches to determine the actual service quality and to evaluate the expectations used.

A special form of the attitude-oriented approach is the **competence-oriented view** of service quality. This special form of the attitude-oriented approach to service quality plays a particularly important role when the quality assessment of services is being focused on, which escape an assessment by the consumer due to their high intangibility. In this case, the skills and abilities of a service company to solve the problems of the consumer are used in this consideration by consumers to assess the service quality. If the service customer has already made experiences with the provider, he can use the **proficiency** documented in connection with the completed transaction of the provider to carry out a competence transfer to similar services. If, however, there is no experience with the specific provider, the provider needs sufficient **skills,** so that the customer can gain

Direct, attitude-oriented quality measurement: (single component approach).

	Very true				Does not apply at all
The service provider's dealer offers a quick repair service.	☐	☐	☐	☐	☐
The local sales partner provides comprehensive support for its customers.	☐	☐	☐	☐	☐

Indirect, attitude-oriented quality measurement: (two component approach).

	Should apply				Should not apply
A service provider should offer different rates for frequent and occasional callers.	☐	☐	☐	☐	☐
My service provider offers different rates for frequent and occasional callers.	☐	☐	☐	☐	☐

Direct, satisfaction-oriented quality measurement: (single-component approach)

	Very satisfied				Very dissatisfied
The period of time between the completion of the contract with the service provider and the activation of the calling card.	☐	☐	☐	☐	☐
Consultation by the local sales partner of the service provider.	☐	☐	☐	☐	☐

Indirect, satisfaction-oriented quality measurement: (two-component approach)

A priori:	Very likely				Very unlikely
Monthly call billing will be on time.	☐	☐	☐	☐	☐
The monthly invoice will be available before the invoice amount is debited.	☐	☐	☐	☐	☐

A posteriori:	Very true				Does not apply at all
Monthly call billing was on time.	☐	☐	☐	☐	☐
The monthly invoice was available before the invoice amount was debited.	☐	☐	☐	☐	☐

Fig. 5.7 Multi-attributive measurement of service quality using the example of a service provider (Source: Based on Hentschel, 1999, p. 300)

confidence in view of the provision of the service, for which no experience exists yet (Weiss, 1992, 59 pp.; Benkenstein, 1993, p. 1102; Golder et al., 2012).

The direct reference of this consideration to the attitude-oriented perspective arises from the fact that the competence assumption towards a service provider is ultimately to be regarded as a predisposition.

Among the multi-attribute methods, a measurement method based on the **GAP model** by Parasuraman et al. has gained a significant position in recent years (see Sect. 4.1). Within the framework of the GAP model, the so-called **SERVQUAL-approach (service quality)** was developed. The so-called GAP 5 represents the central "gap" of the GAP model, which determines the service quality perceived by the customer as a discrepancy between the expected and the perceived service (Parasuraman et al., 1985). With the help of the SERVQUAL approach, GAP 5 is operationalized, that means it has been made measurable. The SERVQUAL approach is a combination of attitude and satisfaction-oriented multi-attribute measurement.

The ten quality dimensions derived by Parasuraman et al. with 97 items could be reduced to five central quality dimensions with 22 items with the help of the data material from two surveys (Parasuraman et al., 1985). The dimensions "convenience of the tangible environment - tangibles", "reliability", "responsiveness", "assurance" and "empathy" are made measurable by the 22 items with a double scale. The 22 items are shown in Fig. 5.8 using the example of a service provider.

In the double scale used, one scale ("expectation scale") records ideal-typical conditions (should profile), while the second scale ("perception scale") records actual conditions (actual profile). Both scales are based on a seven-step subdivision that allows statements from "I completely disagree" (1) to "I completely agree" (7) (see Fig. 5.9).

The difference between the desired and actual state of an item results in an individual value between −6 and +6, with the perceived quality of service increasing with the size of the value for a certain criterion. The center of this continuum finally separates good and bad service quality (Zeithaml et al., 1992).

A **critical examination of the methodology** and the theoretical foundation of SERVQUAL, however, also reveals some difficulties that call the validity of the approach into question. In particular, the central component of the double scalegives reasons for reservations (Brady et al., 2002; Brandon-Jones & Silvestro, 2010). In this context, problems of question answering, claim inflation, discrimination strength and finally the plausibility of the basic assumption are to be mentioned. So the differentiation between expected and experienced service quality can in some cases lead to misinterpretations (Hentschel, 1999; Brady et al., 2002). In addition, the differentiation of the model (i.e. value of the perceived service minus value for the expected service) gives reasons for criticism (cf. e.g. Meffert et al., 2018, p. 211 f.). With regard to the validity of the model, in particular, the low reliability of the method, the convergent validity of the individual indicators and the unstable dimensionality of the SERVQUAL approach are criticized (cf. e.g. Chatterjee & Chatterjee, 2005, p. 479; Tsai & Lu, 2006, p. 641).

	Tangibles
1.	Excellent service providers have modern technical equipment.
2.	The facility of a service provider should be pleasantly eye-catching.
3.	Employees of a service provider should be dressed attractively.
4.	Excellent service providers should make their brochures and communications appealing to customers.
	Reliability
5.	When excellent service providers promise to meet a deadline, the deadline is met.
6.	With excellent service providers, interest in solving a problem should be apparent.
7.	Excellent service providers should do the service right the first time.
8.	Excellent service providers should execute their services at the promised time.
9.	Excellent service providers should have error-free receipts for customers.
	Responsiveness
10.	Employees of excellent service providers can provide information about the timing of a service execution.
11.	Employees of an excellent service provider will serve customers promptly.
12.	Excellent service providers should always be ready to help customers.
13.	At excellent service providers, employees are never too busy to address customer concerns.
	Assurance
14.	With excellent service providers, the behavior of the employees inspires trust in the customers.
15.	Transactions with excellent service providers make you feel safe.
16.	Employees of an excellent service provider are always consistently courteous to customers.
17.	Employees of excellent service providers have the expertise to answer customer questions.
	Empathy
18.	Excellent service providers give individual attention to each of their clients.
19.	Excellent service providers offer their services at times that suit all customers.
20.	Excellent service providers have employees who are personally dedicated to their customers.
21.	Excellent service providers have their customers' best interests at heart.
22.	The employees of outstanding service providers understand the specific service needs of their customers.

Fig. 5.8 Survey of the quality dimensions according to the SERVQUAL approach using the example of a service provider. (Source: Based on Zeithaml et al., 1992, p. 202.)

Example of the double scale (item 16)	I completely disagree						I agree completely
Employees of an excellent service provider are always consistently courteous to customers.	1	2	3	4	5	6	7
Employees of service provider x are always consistently courteous to customers.	1	2	3	4	5	6	7

Fig. 5.9 Double scale for answering the questions according to the SERVQUAL approach

In addition, a constant interpretation with respect to the "This is how it should be" expectations of the subjects is not guaranteed in the SERVQUAL approach. There is thus the danger that the expectation components will be interpreted differently by the test person. The following interpretations are thinkable (cf. also Fig. 2.8 in Sect. 2.4):

- Pure assumption about the level of service quality ("forecasted performance")
- Desired level of service quality ("desired performance")
- Appropriate level of service quality ("equitable performance")

- Minimum level of service quality ("minimum performance")
- Ideal level of service quality ("ideal performance")
- Expression of the importance of this dimension of service quality for the customer ("service attribute importance")

More recent studies also critically point out that a general transferability to any industry is not possible without further ado. Especially because of the very large heterogeneity of services, a modification of the dimensions is often necessary. Individualized measurement concepts are required for individual industries.

Some of these methodological difficulties are often overcome with a suitable, situation-specific survey design. Other problem areas can be avoided through a general adaptation of the methodology. For example, the claim inflation for questions about the importance of quality elements is often countered by specifying a total number of points to be distributed among the various quality features; this forces the surveyed customers to weigh their claims.

The research interest in the SERVQUAL approach is still unbroken— even 30 years after its presentation. A variety of different research projects have dealt over time with proposals for improving or expanding the SERVQUAL approach for the valid conceptualization and operationalization of the concept of service quality (e.g. newer research work by Chatterjee & Chatterjee, 2005; Chu-Mei, 2005; Tsai & Lu, 2006). A prominent **criticism of the SERVQUAL approach** is due to Cronin and Taylor (1992, 1994; Brandon-Jones & Silvestro, 2010). The authors consider the SERVQUAL approach to be only limitedly suitable for conceptualizing service quality. They see no indication that the service quality perceived by customers—as operationalized by SERVQUAL—derived from the satisfaction-oriented "disconfirmation paradigm" as an attitude-like construct from a comparison of expectation and perception of actual performance (GAP 5) arises (Cronin & Taylor, 1992, p. 56). In their opinion, service quality is a purely attitude-oriented construct that is described solely by the performance perceived by the customer (Cronin & Taylor, 1992, p. 58).

As a result of their criticism, the authors present an alternative to SERVQUAL for measuring service quality; the so-called **SERVPERF-approach** (**serv**ice **perf**ormance). Based on the conceptual considerations, the SERVPERF-approach geared towards setting, does away with the separate survey of customer expectations and is limited to the perception component of the SERVQUAL-approach. Transferred to the example in Fig. 5.9, in the context of SERVPERF the test persons would only be asked to assess their opinion on the statement "Employees of service provider x are always consistently polite to customers". According to SERVPERF, the level of perceived service quality then results from the sum of the ratings of all 22 items of the SERVQUAL questionnaire (cf. Fig. 5.8) (Cronin & Taylor, 1992, 1994). A significant advantage of SERVPERF over SERVQUAL lies in the significant **reduction in the amount of data collection.**

In summary, SERVQUAL is characterized by its **completeness and simplicity of ranking.** The popularity of SERVQUAL is due in particular to its high practical relevance. Despite disadvantages such as the possibility of the combination of relevant characteristics while at the same time the design is arbitrary within the scope of the investigation, SERVQUAL has been used as an instrument for measuring the requirements for service quality in a variety of industries with certain modifications.

Measuring service quality in service centers using the example of an airport

The characteristic-oriented approach is also suitable for measuring service quality in so-called service centers. Service centers are an agglomeration of several legally autonomous but economically independent companies ("service partners") that offer their services at a spatially delimited location under the unified management of a center management (Birkelbach, 1993). The measurement and control of quality in service centers is significantly more difficult than in classic services, since the customer comes into contact with different aspects of the service center along a service chain and can form an opinion on them.

Service center quality is defined here as "the quality of all center-specific framework factors (center operation quality) and the entire customer contact points (contact point quality) of the service sequence as perceived by the center user on the basis of individual quality attributes and evaluated with regard to his expectations. It is reflected in the degree to which the customer requirements of the center user are met" (Bruhn et al., 2010, p. 357). The quality assessment by the center user therefore takes place on the basis of individual interactions (in the sense of contact points) with various service partners as well as on the basis of existing institutional framework conditions (i.e. the center operation). The global quality assessment with regards to the service center (DLCQ) of a center user thus represents a function of 'n' contact point quality (KPQ) and the center operation quality. The basis for the partial quality assessments is analogous to the GAP 5 of the GAP model of service quality (cf. Sect. 4.1) the comparison of the perceived center operation quality (WQ_{CB}) and the perceived quality of the contact points used (WQ_{KP}) at individual service partners (SP) with the expected quality (EQ) with regard to various center-specific and contact point-specific quality characteristics (cf. Fig. 5.10).

To measure the quality of service centers, it is therefore necessary to identify relevant quality features with respect to the quality of center operations and contact point quality. In the context of a study by the *University of Basel* on measuring the quality of service centers at an airport, various contact point qualities were identified that are relevant for the perceived quality of service center of airport users and that are based on a typical path or service blueprint of passengers:

- Check-in quality
- Pass control quality
- Security quality

- Shopping quality
- F&B quality
- Waiting area quality

The indicators for measuring the quality of service centers using the example of an airport are shown in Fig. 5.11. The quality characteristics or measurement indicators of the corresponding contact point quality are listed and the exemplary characteristics of the center operation quality for an airport can be taken from there - Fig. 5.12 (Bruhn et al., 2010). ◀

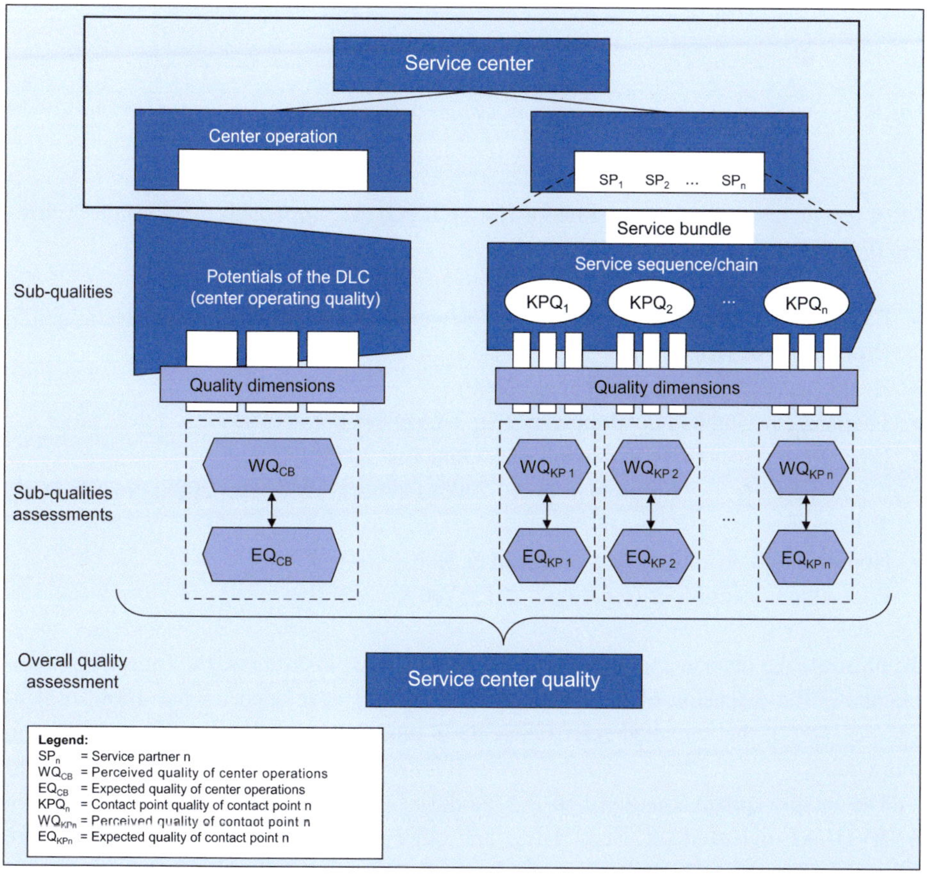

Fig. 5.10 Creation of the overall quality in a service center. (Source: Bruhn et al., 2010, p. 356)

Center operation quality

Quality features

1. The accessibility by car corresponds exactly to my expectations.
2. The accessibility by public transport ...
3. The signs to the airport ...
4. The spaciousness of the airport building ...
5. The temperature in the airport building ...
6. The air quality in the airport building ...
7. The noise level in the airport building ...
8. The perception of security in the airport building ...
9. The variety of destinations served ...
10. The frequency of flight connections ...
11. The airlines represented ...

Fig. 5.11 Survey of quality characteristics of center operation quality of an airport. (Source: Bruhn et al., 2010, p. 373)

Most recent studies and **extensions of the SERVQUAL approach** relate to the following industries:

- Banking and financial services (e.g. Newman, 2001; Mukherjee & Nath, 2005; Olorunniwo & Hsu, 2006)
- Health services (e.g. Hanjoon et al., 2000)
- Tourism, catering and hotel industry (e.g. González et al., 2007)
- Investive services (b2b) in general (e.g. Gounaris, 2005)
- E-commerce (e.g. Parasuraman et al., 2005; Faßnacht & Koese, 2006; Nemati et al., 2012)
- Non-profit organizations (e.g. Vaughan & Shiu, 2001)
- Educational institutions (e.g. Lupo, 2013; Yeo & Li, 2014)

To illustrate the diverse applications of the SERVQUAL instrument, the following example shows the extension of the instrument specifically developed for non-profit organizations (NPO), the so-called **ARCHSECRET model** by Vaughan and Shiu (2001), in more detail.

The quality dimensions used in this model can be divided into the already known **SERVQUAL-oriented** (cf. Sect. 4.1.1) and other **non-profit specific** dimensions (see Fig. 5.13). The **ARCHSECRET-model** is based on several studies on measuring service quality in public and welfare-oriented areas of the city of Glasgow in Scotland, which were carried out from 1995 to 2005.

In the context of a **critical assessment** of the model, its ability to specifically measure the performance quality of NPOs is particularly noteworthy. Here, in particular,

Check-in quality
1. The friendliness of the check-in staff is exactly what I expected.
2. The competence of the check-in staff is exactly what I expected.
3. The layout of the check-in area is exactly what I expected.
4. The speed of check-in is exactly what I expected.
5. The stress-free check-in process is exactly what I expected.
Passport control quality
1. The friendliness of the staff at passport control is exactly what I expected.
2. The setup in the passport control area is exactly what I expected.
3. The speed of the passport control is exactly what I expected.
4. The stress-free passport control process is exactly what I expected.
5. The fair treatment by the officer at passport control is exactly what I expected.
Security quality
1. The friendliness of the staff at the security checkpoint is exactly what I expected.
2. The competence of the staff at the security checkpoint is exactly what I expected.
3. The setup of the security area is exactly what I expected.
4. The speed of the security check is exactly what I expected.
5. The stress-free handling of the security check is exactly what I expected.
Shopping quality
1. The friendliness of the sales staff is exactly what I expected.
2. The competence of the sales staff is exactly what I expected.
3. The diversity of the range of goods is exactly what I expected.
4. The convenience of visiting the store is exactly what I expected.
F&B quality
1. The friendliness of the restaurant/bar staff is exactly what I expected.
2. The competence of the restaurant/bar staff is exactly what I expected.
3. The selection of dishes is exactly what I expected.
4. The taste of the food is exactly what I expected.
5. The relaxation option in the restaurant/bar is exactly what I expected.
6. The convenience of the restaurant/bar visit is exactly what I expected.
Waiting area quality
1. The friendliness of the staff is exactly what I expected.
2. The competence of the staff is exactly what I expected.
3. The comfort of the seating is exactly what I expected.
4. The entertainment options in the waiting area are exactly what I expected.
5. The range of magazines meets my expectations exactly.
6. The atmosphere in the waiting area is exactly what I expected.

Fig. 5.12 Survey of quality characteristics of contact point quality of an airport. (Source: Bruhn et al., 2010, p. 374)

Dimensions of service quality in the ARCHSECRET model	
SERVQUAL oriented dimensions	**Non-profit specific dimensions**
▪ Amenity of the tangible environment (Tangibles) ▪ Reliability ▪ Responsiveness ▪ Competence ▪ Security ▪ Communication	▪ Humanity (Humanness) ▪ Empowerment (Enabling/Empowerment) ▪ Accessibility ▪ Equity
The final ARCHSECRET model contains the following ten dimensions: Access, Responsiveness, Communication, Humanness, Security, Enabling/Empowerment, Competence, Reliability, Equity, Tangibles.	

Fig. 5.13 Dimensions of service quality in the ARCHSECRET-model. (Source: Vaughan & Shiu, 2001, p. 137)

the concrete practical relevance of the instrument is shown, which can support decision-makers in NPOs (Bruhn, 2012, p. 229).

(3) Decomposition methods The decomposition methods proceed in the reverse direction to the multi-attribute approaches (Hentschel, 1999, p. 297). While in multi-attribute methods a global quality assessment results from the evaluation of different sub-services, in decomposition methods a ranking of different services with different feature expressions is formed on the basis of global quality assessments. In a second step, partial qualities are calculated. This can be done, for example, by means of a so-called conjoint analysis.

The **conjoint analysis** method is a decomposition method. It is based on the assumption that the total utility of a product or service consists of the utility of the components (partial utility values). It is mainly used in the context of new product planning. With its help, the relevant properties for a new product or service are to be identified and ranked according to their customer-side importance on the basis of preference judgments of the surveyed customers with regard to different combinations of property expressions (e.g. combinations of price, waiting time and duration of a service) (Backhaus et al., 2018). However, the calculation of partial qualities requires—at least in qualitative terms—the designation of the characteristics. The various services are thus created as bundles of characteristics in each case with different expressions.

The advantage of the conjoint analysis is to be seen above all in the fact that not individual criteria are evaluated by the customers, but a value judgment is made about a complete service bundle. On the basis of the evaluation, partial utility values of the individual service dimensions can then be determined. Complex statistical evaluations are used for the calculation of partial utility values, which are often carried out by means of software programs specially designed for the conjoint analysis.

Conjoint analysis using the example of a hotel stay

A travel organizer wants to use conjoint analysis to determine which service components are most important to customers when choosing a holiday hotel. From preliminary studies it is known that the values price, hotel category, location and travel time are the most important criteria when choosing a hotel. For the four criteria, the possible values shown in Fig. 5.14 are thinkable.

From these criteria with the respective values, it is now possible to create 36 different, partly fictitious, service bundles, which are then to be evaluated by the customers, or to be ranked. In Fig. 5.15 four such fictitious service bundles are shown by way of example.

Depending on the evaluation of the individual stimuli, it is then possible to conclude the relative importance of the individual characteristics. ◀

The **vignette method** is a variant of the decompositional methods that has gained particular importance in the field of service quality measurement (Haller, 1998, p. 113). This method also does not use a global quality assessment to measure service quality. The measurement approach is based on the methodology of conjoint measurement. With the vignette method, it is assumed that the overall quality of a service is composed additively of the perceptions of individual quality dimensions (e.g. friendliness, reliability, etc.) of the service.

Accordingly, the vignette method assumes that quality judgments are based on a relatively small number of factors that are relevant in the customer's perception. A **vignette** represents a fictitious situation that is described by means of certain characteristics. The basic procedure is explained using an example.

A prerequisite for the vignette method is the determination of so-called **"Critical Quality Characteristics"(CQCs),** i.e. those attributes that are relevant for quality assessment from the customer's point of view. For example, the quality assessment of a bank branch can be based on factors analogous to the quality dimensions of the SERVQUAL approach, as shown in Fig. 5.16.

Fig. 5.14 Possible expressions of the service components

Criterion	Expression
Price	▪ 400 EUR ▪ 500 EUR ▪ 700 EUR
Hotel category	▪ 2 stars ▪ 3 stars ▪ 4 stars
Location	▪ Direct beach access ▪ 15 min. walk to the beach
Travel	▪ High season ▪ Low season

Criterion	Stimulus 1	Stimulus 2	Stimulus 3	Stimulus 4
Price	700 EUR	400 EUR	400 EUR	500 EUR
Hotel Category	4 stars	2 stars	3 stars	4 stars
Location	Direct beach access	15 min. walk to the beach	Direct beach access	15 min. walk to the beach
Travel time	Peak season	Peak season	Low season	Low season

Fig. 5.15 Possible expressions of service bundles of a conjoint analysis

Critical Quality Characteristics	Value judgments
Convenience of the tangible environment: e.g. equipment of the business facilities	Appealing
	Not appealing
Reliability: e.g. avoidance of incorrect bookings	High
	Low
Responsiveness: e.g. fast reaction to special requests	Flexible
	Inflexible
Performance Expertise: e.g. return on recommended assets	Competent
	Not competent
Empathy: e.g. consideration of personal sensitivities of customers	Courteous
	Not obliging
Price: e.g. consideration of different price willingness	High
	Low

Fig. 5.16 "Critical Quality Characteristics" and value judgments of a vignette using the example of a bank branch

The individual vignettes are finally formed by assigning one of the associated value judgments to each factor. Each vignette thus represents a combination of different characteristics and value judgments. Figure 5.17 shows an example of a vignette.

Analogous to the conjoint-analysis (Backhaus et al., 2018) the customer is now asked to assess the different vignettes. This is often done on a scale from "very good" to "very bad". As part of the evaluation using a frequency table, the characteristics represent the independent variables and the overall assessments represent the dependent variables. Thus, the influence of the individual attributes on the global quality assessment is expressed by a coefficient.

The vignette method is therefore suitable for **analyzing the ranking** and **weighting of individual quality attributes** of the service and for **determining global quality assessments.** The following example illustrates this procedure.

Evaluation criterion	Value judgment
Tangibility environment	Not responsive
Reliability	High
Responsiveness	Flexible
Performance	Competent
Empathy	Courteous
Price	High
How would you rate a bank branch that fits this description?	
(1) Very good (2) Good (3) Satisfactory (4) Insufficient (5) Bad (6) Very bad	

Fig. 5.17 Example of a vignette

Quality attributes using the example of a Swiss zoo

As part of a study by the *Chair of Marketing and Management at the University of Basel* in cooperation with the *Basel Zoo*, a total of seven quality attributes with two to three different expressions could be identified on the basis of customer surveys carried out on site. These are shown in Fig. 5.18.

Based on the evaluation, the ranking and the weighting of the individual attributes were determined. Based on the response behavior, the importance of the individual properties is calculated—analogously to the conjoint analysis. The customers rated the features "route guidance", "selection of animals" and "admission price" as the most important. Figure 5.19 shows the weighting of the individual quality attributes at a glance. ◀

The high data collection effort is a disadvantage of the procedure; especially since, with a large number of vignettes, the execution of numerous surveys is necessary. Furthermore, there is a risk of information loss, since the diverse quality characteristics of a service are reduced to a few "critical quality characteristics", which on one hand prevents differentiated statements and on the other hand the perception space relevant to the customer cannot be completely captured.

(4) Willingness-to-pay approach The willingness-to-pay approach is a procedure that is based on the **value-oriented concept of quality** (see also Haller, 1998, 40 pp.; Wertenbroch & Skiera, 2002). It is assumed that the customer comes to his quality assessment with regards to a performance by comparing the received performance with the costs incurred in the context of the use of the service, such as financial, time, psychological or physical. These "costs" are usually expressed by the price of the service. The assessment of a service according to the willingness-to-pay approach is often carried out in two ways:

Property	Property expressions
Admission price	▪ 11 CHF ▪ 16 CHF ▪ 21 CHF
Information about animals	▪ Text ▪ Image ▪ Video
Selection of animals at the zoo	▪ Small ▪ Medium ▪ Large
Additional offers e.g. feeds	▪ Petting zoo ▪ Feeding predators ▪ Pony rides
F&B offering	▪ Kiosk ▪ Café ▪ Restaurant
Route guidance	▪ Boring ▪ Varied
Parking facilities	▪ None ▪ Directly in front of the entrance ▪ 10 minutes away

Fig. 5.18 Property expressions of the vignette method using the example of a zoo

- The **benefit** of a service, which consists of the weighted evaluation of individual service features, is related to the price of the service.
- The **price** is included in the list of features in the context of the above-mentioned multi-attribute (price as service quality) or decomposition procedures.

This method is particularly useful if the intention is to vary services as part of the service policy. In such a situation, it is often determined by means of the willingness-to-pay approach whether the extension or improvement of a feature leads to a corresponding higher willingness to pay on the part of the customers (Balcombe et al., 2009; Kang et al., 2012).

Application of the willingness-to-pay approach

Liljander and Strandvik used this approach in an experiment regarding the payment willingness of potential restaurant guests (Liljander & Strandvik, 1992, 15 pp.). These respondents had to imagine that they were visiting a fictional city and had eleven restaurants, described in the form of vignettes, to choose from. To determine the "normal quality level" of a restaurant for the respective person being interviewed, they were

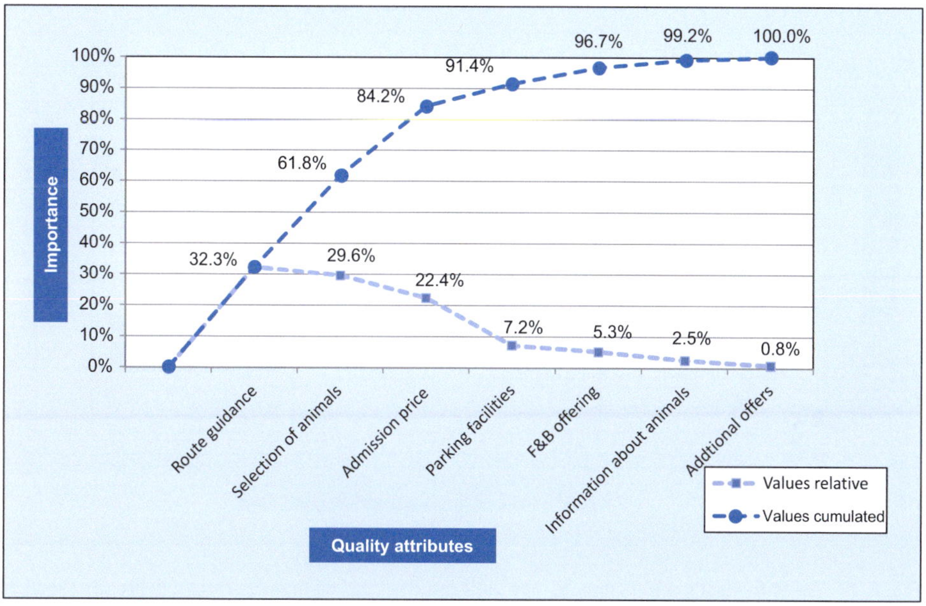

Fig. 5.19 Ranking and weighting of quality attributes using the example of a zoo

asked to think of a restaurant where they regularly have lunch, to name their name and the "normal price" they would pay for a lunch in this restaurant. Subsequently, the eleven fictional restaurants were evaluated by the respondents. For this purpose, they first had to sort the restaurants according to their assessment, secondly to specify the price they would be willing to pay for each restaurant and thirdly to explain which restaurant they would visit at the "normal price". As part of the evaluation, both the payment willingness of the individual (cf. Fig. 5.20) and the aggregated payment willingness were analyzed (cf. Fig. 5.21). ◀

By including the price in the list of attributes of a service, the price of an already existing service is valued from the customer's perspective by means of a multi-attribute procedure. By means of a decompositional procedure, it is possible to vary the price and to determine the relationship between different bundles of attributes, the respective price and the quality of service perceived in this context. This relationship is illustrated in the following example.

Determination of the price willingness by means of a decompositional procedure

The aim of this study was to determine the price willingness of rail customers for certain modalities—in this case the variation of certain service features—with respect to their train ticket and to find answers to the following questions: To what extent is the customer willing to pay a higher price for his train ticket if a certain number of

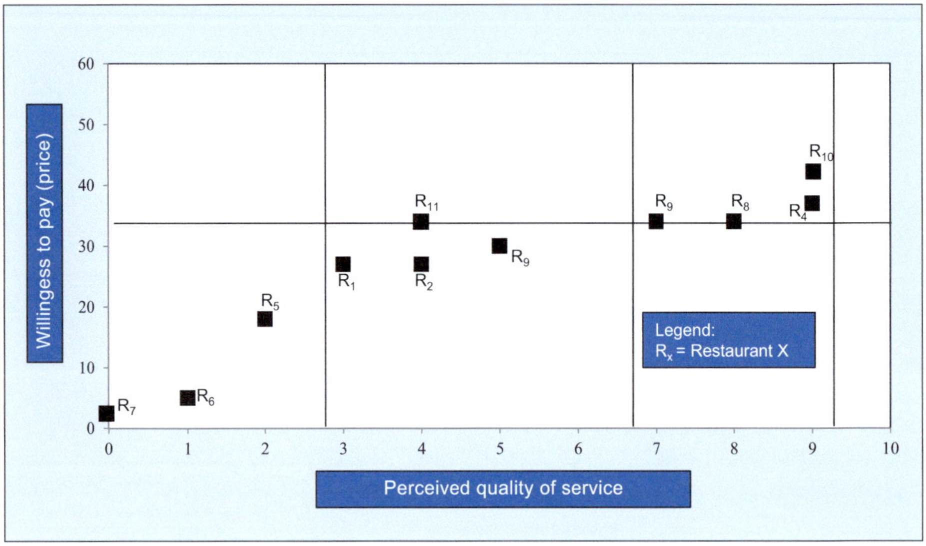

Fig. 5.20 An individual's willingness to pay as a result of the willingness-to-pay approach (Source: Liljander & Strandvik, 1992, p. 20)

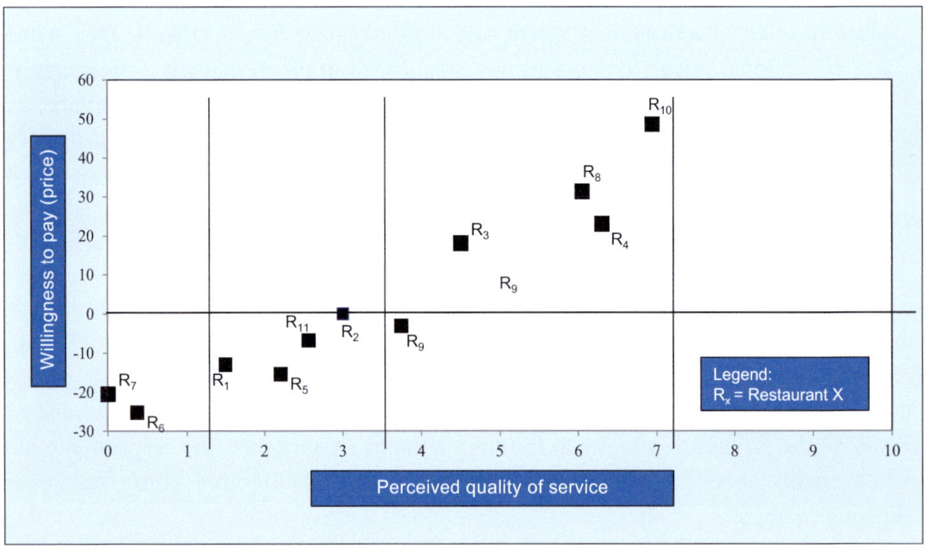

Fig. 5.21 Aggregated payment willingness as a result of the willingness-to-pay approach (Source: Liljander & Strandvik, 1992, p. 21)

people are allowed to accompany him free of charge on his journey? To what extent is the customer willing to pay a higher price for his train ticket if the regional validity of the ticket increases? In order to determine the price willingness by means of a decompositional procedure (here by applying the conjoint analysis), three service features with each three possible variants were used. This resulted in a total of 27 (3 × 3 × 3) possible variants of the service "train ticket". In addition to the preferences of the rail customers, behavior related information of the customers was also collected. This makes it possible to attribute preference differences between the customers to certain customer segments, thus making a target group-specific variation of the service possible. The variants were defined as follows (cf. Fig. 5.22):

- Price: 9 EUR, 12 EUR, 15 EUR
- Price reduction when buying at the machine: 0.50 EUR, 1 EUR, 1.50 EUR
- Regional validity of the ticket: only valid in one federal state, also valid in the adjacent federal states, valid throughout the federal territory

The results of the research showed clear differences between the preferences of different consumer groups with respect to the characteristics of the service "train ticket". Only the frequent travelers are willing to pay a relatively high price for a train ticket, if it is valid throughout the area. The height of the price reduction when buying at the machine is not relevant for the respondents. In addition, the respondents are not generally willing to pay an increased basic price for a train ticket, even if the price reduction when buying at the machine is relatively high. ◀

(5) Penalty-reward-factor approach/Kano method The penalty-reward factor approach is a very similar procedure to the multi-attribute procedures and is based on the relationship between the perception of a performance and the reaction to a performance (see Brandt, 1987; Matzler et al., 2009; Alegre & Garau, 2011). The approach originally assumed that there are quality factors for every service that cause dissatisfaction with the

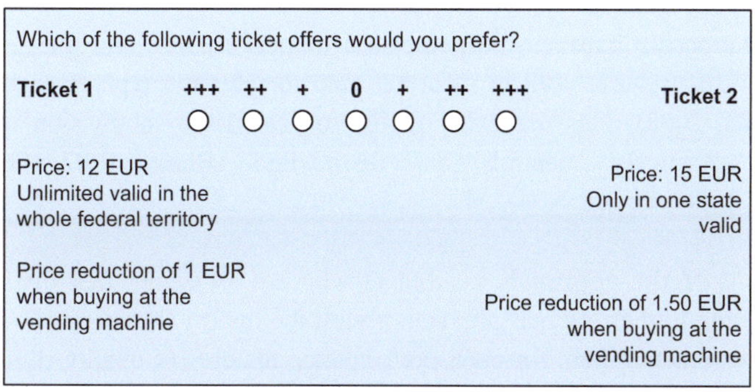

Fig. 5.22 Determining the price willingness of rail customers. (Source: Globalpark, 2004)

customer when they are not fulfilled or confirmed (Brandt, 1987, 61 pp.; Alegre & Garau, 2011). These attributes are referred to as **penalty factors**. In contrast, the **reward factors** represent additional services that create a higher quality perception and therefore a higher satisfaction with the customer. While the customer distributes "bonus points" for the reward factors, he punishes the company for the absence of penalty factors with so-called "demerits" (Brandt, 1987, p. 61). The goal of such a measurement approach is to identify these penalty factors. Therefore, first a general quality assessment of the service is made on a 5-point scale from "very satisfied" to "very dissatisfied". Then the customers are asked to rate the individual attributes of the service, which are given to the subjects, on a scale from "much worse than expected" to "much better than expected". With the help of these data, a multiple regression analysis is carried out using dummy variables (Brandt, 1987, pp. 62 f.). The results of this analysis for individual characteristics are interpreted as follows:

- **Penalty factors** are those attributes of service quality, in which the customer does not give a higher global quality assessment, although the performance was better than expected with regard to the respective attribute. In contrast, the global quality assessment decreases if the quality of the respective attribute was worse than expected.
- **Reward factors** are characterized by the fact that the global quality assessment increases if the quality of the respective attribute is rated better than expected; however, it does not decrease if the performance was worse than expected.

The application of the penalty-reward method in the context of a satisfaction study is illustrated in the following example.

Satisfaction study in public transport

In a study on customer satisfaction in public transport, the penalty and reward factors were determined. The corresponding evaluation showed that the criteria "service" and "comfort" are mainly reward factors, while the dimensions "punctuality", "safety" and "information" can be considered as penalty factors. Figure 5.23 shows the results of the study at a glance.

The relationship between the perception of performance and the reaction to the same performance was used in a further step to infer the type of feature (Gierl & Bartikowski, 2003). This made it possible to extend the subdivision into so-called "satisfiers", "criticals", "neutrals", and "dissatisfiers" (Brandt, 1987, 1988; Brandt & Scharioth, 1998; see Fig. 5.24). ◀

In one **variant of the approach**, customers who are asked are first divided into two groups according to their global satisfaction with the quality of a service—satisfied and dissatisfied customers. Then, for each performance feature (or quality dimension), it is examined how many (globally satisfied/dissatisfied) customers have perceived a positive

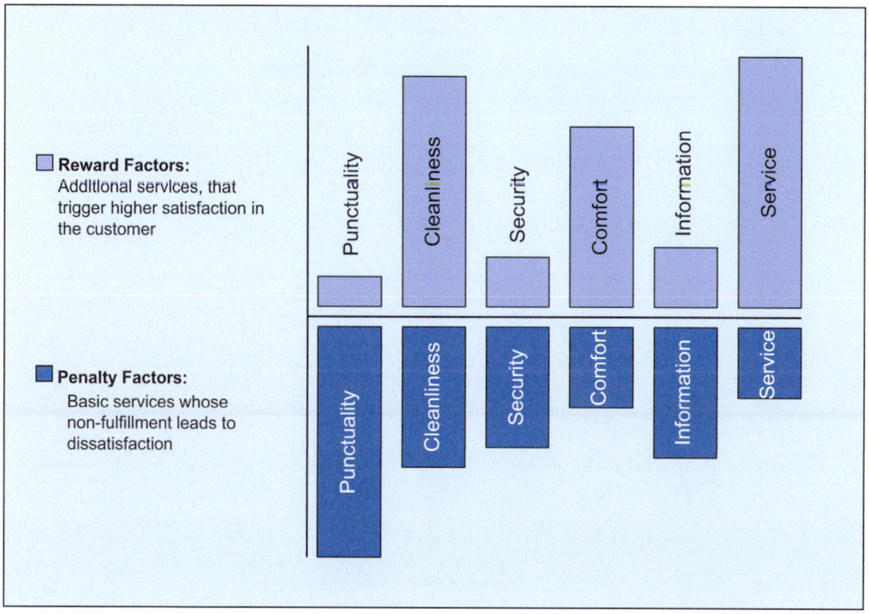

Fig. 5.23 Results of a penalty-reward factor analysis using the example of public transport. (Source: Brandt, 1987, p. 62)

non-fulfillment (feature better than expected), a fulfillment (feature as expected) or a negative non-fulfillment (feature worse than expected) (see also Fig. 5.25).

It is assumed here that a feature is a **"satisfier"** if the proportion of those satisfied can only be increased by more customers perceiving a positive non-fulfillment (or non-confirmation). If a feature has a greater number of customers only confirmed in their expectations, this does not lead to higher satisfaction. **"Dissatisfiers"** are recognized in that the proportion of those dissatisfied increases if more people experience a negative non-fulfillment. If more customers perceive a fulfillment instead of a positive non-fulfillment, the proportion of those satisfied does not increase.

"Criticals" exist if the proportion of those satisfied is highest if a positive non-fulfillment is perceived, and if the proportion of those satisfied is lowest if a negative non-fulfillment exists. Performance features are referred to as **"neutrals"** if the proportions of satisfied customers do not differ from each other. Even if more people experience a positive or negative non-fulfillment, this does not lead to an increase in the proportion of those satisfied or dissatisfied.

By the carried out **division of the quality or performance dimensions**—or from the context between the positive or negative non-fulfillment and the overall satisfaction—characteristics can be divided into two criteria. On the one hand, the question arises as to the potential of a characteristic to cause satisfaction in customers. On the other hand, the question arises as to whether it avoids dissatisfaction. From the perspective of these

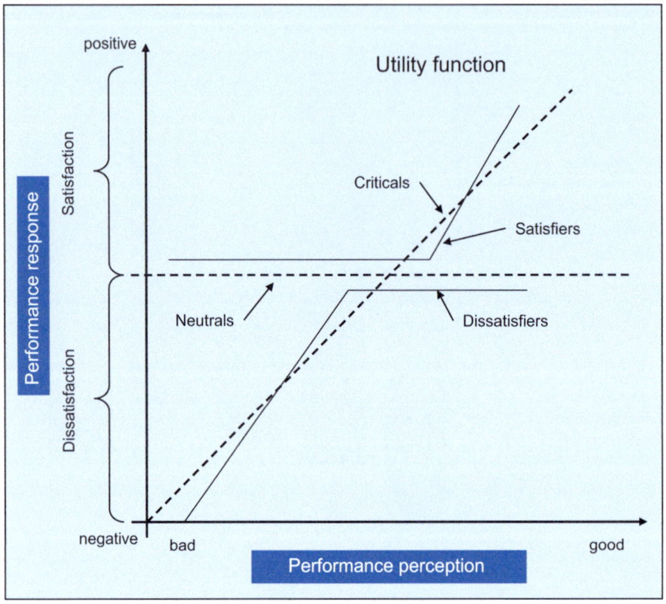

Fig. 5.24 Benefit functions of different feature categories. (Source: Gierl & Bartikowski, 2003, p. 16)

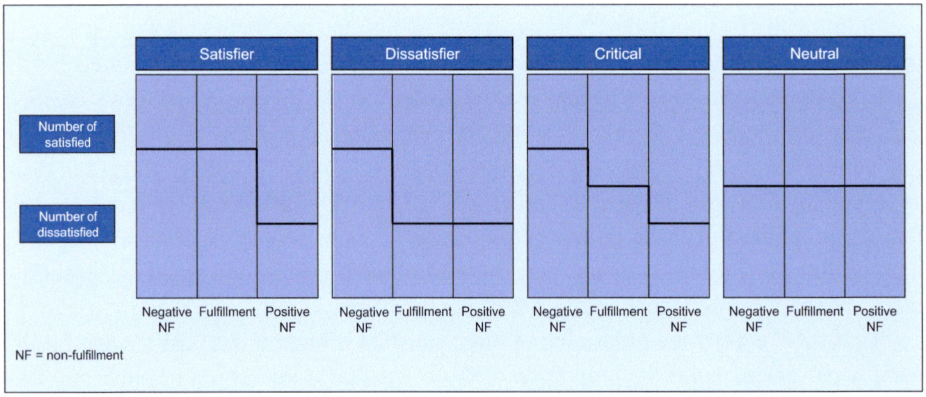

Fig. 5.25 Division of characteristics according to the Brandt method. (Source: Based on Brandt, 1988, p. 38)

criteria, the different characteristic categories can be entered into a characteristic portfolio and corresponding action recommendations can be derived for the different characteristic categories (cf. Fig. 5.26). The relationship between attribute-specific non-fulfillment (NF) and overall satisfaction is illustrated in the corresponding characteristic categories.

Performance response \ Performance perception	No	Yes
Yes	**Satisfiers** Negative NF → Indifference Positive NF → Satisfaction Perform well on some selected characteristics.	**Criticals** Negative NF → Dissatisfaction Positive NF → Satisfaction Perform well on all characteristics
No	**Neutrals** Negative NF → Indifference Positive NF → Indifference Do not perform well on any of these characteristics.	**Dissatisfiers** Negative NF → Dissatisfaction Positive NF → Indifference Provide mediocre rather than good performance on all characteristics.

Fig. 5.26 Feature portfolio of different feature categories and corresponding recommendations for action. (Source: Based on Gierl & Bartikowski, 2003, 15 pp.)

The **advantage** of this approach is that not only quality-relevant results are delivered, which measure the quality of the service from the customer's perspective, but that a targeted use of quality management is also proposed with regard to the different characteristic categories. Consequently, it is up to the company to satisfy these service attributes with the customer first; only then can quality management focus on additional "bonus services".

A similar approach to the classification of quality dimensions or performance characteristics is the **Kano method** (Kano, 1984) (see Sect. 2.4). Here, the surveyed customers are asked directly which category a characteristic falls into. The procedure provides for a functional (satisfaction reaction in the case of a good quality of the feature) and a dysfunctional question (reaction of dissatisfaction in the case of a bad quality of the characteristic) to be asked for each characteristic or quality dimension.

Functional and dysfunctional question according to the Kano method

The question in the Kano method is asked in a functional as well as dysfunctional way. Subsequently, various possibilities for answering the mentioned question types are given. The exemplary questions relate to a fictitious indoor pool "Aquamar". The performance feature in this case corresponds to a catering option that is integrated into the indoor pool building.

Functional form of the question: "What do you think about the fact that there is a catering option in the indoor pool" Aquamar "?"

Dysfunctional form of the question: "What do you think if there was no catering option in the indoor pool" Aquamar "?"

Possible answers to the respective questions:

- “I would be very happy about that.”
- “I take that for granted.”
- “I don’t care.”
- “I could possibly accept that.”
- “That would bother me a lot.”

Figure 5.27 shows an excerpt from a questionnaire. ◀

The characteristics can be assigned to the different categories by combining the answer alternatives. In addition to “satisfiers”, “criticals”, “neutrals” and “dissatisfiers”, this also results in so-called “reverse” and “questionable” categories (cf. Sauerwein, 2000).

Figure 5.28 shows an exemplary characteristic portfolio of a zoo based on the Kano method, which was carried out as part of a study by the *Chair of Marketing and Management at the University of Basel* with *Basel Zoo*.

	That would make me very happy	I assume that provided	I do not care	I could possibly put up with that	That would bother me very much
Functional Question:					
What do you think about the fact that there is a catering facility in the indoor swimming pool "Aquamar"?	○	○	○	○	○
Dysfunctional question:					
How do you think about it, if there was no possibility of catering in the indoor swimming pool "Aquamar"?	○	○	○	○	○

Fig. 5.27 Functional and dysfunctional questions asked according to the Kano method. (Source: Based on Berger et al., 1993)

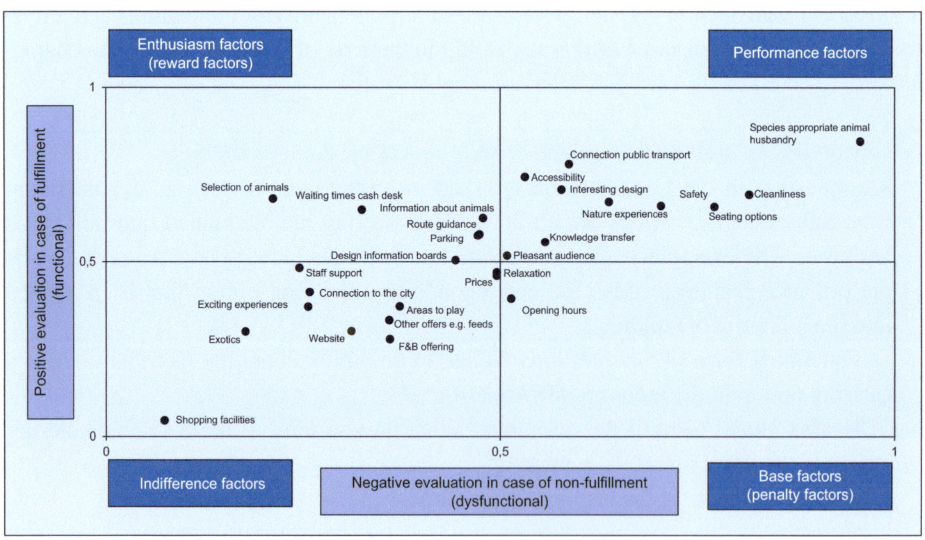

Fig. 5.28 Characteristic portfolio of a Swiss zoo

In a final **appraisal** of all feature-oriented measurement methods, it can be concluded that all approaches have the problem of identifying and selecting the individual relevant attributes. The company provides a limited number of abstractly formulated quality features, which are then assessed by the customers or test subjects in a further step. All methods therefore implicitly assume that a selection of these features is possible taking into account completeness and quality relevance. Therefore, previous studies, such as expert interviews, which increase the effort of the measurement approach, are absolutely necessary.

Feature-oriented methods are therefore only partially suitable for initial quality surveys; however, they enable a valid quality measurement to be carried out if they are combined with other methods, such as complaint measurement. Another criticism of feature-oriented quality measurement is that when using a large number of individual features, respondents can quickly become overwhelmed. Therefore, the number of attributes queried should be kept low. This in turn reduces the informative value of the results.

5.1.2.2 Incident-Oriented Measurement Approaches

Incident-oriented approaches take into account the process-oriented character of service creation in their methodology. They serve to assess this process from the customer's perspective. These methods are based on the so-called story telling, in which service customers are asked to describe their experiences with a service provider relatively unstructured without any specific questions (Scharitzer, 1994, p. 137; Jüttner et al., 2013). A supported incident-oriented measurement is carried out with the following methods:

1. Sequential incident technique (blueprinting)
2. Critical incident technique
3. Critical-to-quality analysis
4. Critical path analysis
5. Root cause analysis

These important incident-oriented methods are described in more detail below.

(1) Sequential event method (blueprinting) In the context of measuring the requirements for service quality, the strengths and weaknesses of existing processes are often determined in a first step. A widely used method for this is the sequential event method, which is based on a decomposition of the service process into several sub-steps. First, a so-called "blueprint" of the interaction between service provider and consumer is to be created, which graphically represents the contact course between provider and customer of a service in a specific situation (Shostack, 1984, 1985; Stauss & Hentschel, 1991; Stauss, 1999; Bitner et al., 2008; see example in Fig. 5.29).

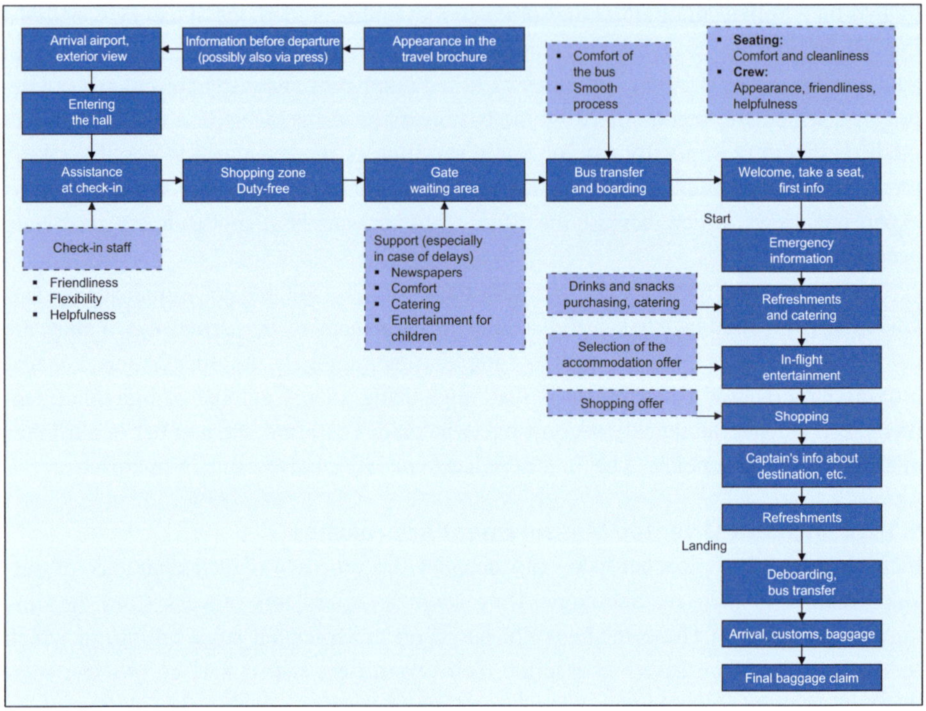

Fig. 5.29 Blueprint using the example of a flight

The so-called "line of visibility" thus illustrates the boundary between the parts of the service visible to the customer and the invisible elements. The customer contact points ("encounter points") can also be determined, which are important for the "moments of truth"; those moments in which there is a direct contact of the customer with the service provider or his offers (Albrecht, 1993, p. 246; Bitner et al., 2008, p. 71; Medberg & Heinonen, 2014).

Over time, the blueprinting method has been supplemented and further developed by the introduction of additional separation lines (Benkenstein & von Stenglin, 2006, p. 58). A comprehensive approach comes from Kleinaltenkamp, who in total—in addition to the "line of visibility"—between further five levels of customer integration in the service process and this by means of corresponding separation lines in the blueprint (e.g. Benkenstein & von Stenglin, 2006, 58 pp.; Medberg & Heinonen, 2014):

The **"line of interaction"** separates customer activities from processes carried out on the provider side. The **"line of perception"** is an extension of the "line of visibility" of visible processes to the overall processes that the customer can perceive (e.g. smells, sounds, etc.). The **"line of internal interaction"** separates the processes directly related to customer integration from the so-called back-office or support processes. The **"line of order penetration"**, the planning line, serves to distinguish between measures that are to be assigned to the customer-induced service process and those that aim at the service potential of the provider. Finally, the **"line of implementation"** is the separation line that separates activities of resource provision from measures of market development (cf. Fig. 5.30).

On the basis of such a modified blueprint, in which the visible elements of the service creation process for the customer are entered in the usual order of contact, the aim is pursued that the surveyedcustomers in personal interviews can mentally-emotionally retrace the phases of the service experience again. With the help of this "re-experiencing", an extensive description of the moments of truth is often achieved in many cases, often also with the help of open questions. Blueprints are also used for processes in which the customer takes over part of the service creation himself. In this way, it becomes clear from the blueprint which measures are taken over by the customer himself and which by the service provider. This shows at which points crossovers between the individual service processes arise.

Blueprint using the example of book production

In Fig. 5.31 the process of creating value is exemplarily shown for the production of a book. The publisher and the author appear as "customers" of the publisher. Here, too, the activities of the publisher are divided into visible and invisible processes. ◀

(2) Critical incident technique On a more differentiated level, the "critical incident technique" deals with the strengths and weaknesses of the service process. The method of capturing and evaluating the so-called "critical incidents" has been the subject of research since the 1950s (Flanagan, 1954). Critical incidents are those events that the customer experiences as particularly satisfying or particularly dissatisfying in the moments of truth. Bitner et al. (1990, p. 73) define these events "as specific interactions between customers and service provider employees that are especially satisfying or especially dissatisfying".

It is precisely these "special" cases that the customer remembers and that also come to mind again and again when the respective service provider or the relevant service is mentioned in the personal environment (Stauss, 1999). The Critical incident technique provides for the capture of these events using standardized open questions in order to encourage customers to remember specific incidents (Mudie & Cottam, 1999):

- Think of an incident in which you experienced a particularly satisfying or particularly dissatisfying service as a customer.
- When did this event take place?
- Describe the specific circumstances that led to this situation.
- How did the employees behave in a specific way (what did they say, what did they do)?
- What causes the feeling that this was a particularly (un)satisfying event?

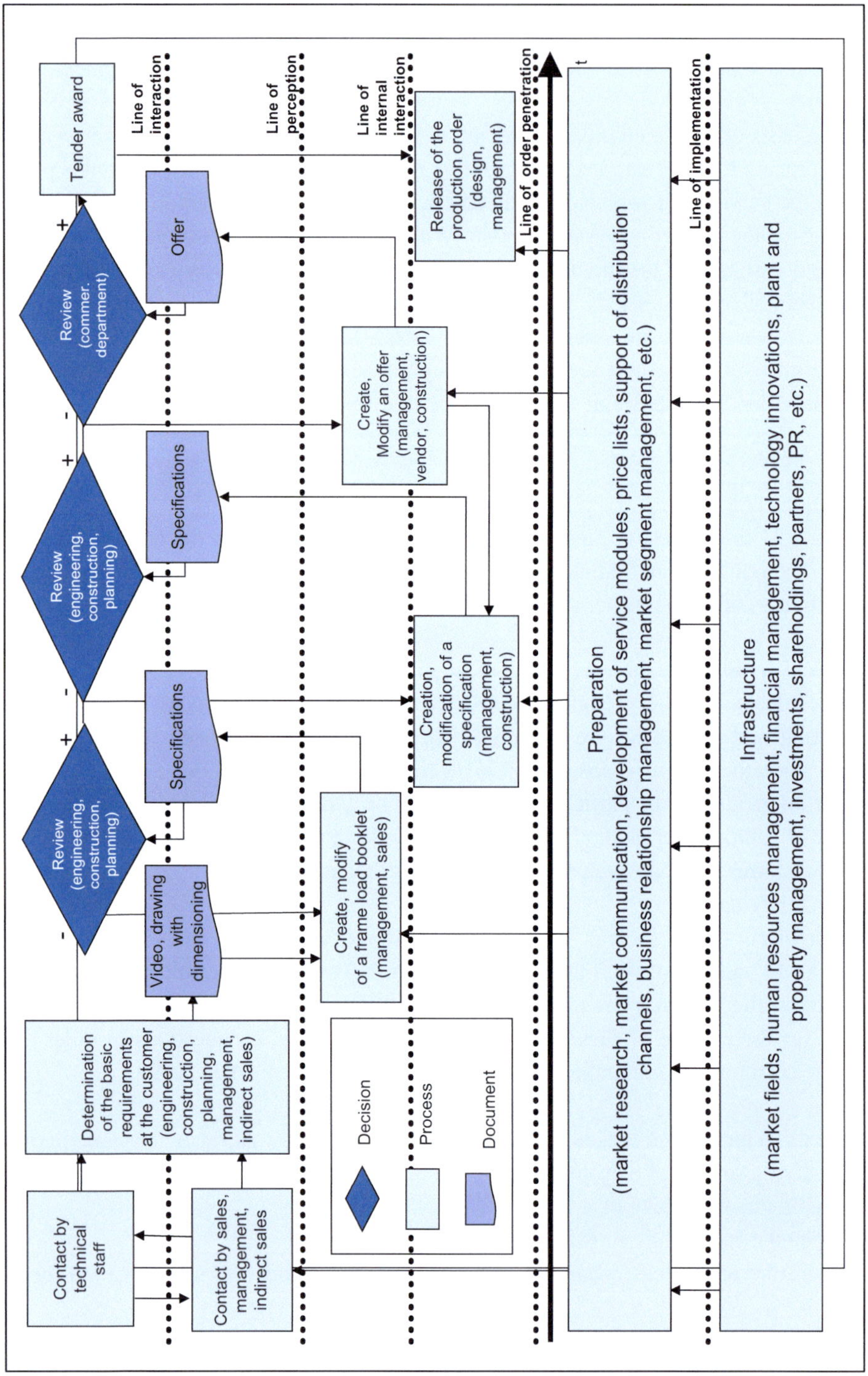

Fig. 5.30 Service blueprint of a sales process. (Source: Benkenstein & von Stenglin, 2006, p. 59)

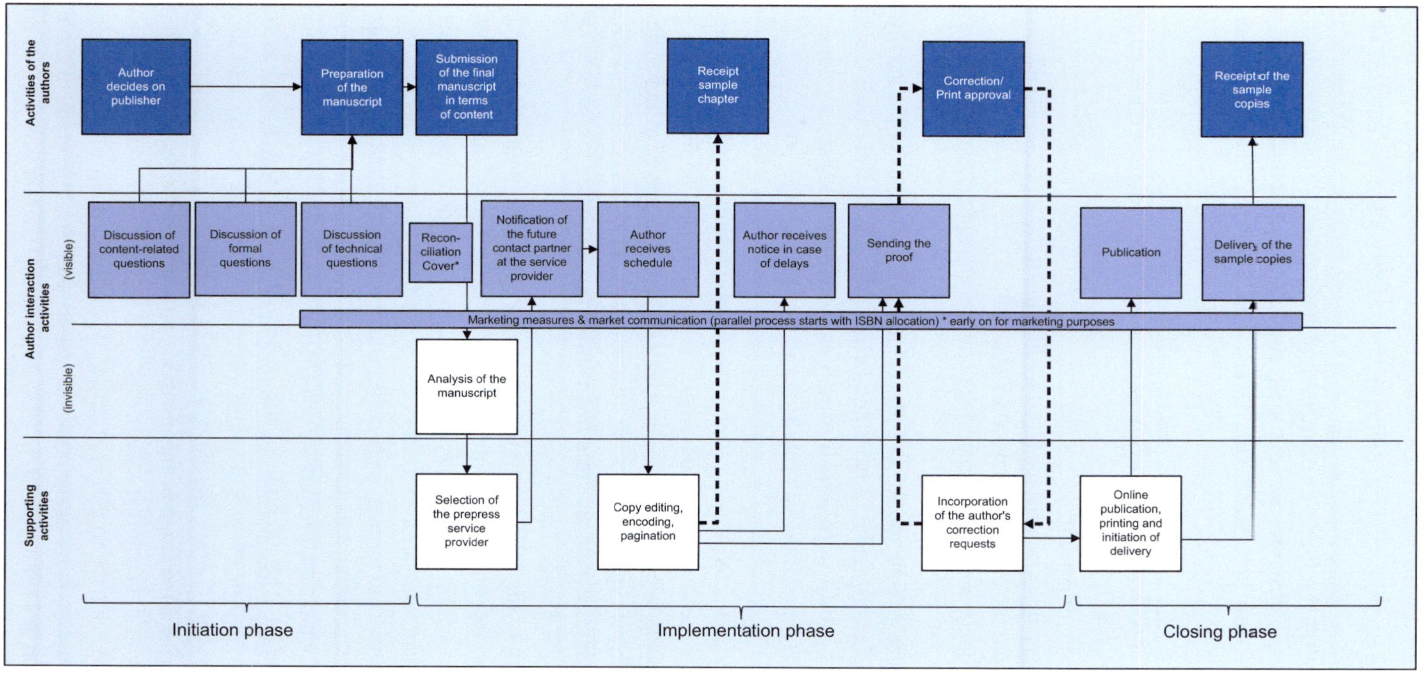

Fig. 5.31 Blueprint using the example of book production. (Source: Bothe, 2013)

An application of the critical incident method is illustrated in the following example using a study by Bitner/Booms/Tetreault (see also Bitner et al., 1990) below.

Application of the critical incident method in industries with high interaction levels

In this study, customer surveys were conducted in restaurants, hotels, and airlines—industries with high interaction levels. The following questions were asked of the subjects:

- Did you remember a particularly (un)satisfying contact with an employee of a restaurant, hotel, or airline?
- When did this occur?
- What specific circumstances led to this situation?
- What did the employee say or do exactly?
- What happened exactly that made you feel the contact was (un)satisfying?

A total of 699 incidents could be determined by means of the survey, of which 347 were satisfactory and 352 were not satisfactory. These were then divided into twelve categories, which were verbally described both for the satisfactory and for the unsatisfactory characteristics. The results are shown in Fig. 5.32. ◀

The opinions collected usually give a fairly comprehensive picture of the perception of the service processes by the consumers. In particular, in addition to the areas in which there is obvious need for action, criteria can also be derived which determine the perception of service quality from the customer's point of view. A requirement profile for service processes developed from this is often helpful when it comes to setting goals for the less positively experienced service aspects.

In analogy to the subdivision of quality characteristics within the penalty-reward-factors analysis, the dimensions of "satisfiers", "dissatisfiers" and "criticals" can also be distinguished within the framework of the critical incident technique. The results of the critical incident analysis shown in Fig. 5.33 show the corresponding proportions of positive or negative statements with respect to the critical incidents. Events in which the proportion of positive statements prevails can be considered as "satisfiers". In the case of events in which the proportion of negative statements prevails, however, a "dissatisfier" is assumed. The attention of the personnel is to be interpreted as "critical", since both positive and negative statements occur relatively frequently and in approximately the same number (Gierl & Bartikowski, 2003, p. 20).

The decisive **advantage** of this method compared to the feature-oriented methods lies in the clarity of the customer statements. This is mainly due to the fact that the surveyed customers do not have to assess abstractly formulated quality features within the framework of a survey, but can describe the events that are important to them in their own words (Bruhn & Hennig, 1993, p. 224; Meffert et al., 2018, 224 pp.). In addition, a high

Event	Satisfactory expression	Not satisfactory expression
Response to customer requests that cannot be fulfilled	My room reservation was lost, but I stayed in the "VIP Suite" for the same price.	Our room reservation was lost – no explanation, no apology, no help finding another hotel.
Response to unreasonably slow service	Although I did not complain about the 1.5 hour wait at check-out, the employee apologized several times.	The employee gave us wrong information, so instead of one hour we had to wait six hours for the flight.
Response to other errors in the creation of the core service	My crab cocktail was half frozen, the waitress apologized and my meal was on the house.	When I complained about damage to my suitcase, the flight attendant said I was lying and trying to trick the company.
Behavior in the event of special customer requests	The flight attendant helped me nurse my travel-sick child.	My child, who was flying alone, was supposed to be supervised by a flight attendant. Instead, she left him standing alone during the transfer.
Behavior with regard to customer preferences	I broke down in the snow, all the hotels were full, one put me up a bed in a social room.	The waitress refused to seat me away from a sunny spot because there was no space available in her area.
Behavior in case of obvious customer error	My glasses, forgotten on the plane, were brought to my hotel.	Due to traffic jams, we missed our flight, and the attendants refused to help us find a flight on another airline.
Reaction to misbehavior of other customers	The manager was watching a suspicious guest so he would not disturb us.	The employees did not speak to the hotel guests who were partying loudly at three in the morning.
Attention given to the customer	The waiter served me royally. He showed that he really cared about me.	The porter paid more attention to his TV than to the hotel guests.
Exceptional behavior of employees	I had overslept and left my hotel room untidy. When I came back, everything was perfectly tidy.	When I took a little longer to order, the waitress said, "read the menu, not the car map."
Behavior with regard to cultural norms	The hotel employee came running after us because my friend had lost a $50 bill.	The waiter at the expensive restaurant treated us like dirt because we were just students at our graduation party.
Positive surprising behavior	The employee was obviously stressed, but he remained calm and reacted very professionally.	
Overall impression	Everything went perfectly.	The flight was terrible – delay, ventilation broken, ...

Fig. 5.32 Example of the critical incident technique. (Source: Bitner et al., 1990, pp. 77–78.)

	Positive statements	Negative statements
Total	347	352
Related to the reaction of the employees when the customer ...		
▪ Complained about the lack of a promised service	6.9 %	8.2 %
▪ Criticized that the service was too slow	4.9 %	15.1 %
▪ Criticized other service features	11.5 %	19.6 %
▪ Help needed	10.4 %	1.7 %
▪ Made a special request	14.7 %	10.5 %
▪ Suffered a mishap	5.8 %	2.3 %
▪ Felt harassed by others	2.0 %	1.1 %
Events not expected by the customer:		
▪ Attention of the staff	13.8 %	13.6 %
▪ Unusual behavior of the staff	6.3 %	11.6 %
▪ Morally impeccable behavior of the staff	4.6 %	11.9 %
▪ Overall impression of the staff	15.9 %	4.3 %

Fig. 5.33 Results of a critical incident analysis. (Source: Bitner et al., 1990, p. 75)

degree of authenticity is ensured by reproducing the original statements of the respondents and not forcing them into a predetermined pattern or category.

The procedure is therefore particularly suitable for service companies with different branches and subsidiaries, as the statements make it possible to identify direct points of attack for improving the individual branches and to create comparisons between the individual branches. This makes it possible to draw conclusions about possible strengths and weaknesses of the individual locations. The main **disadvantage** of this method is its high data collection effort (Stauss, 1999, p. 333). Therefore, the procedure is only limitedly suitable for regular measurement and for checking a sustainable improvement of service quality.

(3) Critical-to-quality-analyse The critical-to-quality-analyse was primarily developed for quality measurement in the area of goods, it is related to various methods of process optimization, such as the continuous improvement process (CIP), the KAIZEN method, the Six Sigma principle, etc. (see the corresponding explanations in the third chapter of this book). However, the analysis can also be used for services. At the center is the consideration (similar to the critical incident technique and the critical path analysis), to bring critical voices about the performance to the attention. However, not only from the customers, but also from other stakeholders, for example from the management and employees. In addition, critical points in the process of performance are recorded by legal or other requirements. All critical voices are summarized in the critical-to-quality-analyse.

From the entire collected critical features, requirements for the creation of the service are developed (similar to the Poka-Yoke method), mostly in the form of indicators. These serve as target specifications for the performance. In this way, the requirements for service quality are formulated more clearly and binding.

As a result, a **critical-to-quality-matrix** can be set up (also possible as a **critical-to-quality-tree**), from which the relevant phases in the creation process of the service can be contrasted with the critical statements, key figures of the critical feature and target specifications for the individual quality features. An example of a critical-to-quality-matrix is shown in Fig. 5.34.

The essential **advantage** of the critical-to-quality-analysis is to be seen in the fact that it tries to identify the critical quality features holistically from quite different perspectives in the process of the service. The **disadvantage** is the high effort involved in collecting the numerous critical points. In addition, a strong quantification of the quality features is carried out by the consistent use of key figures. There are tight limits for certain service processes.

(4) Critical-Path-Analyse This instrument—also known as switching-path-analysis—is a further development of the well-known critical-incident-technique, but with the focus not on the individual transaction, but on the relationship perspective (Roos & Strandvik,

Phases in the service process	Critical statements about incidents/features	Key figure of the critical feature	Goal for the feature
Appointment	Phone frequently busy	Tel. accessibility (in %)	Enable callback within x minutes
	No available appointments on Saturdays	Number of appointments on Saturdays	Saturday appointments beyond normal opening hours
Hair washing	Long waiting times	Waiting time (in min.)	Waiting time max. x min
	Trainees with little experience	Apprentices years of experience	Wash hair only after x months of training
Haircut and blow dry	Hairdresser is distracted with other clients	Focusing exclusively on clients	Create hairstyle in x minutes
	No more time for blow drying by the hairdresser	Time and duration for hair blow dry	Hair blow dry directly after hairstyle by hairdresser in x min
Invoice settlement	Non-transparent invoice	Invoicing with individual partial services	Invoice with partial services available immediately after hairstyle
	Time delay due to limited payment options	Fast payment through all popular options	Invoice settlement within x minutes

Fig. 5.34 Critical-for-quality matrix using the example of a hairdressing salon

1997, p. 623; Roos, 1999, 71 pp.). The critical-path-analysis thus pursues the goal of mapping the entire churn process—starting from a certain trigger to the establishment of a new relationship. This approach to measurement is thus often the starting point for the conception of a customer retention management. Figure 5.35 shows an exemplary churn process from the perspective of an insurance customer.

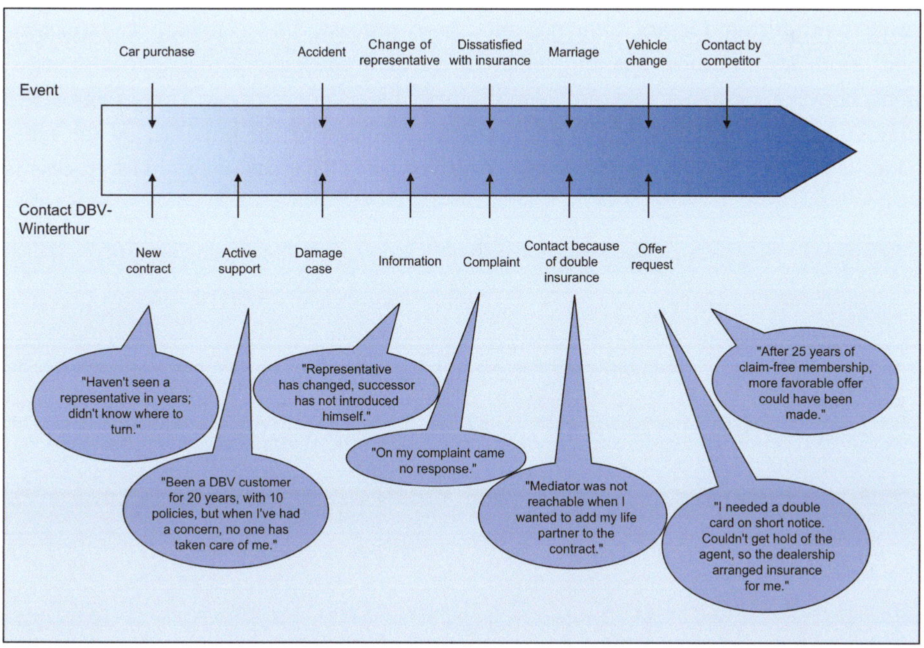

Fig. 5.35 Analysis of the churn process of an insurance customer. (Source: Schröder, 1999, p. 29)

The instrument is based on structured, personal interviews with customers who have migrated. Figure 5.36 provides an overview of the different question areas of such an interview.

In addition to the reasons for customer churn, typical **churn processes** can be identified in the course of these interviews. In the banking sector, for example, six process types can be distinguished (Michalski, 2002, p. 146): reactive churn, short-circuit churn, despair churn, planned churn, forced churn and desired churn. These process types differ in terms of the length of the emigration process. In addition, the process is characterized by specific **customer-related phases** resulting from the mentioned study in the banking sector (cf. Fig. 5.37).

After the so-called latency phase, in which possible negative incident have occurred in customer contact, the perception phase begins. Here, the first thoughts of emigration or a change of provider occur. This phase is followed by a dialogue between provider and customer. On the part of the customer, the dialogue with the provider is sought and complaints are articulated. In the decision-making phase, the customer may begin to inform himself about other provider-external offers and finally make a decision with regard to the termination of the relationship at the end of the phase. The implementation phase is characterized by the termination of the relationship, which marks the end of the churn process (Michalski, 2002, pp. 140–141).

The phases of the churn process are also characterized by corresponding **process characteristics**, such as customer (un-)satisfaction during the latency phase or information search during the decision phase. At the same time, **process-promoting and process-inhibiting factors** influence the churn process. For example, recommendations from third parties regarding other offers or unsatisfactory dialogue or transactions between customer and provider promote the churn process. Convenience and time constraints of customers or the location advantage of the provider, on the other hand, tend to inhibit the churn process. An overview of the customer churn process is

Questions about...	Example
▪ Migration decision	▪ When did you first start thinking about moving?
▪ Churn process	▪ How long did the decision drag on?
▪ Triggers of the migration process	▪ Did a specific event trigger the churn process?
▪ Previous form of business relationship	▪ What was the relationship like before the churn decision?
▪ Corporate behavior after migration	▪ How did the company react to the churn?
▪ Reasons for choosing the new provider	▪ For what reasons was the new provider selected?
▪ Comparison of the new relationship with the old	▪ How does the old business relationship compare to the new one?

Fig. 5.36 Questionnaire for a critical-path analysis. (Source: Roos & Strandvik, 1997, p. 624)

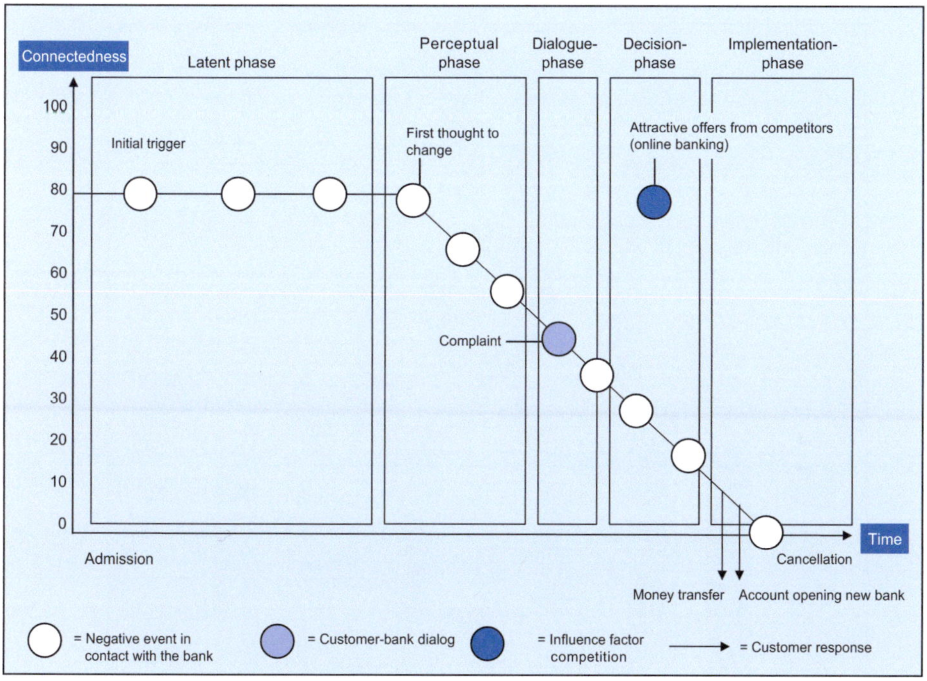

Fig. 5.37 Example of churn phases. (Source: Michalski, 2002, p. 141)

provided by the characteristic-process-type model of customer churn process shown in Fig. 5.38.

Thus, the central task of the illustrated analysis instrument is also to provide decision-relevant information for planning the recovery management. In addition to the benefit of the improved information base for performance deficits and general insights into migration processes, the results can also be used to define suitable indicators for identifying customers at risk of migration.

(5) Root-Cause-Analyse Root cause analysis is an approach to analyzing reasons for migration (Wilson et al., 1993, p. 9; Ammermann, 1998, 52 pp.; Andersen & Fagerhaug, 2006). These are more diverse from the customer's perspective than is often assumed from the company's perspective. The reason for this is the process character of customer migration, which leads to the fact that the termination of the business relationship is usually not due to a single event, but a variety of critical events within a certain period of time determine the customer's decision. Therefore, the traditional methods of market research, such as a written survey of customers about their reasons for migration, are only partially suitable for investigating the "true" causes of migration. The results of such surveys remain very general and specific measures can hardly be derived. What is decisive is therefore not the processing of company-specific migration categories

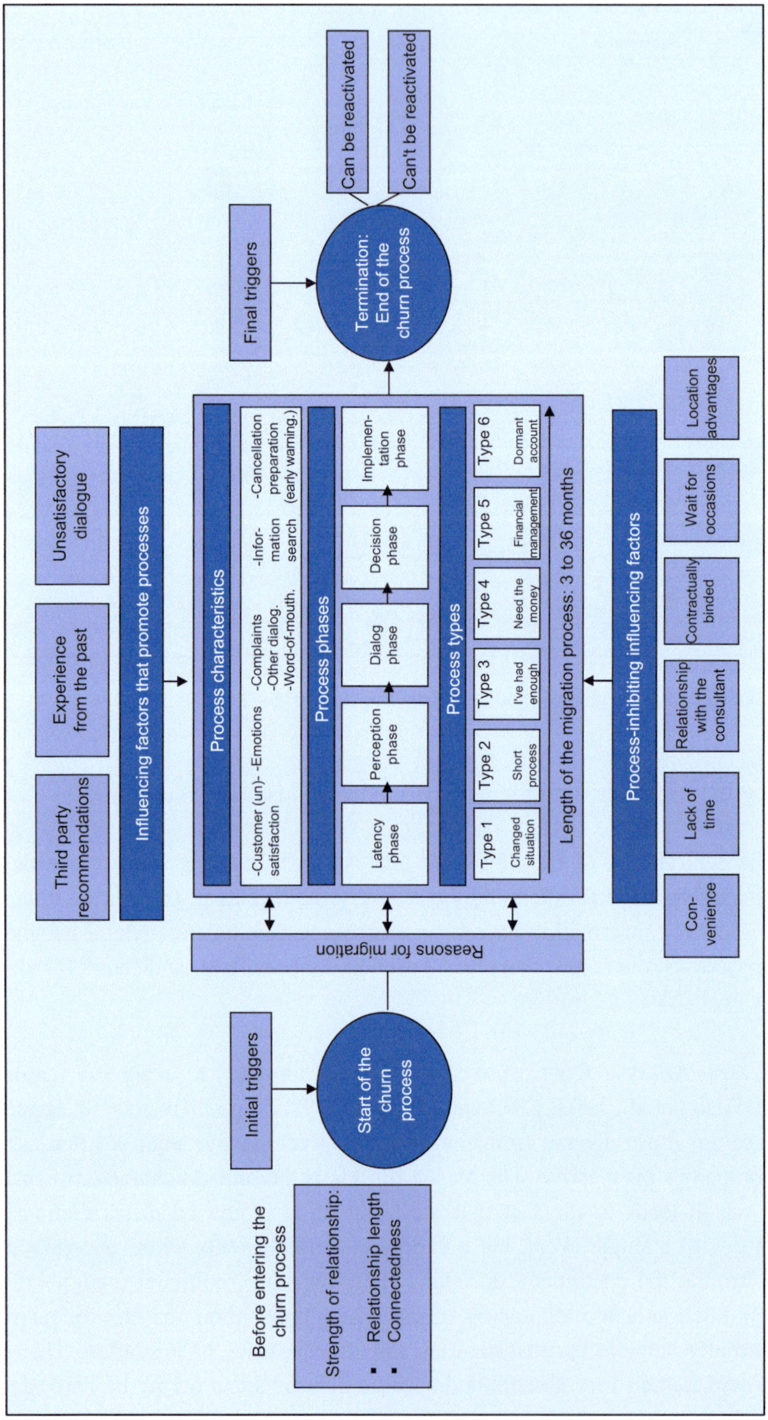

Fig. 5.38 Model of the customer churn process. (Source: Michalski, 2002, p. 109)

(e.g. price, service, performance), but the capture and description of the individual contact points or experiences of a customer within the critical phases of the customer life cycle.

In the root cause analysis, the causes of customer migration are differentiated and identified in a multi-stage process. The starting point of the process are hypotheses about possible reasons for migration, which are described in more detail in the second step in the form of detailed cause trees. This is followed by a telephone survey of former customers based on the story-telling approach. The recording and evaluation of customer conversations is then carried out using a computer-aided survey software (Venohr & Zinke, 1999, p. 160). The result of a root cause analysis is shown in Fig. 5.39 using the example of an insurance company.

In addition to the **company-related reasons for emigration** shown, i.e. reasons caused by the company and thus also influenceable, emigrations can also be based on competition and customer related reasons. Competition related reasons include, for example, headhunting attempts by competitors, triggered by communication policy measures.

For example, *Commerzbank* initiates a change of bank by placing ads and commercials and offering 50 EUR for a change of current account to *Commerzbank* (*Commerzbank*, 2016).

Customer-related reasons for emigration include all motives for emigration that are not caused by dissatisfaction, but by a change in the customer's personal situation (e.g. account closure due to a change of address).

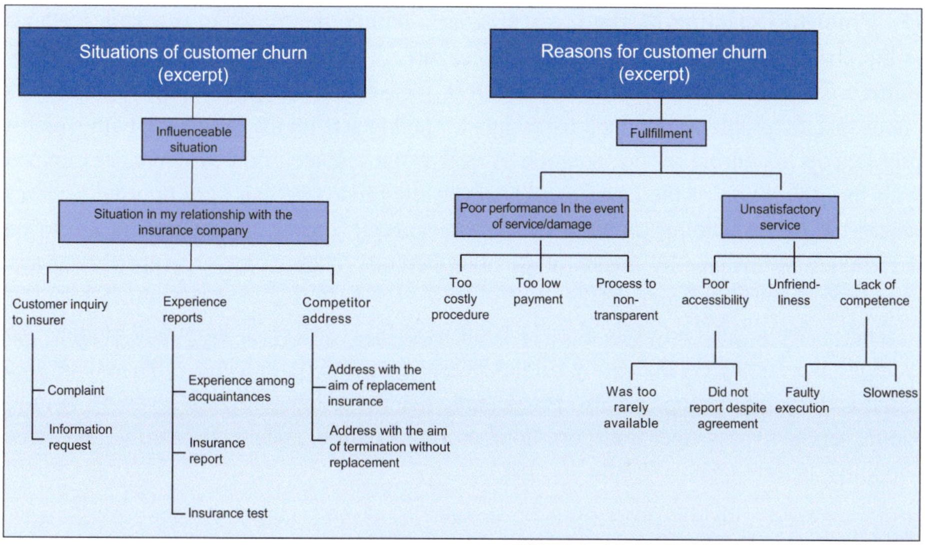

Fig. 5.39 Result of a root cause analysis using the example of an insurance company. (Source: Venohr & Zinke, 1999, p. 160)

The main **advantage** of the incident-oriented methods for measuring service quality presented here lies in the possibility of determining the attributes relevant to the customer of a high service quality. The number of individual features is not limited—in contrast to the feature-oriented methods. Therefore, these methods are used in particular to determine a variety of attributes and to achieve the completeness of the features. **A disadvantage** of the incident-oriented methods is that they cannot be used for quantitative measurement, e.g. by customer-side rating of individual attributes. Therefore, they can only be used to a limited extent for regular measurement of service quality in order to test quality improvements (Meffert et al., 2018, p. 226).

5.1.2.3 Problem-Oriented Approaches to Measurement

In the context of problem-oriented approaches, quality-relevant problem areas are considered from the customer's perspective in the context of service creation. This group of approaches includes:

1. Problem-detecting-method
2. Frequence-Relevance Analysis for Problems (FRAP)
3. Complaint analyses
4. Rating portals on the Internet
5. Customer-to-customer communication in social networks (web 2.0)

The following paragraphs explain the mentioned approaches in more detail and present their possible uses specifically for measuring service quality.

(1) Problem-Detecting-Method An approach within the problem-oriented methods is the problem detecting. The method, which was developed by the advertising agency *Batten, Barton, Durstine & Osborn (BBDO)*, focuses on the questioning of customers about specific problems and their assessment. The focus is on the frequency with which a problem occurs during service creation as well as the valence of the problem is examined with the perception of the customer. However, the procedure can only be used in cases where the corresponding problem classes are already known. These classes are to be determined in advance by means of suitable methods (such as the critical-incident-tecnique) (Stauss, 1999; Pepels, 2013, p. 35).

The problem detecting has already been used in a variety of empirical studies. An example is a survey conducted by Lindqvist among cruise participants. This service area is particularly interesting and also problematic due to the interaction of a variety of individual services (passenger transport, hotel accommodation, restaurant catering, duty-free shopping, entertainment) (Lindqvist, 1987).

Application of the problem detecting method

Lindqvist identified a total of 81 problem areas, which were evaluated with the Lindqvist index, which can be calculated from the following formula:

$$Lindqvist\text{-}Index = \frac{\sum_{i=l}^{n}(a_i + b_i)}{n}$$

The individual variables in this formula have the following meanings:

- a_i indicates how strongly the respondent agrees with the respective statement (i).
- b_i indicates how important it is to the respondent to eliminate the problem addressed (i).
- n is the number of respondents.

Based on the calculation of the indices, Lindqvist sorted the problems according to the Lindqvist index. Accordingly, the problem "There is no place to store luggage after leaving the cabin" led the list of problems with a value of 6182 for the Lindqvist index. Figure 5.40 shows excerpts from the list of problems with the respective Lindqvist index. ◀

(2) Frequence-relevance analysis for problems (FRAP) The method of problem detecting in frequence-relevance analysis for problems (FRAP), which comprises both the determination of the problem classes and the positioning of these classes in an evaluation grid (Stauss, 1999; Woratschek & Roth, 2004; Pepels, 2013, p. 37), found further development. Under the assumption that a service company is all the more urgent to deal with a specific problem of service creation the more often it occurs and the more

Rank	Individual problems	Lindqvist index
1	"There is little room to store luggage after leaving the cabin."	6,182
2	"It sucks when you have to leave your cabin two hours before you arrive in Stockholm."	5,892
3	"It's bad that all the facilities on board close when we arrive in Mariehamm."	5,833
4	"It's impossible to have a conversation over dinner when you're sitting near the band."	5,625
5	"There is a lack of information in the cabins about what to do on board."	5,485
...	...	...
...	...	...
79	"The sales office staff is unfriendly."	2,929
80	"The staff in the terminals are unfriendly."	2,814
81	"It's boring in the cocktail bar without disco music."	2,749

Fig. 5.40 Problem list of the problem-detecting method using the example of a cruise company. (Source: Lindqvist, 1987, p. 18)

significant its appearance is for the customer, customer groups are asked about the occurrence of certain problems, the extent of their upset and their subsequent behavioral reaction. Figure 5.41 shows a two-dimensional matrix with different, condensed scale values of problem relevance and problem frequency values using the example of bank services in the mass business.

This procedure is carried out in several steps. After determining a list of problems, the individual problems thus recorded are condensed into problem clusters. The questionnaire to be created then has three question categories for each problem class:

1. Has the specific problem already occurred?
2. What is the extent of the upset?
3. Which reactions were considered?

The data collection is followed by an evaluation, which converts the customer reactions determined into **frequency and relevance values**. While the determination of the frequency values is largely unproblematic, the relevance values are usually obtained by linking the statements from category two and three. For this purpose, both the extent of

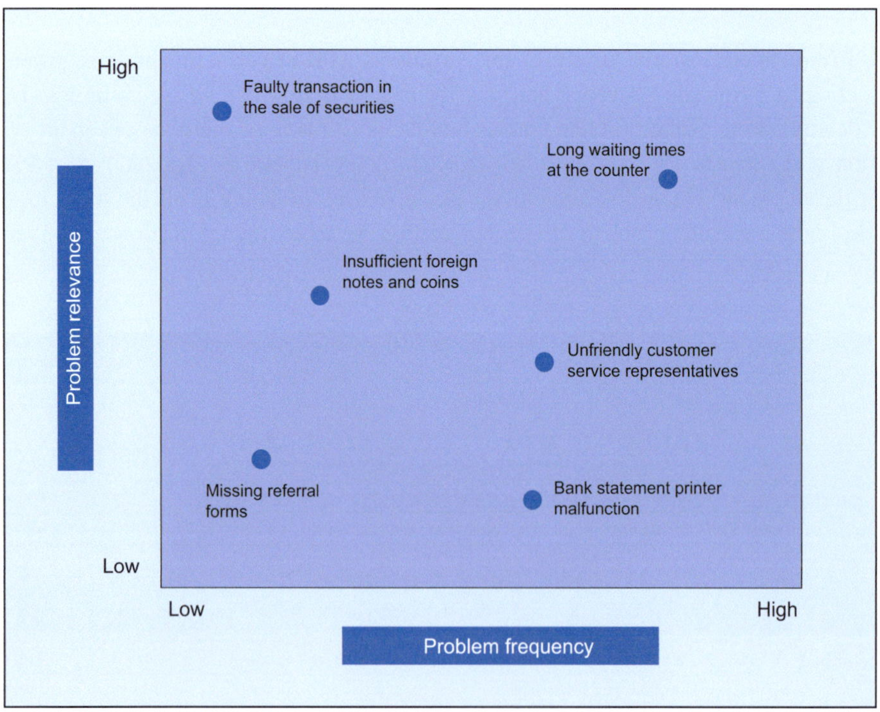

Fig. 5.41 Example of a frequence-relevance analysis for problems (FRAP) in bank services in the mass business

the upset and the individual reaction forms (e.g. "voice" or "exit") are assigned point values, so that the number for the problem relevance results from the multiplication of the two values. The values thus obtained can, for example, be used in the form of the diagram presented here or to form so-called "problem scores".

These **problem value indices** are calculated as the quotient of the sum of the relevance values determined for the individual problem dimensions and the total number of respondents. The values thus determined then give information about the condensation of problems on a few concrete causes and can be transferred to a concentration diagram (Pareto diagram) (Wyckoff, 1988; Haist & Fromm, 1991, pp. 169–179; Stauss, 1999; Stauss & Seidel, 2014; see Fig. 5.42). However, due to the collection of frequency values for the occurrence of problems, the frequency-relevance analysis is only useful in those cases where services are used frequently or over a correspondingly long period of time. Thus, a useful application seems to be largely excluded if an offer is used for the first time or once or only sporadically (Stauss & Hentschel, 1990).

(3) Complaint analyses By effectively using a complaint management (see Sect. 8.4.2) to **measure the volume of complaints** and the **analysis of complaint content**, conclusions can be drawn on acute quality-relevant problem areas of the service offering or the service provider. In addition, service providers can gain an impression of the components of service quality relevant from the customer's point of view from the accumulation of certain complaint situations.

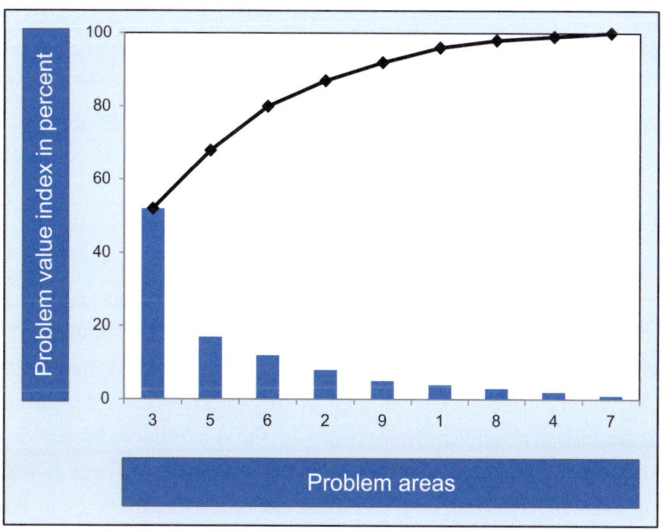

Fig. 5.42 Example of a Pareto diagram

▶ **Complaints** are articulations of dissatisfaction that are expressed towards the company with the purpose of drawing attention to behavior of a provider that is subjectively experienced as damaging, achieving compensation for suffered impairments, and/or causing a change in the criticized behavior (Stauss & Seidel, 2014, p. 28).

Study on the benefits of complaint management

According to a study by *Dortmund University* and software manufacturer *Materna,* in which 220 employees of German companies with a complaint department were surveyed, 40% of companies with systematic complaint management use the complaints received for product and service improvements and quality assurance. Furthermore, the benefits of complaint management are primarily seen in the increase of customer satisfaction, followed by benefits for quality assurance or improvement (Materna, 2012). ◀

For targeted complaint analysis and measurement, a suitable complaint stimulation is recommended. This is primarily achieved by the **removal of complaint barriers** (Riemer, 1986, p. 121; Stauss & Seidel, 2014, p. 131), by making it easier for customers, for example, to express themselves directly on site in the branch using so-called **"comment cards"** or **"complaint boxes"** about the services just used, whether negatively or positively. Direct requests by employees from customers as part of the respective interactions to communicate suggestions, wishes, or deficits often help to remove barriers for customers.

For the purpose of complaint measurement, the establishment of appropriate **complaint channels** is of particular importance. The following types are distinguished here (Stauss & Seidel 2014, 97 pp.):

- Verbal complaint channels (e.g. help desks in railway stations or hotels)
- Written complaint channels (e.g. opinion cards)
- Telephone complaint channels (e.g. hotlines or call centers)
- Electronic complaint channels (e.g. complaint pages or customer forums on the Internet)

Advantages of using complaint analysis are the timeliness and relevance of the problems (customers usually complain very soon if they experience serious quality of service deficiencies) and the relative cost advantage of this procedure, since the complaints are articulated on the initiative and at the expense of the customers (Meyer & Kantsperger, 2005). In complaint analysis, two **problems** in particular arise: There is the difficulty of initiating complaints by the company and there are often deficiencies in complaint capture (Meffert et al., 2018, p. 228).

(4) Rating portals on the Internet With the increasing acceptance of the Internet as an information and communication medium, consumers are opening up new ways of accessing the experiences of other people who have collected them with companies, products and services (Huber et al., 2011). As such a form of interpersonal communication between consumers, **internet-based rating portals** are increasingly gaining attention (Herzog & Luthe, 2010). These offer customers the opportunity to communicate their consumer experiences globally to a potentially unlimited number of Internet users. In this context, the so-called **"Internet-customer-to-customer communication"** takes place mainly in private blogs, opinion forums (Stauss & Seidel, 2014) and increasingly in so-called social networks such as *Facebook* or *Xing* instead.

This is mainly promoted by the change in Internet usage behavior that has been observed for some years, which is mainly discussed under the terms **web 2.0** or "participatory Internet" or "user-generated communication". Under web 2.0, a stronger participation and involvement of Internet users in the Internet is understood, whereby users can generate and share content with others with little effort (O'Reilly, 2006). The Internet is therefore no longer understood as a pure source of information (web 1.0), but as an execution platform to contact other Internet users. Increasingly, Internet users produce and publish their own information on rating portals on the World Wide Web.

The goal of rating portals is generally to make it easier for consumers to make purchasing decisions by providing ratings and experience reports from other consumers (Herzog & Luthe, 2010). Here, resources of various kinds are assessed by users of the system. Resources can be data on objects (e.g. goods, information) or people and their services (e.g. teachers, doctors) (Kartal et al., 2011). Furthermore, a distinction can be made between systems in which the resources are specified by the system operator (e.g. articles in sales portals) and systems in which resources are set by the users (Kartal et al., 2011). In addition to the core functions, numerous rating portals also offer opportunities for social networking. For example, users (open or access-restricted) can join groups or interact with other users in the system, e.g. by adding users to their own friends lists (Kartal et al., 2011). Based on these basic assumptions, object-related and person-related, open and access-restricted, and purchase-oriented and content-oriented rating portals can be distinguished.

First of all, object-related and person-related rating portals can be distinguished. For example, on www.ciao.de experience reports on different products from the categories home & garden, electronics & computer, sport & outdoor and services can be obtained or given, while the rating portal www.jameda.de refers, for example, to doctor recommendations. Here, more than 5 million experience reports on 275,000 doctors can be obtained or given nationwide (Jameda.de).

Furthermore, one can distinguish between open and access-restricted rating portals. For example, only registered ubereats.com-members can rate the performance of a delivery service after ordering. The reason given by ubereats.com is that it wants to ensure a high quality of the rating (ubereats.com). In contrast, on www.yelp.co.uk user contributions and recommendations for top restaurants, shopping opportunities, clubs and entertainment can also be submitted without registration or login. Accordingly, *Yelp* is an open rating portal.

Finally, one can distinguish between purchase-oriented and content-oriented rating platforms. For example, **private blogs** are purely content-oriented rating platforms. Blogs are websites (e.g. www.blogger.com), which are kept in the style of diaries and whose entries are often updated. Private blogs are primarily used for entertainment or personal self-expression, but increasingly also for exchanging experiences about products and services, up to and including a blog's exclusive focus on a company or its services (Stauss & Seidel, 2014). In contrast, **internet opinion forums** are communication platforms that are usually operated by commercially oriented companies and aim at the purchase or use of a service by the customer. Here, for example, *Tripadvisor* (hotels, restaurants, flights, etc.), *Quandoo* (restaurants), *Jameda* and *Weiße Liste* (doctors), *Google my Business* (for businesses), as well as *Amazon* and *Zalando* and many others, where not only services can be bought, but also providers and services can be rated. These opinion and rating portals are now available in many service industries. They offer Internet users the opportunity to read or write experience reports in order to give positive or negative recommendations.

Reviews on *HolidayCheck.de*

The review portal *HolidayCheck* offers customers the opportunity to exchange experiences and assess the quality experienced. Hotels are assessed quantitatively and qualitatively on the basis of six quality criteria hotel, room, service, location & environment, gastronomy and sports & entertainment (cf. Fig. 5.43). In a first step, the experience with the object to be evaluated, in this case the hotel *"Mon Port Hotel & Spa"* on Mallorca, is described in the form of an experience report. For each criterion, a separate text is written. Subsequently, the customer evaluates his experience with the individual quality criteria on a 6-point scale from (1) "very bad" to (6) "very good". From this, the portal automatically calculates an overall assessment of the quality over all ratings that the hotel has offered to the customer. Customers therefore have the great advantage of being able to obtain an assessment of the quality of former guests before booking a hotel. ◄

Accordingly, online rating portals contribute significantly to creating transparency and ensuring a high quality of services to be provided. This compresses the wealth of information and allows for a better overview for assessing a service to be consumed (Kartal et al., 2011). However, rating portals not only offer customers new, extended opportunities to communicate their experiences with providers and their services to other customers, but also offer perspectives for quality management. Based on the rating portals, companies systematically capture and evaluate quality-relevant communication content that customers publish or communicate with other customers on the Internet. A key

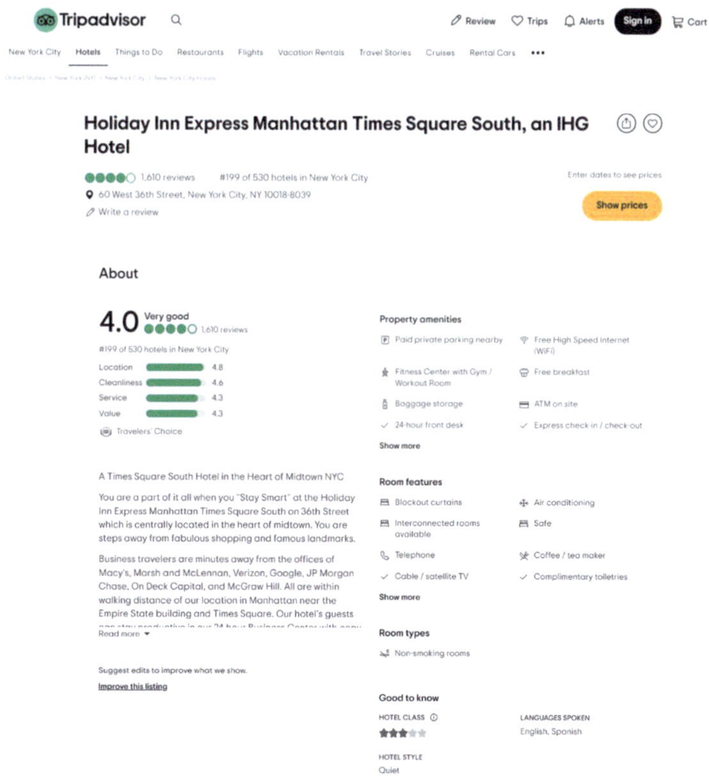

Fig. 5.43 Customer review of a hotel on the review platform *Tripadvisor.* (Source: Tripadvisor, 2023)

advantage of such quality measurement of services lies in the authenticity of the opinions expressed on the Internet (Meffert et al., 2018, p. 232). Analogous to complaint analysis, service providers thus receive customer-initiated hints about the perceived service quality in terms of the overall service and in terms of individual service components without distortion. A **disadvantage** is the high effort involved in reviewing the customer-to-customer communication on the Internet for relevant content that allows for adequate conclusions about the perceived service quality in relation to one's own service offering (Stauss & Seidel, 2014). In order to counter this disadvantage, for example, the rating portal *HolidayCheck* offers hotel managers various functions with which the hotel reviews left on the Internet can be systematically evaluated.

Evaluation of quality judgments with web 2.0 on *HolidayCheck.de*

The rating portal *HolidayCheck,* on which customers can rate hotels and communicate their experiences with them (e.g. in the form of experience reports, self-made videos or photos) to other users, offers hotel operators extensive functionality for evaluating the ratings on *HolidayCheck* as part of their own quality management.

With the help of the offered statistics function, it is possible to check who has rated the hotel in the past. This provides a detailed picture of the guests. In addition, it is possible to observe how the guests' satisfaction has developed in the different categories or over a certain period of time.

In addition, the competition monitor allows the comparison of the own performance indicators with those of up to five self-selected hotels, which can also be filtered by subcategories.

In addition, it is possible to formulate up to six individual additional questions in all languages of the portal, which are added to the rating section of the respective hotel on *HolidayCheck* afterwards. The answers to the additional questions can only be seen by the user of the hotel manager and also evaluated using the statistics functions. This functionality allows hotel managers to ask specific questions about their hotel and supplement and individualize the quality management.

In addition, the quality management can be extended by the manual input option of guest questionnaires. For this purpose, *HolidayCheck* provides a print template of the guest questionnaire, which includes the questions of the *HolidayCheck* hotel rating section. The guest questionnaires can be printed out in the rooms and filled out in writing by the guests. The questionnaires are then entered manually by the hotel manager and evaluated together with the online ratings or separately in the statistics area.

(Source: HolidayCheck, 2013, p. 6 f.) ◀

(5) Customer-to-customer communication in social networks (web 2.0) Social networks are the central channels of web 2.0. The basis of a social network is a platform with the aim of maintaining contacts and mutual exchange. The dialogue of the members joined in the network usually takes place via the so-called "instant messaging". These are message systems that provide users with information about the presence and availability of other users and with the help of which textual messages can be provided to other users in real time. The operators of such networks provide the framework for the various application possibilities (e.g. contact maintenance, data exchange, cooperation, etc.), the content is then created mainly by the users. The most widely used social networks are *Facebook, Twitter, YouTube, MyVideo, Google+, LinkedIn, Xing, Pinterest* and *Instagram* (see Bruhn, 2014, 108 pp. for the individual platforms). The target groups of the individual networks are diverse. There are platforms for singles (e.g. *iLove*), business people (e.g. *Xing*) and platforms without specific targeting of a target group (e.g. *Facebook*). Social networks offer the opportunity to integrate the external factor customer into service creation by obtaining information about and from customers on the individual platforms. This reduces quality uncertainty on the supplier and demand side. In a next step, the information obtained can be used in the internal implementation of quality requirements. Companies also have the opportunity to communicate the results of the service creation process to their customers via the individual social networks and this way reach them as quickly as possible (Weiber & Wolf, 2013).

Pepsi NEXT "Taste Check" on _Facebook_

PepsiCo used a "Taste Check" on *Facebook* to bring the taste of their new variety, "*Pepsi* NEXT", closer to customers. For this purpose, information, including recently experienced events and "likes", from *Facebook* pages of registered users was used to create virtual people. These virtual people corresponded to the profile with which the users represented themselves on their personal *Facebook* page, which was then played by members of the comedy network "Funny or Die" while drinking a new *Pepsi* NEXT. The members simulated the reactions of the individual users based on their virtual profiles while drinking. The videos of the individual reactions were published and sent to the individual *Facebook* users. This way, Pepsi was able to use *Facebook* both as a communication platform for disseminating their new cola and for gaining information about their target group (Source: PepsiCo, 2012). ◄

Contents of customer-to-customer communication in social networks can be praise, requests, mutual help and support, as well as complaints or negative experiences with providers and/or their services, with the latter being the focus (Stauss & Seidel, 2014). Often, unselfish motives predominate in the articulation of public opinion (Hennig-Thurau & Hansen, 2001; Hennig-Thurau et al., 2004); this has a positive effect on the authenticity and credibility of the accounts (Schlosser, 2011).

The **possibilities of customer-to-customer communication in social networks** not only offer customers new, expanded opportunities to communicate their experiences with providers or their services to other customers, but also offer providers perspectives for quality management. This way, the provider openly faces the quality-relevant dialogue with its customers through active participation in customer-to-customer communication in social networks in order to document its own quality requirements in an pro-active way. For example, with *Facebook* it is possible to set up company profiles or groups (fan page) in order to interact directly with customers or to respond to critical content through appropriately trained moderators. In addition, it is possible to route critical customers directly to in-house channels by providing customer service or quality management contact information. Another possibility for providers is to operate their own blogs (corporate blog) on the company's website and to communicate with customers about quality-relevant content. A good example of this is the provider of frozen products *FRoSTA*, which, in addition to its own corporate blog (www.frostablog.de), also maintains a fan page on Facebook (www.facebook.com/frosta.de) and communicates with its followers via short messages with a maximum length of 140 characters via the micromedia platform *Twitter*.

Customer-to-customer communication of the *TUI AG* on the Internet

An example of active participation is the tour operator *TUI,* which, in addition to its own corporate blog (www.blog.tui.at), maintains a fan page on *Facebook* (http://www.facebook.com/TUI.com) and communicates with its followers via short messages with a maximum length of 140 characters via the micromedia platform Twitter (http://twitter.com/tui_travel). The company regularly uses the corporate blog to communicate information from different subject areas, to which customers can leave comments and discussions can arise. Frequently discussed topics are sustainability, travel tips, news, competitions, etc.. The fan page on *Facebook* is used, on the other hand, to present new travel destinations or new *TUI*-owned hotels and to communicate news from the company. Customers use the fan page to exchange experience reports from their holidays with *TUI,* in order to get an overview of the expected quality of the trip and the service providers involved in advance. ◀

However, an important prerequisite for active participation is that the individual channels (e.g. *Facebook* pages) are kept up-to-date. A great danger of such company-initiated platforms is that they are used by customers to express negative opinions about the company and its services. If the company fails to react promptly to negative comments, this can lead to misunderstanding and negative reactions from consumers and negatively affect the quality image through these platforms. Therefore, a person or department is to be named within the company who is responsible for these channels and who is constantly dealing with the customer-created content in these channels.

The customer-oriented methods for measuring service quality presented have in common that the quality of the service is assessed from the customer's perspective. But often it is advisable to supplement the customer-oriented view with an internal perspective in order to obtain a comprehensive quality assessment. For this reason, various company-oriented methods for measuring service quality will be presented below.

5.2 Company-Oriented Measurement Approaches

When using company-oriented measurement approaches, the quality is not assessed from the customer's perspective, but from the company's perspective. When measuring from the company's perspective, it is mainly about translating the expectations of the customers into corresponding performance specifications and ensuring that they are also fulfilled in the process of creating the service. This can be done either by the management or by the employees.

5.2.1 Management-Oriented Measurement Approaches

The management-oriented approaches aim to illuminate from the management perspective the aspects of the service relevant to the customer in terms of quality. The following approaches can be counted among this group:

1. Benchmarking
2. Failure mode and effects analysis (FMEA)
3. Fishbone analysis
4. Statistical process control

The mentioned measurement approaches will be explained in more detail below.

(1) Benchmarking Benchmarking is a management-oriented method with which so-called "best practices"—here in relation to service quality—can be identified. Based on benchmarking, the distance or the **performance gaps** of one's own processes and results on the way to realizing excellent service quality can be determined or measured with respect to the identified "best practice".

At its core, benchmarking aims to relativize and evaluate company-specific processes and results using suitable comparison criteria (Madu & Kuei, 1995, 27 pp.; Siebert & Kempf, 2008; Park et al., 2015). Benchmarking for an internal or external company comparison can relate to the following objects (cf. Sect. 8.2):

- Companies
- Services
- People

Company surveys show that benchmarking is used by the overwhelming majority of companies. This is especially true for the services offered, the staff, the costs and the results of the company's activities.

Benchmarking in higher education practice

The *Centrum für Hochschulentwicklung* (CHE) moderates the self-founded Benchmarking Club Technischer Universitäten (BMC) with the aim of comparing the different areas between the technical universities in Germany. Particular value is placed on quality assurance within the range of classes offered. In order to ensure this, various models, alternatives and experiences are regularly compared within the evaluation of courses among the different members of the BMC. This ensures the quality of the evaluation process and of individual classes and is continuously increased on the basis of "best practices" (Ziegele & Hener, 2014). ◀

Benchmarking is not only suitable for measuring service quality in an enterprise-oriented manner or for identifying the service quality gap: Benchmarking also plays a special role in connection with quality planning (Sect. 8.2) as well as in the context of external customer barometers (see Sect. 14.1) as well as internal service barometers (Sect. 8.4.1).

(2) Failure Mode and Effects Analysis (FMEA) From the perspective of the service provider, error avoidance is to be understood as a central issue of quality management. In particular, the simultaneity of production and consumption of the service makes rectification impossible in most cases. In order to determine the possible weak points in the service process and to determine the resulting consequences, the method of failure mode and effects analysis (FMEA) is suitable (Masing, 1995, p. 252; Estorilio & Posso, 2010; Franklin et al., 2012).

Based on the approach in production, three **forms of FMEA** can also be distinguished for measuring service quality, which build on each other:

- System-FMEA
- Subsystem-FMEA
- Process-FMEA

The system-FMEA is used to examine the cooperation of individual system elements. The subject of such an analysis could be, for example, the cooperation of individual departments in the private customer sector in the banking sector. The subsystem-FMEA, on the other hand, examines whether the structure of individual internal service components meets the defined requirements (in the example, the correct decision on credit card applications). The task of the process-FMEA is finally the analysis of individual internal performance processes, such as the credit worthiness check required for credit card approval.

The **methodology of the three levels of the FMEA** is illustrated in Fig. 5.44 using the example of processing a credit card application in a bank (cf. also the following example). Here, four phases can be distinguished (Masing, 1995; Chen, 2013; Schmitt & Pfeifer, 2015):

FMEA type	Component/ Process	Function/ Purpose	Error effect	Error type	Error cause
System-FMEA	Private customers department	Credit cards "sell"	Resolution bank details	Not granted credit card	Incorrect credit worthiness data
Construction-FMEA	Credit card department	Decision about card applications	Incorrect credit worthiness data	Incorrect credit worthiness data	Reading error on screen
Process-FMEA	Credit worthiness check	Credit rating ensuring	Incorrect credit worthiness data	Reading error on screen	Unclear representation

Fig. 5.44 FMEA methodology using the example of processing a credit card application

- Error description
- Risk assessment
- Measures/solutions
- Result/assessment

As part of the **error description**, all potential sources of error within the service process are to be identified first. This requires the most comprehensive description possible of the overall system, the functions and processes of the service company. This also includes identifying the possible causes and consequences of a potential error. Within this procedure, all, i.e. also “unlikely” errors, must be taken into account in order to ensure the completeness of the procedure.

The task of the **risk assessment** is to quantify the severity of an error, the probability of its occurrence and the probability of its detection; the evaluation is carried out with point values between 1 and 10. The risk priority number (RPZ) results from the product of the three values and is the starting point for further activities. An exemplary application is illustrated in the following example.

FMEA when creating a credit card

a) **Error: Incorrect embossing of the credit cardholder’s name on the card**
Probability of occurrence (A): 3,
Error significance (B:) 9,
Probability of detection (E): 7,
$RPZ = A \times B \times E = 189.$

b) **Error: Unreadability of the magnetic stripe**
Probability of occurrence (A): 4,
Error significance (B): 5,
Probability of detection (E): 1,
$RPZ = A \times B \times E = 20.$

Based on the calculation of the FMEA value, the errors can be categorized and action recommendations can be derived, which errors are to be addressed with the highest priority. Special priority is given to those errors that

- result in a high overall risk priority number
- and/or have very high individual values.

In the example mentioned, error 1 would have very high priority, since the risk priority number of 189 is very high in comparison to error 2, as well as very high individual values in the error significance and the probability of detection. ◀

The example makes it clear that the avoidance of an incorrect name on a credit card has a particularly high priority. For the cardholder, this results in negative consequences (e.g.

unpleasant inquiries in stores, the card is not accepted at all), and the probability of discovery is low (high score).

To introduce **quality improvement measures** or the solution of the underlying problem, four different **approaches** are available (Kersten, 1994, pp. 477–478):

- Avoidance of the error cause(s)
- Reduction of the probability of occurrence of an error
- Reduction of the significance of the error
- Increasing the probability of error detection

From a quality point of view, those measures should be given preference in each case which avoid the occurrence of an error. If, for example, it was found in the course of the error source analysis that insufficient lighting hampers the readability of the application forms, the probability of occurrence of this error can be reduced by a better workplace design.

The **evaluation of success** is finally possible by means of a comparison of the risk priority number in the initial state and after improvements have been made. The actual result of the initiated measures can only be determined after the error-preventing measures or test processes have actually been implemented.

(3) Fishbone Analysis In addition, an identification of problem areas can be achieved by a Fishbone analysis. In a first step, all potential factors that can cause a certain quality deficiency in performance creation are to be written down. The determination of these determinants often take place via the brainwriting technique (Schmitt & Pfeifer, 2015, pp. 243–244; Frehr, 1994, p. 239). The determined factors are transferred to a diagram that is also known as **Ishikawa diagram**. The "head of the fish" is where the problem to be solved is entered, the "main bones" form the central dimensions that influence the problem (Munro-Faure & Munro-Faure, 1992, p. 205; Brüggemann & Bremer, 2012, p. 23).

It becomes clear that the Fishbone analysis is only suitable for problem analysis; specific solutions are not shown. For employees in the company, the Fishbone analysis provides a good basis for discussion when, for example, brainstorming or using other creativity techniques to think about what possible reasons are responsible for the bad image of credit institutions in terms of consulting quality in securities transactions (see Fig. 5.45). In this way, awareness and understanding of specific problems in the provision of services can be created for all participants; Main and secondary influencing variables and thus points of contact for quality improvement measures can be generated from this.

The **advantage** of the cause-effect diagram lies above all in the simple communicative representation of quality defects with the associated influencing variables (Frehr, 1994; Chen et al., 2010, p. 1158). The quality-related main problem is explained by the systematic collection of all conceivable causes and not only suppressed by the assumption of one or two influencing variables. The individual causes are not empirically

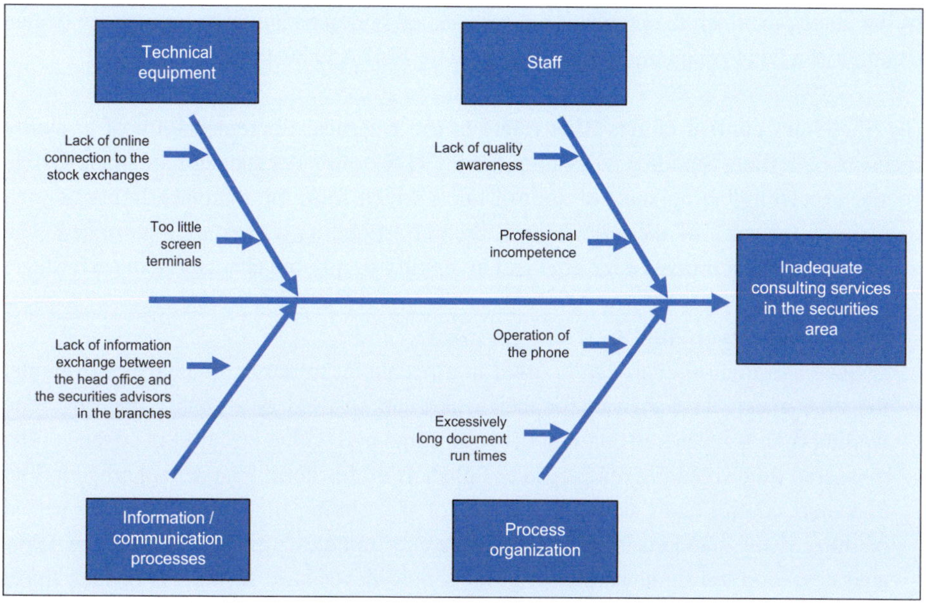

Fig. 5.45 Example of a Fishbone analysis in the field of financial services

determined, but rather developed within the framework of problem perception and solution techniques, such as brainstorming. The central **disadvantage** is above all to be seen in the fact that no specific problem solutions are given. Therefore, the approach is only suitable for analyzing possible sources and causes of a problem.

(4) Statistical Process Control The methods of statistical process control have so far been used predominantly in the area of material goods (Haller, 1998, 134 pp.; Schmitt & Pfeifer, 2015, pp. 226). They are exclusively based on objective data (e.g. transport times, number of complaints, customer waiting times). It is therefore a procedure that enables the service company to standardize processes to a large extent. Deviations and error rates are thereby reduced to a minimum. The goal is to explain whether it is possible to comply with previously defined standards in the current execution of the processes in the company. Another control point is whether the process meets the requirements at any time. If this is not the case, the process flow is often corrected.

According to Wood and Preece (1993), three **core methods** are distinguished from the multitude of statistical methods (Haller, 1998, 184 pp.; Ryan, 2011):

(a) Pareto Analysis This method is based on the fact that errors and mistakes can have different causes, which occur with different frequencies and whose consequences can be more or less severe. The aim of this analysis is to identify and eliminate the few error causes with serious consequences. For this purpose, the error causes are usually visualized in the form of bar charts, in which the problems are represented in descending order of urgency (for an example of a Pareto diagram, see also Fig. 5.39 for FMEA in Sect. 5.2.1).

(b) Shewhart control charts This refers to the graphical implementation of a quality measure over time. Significant deviations from a previously set standard are made visible by the exceeding/falling short of control lines, which form the tolerance limits of such deviations. For each of these deviations, the corresponding causes are determined. The aim is to monitor a process over time and to identify trends and changes at an early stage.

Use of a Shewhart control chart in a restaurant

A Shewhart control chart can be used in a restaurant to ensure quality. For example, the time guests have to wait for their order can be used as a quality measure. The waiting time is measured continuously and compared with a norm set in advance. The measured time should never exceed or fall short of this norm by a certain amount. The Shewhart control chart shows the course of the waiting times over a certain period of time. If the standard is, for example, twenty minutes waiting time and the tolerated deviation ten minutes, then a guest should never have to wait more than thirty or less than ten minutes for his food. If it takes too long for the food to come, the guest will be very angry. If the food is already on the table shortly after the order, the guest might assume that the dishes were not prepared fresh. If a deviation of more than ten minutes is measured, it is to be checked how this error came about. The Shewhart control chart makes it easy to see when and how often an intolerable deviation occurred. ◀

(c) Process Capability Analysis The goal of this analysis is the statistical evidence of how consistent the process is carried out, i.e. whether the result of a process lies within specified requirements. For example, a leisure center has defined a standard (service level) to be adhered to in terms of the reception of telephone calls. According to this, 98% of the calls are to be answered within three rings. It turns out that the actual waiting time significantly exceeded the specified standard (see Fig. 5.46). Only 59% of the calls were answered within a period of up to three rings.

The instruments of process control are presented in more detail in Chap. 13.

5.2.2 Employee-Oriented Measurement Approaches

The external and internal quality perception of individual employees is measured by using employee-oriented measurement approaches. These methods can include:

1. External quality measurement by employee surveys
2. Internal quality measurements

3. Operational suggestion system
4. Poka-Yoke methods

The following section provides a more detailed explanation of employee-oriented measurement approaches.

(1) External quality measurement by employee surveys Similar to customer-oriented, feature-oriented approaches, an external quality assessment can also be made from the perspective of employees using feature-oriented approaches. By adopting the customer perspective through employees, "false" management expectations regarding customer expectations of service quality can be revised (Zeithaml et al., 1992) and indications of the requirements for service quality, for example, by quality-oriented employee surveys from the perspective of customer contact personnel and managers. This way, in particular with regard to customer-employee- interactions, information can be collected that is relevant for quality planning of future services (Haist & Fromm, 1991, 75 pp.). The following **questions** are being focused on:

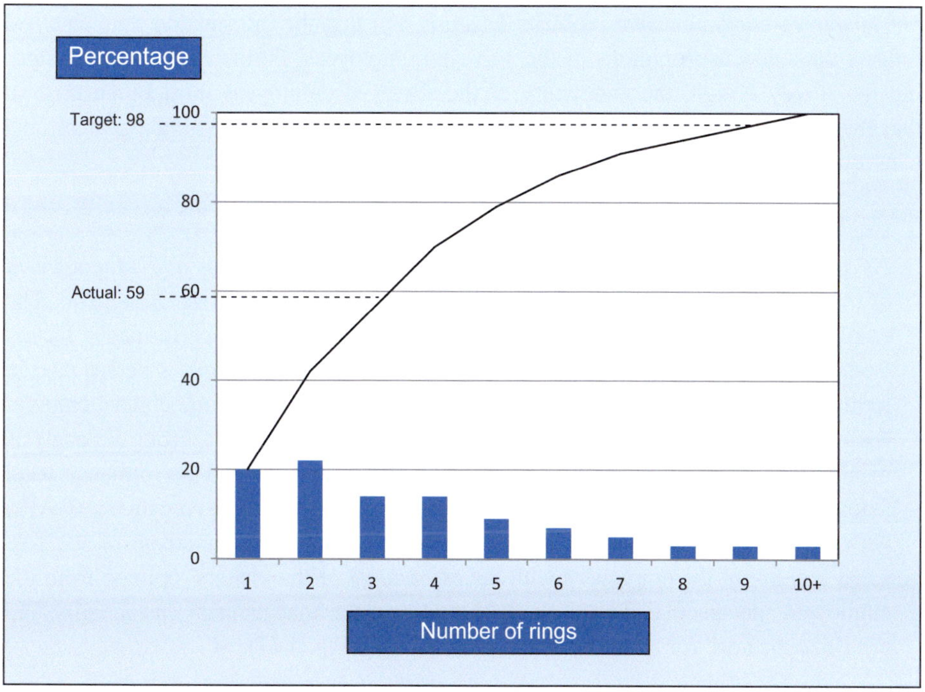

Fig. 5.46 Process capability analysis using the example of a leisure center

- Which aspects of service creation are considered particularly positively (negatively) by employees?
- How is the quality of the service created directly considered by employees?
- What knowledge do employees have on the subject of quality?
- To what extent is there a general awareness of quality among employees?

Such surveys are to be classified as an instrument of employee communication within the upward communication streams (Dotzler & Schick, 1995, p. 281; Domsch & Ladwig, 2013). However, their relevance in the context of inward-looking market research is still underestimated in service companies. To identify problems in the creation of services, Zeithaml et al. (1992) start from two central questions:

- "What is the biggest problem you face in your daily efforts to provide a high-quality offer to your customers?"
- "If you were chairman of your company for one day and were only allowed to make one decision to improve quality, what decision would you make?"

These questions are particularly helpful in terms of the need to prioritize the identification of the most important problem areas. It should be noted that such a procedure does not guarantee comprehensive problem detection and that the information thus obtained reflects the subjective opinions of the surveyed employees. Furthermore, when designing the survey design, the anonymity of the surveyed employees must be ensured so that they can answer the questions openly and without regard to possible professional consequences.

Employee survey of customer contact employees of a zoo

As part of a study on service quality, the *Chair of Marketing and Management of the University of Basel* carried out an employee survey with the *Basel Zoo*. The employees were asked to make a quality assessment of various performance factors from their own perspective as an employee and from the customer's perspective in terms of perspective taking. In this way, assumptions of the customer contact employees regarding the satisfaction of the zoo visitors can be revealed. From the analysis results, management can firstly identify findings on the general performance level from the perspective of the employees. On the other hand, discrepancies (GAPs) between the two perspectives can be determined from the comparison of the personal opinion of the employees with the assessment of the visitors' opinion from the employees' perspective. Based on these findings, the management can identify and prioritize the need for action to improve quality (see Fig. 5.47). ◄

(2) Internal quality measurements In service companies, the quality of internal performance has a significant effect on external quality towards customers. It is assumed

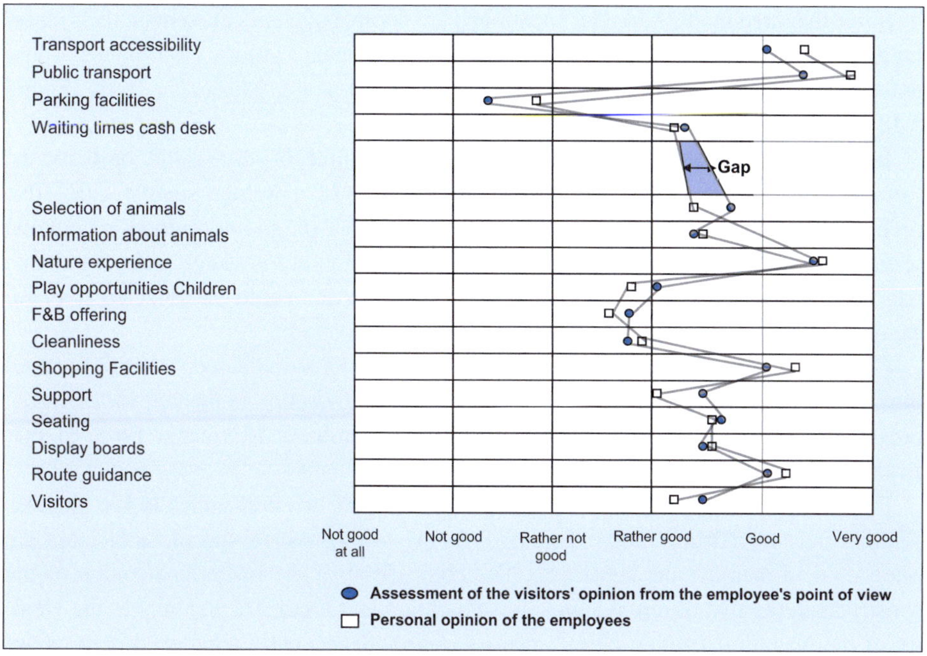

Fig. 5.47 GAP analysis based on a survey of zoo employees

that there are internal customer-supplier relationships within the company. Therefore, it is necessary that regular internal quality measurements, e.g. by means of surveys, are carried out with employees, i.e. internal customers and suppliers (Haist & Fromm, 1991, 75 pp.; Jun & Cai, 2010; Hadwich & Keller, 2015; Braun, 2016).

In the context of these internal customer surveys, the satisfaction of employees with the pre-services of internal suppliers, i.e. "direct" colleagues in their own department or their own office, but also employees of other offices or partner companies, is recorded (see also Sect. 8.4.1).

In order to determine internal service quality, **internal service barometers** are often used as measuring instruments (see also Sect. 14.1). In this context, the principle of the success chain of national customer barometers is transferred to internal services (see also Sect. 14.1). It is assumed that the assessment of internal services by internal customers affects internal customer satisfaction; likewise, that there is internal customer loyalty in the sense of a connection with certain internal providers, which results from the assessment of performance and/or customer satisfaction (Bruhn, 2004; Bruhn & Siems, 2004, pp. 563f.; Bruhn, 2010). Therefore, internal service quality, internal customer satisfaction and internal connection are seen as central model variables, with an effect of service quality on customer satisfaction and an effect of both variables on connection being specified.

Since the three model variables mentioned are complex, latent **constructs**, their measurement—analogous to the procedures within the National Customer Barometer—does not take place directly, but indirectly via indicators (Bruhn & Murmann, 1998). Figure 5.48 illustrates the idea of this approach.

In order to determine suitable **indicators**, which ultimately serve as the basis for the formulation of a questionnaire within the framework of a written survey, qualitative preliminary studies are carried out in which both providers and consumers of internal services were asked which criteria for quality perception are generally relevant and in which a connection with a provider of an internal service could be reflected (see, for example, Homburg & Giering, 1996, p. 11 for this procedure).

For the measurement of **customer satisfaction**, the three indicators "overall satisfaction", "satisfaction compared to expectations" and "satisfaction compared to the ideal" proven in numerous customer barometers are used (Bruhn & Murmann, 1998, 61 pp.; Bruhn, 2003).

Figure 5.49 shows, on the basis of several studies of two companies in the pharmaceutical industry (Bruhn, 2004; Bruhn & Siems, 2004), which specific indicators are often used to measure the constructs. The comparison of the operationalization of the constructs in the two different companies Pharma 1 and Pharma 2 also makes the validity of the operationalization clear, since—despite independently conducted preliminary studies and differences in the organizational structure of both companies—almost identical indicators have been found, which also come into question as possible indicators for all different internal services according to the preliminary studies.

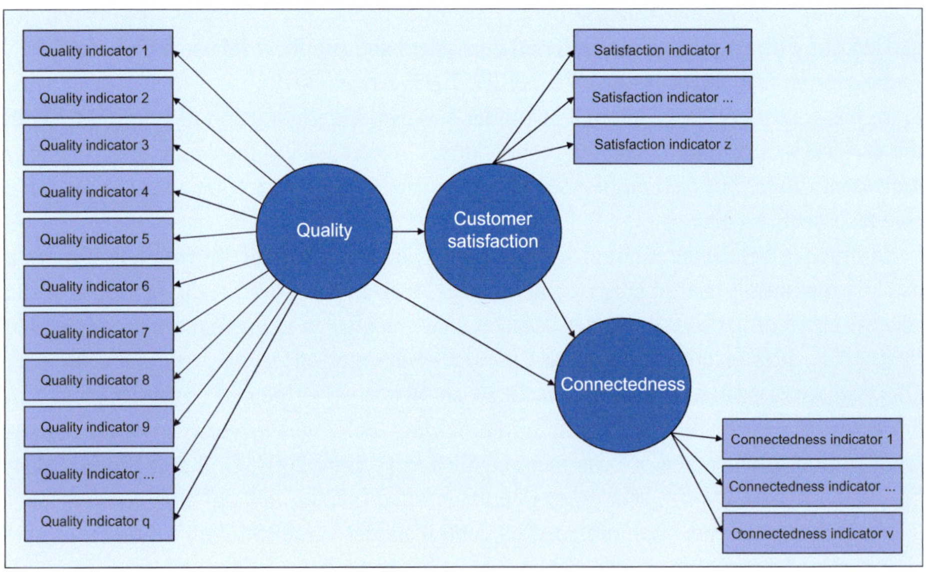

Fig. 5.48 Structure and measurement model of an internal service barometer. (Source: Bruhn & Siems, 2004, p. 563)

	Pharma 1	Pharma 2	
Quality	Competence	Competence	Professional competence
			Social competence
	Accessibility	Accessibility	
	Friendliness	Friendliness	
	Flexibility	Flexibility	
	-	Discretion/confidentiality	
	Reliability	Reliability	
	Speed	Time required for service provision	
		Compliance with deadlines	
	Reaction time	Response time	
	Value for money	Value for money	
	Cost transparency	Cost transparency	
	-	Scope of services offered	
	-	Congruence order/result	
	Individuality	Fulfillment of special customer requirements	
	-	Efficiency	
	Transparency range of services	Transparency Range of services	
	-	Continuity	
	Added value	Utilization of benefits (added value)	
	-	Quality of the service offering	
Customer satisfaction	Overall satisfaction	Overall satisfaction	
	Satisfaction in accordance with expectations	Satisfaction in accordance with expectations	
	Satisfaction in accordance with the ideal	Satisfaction in accordance with the ideal	
Connectedness	Re-election with freedom of choice	Re-election with freedom of choice	
	Recommendation intention	Recommendation intention	
	Thought of other provider or self-fulfillment of the task	Desire for other provider	
	Re-election next time	Re-election next time	
	Change in intensity of cooperation	Change in intensity of cooperation	
	Cross-buying intent	-	

Fig. 5.49 Indicators for the operationalization of the measurement constructs of an internal service barometer. (Source: Bruhn & Siems, 2004, p. 569)

The implementation of an internal service barometer or the measurement of the internal service quality primarily results in the evaluation of performance features of individual internal services. Based on the average values for the indicators, points of attack for improvement measures are defined. The following example illustrates the result of an internal service provider in the pharmaceutical industry, which is relatively poorly rated in comparison to the overall company.

Results of quality measurement of an internal service provider

The result of the internal service provider shows the average values of the evaluated indicators—the so-called Satisfaction Driver—on a scale of one to ten. The term "satisfaction driver" makes the connection between the performance indicators of the internal service and the satisfaction induced by the fulfillment of the respective indicators. For example, it can be seen that the flexibility of the employees is not only rated relatively poorly with 5.55, but also has the largest deviation from the average value for the entire company (7.85) (see Fig. 5.50). Here, urgent action is required, which speaks for training, incentive systems, etc. ◀

Detailed results of such internal service barometers are to be sent exclusively to the departments or branches concerned. The central results of all employee surveys, such as cross-company internal satisfaction indices, are to be published, for example, in the company's internal communication media. Here, a "ranking" of the best departments and branches is often useful for reasons of recognition and role model function.

Of particular importance is also the regular determination of the general employee satisfaction, i.e. the satisfaction of employees with their workplace. Since the satisfaction or dissatisfaction of employees with their daily tasks can significantly affect the quality of service provision to customers, it seems sensible to regularly gain insights from employees about their activities, the relationship with colleagues, etc.. Ideally, all employees of the different branches of the service company are periodically surveyed, but for cost reasons, a restriction to representative surveys at selected branches is possible. However, this is rather to be seen as a minimum requirement in the context of comprehensive quality management. In the long term, it is advisable to introduce a kind of **"Internal Customer Satisfaction Tracking System"** in order to measure, monitor and optimize the quality of internal services on a regular basis. National service companies,

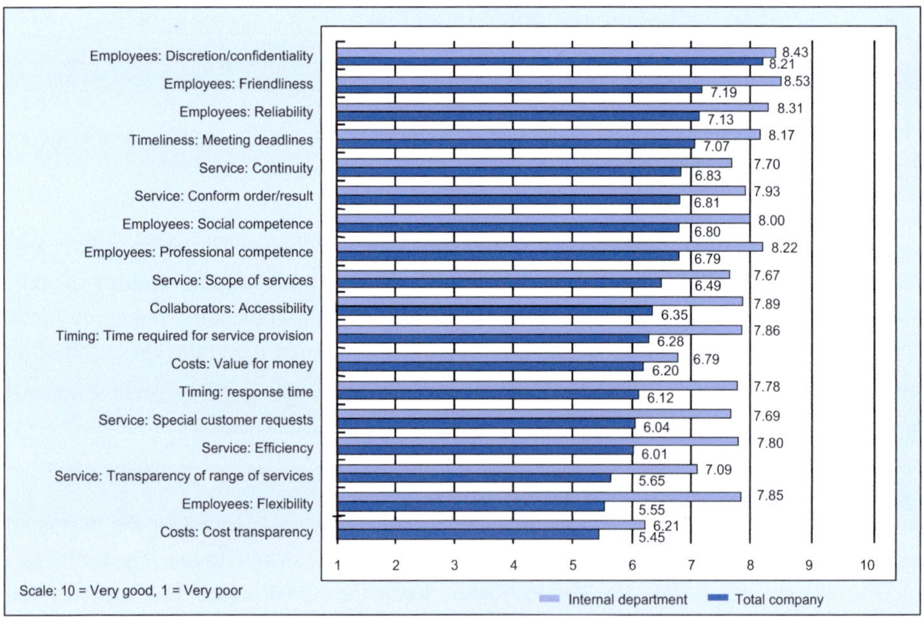

Fig. 5.50 Mean values of the satisfaction drivers for an internal service provider. (Source: Bruhn, 2004, p. 289)

such as *AVIS Deutschland, Deutsche Bahn* or *Deutsche Lufthansa,* already use such systems to continuously capture expectations, attitudes, satisfaction or dissatisfaction of their employees.

(3) Company suggestion system Another employee-oriented approach is the company suggestion system (Merz & Bichler, 1994; Brinkmann & Simon, 1995; Schulte, 2012, 83 pp.), which provides detailed information about company-related problems and solutions from an employee perspective as a central instrument of quality determination. The hearing of employee suggestions by the branch manager with regard to technical and non-technical improvement areas throughout the entire service chain can—in comparison to comprehensive employee surveys—be associated with relatively low personnel and financial expenditure and is therefore generally recommended for service companies.

The company suggestion scheme is often seen as a variant of idea management, in which both the active involvement of employees, the improvement of service quality and rationalizations and minor improvements are in the focus (Anic, 2000). A study by the *Deutsches Institut für Betriebswirtschaft (dib)* showed in 2006 for German companies a participation rate in the business proposal of 22.1%, but with significant fluctuations. Industrial companies were involved with 30%, while non-industrial organizations were only 10% engaged in the proposal (dib, 2006).

To increase motivation of employees to submit suggestions for quality improvement of the service provider, **specific incentives** are to be created and suggestions are to be rewarded accordingly. Furthermore, for practical and credibility reasons, it is necessary to ensure that good employee suggestions are implemented as quickly as possible in the company and announced company-wide as part of internal communication, e.g. in the employee newspaper or in the company intranet (Meffert et al., 2018, 236 pp.).

In principle, various forms of processing of incoming suggestions are conceivable. The *Cooperative for Health and Welfare Work in Germany* distinguishes between central and decentralized assessment processes in the company suggestion system. In a **central assessment process** there is a central commission which collects and evaluates all suggestions. This commission decides whether a suggestion is accepted or rejected. The reward of the employee and the instruction for implementation to the superior is then carried out by the management (cf. Fig. 5.51).

In the case of a **decentralized assessment process**, however, the immediate superior of the employee decides directly whether the improvement proposal submitted is to be accepted and implemented. The representatives in the company are then only informed by the superior of the acceptance of the improvement proposal (cf. Fig. 5.52).

The **advantage** of decentralized processing lies above all in the fact that, through the direct involvement of corporate management, greater importance is attached to the proposal system. The **disadvantage** lies above all in the very time-consuming process. This prevents a quick implementation of improvement proposals. However, with both processing forms it is to be emphasized—especially in the event of negative feedback in the sense of non-implementation of the proposal—that a detailed feedback is given on the

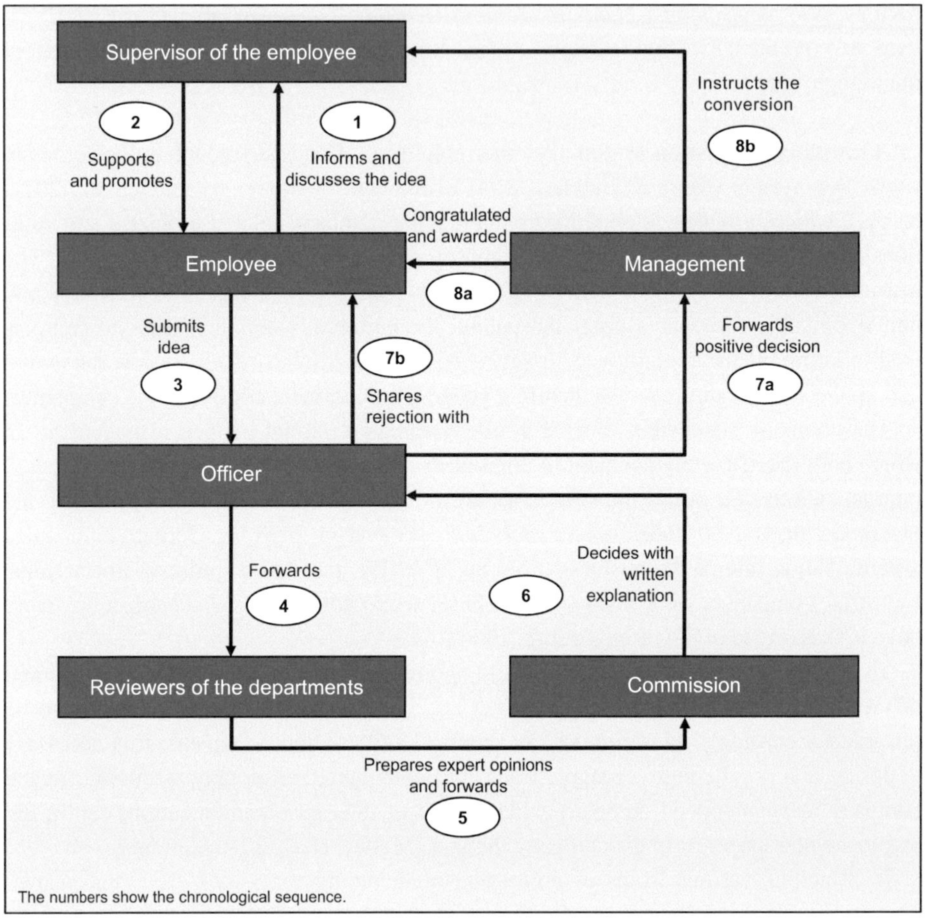

Fig. 5.51 Central processing of the company suggestion system. (Source: BGW, 2008, p. 26)

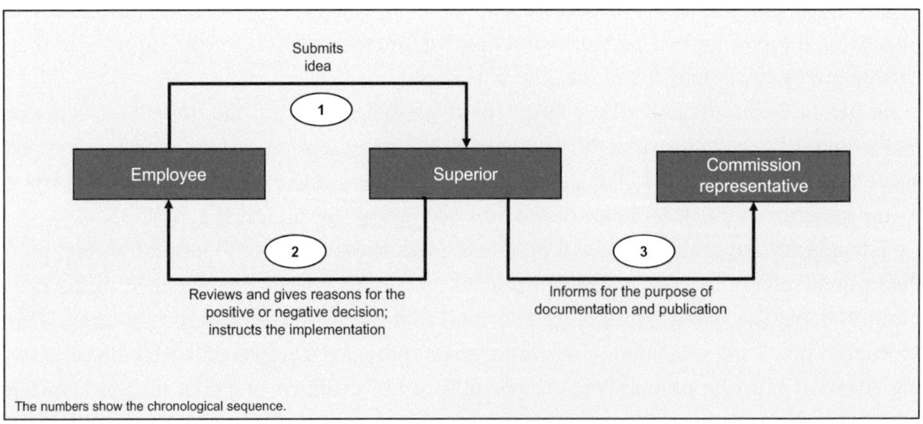

Fig. 5.52 Decentralized processing of the company suggestion scheme (Source: BGW, 2008, p. 27)

improvement proposal. If the rejection is not or only insufficiently justified, there is a risk that this will lead to a demotivation on the part of the employees and thus to a reduction in future improvement proposals.

(4) Poka-Yoke Method Poka Yoke is another one of the employee-related methods. This instrument is related to Kaizen and also has the permanent quality improvement of the services to be provided for sale as its content. The focus is on the **prevention** of intended as well as unintended and random errors with the aim of realizing the zero-error principle (Bühner, 1993, p. 37; Flood, 1993, 28 pp.; Shingo, 1995; Töpfer & Günther, 2009).

The approach developed by Shingo (1991) envisages, in a first step, to analyze and identify each individual phase of service creation where and when errors can occur. The achievement of the claim of a permanent improvement of the quality of the created service requires a source inspection integrated into the individual service creation phases (Bühner, 1993). This error source inspection is required for services in particular due to the Uno-Actu principle, since the rectification of a defective service is usually only possible during the service creation process. A blueprint (see Sect. 5.1.2.2) often provides valuable assistance in the error source inspection to identify customer contact points in which there are interactions between service providers and consumers, and possible errors. Two **types of errors** are distinguished. On the one hand, there are errors that are due to forgetfulness and are only noticed by the employees during their creation, on the other hand, those whose creation the employees overlook (Shingo, 1991; Brinkmann, 1993; Chase, & Stewart, 1994; Shahin & Ghasemaghaei, 2010). In a further step, the possible causes of error are determined and rules for error avoidance, so-called Poka-Yoke, are formulated. Fig. 5.53 shows, using the example of a car repair shop, the Poka-Yoke procedure.

Criticism of this approach is to be criticized in that the social component is both in terms of the relationships between the service users and the employees as well as with regard to the internal processes not taken into account. Thus, the particularly important human factor in service transactions is reduced to purely mechanistic error-avoiding elements (Flood, 1993; Habibbhai & Deshpande, 2014).

The presentation of the approaches to measuring service quality shows that a service company has a variety of methods available to measure the quality of its performance and to identify and avoid quality-reducing errors or problem areas in the creation process. These approaches can be used in the context of quality management of service companies for different purposes. This is clear from the Chap. 6, 7 and 8 in which the concepts and instruments of quality management are described, including the presented measurement approaches.

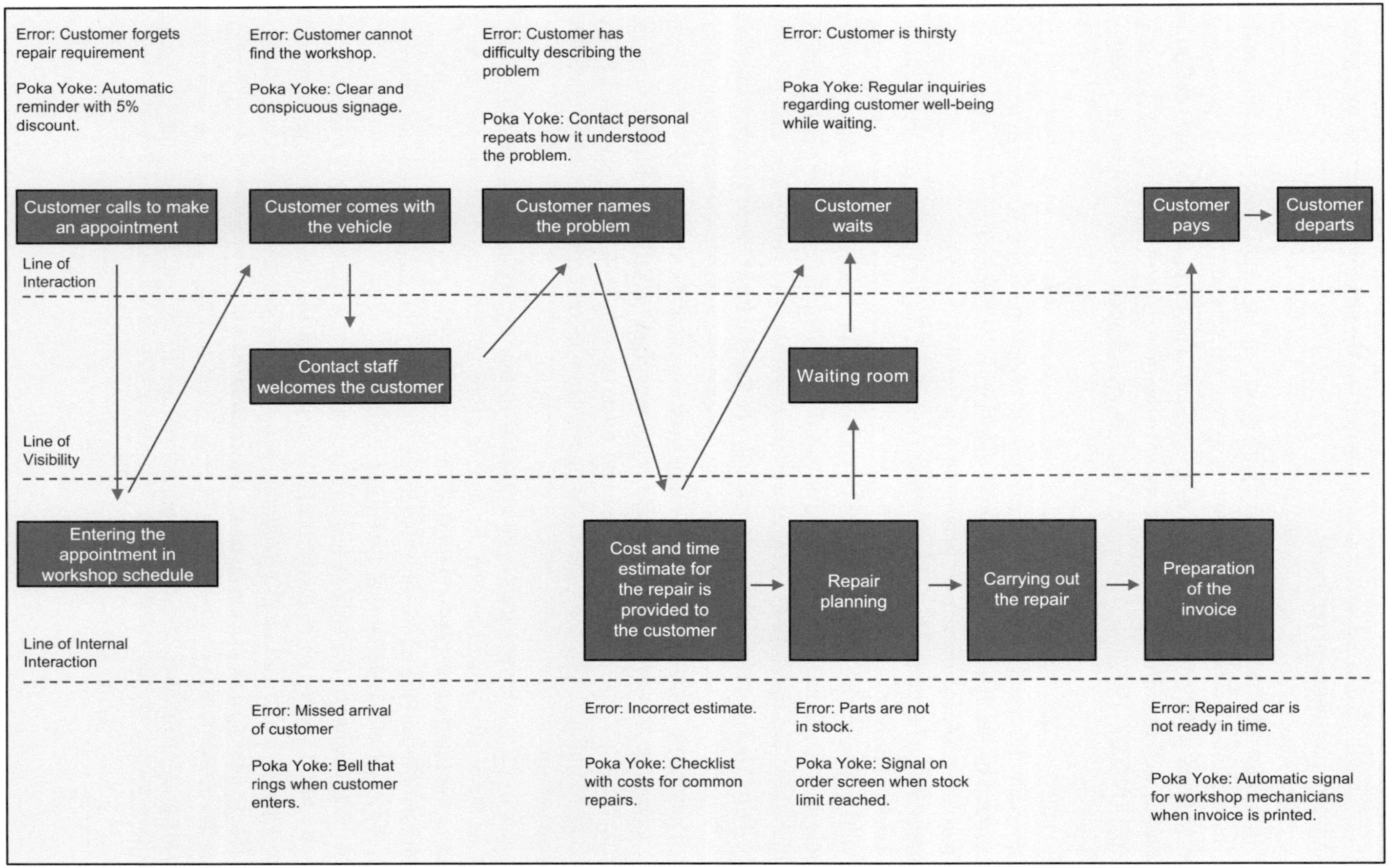

Fig. 5.53 Methodology of the Poka-Yoke method using the example of a car repair shop. (Source: Chase & Stewart, 1994, p. 40 f.)

References

Albrecht, K. (1993). *Total Quality Service. Das einzige, was zählt*. Düsseldorf: Econ.

Alegre, J., & Garau, J. (2011). The factor structure of tourist satisfaction at sun and sand destinations. *Journal of Travel Research, 50*(1), 78–86.

Ammermann, M. (1998). *The root cause analysis handbook. A simplified approach to identifying, correcting and reporting workingplace errors*. New York: Productivity Press.

Andersen, B., & Fagerhaug, T. (2006). *Root cause analysis: Simplified tools and techniques*. Milwaukee: ASQ Quality Press.

Anic, D. (2000). *Ideenmanagement*. Oldenburg: Nomos.

Backhaus, K., Erichson, B., Plinke, W., & Weiber, R. (2018). *Multivariate Analysemethoden. Eine anwendungsorientierte Einführung* (15th Edn.). Berlin: Springer.

Balcombe, K., Fraser, I., & Harris, L. (2009). Consumer willingness to pay for in-flight service and comfort levels: A choice experiment. *Journal of Air Transport Management, 15*(5), 221–226.

BEG. (2019). Qualitätsranking von Bayerischen Eisenbahngesellschaften. http://www.beg.bahnland-bayern.de. Zugegriffen: 4. Apr. 2019.

Benkenstein, M. (1993). Dienstleistungsqualität. Ansätze zur Messung und Implikationen für die Steuerung. *Zeitschrift für Betriebswirtschaft*, *63*(11), 1095–1116.

Benkenstein, M., & von Stenglin, A. (2006). Prozessorientiertes Qualitätscontrolling von Dienstleistungen. In M. Bruhn & B. Strauss (Eds.), *Dienstleistungscontrolling. Forum Dienstleistungsmanagement* (pp. 56–70). Wiesbaden: Gabler.

Berger, C., Blauth, R., Boger, D., Bolster, C., Burchill, G., Du Muchel, W., Pouliout, F., Richter, R., Rubinoff, A., Shen, D., Timko, M., & Walden, D. (1993). Kano's method for understanding customer-defined quality. *Journal of the Japanese Society for Quality Control*, *23*(3), 3–35.

Berufsgenossenschaft für Gesundheitsdienst und Wohlfahrtspflege (BGW) (2008). *Company suggestion scheme als Ideenmanagement*. Hamburg.

Birkelbach, R. (1993). *Qualitätsmanagement in Dienstleistungscentern*. Bern: Lang.

Bitner, M. J., Booms, B. H., & Tetreault, M. S. (1990). The service encounter. Diagnosing favorable and unfavorable incidents. *Journal of Marketing*, *54*(1), 71–84.

Bitner, M. J., Ostrom, A. L., & Morgan, F. N. (2008). Service blueprinting: A practical technique for service innovation. *California Management Review*, *50*(3), 66–94.

Böhm, D. N., Fischl, C., & Rennhak, C. (2006). Kundenzufriedenheitsmessung mittels Mystery Shopping. In C. Rennhak (Eds.), *Herausforderung Kundenbindung* (pp. 87–93). Wiesbaden: Springer.

Bothe, S. (2013). *Servicequalität von Verlagen, Projektbericht*. Hohenheim.

Brady, M. K., Cronin, J. J., & Brand, R. R. (2002). Performance-only measurement of service quality: A replication and extension. *Journal of Business Research*, *55*(1), 17–31.

Brandon-Jones, A., & Silvestro, R. (2010). Measuring internal service quality: Comparing the gapbased and perceptions-only approaches. *International Journal of Operations & Production Management*, *30*(12), 1291–1318.

Brandt, R. D. (1987). A procedure for identifying value-enhancing service components using customer satisfaction survey data. In C. Suprenant (Eds.), *Add value to your service* (pp. 61–65). Chicago.

Brandt, R. D. (1988). How service marketers can identify value-enhancing service elements. *Journal of Services Marketing*, *2*(3), 35–41.

Brandt, D. R., & Scharioth, J. (1998). *Attribute life cycle analysis. Alternatives to the kano method*. ESOMAR Marketing Research Congress, Istanbul.

Braun, C. (2016). *Komplexität interner Dienstleistungen: Konzeptualisierung, Messung und Integration in ein Wirkungsmodell*. Wiesbaden: Springer.

Brinkmann, E. P. (1993). Ausgewählte japanische Methoden des Qualitätsmanagements. Teil 2: Pokayoke mit Fehlerquelleninspektion. In W. Hansen, H. H. Jansen, & G. F. Kamiske (Eds.), *Qualitätsmanagement im Unternehmen. Grundlagen, Methoden und Werkzeuge, Praxisbeispiele, Loseblatt-Sammlung, Kapitel 04.16* (pp. 1–10). Berlin: Springer.

Brinkmann, E. P., & Simon, A. (1995). Qualitätssteigerung durch Partizipation der Mitarbeiter. Teil 2: Company suggestion scheme. Grundzüge. In W. Hansen, H. H. Jansen, & G. F. Kamiske (Eds.), *Qualitätsmanagement im Unternehmen. Grundlagen, Methoden und Werkzeuge, Praxisbeispiele, Loseblatt-Sammlung, Kapitel 03.08* (pp. 1–27). Berlin: Springer.

Brüggemann, H., & Bremer, P. (2012). Total Quality Management (TQM). In H. Brüggemann, & P. Bremer (Eds.), *Grundlagen Qualitätsmanagement. Von den Werkzeugen über Methoden zum TQM* (pp. 178–200). Wiesbaden: Vieweg und Teubner.

Bruhn, M. (1982). *Konsumentenzufriedenheit und Beschwerden. Erklärungsansätze und Ergebnisse einer empirischen Untersuchung in ausgewählten Konsumbereichen*. Frankfurt a. M: Lang.

Bruhn, M. (1999). Verfahren zur Messung der Qualität interner Dienstleistungen. Ansätze für einen Methodentransfer aus dem (externen) Dienstleistungsmarketing. In M. Bruhn (Eds.), *Internes Marketing. Schnittstellen zwischen dem Marketing- und Personalmanagement* (2nd Edn., pp. 537–575). Wiesbaden: Gabler.

Bruhn, M. (2003). Internal service barometers: Conceptualization and empirical results of a pilot study in Switzerland. *European Journal of Marketing*, *37*(9), 1187–1204.

Bruhn, M. (2004). Interne Servicebarometer als Instrument interner Kundenorientierung. Messung und Steuerung der Qualität und Zufriedenheit interner Dienstleistungen. *Marketing ZFP*, *26*(4), 282–294.

Bruhn, M. (2010). Das Konzept des Internen Servicebarometers – Bestandsaufnahme, Methodik und empirische Befunde. In M. Bruhn & B. Stauss (Eds.), *Serviceorientierung im Unternehmen. Forum Dienstleistungsmanagement* (pp. 297–322). Wiesbaden: Gabler.

Bruhn, M. (2012). *Marketing für Nonprofit-Organisationen. Grundlagen, Konzepte, Instrumente* (2nd Edn.). Stuttgart: Kohlhammer.

Bruhn, M. (2013). *Servicequalität – Konzepte und Instrumente für die perfekte Dienstleistung*. München: Beck.

Bruhn, M. (2014). *Integrierte Unternehmens- und Markenkommunikation: Strategische Planung und operative Umsetzung* (6th Edn.). Stuttgart: Schäffer-Poeschel.

Bruhn, M., & Hennig, K. (1993). Selektion und Strukturierung von Qualitätsmerkmalen. Auf dem Weg zu einem umfassenden Qualitätsmanagement für Kreditinstitute. *Jahrbuch der Absatz und Verbrauchsforschung*, *39*(3), 214–238.

Bruhn, M., & Murmann, B. (1998). *Nationale Kundenbarometer. Messung von Qualität und Zufriedenheit. Methodenvergleich und Entwurf eines Schweizer Kundenbarometers*. Wiesbaden: Gabler.

Bruhn, M., & Siems, F. (2004). Interne Servicebarometer zur Messung und Verbesserung von internen Dienstleistungen – Ein Erfahrungsbericht aus der Pharma-Branche. In H. Hippner & K. D. Wilde (Eds.), *Management von CRM-Projekten. Handlungsempfehlungen und Branchenkonzepte* (pp. 559–583). Wiesbaden: Gabler.

Bruhn, M., Batt, V., Hadwich, K., & Meldau, S. (2010). Messung der Qualität in Dienstleistungscentern – Am Beispiel eines Flughafens. *Zeitschrift für Betriebswirtschaft, 80*(4), 351–382.

Bühner, R. (1993). *Der Mitarbeiter im Total Quality Management*. Düsseldorf: VDI.

Chase, R. B., & Stewart, D. M. (1994). Make your service fail-safe. *Sloan Management Review*, *35*(3), 35–44.

Chatterjee, S., & Chatterjee, A. (2005). Prioritization of service quality parameters based on ordinal responses. *Total Quality Management & Business Excellence, 16*(4), 477–489.
Chen, S.-H. (2013). Integrated analysis of the performance of TQM tools and techniques: A case study in the Taiwanese motor industry. *International Journal of Production Research, 51*(4), 1072–1083.
Chen, H. C., Chen, K. S., Chang, T. L., & Hsu, C. H. (2010). An application of six sigma methodology to enhance leisure service quality. *Quality & Quantity, 44*(6), 1151–1164.
Chu-Mei, L. (2005). The multidimensional and hierarchical structure of perceived quality and customer satisfaction. *International Journal of Management, 22*(3), 426–435.
Commerzbank (2016). Initiierung eines Bankwechsels. https://www.google.de/search?q=Wechsel+100+Euro+Commerzbank&ie=utf-8&oe=utf-8&gws_rd=cr&ei=4SuiV-ruhFcWbsAH0p7PICA. Zugegriffen: 22. Jan. 2016.
Cronin, J. J., & Taylor, S. A. (1992). Measuring service quality. A reexamination and extension. *Journal of Marketing*, *56*(3), 55–68.
Cronin, J. J., & Taylor, S. A. (1994). SERVPERF versus SERVQUAL. Reconciling performance-based and perceptions-minus-expectations measurement of service quality. *Journal of Marketing*, *58*(1), 125–131.
dib. (2006). http://dib.de.
Domsch, M. E., & Ladwig, D. H. (2013). Mitarbeiterbefragung – Stand und Entwicklungen. In M. E. Domsch & D. H. Ladwig (Eds.), *Handbuch Mitarbeiterbefragung* (3rd Edn., pp. 3–26). Berlin: Springer.
Dotzler, H. J., & Schick, S. (1995). Systematische Mitarbeiterkommunikation als Instrument der Qualitätssicherung. In M. Bruhn, & B. Stauss (Eds.), *Dienstleistungsqualität. Konzepte – Methoden – Erfahrungen* (2nd Edn., pp. 277–293). Wiesbaden: Gabler.
Drees, N., & Schiller, S. (2003). Mystery Shopping – Ein Instrument zur systematischen Optimierung von Kundenzufriedenheit im Dienstleistungsbereich. In U. Kamenz (Eds.), *Applied Marketing* (pp. 159–172). Berlin: Springer.
Estorilio, C., & Posso, R. K. (2010). The reduction of irregularities in the use of „process FMEA". *International Journal of Quality & Reliability Management, 27*(6), 721–733.
Faßnacht, M., & Koese, I. (2006). Quality of electronic services. *Journal of Service Research, 9*(1), 19–37.
Fishbein, M. (1963). An investigation of the relationship between beliefs about an object and the attitude toward that object. *Human Relations, 16*(3), 233–239.
Flanagan, J. C. (1954). The critical incidence technique. *Psychological Bulletin, 51*(4), 327–358.
Flood, R. L. (1993). *Beyond TQM*. Canada: Wiley.
Franklin, B. D., Shebl, N. A., & Barber, N. (2012). Failure mode and effects analysis: too little for too much? *BMJ Quality and Safety, 21*(7), 607–611.
Frehr, H.-U. (1994). *Total Quality Management. Unternehmensweite Qualitätsverbesserung* (2nd Eds.). München: Hanser.
Gierl, H., & Bartikowski, B. (2003). Ermittlung von Satisfiers, Dissatisfiers und Criticals in der Zufriedenheitsforschung. *Der Markt, 42*(164), 13–34.
Globalpark. (2004). Bestimmung der Preisbereitschaft von Bankkunden. http://www.globalpark.de. Zugegriffen: 16. Juli 2012.
Goerlich, B. (1994). Gebot der Stunde. *Wirtschaftswoche*, *48*(10), 95–96.
Golder, P. N., Mitra, D., & Moorman, C. (2012). What is quality? An integrative framework of processes and states. *Journal of Marketing, 76*(4), 1–23.
González, M. E. A., Comesaña, L. R., & Brea, J. A. F. (2007). Assessing tourist behavioral intentions through perceived service quality and customer satisfaction. *Journal of Business Research, 60*(2), 153–160.

Gounaris, S. (2005). Measuring service quality in B2B services. An evaluation of the SERVQUAL scale vis-à -vis the INDSERV scale. *Journal of Services Marketing, 19*(6), 421–435.

Habibbhai, A. A., & Deshpande, V. A. (2014). A review of various tools and techniques for lead time reduction. *International Journal of Engineering Development and Research, 2*(1), 1159–1164.

Hadwich, K., & Keller, C. (2015). Interne Servicequalität in Unternehmen: eine empirische Untersuchung der Einflussfaktoren und Auswirkungen. *Zeitschrift für betriebswirtschaftliche Forschung, 67*(2), 170–205.

Haist, F., & Fromm, H. (1991). *Qualität im Unternehmen. Prinzipien, Methoden, Techniken* (2nd Edn.). München: Hanser.

Haller, S. (1998). *Beurteilung von Dienstleistungsqualität. Dynamische Betrachtung des Qualitätsurteils im Weiterbildungsbereich* (2nd Edn.). Wiesbaden: Gabler.

Hanjoon, L., Delene, L. M., Bunda, M. A., & Chankon, K. (2000). Methods of measuring health-care service quality. *Journal of Business Research, 48*(3), 233–246.

Hennig-Thurau, T., & Hansen, U. (2001). Kundenartikulation im Internet. *Die Betriebswirtschaft, 61*(5), 560–580.

Hennig-Thurau, T., Gwinner, K. P., Walsh, G., & Gremler, D. D. (2004). Electronic-word-of-mouth via consumer-opinion platforms: What motivates consumers to articulate themselves on the internet. *Journal of Interactive Marketing, 18*(1), 38–52.

Hentschel, B. (1999). Multiattributive Messung von Dienstleistungsqualität. In M. Bruhn, & B. Stauss (Eds.), *Dienstleistungsqualität. Konzepte – Methoden – Erfahrungen* (3rd Edn., pp. 289–320). Wiesbaden: Gabler.

Herzog, L., & Luthe, M. (2010). Liebe auf den zweiten Blick – Vom souveränen Umgang mit Hotelbewertungen. In D. Amersdorffer, F. Bauhuber, R. Egger, & J. Oellrich (Eds.), *Social Web im Tourismus. Strategien – Konzepte – Einsatzfelder* (pp. 161–168). Berlin: Springer.

Tripadvisor. (2023) https://www.tripadvisor.com/Hotel_Review-g60763-d3529145-Reviews-Holiday_Inn_Express_Manhattan_Times_Square_South_an_IHG_Hotel-New_York_City_New_York.html

HolidayCheck. (2019). Kundenbewertung eines Hotels auf der Bewertungsplattform HolidayCheck. https://www.holidaycheck.de/hr/bewertungen-mon-port-hotel-spa/3b28babd-7425-33f6-9015-8adf9cab38f8. Zugegriffen: 4. Apr. 2019.

Homburg, C., & Giering, A. (1996). Konzeptualisierung und Operationalisierung komplexer Konstrukte. Ein Leitfaden für die Marketingforschung. *Marketing ZFP, 18*(1), 5–24.

Huber, F., Krönung, S., Meyer, F., & Vollmann, S. (2011). Akzeptanz von Bewertungsportalen als Basis von Electronic Word-of-Mouth. *Marketing Band, 54*, 1–2.

Jun, M., & Cai, S. (2010). Examining the relationships between internal service quality and its dimensions, and internal customer satisfaction. *Total Quality Management & Business Excellence, 21*(2), 205–223.

Jüttner, U., Schaffner, D., Windler, K., & Maklan, S. (2013). Customer service experiences: Developing and applying a sequentialincident laddering technique. *European Journal of Marketing, 47*(5/6), 738–769.

Kang, K. H., Stein, L., Heo, C. Y., & Lee, S. (2012). Consumers' willingness to pay for green initiatives of the hotel industry. *International Journal of Hospitality Management, 31*(2), 564–572.

Kano, N. (1984). Attractive quality and must-be quality. *Journal of the Japanese Society for Quality Control, 14*(2), 39–48.

Kartal, A., Doerfel, S., Roßnagel, A., & Stumme, G. (2011). Privatsphären- und Datenschutz in Community-Plattformen: Gestaltung von Online-Bewertungsportalen. *Informatik, 412*, 1–15.

Kersten, G. (1994). Fehlermöglichkeits- und -einflussanalyse (FMEA). In W. Masing (Eds.), *Handbuch Qualitätsmanagement* (3rd Edn., pp. 469–490). München/Wien: Hanser.

Kroeber-Riel, W., & Gröppel-Klein, A. (2013). *Konsumentenverhalten* (10th Edn.). München: Vahlen.

Liljander, V., & Strandvik, T. (1992). *Estimating zones of tolerance in perceived service quality and perceived service value, Meddelanden Working Papers No. 247*. Helsinki.

Lindqvist, L. J. (1987). Quality and service value in the service consumption. In C. Suprenant (Eds.), *Add value to your service* (pp. 17–20). Chicago.

Lupo, T. (2013). A fuzzy SERVQUAL based method for reliable measurement of education quality in Italian higher education Area. *Expert Systems with Applications, 40*(17), 7096–7110.

Madu, C. N., & Kuei, C.-H. (1995). *Strategic total quality management. Corporate performance and product quality*. Westport: Praeger.

Masing, W. (1995). Planung und Durchsetzung der Qualitätspolitik im Unternehmen. Zentrale Prinzipien und Problembereiche. In M. Bruhn, & B. Stauss (Eds.), *Dienstleistungsqualität. Konzepte – Methoden – Erfahrungen* (2nd Edn., pp. 239–253). Wiesbaden: Gabler.

Materna. (2012). Studie zum Nutzen des Beschwerdemanagements. http://www.pressetext.com/news/20051108012. Zugegriffen: 16. Nov. 2012.

Matzler, K., Sauerwein, E., & Stark, C. (2009). Methoden zur Identifikation von Basis-, Leistungs- und Begeisterungsfaktoren. In H. H. Hinterhuber & K. Matzler (Eds.), *Kundenorientierte Unternehmensführung. Kundenorientierung – Kundenzufriedenheit – Kundenbindung* (6th Edn., pp. 319–344). Wiesbaden: Gabler.

Medberg, G., & Heinonen, K. (2014). Invisible value formation: A netnography in retail banking. *International Journal of Bank Marketing, 32*(6), 590–607.

Meffert, H., Bruhn, M., & Hadwich, K. (2018). *Dienstleistungsmarketing. Grundlagen, Konzepte, Methoden* (9th Edn.). Wiesbaden: Springer.

Merz, E., & Biehler, B. (1994). *Company suggestion scheme – Professionell und wirksam* (2nd Edn.). Landsberg am Lech: Moderne Industrie.

Meyer, A., & Kantsperger, R. (2005). *Mitarbeiterperspektive: Entwicklung, Anwendung, Implikationen. Call Center Benchmarking*. Wiesbaden: Gabler.

Michalski, S. (2002). *Kundenabwanderungs- und Kundenrückgewinnungsprozesse. Eine theoretische und empirische Untersuchung am Beispiel von Banken*. Wiesbaden: Gabler.

Mudie, P., & Cottam, A. (1999). *The management and marketing of services* (2nd Edn.). Oxford: Routledge.

Mukherjee, A., & Nath, P. (2005). An empirical assessment of comparative approaches to service quality measurement. *Journal of Services Marketing, 19*(3), 174–184.

Munro-Faure, L., & Munro-Faure, M. (1992). *Implementing total quality management*. Boston: Trans-Atlantic.

Nemati, B., Gazor, H., MirAshrafi, K. N., & Ameleh, K. N. (2012). Analyzing e-service quality in service-based website by E-SERVQUAL. *Management Science Letters, 2*(2), 727–734.

Nerdinger, F. W. (1994). *Zur Psychologie der Dienstleistung. Theoretische und empirische Studien zu einem wirtschaftspsychologischen Forschungsgebiet*. Stuttgart: Schäffer-Poeschel.

Newman, K. (2001). Interrogating SERVQUAL. A critical assessment of service quality measurement in a high street retail bank. *International Journal of Bank Marketing, 19*(3), 126–139.

Nürnberg, V., & Schneider, B. (2014). Analyse der Kundenorientierung in Krankenhäusern. In V. Nürnberg, & B. Schneider (Eds.), *Kundenmanagement im Krankenhaus: Service – Qualität – Erreichbarkeit* (pp. 55–67). Wiesbaden: Springer.

O'Reilly, T. (2006). Nutzergenerierte Kommunikation. http://www.oreilly.de/artikel/web20.html. Zugegriffen: 22. Jan. 2010.

o. V. (1. Dez. 1994). Die Stiftung Warentest will künftig mehr Dienstleistungen prüfen. *Frankfurter Allgemeine Zeitung*, 19.

Oliver, R. L. (1980). A cognitive model of the antecedents and consequences of satisfaction decisions. *Journal of Marketing Research, 17*(4), 460–469.

Olorunniwo, F., & Hsu, M. K. (2006). A typology analysis of service quality, customer satisfaction and behavioral intentions in mass services. *Managing Service Quality: An International Journal, 16*(2), 106–123.

Parasuraman, A., Zeithaml, V. A., & Berry, L. L. (1985). A conceptual model of service quality and its implications for future research. *Journal of Marketing, 49*(1), 41–50.

Parasuraman, A., Zeithaml, V. A., & Malhotra, A. (2005). ES-QUAL. A multiple-item scale for assessing electronic service quality. *Journal of Service Research, 7*(3), 213–233.

Park, H., Geum, Y., & Park, Y. (2015). A dual quality function deployment approach for benchmarking service quality. *Total Quality Management & Business Excellence, 26*(5–6), 569–582.

Pepels, W. (2013). Qualitäts- und Zufriedenheitsmessung als CRM-Basis. In S. Helmke, M. F. Uebel, & W. Dangelmaier (Eds.), *Effektives Customer Relationship Management. Instrumente – Einführungskonzepte – Organisation* (5th Edn., pp. 23–49). Wiesbaden: Springer Gabler.

PepsiCo. (2012). Pepsi NEXT „Facebook taste test". http://www.pepsico.com/PressRelease/Pepsi-NEXT-Unveils-First-Internet-Taste-Test-to-Inspire-Cola-Lovers-to-Drink-It-04032012.html. Zugegriffen: 6. März 2013.

Richardson, B. A., & Robinson, C. G. (1986). The impact of internal marketing on consumer service in a retail bank. *International Journal of Bank Marketing, 4*(5), 3–30.

Riemer, M. (1986). *Beschwerdemanagement*. Frankfurt a. M: Campus.

Roos, I. (1999). Switching processes in customer relationships. *Journal of Service Research, 2*(1), 68–86.

Roos, I., & Strandvik, T. (1997). *Diagnosing the termination of customer relationships*. Proceedings of new and evolving paradigms: The emerging future of marketing-conference, Dublin.

Rosenberger, G. (1999). Messung der Dienstleistungsqualität durch die Stiftung Warentest. In M. Bruhn, & B. Stauss (Eds.), *Dienstleistungsqualität. Konzepte – Methoden – Erfahrungen* (3rd Edn., pp. 341–359). Wiesbaden: Gabler.

Rosigkeit, A. (1997). Hochschul-Ranking, Hintergründe und kritische Anmerkungen zu einem modernen Bewertungsverfahren. *Beiträge zur Hochschulforschung, 1*, 23–49.

Ryan, T. P. (2011). *Statistical Methods for Quality Improvement* (3rd Edn.). New Jersey: Wiley.

Sauerwein, E. (2000). *Das Kano-Modell der Kundenzufriedenheit. Reliabilität und Validität einer Methode zur Klassifizierung von Produkteigenschaften*. Wiesbaden.

Scharitzer, D. (1994). *Dienstleistungsqualität – Kundenzufriedenheit*. Wien: Service Fachverlag.

Schlosser, A. E. (2011). Can including pros and cons increase the helpfulness and persuasiveness of online reviews? The interactive effects of ratings and arguments. *Journal of Consumer Psychology, 21*(3), 226–239.

Schmitt, R., & Pfeifer, T. (2015). *Qualitätsmanagement: Strategien – Methoden – Techniken* (5th Edn.). München: Hanser.

Schröder, K. (1999). Storno. Mit Service den Kunden binden. *Aspekte. Nachrichten-Magazin der DBV-Wintherthur-Gruppe, o. V.*(3), 22–23.

Schulte, C. (2012). *Personal-Controlling mit Kennzahlen* (3rd Edn.). München: Vahlen.

Shahin, A., & Ghasemaghaei, M. (2010). Service Poka Yoke. *International Journal of Marketing Studies, 2*(2), 190–204.

Shingo, S. (1991). *Poka-Yoke. Prinzip und Technik für eine Null-Fehler-Produktion*. St. Gallen: Hanser.

Shingo, S. (1995). *Zero Quality Control. Source Inspection and the Poka-Yoke System*. Portland: Routledge.

Shostack, G. L. (1984). Designing services that deliver. *Harvard Business Review, 62*(1), 133–139.

Shostack, G. L. (1985). Planning the service encounter. In J. A. Czepiel, M. R. Solomon, & C. F. Suprenant (Eds.), *The service encounter. Managing employee/customer in service businesses* (pp. 243–253). Lexington: Lexington Books.

Siebert, G., & Kempf, S. (2008). *Benchmarking. Leitfaden für die Praxis* (3rd Eds.). München: Hanser.

Splendid Research (2019). Mystery shopping. http://splendid-research.com. Zugegriffen: 3. Apr. 2019.

Stauss, B. (1999). „Augenblicke der Wahrheit" in der Dienstleistungserstellung. Ihre Relevanz und ihre Messung mit Hilfe der Kontaktpunkt-Analyse. In M. Bruhn, & B. Stauss (Eds.), *Dienstleistungsqualität. Konzepte – Methoden – Erfahrungen* (3rd Edn., pp. 321–340). Wiesbaden: Gabler.

Stauss, B., & Hentschel, B. (1990). Verfahren der Problementdeckung und -analyse im Qualitätsmanagement von Dienstleistungsunternehmen. *Jahrbuch der Absatz- und Verbrauchsforschung, 36*(6), 232–259.

Stauss, B., & Hentschel, B. (1991). Dienstleistungsqualität. *Wirtschaftswissenschaftliches Studium, 20*(5), 238–244.

Stauss, B., & Seidel, W. (2014). *Beschwerdemanagement: unzufriedene Kunden als profitable Zielgruppe* (5th Edn.). München: Hanser.

Stiftung Warentest. (2012). Banken im Test – Die Blamage geht weiter. http://www.test.de/Bankenim-Test-Die-Blamage-geht-weiter-4113924-4114315/. Zugegriffen: 22. Apr. 2013.

Stiftung Warentest. (2016). Dienstleistungstests. http://www.test.de/unternehmen/zahlen/. Zugegriffen: 22. Jan. 2016.

Sutton, R. I., & Rafaeli, A. (1988). Untangling the relationship between displayed emotions and organizational sales. The case of convenience stores. *Academy of Management Journal, 31*(3), 461–487.

Tagesspiegel. (2011). Qualitätsrankings. http://www.tagesspiegel.de/medien/qualitaets-ranking-tagesspiegel-belegtspitzenplatz/4453572.html. Zugegriffen: 22. Jan. 2016

Töpfer, A., & Günther, S. (2009). Mehrere Wege zu verschwendungsfreien Prozessen und Null-Fehler-Qualität: Einführung und Überblick über die Beiträge. In A. Töpfer (Eds.), *Lean Six Sigma* (pp. 3–23). Berlin: Springer.

Trommsdorff, V. (1975). *Die Messung von Produktimages für das Marketing. Grundlagen und Operationalisierung*. Stuttgart: Heymann.

Trommsdorff, V., & Teichert, T. (2011). *Konsumentenverhalten* (8th Edn.). Stuttgart: Kohlhammer.

Tsai, H.-H., & Lu, I.-Y. (2006). The evaluation of service quality using generalized choquet integral. *Information Sciences, 6*(176), 640–663.

van der Wiele, T., Hesselink, M., & van Iwaarden, J. (2005). Mystery shopping: A tool to develop insight into customer service provision. *Total Quality Management & Business Excellence, 16*(4), 529–541.

Vaughan, L., & Shiu, E. (2001). ARCHSECRET: A multi-item scale to measure service quality within the voluntary sector. *International Journal of Nonprofit & Voluntary Sector Marketing, 6*(2), 131–144.

Venohr, B., & Zinke, C. (1999). Kundenbindung als strategisches Unternehmensziel. Vom Konzept zur Umsetzung. In C. Homburg & M. Bruhn (Eds.), *Handbuch Kundenbindungsmanagement. Grundlagen, Konzepte, Erfahrungen* (3rd Edn., pp. 151–168). Wiesbaden: Gabler.

Weiber, R., & Wolf, T. (2013). Der Qualitäts-Dreisprung – Ein konzeptioneller Ansatz zur Verbesserung des Qualitätsmanagements bei Dienstleistungen durch Social Media. In M. Bruhn, & K. Hadwich (Eds.), *Dienstleistungsmanagement und Social Media. Forum Dienstleistungsmanagement* (pp. 393–418). Wiesbaden: Springer Gabler.

Weiss, P. A. (1992). *Die Kompetenz von Systemanbietern. Ein neuer Ansatz im Marketing für Systemtechnologien*. Berlin: Schmidt.

Wertenbroch, K., & Skiera, B. (2002). Measuring consumers' willingness to pay at the point of purchase. *Journal of Marketing Research, 39*(2), 228–241.

Wilson, A. M. (1998). The use of mystery shopping in the measurement of service delivery. *Service Industries Journal, 18*(3), 148–163.

Wilson, P. F., Dell, L. D., & Anderson, G. F. (1993). *Root cause analysis. A tool for total quality management*. Milwaukee: American Society for Quality.

Wood, M., & Preece, D. (1993). *The use of statistical process control for service processes*. Proceedings of the workshop on quality management in services III. Helsinki.

Woratschek, H., & Roth, S. (2004). Informationsökonomischer Erklärungsansatz der Markenführung. In M. Bruhn (Eds.), *Handbuch Markenführung* (2nd Edn., pp. 347–370). Wiesbaden: Gabler.

Wyckoff, D. D. (1988). New tools for achieving service quality. In C. Lovelock (Eds.), *Managing services* (pp. 226–239). London.

Yeo, R. K., & Li, J. (2014). Beyond SERVQUAL: The competitive forces of higher education in Singapore. *Total Quality Management, 25*(2), 95–123.

Zeithaml, V. A., Parasuraman, A., & Berry, L. L. (1992). *Qualitätsservice. Was Ihre Kunden erwarten – Was Sie leisten müssen*. Frankfurt a. M: Campus.

Ziegele, F., & Hener, Y. (2014). Benchmarking in der Hochschulpraxis – Effizienzsteigerung und Prozessoptimierung durch kooperatives Handeln. In W. Benz (Eds.), *Handbuch Qualität in Studium und Lehre* (3rd Edn., pp. 1–20). Berlin: Raabe.

Part III

Planning the Quality Management for Services

6 Strategic Orientation of Quality Management for Services

Abstract

The increasing quality requirements of the various stakeholder groups of service companies have not only brought about new methods for measuring service quality, but above all have changed the companies' perspective towards a stronger strategic and comprehensive quality orientation. Such a holistic orientation of the entire service company towards quality requires the development and implementation of a company-specific quality management concept that meets the specific features of the company. Accordingly, various requirements, goals, principles and instruments of quality management are discussed in this chapter. Based on this, different types of quality management strategies are discussed, which relate to the expectation of the service quality as well as the creation of the service quality.

The rising quality requirements of the various stakeholder groups of service companies have not only brought about new methods of measuring service quality, but above all have changed the companies' view towards a stronger **strategic and comprehensive quality orientation**.

Such a holistic orientation of the entire service company towards quality requires the development and implementation of a "suitable" quality management concept, i.e. one that takes into account the individual needs and resources of the company. Accordingly, various requirements, goals, principles and instruments of quality management are discussed below.

M. Bruhn, *Quality Management for Services*,
https://doi.org/10.1007/978-3-662-67032-3_6

6.1 Requirements for Quality Management for Services

To systematically develop and successfully implement quality management for services, various requirements must be met. The following—pragmatically and plakativ formulated—**principles of quality management (10 Cs)** for services are in the focus (see Fig. 6.1; Bruhn, 1998, 33 pp.):

- customer orientation
- consequence
- competitive differentiation
- consistency
- congruence
- coordination
- communication
- completeness

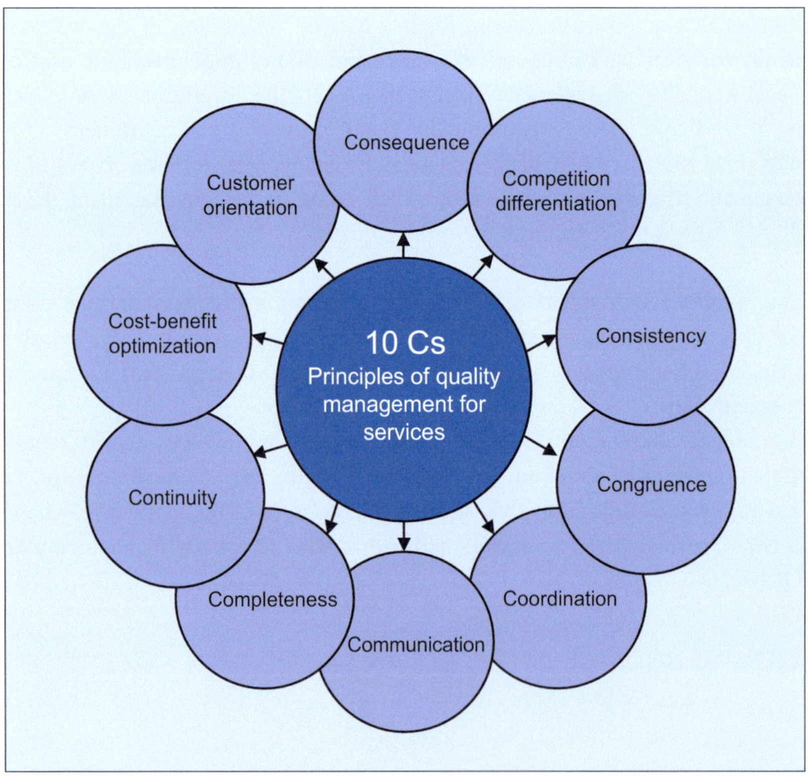

Fig. 6.1 Principles of quality management for services

- continuity
- cost-benefit orientation

Customer orientation has a central importance. Since the achievement of a high perceived service quality and customer satisfaction is decisive for the success of a service company, the entrepreneurial activities are to be oriented sustainably to the customer requirements. Customer orientation is the comprehensive, continuous determination and analysis of customer expectations as well as their internal and external implementation in entrepreneurial services and interactions with the aim of establishing long-term stable and economically advantageous customer relationships (Bruhn, 2012, p. 15). A high customer orientation is, for example, the case in the following cases:

All customer-related information is recorded and disseminated throughout the company. Regular customer satisfaction and customer loyalty studies as well as a comprehensive database are a given in this case.

With regard to all activities, in comparison, most measures are taken to change the corporate culture, such as workshops and seminars about different culturs, development of corporate mission statements, etc.. The reason for this is that the central common value concepts and norms within a company are concretized in the corporate culture, which influence the thinking and behavior of the employees (Schein, 1995; Homburg et al., 2003). Accordingly, there is agreement in the scientific literature that a central anchoring of the value of customer orientation in the corporate culture is necessary in order to be able to realize external and internal customer orientation (Homburg et al., 2003; Stauss & Bruhn, 2010, p. 21).

In order to create customer orientation within the service company, the **consequence** of the orientation of the behavior of all members of the company is required. Both the managers and the employees of lower hierarchical levels must be aware of the need for customer and quality orientation. In addition to creating a "service culture" in the company, which refers to the behavior towards both external and internal customers, it is necessary to develop and define a uniform customer-oriented quality concept within the company itself.

In order to survive in the service competition, a quality-related **competitive differentiation** is necessary through a corresponding positioning in the market, to which the quality measures of the service provider have to be oriented. The planned positioning of the service company determines the target image towards the different target groups and thus represents a quality platform for the entire company, which allows for a consistent service orientation internally as well as externally. The strategic anchoring of quality management in the company is required to achieve this, which goes beyond the implementation of individual operational measures.

The various quality management measures are to be designed in such a way that they can be perceived without contradiction or uniformly by the customers and employees of the service company. This **consistency** avoids that irritations of the customers, caused by inconsistent service provision, for example in terms of the behaviour of the employees or the technical equipment of individual branches or offices, lead to credibility and acceptance problems. Therefore, it is the task of quality management to ensure that no contradictions in content or form occur when creating the service quality.

A quality management system can only be successfully implemented within the service company if the behaviour of the employees towards each other in the departments, business areas and subsidiaries corresponds to the behaviour towards the external customers. This **congruence** of internal and external quality and service orientation is a significant success factor for creating a comprehensive customer orientation (Anosike & Eid, 2011). In this context, it is also necessary to create unity from the differentiated quality-related measures and instruments within the framework of an integrated quality management, which includes the orientation of all company activities towards the needs and requirements of both external and internal customer groups.

Due to the relevance of all activities of a service provider for the perceived service quality, the **coordination** of activities with a view to high quality is to be focused on at a corporate level. It is not only the activities with an explicit quality character, but also the measures with apparently low quality importance that have to be coordinated with each other (Parente et al., 2002).

Furthermore, the **communication** of the service company is to be oriented towards the quality requirements. By means of internal communication, it is necessary to create a uniform and company-wide customer orientation. With regard to external communication, in particular for employees in customer contact, the awareness of their role as communication carriers and quality signals is of increased importance.

To ensure the consistent implementation of customer orientation in service companies, it is important to pay attention to the **completeness** of its orientation when developing a quality management. It is necessary that a holistic quality concept is chosen in order to involve employees of all hierarchical levels in the customer orientation of the entire company.

A comprehensive quality management requires a medium to long term **continuity** in the use of quality-related partial concepts and instruments. If a future-oriented focus of the developed quality-related measures is lacking, such as customer and employee surveys, the success of a consistent quality orientation will only be partially achieved. Without the continuous orientation of quality management, a quality awareness can only be created to a limited extent with the employees of service companies.

Finally, the development and design of quality management must take place under economic aspects. Only a comprehensive **cost-benefit orientation** of quality management can contribute to the achievement of high quality and satisfaction and to the long-term economic success of the service provider.

Based on these principles, the integrated design of a quality management system is to be carried out for the systematic implementation of customer orientation and quality awareness. The mentioned aspects are the central requirements for the design of a quality management system in service companies.

6.2 Tasks and Instruments of Planning a Quality Management System for Services

When planning a quality management system for services, it is necessary to define the basic framework of quality management and thus the quality-related strategic orientation of the service company in line with the corporate strategy. Thus, the strategic planning of a quality management system for services has five basic **tasks**:

1. Determining the strategic quality position
2. Determining the quality strategy
3. Determining the company's vision, mission and quality principles
4. Specifying quality requirements
5. Determining quality objectives

These tasks are explained in detail below.

(1) Determining the strategic quality position Determining the strategic quality position of the service company is essential for designing a quality management concept, because, depending on the current and future desired quality position in regards to the competitors in the market, different priorities have to be set within the framework of quality management (Kim et al., 2012; Brown et al., 2013).

As part of a general **analysis process of strategic planning**, it is up to service companies to find an **optimal quality position** against the background of the current market and competitive situation of the supplier. In principle, as postulated in the success chain of quality management for services, a positive economic success effect results from quality management for services or from superior service quality (cf. Sect. 1.2). Nevertheless, a **differentiated consideration** is required when defining the strategic quality position, taking into account the following aspects:

- Quality relevance in the market environment
- Complexity of the service quality perceived by customers

The determination of the desired **competitive advantage** plays a central role in the strategic orientation of a service provider. In order to secure their market position, companies strive for differentiation, cost and time advantages; often, companies pursue several competitive advantages in the sense of a combination of these competitive advantage dimensions (Meffert et al., 2018, pp. 166–167). Figure 6.2 shows the dimensions of competitive advantages for the implementation of competitive advantage strategies.

If competitive advantages are based on a cost leadership over relevant competitors, they are **cost advantages** (e.g. outsourcing of IT services to low-wage countries). **Time advantages** are characterized by a competitive advantage in terms of the duration or the

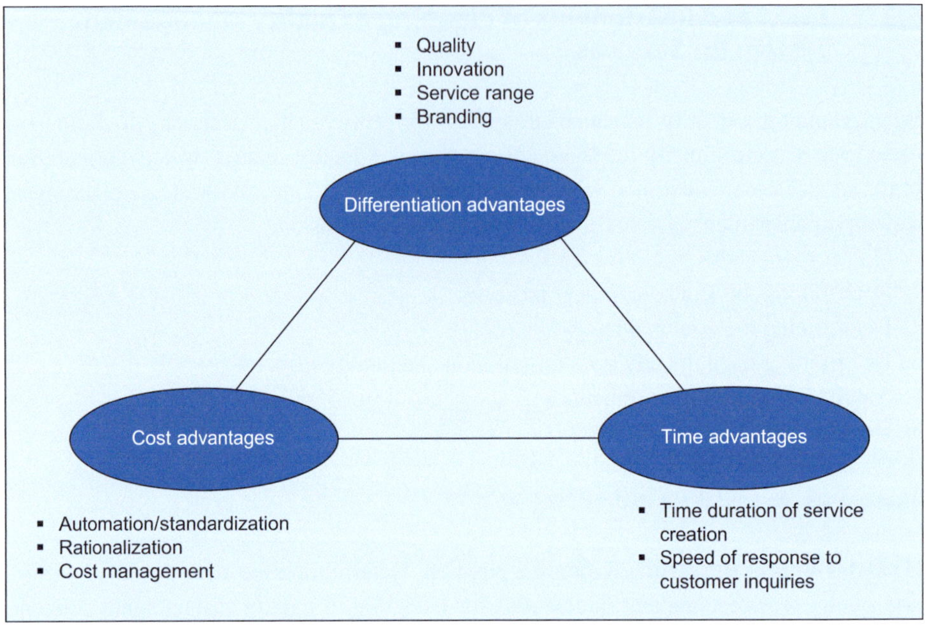

Fig. 6.2 Dimensions of competitive advantages. (Source: Meffert et al., 2018, p. 167)

speed at which customer wishes are fulfilled. **Differentiation advantages** are characterized by the fact that they enable service providers to stand out from the competition in terms of the services offered. Achieving a **superior quality position** from the customer's point of view represents a possible differentiation advantage for service providers over competitors (Meffert et al., 2018, 166 pp.).

In addition to the fact that a certain quality level perceived by the customer is essential for the success of a company, it is also necessary for a service provider to define, against the background of the relevant market environment, on the basis of which competitive advantage (or on the basis of which combination of different competitive advantages) a strategic positioning is to be pursued. It is advisable to make the extent of the strategic quality position ultimately dependent on the importance or relevance of quality for the success of a service provider in its specific market environment. If it is rather a quality-sensitive market environment (e.g. management consulting, asset management, etc.), in which a supplier operates, a competitive advantage strategy based on a superior quality position is more suitable than, for example, in a strongly cost-oriented competitive environment (e.g. telecommunications services). In any case, a differentiated consideration according to the **quality relevance in the specific market environment** is required in order to avoid a misallocation of resources by an inadequate strategic orientation.

In addition, determining a strategic quality position for service providers turns out to be a **complex, multidimensional optimization problem** (Meffert et al., 2018, 166 pp.). The multidimensionality results from the fact that service quality is judged subjectively by customers on the basis of different quality dimensions (cf. Chap. 4). This can lead to individual quality features being weighted or judged differently by customers. For example, it is conceivable that customers of a car rental company would attribute greater importance to the quality dimension "competence of the supplier" (assurance) than, for example, to the "empathy" of the customer contact employees. This makes it difficult to determine the strategic quality position and requires intensive and differentiated consideration of customer requirements.

In view of the special features outlined here in connection with finding the strategic quality position of service providers, it is necessary to determine and define them as part of a situational analysis. For this purpose, so-called **quality portfolios** can be used, which show the strategic position of the company in terms of the service quality of individual strategic business areas (Wilken, 2013). If, for example, a business area of the service company is positioned on the basis of the dimensions "relative quality position of the company" and "importance of quality in the service sector", the quality portfolio shown in Fig. 6.3 can result. This way, general directions can be shown in terms of the actual and desired situation of various business areas of a service company. However,

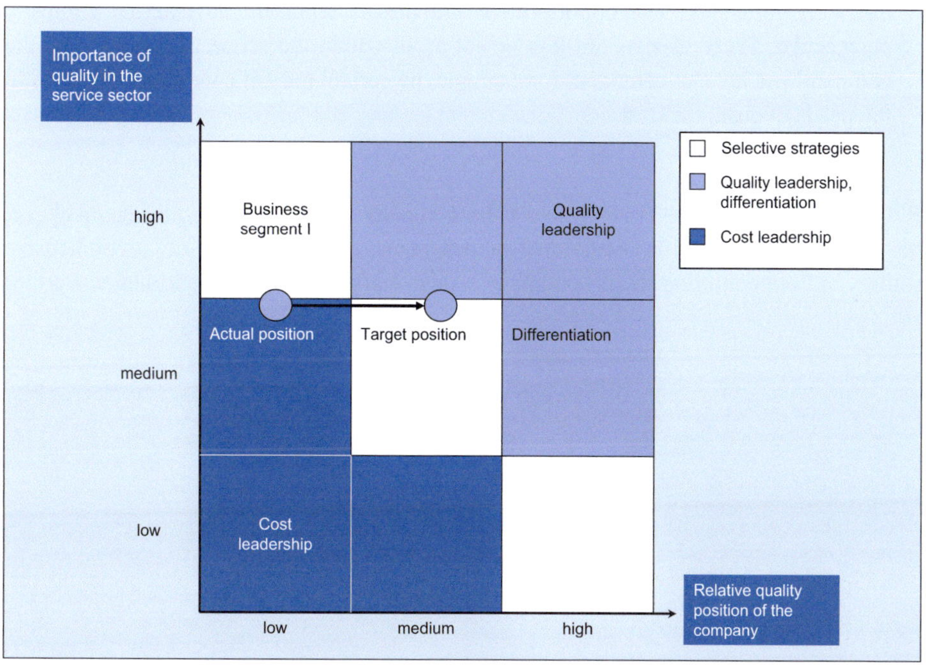

Fig. 6.3 Example of a quality portfolio for service companies

more detailed information can only be obtained with the help of extensive customer surveys using quality planning instruments (cf. Sect. 8.2).

A concretization of the current quality position of the service company and an indication of approach points for achieving the target position can be made using the quality-related **SWOT analysis**.

While the **opportunities and threat analysis** can be used to identify and anticipate those environmental forces that will be of importance in the context of strategic planning processes for service companies in the future, the **strengths and weaknesses analysis** attempts to determine which activities the service company can take in a strategically meaningful way in view of the current and future resource situation (Meffert et al., 2018, p. 133).

As a rule, **strengths - weaknesses and opportunities - threats analyses (SWOT analyses)** are carried out in parallel on the basis of the company-specific context, and the strategic key factors are summarized in a matrix, as exemplarily shown for a bank in the following example. The comparison of the two analyses can further specify the strategic decision-making field, and concrete strategic directions can be derived with regard to the desired competitive advantages.

Quality-oriented SWOT analysis using the example of a bank

From an internal company perspective, the strengths and weaknesses of the bank are set out (cf. Fig. 6.4). Here, aspects are considered that the company itself can significantly influence. The opportunities and threats represent an external company perspective. These aspects are less or not at all within the scope of influence of the company and are characterized, for example, by certain market conditions or trends in the needs of customers. ◄

Responsible for the SWOT analysis of the company is basically the management, i.e. the executive board and the individual branch managers, who carry out comprehensive studies in cooperation with the company's own market research departments. Further

Strengths	Weaknesses
▪ High penetration in rural areas (branch network) ▪ Allfinanz offers of the Savings Banks Finance Group ▪ Decentralized sales structure ▪ Regional competencies/location advantage	▪ Image deficits amongst wealthy consumers ▪ Collaboration in the network ▪ Dominance of stationary sales systems ▪ Inertia/lack of strategic flexibility
Opportunities	**Threats**
▪ Increased demand for complete solutions/ bancassurance offers ▪ Success potentials in micro marketing/local market	▪ Increasing competition among bancassurance providers ▪ Dynamics of technological developments ▪ Internationalization of the banking market

Fig. 6.4 Quality-oriented SWOT analysis using the example of a bank

findings for the determination and achievement of the optimum quality position are provided by the implementation of an **importance-performance analysis**. With the help of such an analysis, it can be shown which importance a quality feature has for the overall satisfaction of the service recipient and which improvement potential still exists with regard to this feature. These findings thus enable optimum control of improvement measures. Depending on the positioning of the quality features in the importance-performance portfolio, corresponding norm strategies can then be derived (cf. for the Importance-Performance analysis Sect. 5.1.2.1).

(2) Determination of the quality strategy Based on the quality-related strengths, weaknesses, opportunities and threats, the **quality strategy** is determined with which the desired quality position is to be achieved. Competitive quality strategies show the basic orientation of a service company and its quality management towards the market. These have the goal of a clear positioning of the company on the market and at the same time the exploitation of a profitable market potential (Heskett, 1988, p. 47; Hoang et al., 2010).

In which **business areas or segment areas** the service company can realize quality and/or cost advantages depends on the requirements of the customers for the various products and services as well as on the offers of the competitors on the market. In comprehensive customer surveys and market research, the different dimensions of service quality are to be determined in each case, which in a further step serve as a basis for the planning of performance advantage or cost advantage strategies in the respective business areas. Clearly formulated internal and external quality strategies show different directions for creating service quality (Bruhn, 1999, p. 34) and thus derive specific tasks for quality control and quality assurance (see Sect. 6.3).

(3) Determination of corporate vision, mission and quality principles Based on the desired quality position and the chosen quality strategy of the service company, a corresponding visualization of the quality strategy must be ensured. Suitable instruments for this are the formulation and communication of the **vision, mission and principles with a quality reference**. These are to be communicated appropriately internally to the employees and externally. The **vision** stands as a future-oriented, holistic image of quality management for the company. The **mission**, on the other hand, is mainly directed at the external stakeholder groups and primarily at the customers. With the company mission, the company signals which quality claim it has on itself and its services in order to be perceived by the customers as high quality. The **quality principles**, on the other hand, provide detailed statements on how the company's vision and mission are to be implemented in daily operations and are primarily aimed at employees, but are often also made accessible to external stakeholder groups, e.g. via the company homepage. In Fig. 6.5 examples of the company vision, mission and quality principles for two different service companies are shown.

	Five star hotel	Real estate developer
Vision	As a five-star hotel, we want to be among the top 3 hotels in the relevant market and always be shortlisted from the perspective of our target groups.	As a real estate developer, we want to be a competent contact partner in all project phases and be perceived and accepted as such by the individual contractual partners.
Mission	We have set ourselves the goal of convincing guests and customers again and again through consistently high quality. To ensure this, all employees live our quality promise.	We aim to provide the highest level of quality in all issues through the most innovative technical equipment and our competent staff.
Quality principles	1. Quality and the well-being of our guests are our central corporate goals.	1. The realization of high-quality projects is one of our central corporate goals.
	2. We promote quality awareness in all areas of our hotel.	2. We take care to create added value for our customers and the respective communities with our projects.
	3. Suggestions and ideas are important to us and contribute to the continuous improvement of our quality standards.	3. Our business activities are based on partnership and sustainable cooperation with municipalities and contractual partners.
	4. We maintain the equipment and furnishings of our hotel to always offer a quality ambience.	4. Not only the quality of the houses and apartments, but the cooperation in the individual project phases decides on the future-oriented development.
	5. Our employees are motivated to fulfill all the guests' wishes and make their stay pleasant.	5. In doing so, we are committed to continuous improvement processes in the individual project phases.
	6. We want to meet and even exceed the demanding, constantly changing expectations of our customers and guests with our quality.	6. We want to offer our customers the best possible services at an attractive value for money.
	7. Service quality is teamwork – we foster a work environment that promotes employee motivation and competence.	7. In doing so, we aim to achieve sustainable development for our diverse customers (e.g. municipalities, investors, tenants).
	8. Our appearance is friendly and courteous. Every employee is the contact person for your questions and wishes.	8. Our professionalism and the competence of our employees guarantee that we implement high quality in the long term.
	9. We are aware that the responsibility for our quality lies with each and every employee, and therefore we all want to be good role models.	9. The satisfaction of all our project partners is the highest goal of our daily actions.
	10. For us, quality management is a long-term measure for the continuous optimization of our service quality. To this end, we regularly review our standards.	10. We feel obliged to contribute to the social, ecological and economic development of society through our activities.

Fig. 6.5 Examples of vision, mission and principles with a quality reference

The formulation of binding quality principles thus forms the foundation, so to speak, for the quality management and quality improvement measures to be carried out in the service company. Accordingly, it is the **task of the management**, i.e. the respective managing director and branch manager, together with managers of different hierarchical levels, to develop and specify concrete quality principles in the company's guiding principles (Reimann, 2012; Hinsch, 2014).

In this context, it should be noted that it generally appears to be necessary to specify in detail for all activities in service companies which employees are to carry out which quality measures at a certain time or within a specific period of time when developing and implementing quality management.

The formulation of binding quality principles also imparts the **linguistic regulation for efficient internal and external communication** and thus represents the basis for a uniform approach. Its binding character is often underscored by the written documentation. A quality principle book or posters accessible to all employees—at least in the back-office area of the individual offices or branches—are useful tools for making quality guidelines transparent to the inside and outside. It is particularly important for the management and all the managers of the company to be constantly striving to communicate the quality principles set out horizontally and vertically, but above all precisely in their daily work by, for example, introducing plans and results of quality control as part of important meetings (Berry et al., 1989, p. 91; Russell & Stone, 2002).

Quality principles of *Shell Deutschland Oil GmbH*

Figure 6.6 shows the quality principles of *Shell Germany,* as they were published in 2014, with which the company wants to document that the customer is always at the center of all company activities. The quality principles also contain the importance of environmental protection as well as a certain basic attitude of the employees, which ultimately reflects the corporate culture prevailing at *Shell* (Shell, 2019). ◀

Quality principles are to be formulated in service companies in such a way that they are valid for **all company areas** and are therefore relatively general in order to ensure uniformity and the necessary sense of community among employees. However, specific derivations and modifications for the specific tasks of individual departments may be necessary and are to be carried out if necessary by the management (Frehr, 1999, p. 43; Reimann, 2012).

When defining quality principles, it is necessary to take into account the basic statements or the **mission statement of the service company**. Only if the central entrepreneurial task is reflected in the principles, a uniform appearance and thus a clear positioning in the market can be possible.

In order to ensure a uniform, high and continuous level of quality, it is advisable to define quality requirements for the entire service chain, but it is not possible or necessary to specify detailed "standards" for all situations and individual processes. It is

SHELL IN GERMANY
QUALITY PRINCIPLES

Shell's products in Germany make a fundamental contribution to mobility and to meeting our society's heating and energy needs.

We want to satisfy each and every one of our customers with our products and service. Therefore, we always strive to meet the demands of consumers and the expectations of the public for our products, services and work processes.

The focus is not only on the quality and cost-effectiveness of our products, but also on occupational safety and the protection of health and the environment. On this basis, it is our goal to consolidate and further expand our market position.

THESE PRINCIPLES SUMMARIZE OUR COMMITMENT TO QUALITY ASSURANCE:

- We are in close dialog with our customers. Only this way can we understand the requirements of our customers and recognize new needs in due time. We attach great importance to reacting to special needs with creativity and flexibility.
- We work continuously to improve our work processes, products and services. Together with the Shell Group's research and product technology centers, we are constantly looking for innovative solutions to meet the demands of the market.
- Strict cost control ensures the profitability of our business and modern technologies guarantee that our products meet the high quality requirements of our customers. We are always a reliable supplier and offer competitive value for money.
- We inform our customers in an appropriate way about the quality, environmental, health and safety aspects related to the handling, use, storage and disposal of our products.
- We promote a sense of responsibility, action competence and customer orientation among our employees through extensive communication and regular training. All employees are involved in the implementation of these principles. They are provided with the means and skills to perform their tasks competently and with a sense of responsibility. Our commitment as a supplier to our customers also applies in full to processes within our company.
- We expect high quality from our suppliers. The delivered products and services are checked accordingly and the suppliers are evaluated on the basis of an evaluation system.
- We want to achieve continuous improvement in quality assurance. To this end, we have introduced quality management systems that are permanently developed and maintained.
- We set ourselves targets for improvement and determine the steps necessary to implement them. The results of this continuous improvement process are regularly measured and assessed according to a standardized system and presented in internal and public reports.

Peter Blauwhaff
Country Chairman Germany, Februar 2010

Fig. 6.6 Quality principles of *Shell Germany*. (Source: Shell, 2014)

particularly difficult to specify comprehensive specific **behavioral requirements** for employees in customer contact and then to check them. Here it is rather recommended to check the compliance with general principles of behavior.

The **quality requirements** can be given in the form of "key performance indicators" (e.g. number of processing days, number of incorrectly completed forms per unit of time; Haist & Fromm, 1991; Balck, 2004; Hernandez, 2010) or in the form of detailed descriptions (e.g. checklists, processing guidelines). The form of the quality requirements is ultimately of minor importance, as long as they represent concrete instructions for employees that allow the fulfillment of the identified customer expectations. The prerequisite for the acceptance and implementation of the quality requirements is the involvement of employees of the affected departments in the process of specifying the requirements (Berry et al., 1989, p. 120; Rothlauf, 2014).

When specifying the quality requirements, it is also important to take into account the quality expectations of the external as well as the **internal customers** (Chase & Bowen, 1991, p. 160; Jun & Cai, 2010; Nazeer et al., 2014; Hadwich & Keller, 2015). Since every employee in the company can act as a supplier and a customer of services, the quality requirements are to be defined in such a way that they are also valid throughout the entire service chain.

In companies, **quality circles** or **quality teams** are often responsible for developing quality requirements, which collect the experiences of all employees and managers of the service company in the respective areas of responsibilities, condense them and translate them into realistic requirements. The main responsibility for setting the quality requirements lies primarily with the company's management, as only this way acceptance amongst all employees, a uniform determination of the requirements as well as a later comparison of the different locations with regard to the compliance with the quality requirements can be achieved.

For example, in the context of the service chain of credit institutions, the following areas could be **customer-oriented quality requirements** set and controlled:

- Waiting times at the counter
- Telephone availability
- Average waiting time for accepted telephone calls
- Response time for written customer inquiries
- Processing time for routine transactions (transfers, currency exchange, other bookings)
- Key data of the consulting process (behavior, technology, etc.)
- Error rates, frequencies in transactions
- Transmission times in online systems, etc.

The following example shows application-related quality requirements of a supplier in the construction industry.

Project-related quality standards

Figure 6.7 shows an exemplary tabular overview of quality standards in the service process of *HOCHTIEF Construction AG*. The systematization takes into account criteria for the orientation (internal/external) and the type of measurement (quantitative/qualitative) of the quality standard. In addition, criteria are added with regard to the subject to which the quality standard refers (activity, document, behavior) as well as with regard to the content dimension (formal/informal) of the standard (Fliess et al., 2006, p. 83). ◀

There are different names for **quality requirements**, such as quality standards, -norms, regulations or specifications, which are not used here in order to avoid the difficulties in a clear delimitation within this variety of terms.

(4) Specification of quality requirements The established quality principles must be further concretized for consideration in the daily work of the service company in order to give the employees precise instructions and assistance for their tasks so that the desired quality objectives can be achieved. In order to ensure this uniform, high and continuous quality level, the definition of **quality requirements**—based on the quality principles—must be made for the entire chain of service creation. However, it is not possible and necessary to specify detailed standards for all situations and individual processes. The difficulty lies in specifying comprehensive and general **behavioral requirements** for the employees in contact with the beneficiaries and subsequently checking them, especially in very personnel-intensive situations. In these cases, the specification and control of general principles of behavior are recommended. **Quality requirements** are, for example, given in the form of key performance indicators (e.g. number of processing days, number of incorrectly completed forms per unit of time; Haist & Fromm, 1991; Hernandez, 2010) or in the form of detailed descriptions (e.g. checklists, processing guidelines).

The form of quality requirements and instructions is of minor importance in principle. What is decisive here is that the instructions lead accordingly to the fact that the identified customer requirements are met to the best possible extent. As a prerequisite for the acceptance and implementation of quality requirements, it is necessary to involve the corresponding employees in the **process of defining requirements**. When formulating the quality requirements, it is also important to ensure that the quality expectations of the **external** as well as the **internal customers** are taken into account to the same extent (Parente et al., 2002; Nazeer et al., 2014). Since every employee of the service company often acts as both a supplier and a customer of services, the quality requirements must be defined in such a way that they are also valid within the entire value creation chain and the internal customer requirements are sufficiently taken into account.

No.	Quality standard	Quality dimension	Subject	Content dimension	Alignment	Measurement
1	Protocol standard (content)	Customer engagement	Document	content	external/ internal	qualitative
2	Kick-off workshop	Customer engagement	Activity	formal/content	external	qualitative
3	Presentation standard (design)	Customer engagement	Document	formal	external	qualitative
4	Final discussion (follow-up)	Customer engagement	Activity	formal/content	external	qualitative
5	Customer inquiries are to be answered within one day.	Speed	Activity	formal	external	quantitative
6	Interim reports of the project are to be prepared within 5 days.	Speed	Activity	formal	external	quantitative
7	Minutes are to be taken within 5 days.	Speed	Activity	formal	external	quantitative
8	Check lists/task lists for project team meetings.	Speed	Document	content	internal	quantitative/ qualitative
9	100% accessibility of employees during working hours: on duty arrangements are in place outside working hours.	Accessibility	Behavior	formal	external	quantitative
10	For the individual customer, there is a fixed and consistent contact person in the company.	Accessibility	Behavior	formal	external	quantitative
11	Within the company, a permanent employee is responsible for answering customers' questions via the Internet.	Accessibility	Behavior	formal	external	quantitative
12	100% customer focus: customer requested dates and locations have priority.	Accommodation	Behavior	content	external	qualitative

Fig. 6.7 Systematization of project-related quality standards of the *HOCHTIEF Construction AG*. (Source: Fliess et al., 2006, p. 84)

Quality requirements in the form of key performance indicators in the example of a hair salon

A large number of quality requirements in the form of key performance indicators can be identified in order to increase customer loyalty in a hair salon. The following is a selection of quality requirements for customer loyalty in a hair salon with the corresponding target direction:

- Faster processing time for customers
- Minimization of processing costs
- Increasing customer satisfaction
- Minimization of complaints
- Increase in positive feedback
- Increasing the recommendation rate
- Improving communication
- Increasing the image value
- Increasing the number of employees in customer service
- Improving the knowledge of customer by the employees
- Increasing the usage of the customer database
- Improving relationship competence
- Increasing contact intensity, etc. ◀

(5) Determination of quality objectives The general quality principles and requirements presented are often concretized in companies within the framework of the strategic quality planning of the service company for the various branches, departments and functional areas, i.e. determined by the management as **long and short-term quality objectives**. In Fig. 6.8 different short-term quality objectives are exemplarily shown, which can be important for service companies.

As **quality management objectives** the imperatives (preferred states) set in the company concept are generally characterized, which are to be achieved by the use of quality-related instruments.

The formulation of clear, long-term oriented goals is necessary in order to avoid a purely reactive adaptation of the company to environmental changes with the danger of only "muddling through" (Raffée, 1984, p. 67). Given the variety of possible quality management goals, a systematization of the different intentions and classification into the **goal level hierarchy of the service company** is necessary, as shown in Fig. 6.9 and examined in more detail below.

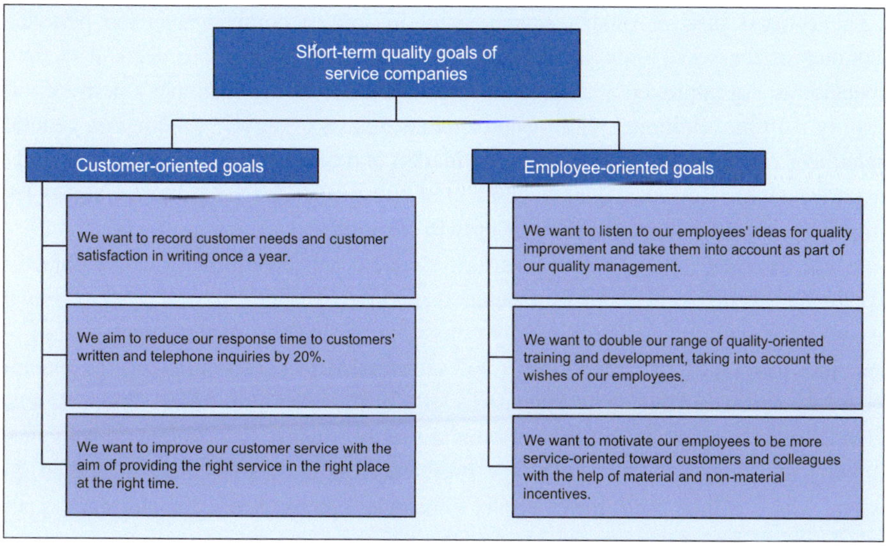

Fig. 6.8 Examples of short-term quality management objectives for service companies

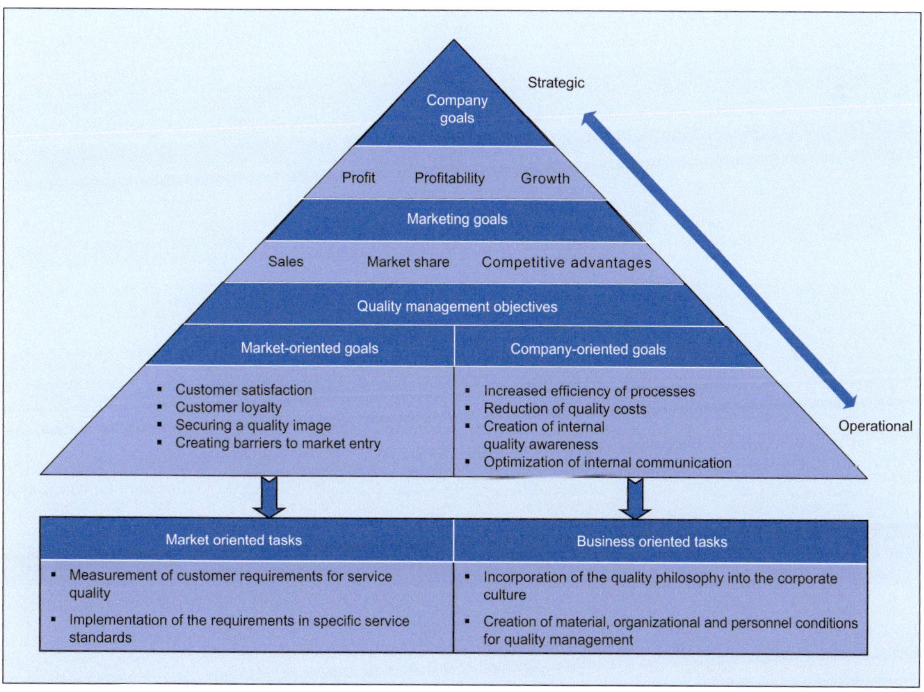

Fig. 6.9 Goal system of quality management of service companies

The **central task** of quality management in service companies is the permanent assurance of the service quality defined by the supplier, taking into account customer expectations. An improved quality of services offered from the customer's point of view, possibly with the additional realization of reduced costs of service production, generally strengthens the company's position in the market and often creates comparative competitive advantages. However, quality goals must always be derived from the higher-level company goals, such as profitability or growth security.

As can be seen in Fig. 6.9, the various strategic and operational quality objectives have to be taken into account in a detailed quality management goal system. For the purpose of structuring and analytically consistent application of the target system, the procedure developed by Saaty (1980) and attributable to decision-making or planning theory, the **Analytic Hierarchy Process (AHP)**, offers itself (e.g. Ahlert, 2003; Blockus, 2010). The AHP approach often represents a tool for quality responsible persons in the company to bring their own goal type preferences into a hierarchically ordered goal system, which is then used, for example, to calculate preference values for each quality goal. These values make it clear whether the task or areas of responsibility located on the lower levels of the goal hierarchy adequately take into account the desired top-level objectives. Fig. 6.10 provides an overview of the quality management objectives of service companies.

The **market oriented objectives** basically include, in addition to classical economical objectives, the **psychological objectives** such as the increase in customer satisfaction

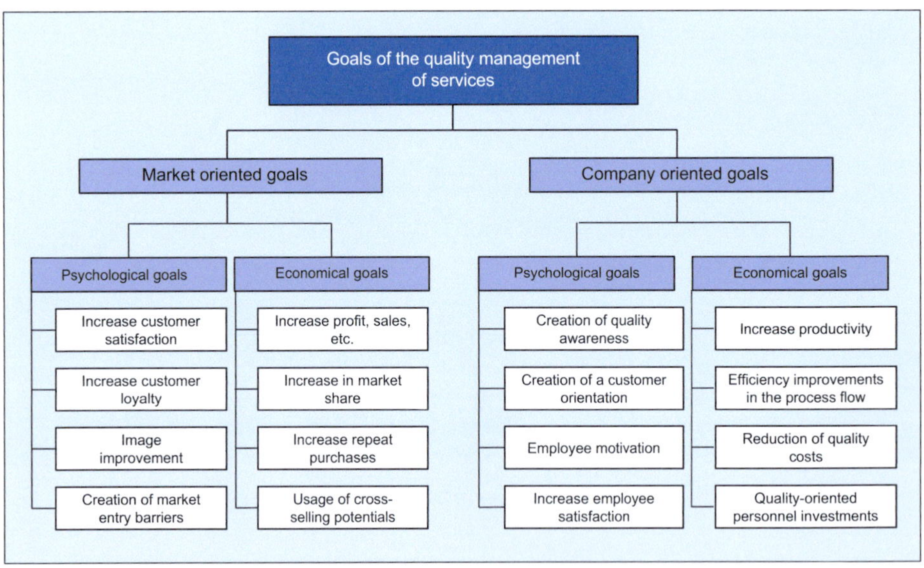

Fig. 6.10 Quality management objectives of service companies

and loyalty, image improvements and the creation of market entry barriers to potential competitors.

Even with the **company oriented objectives** of quality management, economical and psychological objectives can be distinguished. The **economical objectives** include productivity improvements, efficiency improvements in internal service processes, the reduction of quality costs and quality-oriented personnel investments. The **psychological objectives** include the creation of quality awareness and customer orientation among employees of all hierarchy levels as well as the increase of employee motivation and satisfaction. After the various quality management objectives have been determined, the specific market and company oriented tasks have to be derived which serve to achieve the previously determined objectives (Meffert et al., 2018, 151 pp.).

While the market oriented tasks in addition to measuring customer requirements in terms of the quality of services also include the implementation of the requirements for specific quality standards, the primary **company oriented tasks** are mainly concerned with the incorporation of the quality philosophy into the culture of the service company and the creation of factual, organizational and personnel conditions for quality management.

The determination of the individual goals and tasks of quality management always requires a specification in terms of **content, extent and time reference,** so that there are clear measuring instructions, based on which the achievement of the target is controlled in turn. Of particular importance is the clear formulation of the market and company oriented quality objectives, so that they are understandable and comprehensible for

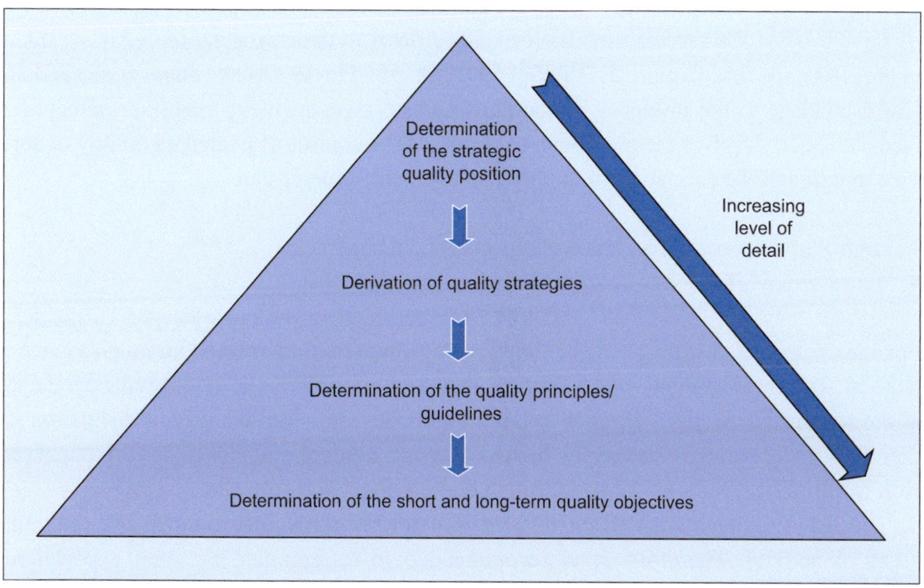

Fig. 6.11 Pyramid of the planning of a quality management for services

all groups of stakeholders, i.e. employees and customers, both in terms of content and extent. It is important to formulate these in a measurable and realistic manner and to differentiate them according to the time reference, i.e. short-term (e.g. for one year) or long-term (e.g. for five years), so that their fulfilment can be demonstrated.

In Fig. 6.11 the relationship between the different elements of the planning of a quality management is summarized.

In general, these different levels are always to be set by the service company's management, i.e. the management and the branch managers, but it is recommended to involve the suggestions and ideas of employees at lower hierarchical levels. A strategic embedding is required for the systematic planning of specific quality management measures (see Chap. 7 and 8). This is done by specifying quality management strategies for which service companies have numerous options.

6.3 Quality Management Strategies

6.3.1 Significance and Types of Quality Management Strategies

Based on the determination of the strategic quality position of a service company, the strategies for quality management can be derived. The **increase in the perceived quality of service** is the central basis for the formation of the strategies.

The central dimensions in the assessment of the quality of services or the perceived quality of service are the **delivered performance** and the **customer expectations** (Boulding et al., 1993; see also Chap. 2), because customers then perceive a high quality of service when the service used corresponds at least to their expectations (Parasuraman et al., 1988, p. 16; Cronin & Taylor, 1992, p. 56; Hoyle, 2009; Zhao et al., 2012). Corresponding to the influence of the performance assessment by customer perception (see also Sect. 2.4) two central points to approach the control of perceived quality of service in terms of the formation of quality management strategies are:

- control of customer expectations (expectation management)
- improvement of the service (classic quality management)

For the control of customer expectations, various **expectation related strategies** can be derived. The improvement of the service, the main goal of classic quality management, is the subject of the **performance related strategies**. In a similar way, a distinction is therefore made between **strategies for expectation control** and **strategies for expectation fulfillment,** which is realized through the improved performance.

The listed strategies are to be divided into further steps according to the different areas of activity. Therefore, it is recommended to differentiate between expectation related and performance related strategies in each case with regard to the dimensions of **market, customer** and **competition related strategies** (Bruhn, 2000).

Generally, **market oriented strategies** are basic strategic decisions regarding market choice and market management of a service provider. In addition to the basic market oriented decisions, **customer oriented strategic directions** are to be set according to the customer needs—according to the different phases of the customer life cycle—and according to the customer value of the respective customer groups. **Competition oriented strategies** finally define the specific behaviour of a provider towards its competitors (Meffert et al., 2018, 163 pp.).

An overview of the different characteristics of expectation oriented and performance oriented strategies is shown in Fig. 6.12.

6.3.2 Expectation Oriented Strategies

6.3.2.1 Strategic expectation management

The control of customer expectations is often a suitable way to increase the perceived quality of service, since improvement potentials of the actual performance itself are limited or only marginally achievable in some industries, even with a professional quality management system. In addition, companies often face the danger of the so-called **"expectation spiral"** in this context (Bruhn & Georgi, 2000, p. 185). This phenomenon is characterized by the fact that an improved performance leads to continuously higher expectations on the part of the customers. As a result of such findings, it is therefore sensible to use customer expectations as a control parameter of the perceived quality

	Expectation based strategies			Performance related strategies		
	Strategies of expectation management			Strategies for meeting expectations		
	Normative expectations		Predictive expectations			
Market related strategies	Isolated strategy		Integrated strategy	Isolated strategy		Integrated strategy
				Market penetration		Service Development
				Market development		Diversification
Customer related strategies	Differentiated		Undifferentiated	Differentiated		Undifferentiated
	Control of the purchase intention	Control of the performance appraisal	Customer loyalty management	Potential oriented strategy	Process oriented strategy	Result oriented strategy
	Acquisition related expectation strategy	Commitment related expectation strategy	Recovery related expectation strategy	Standardization strategy		Individualization strategy
Competitive strategies	Conflict strategy		Cooperation strategy	Evasion strategy		Cooperation strategy
				Conflict strategy		Adaptation strategy

Fig. 6.12 Expectation and performance oriented strategies of quality management

of service and to implement them with the customers by means of a **systematic expectation management**. The aim is to establish a realistic level of expectations with the customers in order to increase the perceived quality of service through their own performance (Bruhn, 2000). In this context, the **determinants of customer expectations** are to be used to identify the approach for the expectation management. In addition to measures taken by the company, such as corporate communication, a variety of non-controllable factors, such as situational influences or personal needs of the customers, are attributed to the expectation determinants (Bruhn & Georgi, 2000, p. 186).

From a strategic perspective, the question arises as to whether customer-related expectations are to be controlled downwards or upwards in general or to be stabilized at an existing level of expectations (Richter, 2005, pp. 181–182). From the perspective of the consumption phases of a service, the control of expectations is to be taken into account **before, during** and **after service utilization**. From this perspective, the following areas can be distinguished:

- Control of purchase intention
- Control of performance assessment
- Control of customer loyalty

Figure 6.13 shows the consumption-phase-related goals of an expectation management as well as the corresponding influencing factors on the expectation control.

The **construct of expectations** is used in numerous scientific disciplines, such as learning theory, decision theory or social science, to explain human behavior. The expectation of an individual always represents a psychological state that refers to future behavioral consequences for the individual (van Raaij, 1991, pp. 401–402). Based on this general definition of expectation, a variety of interpretations of the concept of customer expectations can be found in the literature on perceived service quality and customer satisfaction, within which it is primarily a matter of establishing a realistic level of expectations in customers in order to increase the perceived service quality through one's own performance (see Bruhn & Georgi, 2000, p. 188 for an overview). The perceived service quality is influenced by the **normative expectations** and thus expresses the quality of a performance from the customer's perspective according to his requirements for a performance of the corresponding performance category. However, even if the perceived service quality is weak, the customer can be satisfied with a performance if the level of service quality is as planned, i.e. corresponds to his predictive expectations. In this context, the relevance of **predictive expectations** becomes clear with regard to the perceived service quality. In the development of expectation-related strategies, it is therefore necessary to base the working of a systematic strategic expectation management and to differentiate between normative and predictive expectations (see Sect. 2.4).

In addition, the influence of expectations as a determinant of perceived service quality in the operational implementation of expectation management must be taken into account. Only those expectation determinants provide a starting point for the systematic

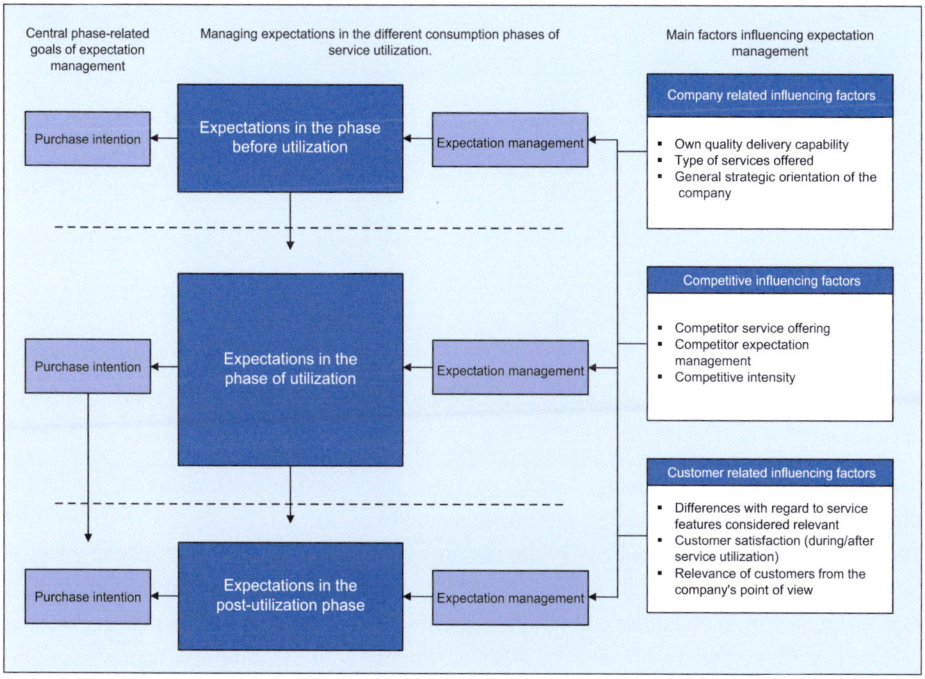

Fig. 6.13 Consumption phase related goals of an expectation management and main influencing factors on the expectation increase. (Source: Richter, 2005, p. 192)

control of customer expectations that can also be influenced by the service company. These determinants can be divided into the determinants service promises (P), service information (I) and service performance (P) according to a so-called **PIP model of expectation management** (see Fig. 6.14).

The **service promise** includes all statements of the service company that emphasize a certain service or quality level of the services offered to customers. Direct service promises (e.g. fast food restaurants promise that a customer's waiting time for an order will not exceed five minutes) relate to those service features with respect to which customer expectations are to be influenced. In contrast, indirect service promises (e.g. price promises) are not primarily used to control expectations, but rather influence customer expectations in the form of quality signals (Bruhn & Georgi, 2000). **Service information** has an explanatory character in that the overall service or individual service aspects are made tangible. In contrast to service promises, these can also contain negative aspects. In the case of direct service information, the supplier itself is the source of information (e.g. information on successful test reports, which are displayed in company brochures or in sales rooms). In contrast, indirect service information is transmitted by independent institutes (e.g. impersonal recommendation by the report on service tests of a consumer

Type of expectation control / Measures	P Service promise	I Service information	P Service performance
Direct expectation control	**Direct service promise** Example: One-sided corporate communication	**Direct service information** Example: Two-way communication	**Direct service performance** Example: Satisfying current customers
Indirect expectation control	**Indirect service promise** Example: Tangible performance elements, price	**Indirect service information** Example: Impersonal word of mouth communication through public channels	**Indirect service performance** Example: Positive word of mouth communication through satisfiying current customers

Fig. 6.14 PIP model of expectation management. (Source: Bruhn & Georgi, 2000, p. 186)

organization) (Bruhn & Georgi, 2000, p. 186). The **service performance** requires a direct contact between supplier and customer and can in turn be divided into a direct and indirect service performance. While the direct service performance relates to the customers' own experiences with the provider (e.g. experiences during previous visits to a hair salon or during consultations with a financial advisor at a bank), the indirect service performance includes experiences of other customers with the supplier, which are communicated to the respective customer (e.g. recommendation of a hotel) (Bruhn & Georgi, 2000).

As a result, the basis for successful expectation management is the analysis of the expectation-related control options and their impact on customer purchasing behavior.

6.3.2.2 Market Oriented Strategies

In terms of market oriented strategy, it is important for service companies to check the clarity of the market delimitation processed by the company's performance (see, for example, Abell, 1980; Backhaus & Voeth, 2014). The delimitation of the market takes place in this sense in geographical as well as product and service related terms. If a clearly delimited market is available in this context, the service provider is only confronted with the expectations of the customers who are to be found in this market. From this situation results the pursuit of an **isolated strategy** (Bruhn, 2000, p. 1041).

An isolated strategy thus deliberately neglects the diversity of circumstances in the different markets. This is based on the consideration that these do not have any influence on the company's customers.

However, the market of a company is often closely related to other markets. As a result, customer expectations are also influenced by other markets. In this case, the expectation management also uses a control of the expectations existing in other markets as part of an **integrated strategy**.

6.3.2.3 Customer Oriented Strategies

In terms of customer oriented strategy formation, the **heterogeneity of customer needs** plays a particularly important role (Bruhn, 2000, pp. 1036–1037). If the company's target groups have relatively homogeneous customer needs, an **undifferentiated strategy** is to be pursued as part of expectation management. If customer needs are extremely heterogeneous, it is no longer possible to control expectations of all customers in a uniform manner. In this case, a **differentiated strategy** is to be pursued. This means, for example, that a restaurant chain with different customer segments or customer types, which have different expectations, should strive for a differentiated design of the communication instruments in each case in order to address the respective segments.

A concept that structures customer needs is the so-called **customer relationship life cycle**. This includes, for example, the differentiation of the respective needs of customers based on the corresponding life cycle phases. In this context, customer related expectations management strategies can be divided into three **phases** according to the customer relationship life cycle (Bruhn, 2016, 53 pp.):

1. Customer acquisition (Recruitment)
2. Customer loyalty (Retention)
3. Customer recovery (Recovery)

This results in acquisition related, retention related and recovery related expectations strategies.

(1) Acquisition related expectation strategies In the context of an acquisition related expectation strategy, it is primarily a matter of distinguishing between **desirable and undesirable customers** in order to achieve a corresponding self-selection among customers by specifying expectations (Richter, 2005, p. 197). For example, families are desirable customers for ski resorts in Switzerland and Austria. With regard to desirable customers, it is desirable to pursue high predictive expectations within the framework of new customer acquisition (e.g. by advertising for children's slopes, ski schools and ice castles in family ski resorts) on the basis of the positive relationship between predictive expectations and the intention to purchase. At the same time, relatively high normative expectations represent a desirable state in order to make it more difficult for the competition to satisfy the potential customers from the perspective of the provider (e.g. the requirement that children's ski lifts in family ski resorts are actually usable). In the case of undesirable customers, on the other hand, expectation management is specifically used to reduce the intention to purchase (e.g. the lack of interest of sports skiers in family ski resorts).

A differentiation of the desired and potential customers often takes place on the basis of the criterion of customer value (cf. e.g. Krafft, 2007; Reinecke & Janz, 2007) if possible. On the assumption of a high customer value, a differentiated expectation management is to be pursued in the case of correspondingly attractive customers or target groups (e.g. young families in a family-friendly ski resort) in order to increase the willingness to buy (e.g. family-specific service promises and offers of the ski resort). In contrast, for less attractive customers or target groups (with low customer value), a standardised expectation management is to be pursued in order to increase the willingness to buy (e.g. indication of piste kilometres) (cf. Fig. 6.15).

(2) Expectation strategies related to loyalty Within the context of expectation strategies related to loyalty, the distinction between satisfied and dissatisfied customers must be taken into account. In order to continue to make satisfied customers loyal to the supplier or company, an increase in predictive expectations (e.g. no advertising with information on prices for children's ski courses and ski lifts for children) is not generally required due to the high perceived service quality. However, to a limited extent, it is necessary to reduce normative expectations in order to secure future customer satisfaction on the expectation side (e.g. reduction of ski pass prices for repeated visits to the ski area).

As for creating loyalty of dissatisfied customers, high predictive expectations are required within the context of **compensation strategies related to loyalty** in order to compensate for possible negative behavioural intentions due to the negative perceived service quality (e.g. no advertising for number of children's lifts). At the same time, it is necessary to reduce the normative expectations of such customers (e.g. discounts for the next visit to the ski area after the children's lift is out of order/defective, in order to

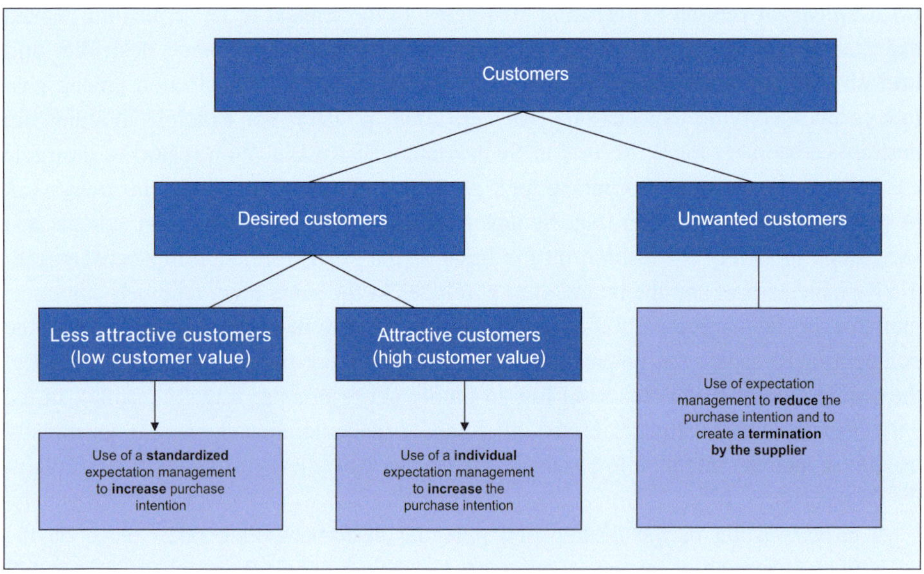

Fig. 6.15 Decision tree between desirable and undesirable customers and corresponding courses of action. (Source: Based on Richter, 2005, p. 197)

reduce the claim character of the customers' expectations), in order to achieve a higher perceived service quality at the next service transaction within the context of a **prevention strategy related to loyalty**.

(3) Expectation based recovery strategies Analogous to customer based expectation strategies, it is necessary to differentiate between satisfied and dissatisfied customers when using expectation based recovery strategies. With satisfied customers who have migrated or are at risk of migrating, the positive perceived service quality does not lead to positive behavioral intentions (e.g. no return visit to the ski resort despite satisfaction). Therefore, it is necessary to increase the potential of customer satisfaction in the event that the customer uses the supplier's services again through high predictive expectations (e.g. advertising with expansion of children's ski slopes; construction of castles made of ice or family-friendly hotels). Similarly, relatively high normative expectations should be pursued (e.g. expansion of children's ski lifts and slopes or increase in ski lessons for children), as these—analogous to the acquisition based expectation strategy—can be fulfilled by the supplier's own performance and often represent a competitive barrier for potential new competitors.

In order to regain dissatisfied customers, high predictive expectations should be pursued in particular (e.g. advertising for expansion of children's ski lifts and slopes as well as other attractions). However, normative expectations should be kept at a low level, as it is assumed that these could not be fulfilled during the last contact (e.g. one-time discounts for the next visit to the ski resort after the failure/defect of the children's lift).

In a final consideration of phase related strategies, it becomes clear that the demand for high predictive and low normative expectations is not consistently valid throughout the customer relationship life cycle. The analysis of satisfied and dissatisfied or attractive and unattractive customers underlines the fact that the desired strategies vary to a great extent depending on the life cycle phase and the degree of customer satisfaction and, in particular, differ with regard to normative expectations.

6.3.2.4 Excursus: Termination of Unwanted Customers (Termination Quality)

The customer based expectation strategies described relate to the acquisition, retention and recovery **of desired customers**. Expectation strategies can also relate to the management **of unwanted customers**, to whom a provider does not want to maintain customer relationships. For this purpose, supplier-side termination strategies are particularly suitable for reducing both predictive and normative expectations.

In cases where continuing the customer relationship is not desirable from the provider's perspective, it is advisable for the provider to pursue a so-called **relationship termination strategy** (see also Lucco, 2009; Bruhn, 2016, 139 pp.). With this type of strategy, the following **goals** are pursued:

- Increasing economic efficiency by terminating unprofitable relationships
- Adapting to changed legal framework conditions to which the relationship of certain customer groups is not compatible
- Avoiding the endangerment of "uncritical" customer relationships by the untenable (criminal or unpleasant) behavior of certain customer groups

Strategy options for relationship termination can be differentiated according to whether the relationship termination is carried out by the provider "openly" or "covertly" (Alajoutsijärvi et al., 2000; Bruhn et al., 2008; Günter et al., 2017).

A total of four **types of hidden strategy options** can be distinguished, which result from the dimensions "orientation of the strategy" (customer oriented versus selforiented) and "activity level of the strategy" (passive vs. active) (see Fig. 6.16):

- The **strategy of de-escalation** aims for a hidden, passive withdrawal from the customer relationship, which, however, is carried out in a customer oriented manner. This strategy is for example conceivable if the continuation of the customer relationship no longer corresponds to the strategic orientation of a company, but the continuation also—at least in the short term—does not cause any damage (e.g. pseudo-continuation of the relationship at low relationship intensity).
- In contrast to de-escalation, **escalation** is a self-oriented termination strategy in which the supplier tries to end the customer relationship in such a way that this is not rated as an active cancellation attempt from the customer's point of view (e.g. by increasing the account maintenance fee).
- The strategy of the **"let it expire"** pursues an active, but customer oriented plan of action for the termination of the relationship. By deliberately reducing customer specific investments, the relationship is to be brought to a standstill step by step (e.g. by reducing the dispatch of current offers and discount campaigns).

Orientation / Activity level	Customer orientation	Self orientation
Passive withdrawal	**De-escalation strategy** Example: Pseudo continuation in case of low relationship intensity	**Escalation strategy** Example: Re-profitabilisation through price increases
Active withdrawal	**Let it expire "strategy"** Example: Reduction of investment in the relationship	**Withdrawl strategy** Example: Reduction of trust and familiarity

Fig. 6.16 Strategies and measures of the hidden relationship termination. (Source: Based on Alajoutsijärvi et al., 2000, p. 1274)

- The fourth hidden withdrawal strategy aims for the **active, self-oriented withdrawal** from the customer relationship with the goal of the termination of the relationship, for example by deliberately reducing trust and familiarity (e.g. by exchanging the long-standing customer advisor).

In contrast to the hidden termination of the relationship strategies, four generic **types of open termination strategies** can be distinguished on the basis of the dimensions "orientation of the strategy" and "finality of termination" (cf. Fig. 6.17):

- The **strategy of declared termination** aims for an irrevocable termination of the relationship and explains the reasons for this to the customer (e.g. account cancellation for private customers, as the bank's focus is now on business customers).
- Under the **fait accompli strategy**, the customer is openly informed of the end of the relationship (e.g. written notice to the customer of the termination of his account).
- The **strategy of consensual conversation** signals the provider's willingness to compromise or the conditions for the continuation of the relationship, thus leaving open the option of resuming the customer relationship (e.g. reduction of the bank's investments in the relationship with the customer through reduction of personalized advertising).
- A **strategy of unproductive conversation** also provides for the possibility of resuming the relationship. However, the resumption is not linked to the condition of a compromise (e.g. resumption of the customer without compromises on the part of the bank).

Orientation / Activity level	Customer orientation	Self orientation
Final termination	**Strategy of declared termination** Example: Interview to explain the reasons of the company	**A fait accompli strategy** Example: Written notification to the customer
Revocable termination	**Strategy of the consensual conversation** Example: Reduction of investment in the relationship	**Strategy of the inconclusive conversation** Example: Resumption without compromise on the part of the company

Fig. 6.17 Strategies and measures of the open relationship termination. (Source: Based on Alajoutsijärvi et al., 2000, p. 1274)

In connection with the strategy selection with regard to the supplier-side relationship termination, both the **termination reasons** and the possible **public consequences** of the implementation of a relationship termination strategy are to be differentiated. There is often the risk of damage to reputation if an open relationship termination strategy is chosen without the public being able to understand the reasons for the procedure. Furthermore, in connection with the relationship termination, the **perceived termination quality from the customer's perspective** is to be taken into account:

Termination quality is the ability of a company to terminate a business relationship with an existing customer on the basis of the transaction oriented expectations that the customer will have of a termination at a certain level of requirements. It is determined by the sum of the characteristics of the entire termination process, to meet certain requirements (Lucco, 2009, p. 13).

Based on this, the **perceived termination quality** is understood as the judgment of a customer with regard to individual interactions and transactions, the process and the personal balance of the termination. The perceived termination quality is therefore a multi-stage construct that covers various transactions and includes the complexity of the supplier-side termination (Lucco, 2009, p. 77). In terms of content, the construct includes the distribution, procedural and interaction quality, with the interaction quality being the strongest content-related driver of termination quality (Lucco, 2009, p. 169). While the distribution quality is mainly related to the result of the termination in the form of fair compensation, the procedural quality includes the reaction speed of the provider and the perceived fairness in the context of the termination process. The interaction quality finally includes the understanding and respectful treatment brought about and is therefore mainly related to relationship aspects between supplier and customer (Lucco, 2009).

Customer emotions are central **consequences of the termination of the relationship.** These can be both positive and negative. For example, Lucco (2009, p. 169) shows that the perceived quality of termination can lead to an increase in customer satisfaction with termination, but also causes the customer to feel the same degree of negative experience. Finally, it becomes clear that a high quality of termination is crucial to reducing negative emotions (2009, p. 169). Since the intensity of customer emotions expresses the customer's level of activation through the experienced termination, it is necessary to achieve the lowest possible activation through the termination in order to maintain a positive attitude towards the supplier despite the termination. Consequently, the study by Lucco (2009) shows that both the perceived quality of termination and the resulting intensity of emotions have an impact on the customer's attitude towards the supplier. However, it also becomes clear that it is not only the perceived quality of termination that shapes the customer's attitude, but rather the perceived quality of the relationship during the active business relationship. In turn, this is also influenced by the perceived quality of termination (2009, p. 169).

The findings regarding the determinants and effects of the perceived quality of termination result in central **implications** for service providers in the context of their relationship termination strategy. Companies should pay particular attention to how they can preserve a positive attitude in the customer—despite the termination by the supplier—and reduce the risk of image-damaging consequences, for example through negative word-of-mouth communication of terminated customers.

6.3.2.5 Competition Related Strategies

Relevant determinants of competition related strategies are the **intensity of competition** within a market and the company's own **position within the competition** (Bruhn, 2000, pp. 1038–1040). The fact that companies rarely act independently of competition in the market leads to the two basic strategies of **conflict** and **cooperation** (2014).

When one's own company is superior to the competition or to competing suppliers, high normative expectations in the sense of an expectation related **conflict strategy** are to be pursued. However, if there is also high competition intensity, there is the danger that the competing companies will increase the (normative) customer expectations very strongly by means of an **"expectation struggle"** and thus make it increasingly unlikely that the expectations will be met. Possible examples of an expectation struggle are often associated with statements such as "… the best in its class" or also "… unbeatable" or "… better than competitor XY".

Conflict strategy of telecommunications providers

The conflict strategy is often pursued by telecommunications providers. By means of aggressive advertising measures, price offers, etc., they try to win new customers and to poach them from the competition. Many providers lure customers with the waiver of the basic fee for the first months, with the complete assumption of incurred switching costs and the administrative tasks of a provider change (e.g. termination with the current provider). In addition, it can be observed on the market that so-called flat-rate offers for Internet and landline use are continuously undercut by the individual providers. ◀

Therefore, it is also advisable to consider an **cooperation strategy** of expectation management in case of high competition intensity. This should also be taken into account as a result of a conflict strategy.

Cooperation strategy between *Deutsche Post* and *Lufthansa*

Forms of so-called "co-opetition", i.e. cooperation between competitors, can be observed in international transport and logistics companies. A co-opetition usually serves to jointly serve markets, i.e. to expand the respective areas of activity and to better utilize the existing capacities. Under the name "Global Match", *Deutsche Post World Net* and *Deutsche Lufthansa AG* have established a close cooperation between their airfreight subsidiaries *Lufthansa Cargo* and *DHL* International. Airline alliances in passenger transport, on the other hand, are already classic cooperation. In addition to a common range of services, they often also combine their customer loyalty programs (e.g. *"Miles and More"*) under one roof. ◀

Ultimately, expectation related strategies—regardless of the specific strategies pursued—form the framework of expectation management and determine its **operational design** (cf. Chap. 7).

6.3.3 Performance Related Strategies

While expectation related strategies are focused on the control of customer expectations, performance related strategies focus on the quality level of the company's services, i.e. on the **fulfillment of customer expectations.** As a result, both types of strategies are closely linked. Performance related strategies can be dimensioned in the same way as expectation related strategies, according to the areas of market, customer and competition.

6.3.3.1 Market Related Strategies

As with the principles of expectation related strategies, the clarity of the distinction in relation to the market processed by the company's performance is also checked here. In the case of a clearly delineated market, an **isolated strategy** comes into force. Here, forms of market field strategies are to be pursued in particular in the sense of the strategy of **market penetration**.

Due to the often close connection of the company to other markets, market field strategies form the starting point in particular as strategies of **service development**, **market development** and **diversification** (cf. e.g. Meffert et al., 2018, 165 pp.). For example, the company *American Express* offers, as part of a service development, a booking service or life and travel insurance in connection with the credit card. The expansion of the offer of a scheduled airline to the segment of "charter flights" or "low-cost flights" can be cited as an example of the strategy of market development. An example of a diversification strategy is finally the inclusion of additional financial services in the offer of building societies. The fact that the fulfilment of customer expectations is also influenced by the performance of other markets leads to a greater focus on an integrated view of all services or service quality of other markets in the form of **integrated strategies**. An integrated strategy of a supplier of financial and insurance services is the consideration of performance standards within the financial and insurance industries. The integration of the requirements from both areas thus leads to the application of an integrated strategy.

6.3.3.2 Customer Related Strategies

The heterogeneity of customer needs is also the most important determinant in the area of performance related customer strategies. This results analogously in **undifferentiated performance related strategies** and **differentiated performance related strategies**.

Within undifferentiated performance-related strategies, a specific adaptation of the services to the customer needs is omitted. A differentiated performance related strategy of quality management is characterized in particular by a differentiation of the service quality in relation to the customer needs.

In a similar way to customer oriented expectation strategies, customer needs are distinguished according to the classical phases of the customer life cycle (customer acquisition, customer retention and customer reacquisition) as well as the strategies of relationship termination (see excursus in Sect. 6.3.2.3). Performance related strategies can be differentiated in the various phases according to the **degree of standardization vs. individualization** in relation to the quality of the service. This results in customer related strategy types being defined in the phases of customer acquisition, customer retention and customer reacquisition.

In the **phase of the customer acquisition**, it must be decided whether the service will be specifically adapted to the needs of the customers to be acquired (individualization strategy) or not (standardization strategy). In the first case, the quality of the service is differentiated according to the expected profitability of different customer groups (e.g. differentiation of the offer for the trial use of a service), whereas in the second case the quality of the service does not vary with respect to different potential customer groups.

In the **phase of customer acquisition**, the decision option of whether the quality of the service is standardized or differentiated for all loyal customers also arises. The consideration of the customer value of different customer groups also offers itself as a decision criterion here. This means that the quality of the service is more aligned with the individual needs of customer groups with a high customer value than of loyal customer groups with a lower customer value. Studies have shown that more than 60% of customer relationships in banks' "broad" private customer business do not generate a positive contribution margin. Banks are therefore increasingly focusing on retaining wealthy private customers and customers with high income expectations (Benkenstein et al., 2006).

With regard to the **phase of customer recovery**, the general aim is to avoid customer churn or to reacquire customers who have already left. As part of the customer related quality management strategies, it must be decided whether the measures used for reacquisition (e.g. credits, discounts, complaint responses, repairs) are used in the same way for all customers (standardization) or whether the measures and instruments are designed differently for different customer groups (e.g. compensation payments for less important customers vs. innovations according to the needs of important customers).

In terms of customer expectations being met through the provision of a service by the provider, another service-specific focus can be used: the subdivision of the dimensions of service quality in relation to the potential, process and result of a service. From this perspective, strategic priorities can be established for quality management. Under the heading of performance related strategies, these—in the sense of a focus of quality management activities on the service dimensions—are recorded as follows:

- Potential oriented strategies
- Process oriented strategies
- Result oriented strategies

Potential oriented strategies focus quality related activities on the service potentials of the service provider (cf. Sect. 2.4). This is particularly recommended for potential oriented services, in which the service potentials play a more important role in the provision of the service than the service processes or the service result. Examples of this are thermal baths or amusement parks. In both cases, the service potentials play an important role in the quality assessment. Furthermore, the service potentials in general play an important role in the evaluation of the service quality, since only these are equipped with so-called search properties, based on which service customers can at least partially assess the service quality before using the service.

If, on the other hand, the service process is in the focus of the quality management activities, this is referred to as a **process oriented strategy**. This is particularly suitable for services in which the service process is predominantly important. Examples of this are concert events. In this case, the synchronous contact between supplier (concert promoter) and customer (concertgoer) is emphasized. Accordingly, so-called experience properties of the service play a special role in the customer-side quality assessment of the concert.

If a service is primarily about the service result, an **outcome oriented strategy** might be suitable. The focus of quality efforts is then on ensuring a high-quality service result, as is absolutely necessary, for example, in the context of medical services or insurance benefits. As a result, an outcome oriented strategy is always to be chosen when services are based primarily on so-called trust properties. In this case, it is then necessary to generate trust, for example in the form of strong brands, in order to make intangible offers tangible through information substitutes and quality surrogates (Meffert et al., 2018, p. 329).

The differentiated analysis of customer needs often also results in the need to link different approaches of the types of strategies.

6.3.3.3 Competition Related Strategies

In terms of competition related strategies, both the behavior of the competition and the intensity of competition within a certain market serve as a decision making basis. In the area of performancerelated strategies, four **strategy directions** can be described in dealing with the competition:

- Avoid strategy
- Cooperation strategy

- Conflict strategy
- Adaptation strategy

The principle of an **avoid strategy** represents the fulfillment of customer expectations for service quality by means of services that prevent direct competition with competing companies. An avoid strategy can be realized, for example, by means of innovative services or service processes as well as by pronounced marketing efforts (Meffert et al., 2018, pp. 184f.). An example of this are online video libraries, in which a customer obtains the rental films by download via the Internet. In the context of a **cooperation strategy**, on the other hand, the cooperation with the competition is deliberately sought at the level of quality management. An example of this is the agreement and fixation of processes with low expenditure and error potential for the processing of customer changes between competitors, for example in the field of energy supply or in the telecommunications sector. In the case of industry related high normative expectations and low potential for the formation of predictive expectations, the perceived service quality is primarily based on the objective service quality. In this context, the **conflict strategy**, which is in contrast to the cooperation strategy, is to be mentioned. Here, a "performance struggle" with competing companies is sought in the sense of service quality, for example by means of specifying own quality standards. A so-called **adaptation strategy** is characterized in that companies, for example, adapt to given market forms or to existing quality guidelines. This corresponds, for example, to the imitation of the existing industry related service quality or to the fulfillment of generally valid market standards.

In the context of a **critical assessment** of existing strategies, the focus is on the differentiated treatment of normative and predictive expectations in the context of expectation based strategies. In this context, it is necessary to capture the level of predictive and normative customer expectations with respect to the potential, process and result features of the service offering identified as relevant (absolute expectation analysis) in an analysis phase preceding the formulation of strategy. It's also necessary, in the context of a relative expectation analysis, to identify those service features for which there is a need for action. Furthermore, a competitive analysis is required in order to obtain the predictive expectations with respect to the services of the main competitor and to compare them with the expectations related to the company's own service offering. Finally, it is also useful in this context to observe the expectations management measures of the competition and their effects on the company's own expectations management and to incorporate them into the strategy formulation process (Richter, 2005, 184 pp.).

The formulation of **performance based strategies**, requires a knowledge of and intensive engagement with the quality requirements relevant to customers in order to efficiently improve the quality of the service actually delivered or to optimally meet the expectations of the service quality. This is where methods for measuring service quality come into play (see Chap. 5).

In addition, expectation based and performance based strategies are focused on quality management in this case. Accordingly, corporate strategies that are subordinate to quality management must be taken into account. Finally, in the context of the **overall corporate strategy**, the strategic quality position is to be determined. This, in turn, defines the cornerstones for the strategic orientation of quality management.

The definition of the strategic orientation of quality management forms the basis of the **operational orientation** of quality management within the framework of planning and control. In this context, activities in the field of expectation management and performance-related quality management are distinguished—analogous to the formation of quality management strategies. In the following Chap. 7 therefore first the operational planning of expectation management is presented, while in Chap. 8 the operational planning of performance related quality management takes place.

References

Abell, D. F. (1980). *Defining the business. The starting point of strategic planning*. Englewood Cliffs: Prentice Hall.

Ahlert, M. (2003). *Einsatz des Analytic Hierarchy Process im Relationship Marketing. Eine Analyse strategischer Optionen bei Dienstleistungsunternehmen*. Wiesbaden: Gabler.

Alajoutsijärvi, K., Möller, K., & Tähtinen, J. (2000). Beautiful exit: How to leave your business partner. *European Journal of Marketing*, *34*(11/12), 1270–1289.

Anosike, U. P., & Eid, R. (2011). Integrating internal customer orientation, internal service quality, and customer orientation in the banking sector: An empirical study. *The Service Industries Journal, 31*(14), 2487–2505.

Backhaus, K., & Voeth, M. (2014). *Industriegütermarketing* (10th Edn.). München: Vahlen.

Balck, H. (2004). Wertschöpfungsketten im Industriellen Facility Management. In U. Lutz & K. Galenza (Eds.), *Industrielles facility management* (pp. 9–26). Berlin: Springer.

Benkenstein, M., Stuhldreier, U., & Uhrich, D. S. S. (2006). Customer Lifetime Value durch Beziehungsmarketing im Privatkundengeschäft von Universalbanken. In B. Günter, & S. Helm (Eds.), *Kundenwert: Grundlagen – innovative Konzepte – praktische Umsetzungen* (pp. 709–730). Wiesbaden: Gabler.

Berry, L. L., Bennett, D., & Brown, C. (1989). *Service quality. A profit strategy for financial institutions*. Homewood: Irwin Professional Pub.

Blockus, M. O. (2010). *Komplexität in Dienstleistungsunternehmen: Komplexitätsformen, Kosten- und Nutzenwirkungen. Empirische Befunde und Managementimplikationen*. Wiesbaden: Gabler.

Boulding, W., Kalra, A., Staelin, R., & Zeithaml, V. A. (1993). A dynamic process model of service quality. From expectations to behavioral intentions. *Journal of Marketing Research,30*(1), 7–27.

Brown, S., Bessant, J. R., & Lamming, R. (2013). *Strategic operations management*. London: Routledge.

Bruhn, M. (1998). *Wirtschaftlichkeit des Qualitätsmanagements. Qualitätscontrolling für Dienstleistungen*. Berlin: Springer.

Bruhn, M. (1999). Qualitätssicherung im Dienstleistungsmarketing. Eine Einführung in die theoretischen und praktischen Probleme. In M. Bruhn & B. Stauss (Eds.), *Dienstleistungsqualität. Konzepte – Methoden – Erfahrungen* (3rd Edn., pp. 21–48). Wiesbaden: Gabler.

Bruhn, M. (2000). Kundenerwartungen. Theoretische Grundlagen, Messung und Managementkonzept. *Zeitschrift für Betriebswirtschaft, 70*(9), 1031–1053.

Bruhn, M. (2012). *Marketing für Nonprofit-Organisationen. Grundlagen, Konzepte, Instrumente* (2nd Edn.). Stuttgart: Kohlhammer.

Bruhn, M. (2016). *Relationship Marketing. Das Management von Kundenbeziehungen* (5th Edn.). München: Vahlen.

Bruhn, M., & Georgi, D. (2000). Kundenerwartungen als Steuerungsgröße. *Marketing ZFP, 22*(3), 185–196.

Bruhn, M., Lucco, A., & Wyss, S. (2008). Beendigung von Kundenbeziehungen aus Anbietersicht-Wirkung der wahrgenommenen Gerechtigkeit auf die Zufriedenheit und Verbundenheit ehemaliger Kunden in unterschiedlichen Beendigungsszenarien. *Marketing ZFP, 30*(2), 221–237.

Chase, R., & Bowen, D. (1991). Service quality and the service delivery system. In S. W. Brown, E. Gummesson, B. Edvardsson, & B. Gustavsson (Eds.), *Service quality. Multidisciplinary and multinational perspectives* (pp. 157–176). Lexington: Lexington.

Cronin, J. J., & Taylor, S. A. (1992). Measuring service quality. A reexamination and extension. *Journal of Marketing, 56*(3), 55–68.

Deutsche Gesellschaft für Qualität e. V. (1995). *Begriffe zum Qualitätsmanagement* (6th Edn.). Berlin: Beuth.

Fliess, S., Lasshof, B., & Willems, G. (2006). Qualitätsstandards im Dienstleistungsprozess. In M. Bruhn & B. Stauss (Eds.), *Dienstleistungscontrolling. Forum Dienstleistungsmanagement* (pp. 71–88). Wiesbaden: Gabler.

Frehr, H.-U. (1999). Total-Quality-Management. In W. Masing (Eds.), *Handbuch Qualitätsmanagement* (4th Edn., pp. 31–48). München: Hanser.

Günter, B., Helm, S., & Eggert, A. (2017). *Kundenwert. Grundlagen – Innovative Konzepte – Praktische Umsetzungen* (4th Edn.). Wiesbaden: Gabler.

Hadwich, K., & Keller, C. (2015). Interne Servicequalität in Unternehmen: eine empirische Untersuchung der Einflussfaktoren und Auswirkungen. *Zeitschrift für betriebswirtschaftliche Forschung, 67*(2), 170–205.

Haist, F., & Fromm, H. (1991). *Qualität im Unternehmen. Prinzipien, Methoden, Techniken* (2nd Edn.). München: Hanser.

Hernandez, H. (2010). Quality audit as a driver for compliance to ISO 9001:2008 standards. *The Total Quality Management Journal*, 22(4), 454–466.

Heskett, J. L. (1988). *Management von Dienstleistungsunternehmen. Erfolgreiche Strategien in einem Wachstumsmarkt*. Wiesbaden: Gabler.

Hinsch, M. (2014). Verantwortung der obersten Leitung. In M. Hinsch (Eds.), *Qualitätsmanagement in der Luftfahrtindustrie* (pp. 41–49). Berlin: Springer.

Hoang, D. T., Igel, B., & Laosirihongthong, T. (2010). Total quality management (TQM) strategy and organisational characteristics: Evidence from a recent WTO member. *Total Quality Management & Business Excellence*, *21*(9), 931–951.

Homburg, C., Fassnacht, M., & Guenther, C. (2003). The role of soft factors in implementing a service-oriented strategy in industrial marketing companies. *Journal of Business to Business Marketing, 10*(2), 23–51.

Hoyle, D. (2009). *ISO 9000 quality systems handbook. Using the standards as framework for business improvement* (6th Edn.). Oxford: Butterworth-Heinemann.

Jun, M., & Cai, S. (2010). Examining the relationships between internal service quality and its dimensions, and internal customer satisfaction. *Total Quality Management & Business Excellence, 21*(2), 205–223.

Kim, D. Y., Kumar, V., & Kumar, U. (2012). Relationship between quality management practices and innovation. *Journal of Operations Management, 30*(4), 295–315.

Krafft, M. (2007). *Kundenbindung und Kundenwert* (2nd Edn.). Heidelberg: Physica.

Lucco, A. (2009). *Anbieterseitige Kündigung von Kundenbeziehungen: Empirische Erkenntnisse und praktische Implikationen zum Kündigungsmanagement*. Wiesbaden: Springer.

Meffert, H., Burmann, C., & Kirchgeorg, M. (2014). *Marketing. Grundlagen marktorientierter Unternehmensführung* (12th Edn.). Wiesbaden: Springer Gabler.

Meffert, H., Bruhn, M., & Hadwich, K. (2018). *Dienstleistungsmarketing. Grundlagen, Konzepte, Methoden* (9th Edn.). Wiesbaden: Springer.

Nazeer, S., Zahid, M. M., & Malik, F. A. (2014). Internal service quality and job performance: Does job satisfaction mediate? *Journal of Human Resources, 2*(1), 41–65.

Parasuraman, A., Zeithaml, V. A., & Berry, L. L. (1988). SERVQUAL: A multiple-item scale for measuring consumer perceptions of service quality. *Journal of Retailing, 64*(1), 12–40.

Parente, D. H., Pegels, C. C., & Suresh, N. (2002). An exploratory study of the sales-production relationship and customer satisfaction. *International Journal of Operations & Production Management, 22*(9), 997–1013.

Raffée, H. (1984). Strategisches Marketing. In E. Gaugler, O. H. Jacobs, & A. Kieser (Eds.), *Strategische Unternehmensführung und Rechnungslegung* (pp. 61–81). Stuttgart: Schäffer-Poeschel.

Reimann, G. (2012). *Erfolgreiches Qualitätsmanagement nach DIN EN ISO 9001:2008. Lösungen zur praktischen Umsetzung* (2nd Edn.). Berlin: Beuth.

Reinecke, S., & Janz, S. (2007). *Marketingcontrolling: Sicherstellen von Marketingeffektivität und -effizienz*. Stuttgart: Kohlhammer.

Richter, M. (2005). *Dynamik von Kundenerwartungen im Dienstleistungsprozess. Konzeptionalisierung und empirische Befunde*. Wiesbaden: Gabler.

Rothlauf, J. (2014). *Total Quality Management in Theorie und Praxis. Zum ganzheitlichen Unternehmensverständnis* (4th Edn.). München: Oldenbourg.

Russell, R. F., & Stone, A. G. (2002). A review of servant leadership attributes: Developing a practical model. *Leadership & Organization Development Journal, 23*(3), 145–157.

Saaty, T. L. (1980). *The analytic hierarchy process. Planning, priority setting, resource allocation*. New York: McGraw-Hill.

Schein, E. H. (1995). *Unternehmenskultur: Ein Handbuch für Führungskräfte*. Frankfurt a. M: Campus.

Shell (2014). Qualitätsgrundsätze von Shell. http://www.shell.de/content/dam/shell-new/local/country/deu/downloads/pdf/shell-de-qualitaetsgrundsaetze-2014.pdf. Zugegriffen: 20. Mai 2016.

Shell (2019). www.shell.de/ueber-uns/our-values.html. Zugegriffen: 26. Apr. 2019.

Stauss, B., & Bruhn, M. (2010). Serviceorientierung im Unternehmen – Eine Einführung in die theoretischen und praktischen Problemstellungen. In M. Bruhn & B. Stauss (Eds.), *Serviceorientierung im Unternehmen. Forum Dienstleistungsmanagement* (pp. 4–32). Wiesbaden: Gabler.

van Raaij, F. W. (1991). The formation and use of expectations in consumer decision making. In H. H. Kassarjian, & T. S. Robertson (Eds.), *Handbook of consumer behavior* (pp. 401–418). Englewood Cliffs: Prentice Hall.

Wilken, C. (2013). *Strategische Qualitätsplanung und Qualitätskostenanalysen im Rahmen eines Total Quality Management*. Heidelberg: Springer.

Zhao, L., Lu, Y., Zhang, L., & Chau, P. Y. (2012). Assessing the effects of service quality and justice on customer satisfaction and the continuance intention of mobile malue-added services: An empirical test of a multidimensional model. *Decision Support Systems, 52*(3), 645–656.

7 Operational Design of Expectation Management for Services

Abstract

In addition to the service perceived by the customer, customer expectations form a second decisive determinant of perceived service quality. Therefore, it is not only the task of a comprehensive service quality management system to control the quality of the services provided. Rather, the customer-side expectations also form a decisive control variable for the management of service quality. Accordingly, this chapter provides a consideration of the tasks of a so-called expectation management. In addition, the most important influencing factors on customer expectations are analyzed. Based on this, a measurement and control of customer expectations is possible. The approaches discussed in this context are distinguished by direct (e.g. guarantees) and indirect (e.g. service promises) measures of "expectation management" and presented in detail.

7.1 Tasks of Expectation Management

The **customer expectations** are, in addition to the performance perceived by the customer, the second determinant of the perceived service quality (cf. Chap. 2). This means that it is not only the task of a comprehensive quality management system to control the quality of the services delivered (cf. Chap. 8). Rather, the customer expectations also represent a control variable of a comprehensive quality management system.

This is all the more true because the desired control direction of the customer expectations is not clear. First, there are **conflicting relationships between different types of expectations.** Predictive customer expectations, i.e. the level of performance considered probable by the customer, have a positive effect on the perception of quality. In contrast, the normative customer expectations, i.e. the requirements of the customers for the

M. Bruhn, *Quality Management for Services*,
https://doi.org/10.1007/978-3-662-67032-3_7

performance level of a provider, represent a negative determinant of the quality perception (cf. Chap. 2). Since predictive and normative expectations also interact to a certain extent, measures of expectation management can also have a counterproductive effect. This is the case when the predictive expectations are, as ideally desired, increased, the normative expectations also experience an increase and, as a result, no improvement in perception can be achieved due to the opposite effects on the perceived service quality.

Complicating matters is the fact that **activities of other corporate functions** in addition to expectation management also—and to some extent always have—made customer expectations the subject of control. Thus, measures of classical marketing (e.g. classical advertising, PR) serve to increase predictive expectations. Service companies strive, through advertising or also an appropriate pricing, that customers receive a positive image of the company's services, i.e. expect a high quality of performance (in terms of predictive customer expectations). The goal of these customer oriented effects is to achieve a purchase decision of the customer with respect to their own services. In addition to an increase in predictive expectations, however, the corresponding measures also have the effect of increasing normative expectations. Customers choose a supplier, amongst other reasons, on the basis of a high price, since the price serves as a quality indicator. If they pay a high price, they expect (in terms of normative expectations) a high quality of performance in return.

Thus, a central task of expectation management is to see the **alignment of predictive and normative customer expectations**. Many expectation determining measures affect both types of expectations, so it is important for companies to weigh up exactly what the overall effect of the respective measures on the perceived service quality will be. It is even more effective to coordinate the two aspects in such a way that the normative expectations are simultaneously reduced and the predictive expectations are increased. The following example shows the practical implementation of these considerations.

Expectation management at an airline

It generally appears to be difficult to control normative and predictive expectations separately. However, this would be necessary to achieve the desired effects on perceived service quality.

A few years ago, an American airline chose an interesting approach in this regard: In an advertisement, it was proclaimed that all airlines have problems keeping to their scheduled arrival times (lowering of normative expectations), but that it copes best with this problem itself (increasing of predictive expectations) (Boulding et al., 1993).

This advertisement therefore addresses the quality characteristic "punctuality" of an airline. The measure serves to control the normative expectation type of the product-type level, i.e. the average quality level in an industry. If the customer knows that all companies have these problems, he or she will not deal with it too critically at his or her own airline. In addition, predictive expectations are controlled by assuring the customer that one copes best with this problem oneself. This generally leads to a more positive perception of the actual punctuality in the sense of a "self-fulfilling prophecy". As a result, the customer's expectations are controlled by the advertisement in such a way that the customer perceives the punctuality of the airline more positively. ◀

Furthermore, it must be taken into account in expectation management that neglecting expectation management carries the **risk of an expectation spiral** (Bruhn, 2000). The basic principle of quality management is to meet or even exceed normative expectations in order to achieve a positive perception of quality as a result. Experience shows that exceeding customer expectations in particular can lead to increased expectations. If the customer realizes that the provider can even do more than he originally expected, he will adjust his expectations upwards as a result of this experience. In turn, the provider tries to improve his performance, thereby increasing customer expectations again. A continuation of this thought experiment has the expectation spiral as a result. So it is another task of expectation management to ensure a **moderate level of normative expectations** (Bruhn & Georgi, 2000).

It becomes clear that customer expectations are a much more complex control variable than other marketing target variables. Variables such as sales, customer satisfaction or the perceived quality of service are basically, as long as it makes economic sense to increase them, but in the rarest of cases to reduce. In contrast, there are **conflicting goals** in expectation management, so it is very surprising that approaches to systematic expectation management are hardly known in literature and practice.

Expectation management requires certain levers by which customer expectations can be controlled. The determinants of customer expectations represent starting points for the identification of such levers.

7.2 Determinants of Customer Expectations as a Starting Point for Expectation Management

In particular, due to the complexity of customer expectations, service providers usually start with the **determinants of customer expectations** as levers for their control. At a higher level, customer related, supplier related and competitor related determinants can be differentiated as determinants of customer expectations (cf. Fig. 7.1; see for an overview Zeithaml et al., 1993).

The **demand related determinant** subsumes those aspects which are derived from the customer's personality or the customer's situation (Clow & Beisel, 1995, p. 35). These aspects cannot be controlled by the supplier. The personality related determinants (Thompson and Kaminski, 1993) include the customer's personal needs (e.g. money and time savings through online check-in), his personal service philosophy as well as his own performance capability (e.g. competence for carrying out online check-in). Especially for services for which intensive involvement of the external factor is required for their creation (e.g. advisory services), this determinant affects the expectations. If a

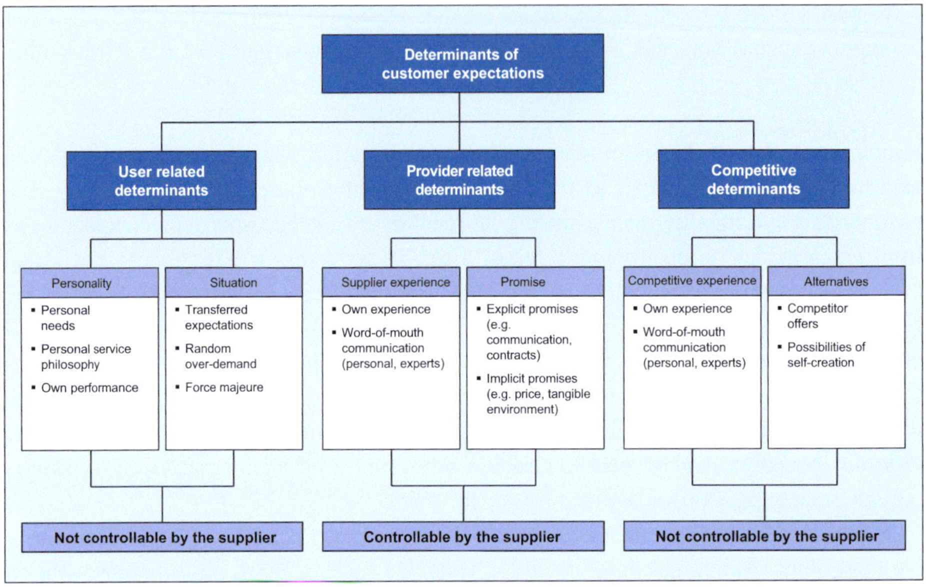

Fig. 7.1 Determinants of customer expectations. (Source: Based on Zeithaml et al., 1993, p. 8)

customer is aware that he cannot contribute to the creation of the service to the same extent (e.g. provision of the required information for the provision of the advisory service) as is usual or agreed, he will not demand an outstanding service from the supplier. The situation related determinants are attributed to higher powers (e.g. bad weather), a random over-demand and so-called transferred expectations. Transferred expectations refer to expectations which a customer has of a supplier in order to fulfil the expectations of other persons (e.g. his own customers or his superiors).

Those determinants which are generated in the contact between customer and supplier and which can thus be controlled by the supplier are referred to as **supplier related determinants**. Here, experience with the supplier (e.g. hotel stay) and promises made by the supplier (e.g. service promises of the hotel) can be distinguished. Experience with the supplier refers to the assessment of the services received by the supplier from the customer's point of view up to now (Kopalle & Lehmann, 1995; Goldstein et al., 2002). Here, both the customer's own experience (e.g. previous stays in the hotel) and the experience of other customers conveyed to him by word of mouth from personal friends and acquaintances or experts (e.g. performance tests in magazines; experience reports from the Internet) are taken into account (Webster, 1991; Anderson, 1994; Helm and Günter, 2000). Promises made by the supplier are activities of the supplier which point to the level of the service to be used. In explicit service promises (e.g. communication, contracts, hotel website), this level of service is specified in concrete terms (Strader & Katz, 1990; Hill & Gandhi, 1992, p. 70; Kopalle and Lehmann, 1995, p. 281). In implicit promises of performance (e.g. tangibles environment, price of hotel stay), the customer infers the expected level of performance from observable quality indicators (Sweeney et al., 1992; Kopalle & Winer, 1996; Eboli and Mazzulla, 2011).

Finally, **competitive determinants** refer to those aspects that are related to other suppliers of the same services. Like the demand related determinants, the competitive determinants are also not controllable by the supplier himself. Competitive determinants include experience with other suppliers (e.g. other hotels) and perceived service alternatives (e.g. accommodation offers from other hotels). With regard to experience with other suppliers, both one's own experience (e.g. already made overnight stays) and experiences of others, which are known through word of mouth communication (e.g. experience reports on hotels on the Internet), can be differentiated. The perceived service alternatives concern the customer's ability to obtain the suppliers services in other ways (e.g. booking an alternative hotel with the same or similar properties). In addition to the offers of the competition, the possibility of self-production, in which the customer competes with the supplier, can also affect customer expectations.

In the light of the determinants of customer expectations as potential levers for expectation management, service companies develop a systematic expectation management.

7.3 Instruments of Expectation Management

Expectation management is primarily about influencing customer expectations in a productive direction. For this it is necessary, firstly, to know the customer expectations and, secondly, to identify means by which they can be influenced. Accordingly, there are two groups of instruments in the context of expectation management:

1. Instruments for the analysis and measurement of customer expectations
2. Instruments for the control of customer expectations

7.3.1 Measuring and Analyzing Customer Expectations

The analysis phase of the customer expectation management process has the analysis of customer expectations as its subject. This means first of all that a **measurement of customer expectations** is carried out in order to make decisions on this basis later. For the measurement of customer expectations, basically numerous of the methods discussed in the quality measurement procedures (see Chap. 5) can be used. From a perspective of expectation management, four **dimensions** can be used to differentiate between different measurement approaches (Bruhn, 2000; see Fig. 7.2):

(1) Dimensionality of expectation measurement In one-dimensional measurement, expectations are measured in relation to overall performance (global measurement)

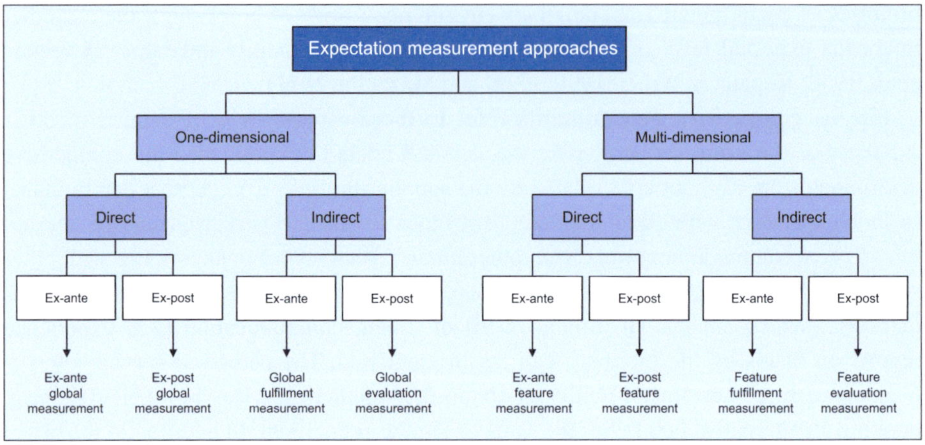

Fig. 7.2 Approaches to measuring customer expectations

(Liljander & Strandvik, 1993, p. 121; Imrie et al., 2002), while multidimensional measurement focuses on individual performance characteristics (e.g. waiting time at the bank counter) (feature measurement). Compared to global measurement, feature measurement has the advantage that it can be used to draw specific implications for expectation management measures. The disadvantage of feature measurement is that there are performance characteristics that are difficult for customers to assess, for which they do not form expectations or which they do not even perceive as a performance characteristic before using the service. For certain features (e.g. color for products, waiting time in line for services), expectation concretization is more possible from the customer's perspective than for others (e.g. friendliness of employees for services). Accordingly, not all features are necessarily captured by a feature measurement and, as a result, incorrect priorities may be derived with regard to the features most relevant to the customer. The following example illustrates this.

Implicit customer expectations in restaurants

The manager of an English restaurant was wondering how he could improve the design and offer of the restaurant and turned to the *City College Norwich* in England. This designed a survey to measure the quality of the restaurant and the (predictive) expectations with regard to the restaurant services. The visitors of the restaurant were asked to name the aspects that had struck them positively or negatively in the restaurant. Passers-by were asked about their positive or negative assessment with regard to the services to be expected in the restaurant (predictive expectations). Subsequently, it was compared which aspects were expected by passers-by and perceived by customers. Figure 7.3 shows the corresponding results.

For example, 5% of customers noticed that the staff was friendly. None of the customers (0%) had explicitly formulated this as an expectation. With regard to negative aspects, 23% of customers were put off by the price, while none of the passers-by expected this as a negative point.

These results illustrate that it is quite realistic that customers perceive quality features in contact with the service provider, about which they had not previously thought and for which they had not formed any expectations.

(Source: Johns and Howard, 1998) ◄

(2) Timing of expectation measurement Expectations can be measured before (ex-ante measurement) or after service consumption (ex-post measurement) (Swan, 1977; Swan & Trawick, 1980; Cadotte et al., 1987; Liljander & Strandvik, 1993, p. 121; Hawkins et al., 2015). If expectations are measured ex-post, there is a risk of retrospective adjustment of expectations. As a result, expectations are not measured as they were previously. In contrast, ex-ante measurement is associated with a higher effort; in particular,

	Expressed as expectation (%)	Evaluation at the restaurant (%)
Positive features that were not expected by the customer:		
Food		
Traditional/authentic	0	5
Consistent	0	0
Climate		
Own Table	0	9
Spacious	0	5
Artwork	0	14
Employees		
Competent	0	0
Friendly	0	5
Atmosphere		
Friendly/hospitable	0	0
Negative features that were not expected by the customer:		
Food		
Poorly cooked	0	14
Price		
Service fee	0	23
Employees		
Young/inexperienced	0	9
Inappropriately dressed/shabby	0	9

Fig. 7.3 Implicit customer expectations with regard to a restaurant. (Source: Johns & Howard, 1998, p. 256)

if in addition to expectations, quality perception (ex-post) is measured. Furthermore, the ex-ante measurement is associated with the problem described above that customers may not be able to express their expectations before service consumption. Fig. 7.4 shows an example of how predictive expectations were measured before the first visit to the language school.

(3) Directness of expectation measurement In direct measurement, customers are specifically asked about their expectations, while in indirect measurement, expectations are determined by measuring quality assessments (Berry & Parasuraman, 1998; Ladhari, 2009). Direct measurement places high demands on the subjects to express such a complex construct as expectations. The subject is expected to be able to retrospectively separate expectation and perception in an ex-post measurement. An indirect measurement, in which inferences are made about expectations indirectly through the importance weights of individual quality features from the customer's perspective, can only take place ex-post. In addition, it can not determine expectation levels, but only the relative importance of quality features.

What do you think **will** the **equipment** and **staff** of Gamma language school **be like**?

Please check only one box!	very good	good	rather good	medium	rather bad	bad	very bad
2. Equipment and personnel will be	☐	☐	☐	☐	☐	☐	☐

What do you think the **course of** the language course at Gamma will be like?

Please check only one box!	very good	good	rather good	medium	rather bad	bad	very bad
3. The course of the language course will be	☐	☐	☐	☐	☐	☐	☐

What do you think **will be** the **personal impact/outcomes** achievable for you through the language course at Gamma?

Please check only one box!	very good	good	rather good	medium	rather bad	bad	very bad
4. The for me personal impact/outcomes at Gamma will be...	☐	☐	☐	☐	☐	☐	☐

What do you think it **will be like** at Gamma language scholl **overall**?

Please check only one box!	very good	good	rather good	medium	rather bad	bad	very bad
5. Overall it will be at Gamma language school...	☐	☐	☐	☐	☐	☐	☐

Fig. 7.4 Measurement of predictive expectations in relation to a language school. (Source: Richter, 2005, p. 269)

In general, there are different methods available for measuring the **importance of quality features**. The range extends from relatively simple methods (e.g. paired comparisons, ranking methods, ratings, constant sum methods) via certain standardized methods (e.g. Kano model, price sensitivity analyses, scoring model, portfolio analyses) to the relevant statistical methods (e.g. regression analyses, conjoint analyses). Reference is made to the corresponding standard literature of market research and statistics.

(4) Explicitness of expectation measurement This criterion is a subdimension of indirect expectation measurement. In the case of explicit measurement, expectations are explicitly referred to in the statement formulation (fulfillment measurement), while in the case of implicit measurement, expectations are analytically determined on the basis of performance assessments (evaluation measurement) (cf., for example, Parasuraman et al., 1988, 31 pp.; Rosen and Karwan, 1994; Ladhari, 2009). In the case of implicit measurement, the requirement to derive customer expectations from the evaluation measurement is even greater.

Depending on the respective expectation measurement a different **operationalization for the capture of expectations** is carried out. Figure 7.5 shows different statement formulations as they are used in customer surveys of service companies. With regard to the two types of expectations, it must be taken into account here that the measurement of normative expectations is carried out ex-ante and ex-post in the same way, as they have a time-independent character. This does not apply to the predictive expectations, which explicitly relate to future service utilization. Therefore, an ex-post measurement is more

Expectation type / Measurement approach	High normative expectation	Low normative expectation	Predictive expectation
Ex-ante global measurement	"An excellent supplier always delivers a performance without decency."	"A supplier should at least deliver the service in a way that is usable."	"I think supplier X will deliver an outstanding performance."
Ex-post global measurement			"When I chose supplier X, I thought he would deliver outstanding performance."
Global fulfillment measurement	"Supplier X's performance was like an optimal vendor."	"Supplier X's performance was acceptable for a provider in this category."	"Supplier X's performance was as I had previously imagined."
Global assessment measurement	"I am very satisfied with the services provided by supplier X."		
Ex-ante feature measurement	"The staff of an excellent supplier is always friendly."	"Supplier X's performance was acceptable for a provider in this category."	"I think that the employees of supplier X will be very friendly becoming."
Ex-post feature measurement			"When I chose supplier X, I thought his staff would be very friendly."
Feature fulfillment measurement	"Supplier X's staff was as friendly as an optimal supplier's staff."	"Supplier X's staff was as friendly as can be expected, at least."	"The staff at supplier X were as friendly as I thought they would be."
Feature assessment measurement	"I am very satisfied with the friendliness of supplier X's staff."		
The statements are to be rated on a scale from "strongly agree" to "strongly disagree".			

Fig. 7.5 Question formulations in measuring customer expectations. (Source: Bruhn, 2000, p. 1041)

difficult. Furthermore, problems can arise if the two types of expectations are measured at the same time (Oliver, 2010, pp. 81–82). However, there are studies in which, for example, a separate measurement of predictive and normative expectations was carried out as part of experiments (see, for example, Boulding et al., 1993; Spreng et al., 1996; Fiedler et al., 2010).

The results of the expectation measurement are further analyzed by the service provider in order to derive the basis for systematic measures of expectation management. Here, two essential **analysis approaches** can be distinguished:

- In the **absolute expectation analysis** it is about measuring and analyzing the level of customer expectations. This is done primarily against the background that predictive expectations have a positive and normative expectations have a negative effect on quality perception.
- In the **relative expectation analysis**, the expectations are analyzed differently for customer groups or types of services. The results then allow the derivation of marketing measures, differentiated according to customer groups and/or differentiated according to services.

Ideally, the analysis results flow into the planning of expectation management according to the planning process of expectation management.

7.3.2 Control of Customer Expectations

Based on the analysis results, service companies try to steer customer expectations in the desired direction. Expectations are not stable over time, so measures of expectation management can be carried out both before and after the use of a service. For services that are created in contact with the customer, measures of expectation management are also necessary during the purchase or creation phase, as expectations can also change during the creation process or are controllable.

The already identified **determinants of customer expectations** (cf. Fig. 7.1) act as starting points for a control of expectations. The measures of expectation management can be set at these determinants. At an aggregated level, these determinants can be condensed into three essential **control levers of expectation management** (Richter, 2005, pp. 213 f.):

- Expectation relevant service information which is imparted to the customer by the company implicitly and explicitly.
- The quality of the customer's perceived previous experiences with the supplier and its services.
- Expectation relevant information about the supplier in the form of recommendations imparted to the customer by other persons.

Accordingly, expectation management for services can be divided into areas of **information control**, **experience control**, and **recommendation control**. The influence of the determinants on customer expectations also differs in terms of directness, so that direct and indirect expectation management can be differentiated as forms of expectation management (cf. Fig. 7.6).

7.3.2.1 Direct Expectation Management

In direct expectation management, measures are used that have an immediate effect on customer expectations. For this reason, the term expectation management in the narrower sense can also be used in this context (see also Richter, 2005, pp. 220 ff. below). Within the framework of direct expectation management, expectations are controlled by means of **information control**. By means of information control, an increase or stabilization or reduction of predictive and normative expectations is possible. Measures of information control can be divided into explicit and implicit service information. The group of **explicit service information** includes:

1. Guarantees
2. Service statements
3. Presentation of test results or performance evidence

The different forms of explicit service information are briefly explained below.

(1) Guarantees Guarantees have a binding character for companies with regard to the compliance with the guarantees given to the outside (Tucci & Talaga, 1997). This leads to a high credibility of guarantees and thus to a high expectation steering effect (Wu et al., 2012).

Guarantees are obviously only suitable for expectation enhancement and not for expectation reduction. An example of a service guarantee is a commitment regarding the maximum waiting time for an order in a restaurant. For example, many pizza delivery services advertise that a delivery is made within a certain radius within 30 minutes. If the delivery takes longer, the customer does not have to pay for the delivery.

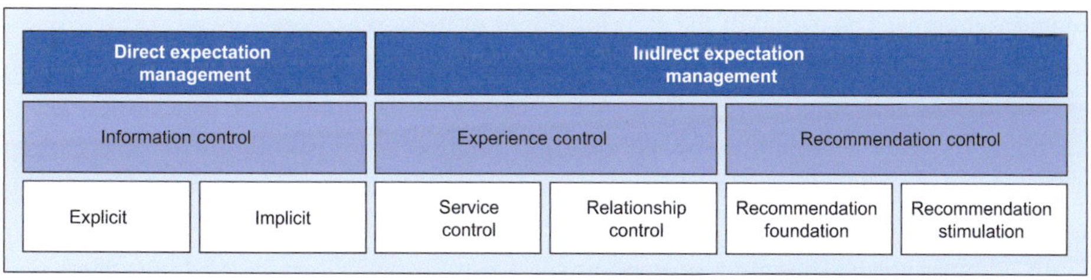

Fig. 7.6 Direct versus indirect expectation management for services. (Quelle: Richter, 2005, S. 183)

(2) Service statements In contrast to guarantees, service statements are suitable for both increasing and decreasing customer expectations. A language school can point to the excellent reputation of the certificate of completion. In this case, the service statement is a service promise that increases the customer's expectations. Vice versa, expectations can also be lowered by service statements. Using the example of the language school, this could be done by pointing out that only the basics of the language can be taught in a basic language course and that further follow-up courses are necessary for greater learning progress. However, when it comes to lowering expectations, it is important to take into account the strategic guideline that is usually in place to lower normative expectations more than predictive expectations. Using the example of the language school, this could be implemented by the service statement that while only the basics are taught in the basic course, the language progress will be noticeably better than that of competing offers. The success of service statements in changing expectations depends primarily on the credibility of the service statements.

In a close relationship to service statements as part of a systematic management of expectations for services are the aspects of communication policy and the importance of personnel policy. In principle, it is important to critically examine all communication messages of the supplier for their potential influence on expectation management and to adjust them if necessary in order to exclude unintended, negative influencing effects.

In the service creation process, customer contact employees play a key role in relation to expectation management (Sirianni et al., 2013; Bitner & Wang, 2014). The success of expectation management depends largely on the motivation and ability of employees to implement the planned expectation management strategies. This is to be ensured by the use of employee related quality management instruments (see Sect. 8.3.1).

(3) Presentation of test results or performance evidence The presentation of test results (e.g. seals of quality, stars in a restaurant, test results of goods or services, etc.) serves primarily to credibly support service statements. The presentation of these test results can be carried out via all communication channels in principle. Test results are often integrated into television advertising, published on the homepage, or displayed directly at the point of sale (branches, business premises, etc.). Due to their neutral and objective character, test results have a high degree of credibility. The predictive expectations of customers can be increased by an active communication of test results or performance evidence. The extent of the steering effect of these measures depends primarily on the awareness and reliability of the neutral institution issuing the test results.

The use of test results or performance evidence in the context of expectation management requires recognition by an external neutral test institution. To obtain this, a high level of performance quality must be ensured by means of a systematic quality management.

In addition to the measures of explicit service information presented here, there is a variety of **implicit service information.** This includes, in particular, the price of the service, the tangible environment, the sales system and the reputation of the supplier or the service. A high degree of these categories of implicit service information acts as a positive quality indicator, which increases customer expectations. In particular, when customers have no other credible information or experience with a service provider—which is often the case with services due to a high proportion of experience and belief properties—customers will form their preliminary expectations on the basis of implicit service information. Nevertheless, it must be taken into account that the design of implicit service information for targeted expectation management—with the exception of the price—is a lengthy and difficult process. Fig. 7.7 shows, by way of example, measures for explicit and implicit information management.

7.3.2.2 Indirect Expectation Management

In contrast to direct expectation management, measures of indirect expectation management do not have an immediate effect on customer expectations. At an aggregated level, two essential control levers can be identified (see also Richter, 2005, 213 pp.). These are the expectation management by

1. Experience control
2. Recommendation control

The two approaches are presented below.

(1) Expectation management by experience control Experience control does not act directly on customer expectations, but indirectly via a contact situation with the supplier that is perceived as positive or negative. The expectation management by **experience control** is based on the consideration that the customer's expectations are all the higher the more positive the customer assesses his previous experiences with the provider. Experience control is attributed to both an influence on predictive expectations (Boulding et al., 1993) and—to a lesser extent—to normative expectations. With regard to normative expectations, positive experiences in particular influence the customer's

Information control measures			
Phase / Task	Before using the service	During the service utilization	After the service has been used
Explicit service information	E.g. ▪ Formulation of advertising promises ▪ Incorporation of test results in the communication messages	E.g. ▪ Service statements on the quality of results in the context of personal communication between customer and employee	E.g. ▪ Issuing of guarantees ▪ Apology for any service deficiencies
Implicit service information	E.g. ▪ Price fixing ▪ Design of the distribution system	E.g. ▪ Design of the tangible environment ▪ Internalization/externalization	E.g. ▪ Sending an image brochure about the supplier ▪ Sending exclusive gifts for the birthday of customers

Fig. 7.7 Example measures of information control management. (Source: Richter, 2005, p. 228)

expectation level (e.g. discounts received when booking a holiday). In general, a customer's expectations of a supplier or a service increase if he has already had positive experiences with a service provider or a service offering (e.g. repeated desire for discounts if the holiday is booked again with the same supplier). Vice versa, expectations decrease in the event of negative experiences (e.g. after an unsuccessful hotel stay, expectations decrease when staying overnight in this hotel again).

The central tasks of experience management relate to the **control of performance and the relationship** with customers. Experience management is particularly suitable for increasing expectations. This can be achieved, for example, by a gradual and consistent improvement in performance, which ideally ends in setting an industry standard (e.g. the *Ritz-Carlton hotel group* sets decisive service standards in the hotel industry). In the context of relationship management, for example, measures to maintain individual business relationships with the customer in the sense of relationship marketing are suitable. Measures of both performance and relationship management can be taken in the pre-consumption, consumption and post-consumption phases. Figure 7.8 gives an overview of possible measures of experience management depending on the consumption phase of a service.

(2) Expectation management by recommendation control Finally, in the context of holistic expectation management, a recommendation control is also recommended, as recommendations—from friends and acquaintances as well as from independent experts—can influence customer expectations in a positive and negative way. They determine the predictive customer expectations, because the customer can form a more concrete image of the service offering of a company on the basis of statements to which he trusts more than company information (Boulding et al., 1993). In addition to that, recommendations can control the normative expectations by customers demanding aspects that they experience through word of mouth communication about certain suppliers.

Experience management measures			
Task \ Phase	Before using the service	During the service utilization	After the service has been used
Service control	E.g. ▪ Trial use of the service ▪ Offer of customized services	E.g. ▪ Adding innovative service components ▪ Optimizing service components	E.g. ▪ Offer after-sales services ▪ Complaint management
Relationship control	E.g. ▪ Use of free service numbers ▪ Offer of information options	E.g. ▪ Empowerment of employees (to give them the opportunity to consciously shape the relationship with customers)	E.g. ▪ Sending congratulatory cards for client's birthdays or similar personal occasions

Fig. 7.8 Example measures of experience management. (Source: Richter, 2005, p. 216)

Due to the immateriality of services, the recommendation by third parties often plays an important role in the purchase decision. For this reason, a large number of evaluation platforms (e.g. *Tripadvisor*) have been formed on the Internet. There, consumers have the opportunity to rate services that they have already experienced, such as hotels or whole holiday trips. Since these assessments are written by independent third parties, they are often considered to be highly credible.

The great danger of (negative) word of mouth communication is that it is hardly influenceable and controllable by companies. Especially because of the rapid spread of word of mouth communication via the Internet, the central **task of recommendation control** is to encourage one's own customers to engage in positive word of mouth communication (see also Richter, 2005, 216 pp. below). A prerequisite for positive customer communication is the realization of a high level of service quality. After all, a customer will report on his positive experiences with a supplier, especially if he has had positive experiences. The realization of a high level of service quality is therefore a essential foundation for the recommendation management. In this context, one can also speak of a recommendation foundation (Bruhn, 2000, p. 1045).

In addition to the recommendation foundation, the offensive **stimulation of word of mouth communication** is another task of the recommendation control. This is about activating the recommendation potential of customers. This can be done on the basis of financial incentives (e.g. launch of "customer referral campaigns") or also intrinsic or altruistic motives. An example of this is the establishment of customer forums in which customers report on their experiences with a service provider from unselfish motives (Gummerus et al., 2012; see Sect. 5.1.2.3). Furthermore, the use of so-called "professional rumor managers" would be conceivable, who, on behalf of the service provider, influence the expectations of potential or current customers, for example, by spreading positive word of mouth communication in online customer forums, at the supplier itself or in public places (Richter, 2005, pp. 218–219).

Customer referral programs

- Banking services: When customers recommend the "*Postbank Giro plus*" at the *Postbank* to friends and acquaintances, they secure 50 EUR for themselves and the newly recruited customers.
- IT services: Customers receive bonus storage space at *Dropbox* when friends or acquaintances install *Dropbox* on their computers.
- Mobile phone services: If the mobile phone provider *Simyo* is recommended to friends, the existing customer will receive a bonus credit of up to 50 EUR.
- Insurance services: If friends or acquaintances join *ADAC*, they can consume prizes from different categories (e.g. car & tools, cooking & living, house & garden, etc.).
- Electricity services: If friends, relatives and acquaintances are recruited for *Naturstrom*, they can choose between material and money prizes. ◀

The latest technological development of the Internet as well as the resulting change in user behavior towards an increasing design of content by users ("Web 2.0", see Sect. 5.1.2.3) open up new possibilities for **expectation management by recommendation control.** In particular, so-called social networks (e.g. *Facebook, LinkedIn*) allow Internet users to forward content to people connected in a virtual network in a few steps. In addition to social networks, there are also other social media forms such as micromedia (e.g. the short message service *Twitter*) or blogs that offer the possibility—usually in the form of recommendation buttons above or below the content on the website—to recommend the content to other Internet users. This offers companies the opportunity for active recommendation control via the Internet. The first step for companies is to provide content (e.g. an advertisement) on the Internet (e.g. on the provider's homepage) in the hope that it will be recommended or forwarded to other Internet users as quickly as possible. The goal of a so-called "viral marketing campaign" is that the company message infects the Internet users and spreads like a virus on the Internet and thus positively influences the expectations of the customers.

Recommendation on web 2.0

- Expectation control through recommendations: Recommendation of hotels by experience reports on Tripadvisor.com.
- Expectation control through advertising: With the help of the "tell a friend" function, companies can send recommendation e-mails to friends or acquaintances of the customer via online marketplaces.
- Expectation control through social networks: On *LinkedIn* networked contacts can be specified as references. They are a credible proof of the quality of one's own work and more personal than the formal certificate of the last employer. ◀

Figure 7.9 shows exemplary measures of recommendation control depending on the respective phase of use of a service.

In summary, a service provider has a large number of different measures of information, experience and recommendation control in the context of expectation management to influence customer expectations. However, these measures cannot be used independently of each other. Rather, in terms of an **integrated concept**, coordination between the instruments is required. Only by coordinated use can synergy effects be realized in terms of a higher control effect of the measures and compensatory effects of the individual instruments be avoided (Richter, 2005, pp. 228–229). In addition, it becomes clear that expectation management cannot be considered independently of classical quality management, since the perceived service can have a significant influence on expectation formation—as has been shown in particular in the context of experience and recommendation control.

Recommendation control measures			
Task \ Phase	Before the service utilization	During the service utilization	After the service utilization
Recommendation foundation	E.g. ▪ Service description with detailed information about special features of the provider ▪ Prize competition	E.g. ▪ Showing reference lists with well-known customers who are convinced of the performance or the service provider	E.g. ▪ Sending special information about the advantages of services the participants of the customer recruit customer offer
Offensive recommendation simulation	E.g. ▪ Lobbying ▪ Public relations measures ▪ Viral marketing campaign	E.g. ▪ Personal request for recommendation ▪ Draw attention to customer recruit customer offer ▪ Viral marketing campaign	E.g. ▪ Setting up interactive customer forums ▪ Sending documents for newly acquired customers ▪ Viral marketing campaign

Fig. 7.9 Exemplary measures of recommendation control. (Source: Richter, 2005, p. 220)

The following example shows the experiment of a hotel with regard to the different effects of expectation control.

Control of customer expectations in a hotel

The general manager of a hotel has the habit of spontaneously questioning the guests of his hotel in the lobby about their well-being and expectations of the hotel. It turned out that the customer expectations are partly very heterogeneous and the ideas of the expected performance differ significantly. Against the background of the very standardized performance of a hotel, the manager saw the danger of not being able to meet all customer expectations. Based on this, an investigation was carried out into the effects of measures of expectation management in the hotel. The influence of the three **expectation determinants**

- Service promises (exaggerated promises with regard to the expected performance)
- Service information (information about a negative situation during the stay, namely a disturbance of rest due to a construction site)
- Service performance (the actual perception of the guest of the performance and the hotel itself)

on the two **expectation types**

- predictive and
- normative expectations

was examined.

The following, partly surprising **results** were achieved:

- The predictive expectations were only influenced by the service information (negative). In contrast, they were neither influenced positively by the service promises nor by the service performance.
- The normative expectations, which are rather considered to be resistant to control, were significantly influenced by all three determinants, namely negatively by the service information and positively by the service promises and the service performance.

This allows various **conclusions** to be drawn for expectation management:

- The service promises only affect the normative expectations. This makes it more difficult to satisfy the customers. Exaggerated representations are therefore not useful in the context of expectation management.
- The more realistic service information has a negative effect on both types of expectations. After all, the normative expectations are also lowered and the probability of satisfying the customers is increased. However, the predictive expectations also fall. Too negative representations can therefore lead to a negative assessment of a service provider and, for example, deter potential customers.
- The service performance only has a positive effect on the normative expectations. This is a typical indication of the expectation spiral. In this case, it could be important for the hotel to also confirm guests' opinions by moderate communication.

(Source: Bruhn & Georgi, 2000, 191 pp.) ◀

After defining and designing expectation management, the next step is to plan the operational design of a performancerelated quality management. The control loop of quality management serves as the basis, and the implementation of quality management is oriented towards its individual phases.

References

Anderson, E. W. (1994). *Customer satisfaction and word-of-mouth. Working paper*. Michigan.

Berry, L. L., & Parasuraman, A. (1998). Wie Servicewünsche genau erfasst werden. *Harvard Business Manager, 20*(3), 80–91.

Bitner, M. J., & Wang, H. S. (2014). Service encounters in service marketing research. In R. T. Rust, & M. H. Luang (Hrsg.), *Handbook of service marketing research* (pp. 221–243). Northampton: Edward Elgar Pub.

Boulding, W., Kalra, A., Staelin, R., & Zeithaml, V. A. (1993). A dynamic process model of service quality. From expectations to behavioral intentions. *Journal of Marketing Research, 30*(1), 7–27.

Bruhn, M. (2000). Kundenerwartungen. Theoretische Grundlagen, Messung und Managementkonzept. *Zeitschrift für Betriebswirtschaft, 70*(9), 1031–1053.

Bruhn, M., & Georgi, D. (2000). Kundenerwartungen als Steuerungsgröße. *Marketing ZFP,22*(3), 185–196.

Cadotte, E. R., Woodruff, R. B., & Jenkins, R. L. (1987). Expectations and norms in models of consumer satisfaction. *Journal of Marketing Research, 24*(3), 305–314.

Clow, K. E., & Beisel, J. E. (1995). Managing consumer expectations of low-margin, high-volume services. *Journal of Services Marketing, 9*(1), 33–46.

Eboli, L., & Mazzulla, G. (2011). A methodology for evaluating transit service quality based on subjective and objective measures from the passenger's point of view. *Transport Policy, 18*(1), 172–181.

Fiedler, M., Hossfeld, T., & Tran-Gia, P. (2010). A generic quantitative relationship between quality of experience and quality of service. Network. *IEEE, 24*(2), 36–41.

Goldstein, E. B., Ritter, M., & Herbst, G. (2002). *Wahrnehmungspsychologie* (2nd Edn.). Heidelberg.

Gummerus, J., Liljander, V., Weman, E., & Pihlström, M. (2012). Customer engagement in a facebook brand community. *Management Research Review, 35*(9), 857–877.

Hawkins, T. G., Gravier, M. J., Berkowitz, D., & Muir, W. A. (2015). Improving services supply management in the defense sector: How the procurement process affects B2B service quality. *Journal of Purchasing and Supply Management, 21*(2), 81–94.

Helm, S., & Günter, B. (2000). Kundenempfehlungen. Resultat und Ausgangspunkt des Kundenbindungsmanagements im Dienstleistungsbereich. In M. Bruhn, & B. Stauss (Ed.), *Kundenbeziehungen im Dienstleistungsbereich. Jahrbuch Dienstleistungsmanagement 2000* (pp. 103–130). Wiesbaden: Springer.

Hill, D. J., & Gandhi, N. (1992). Services advertising. A framework to its effectiveness. *Journal of Services Marketing, 6*(4), 63–76.

Imrie, B. C., Cadogan, J. W., & McNaughton, R. (2002). The service quality construct on a global stage. *Managing Service Quality: An International Journal, 12*(1), 10–18.

Johns, N., & Howard, A. (1998). Customer expectations versus perceptions of service performance in the foodservice industry. *International Journal of Service Industry Management, 9*(3), 248–265.

Kopalle, P. K., & Lehmann, D. R. (1995). The effects of advertised and observed quality on expectations about new product quality. *Journal of Marketing Research, 32*(3), 280–290.

Kopalle, P. K., & Winer, R. S. (1996). A dynamic model of reference price and expected quality. *Marketing Letters, 7*(1), 41–52.

Ladhari, R. (2009). A review of twenty years of SERVQUAL research. *International Journal of Quality and Service Sciences, 1*(2), 172–198.

Liljander, V., & Strandvik, T. (1993). Different comparison standards as determinants of service quality. *Journal of Consumer Satisfaction, Dissatisfaction and Complaining Behavior, 6*, 118–132.

Oliver, R. L. (2010). *Satisfaction. A behavioral perspective on the consumer* (2. Edn). New York.

Parasuraman, A., Zeithaml, V. A., & Berry, L. L. (1988). SERVQUAL: A multiple-item scale for measuring consumer perceptions of service quality. *Journal of Retailing, 64*(1), 12–40.

Richter, M. (2005). *Dynamik von Kundenerwartungen im Dienstleistungsprozess. Konzeptionalisierung und empirische Befunde*. Wiesbaden: Gabler.

Rosen, L. D., & Karwan, K. R. (1994). Prioritizing the dimensions of service quality. *International Journal of Service Industry Management, 5*(4), 39–52.

Sirianni, N. J., Bitner, M. J., Brown, S. W., & Mandel, N. (2013). Branded service encounters: Strategically aligning employee behavior with the brand positioning. *Journal of Marketing, 77*(6), 108–123.

Spreng, R. A., MacKenzie, S. B., & Olshavsky, R. W. (1996). A reexamination of the determinants of consumer satisfaction. *Journal of Marketing, 60*(3), 15–32.

Strader, M. K., & Katz, B. M. (1990). Effects of a persuasive communication on beliefs, attitudes, and career choice. *Journal of Social Psychology, 130*(2), 141–150.

Swan, J. E. (1977). Consumer satisfaction with a retail store related to the fulfillment of expectations on an initial shopping trip. In R. L. Day (Ed.), *Consumer satisfaction, dissatisfaction and complaining behavior* (pp. 10–17). Bloomington: Indiana University.

Swan, J. E., & Trawick, F. I. (1980). Satisfaction, disconfirmation and comparison of alternatives. In K. H. Hunt & R. L. Day (Eds.),*Refining concepts and measures of consumer satisfaction and complaining behavior. Papers from the fourth annual conference on consumer satisfaction, dissatisfaction and complaining behavior* (pp. 17–24). Bloomington: Indiana University.

Sweeney, J. C., Johnson, L. W., & Armstrong, R. W. (1992). The effect of cues on service quality expectations and service selection in a restaurant setting. *Journal of Services Marketing, 6*(4), 15–22.

Thompson, A. M., & Kaminski, P. F. (1993). Psychographic and lifestyle antecedents of service quality expectations. A segmentation approach. *Journal of Services Marketing, 7*(4), 53–61.

Tucci, L. A., & Talaga, J. (1997). Service guarantees and consumers' evaluation of services. *Journal of Services Marketing, 11*(1), 10–18.

Webster, C. (1991). Influences upon consumer expectations of services. *Journal of Services Marketing, 5*(1), 5–17.

Wu, C. H. J., Liao, H. C., Hung, K. P., & Ho, Y. H. (2012). Service guarantees in the hotel industry: Their effects on consumer risk and service quality perceptions. *International Journal of Hospitality Management, 31*(3), 757–763.

Zeithaml, V. A., Berry, L. L., & Parasuraman, A. (1993). The nature and determinants of customer expectations of service. *Journal of the Academy of Marketing Science, 21*(1), 1–12.

Operational Design of Quality Management for Services

8

Abstract

For the operational implementation of a Total Quality Management, a systematic use of specific instruments of quality management of services is necessary to ensure a high quality in the various phases of the service creation process. For this purpose, the control loop of quality management, consisting of the four phases of quality planning, control, auditing and management documentation, is used in this chapter. Based on this concept, the various instruments for the implementation of quality management are explained in detail and their possible uses are shown. The chapter closes with possibilities for coordinating and integrating the individual phases into a comprehensive quality management system for services.

8.1 Quality Management Loop

For the implementation of a Total Quality Management (TQM) and a systematic quality management system for service companies, the use of specific quality management instruments is necessary to ensure the quality of the service in the different phases of the service creation process. In this context, most of the time it is spoken of as a **quality management loop** (Lehmann, 1995, p. 173; Schreiber, 2007; Seifert, 1998). Based on this concept, companies with an implemented quality management system have built their own systems according to their specific needs in the past (see the quality management systems of various service companies in Bruhn & Meffert, 2002; Schlick & Rochhold, 2007).

M. Bruhn, *Quality Management for Services*,
https://doi.org/10.1007/978-3-662-67032-3_8

Ideally, a quality management system can be oriented towards the classic management functions of planning, implementation and control according to the loop concept. Here, four **phases** can be distinguished (DIN ISO 9001, 2015):

- Quality planning
- Quality control
- Quality auditing
- Quality management documentation

These ideal typical phases are shown in Fig. 8.1 as a control loop.

How the different **phases of the quality management system for service companies** are to be designed and which individual quality-related measures necessarily have to be used in the respective phases will be shown below. It is necessary here to specify for the respective quality measures which employee groups of the service company are mainly responsible for which concepts. In this context, four **levels of responsibility for quality measures** are to be distinguished in principle (cf. Fig. 8.2). Deviating structures may be necessary due to the different sizes and organizational structures of service companies.

In order to achieve a goal-oriented pursuit of the tasks of quality management and thus an efficient implementation of the quality objectives, a **holistic, integrated implementation of the quality management system** is necessary within the company. When designing an integrated quality management system, two integration levels can be distinguished:

- Level of intraphase integration
- Level of interphase integration

On the **level of intraphase integration**, it is about aligning the instruments within one phase. In this context, it is first necessary to examine for each phase which networking

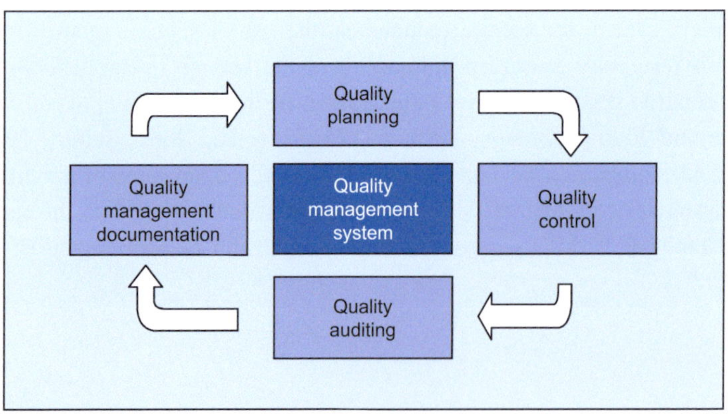

Fig. 8.1 Ideal typical phases of a quality management system

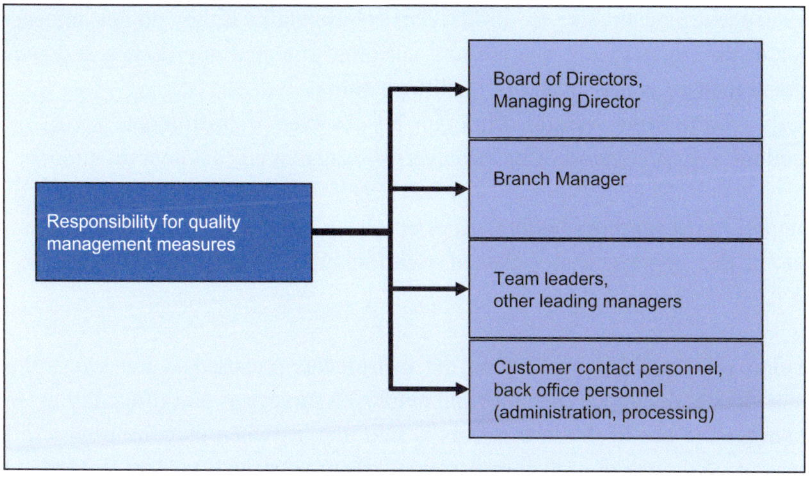

Fig. 8.2 Levels of responsibility for quality management measures

potential exists with the individual instruments in order to achieve synergy effects through the integrated use of the instruments of the respective phase in the realization of quality objectives. Within the framework of this interdependence analysis of quality instruments, **functional, temporal and hierarchical relationships** are analyzed. While the temporal or hierarchical relationship analysis examines intertemporal or hierarchical relationships, the functional relationships are concerned with the contextual and factual interdependency of effects with regard to a quality objective. (Bruhn, 2014, pp. 123–134).

In **functional relationships** five **types** can be distinguished:

- If two or more instruments show **complementary relationships** to each other, they supplement or support each other in their effect. Lead and follow-up instruments can be distinguished, with the latter taking on a support function for the lead instruments. For example, in the context of quality management, the work of quality circles, which, among other things, is to find solutions to quality problems, can be supported by measures of personnel development in order to convey quality related know-how to employees.
- In **conditional relationships** between instruments, the use of one (or more) instrument(s) is necessary for the effect of other instruments, e.g. in the context of quality assurance before the Problem-Detecting-Method is carried out, the determination of potential problem areas, e.g. by means of incident oriented methods, is necessary.
- **Substituting relationships** are then established if the same effect can be achieved with two or more instruments, e.g. cross-departmental and non-position related personnel development.

- Between instruments there are **indifferent relationships** if they do not influence each other in their effects, e.g. feature and customer oriented approaches and sequential incident method in the context of quality planning.
- Finally, **competitive relationships** can be observed if instruments act in opposite directions, e.g. strong cost orientation versus exaggerated customer orientation.

With regard to the **temporal effect**, an effect delay can be distinguished from an effect transfer. In this context, four **types of relationships** are distinguished (Bruhn, 2014, 117 pp.):

- **Parallel relationships** exist when the instruments are used at the same time, e.g. simultaneous conduct of customer and employee surveys as part of quality assurance.
- An **successive use** of the instruments is said to exist when they are used at different times, e.g. determination of problem areas using the critical incident technique before the FRAP is carried out as part of quality planning.
- **Intermittent relationships** exist when part of the instruments is used continuously, while other instruments are used on an ad hoc basis, e.g. incentive systems and ad hoc training as part of quality management.
- By replacing one instrument with another over time, **replacing relationships** are characterized, e.g. replacement of the problem detecting method by the FRAP.

If you look at the **hierarchical relationships** between instruments of quality management, then part of the instruments has strategic importance (e.g. features and customer oriented approaches as part of quality planning), while other instruments have operational character (e.g. Fishbone analysis as part of quality planning).

In addition to the networking analyses within the individual phases, it is necessary to plan, on the **level of interphase integration**, the use of quality instruments holistically and across phases in order to exploit as many synergy effects as possible to achieve the quality objectives.

The following analyzes which instruments can be used in each of the four phases of the quality management control loop. The instruments to be examined are, in part, the approaches to measuring service quality presented in Chap. 5. When dealing with these instruments, this chapter will therefore not provide a detailed description, but rather focus on their concrete possibilities for use, their use and networking possibilities within an integrated quality management. In addition to the approaches to measuring service quality, however, further instruments, in particular those of quality control and management, will be added in the following, to which correspondingly more detailed explanations will be given.

After each phase, the instruments of the respective phase will be characterized in a systematic assessment grid with regard to their possibilities of use within quality management according to the following **features**:

- Strategic goal
- (Operative) goals for action
- Subject of investigation
- Individual measures
- Success drivers
- Main contact person in the company
- Resources required
- Frequency of implementation
- Responsible person
- Cooperation with other departments

The assessment according to these criteria should provide indications for the **suitability of quality management instruments** for companies in their specific situation. Based on the control loop of quality management, the individual phases of planning, controlling, auditing and documentation of quality are explained below.

8.2 Instruments of Quality Planning

The **Quality Planning** as the first phase of a systematic quality management can be defined as follows (EN ISO 9000, 2014):

▶ **Quality Planning** is that part of quality management that is directed towards specifying quality objectives and the necessary execution processes as well as the associated resources for achieving the quality objectives.

A short definition describes quality planning as "planning and further developing the quality requirements for the unit under consideration" (Deutsche Gesellschaft für Qualität e. V., 1995a, p. 95). Accordingly, this first phase of quality management for service companies includes the planning and further development of the quality requirements for the various services of the company. Not the quality of the services themselves, but the various quality requirements are to be planned.

8.2.1 Overview of the Instruments

The **tasks of quality planning** are basically the following activities:

- Determination of customer expectations with regard to product and service quality
- Setting specific quality objectives
- Development of concepts for their implementation

A variety of instruments are available to the service company to implement these tasks, as shown in Fig. 8.3. Within the **quality planning**, the specific requirements for service quality from the customer's and supplier's perspective are determined in order to be able to develop and offer products and services of the company in accordance with the expectations of the relevant stakeholder groups. As generally the use of quantitative and qualitative **measurement methods for the capture of quality expectations and perceptions of service customers** could be carried out, is shown in Fig. 8.4 in overview. The management of the service company will be primarily responsible for ensuring that comprehensive customer surveys are carried out at least once a year. It is advisable if each branch office receives a detailed evaluation of its results, whereby the customer

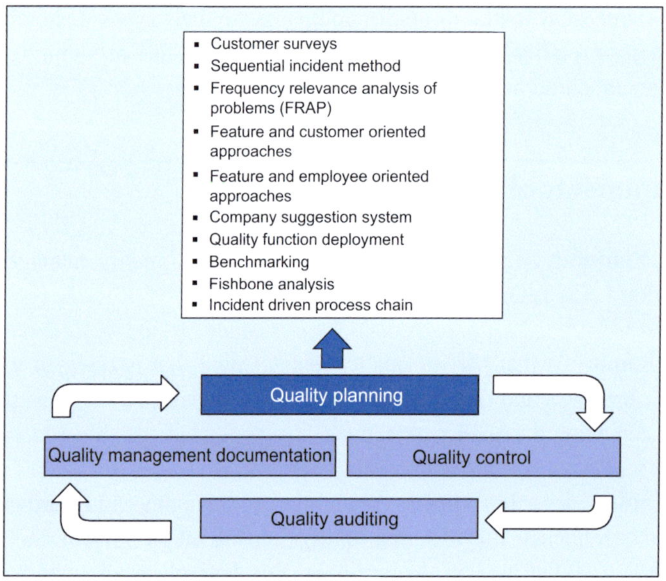

Fig. 8.3 Instruments of quality planning of service companies

Quality instrument	Quantitative/qualitative measurement methods (SERVQUAL, Blueprint, FRAP analysis)
Objective of the assignment	Capture customers' quality expectations/perceptions
Areas of investigation	All aspects of the service chain in the company
Responsible parties	Executives of the service company, market research departments
Responsible parties	Regular customer surveys and data evaluations
Activities period/time	At least once a year

Fig. 8.4 Exemplary planning scheme for measuring service quality

judgments are to be prepared in a comprehensible manner in relation to the different individual criteria, but also global values.

In view of the numerous quality measurement methods (cf. Chap. 5) the following instruments will be briefly introduced, which appear to be particularly suitable for the **recording of specific quality criteria** (Benkenstein, 1993; Haller, 1998; Brüggemann, 2012). These are the instruments listed below, some of which have already been discussed in detail in the context of measuring service quality (cf. Chap. 5):

a) Classical customer surveys
b) Sequential incident method (SIM)
c) Frequency relevance analysis of problems (FRAP)
d) Feature and customer oriented approaches
e) Feature and employee oriented approaches
f) Company suggestion system
g) Quality function deployment (QFD)
h) Benchmarking
i) Fishbone analysis
j) Incident driven process chain

The listed quality planning instruments will be explained in more detail below.

(a) Classical customer surveys Customer surveys are used to determine the quality requirements for a service from the customer's perspective. Often, customer satisfaction with the quality of the service is queried. This is done by querying the global overall satisfaction with the company or by querying the satisfaction with individual quality features. The subject of the investigation is the service perceived by the customer. Questionnaires are often used as instruments for customer surveys. The questionnaire can be created and sent in writing, verbally or online by e-mail. The advantage of standardized questionnaires lies in the uniformity and high standardization of the survey. A large number of customers can be surveyed and their answers compared by using standardized questionnaires. The disadvantage is a often quite low return rate and the limited possibility to query customer specific features, since the questions and question categories have been defined in advance and can not be changed. Customer surveys are regularly carried out, at least once a year.

Customer survey using the example of a painting company

Fig. 8.5 shows the example of a customer questionnaire of a painting company, which is sent standardized to each customer after the service has been provided. This should give the painting company as comprehensive a picture as possible of the most important features from its point of view, such as punctuality, friendliness, etc.. The information on the assessed painter and the date of the service creation also makes it possible to identify valuable information on the respective performance level of the individual employees and possible weaknesses. ◀

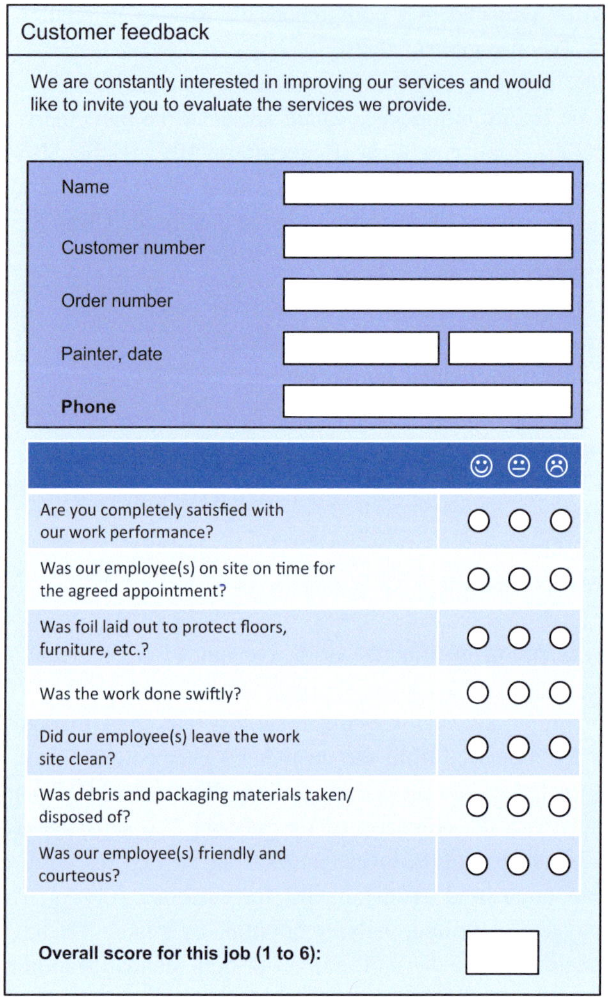

Fig. 8.5 Example of a customer questionnaire for a painting company

(b) Sequential Incident Method (SIM) In measuring the requirements for service quality based on specific key experiences of customers, the Sequential Incident Method results in a systematic representation of the service process using a graphical flowchart (Stauss & Hentschel, 1991, p. 242; Meffert et al., 2018, pp. 224 f.). This so-called **Blueprint** enables the complete capture and evaluation of various contact situations from the customer's perspective (see Sect. 5.1.2.2).

(c) Frequency Relevance Analysis of Problems (FRAP) Furthermore, the Frequency Relevance Analysis of Problems (FRAP) is often used in the context of operational quality planning. Visual measurement instruments such as blueprinting and FRAP analysis are often used by service companies to supplement the capture of various components of the service process and to identify obvious problems or areas for action (see Sect. 5.1.2.3).

(d) Feature and customer oriented approaches With the help of feature and customer oriented approaches, especially through the multi-attribute measurement methods, in particular the performance criteria relevant from the customer's point of view as well as general satisfaction measures and internal determinants of service quality are determined (Parasuraman et al., 1985, 1988). In practice, the SERVQUAL or SERVPERF approach is often used for this purpose, which represent extremely practical methods for service companies from different industries (cf. Sect. 5.1.2.1). When using these instruments in connection with quality planning, companies have to make sure that they first test in focus group interviews with customers, expert interviews and pretests which quality criteria are particularly relevant for their company or industry. In regular, i.e. at least annual, customer and employee surveys, it is then to be determined how the stakeholder groups assess the performance criteria, what importance they attach to them and in which areas the causes of poor quality can be found within the company. In this context, the comparison of the assessment of performance criteria with their respective importance for customer satisfaction is of increased relevance. This relationship is illustrated in the following example.

Action relevance matrix for performance criteria in the example of online banks

The results of feature and customer oriented approaches using multi-attribute methods can be used to quantify the evaluation of the performance features of a service as well as their importance from the customer's perspective. The evaluation is shown in the example on a scale from one (very satisfied) to four (very dissatisfied), the importance is represented by the correlation value with the overall satisfaction of the service. The satisfaction of the features is set against each other in a coordinate system—a so-called action relevance matrix—of the respective importance (cf. Fig. 8.6). This allows implications of action in the sense of strategic directions for quality management to be derived. For example, the need for action lies particularly with those performance features that have a high positive impact on overall satisfaction, but are rated poorly by customers (quadrant bottom right, "Start here"). The action relevance matrix is also called importance relevance matrix. ◀

(e) Feature and employee oriented approaches Comprehensive employee surveys (Domsch, 1991; Borg, 2003; Brüggemann & Bremer, 2012; Homburg & Stock-Homburg, 2012), which can be evaluated using feature and employee oriented approaches, are of particular importance for the development and implementation of a quality management system in service companies (see Sect. 5.2.2).

Regularly, i.e. preferably at intervals of no more than twelve months, the market research departments of the service company are to carry out surveys with all employees to record the perceived quality of services provided externally and internally. It is

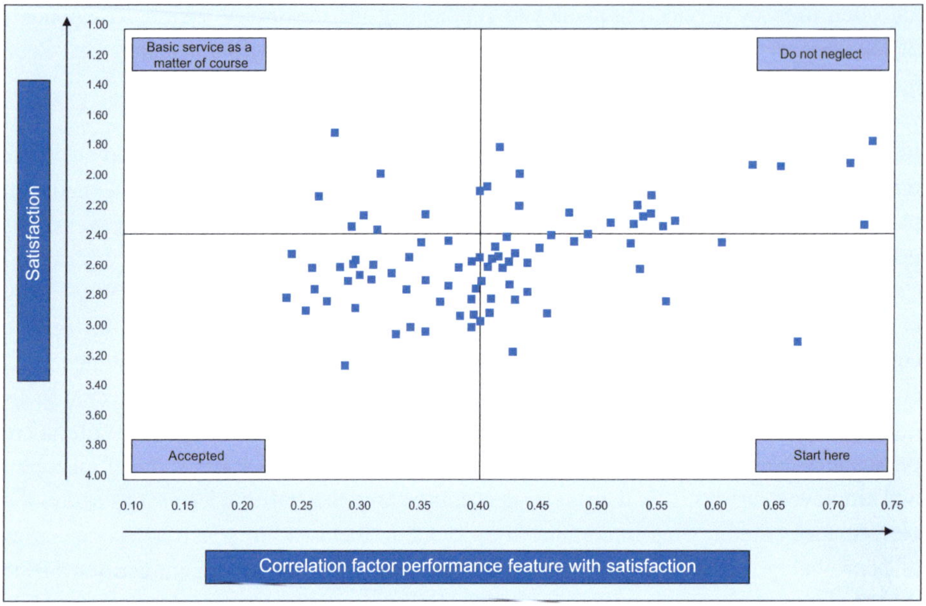

Fig. 8.6 Action relevance matrix of a service provider

important to determine how the employees assess the quality of the products and services from their personal point of view (Haist & Fromm, 1991; Homburg & Stock-Homburg, 2012), which customer expectations they accept. It is important to determine which quality criteria they consider to be important. Differences between the views of managers and employees in customer contact as well as differences between individual locations are particularly recorded here.

The management of the service company is generally always responsible for the comprehensive employee surveys, so that the surveys are accepted company-wide.

(f) Company suggestion system In addition to the necessary employee surveys, the company suggestion system (Merz & Biehler, 1994; Brinkmann & Simon, 1995; Thom, 2003; Kummer & Genz, 2004; Schulte, 2012; see Sect. 5.2.2) can be used to discover quality relevant problems that need to be considered when defining quality requirements as part of quality planning. For the sake of employee motivation, it is important to reward (both materially and immaterially) and implement improvement suggestions. In addition, it may be appropriate to publish rewarded suggestions, for example in employee newspapers or in the company's intranet, in order to give colleagues incentives to participate in the suggestion program (Haist & Fromm, 1991; Homburg & Stock-Homburg, 2012).

(g) Quality Function Deployment The concept of quality function deployment (QFD) represents a comprehensive systematics for customer oriented quality planning (Sondermann, 1994, 238 pp.; Gogoll, 2003; Saatweber, 2011, p. 72; Schmitt & Pfeifer, 2015, 709 pp.).

The methodology for ensuring customer orientation in all phases of planning and realization of products or services, developed in Japan at the end of the 1960s and then adopted by Europeans in the mid-1980s, is based on one (or more) matrix arrangements (see, for example, Saatweber, 2011), as schematically shown in Fig. 8.7.

Due to its specific form, this matrix is also referred to as **"House of Quality"** (Hauser & Clausing, 1988; Bruhn, 2002b, p. 168; Schmitt & Pfeifer, 2015); in it, the different analysis and planning steps of product or service concept and development planning are documented.

With regard to the specific procedure in the QFD method, the following **central process steps** can be identified (Saatweber, 1994, 449 pp., 2011; Gogoll, 2003, 5 pp.; Schmitt & Pfeifer, 2015, pp. 711-712):

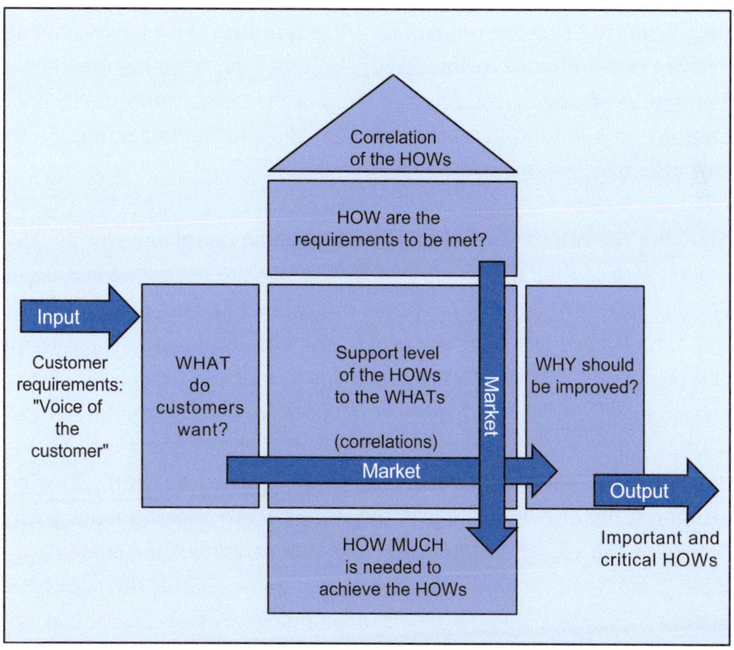

Fig. 8.7 Basic concept of quality function deployment (QFD) as "House of Quality". (Source: Saatweber, 2011, p. 72)

Step 1: Determination of the relevant customer groups The QFD process starts with an **identification of the target groups,** i.e. the relevant customer groups of the service company **(WHO?).** Here the internal and external recipients of the services are determined.

Step 2: Recording of customer requirements and needs Subsequently, the—often arranged in a tree structure and weighted—**wishes, needs** and **specific requirements** of the customers **(WHAT?)** are recorded for a service; data from primary and secondary market research can be used here.

Step 3: Derivation of quality features Furthermore, the performance features **necessary to meet customer requirements (HOW?)** are to be specified. Here, within the framework of the QFD project, a mental transformation process is necessary, which requires accuracy and creativity to develop different solution alternatives for often globally formulated customer wishes (Saatweber, 1994, p. 454).

Step 4: Determination of target values in relation to quality features In this work step, measurable or quantifiable target values and units **(HOW MUCH?)** are assigned to the quality features. The determination of the **quantitative expressions of the performance features** is followed by the determination of the optimization direction for the individual feature expressions (i.e. determination of the target value and its maximization or minimization). In addition, the technical difficulties in implementing the service, i.e. the degree of difficulty, are to be determined.

Step 5: Checking for interactions In addition, possible correlations or positive or negative interactions between the individual quality features are examined and documented in the correlation matrix, the "roof" of the house of quality. For example, the quality criterion "advice from employees" can correlate negatively with the feature "waiting times at the counter" and positively with the feature "behavior in case of complaints". Negative interactions provide clues for required changes or the need for completely different solutions.

Step 6: Performance comparisons with the services of competitors Finally, own and external customers, but also employees of the company are asked to judge **their own services** and **those of the closer competitors**. The comparison of the answers to questions such as "How good is our own performance?" and "How good is the performance of the main competitor?" provides clues to development potentials and competitive advantages.

All results of the activities described are then recorded in the so-called **"House of Quality",** often with the results of a QFD diagram serving as input variables for another QFD project at the same time (Schmitt & Pfeifer, 2015, pp. 713–715). The number of necessary QFD planning steps depends on the complexity of the service considered. In the following example, a section of a "House of Quality" for asset consulting by a credit institution is shown.

Excerpt from a "House of Quality" of a credit institution

Investment advice is part of the services offered by a credit institution. In this respect, the column **"WHAT"** lists the customer requirements with regard to the investment business. The column **"HOW"** lists the performance specifications of the investment business. These reflect the company's own resources that are made available to meet customer requirements. The comparison of customer requirements and performance specifications shows the causal relationship of the company's activities. This allows performance specifications to be differentiated that have a stronger or weaker influence on the fulfilment of customer requirements. At the same time, positive and negative interactions between the performance specifications are represented in the "roof" of the "House of Quality". Fig. 8.8 shows an example of the house of quality using the example of a credit institution. ◀

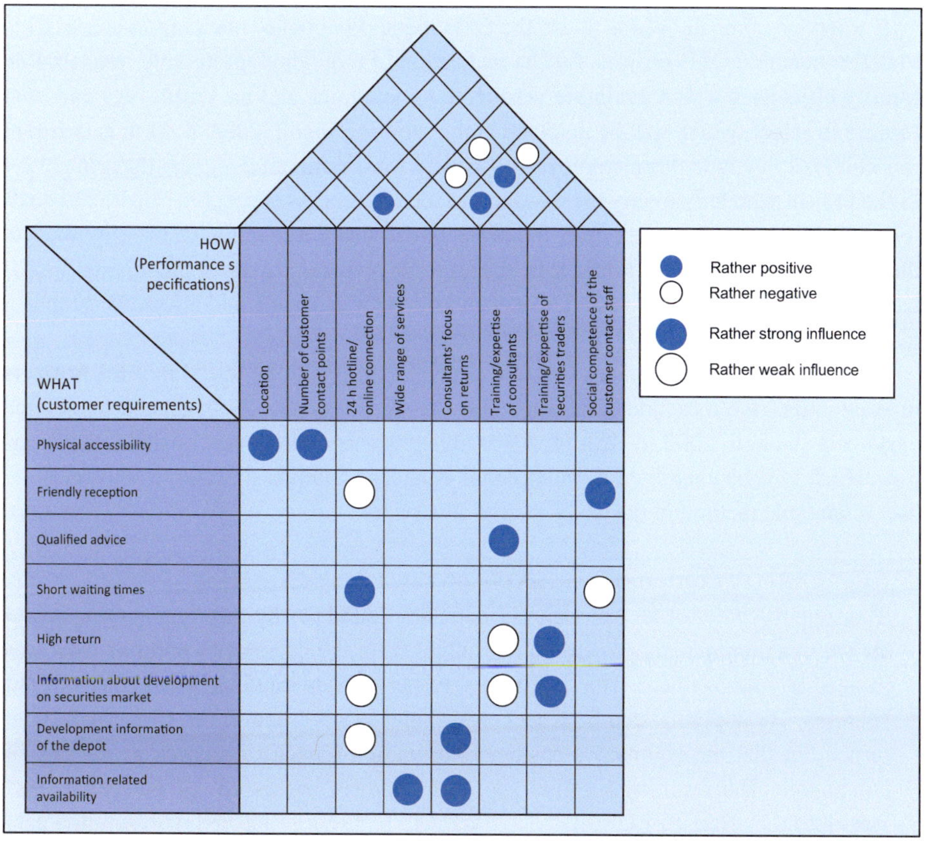

Fig. 8.8 Excerpt from a "House of Quality" using the example of investment advice

A problem with the traditional QFD method is that the derivation of quality features is too strongly based on subjective assessments and does not take into account interactions between the selected quality feature (Chun-Lang, 2006, p. 294). In addition, there are still some **difficulties** to be mentioned which can occur when using the QFD method in the service sector. First of all, it is already to be noted in the context of the capture of customer requirements that no detailed specification is made of the market research methods and procedures to be used (Gogoll, 2003, p. 6). Proposals for the use of the sequential incident method, information from customer complaints or the FMEA method also include the problems associated with these methods of data capture and analysis. In addition, the **phases of analysis of core services and optimization of processes** are also fraught with problems. For the use of the QFD method or the house of quality in the service sector, it is necessary to modify the concept. The immateriality of the services and the integration of the external factor make it difficult to create a clear separation of the dependency matrices. Nevertheless, the QFD method allows a **focusing of the entire quality related planning process** on the wishes of the customer.

If a service company wants to use the QFD method in quality planning, it is advisable to define concrete QFD projects for the realization of individual, preferably **quantifiable quality objectives** which **available resources** to determine and the **employees and customers to select,** which will be involved in the project (Gogoll, 2003, p. 5). It is useful in this context to include those employees—usually from different company departments—in the project who are directly involved in the service processes concerned with the quality or project goal. If, for example, the project goal were "Reduction of waiting times at the bank counter by 10%", it might be useful to involve employees at the counter, in the back office and in the IT department, as well as branch managers and affected customers, in the QFD project.

After carrying out the quality function deployment, it is to be tested in **pilot projects** to what extent the measures developed are accepted or positively evaluated by service customers (Gogoll, 2003, p. 16). However, this phase is already part of the quality control, which will be examined in more detail later. The following example illustrates the use of the QFD method in the development of services.

Development of a customer feedback system based on a QFD analysis

The support center of an American IT company found out by means of a QFD analysis that the communication with the customer from the customer's point of view was in need of improvement. This included services such as training, availability or consultation. By applying step 5 "Checking of interactions" within the QFD analysis, it turned out that the company's customers primarily demanded a quick processing of complaints. The resulting solution approach "Percent processed software problems" went into the "House of Quality" as a "WHAT" and was connected with solution possibilities, the "HOWs". This resulted in a comprehensive customer feedback system that should improve communication with customers. In the form of a pilot project, it was first tested in a local market and, due to the positive response of the service customers, later introduced worldwide (Saatweber, 2011, pp. 272–279). ◀

(h) Benchmarking The instrument of so-called benchmarking can also be used within operational quality planning. Here, company internal processes and results are relativized and evaluated on the basis of certain comparison criteria (Madu & Kuei, 1995, pp. 27–34). The rough **procedure** of benchmarking in the context of quality planning comprises the following steps:

1. Determination of the benchmarking object (e.g. sub-service, single service, entire company)
2. Selection of the benchmarking partner(s)
3. Comparison of the benchmarking object with the corresponding object of the benchmarking partner

In this context, companies, services and/or individuals can be considered as objects of an internal/external comparison (cf. Fig. 8.9). The application of benchmarking as an instrument of quality planning thus requires the capture of specific information regarding the areas to be compared (e.g. in the area of external services, the collection of a customer satisfaction index). The basis for this are often upstream, customer oriented measurements.

In the case of **benchmarking of company units**, the results of a subsidiary, branch, department or service unit are compared with certain key figures of other internal service units. For example, the performance of a large corporation's patent department can

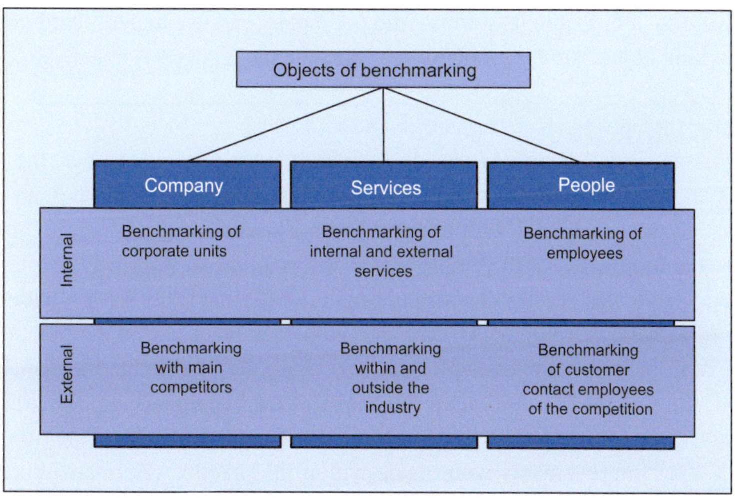

Fig. 8.9 Objects of benchmarking in service companies (with examples)

be compared with the performance of other patent departments in the company's branch offices.

If the **benchmarking of one's own company with major competitors** takes place, the whole company or a smaller unit, for example in terms of market selection or market development, is compared with performance units of competitor companies. In the case of **benchmarking of external and internal services**, multi-attribute, feature oriented measurement methods, such as the SERVQUAL approach, are the starting point. A comparison of services offered by a company to external customers or those services created within the company's internal customer-supplier relationships (e.g. measured by internal service barometers or satisfaction indices) provides important starting points for detailed improvement measures.

Benchmarking within and outside the industry is the classic case of benchmarking. The performance of a company is compared with the performance of other companies (e.g. through competitor analysis or national customer barometers). Examples of such benchmarks are values for customer satisfaction, customer loyalty or the productivity of a service provider. However, this form of benchmarking, which only makes sense when aligned with "honest" comparison values, is made more difficult by the reservations of competing companies when it comes to sharing sensitive company data. A significant reduction in these reservations is possible, for example, if the benchmarking is carried out by a reputable external provider, such as specialized benchmarking consulting firms, which ensure the anonymization of information used in benchmarking. However, this approach can be associated with a significant financial expenditure.

The data basis of a **non-industry benchmarking** (best in class) is usually easier to establish. Here, too, the same services—for example the internal accounting of documents or the telephone behaviour when receiving calls—are compared with each other, but the belonging to different industries allows for a more intensive—and open—exchange of data. For example, a forwarding company can use a credit card company as a benchmarking object for the accounting of documents.

Benchmarking between a doctor's office and a hotel

An example of cross-industry benchmarking is provided by the German Benchmarking Centre. In the context of a "best in class" benchmarking, an orthopaedic practice was compared with a 5-star hotel. The aim of the benchmarking was to increase the motivation of the employees in the practice, to improve the internal communication flow and to significantly increase the service level for the patients. First, similar tasks were identified and compared with each other within the practice and the hotel. These included, for example, the way customers are treated and the presence at the workplace. As one of several initiatives of the benchmarking, the practice staff took part in the service training of the hotel. As a result of the benchmarking project, the following improvements were made within the practice (German Benchmarking Centre, 2016):

- Wearing uniform work clothing increases the identification of employees with the practice
- Patients are picked up personally from the waiting area
- Establishment of a shift leader who regulates the practice organization in case of high patient volume ◀

In **benchmarking of employees**, managers and employees within individual units (e.g. departments) or between different units (e.g. employees of branch A with those of branch B) are compared. Examples of such benchmarks are indicators of employee satisfaction , employee loyalty or performance indicators such as the number of new customers acquired within a quarter.

The external comparison of employees with those of competitor companies represents the **benchmarking of customer contact employees** and is limited to employees who are in customer contact due to the difficulty of data acquisition. Only with customer contact employees is it possible to obtain data on employees of other service companies through customer surveys or “mystery shopping”. Since customer interaction is of high importance for service quality and is used to build up decisive competitive advantages in many industries, this form of benchmarking plays an important role that is still little noticed in corporate practice today.

In analogy to the preceding categorization according to objects of benchmarking, five **forms of benchmarking** can be distinguished (Madu & Kuei, 1995; Mertins & Kohl, 2009, pp. 31 et seq.), which are shown in Fig. 8.10:

- Product benchmarking
- Process benchmarking
- Best practice benchmarking
- Strategic benchmarking
- Systemic benchmarking

While with **product benchmarking** the service result is examined in detail, with **process benchmarking** the performance creation process or the service chain is the subject

Benchmarking form	Exemplary benchmark
Product benchmarking	Punctual compliance with arrival times
Process benchmarking	Satisfaction with the courtesy of stewardesses
Best practice benchmarking	Employee performance of an airline paying higher salaries
Strategic benchmarking	Fluctuation rate
Systemic benchmarking	Environmental commitment

Fig. 8.10 Forms of benchmarking (Source: In analogy to Madu & Kuei, 1995, p. 118)

of the investigation. **Best practice benchmarking** focuses on specific management methods. For example, the performance of employees is examined in relation to reward and incentive systems or the design of the work environment.

A little further away from the classical benchmarking idea are the two last expressions. Thus, **strategic benchmarking** tests the compatibility of the results of individual benchmarks with the company's strategy; individual key figures are analyzed with regard to their consistency in the overall entrepreneurial system. **Systemic benchmarking** finally places the individual company in the context of its environment and thus has a more social dimension. For example, the contribution of a company to environmental protection or the consideration of social issues in the light of the needs of all people—including non-customers—are subjected to closer scrutiny. The achievement of customer satisfaction receives a—still neglected—further facet in this form of entrepreneurial comparison.

In a framework concept specially tailored to the **benchmarking of quality objectives of service companies**, the following procedure can be followed (Chen, 1998, pp. 123 f.):

1. Identification of the trigger for the quality problem
2. Identification of partners for benchmarking
3. Carrying out the benchmarking and prioritizing quality objectives

First of all, it is necessary to **identify the triggers for the quality problems** in the provision of services. Both customer contact and back office processes are to be examined with a focus on customer contact employees, time, place, tangible and intangible elements of the service creation process with regard to problem causes. In particular, the SERVQUAL approach is suitable for measuring service quality with regard to the intangible elements. It is recommended to collect and analyze quality related data via customer surveys and by using the sequential event method and process observation.

The next step requires the **identification of suitable benchmarking partners.** These can be found, for example, in so-called best practice databases, in published studies or through personal contacts. In preparation for the first meetings with potential partners, it is necessary to do thorough preparatory work. In addition, it is not advisable to primarily align the partner selection with the world class company. Rather, the selection of partners should be realistic in view of the company's own quality potential.

When carrying out the **benchmarking and the prioritization of quality objectives**, it is recommended to compare the potential, the creation process and the result of a service with the practices of the benchmarking partner. The quality improvement programs derived from the benchmarking are to be prioritized in line with the company's overall quality strategy and resources.

As part of a **critical assessment** of the benchmarking, the favorable cost factor and the aspect of relative standards are in the focus; this facilitates the comparability of the benchmarking objects. In addition, benchmarking provides important impulses to exploit existing potentials and promote innovations, e.g. by adopting new ideas within internal or external comparison objects. As a disadvantage, the different strategies of the benchmarking objects underlying—especially in cross-industry benchmarking—are to be mentioned (e.g. between a food and a service provider). Possible improvement impulses are to be considered accordingly.

With regard to the **definition of responsibility** for a benchmarking project within the framework of quality planning, the selection of benchmarking objects and benchmarking partners is preferably to be carried out by the management. The actual performance comparison lies more in the area of responsibility of line or staff employees with quality knowledge.

(i) Fishbone analysis As another instrument of operational quality planning, the various causes of a specific quality problem from a supplier perspective can be systematically and comprehensively determined and graphically represented using a fishbone analysis on the basis of an Ishikawa diagram (cf. Sect. 5.2.1).

(j) Incident driven process chain Currently, process models are increasingly being discussed as part of quality planning. A process is understood to be a temporal logical sequence of activities in the context of service creation (cf. also Scheer et al., 2006, pp. 37 pp. below). The incident driven process chain method from business informatics can be used to model processes.

In incident driven process chains, processes are represented as **sequences of functions**. For each function, the start and end incidents can be specified. Incidents are triggers of functions and results of functions. An incident is said to have occurred when a change of state (e.g. postal arrival of a customer enquiry) has taken place, which in turn triggers subsequent company activities. In contrast to a function, which represents a time consuming incident, an event is related to a point in time.

The connections in incident dirven process chains are represented by **linking points** which contain so-called linking operators. These contain rules and conditions and thus represent means of controlling the service process. Possible values are conjunctive ("and"), adjunctive ("or") and disjunctive ("exclusive or") linking operators.

Functions, incidents and linking operators form the **basic elements** of an incident driven process chain. Additional display objects make it possible to take into account further aspects, such as the mapping of organizational units.

Fig. 8.11 shows an **example of an incident driven process chain.** The starting event represents the reservation request in a hotel. In the downstream function, the hotel employee checks whether there are still rooms available in the period specified by the customer. This can result in a positive or negative result. This event leads to the function "Feedback" etc.. This simple example makes it clear that the incident driven process chain is a method that allows for systematic and complete planning of customer contact points and the processes behind them.

In summary, Fig. 8.12 shows the individual instruments of quality planning and their possible uses in an overview.

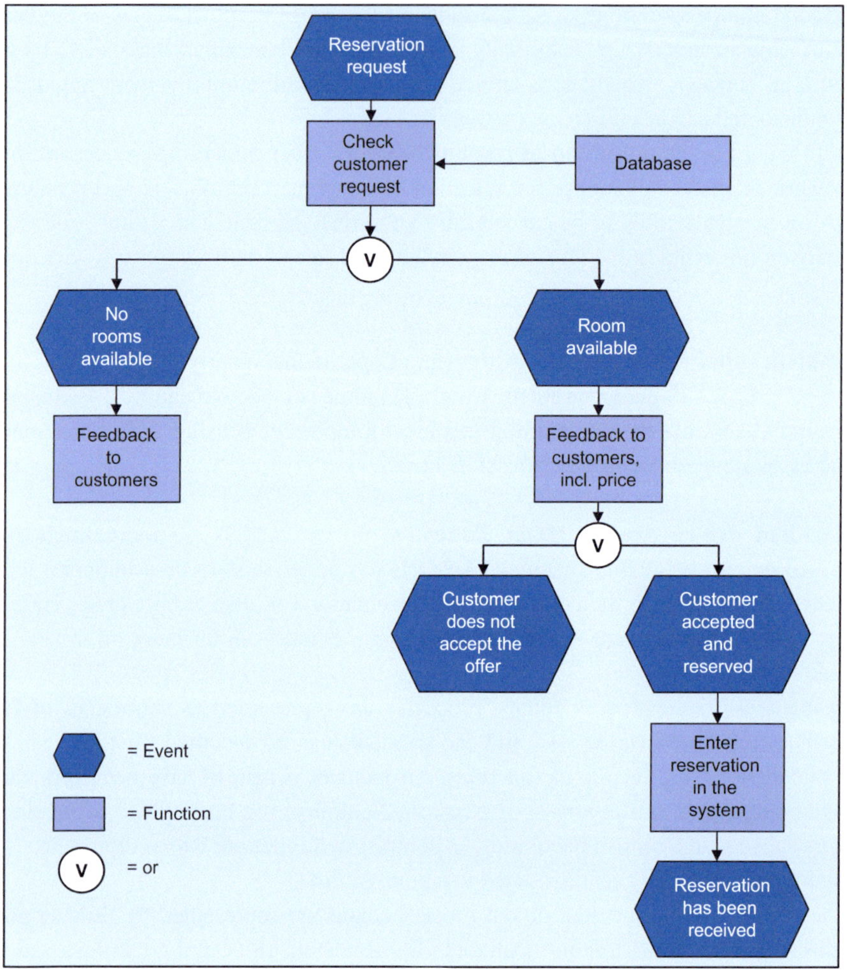

Fig. 8.11 Example of an incident driven process chain for a reservation request

8.2.2 Integration of Instruments for Quality Planning

In order to consider the integration of quality planning instruments, it is first necessary to analyze the **networking potential**. This requires investigating the relationships between the individual instruments in terms of their functional, temporal, and hierarchical character. Fig. 8.13 shows such a **relationship analysis** for all quality planning instruments. The example of the feature and customer oriented approaches is used to explain this relationship analysis.

When using **feature and customer oriented approaches**, the following **linking options** with other instruments can be identified within the scope of quality planning.

	Customer surveys	Sequential Incident Method (Blueprint)	Frequency-relevance analysis of problems	Feature- and customer oriented pproaches	Feature- and employee oriented approaches	Company suggestion system	Quality Function Deployment	Benchmarking	Fishbone analysis	Incident-driven process chain
Strategic goal	Analysis of service quality from the customer's point of view	Analysis of service quality from the customer's point of view	Identification of obvious problems	Analysis of relevant performance criteria from the customer's point of view	Analysis of relevant performance criteria from the employee's perspective	Internal analysis of service quality	Systematics for customer-oriented quality planning	Comparative analysis of service quality	Analysis of service quality problems	Internal analysis of service quality
Goals for action (operational goals)	Query custome expectations; Determination critical success driver	Mapping of the service creation process from the customer's point of view	Mapping of occurring problems from the customer's point of view	Multiattribute measurement of service quality from the customer's perspective	Multiattribute measurement of service quality from the employee's perspective	Utilization of employee know-how for quality improvement	Ensuring customer orientation within the framework of planning and realization of services	Comparison of service quality with other companies/departments	Identification of causes for poor service quality from the provider perspective	Mapping of all internal and external processes within the scope of service creation
Subject of investigation	Sales and Customer Support	Processes within the scope of service provision	Processes within the scope of service provision	Attributes of the quality of service	Attributes of the quality of service	All areas	All areas	Total company	Total company	All processes in the company
Individual measures	Customer questionnaires, comment cards, interviews, etc.	Mapping of the service creation process together with the customers	Mapping of occurring problems of the service production process together with the customers	Customer questionnaires	Employee questionnaires	Collecting, reviewing and implementing suggestions for improvement from employees	Analysis of core services and optimization of quality-related planning processes	Evaluation of company performance and comparison with other companies	Graphical representation of causes of poor service quality	Analysis of all internal and external process steps
Result/Output	Know and meet customer expectations, increase customer satisfaction	Designing optimal service processes from the customer's point of view, increasing customer satisfaction	Identification of relevant problem areas in the service process	Know and meet custome expectations, increase customer satisfaction	Knowing customer expectations from the employee's perspective	Elimination of weak points, error reduction, increase in efficiency	Focusing the entire quality-related planning process on the wishes of the customer	Adoption of "best practice" concepts, implementation of optimal processes	Systematic mapping of causes for a concrete quality problem	Design of optimal service processes, avoidance of process errors, increase in efficiency
Main contact partner in the company	Marketing, market research	Sales and marketing	Sales and marketing	Market research and marketing	Market research	Quality Management, Department Manager	Quality Management, Department Manager	Corporate Governance	Corporate Governance	All department heads
Resources-effort	Medium	Medium	Medium	Medium to high	Medium	Low	High	Medium	Medium	High
Frequency of implementation	Regularly, once a year	Regular	Regular	Regular	Regular	Regular, ongoing	One time per planned service	Regularly, if necessary once a year	Regular	If required
Responsibility carrier	Marketing	Sales and Marketing	Sales and Marketing	Sales and Marketing	Corporate governance	Corporate governance	Corporate governance	Corporate Governance	Corporate management	Corporate management
Cooperation with other agencies	Sales and Customer Support	Sales and Marketing	Sales and Marketing	Sales and Marketing	Market research and marketing	All employees	Different departments and customers	Areas to be benchmarked	All departments	All departments

Fig. 8.12 Instruments of quality planning at a glance

Integrated quality planning	Customer surveys	Sequential incident method (Blueprint)	Frequency-Relevance-Analysis of problems	Feature and customer-oriented approaches	Feature and employee-oriented approaches	Company proposal system	Quality function deployment	Bench-marking	Fishbone analysis	Incident driven process chain
Customer surveys		Conditional successive	Conditional parallel	Conditional parallel	Comple-mentary successive	Conditional parallel	Conditional parallel	Comple-mentary parallel	Conditional successive	Comple-mentary successive
Sequential incident method (SIM) (Blueprint)			Conditional successive	Conditional successive	Conditional successive	Comple-mentary parallel	Conditional parallel	Comple-mentary parallel	Conditional successive	Comple-mentary parallel
Frequency-Relevance Analysis of Problems (FRAP)				Comple-mentary parallel	Comple-mentary parallel	Conditional parallel	Indifferent parallel	Comple-mentary parallel	Conditional successive	Comple-mentary parallel
Feature and customer oriented approaches					Comple-mentary parallel	Conditional parallel	Conditional parallel	Comple-mentary parallel	Conditional successive	Comple-mentary successive
Feature and employee oriented approaches						Conditional parallel	Conditional parallel	Comple-mentary parallel	Conditional successive	Comple-mentary successive
Company suggestion system							Indifferent parallel	Indifferent parallel	Conditional parallel	Comple-mentary successive
Quality function deployment								Indifferent parallel	Indifferent parallel	Comple-mentary successive
Benchmarking									Indifferent parallel	Indifferent parallel
Fishbone analysis										Comple-mentary parallel
Incident driven process chain										

Fig. 8.13 Relationship analysis as part of an integrated quality planning

- In **functional terms**, these approaches have conditional relationships to those strategic quality planning instruments for which the collection of quality requirements is necessary, e.g. quality objectives and standards. Regarding the operational quality planning instruments, the use of feature and customer oriented approaches requires, in particular, the results of investigations with instruments that provide indications of possible relevant quality requirements from the customer's perspective (e.g. Sequential incident method, company suggestion system). Complementary relationships of feature and customer oriented approaches are mainly given with those planning instruments that also systematically collect certain quality criteria (e.g. feature and employee oriented approaches).
- The **temporal links** with other planning instruments concern the feature and customer oriented approaches, which are to be carried out before those procedures that serve a deeper analysis of individual quality criteria or problem areas (e.g. FMEA, Fishbone analysis). On the other hand, such procedures are to be carried out before those that contribute to the determination of the criteria to be queried (e.g. Sequential incident method).
- With regard to the **hierarchical relationships**, the feature and customer oriented approaches have a more strategic character than the other planning instruments because these approaches systematically and completely capture customer requirements.

To illustrate the **hierarchical and temporal character** of the quality planning instruments, the use of different planning instruments in the context of quality management is exemplarily shown in Fig. 8.14. First, stations in the service creation process are to be determined by means of the sequential incident method, which are particularly strongly perceived by customers and in which potentially occurring problems can thus have a great influence on the quality perception. These criteria can be used as a starting point for developing a questionnaire for the implementation of multi-attribute measurement methods. With the FRAP, FMEA and Fishbone analysis, problem areas possibly recognized by the sequential incident method are to be analyzed in more detail. The instrument of the company suggestion system can also provide criteria whose characteristics are to be recorded in customer and employee surveys.

8.3 Instruments of Quality Control

The **quality control** phase is basically based on the results of quality planning. The content of quality control is to be described as follows (EN ISO 9000, 2014):

▶ The **quality control** is that part of quality management that is oriented towards the fulfillment of the quality requirements. Quality control includes work techniques and activities for both monitoring a process and eliminating the causes of unsatisfactory

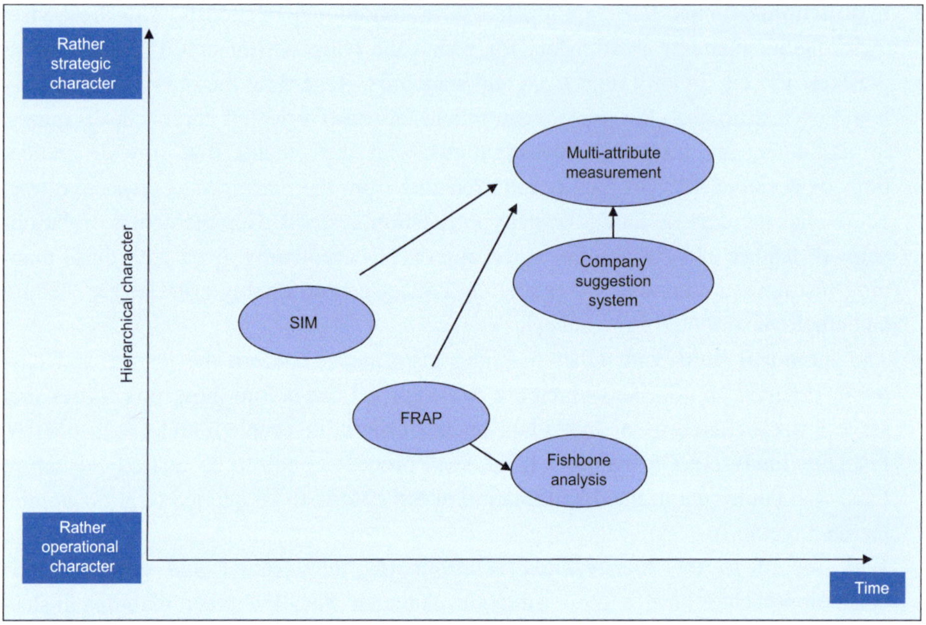

Fig. 8.14 Exemplary use of quality planning instruments

results. Measures of quality control and quality auditing/quality management documentation are interrelated.

This phase therefore includes all activities that serve the realization of the requirements for the quality of the services from the customer's and the company's point of view.

The following **instrument groups** can be used to systematize the instruments to be used in this phase:

- Employee related instruments
- Culture related instruments
- Organization related instruments

Fig. 8.15 shows an overview of the quality management instruments.

8.3.1 Employee Related Instruments of Quality Control

Services are usually carried out between an company's employees—the contact personnel—and the customer as part of a so-called **interaction oriented process** (Schulze, 1999, p. 263; Ford & Mouzas, 2012). The importance of the interactive event as an

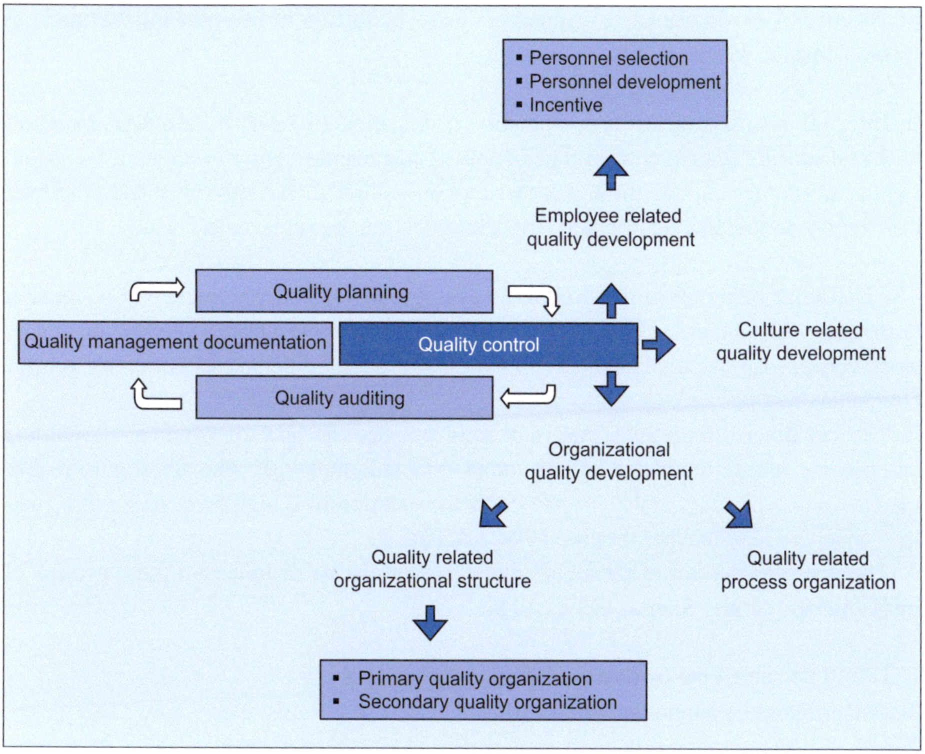

Fig. 8.15 Instruments of quality control of service companies

element of service quality can differ depending on the duration, intensity and integration of the external factor. Both in **problem oriented interactive** and in **personally interactive** services (cf. the classification by Mills & Margulies, 1980) the personal contact between the consumer and the contact personnel plays an important role in the **service encounter**. Thus, the seller or consultant often acts as a personification of the entire service offered or the service company. Consequently, it is of decisive importance for the perceived consulting quality of the company how well it succeeds in adapting the qualification profile of the customer contact person to the requirement profile considered relevant by the customer in the consulting situation.

Due to the importance of employee behavior as a key quality feature of services, the customer orientation coming from customer contact employees has a significant impact on the service quality perceived by the customer. This highlights the importance of implementing a holistic customer oriented corporate management. In this context, the concept of **internal marketing** is important (e.g. George, 1977; Berry, 1984; Grönroos, 1985; Gummesson, 1987; Gleitsmann, 2007; Ahmed & Rafiq, 2012; Pepels, 2016, pp. 342–352). In the literature, the term internal marketing has different interpretations with regard to the extent of the marketing orientation focused on an internal company perspective. A comprehensive, original inclusive **definition of internal marketing** is as follows (Bruhn, 1999a, p. 20):

▶ **Internal marketing** is the systematic optimization of internal company processes with instruments of marketing and personnel management in order to implement marketing as an internal attitude through consistent and simultaneous customer and employee orientation, so that market oriented corporate goals can be achieved efficiently.

The **customer orientation** is therefore an essential part of internal marketing (George & Grönroos, 1999; Stauss, 1999; Pantouvakis, 2012) and represents an appropriate framework concept for the implementation of a quality oriented personnel policy for service companies.

The customer oriented internal marketing pursues the goal of winning, developing and binding highly motivated and customer oriented employees who are able to realize the external marketing goals—in particular the creation of a high level of service quality—in an efficient manner (Stauss, 1999, p. 210).

The instrumentarium of personnel management can be divided into three **groups of instruments** (cf. also Stauss, 1999, 210 pp.):

1. Use of personnel policy instruments oriented towards sales
2. Use of internal communication instruments oriented towards sales
3. Use of external marketing instruments oriented towards personnel

The **use of personnel policy instruments oriented towards sales** is about the use of the classical personnel policy action areas such as personnel selection, deployment, development, release as well as wage policy under the aspect of sales market policy requirements.

The **use of internal communication instruments oriented towards sales** is about the application of communication measures directed towards the staff, such as employee newspapers or individual training.

Under the **use of external marketing instruments oriented towards personnel**, primarily customer oriented marketing instruments such as advertising and public relations, which at the same time motivate the employees to pick up a sales or customer oriented behavior, fall. An example of this is a guarantee promise articulated to the customer (e.g. maximum waiting times), which implicitly obliges the employees to a certain quality orientation.

Accordingly, **personnel measures** are to be taken by the management in the sense of internal marketing in order to ensure and improve both the ability and the willingness of employees to create a service quality at a high—the required—level. Different instruments are available for this (see, for example, Stock-Homburg, 2013 as an overview), of which the following quality-related instruments are presented here:

- Recruitment
- Staffing
- Personnel development
- Incentive systems for employees

8.3.1.1 Quality Oriented Recruitment

The process of **recruitment** can be divided into the phases of personnel acquisition and selection, to which certain instruments can be assigned. The primary goal of recruitment is to advertise and select quality oriented or suitable employees who are able to meet the complex requirements of comprehensive quality management (Stauss, 1999, p. 211). The following presents the individual instruments of personnel acquisition and selection.

(1) Instruments of personnel acquisition The **personnel acquisition** or advertising to win new, motivated and qualified employees for the successful implementation of a quality management in service companies is to be given a comparable importance to the advertising for the actual service offer. It is recommended to also use instruments of advertising for the purpose of personnel acquisition. These are also to be designed and placed specifically for target groups (Stauss, 1999, p. 211).

Personnel acquisition is the sum of measures with which contact is made with potential employees in the context of personnel acquisition (Weber et al., 1993, p. 197).

The variety of instruments for personnel acquisition can be divided into direct and indirect instruments. Unlike indirect instruments, the company uses direct instruments to attract potential employees without the intervention of other institutions. When designing direct instruments, it is important to consider whether specific groups of potential employees (e.g. university graduates) or the entire job market are addressed (Fröhlich & Langecker, 1989, pp. 154–159; Stock-Homburg, 2013, pp. 159–166).

The **direct instruments** include:

- Personnel image advertising
- Job advertisements in print media
- Job advertisements on the Internet
- Personnel acquisition via social media
- Direct approach

The **personnel image advertising** serves to create an image as an attractive and employee oriented employer. This effect is achieved by addressing potential employees using instruments of external communication (e.g. advertising, public relations, sponsoring) (Fröhlich & Sitzenstock, 1989, pp. 134-141; Nagel, 2011, p. 42). For services, the importance of employees as potential factors in the creation of services plays a central role. This is not only to be discussed in the communication of work requirements, but offers a way to highlight specific personality traits and the opportunity for personal development. Accordingly, it is important to consider personality traits such as helpfulness or empathy, for example (Meffert et al., 2018, p. 415).

With the help of **job advertisements in print media**, a relatively large number of potential employees can be reached. The success of a job advertisement depends on how cognitive effects (e.g. creating attention due to the advertisement), affective effects (e.g. positive attitude towards the offered position) and conative effects (e.g. making an application) are achieved (Meffert et al., 2018, p. 411).

Increasingly, **job advertisements on the Internet,** e.g. on career pages on the company homepage or on so-called job portals of external providers, are becoming the standard, which offer companies advantages of more efficient processing processes such as better archiving and more efficient comparisons among applicants and thus, among other things, a shorter processing time (cf. to the opportunities and risks of internet-based applications Stock-Homburg, 2013, pp. 164–165). This results in savings potentials of up to 50% of the hiring costs (Kupitz, 2006). It should be noted that both the design of the job advertisement and the entire Internet presence serve to build an image and position the company on the recruitment market.

Personnel acquisition via social media

Many companies now also use various **social media platforms** to address potential applicants. The company *Google*, for example, has published a recruiting video on *YouTube.com*. There employees, executives, etc. report on the advantages and everyday working life at *Google.* Other companies such as *Bertelsmann* and *Bayer AG* have complete recruiting pages on *Facebook*, on which regular information on recruiting events, trainee programs and other contact and entry opportunities of the company are presented. LinkedIn is another platform that is used frequently for personnel acquisition worldwide. ◀

Finally, there is the possibility of **direct approach** by addressing potential employees personally. This is done by recruiting employees from the competition (e.g. at trade fairs), from suppliers (e.g. recruiting a company consultant by a bank) and by recruiting customers (e.g. recruiting a club visitor as a waitress).

When using **indirect instruments** of personnel acquisition, customer approach takes place through an intermediary institution. The following alternatives are possible here:

- Personnel consulting
- Executive search/employment agencies
- Use of employees in temporary work

The use of **personnel consulting** serves the targeted search for employees, according to a specification of the service company in terms of the relevant professional and personal characteristics of the potential job applicant. Advantages of using a personnel consultant are the good overview of the labor market and the time savings for the company's own employees. However, it is important to ensure that the personnel consultant knows the processes of service delivery and the quality requirements at the respective service company or is extensively informed about the requirements of the position.

Executive search and **employment agencies** have a great overview of the labor market in common, but usually do not offer a qualified selection of applicants in contrast to personnel consultants.

Under the **temporary work**, the working time of employees is temporarily used, which are mediated by personnel leasing companies. In addition to the use of temporary work in the context of interim management (e.g. during pregnancy, departure of executives), it is particularly suitable for services with time-dependent, predictable demand fluctuations (e.g. catering, tourism). The disadvantage of temporary work is that, due to the often short period of employment, the use of training and further education measures is not worthwhile. Therefore, it is particularly important for part time employees with direct customer contact to convey an understanding of quality and to clarify the high influence of their behavior on the customer's quality perception of the overall company.

(2) Instruments of personnel selection The use of **measures of personnel selection** is necessary when there are several applications for a position. The selection criteria for employees to achieve corporate goals must be derived from the respective defined service standards of the service company and be appropriately job specific (Stauss, 1999, p. 212). This way, an intended external customer focus is made possible by the definition of relevant service or customer oriented selection criteria in the context of personnel selection. In addition to **formal criteria** (e.g. age, education), **job specific requirements** (e.g. willingness to help or empathy) are to be taken into account in the selection of applicants (see for an overview of the general and specific requirements for customer contact personnel in service companies Becker & Wellins, 1990, p. 49; Schneider & Schechter, 1991, p. 223). In particular, it is necessary to determine selection criteria which, in addition to the professional qualifications, also test the social competence of the potential employees (Hoffmann, 1989, p. 412; Ahmed & Rafiq, 2012).

Accordingly, in addition to written aptitude tests and interviews, for example, the conduct of behavior tests (role-plays, group discussions, etc.) is to be recommended, in which the applicants are confronted with "real" customer contact situations and observed with regard to their communication skills (Schneider & Schechter, 1991, p. 225; Nerdinger et al., 2014, 249 pp.). In this context, it is often necessary to also define precise requirements or job descriptions of potential employees, as illustrated in the following example.

Requirements and job description for flight attendants of the *Deutsche Lufthansa AG*

The requirements for flight attendants of the *Lufthansa* includes a high degree of customer orientation, sociability, communication skills and team spirit, as well as personal, friendly and well-groomed appearance. Social competence is also an essential quality of future employees (cf. Fig. 8.16). ◄

The relevance to reality in the context of aptitude tests and job interviews can be further increased if the applicants have to prove their skills in an authentic service encounter under observation. This is often observed, for example, in the catering industry by a one to multi-day trial work (Meffert et al., 2018, p. 415).

If the potential employee shows little or no service mentality, sociability and empathy during the **job interviews**, he will later find little satisfaction in his work as a customer contact person and will also be of little help to the satisfaction of the customers. Specialist knowledge, dealing with technology, etc. can usually be imparted or improved by means of training; the basic service and communication or contact readiness **(Service Minded)** however, are part of human personality and can therefore only be learned to a limited extent in training. The management responsible for the recruitment of new employees in the personnel and specialist departments are therefore to be made aware that they pay more attention to the service and communication readiness of the applicants. This situation is consistently taken into account, for example, in the personnel

Age	▪ Minimum age: 18 years
Education	▪ Completed school education ▪ German and English fluently
Social competence requirements	▪ High service orientation ▪ Sociability, strong communication skills and the ability to work in a team ▪ Friendly appearance and a well-groomed appearance ▪ Intercultural competence ▪ Willingness to work even under stress, diplomacy even in difficult situations ▪ Self-reliance, commitment and initiative ▪ Sense of responsibility and reliability
Other requirements	▪ Minimum height 1.60 m and a reasonable body weight ▪ Passport without restriction ▪ Visual acuity between +/- 5 diopters ▪ Willingness to work shifts and assignments abroad
Perks	▪ Retirement benefits provided by Lufthansa in addition to pension insurance ▪ Numerous training offers for addtional qulifications ▪ Special discounts at many hotels and rental car companies
Remuneration	▪ Starting salary: EUR 1,533.23 (+ 16.3 % shift bonus) ▪ Lump-sum vacation allowance of EUR 1,022.85 ▪ Absence allowance per day: 42.00 EUR continental and 48.00 EUR intercontinental (partially subject to tax and social security contributions)

Fig. 8.16 Requirements and job description for flight attendants of the *Lufthansa*. (Source: Lufthansa, 2012)

selection of the American department store chain *Nordstrom*: The so-called *Nordstrom* principle—"Nordstrom hires the smile and trains the skill"—postulates that service oriented employees are to be recruited and taught sales skills rather than the other way around (Spector & McCarthy, 1995, p. 181; Schneider & Bowen, 2010).

8.3.1.2 Quality Oriented Personnel Deployment

In connection with quality oriented personnel deployment, it is necessary to distinguish between task oriented and relationship oriented tasks, which employees have to take over in the context of a consistent implementation of a quality management, depending on the degree of interaction in customer contact (Stauss, 1999, p. 212).

▶ The **personnel deployment** in the wider sense subsumes all measures which are connected with the determination and design of the work organisation, the workplace as well as the working time.

The following **categories of instruments** can be distinguished on the basis of the definition of personnel deployment:

1. Workplace related instruments
2. Work time related instruments
3. Work organisation related instruments
4. Activity independent measures

The various instruments and measures are explained in more detail below.

(1) Workplace related instruments The workplace related instruments of personnel deployment relate to measures that are directly related to the individual employee's workplace. Therefore, the extent of the leeway granted to them is significant for the quality of their performance. A distinction is made between the extension of the task, the enrichment of the task and the limitation of the task (Bokranz, 1989, p. 429).

An **extension of the task** means the expansion of the scope of activity (job enlargement) (Stock-Homburg, 2013)—for example, if a porter in a hotel is additionally responsible for room service or the fulfilment of special requests from guests—as well as by task rotation (job rotation), the systematic change of workplace at regular intervals (Thommen & Achleitner, 2012, p. 759). In addition to the often resulting increase in motivation, the employee is also sensitized to different processes of performance creation during customer contact and enabled to gain an overview of the performance creation process and to pay particular attention to critical process steps for the customer.

A **enrichment of the task** (job enrichment) leads to an increase in the scope of decision-making within the framework of a qualitative expansion of the task field (Stock-Homburg, 2013, p. 242). The concept of "empowerment" (Bowen & Lawler, 1995) is associated with this. Employees are given more leeway to meet customer needs and respond to complaints. As a result, employee motivation and satisfaction are usually increased and, at the same time, customer satisfaction is increased.

A **limitation of the task** leads to a reduction in the scope of activity and/or the scope of final decision making. These measures represent a form of "punishment" for the creation of poor performance quality. Such measures are also conceivable in order to promote quality orientation of the employee through specialization.

(2) Work related instruments The work related instruments can be categorized according to the times of performance creation. The following dimensions are significant:

- **Duration of the customer-employee contact:** With services that have a high degree of integration of the external factor, the duration of the customer-employee contact is relatively long. Therefore, the awareness must be created in the employees that the customer contact time is not considered wasted time.
- **Time flexibility:** Some services are necessarily associated with a certain time from the customer's point of view. This is based on typical demand behavior (lunch in a restaurant) or customer related time constraints (passenger does not want to wait two hours for a bus). The more time flexibility there is with regard to performance creation, the more likely it is that there will be special features for working time design, for example in the form of shift work or the use of temporary workers.

(3) Work organization related instruments A central work organization related instrument is **team work**. A quality oriented way of working often requires close cooperation of various employees in teams during the performance creation process in order to ensure a satisfactory performance creation for the customer (Wirtz & Lovelock, 2016, p. 432; Noé, 2016, 12 pp.). The use of teams offers employees a more positive working environment and the performance creation processes can be designed more efficiently. This leads, in the next step, to a higher willingness to perform and a greater motivation of employees (Berry et al., 1990, p. 33).

Studies by Berry and Parasuraman come to the result that customer contact employees, who, according to statements by colleagues and superiors, do not meet the quality standards, in the context of surveys, answer negatively to statements to a positive perception of team orientation of the working environment (e.g. "I feel as a member of a team in my department") (Berry & Parasuraman, 1999, p. 99).

These findings illustrate the need for a team oriented work climate within the company and individual departments. The management has a special responsibility for the climate and motivation of the employees. After all, employee satisfaction usually affects the lower levels of the hierarchy, i.e. the customer contact personnel, to a high degree of customer satisfaction. For **successful teamwork** the following requirements can be named:

- Permanent membership in the team
- Regular contacts in the team
- Team oriented personnel management
- Common goals
- Team related performance measurements and rewards

Another instrument in terms of personnel deployment is the **type of employee leadership.** A participatory leadership style is necessary for the employee to feel involved in decision making processes of the company and thus appear as a responsible service provider to the customer. The advantage of early involvement of employees in decision making processes lies mainly in the fact that the employees are directly involved in the ideation and decision making process and thus the probability increases that they identify with the decision and are committed to a successful implementation. The disadvantage of intensive participation of employees is the increase in complexity and the high time requirement (Homburg & Stock-Homburg, 2012; Stock-Homburg, 2013).

(4) Activities independent measures Building on the knowledge that satisfied employees also tend to act more quality oriented and that employee satisfaction also directly affects customer satisfaction (Homburg & Stock, 2001; Malhotra & Mukherjee, 2004), more and more companies are offering their employees special benefits that are not related to their work in the company (e.g. offer of an internal company gym).

Studies have shown that employee satisfaction not only positively affects the quality of service provision, but that in addition, sick leave caused by stress at work decreases significantly. Since absences of this kind account for up to 50% of all work stoppages, the increase in employee satisfaction also makes it possible to reduce personnel costs. In addition, it is assumed that the employees' loyalty to the company increases (Lee & Bruvold, 2003). Through a lower turnover rate, the costs for the training of new employees are also reduced, so that the increase in employee satisfaction also makes sense from this point of view.

8.3.1.3 Quality Oriented Personnel Development

After personnel acquisition, it is the task of personnel management to continusouly educate the available personnel according to their knowledge and the quality requirements of the workplace.

▶ The **personnel development** includes all measures for the transfer of qualifications that increase the current and future performance of managers and employees (education), as well as measures that support the professional development of managers and employees (promotion) (Stock-Homburg, 2013, p. 205).

Measures of quality oriented personnel development, in particular the **quality oriented training and education** (cf. e.g. Methner, 1999; Bullinger & Moll, 2003; Wesselmann &

Hohn, 2012, p. 235), play an important role in the context of quality management in order to be able to realize the claim of a company-wide implementation of the quality management approach and the desired involvement and participation of all employees. In this context, the basic task of personnel development is to increase the action competence of employees. Four **types of action competence** can be distinguished (Becker, 1999, p. 276; Agarwala, 2003, pp. 191 f.; Nerdinger et al., 2014, p. 210):

- **Professional competence and expertise:** Expert knowledge for the respective position, e.g. knowledge of the insurance market for an insurance broker.
- **Methodological or conceptual competence:** Ability to apply different solution methods to a problem, selection and learning ability, e.g. treatment of specific customer problems.
- **Social competence:** Teamwork, responsibility, communication skills, customer orientation.
- **Psychological competence:** Motivation, attitudes, willingness to work, e.g. willingness to realize a customer-oriented corporate culture.

In the context of a company-wide quality oriented personnel development with the goal of implementing the quality management system, it makes sense to expand or improve the existing personnel development program of the service company to the effect that, in addition to a continuous imparting of subject specific and methodological competence, social and psychological competence is also brought closer to the employees. The regular conduct of employee orientation interviews plays a relevant role in this context. An corresponding criteria catalogue is presented in the following example.

Criteria catalog of an employee orientation conversation of the hotel *Schindlerhof*

The *Schindlerhof* is one of the top-class hotels of the German conference hotel industry and has an international reputation that has been extended to a wide German-speaking public. By consistently implementing a TQM approach, the *Schindlerhof* also pursues a quality oriented personnel development. As part of the annual employee orientation conversation, the employee evaluates himself according to certain criteria. Subsequently, the supervisor evaluates the assessment according to predetermined criteria. Figure 8.17 shows the criteria catalog of the employee orientation conversation in this regard. ◄

For example, if the focus of a **transactional analysis oriented personnel training** (Schulze, 1999, 275 pp.) is the provision of psychosocial skills, the following sub-skills can be distinguished, which need to be learned. In the context of transactional analysis (Berne, 1961; Schulze, 1999, p. 268; Mohr, 2012), any form of attention, i.e. all signs of social exchange, is referred to as a transaction:

▪ Flexibility	▪ Entrepreneurial thinking
▪ Appearance	▪ Empathy
▪ Willingness to work overtime	▪ Punctuality
▪ Ability to work in a team	▪ Creativity
▪ Interest in further training	▪ Work quality
▪ Reliability	▪ Diligence
▪ Ability to work systematically	▪ Consistency of implementation
▪ Ability to accept criticism	▪ Compliance with the team concept
▪ Willingness to innovate	▪ Loyalty

Fig. 8.17 Criteria catalog of an employee orientation conversation at the hotel *Schindlerhof.* (Source: Bruhn et al., 2002, p. 153)

- Recognizing different transaction processes in interaction episodes during the creation of a service
- Active design of these interaction episodes based on the findings of transactional analysis
- Coping with difficult and negative customer service quality interaction processes
- Understanding one's own basic attitude and preferred roles in dealing with people
- Recognizing "psychological games", i.e. transactions that repeat themselves, and learning ways to break them
- The (social) ability to recognize, accept and deal appropriately with one's own and other people's emotions (emotional intelligence)

The approach to emotional intelligence developed by Salovey and Mayer (1990) and extended by Goleman (2015) is based on the concept of social intelligence, which differentiates between intrapersonal and interpersonal skills. The core capacity of intrapersonal skills is access to one's own emotional life. The ability to perceive one's own emotions and emotional states (self-perception), to control perceived emotions (self-control), and to make use of perceived emotions, for example, not to be negatively influenced by negative emotions (self-motivation). As the opposite of the perception of one's own emotions, interpersonal skills include the perception of emotions of other people. Empathy, i.e. the ability to emotionally adjust to others, plays a central role. This ability is, in addition to intrapersonal skills, a prerequisite for adequate social competence (cf. Fig. 8.18; Salovey & Mayer, 1990; Finsterwalder & Tomczak, 2001; Goleman, 2015).

In **employee surveys**, it is also necessary to determine in which areas the employees themselves see the need for training for themselves and their colleagues, so that they are not "trained beyond the need". Furthermore, feedback interviews are useful for

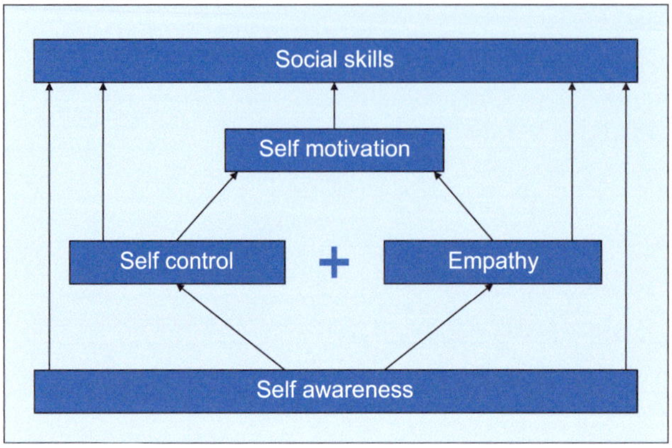

Fig. 8.18 Emotional intelligence hierarchy model (Quelle: Salovey und Mayer, 1990, S. 200)

evaluating training content, forms and times in order to permanently improve the quality of training and further education measures.

The aim of changing the various competencies is to prepare the employees for future tasks, i.e. to qualify them. The following **qualification** directions can be differentiated (Becker, 1999, p. 276; Oechsler, 2011):

- **Qualification for the professional extension of the activities,** in which the extent of the action competence is to be increased without intending a change of position, e.g. training of a credit processing clerk in a new method for assessing creditworthiness.
- **Qualification for new professional tasks,** in which the qualification of the employee is to be adapted to external or internal company developments, e.g. computer course for bank employees.
- **Qualification for career advancement,** in which an employee is to be prepared for a new position, e.g. "creativity techniques" course for an employee who is to be used as a project manager in the future.

The large number of instruments of personnel development can be classified according to the activity relatedness of the measure and the continuity of its use according to Fig. 8.19 (see, for example, Becker, 1999; Paschen, 2004; Stock-Homburg, 2013):

- **On the job training (On the Job)**
  This is where continuous qualification takes place at the workplace (Stock-Homburg, 2013, pp. 239–247). In the service sector, this form of employee development is of particular importance for positions that involve customer contact. It should be noted that in "sensitive" service sectors (e.g. banks), on the job training may lead to customer irritation.

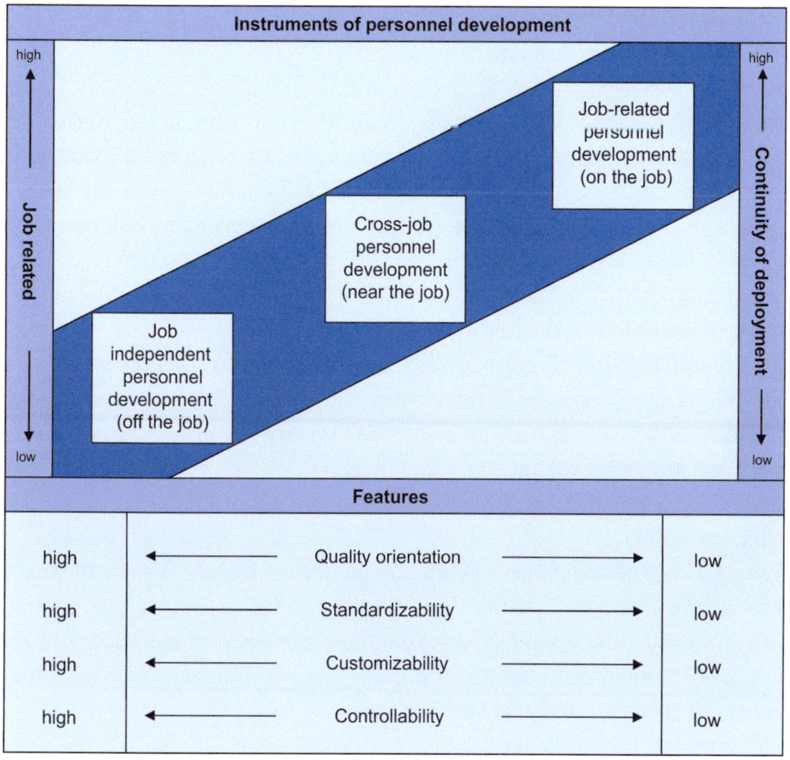

Fig. 8.19 Instruments of personnel development

- **Cross-position employee development (Near the Job)**
  This refers to measures that are close to the workplace in terms of space, time and content, e.g. quality circles, project group work (Stock-Homburg, 2013, pp. 245-247). Such cross-position groups are particularly suitable for anchoring customer orientation within the company, i.e. getting to know the needs of internal customers and realizing their importance for external customers.
- **Unattached employee development (Off the Job)**
  Such measures are carried out outside the workplace, e.g. lecture, conference, courses, without being related to the employee's field of activity. This type of training is suitable, amongst other things, for preparing for dealing with such customers and customer segments in which there are clear differences compared to the familiar environment.

For a more detailed description of the instruments, further criteria can be used:

- Quality orientation of the measures
- Standardizability of the measures

- Individualizability of the measures
- Controllability by the management

In this context, the independent and cross-sectional instruments in the service sector are of particular importance. Since the service creation process often has a character of interactivity between employees and customers, measures of training near the job (e.g. role playing) can be used to **simulate the service creation processes** (Bieberstein, 2006). The employee has to serve a "Exemplary customer" (e.g. psychologist, coach), so that an instant assessment of the exemplary employee performance is possible and he can be made aware of possible weaknesses immediately.

To the central **building blocks quality related training** for employees of service companies target group specific various areas are to be carried out, with the following three modules, which are already in successful service companies (Bruhn & Meffert, 2002), appear particularly useful:

- **Quality seminars**
  The target group are employees of all company departments. The seminar can serve as a forum for discussion of the most diverse quality issues, such as customer and employee surveys, the optimization of internal processes or the quality of customer advice. It is particularly important to convey the willingness and the pronounced awareness of customer or service orientation.
- **Service training for customer contact personnel**
  The target group are all employees in customer contact (eg receptionists, telephone operators, customer service). The training can serve the further development of personal performance skills by expanding the repertoire of behavior, by conversation and question techniques as well as sales psychological knowledge are imparted. Employees in customer contact have to learn above all, to pay more attention to the wishes and needs of the customers.
- **Coaching of executives**
  The target group are managers of service oriented areas in business units and head offices. This seminar module can align the serviceoriented management of the service company by further developing the communicative and motivational skills as a prerequisite for improved leadership quality and thus also to ensure the quality of service better. To induce attitude and behavior changes in employees with regard to a stronger service or customer orientation, all managers are to be trained in psychological leadership and communication techniques as well as assistance in the design of employee interviews.

Contents of complaint seminars

There are numerous seminars on complaint management offered by various institutes (Semigator, 2019). These one to two day seminars are offered both in general and industry specific form and include numerous content, such as: importance of complaints, relationship between customer (un)satisfaction, complaints and customer loyalty, causes of complaints, complaints as a process of stimulation, acceptance, processing, reaction and processing, telephone conversation, process oriented processes, complaints on the Internet, analysis of rating portals, legal questions, etc. ◀

Finally, it should be pointed out that the central aspect of the **applicability and implementability of the training content** is independent of the scope and design of the qualification measures (Schildknecht, 1992, p. 154; Sokolakis, 2011). A prerequisite for a positive learning effect and thus an indispensable prerequisite for the successful implementation of a quality management concept is the actual feasibility of customer oriented behavior in the company, which is usually dependent on corresponding action and decision making responsibilities.

In order for employees, in customer contact, to use their action competence customer and quality oriented, it is the task of management to create the organizational conditions for this and to give the employees the necessary room for action. It is therefore necessary to ensure sufficient **empowerment** of the customer contact employees (e.g. Berry, 1995; Bowen & Lawler, 1995; Rafiq & Ahmed, 1998; Ahmed & Rafiq, 2012). In its core, empowerment means giving the customer contact employees a certain degree of **discretion and room for action** within the framework of the service creation process (Rafiq & Ahmed, 1998, p. 380). Greater room for action is particularly necessary when—as is typical for most services—the service production coincides with the service consumption at the same time (Uno-Actu-Principle) and only limited opportunities for the service personnel to influence the service result or the service quality subsequently.

Example of empowerment of the *Ritz-Carlton Hotel Company*

An example of empowerment are the discretionary powers granted to service employees by management in terms of an adequate response to a complaint. Thus, for example, each employee of the *Ritz-Carlton* hotel chain has a sum of up to 2,000 USD at their disposal to respond to a complaint and rectify errors or deficiencies (Braun & Burger, 2007, p. 84). ◀

Empowerment is used to pursue various **objectives**: For example, employees are granted expanded sales options (up- and cross selling), which can lead to increased revenue. Furthermore, empowerment strives for higher employee motivation and satisfaction, higher employee efficiency and responsiveness to meet customer needs. the creation of customer specific services, and faster rectification of complaints (e.g. Rafiq & Ahmed, 1998, pp. 381–382; Ahmed & Rafiq, 2012).

However, empowerment also comes with certain **costs** that, for example, arise from qualification measures that adequately prepare employees for the increased scope of tasks brought about by empowerment. In addition, there is the **risk** that, through empowerment of customer service employees, the service creation process slows down (Rafiq & Ahmed, 1998, p. 382; Ahmed & Rafiq, 2012). Against this background, it is up to management to find the **optimal level of empowerment** depending on the type of service offered, so that customer service employees can make use of the competencies acquired through quality-oriented personnel development in the best possible way.

8.3.1.4 Quality Oriented Incentive Systems for Employees

In order for employees to be willing to implement the customer orientation knowledge imparted in quality related training and to meet the quality standards set out in the quality principles, it is necessary for the management to develop and implement suitable motivational measures (von Diemer, 1999; von Rosenstiel, 2010). In particular, when developing and introducing the quality management system, **extrinsic motivation,** i.e. incentives that lie in the work environment or in the consequences of the activity, are of particular importance in order to reward the commendable performance of employees. However, in order to ensure that employees behave customer oriented in accordance with quality standards in the company in the long term, it is also necessary to create **intrinsic motivation,** i.e. incentives that lie in the activity itself. The working conditions and content are to be designed in such a way that employees are willing and convinced to provide high-quality services from an inner motivation.

In order to motivate employees to act in a quality oriented manner, so-called **customer oriented incentive and compensation systems** are particularly suitable (Tuzovic, 2004). The special feature compared to traditional incentive and compensation systems is their focus on customer oriented behavior. While traditional incentive and bonus systems are based on financial variables (e.g. sales), customer oriented incentive and compensation systems use customer related variables such as customer satisfaction as the basis for calculation. When introducing a customer oriented incentive and compensation system, two basic **phases** must be carried out (see Fig. 8.20):

- As part of the **conceptualization of the system**, for example, the basis for calculation and the incentives are defined.
- As part of the **institutionalization of the system**, networking with other management systems takes place, without which the incentive and compensation system would be detached from the company's management model and therefore be ineffective.

The effectiveness of a customer oriented incentive and compensation system is essentially determined by the incentives chosen. In Fig. 8.21 **individual incentive forms** are listed with the goal of motivating employees and encouraging a customer oriented behavior. Also **group related incentive systems,** which include team premiums, recognition of service teams, etc., are appropriate in order to motivate project groups or departments to have a greater quality focus.

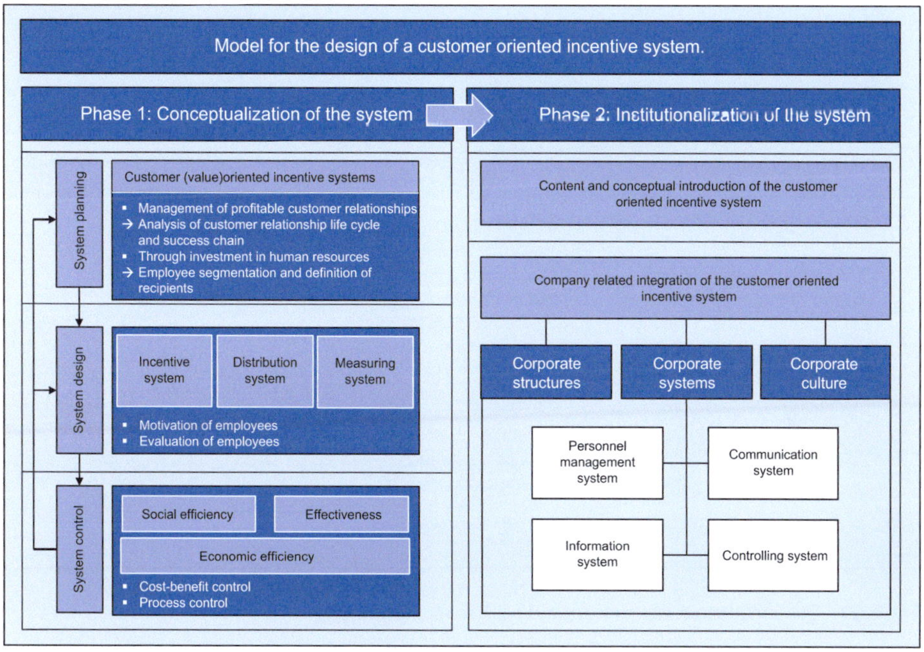

Fig. 8.20 Procedure for introducing customer oriented incentive and compensation systems. (Source: Tuzovic, 2004, p. 58)

In the case of customer oriented incentive systems, the involvement of customers in the performance assessment of an employee is of particular importance. In this case, the incentive systems can be divided into **direct** and **indirect compensation models**.

In a **direct compensation system** the customer has direct influence on the variable compensation. For example, the airline Northwest Airlines distributes checks for 50 USD to especially friendly flight attendants. The direct specification of customer oriented compensation is often problematic due to labor and tariff law restrictions, but also due to cultural differences in the individual assessment of customer orientation. In addition, such compensation systems appear to be problematic against the background of situational factors (e.g. weather, delays, other passengers as additional influences on a flight), which do not allow for uniform budgeting and allocation within the compensation system.

An **indirect compensation system** involves a more comprehensive inclusion of customers and provides better opportunities for setting variable compensation. First, a standard must be defined which is to be recognized as "normal performance". Furthermore, an upper limit and a lower limit are to be set, which specify a range for the variable share of the compensation. The basis for the amount of the compensation is often the customer satisfaction in customer oriented compensation systems, which serves as a criterion for

Quality oriented forms of incentives in service companies		
Incentives to maintain and increase extrinsic motivation		Incentives to maintain intrinsic motivation
Material incentives	Non-material incentives	
▪ Premiums for customer oriented consulting ▪ Performance-related salary components (success from the customer's point of view) ▪ Wage increase for customer oriented behavior ▪ Right to attend seminars ▪ Incentive travel	▪ Individual award ▪ Announcement of performance in the company's own media (employee newspaper, bulletin board) ▪ Transfer of responsibility ▪ Assignment of project management tasks (e.g. for quality groups) ▪ Offering opportunities for promotion	▪ Expressing personal praise ▪ Improvement of working conditions ▪ Modification of the design of work content

Fig. 8.21 Quality oriented incentive forms in service companies

customer oriented behavior of the employees and for which different possibilities for its collection exist. Since customer satisfaction is subject to subjective assessment, a method is to be used which always sets the same standards despite the inevitable subjectivity in order to ensure a fair compensation system.

Of particular importance is the **coordination of possibly different incentive systems** of the different company departments and divisions, so that disadvantages of individual groups of employees can be avoided despite identical performance or level related discrepancies between the incentive systems.

In summary, customer oriented compensation systems represent an effective instrument for the long-term anchoring of quality orientation in the company, but the involvement of all employees in the system is essential for this. This is because only this way can it be demonstrated that internal processes have a decisive influence on external performance assessment or customer satisfaction (Zeithaml et al., 2017) and customer oriented compensation systems therefore also have to include areas without customer contact.

The use of incentive systems in practice

The management of the courier and logistics company *FedEx* which operates worldwide is given targets regarding customer and employee satisfaction. The achievement of these targets is a prerequisite for the management to be included in the company's bonus pool.

At the greeting card manufacturer *Hallmark* the variable compensation of each employee depends on their position and is between 5 and 10%. In addition to many other indicators, customer satisfaction is used as a basis for assessment.

At the branch managers of the fast food chain *Pizza Hut* 50% of the total quarterly bonus payments are linked to the result of customer satisfaction surveys and a "Customer Loyalty Index". To determine this index, 50,000 customers are called each week.

The retail chain *Kmart* has introduced a variable compensation system in connection with its "Super Service Initiative" bonus program. Based on customer surveys, a super service index (SSI) is formed. This serves as the basis for calculating quarterly bonuses for associates. ◀

However, the use of incentive systems can also have negative consequences. Often, incentives that are intended to lead to increased performance are judged by employees who already showed a high level of commitment before the introduction of the incentive system, as unfair. These employees criticize that behavior which is considered self-evident is rewarded and thus tends to benefit less engaged employees.

Negative effects of incentive systems

A study by *Harvard Business School* has shown that incentive systems can have a negative effect on the overall productivity of the company (Gubler et al., 2013). The company introduced a lottery to increase the punctuality of employees. All employees who always arrived on time were eligible to participate in the lottery. The lottery prize was worth 75 USD.

At first, the goal was achieved by the proportion of employees who were late for work by 40%. In the long run, however, it was found that the work productivity of the committed employees, who had previously started work without incentives, was declining. They considered the lottery to be unfair and therefore reduced their work.

In addition, the lottery had another effect. Instead of appearing late, some workers preferred to call in sick and stay at home to continue participating in the lottery. As a result, the number of sick days increased by almost 50%. Due to these negative side effects, the lottery was discontinued shortly after its introduction

(Source: Gubler et al., 2013). ◀

8.3.2 Culture Related Instruments of Quality Control

In addition to the personnel policy aspects, the "climate" of cooperation between employees plays an important role in the development and implementation of quality management in the company (Schneider & Bowen, 1999). An adaptation of the **corporate culture** is necessary in the sense that the system of shared value and norm concepts as well as shared ways of thinking and behavior, which shape the decisions, actions and activities of members of the organization (Heinen & Dill, 1990, p. 17; Homburg & Pflesser, 2000), develops in the direction of stronger customer and quality orientation. The service culture serves employees in special service situations as an orientation (Hartmann, 2012, p. 611). A clear common value orientation of the company thus makes it possible to increase performance in addition to stronger employee motivation and identification with the company. By specifically aligning corporate values with customers, employees are given an everyday focus and overall more efficient resource allocation is achieved. Strong competition, complexity and dynamic change also require a flexible adaptation of corporate culture, so that organizational ability is seen as a major challenge for companies. Workshops for employees are just one example of how to optimize corporate culture with regard to a quality orientation. In order to implement a successful corporate culture, it is also essential to reduce rivalry and departmental thinking. The mediation of a uniform understanding of quality and increased quality awareness as well as the involvement of employees in the collective learning and change process are necessary adjustments for a successful quality oriented corporate culture (Homma & Bauschke, 2010, pp. 20–29).

There are four **dimensions of corporate culture** in particular: a flexible adaptation (e.g. to customer needs), a shared vision, the building of an emotional bond to the company and a closeness within the company (Nerdinger & Wilke, 2009, p. 30). Only if all managers and employees in customer contact and back-office departments appreciate good service quality and see internal and external customer orientation as "natural way of life", that is, as self-evident task in everyday work, can a cross-departmental **service culture** arise in the various departments, branches and central offices of the company (Grönroos, 2007, p. 415; Meffert et al., 2018, pp. 467–484).

However, the change in corporate culture is a very long-term and difficult process. Some employees of the service company may react negatively to the development of a quality management system if, in their opinion, it represents changes that are in contradiction to their values and norms, which have been grown and firmly anchored over the years. In contrast to the restructuring of organizations or the introduction of certain quality activities, it is only very limited possible to implement fundamental cultural changes exclusively through formal changes (Meffert et al., 2018, p. 471).

Against the background of these potential corporate cultural barriers, the management of the service company is primarily challenged to live the quality idea through their decisions, actions and activities towards employees and market partners and thus to develop a service culture. With regard to the numerous points of contact at which the management and business unit managers, but also the department managers can take on a **role model function** (e.g. Schneider & Bowen, 1999, p. 273; Bayer, 2009, p. 106), the culture aspects mentioned in Fig. 8.22 are of particular importance.

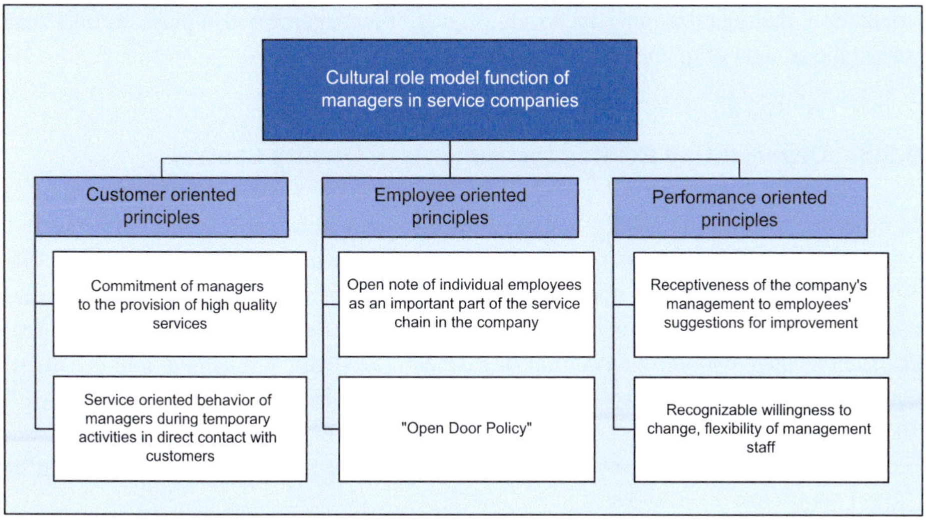

Fig. 8.22 Principles of the cultural role model function of managers in service companies

Role model function of executives at *Stadtwerke Düsseldorf AG*

The importance of the role model function of executives is particularly in focus at *Stadtwerke Düsseldorf*. The focus of the role model function lies in the daily living and implementation of the quality concept at the management level. The requirements for quality oriented role model function of *Stadtwerke Düsseldorf* are shown in Fig. 8.23. ◀

In addition to the employee and culture related instruments of quality management, which are mainly intended to change the company specific values and ways of thinking and behaving of employees in the direction of a stronger customer and service

Managers act as role models: They identify with the company's goals, are reliable, open, trustworthy, show appreciation and are willing to change.	
Managers focus their actions on the continuous improvement of quality, processes, resource utilization and costs.	Managers encourage the introduction of new ideas and support the implementation process.
Managers have a constructive attitude towards errors, which they analyze with regard to the cause. Managers take measures to avoid errors in the future.	Managers define criteria against which quality, process optimization and resource utilization can be measured.

Fig. 8.23 Requirements for the role model function of executives of *Stadtwerke Düsseldorf*

orientation, the organizational anchoring of quality management also plays an important role in the context of quality management.

8.3.3 Organization Related Instruments of Quality Control

In order for the specific quality management measures to be successfully implemented, various **structural and procedural organizational requirements** (e.g. Zeller, 1994; Schneider & Bowen, 1999; Wirtz & Lovelock, 2016) must be created in service companies. In particular, a comprehensive team structure, restructuring of the hierarchy levels, such as the change from a central to a decentralized structure, and a high density of communication are necessary for the realization of quality management (Haller, 2017, 300 pp.). Thus, the problem of **institutionalizing quality management** in the organization of the service company arises, so that a fast and flexible quality control can be carried out.

8.3.3.1 Quality Related Organizational Structure

The need for organizational anchoring and shaping of quality management in the enterprise arises directly from the organizational and structural barriers mentioned above. The design measures have to meet the **requirements** shown in Fig. 8.24.

The organizational design of quality management in service companies has to include not only the many and heterogeneous departments, but also sometimes subsidiaries and affiliated companies in the organizational design. This not only affects service and marketing departments, but all employees of the service chain of a company, i.e. personnel in customer contact areas and back-office departments, as well as department heads and managing directors.

It appears to be expedient to carry out both **primary**, i.e. permanently anchored in the company organization, and **secondary**, i.e. temporarily created, **measures of quality organizations**.

(a) Primary quality organization It is necessary to transfer **quality responsibility** to **line management** or to oblige each manager to ensure that she/he meets the quality and customer oriented requirements of her/his subordinates and achieves the set quality goals. In addition to the decentralized, operational units of quality management in the respective branches or offices, it is advisable to set up a **central quality management office** which performs coordination and communication functions. For example, a **quality committee** can be set up with **quality subcommittees** which are responsible for the conceptual (further) development of quality management as well as for generating quality goals and conducting customer and employee surveys. The conception and implementation of quality training programs and quality audits are also to be carried out by the employees of this central department in principle, in order to achieve uniform

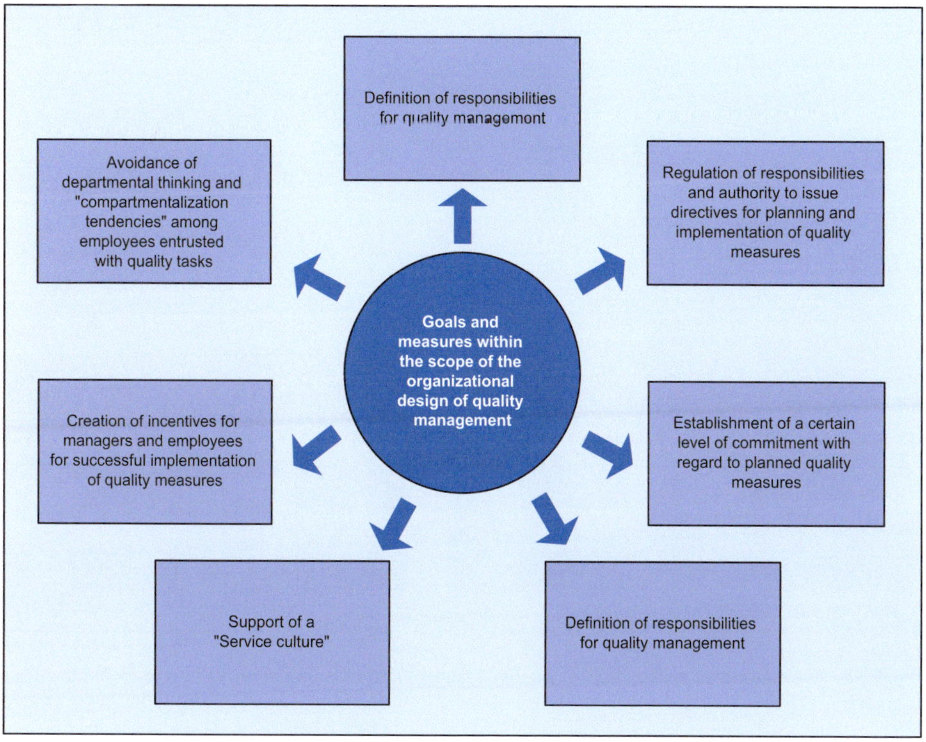

Fig. 8.24 Goals and measures in the context of organizational design of quality management

quality standards and to enable comparisons between departments and offices. Fig. 8.25 shows an example of the organizational embedding of quality management in the form of a central quality staff unit (quality management commissioner) using the example of an engineering office.

In highly branched service companies, it may be expedient to set up a so-called **quality manager** (Kamiske & Krämer, 1994; Hobel & Schütte, 2006; Biel, 2008; Sommerhoff, 2008; Gruhlich, 2016) for each region, who is responsible for the cross-departmental coordination and control of quality. The quality manager thus takes on the role of consultant and coordinator and can contribute to test methods through the quality management specific expertise and overall shape and develop the quality management system (Hobel & Schütte, 2006; Biel, 2008; Sieber, 2008; Sommerhoff, 2008; Gruhlich, 2016). Although the definition of a manager responsible for quality is very important for the implementation of a systematic quality management system, it seems that numerous companies are not aware of this. A survey of Austrian credit institutions showed that 23% of companies have transferred quality responsibility to line manager positions, 13% of companies to staff positions, but 63% have not transferred quality responsibility (Breyer & Zinnagl, 1995, p. 407).

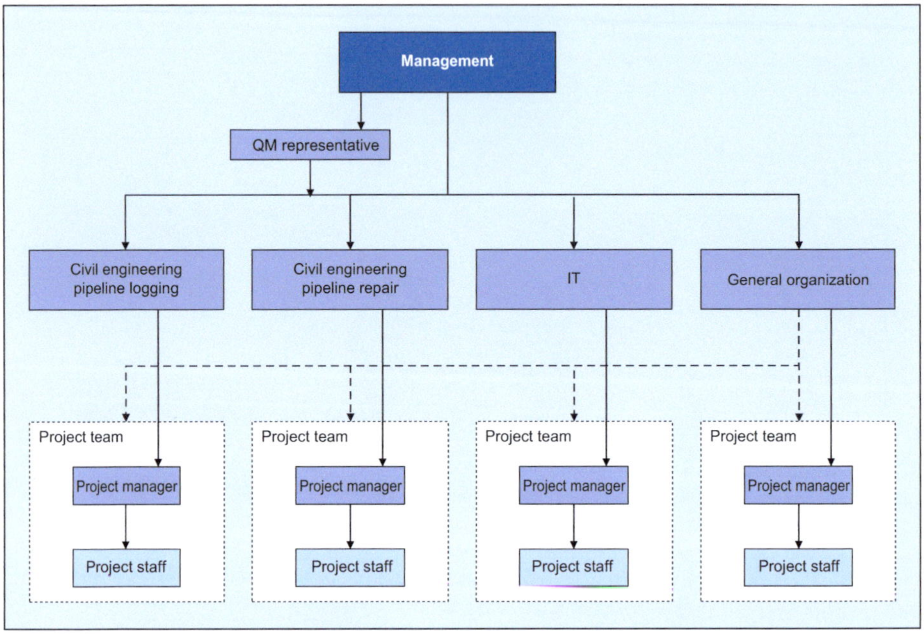

Fig. 8.25 Organizational embedding of quality management as a staff unit using the example of an engineering office

(b) Secondary Quality Organization The **quality circle concept** can be seen as a complementary or secondary form of quality management (cf. e.g. McKinsey, 2007; Evans & Lindsay, 2011; Kamiske & Brauer, 2011). Quality circles are permanently established discussion groups in which five to ten employees of one or more work areas of the lower hierarchy level of a business area meet on a voluntary basis at regular intervals (at least once a month) during or outside regular working hours to discuss self-selected quality problems of their own work area (Beriger, 1995, p. 121; Gabler Wirtschaftslexikon, 2013). Under the guidance of a trained moderator, solutions are to be developed and their implementation at least initiated and then also controlled with the help of special problem solving techniques, with the improvement proposals to be incentivized within the framework of the legal and operational provisions (Huber, 1992, p. 989; Brehm, 2001). It is important that each employee has what is known as an end-to-end understanding. This means that quality improvement begins with the collection of customer opinions and ends with the realization of the quality improvements perceived by the customer (McKinsey, 2007, pp. 58-59). The following example shows the structure of a quality circle system in a simplified way.

Structure of a quality circle system

The quality circle system shown in Fig. 8.26 shows the levels of the corporate hierarchy. Here, the hierarchy is divided into four areas. This can be adapted according to the company situation or the complexity of the service process. The respective hierarchical levels are assigned a control, coordination, moderation or active participation function in different ways. ◀

Based on the quality circle structure, various quality management objectives can be pursued within this secondary quality organization. Steinle et al. (1995) conducted a survey of companies that included service companies in order to gain insights into the primary **objectives of quality circle concepts** in practice. Fig. 8.27 provides an overview of the objectives that the surveyed companies considered to be significant. It is evident that the motivation and quality of service are at the forefront of quality circles.

With regard to the **tasks and functions of quality circle employees**, the control group, the coordination team, circle moderators, members of the quality circle, as well as supporters and experts can be distinguished in principle (see, for example, Brehm, 2001; Bechmann, 2010).

The **control group** represents the entrepreneurial will to actively support quality enhancing measures. This control group is to be formed from members of the board or management and additionally from selected members of regional quality circles (e.g. branch managers and the associated departmental manager circle). The most important **tasks of this control team** include:

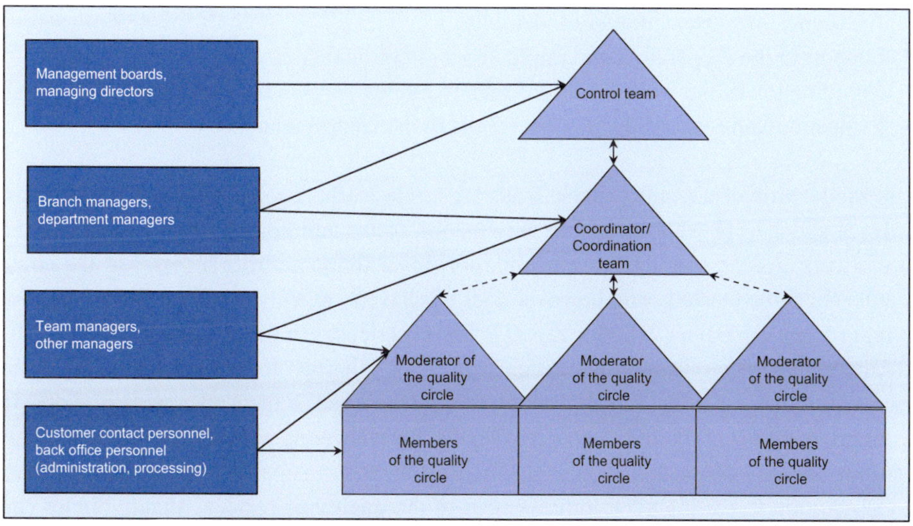

Fig. 8.26 Structure of a quality circle system

Goals of quality circles	Importance from the perspective of companies (in percent of companies surveyed)
Motivation/employee satisfaction	58.8
Product/service quality	52.9
Costs	47.1
Productivity	35.3
Communication/cooperation	27.5
Working conditions	23.5
Human resources development	21.6
Innovation	11.8
Other	7.8

Fig. 8.27 Objectives of quality circles. (Source: Steinle et al., 1995, p. 38)

- Development of a company specific quality strategy
- Provision of required work and auxiliary means
- Setting of training programs, schedules and general procedures
- Material and immaterial promotion of quality circle work

The **coordination team** of the quality circle is to consist of management personnel of the service company who are to fulfill the following tasks on a full-time or part-time basis:

- Assistance in the selection of moderators
- Support in the preparation and implementation of quality circle work
- Coordination of the work of various quality circles
- Evaluation and forwarding of work results to the control team

The **moderator of a quality circle** leads the circle work, organizes the group meetings, informs about work techniques and presentation forms and documents the results of the circle work. As the employee responsible for this activity, a direct superior of the circle members can be elected, who however does not have to exercise his function as superior here, or a circle member, who for example has experiences in group work techniques.

The **group of circle members** then works out solutions together for obvious problems within the company, which have preferably occurred in their own areas of work. In addition, experts can be involved occasionally for answering subject specific questions, but they are only allowed to support the moderators and/or superiors in the quality circle.

In addition to the intended improvement of the quality of service, quality circles can contribute to the promotion of the quality awareness and the quality responsibility of the involved employees, to the identification of quality related problems that are not limited to a certain area and to a significant improvement of the quality of internal communication in the branch or headquarters. The possibilities of exchanging experiences and personality development document the importance of quality circles as a management tool. Consequently, various **subject areas** have to be discussed and analyzed in the quality circles. In Fig. 8.28 different topics for quality circles are listed as an example.

In order for the findings and suggestions for service improvement gained in the quality circles to be taken seriously and implemented concretely, it is necessary to convince all management levels of the service company of the quality circle work and, for example, to institutionalize a presentation of the results in front of the members of the management board.

8.3.3.2 Quality Related Process Organization

Finally, an adaptation of the **information and communication systems** is also required for successful quality management in service companies.

One reason for this is that, in addition to the professional and personal qualities of the contact personnel required by the customers, the performance of the information and

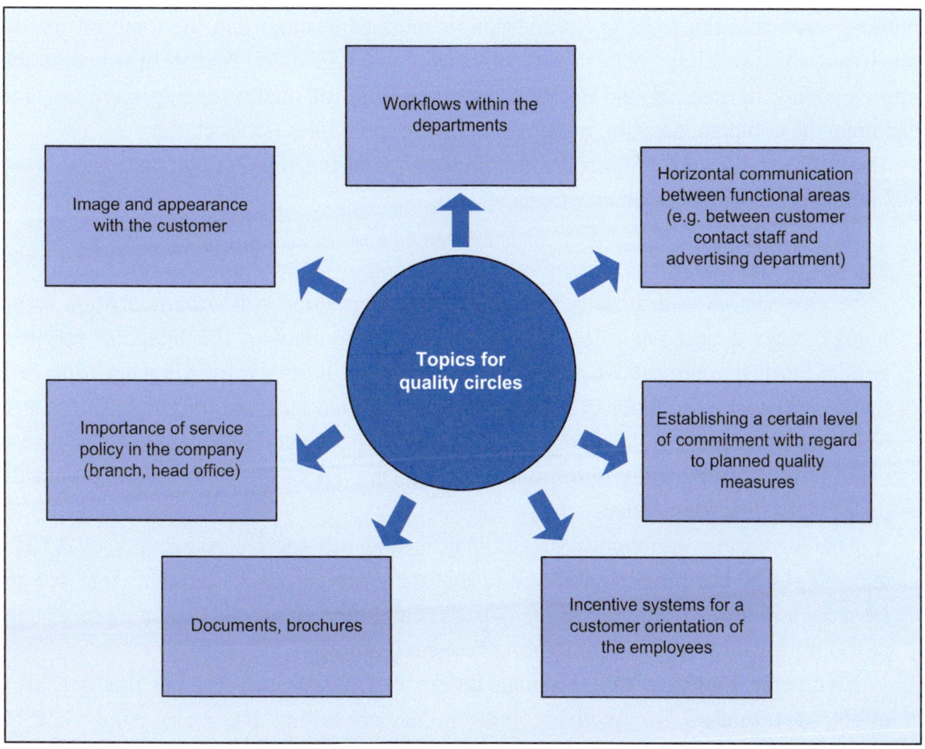

Fig. 8.28 Exemplary topics for quality circles

communication technologies used increasingly determines the quality of services. The task of the **information systems** is to collect quality relevant information from the individual departments, which serves as the basis for deriving quality measures. For example, the results of customer surveys on customer satisfaction from the market research department are the basis for identifying quality gaps. The direct contact between the employee and the customer also allows the sales staff to obtain further information from the customers first-hand. A key task of the information systems is to systematically capture the required information from the various sources, prepare it and make it available for quality management.

The **communication systems** of the service company are to be designed in such a way that an optimal implementation of quality management is ensured by consistent internal and external communication. Due to the special importance of the employees, internal employee communication plays an important role (Meffert et al., 2018, p. 463). This ensures that employees are always kept informed of news regarding quality management (work manuals, guidelines, etc.) in order to ensure a uniform and quality compliant appearance of the company. External communication refers to communication with stakeholder groups outside the company, such as customers, suppliers, distribution partners, etc. It is necessary here that the communication of quality is coordinated consistently across the various instruments (e.g. field service, media advertising) so that a uniform message arrives at the receiver. In addition, it must be ensured that the topic of quality receives a corresponding focus in the company's communication policy. Instruments for implementing the internal and external communication of quality management are, for example, the company's vision, mission and quality principles (cf. Sect. 6.2).

The following example illustrates the importance of information and communication technology instruments of quality management.

The role of information and communication technology in the financial services sector

In service sectors with high transaction rates, information and communication technology plays a decisive role. This is particularly evident in the financial services sector: From the perspective of credit institutions, it is necessary to counter the fact that of 100 customer contacts at the counter, more than 90% are routine business and less than ten percent are advisory (Ambros, 1991, p. 10) and service personnel invest about a third of their time in the issuance of forms. This aspect must also be considered from a cost perspective.

The increasing importance of automation and self-service, especially at **ATMs** and **telephone banking,** clearly shows that there are customer segments that see an advantage in terms of quality in the paperless and time economic distance processing of routine business without limitation of opening times.

Electronic banking, i.e. computer based banking services via the Internet, also offers opportunities for quality control in the consulting phase, as employees in customer contact can retrieve current market, customer and product information online, which is helpful or even necessary in individual, comprehensive advice. Simulations of alternative solutions in money investment or financing as well as model calculations in computer dialogues can also be used as part of expert systems to serve the information needs of customers and advisers. ◀

The expansion of the information system and data processing necessary for the quality management system of service companies mainly includes four tasks. An expansion of the information system is to be ensured in order to enable **market analysis and observation** in the individual branch offices. In addition, functional **customer information systems** of the service provider as well as **database and information systems** for daily use in consulting and sales need to be implemented. Additionally, it is the task of the information system to provide the basis for **control and comparison options with other branches** and their sales performance. The exemplary implementation of these tasks using the example of a car dealership is presented in the following example.

Quality oriented information and communication systems using the example of a car dealership

In order to ensure a high level of service quality, various information and communication systems are used in car dealerships (see the overview in Fig. 8.29).**Market analyses** include the classic key figures such as market share development and sales figures. In addition, a car dealership always has to keep up to date with trends and developments in the automotive market. Information on average selling prices and equipment packages of new cars provides information on the quality standards that need to be met in the relevant segment. **Customer information systems** can be mapped at both a global and an individual level. Global customer information systems provide information about the entire customer base. Individual customer information, on the other hand, makes it possible to map the entire purchase history of a customer. This makes it possible for advisers to better understand the preferences and needs of the customer and, based on this, to provide individual and thus high-quality advice. **Data bases and information systems** in consultation and sales also support high-quality advice by making all relevant information available and thus making reliable statements (e.g. with regard to the range of used cars available, delivery times) possible. This also includes information on test reports, quality rankings (e.g. *ADAC* breakdown statistics) etc. If such information is lacking, there is a risk of incorrect advice, resulting in the customer not receiving the optimal service for him. **Control and comparison options** with other branches make it possible to check one's own quality and to identify any weaknesses. In addition, in this context, opportunities for benchmarking and mutual learning are identified. ◀

Market analyses and observations	▪ Market share developments of various car dealerships and brands in the sales area ▪ Average selling price of new cars in the target area ▪ Average quality level of new cars sold in the target area ▪ Information on developments in market trends and customer needs
Customer information systems	*Global customer information (entire customer base)* ▪ Number of customers in the customer base ▪ Number of new customers ▪ Sales development of the customer base *Individual customer information (at individual customer level)* ▪ Customer contribution margin ▪ Customer history (date of the last transaction, information about all transactions and services with the customer) ▪ Information about potential needs of the customer (e.g. next inspection date, need for winter tires, etc.)
Database and information systems in consulting and sales	▪ Database of all available used cars ▪ Information about available spare parts and accessories for all models ▪ Price information and information about delivery times ▪ Information about current trends in the automotive market (e.g. legal changes, test reports, etc.)
Control systems with other business units	▪ Comparison of sales figures with other branches ▪ Comparison of quality related customer satisfaction surveys ▪ Comparison of profitability with other branches ▪ Comparison of structures with other subsidiaries (e.g. number of employees in workshop, consulting/sales, administration, etc.)

Fig. 8.29 Quality oriented information and communication systems using the example of a car dealership. (Source: Bruhn, 2013, p. 151)

A final overview of the individual instruments of quality planning is provided in Fig. 8.30. Based on the criteria, it can be checked which instruments are particularly suitable for a company under which conditions.

If the companies carry out the preventive and accompanying measures of quality management described above, which are necessary for the creation of services, it is essential that the quality of the services and products is subsequently or checked and controlled during the process.

8.3.4 Integration of Instruments for Quality Control

In order to carry out an integrated use of the instruments of quality management, an **relationship analysis** between the individual steering instruments must be carried out at the level of intra-phase integration. Fig. 8.31 shows the results of a relationship analysis for all instruments of quality management. The possible links between **personnel development** and the other instruments of quality management are to be shown by way of example. The instrument of personnel development can have the following functional, temporal and hierarchical relationships with other steering instruments:

	Quality-oriented personnel management	Quality-oriented incentive systems	Role model function of managers	Vision, mission, quality principles	Organizational anchoring/ Staff unit for quality	Quality Circle	Information and communication systems
Strategic goal	Building a qualified workforce	Motivation of employees to act in a quality-conscious manner	Exemplifying the qualit philosophy	Establishment and visualization of the quality philosophy	Organizational anchoring of quality	Organizational anchoring of quality	Optimal dissemination of quality-relevant information within the company and externally
Goals for action (operational goals)	Optimal personnel selection, -development and -definition under quality aspects	Development of monetary and non-monetary incentive systems to promote quality-oriented behavior.	Teaching quality-oriented behavior and motivating employees to behave in this way.	Communication of the quality vision and mission internally and externally, formulation of binding principles for all employees as a basis for daily work	Anchoring of quality management in the daily work of the company, establishment of a contact partner for quality issues.	Establish a quality circle of 5-10 employees to discuss quality issues at regular intervals.	Establishment of communication channels, continuous information of employees and customers, development of information and data processing systems.
Subject of investigation	Employee qualification	Compensation systems, employee behavior	Employee behavior, interaction with each other	Employee behavior, external image of the company	Company organization	General quality topics	Information and communication system
Individual measures	Personnel acquisition, training, continuing education measures	Quality-related bonus payments, praise, privileges, responsibility, etc.	Daily example by superiors, dialog with employees	Documentation and communication of the vision, mission and principles, regular dialog with employees and customers.	Establishment of a quality staff unit with all necessary competencies and responsibilities, introduction of a quality committee consisting of employees from different departments.	Developing solutions for quality problems, ensuring a permanent focus on quality	Set up a quality-related database with access for all employees, regular quality-related communication, e.g. through employee magazine, newsletter, etc.
Result/Output	Qualified employees, quality-related knowledge	Motivated employees	Quality-oriented behavior among all employees	Quality-oriented behavior among all employees, uniform, competent external appearance, high quality image	Quality issues are discussed and dealt with at regular intervals; quality is given high priority in the company; there is a contact person who can be contacted in the event of quality issues.	Improvement of internal performance quality, cost reduction, employee satisfaction	Optimal level of information for all employees regarding quality issues, fast and regular communication on quality-relevant topics within the company and externally
Main contact person in the company	Personnel management	Personnel management	Management, supervisors	Management, Marketing/ Communication	Quality Representative	Circle members, quality representative	Communication department, quality representative
Resources	Medium to high	Low	Very low	Low to medium	Medium	Medium	Medium to high
Frequency of implementation	Regular, continuous	Regular, continuous	Continuous	One-time formulation to ensure continuity, but ongoing communication	If needed	Continuous	Continuous
Responsible parties	Human Resources Management	Head of department, management	Head of department, management	Management	Management	Management, Qualit Representative	Management, Quality Representative
Cooperation with other agencies	Head of Department	Head of department, management	All employees	All internal and external stakeholders	Head of department	Circle members from different departments and hierarchy levels	Management, Communication, Sales, Human Resources

Fig. 8.30 Quality control instruments at a glance

Integrierte		Hiring new employees	Human Resources Development			Incentive systems	Role model function	Mission, vision, principles	Organizational anchoring	Quality circle	Information communication systems
			Job linked	Job independent	Cross job						
Hiring new employees			complementary successive	complementary successive	complementary successive	indifferent parallel strategic	indifferent parallel operative	indifferent parallel strategic	indifferent parallel operative	indifferent parallel strategic	indifferent parallel operative
Human resources development	Job linked			complementary parallel operative	complementary parallel operative	complementary parallel	conditional successive operative	conditional successive strategic	conditional successive operative	conditional successive strategic	indifferent parallel operative
	Job independent				complementary parallel	complementary parallel	conditional successive operative	conditional successive strategic	conditional successive operative	conditional successive strategic	indifferent parallel operative
	Cross job					complementary parallel	conditional successive operative	conditional successive strategic	conditional successive operative	conditional successive strategic	indifferent parallel operative
Incentive systems							indifferent parallel operative	indifferent parallel operative	indifferent parallel operative	indifferent parallel operative	indifferent parallel operative
Role model function								conditional successive strategic	indifferent parallel operative	indifferent parallel operative	conditional successive strategic
Mission, vision, principles									indifferent parallel operative	conditional successive strategic	conditional successive strategic
Organizational anchoring										complementary parallel strategic	complementary parallel
Quality circle											indifferent parallel operative
Information and communication systems											

Fig. 8.31 Relationship analysis as part of integrated quality control

- In **functional terms**, personnel development has a complementary relationship to quality oriented personnel selection, because both groups of measures ensure that the company's employees implement quality awareness and customer orientation. Conditional relationships exist with those instruments for which quality related know-how of the employees is required (e.g. quality responsibility, quality circles).
- If you look at the **time relationships** of personnel development, quality oriented personnel development follows personnel selection. Due to the desirable continuous use of personnel development, there is a parallel relationship to the other instruments.
- With regard to the **hierarchical relationships**, personnel development has a more operational character than the distribution of quality responsibility and the information and communication systems, and a more strategic character than feedback interviews and quality circles. Compared to incentive systems and the recruitment of new employees, the hierarchical relationship type depends on the design of the personnel development measures.

In Fig. 8.32 the use of quality management instruments taking into account the hierarchical and temporal relationships between the instruments is shown by way of example. It can be seen here that the quality oriented design of the information and communication systems of the service provider is to be used on a more strategic level and the implementation of quality related incentive systems on a more operational level is to be carried out

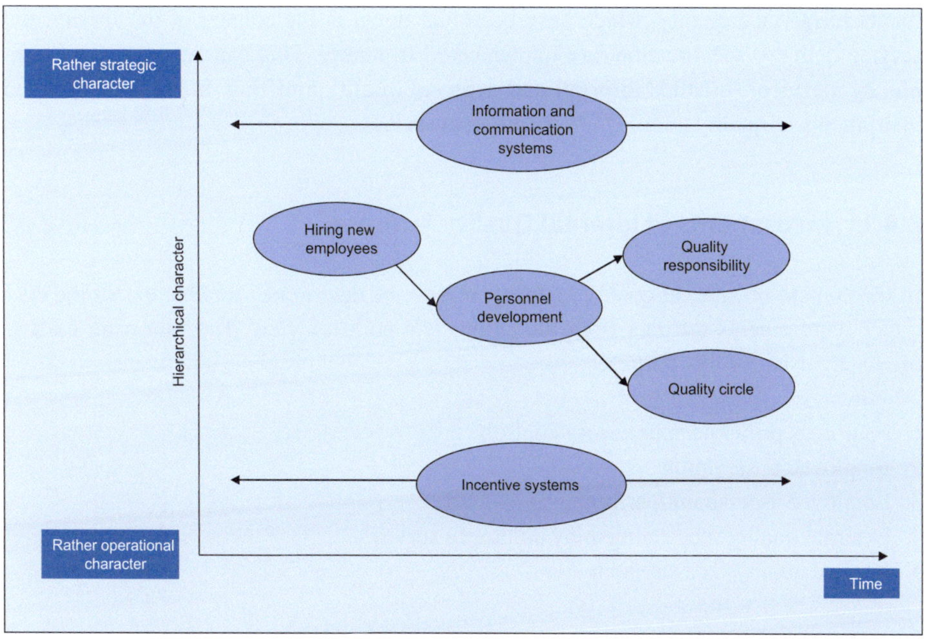

Fig. 8.32 Exemplary use of quality control instruments

in parallel with the other control instruments. When recruiting new employees, who are to be positioned in time before the personnel development measures from an individual point of view, attention is to be paid to the ability of the new employees to act in a quality and customer oriented manner. These abilities are then to be strengthened by means of personnel development in order to enable the employees to also assume responsibility for quality and to participate in quality circles.

Following quality management is the phase of quality auditing. The task of this phase is to check to what extent the goals defined in the planning phase were actually achieved by the quality management measures.

8.4 Instruments of Quality Auditing

In addition to planning and controlling service quality, in a quality management system in the **quality auditing phase** the actual fulfillment of the quality requirements is to be determined:

▶ The **quality auditing** is that part of quality management that makes an assessment of conformity by observing, judging, measuring or comparing (DIN EN ISO 9001, 2015, p. 10).

It is therefore to be checked by suitable methods whether and to what extent the requirements for service quality, which have been laid down in the context of the service and service delivery specification, are implemented in reality. This can be done, for example, by means of so-called **internal and external quality auditing**. In Fig. 8.33 selected instruments of quality auditing are shown in overview.

8.4.1 Instruments of Internal Quality Auditing

In the context of internal quality auditing, it must be determined to what extent the customer requirements are met from the company's point of view. The following **instruments** are used for this purpose:

1. Four-eyes principle/supervisory controls
2. Employee observation
3. Employee assessment/performance reviews
4. External quality measurement by employee survey
5. Internal quality measurement

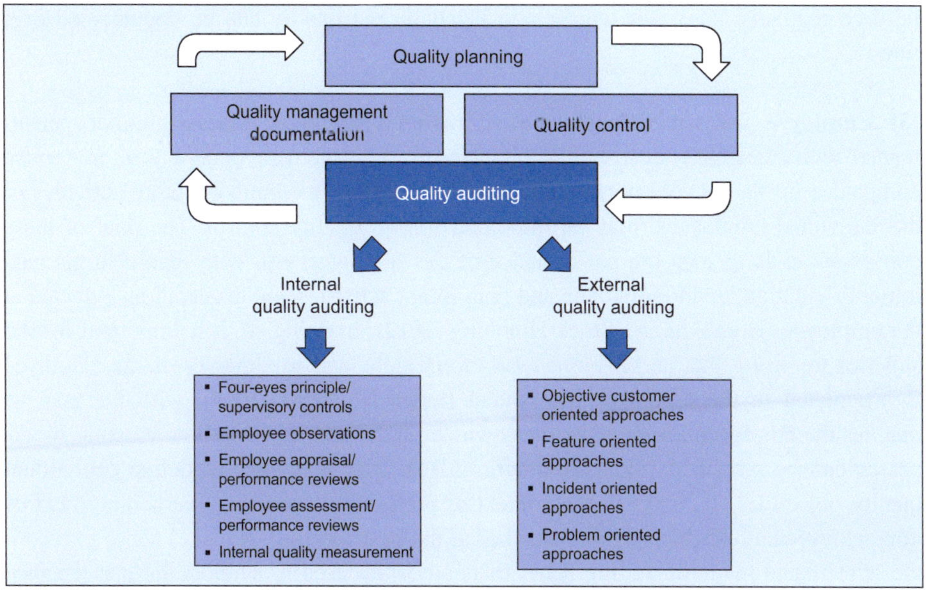

Fig. 8.33 Instruments of quality auditing of service companies

The quality auditing instruments are explained in more detail below.

(1) Four-eyes principle/supervisory controls First, the four-eyes principle can be used to carry out an internal quality check. It should be noted that this method can only be used to check the external quality if all employees have a pronounced awareness of quality and are aware of the customer requirements. In addition to the four-eyes principle, supervisory controls and computer based data backups can be used as control instruments against abusive behavior, as well as serve to comply with predetermined quality standards. The four-eyes principle is used across industries, mainly to secure "critical" processes that can lead to severe negative consequences in the event of an error. A process is considered critical if there is a risk of personal injury or significant financial losses. In addition, the four-eyes principle is used to prevent corruption (e.g. bribery, favoritism, abuse of office, etc.).

(2) Employee observation As part of the internal quality check, the supervisor occasionally takes part in the customer consultation conversation in a passive, observing capacity. Following employee observation carried out this way, a feedback conversation must always take place in which the strengths and weaknesses of the customer contact person in terms of service oriented behavior are discussed. However, the danger of employee observation is that the employee feels uncomfortable in the observation situation and adapts his behavior accordingly. The instrument of employee observation is to be used regularly. This way, changes in behavior and quality can be documented over time.

(3) Employee assessment/performance reviews Employee assessments or performance reviews, which are usually conducted by management once a year in service companies for the general **performance assessment** and planning of future activities of the individual employees, play an important role in quality control. The goal of these conversations is to map the performance of the employee, e.g. with regard to his customer orientation, and to articulate and plan future activities and expectations directed at the employee (Homburg and Stock-Homburg, 2012). In doing so, it is important for the superior to ensure that an balanced ratio of strengths and weaknesses of the employee is articulated in the employee assessment (Stock-Homburg, 2013), with the goal of making the employee aware of his/her own strengths and showing how existing weaknesses can be remedied (Stock-Homburg, 2013). It is necessary to define operational quality objectives and tasks in the context of performance reviews, the achievement or non-achievement of which will be checked at the next meeting. It makes sense to record the agreements made in writing. Checklists are often used to support the performance reviews. The most important points, e.g. to ensure a high level of customer orientation, are noted on the checklist, which are to be worked through during the joint conversation between the employee and his/her superior. Fig. 8.34 shows an exemplary protocol for a performance reciew.

Causes of deficiencies in service delivery can include, for example, incorrect job descriptions and requirements profiles that can be captured and improved during these conversations (Horváth & Urban, 1990, p. 182; Bolton & Saxena-Iyer, 2009). Since there is generally a strong relationship between employee and customer satisfaction, it makes sense to carefully analyze the problems and ill-feelings articulated by employees to their supervisors, and possibly change the areas of activity and competence of employees. Within the **performance reviews**, it is also important to exchange information about the individual employee's contribution to ensuring and improving the quality of the service perceived by customers within the quality management system. Only if each employee is credibly made aware of the great importance of his or her individual performance in external or internal customer contact for the entire service chain and the image of the service company as a whole will his or her motivation for customer orientation ultimately determine the success of the quality management system.

(4) External quality measurement by employee surveys This employee oriented process can be used to determine compliance with external customer requirements for high service quality from the perspective of employees. The employees evaluate their own performance for the customers by putting themselves in the customer's perspective (perspective change) and forming an image of themselves by comparing their expectations of their own performance with their perception of their own performance. The **self-image** is of particular importance in the comparison to the so-called **external image,** which

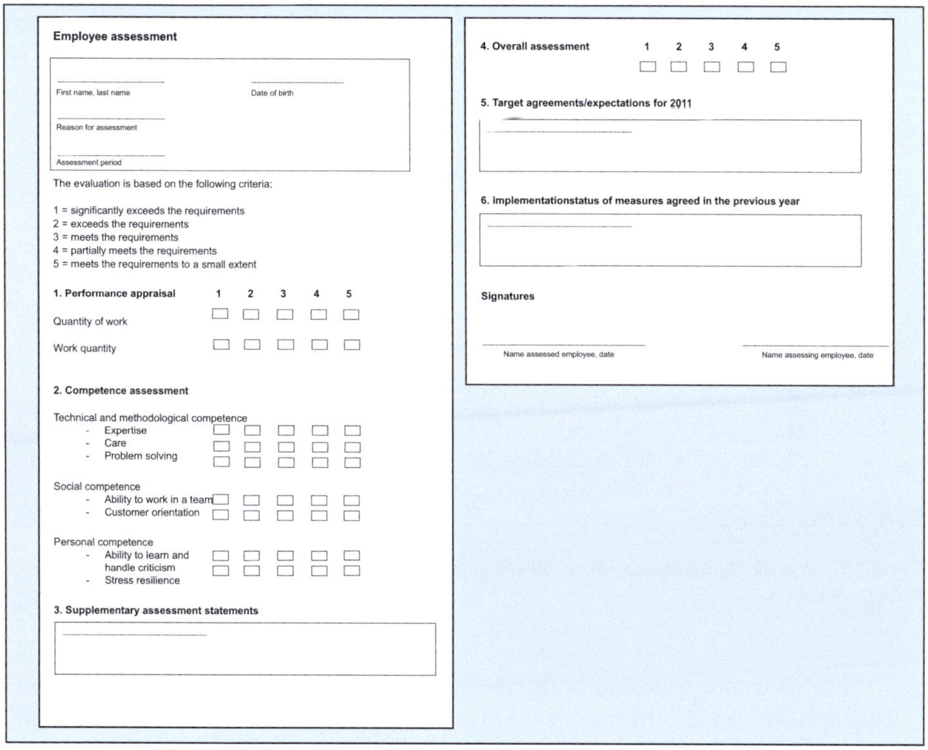

Employee assessment

First name, last name .... Date of birth ....

Reason for assessment ....

Assessment period ....

The evaluation is based on the following criteria:

1 = significantly exceeds the requirements
2 = exceeds the requirements
3 = meets the requirements
4 = partially meets the requirements
5 = meets the requirements to a small extent

1. Performance appraisal 1 2 3 4 5

Quantity of work ☐ ☐ ☐ ☐ ☐

Work quantity ☐ ☐ ☐ ☐ ☐

2. Competence assessment

Technical and methodological competence
- Expertise ☐ ☐ ☐ ☐ ☐
- Care ☐ ☐ ☐ ☐ ☐
- Problem solving ☐ ☐ ☐ ☐ ☐

Social competence
- Ability to work in a team ☐ ☐ ☐ ☐ ☐
- Customer orientation ☐ ☐ ☐ ☐ ☐

Personal competence
- Ability to learn and handle criticism ☐ ☐ ☐ ☐ ☐
- Stress resilience ☐ ☐ ☐ ☐ ☐

3. Supplementary assessment statements

4. Overall assessment 1 2 3 4 5

☐ ☐ ☐ ☐ ☐

5. Target agreements/expectations for 2011

6. Implementationstatus of measures agreed in the previous year

Signatures

Name assessed employee, date .... Name assessing employee, date ....

Fig. 8.34 Protocol for a performance review

results from the customer's quality assessment. The comparison of the service provider's image of oneself and the customer's external image can show the different quality assessments of the interaction partners and thus (perception) discrepancies in the customer-supplier relationship. The goal is to check the ability of employees to change perspectives, which is necessary for high service quality. The following example shows the comparison of image of oneself and external image in the quality measurement by an employee survey.

Perception discrepancy between self-image and external image

In the internal analysis of the service quality of a human resources department, employees of the entire company were asked for an assessment. From this, an external image was determined, which was then compared to the self-image according to the employees' own assessment. Fig. 8.35 shows the perception discrepancy between self-image and external image on a scale of 1–5. It is clear that only the criteria of employees' qualifications, reliability and awareness of services show consistent assessment. It is clearly visible that the self-image has significantly higher values than the external image.

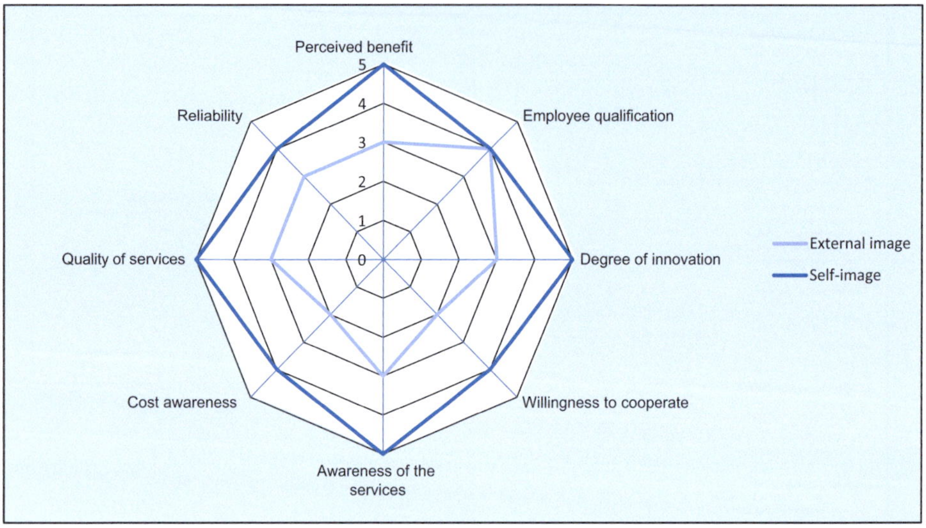

Fig. 8.35 Perception discrepancy between self image and external image among employees (Source: Ladwig, 2006, pp. 257-258.)

Such an employee survey with criteria such as quality of services or employees' qualifications can also be transferred to external quality measurements. For example, bank customers could be asked about the service quality of customer service employees in order to determine an external image. Subsequently, a self-image would have to be created by an assessment of the employees regarding their own service quality. When comparing, it could then be determined where action is needed to ensure high service quality. ◀

(5) Internal quality measurement The internal quality measurement process can be used to determine compliance with internal customer requirements.

It makes sense to transfer the concept of the national customer barometer to internal company or group questions in a modified form (cf. Sect. 5.2.2). According to the barometer concept, **internal service barometers** are to be understood as cross-service measurements of customer satisfaction, their determinants and their effects, which are carried out regularly by a neutral institution in a company or group.

In addition to capturing internal quality issues, internal customer surveys are used to identify improvement proposals and ideas from the surveyed employees (Hadwich & Keller, 2015). For reasons of survey effort, it does not make sense to have all internal services and service providers evaluated by all employees, but rather to select a sample of the people and departments to be evaluated. The implementation of the employee surveys is to be carried out by a central information and data processing center of the overall company or the market research departments of the business units in order to realize cost reduction potentials and comprehensive control of this instrument.

The special advantages of an internal barometer approach result from the continuity of the survey of comparable satisfaction data across the company, as well as—depending on the design—from the potential use of partial satisfaction or partial quality indices as control instruments for quality management (cf. also Kövesi-Grafinger and Schmoll, 1998, pp. 134–144; Parente et al., 2002). Depending on the size of the company and the company or group structure, in addition to an overall index, sub-indices are often formed at different company levels.

As examples, here are companies with their own legal entity, profit center or even individual service departments and specific services. In Fig. 8.36 the respective indices of an internal service barometer of individual service areas (labeled SU1 to SU24 in the figure) of a pharmaceutical company are shown. In this context, the necessary action needs in relation to the internal service areas become apparent. This results primarily from the comparison of the area related index with the company wide average of the indices.

8.4.2 Instruments of External Quality Auditing

In the context of external quality assurance, it is about checking the fulfilment of customer requirements from the customer's perspective. In doing so, some of the

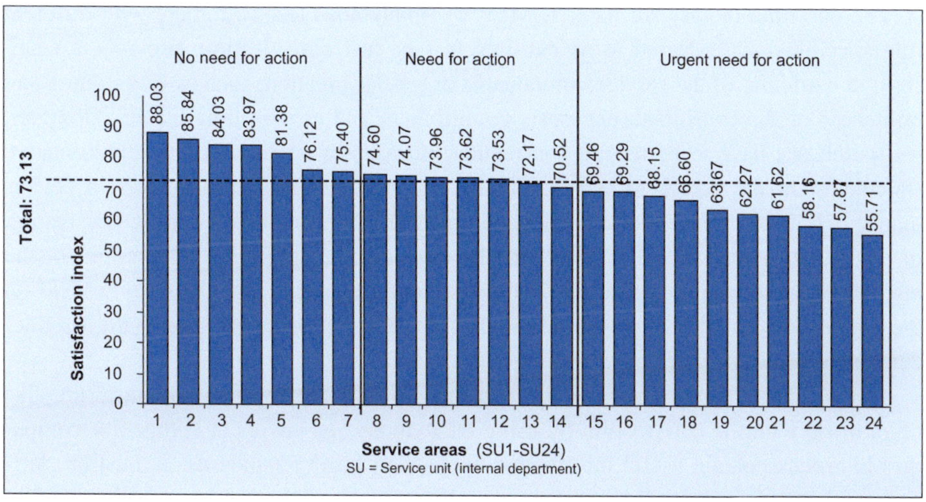

Fig. 8.36 Results of an internal service barometer of a pharmaceutical company. (Source: Bruhn, 2004, p. 288)

approaches to measuring customer expectations and perceptions in terms of the quality of services can be used. The preferred methods include:

1. Objective customer oriented approaches
2. Feature oriented approaches
3. Problem oriented approaches

The different approaches to external quality assurance are explained below.

(1) Objective customer oriented approaches For some years now, service companies, especially banks, have been increasingly concerned with so-called **test purchase methods,** whereby the observation of a real customer relationship, the maintenance of test accounts and the conduct of test consulting conversations can be distinguished (Drewes & Schneider, 1999).

In these forms of so-called "mystery shopping" or "silent shopping", an agreement made with the customer for the temporary observation of his accounts serves essentially to detect weaknesses in standard services from the customer's point of view. For this purpose, the account documents are checked at regular intervals and the customer's experience reports, for example on the clarity of the account statements, the comprehensibility of the forms, the booking periods or the postal delivery times, are obtained.

The management of fictitious **test accounts** without the knowledge of the account-holding institution, but in coordination with the auditing departments, is a more controllable instrument of quality control, since specific bookings and business transactions from the investment or credit business can be initiated and subsequently be checked.

The one-time inquiry for bank services by anonymous test customers with different role specifications, referred to as **counter test** or **test consultation**, provides a snapshot, in particular of the service components of service creation, such as friendliness and politeness of the consultant, expertise, commitment and cross-selling efforts. However, test purchases have to be repeated at regular intervals in order to obtain representative statements on the perceived service quality and the obvious improvements in performance over time. It is also necessary to design detailed questionnaires for the evaluation in order to systematize and at least partially objectify the subjective impressions of the test persons after the conversation. Test consultation conversations can also be held on the phone, which will become increasingly important in the future due to the stronger demand of customers for home banking services.

Test purchases can not only be made in one's own company. Test purchases are also often made with the competition by using anonymous test buyers at competitor companies in order to obtain useful information about their quality standards. It must be clarified in advance which quality features are to be assessed by the test buyer and according to which criteria these are to be evaluated.

Test purchases from competing companies in the retail and specialist trade

The retail and specialist trade is particularly suitable for carrying out test purchases from competitors, as it is relatively easy to collect an understanding of the quality of the services offered on site anonymously and undisturbed. The managing director of the *Tengelmann Group* explained in an interview that he regularly shops at competitors such as *Aldi, Rossmann* etc. in order to obtain information about the product range, the store design, etc.. In general, the following information is of importance for test purchases in commercial enterprises of the competition (see, for example, Weiss, 2009, pp. 31 et seq.):

- **Range of services:** Which products and services are offered? How are the services arranged and offered?
- **Value for money:** Which highlights are set? What special offers are there? Which quality products are offered at what prices? Which additional services are included in the price? Is a guarantee given for various products? Is a free return policy offered? Is there free delivery and assembly? If so, is this linked to the purchase price?
- **Customer service/employees:** Is trained and competent staff available? Is the customer given comprehensive advice? Does the staff behave friendly towards the customer? How are (special) wishes dealt with? How are questions answered? How are objections dealt with? How is the sales conversation conducted?

These criteria can be transferred to all possible industries with small adjustments. If, for example, it is a specialist shop, the focus may be on the specialist knowledge of the employees and the consulting service, while in hotels, on the other hand, the cleanliness of the facility and the well-groomed appearance of the employees are of high relevance. ◀

(2) Feature oriented approaches If the customer surveys carried out as part of the feature oriented approaches are not only carried out once, but continuously as part of so-called "**Customer Satisfaction Tracking Systems**", it is possible to obtain comparative data with regard to time comparisons, department comparisons, branch comparisons, etc. in order to show positive and negative changes in the quality of service perceived by customers.

It is important for service companies to carry out comprehensive customer surveys on a regular basis, not only for the already discussed reasons of capturing relevant quality criteria as part of quality planning, but above all for the control of the quality measures implemented. In order to be able to determine, for example, which individual instruments of quality management prove to be particularly effective, it is necessary to ask service customers before, during and after the implementation of the quality management system designed for their opinion on the products and services of the company.

However, it should be noted that quality measures often only take effect with a time delay (time lag) and in the long term; accordingly, direct cause-effect relationships can only be derived to a limited extent. However, positive or negative trend developments can be seen in regular customer surveys, which can indicate quality improvements or deteriorations and, in a comparison of data from different departments or branches, also "outliers", i.e. branches with operational implementation difficulties.

(3) Problem oriented approaches The central instrument of external quality assurance for service companies is a systematic complaint management (Stauss, 1989; Hoffmann, 1989; Rothlauf, 2014; Stauss & Seidel, 2014). The global goal of complaint management is to increase the profit and competitiveness of the company. However, a number of subgoals for complaint management have to be achieved in order to achieve this global goal, which are shown in Fig. 8.37.

The information from the complaints articulated by customers verbally or in writing can, provide indications of possible quality deficiencies in service provision and reveal potential for improvement for the service to be provided in the future. However, a complaint handling that is satisfactory for the customer (e.g. apologies, quick research on documents, compensation for damages) can lead to even higher customer loyalty if the customer feels "in good hands" at this branch in case of problems.

In order to systematically record and process different complaints from service customers, it is necessary to institutionalize or implement various **steps of a complaint management** in the company. A modern systematic complaint management system can

Global goal	▪ Increase profit and competitiveness
Subgoals relevant to customer relations	▪ Stabilization of endangered customer relationships or avoidance of customer losses by creating (complaint) satisfaction ▪ Achieving additional purchases by increasing purchase intensity and frequency and promoting cross-buying behavior ▪ Promotion of a customer oriented corporate image ▪ Creation of additional advertising effects by means of influencing verbal communication
Quality relevant subgoals	▪ Improving the quality of products and services by using the information contained in complaints ▪ Avoid external error costs ▪ Avoid internal error costs
Productivity relevant subgoals	▪ Efficient task fulfillment ▪ Detection of cost reduction potentials

Fig. 8.37 Goals of complaint management. (Source: Stauss and Seidel, 2014, pp. 63)

be divided into a **direct** and **indirect** complaint management process (Stauss & Seidel, 2014, 77 pp.).

The **direct complaint management process** must therefore include the following sub-steps, the fulfillment of which is directly related to the contact with the complainant (based on Stauss and Seidel, 2014, 77 pp.):

1. Complaint stimulation
2. Complaint acceptance
3. Complaint processing
4. Complaint reaction
5. Complaint handling

Figure 8.38 shows the individual phases of direct complaint management.

(1) Complaint stimulation In order to record complaints on as broad a basis as possible, the use of **complaint stimulating measures** is particularly to be considered. Companies have a wide range of options available. First and foremost, appropriate complaint channels are to be set up. The personal complaint process is particularly suitable for services with a high degree of interaction of the external factor. The media complaint process is divided into the following **forms**:

- Active written complaints (e.g. complaint letters and e-mails) and passive written complaints (e.g. comment cards in hotel rooms)
- Telephone complaints (e.g. by setting up toll-free telephone numbers)
- Online complaints (e.g. via forms on the company's website)

However, various studies show that a large proportion of dissatisfied customers do not contact the company with a complaint (see, for example, Goodman et al., 2000).

Therefore, it is first necessary to analyze the reasons that have so far prevented customers from complaining, and in addition to identify criteria that determine the complaint behavior of customers (Bruhn, 1982, 158 pp.; Stauss & Hentschel, 1991; Eder et al., 2010).

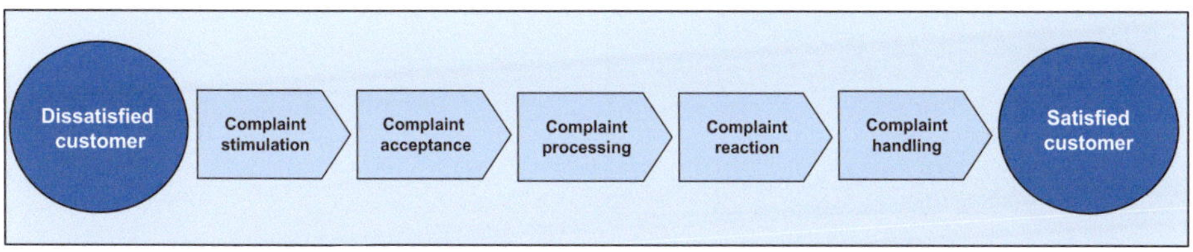

Fig. 8.38 Phases of direct complaint management. (Source: Based on Stauss & Seidel, 2014, p. 67)

For example, studies amongst dissatisfied car customers showed that the expected **costs and risks of making a complaint** are compared to the expected benefits or success, whereby both economic and non-economic criteria are used to assess success. Accordingly, complaints are articulated mainly for two reasons (Bruhn, 1982; Hansen & Jeschke, 1999; servmark, 2012):

- Customers consider the costs of complaint to be too high
- The probability of success of a complaint is considered to be low

In this context, Fig. 8.39 provides an overview of the reasons why **dissatisfied car workshop customers** have refrained from making a complaint (Bruhn, 1982; Hansen & Jeschke, 1999) as well as cited complaint barriers from a current study of the banking sector (servmark, 2012).

The complaint behavior of customers also depends on **service specific factors**. For example, a high—subjectively perceived—problem relevance leads to an increased complaint frequency. In contrast, problems with personal services tend to have lower complaint rates, as customers feel partly responsible for the poor service quality. In particular, deficiencies in the process properties of personal services are more likely to result in immediate churn, passivity or negative word of mouth.

Other determinants of complaint behavior are **person and situation related factors**. Both socio-demographic characteristics and psychological factors influence the complaint behavior of customers, with the final decision to file a complaint finally also determined by situational factors—such as the complaint image of the service provider (Hansen & Jeschke, 1999; Brock, 2009).

Reasons given by car workshop customers (multiple answers possible)	
Financial expenditure to remedy the defect was insignificant	22 %
Functionality of the product was not impaired	12 %
Problem not serious enough	36 %
Time spent on complaints too high	19 %
Complaint costs too high	15 %
Nothing achievable, because trade and manufacturer stick together	11 %
Legal requirements for complaint not known	3 %
Unknown complaint location	4 %
Ignorance about complaint handling	1 %
Complaints are unsuccessful according to experience	16 %
Complaints bring only trouble	12 %

Reasons given by respondents to the servmark study (2012) (multiple answers possible)	
Too much effort	39 %
No prospect of success	34 %
Churn to the competition	9 %
No contact person reached	7 %
Problem solved elsewhere	3 %

Fig. 8.39 Reasons for the non-filing of complaints by dissatisfied car workshop customers. (Sources: Bruhn, 1982, p. 66; Hansen & Jeschke, 1999, p. 443; servmark, 2012)

(2) Complaint acceptance The tasks of complaint management in the context of complaint acceptance primarily relate to the aspects of organization of the complaint intake and capture of complaint information (Stauss and Seidel, 2014, 127 pp.). It is important to systematically organize the complaint intake and forwarding within the company. In this context, the term **complaint ownership** is often used. This means that the employee who receives the complaint is the "owner" of the complaint and is responsible for ensuring that the problem is captured, processed and completed. Employees who receive complaints in direct or indirect customer contact must be prepared for this situation and trained in dealing with customer complaints.

In a next step, all relevant information on customer dissatisfaction contained in the complaints, in particular with a view to efficient processing of the complaint, is to be recorded. In this context, the focus is on the development of a suitable categorization system for the complaint information. A possible categorization of the **complaint information** is shown in Fig. 8.40.

Closely linked to the organization of the complaint intake is also the form in which the complaint content is captured. Often, the capture is carried out using standardized form sheets and complaint software programs in order to enable a fast, comprehensive and structured recording of the complaint.

(3) Complaint Processing Complaint handling includes all internal processing steps in connection with a complaint (Niefind & Wiegran, 2010, 49 pp.; Stauss and Seidel, 2014, 167 pp.). In particular, it is important to define the processing of the complaint. In this sense, responsibilities during complaint processing and handling deadlines are to be set. Likewise, mechanisms for monitoring complaint processing are to be implemented and communication between the processing departments is to be ensured.

(4) Complaint Reaction The **complaint reaction** includes all activities of complaint management that the customer experiences during the complaint process (Niefind and

Complainant information	Complaint issue information	Complaint object information
▪ Identity of the complainant - Information about the person/ organization - Accessibility - Internal or external customer	▪ Exact circumstances of the complaint incident - Affected organizational unit - Time of the problem occurrence - Specific situation of the incident	▪ Products and/or services
▪ Role of the complainant in the complaint process	▪ Type of issue	▪ Marketing aspects
▪ Exasperation level and behavioral consequences - Extent of upset - Intention to act or behavioral consequences	▪ Problem cause ▪ Initial or follow-up complaint	▪ Socio-political behavior

Fig. 8.40 Categorization of complaint information

Ratajczak, 2010; Stauss & Seidel, 2014, 197 pp.). This includes the immediate interaction with the complaining customers and the possible solutions or compensation measures in this context. The importance of an adequate complaint reaction is shown by the results of empirical studies on customer reactions in connection with complaint handling: The proportion of customers who are satisfied with the complaint reaction is between 30 and 60% in the majority of studies. Furthermore, even 70 to 90% of the satisfied complainants have an increased customer loyalty due to the complaint reaction (Homburg & Fürst, 2005, pp. 51 f.).

(5) Complaint Handling Structural, informational and personnel measures are available for complaint handling. **Structural measures** include, for example, the implementation of a complaint culture, in which employees are made aware of complaints. A complaint is therefore not to be seen as a burden, but as an opportunity to improve the company's performance. Other structural measures include the creation of responsible organizational units for complaint management. One part of this unit is then taken over by the process owner, who is responsible for the entire complaint management process and ensures that this process runs smoothly.

Under **informational measures**, the implementation of complaint systems (e.g. *SAP* program) and complaint management tools (complaint form) are summarized. In addition, it is ensured that all those involved in the complaint process always have access to the relevant information. Regular feedback loops during the handling process ensure that the customer as the complainant is continuously informed about the status of his or her pending complaint.

By **personnel measures**, a corresponding employee structure is created within the company, which enables a smooth handling of the complaint process. The employees are trained by means of appropriate training. In this context, it is important to ensure that the employees appear polite, friendly and understanding. The role model function of the management also plays an important role here. The complaint culture can only be successfully implemented if it is actively lived by the management. Therefore, the complaint management needs sufficient appreciation from the superiors in order to be carried by the employees accordingly.

In addition to the direct, there is an **indirect complaint management process.** This is characterized by the fact that the customer is not directly involved. The individual steps of the process therefore take place within the company. In this context, one speaks of an internal, company oriented complaint management process. The following tasks fall under the indirect complaint management process (Stauss & Seidel, 2014, pp. 69-):

- Complaint analysis and measurement
- Complaint management controlling
- Complaint reporting
- Complaint information use

The **analysis and measurement of complaint behavior** of consumers provides information about problematic areas within the service creation process itself, even when the business relationship with the customer is rather **aperiodic in character**. The accumulation of complaints in a certain functional or object context can support the specification of the later areas to be examined in more detail (cf. Sect. 5.1.2.3)

The **complaint management control** has the goal of controlling and monitoring complaint management. As part of an evidence control, it is about monitoring the ability of complaint management to detect dissatisfaction spread among the company's customers in the form of complaints. The task control is aimed at monitoring the fulfillment of the individual task areas of complaint management, whereas the cost-benefit control has the function of assessing the cost and benefit effects of the complaint management system (Brachetti & Wiegran, 2010; Stauss & Seidel, 2014).

In **complaint reporting** it is about making the findings of complaint analysis and complaint management control available to the various internal target groups of the company (Rothlauf, 2014, p. 217; Stauss & Seidel, 2014, pp. 393–394).

The final sub-step—the **complaint information use**—aims to steer the gained complaint information into quality management. This is possible using classical instruments of quality planning, as well as for example through a systematic inclusion of complaint information in quality circles (Schöler, 2009; Rothlauf, 2014; Stauss & Seidel, 2014). Ultimately, the consistent use of complaint information is an important prerequisite for the success of a quality management system.

The **further development of the Internet** and the associated technological possibilities (Web 2.0, see Sect. 5.1.2.3) open up new perspectives in connection with the complaint management of companies. This is particularly true for the stimulation of complaints by the establishment of new complaint channels, which can significantly increase the likelihood that customers will articulate their dissatisfaction in the form of complaints to the company and give them the opportunity to take immediate countermeasures.

For example, the use of web analysis tools such as blog search engines (e.g. technorati.com) or the *Google* search can be used to identify forums and blogs for critical statements about a service provider. However, such an approach requires the willingness of the supplier to engage in an open dialogue with critics in the relevant blogs or forums. Another possibility is the establishment of a **complaint channel** via social media platforms such as *Facebook* or via own chat applications on company websites. This not only accelerates the reaction times to complaints, but also simplifies the documentation and electronic further processing of complaints considerably. The *Deutsche Telekom AG*, for example, has its own *Facebook* page on which customers can submit complaints and questions, which are then processed by employees (see Fig. 8.41).

The German software company *SAP* has set up a chat on its homepage specifically for questions and problems, through which customers can directly contact employees. This allows for the immediate processing of questions and problems (cf. Fig. 8.42).

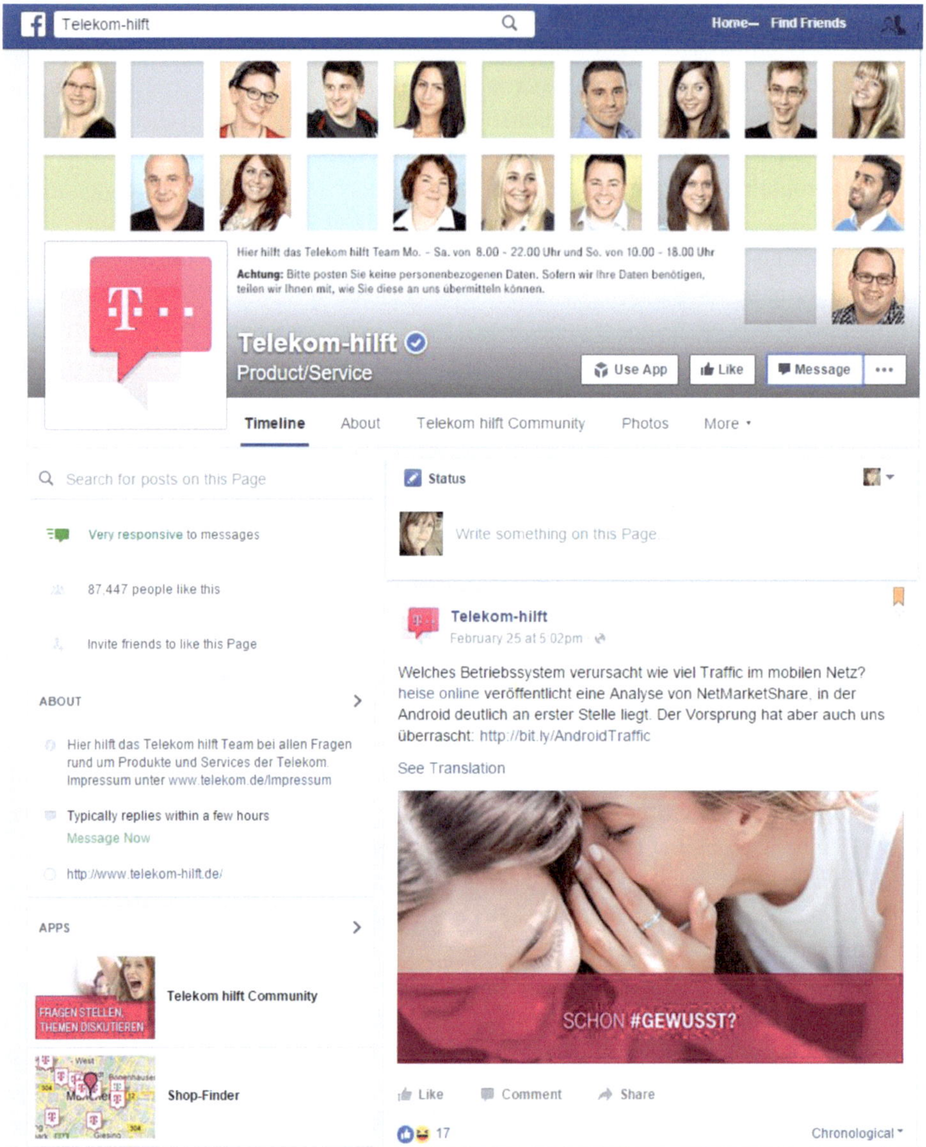

Fig. 8.41 Facebook as a complaint channel of the *Deutsche Telekom AG*. (Source: Facebook, 2016)

The instruments of quality planning are summarized in Fig. 8.43 and characterized in more detail with regard to various characteristics. On this basis, it can be checked which instrument is suitable for companies to be used for quality control under which conditions.

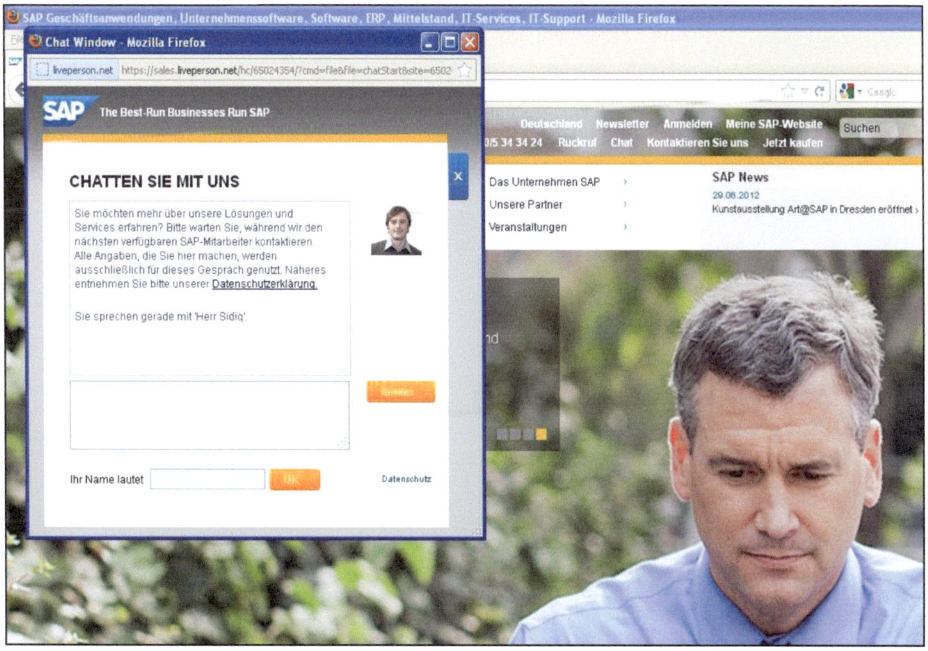

Fig. 8.42 Live chat as a complaint channel at *SAP AG*. (Source: SAP, 2016)

8.4.3 Integration of Instruments for Quality Auditing

In order to implement a holistic, enterprise-wide quality management, it is necessary, as in the phases of quality planning and quality control, to carry out a networking analysis between the individual instruments. Fig. 8.44 shows an overview of a **relationship analysis** of possible instruments for quality control.

For example, the internal testing instrument for internal quality analysis and the external instrument for complaint analysis are to be explained with the respective networking with other quality control instruments. For the **internal quality measurement**, the following functional, temporal and hierarchical relationships are highlighted in this context:

- In **functional terms**, there are conditional relationships with the four-eye principle and employee observation, because the results of the latter instruments flow into the performance assessment as part of the internal quality measurement. However, there is a complementary relationship with respect to employee assessment, because both procedures serve to assess the performance of employees. Since this is done from the management's point of view in the case of employee assessment and also from the colleagues' point of view in the case of internal quality measurement, the joint use of both instruments can lead to higher validity of the measurement results. There are

	Four-eyes principle/ Internal affairs controls	Employee observation	Employee assessment/ employee reviews	External quality measurement through employee surveys	Internal quality measurement	Test purchasing methods	Customer satisfaction tracking systems	Complaint management
Strategic goal	Control of the employee behavior	Control of the employee behavior	Assessment of the performance of the employees	Fulfillment analysis of external customer requirements from the employees point of view	Fulfillment analysis internal quality requirements	Observation of real customer relations	Identifying the development of the service quality over time	Identification of indications of possible quality deficits
Goals for action (operational goals)	Verification of compliance with standards; control against abusive Behavior	Identifying the strengths and weaknesses of employees through Observation in daily situations (especially during counseling interviews)	General performance evaluation and planning of future activities; Definition of measurable Quality objectives	Evaluation of own performance by the employee from the customer's point of view perspective, counter Cross-referencing of self- perception and external perception of performance quality	Regular survey of the determinants and effects of internal Customer Satisfaction, Deriving suggestions for improvement from the employees	Analysis of employee behavior in customer contact under real conditions	Time comparisons, departmental comparisons, branch office comparisons with regard to the service quality. Analysis of positive and negative changes	Recognition of weak points in the performance and implementation of implementation of measures; satisfaction and retention dissatisfied customers
Examination object	Employee behavior, process of performance creation	Current, daily behavior of the employee	Quality related performance of the employee and future goals	Fulfillment of customer expectations from the employees' point of view	Internal customer satisfaction, analysis of internal customer-supplier relationships	Consultations, employee behavior, process of performance creation	Service quality from the customer's point of view	Satisfaction and retention of dissatisfied customers
Individual measures	Control of the employee behavior and the service provision by several persons, possibly independent institutions	Silent observation of the Employees through Supervisors, human resources department, experts	Conversation between employees and Supervisor; analysis of past performance and joint definition of future goals.	Employee surveys, Comparison of the own and external image of the employees in the context of service qualitya	Internal Service Barometer Purchases or Employee surveys	Anonymous Test Test consultations through external Institutes	Regular customer surveys to check the measures carried out	Complaint- - stimulation - acceptance - processing - reaction and - handling
Result/Output	Error reduction, consistently high quality	Error reduction, consistently high quality, improvement of employee behavior	Increasing motivation, setting common goals, employee satisfaction, performance improvement	Merging external and internal view regarding the performance quality, create awareness among employees for customer viewpoint	Quality improvement, error prevention, eliminate weaknesses	Improvement of service quality, error reduction	Recognize the development over time of the service quality; thereby deriving the increase of the image Contribution of individual measures to success (cause-effect relationships)	Error reduction, increase in customer satisfaction image enhancement
Main contact partner in the company	Quality representative	Human Resources	Head of Department	Market research department, Head of Department	Market research department	Market research	Market research	Complaints department
Resources required	Medium	Low	Low	Medium	Medium	Low	Medium to high	Medium
Frequency of implementation	If needed	Regularly, when needed	Regularly, at least annually	Regular, annual	Regularly, at least annually	Regularly or when needed	Regularly, at least annually	Continuous
Responsibility carrier	Human Resources Department, Head of Department	Human Resources Department, Head of Department	Human resources department, Head of Department	Market research department, Head of Department	Market research department, Head of Department	Market research, sales	Market research	Complaint department, customer service
Cooperation with other agencies	Human Resources Department, Head of Department	Human Resources Department, Head of Department	Personnel department, department manager, employee	All departments, especially sales	All employees	Sales, market research, external institutes	All departments to be "tracked"	All departments where complaints occur

Fig. 8.43 Instruments of quality auditing at a glance

Integrated quality auditing	Four-eyes principle/ supervisory controls	Employee observation	Employee assessment/ performance reviews	External quality measurement through employee surveys	Internal quality measurement	Test purchase methods	Customer satisfaction tracking system	Complaint management
Four-eyes principle/ supervisory controls		complementary parallel	indifferent parallel operative	indifferent parallel operative	conditional parallel operative	indifferent parallel	indifferent parallel operative	indifferent parallel operative
Employee observation			conditional successive operative	indifferent parallel operative	conditional successive operative	indifferent parallel	indifferent parallel	indifferent parallel
Employee appraisal/ performance reviews				indifferent parallel operative	complementary parallel operative	indifferent parallel	indifferent parallel	indifferent parallel
External quality measurement through employee surveys					indifferent parallel	conditional successive operative	complementary parallel	complementary parallel
Internal quality measurement						indifferent parallel strategic	indifferent parallel operative	indifferent parallel operative
Test purchasing methods							complementary parallel	indifferent parallel
Customer satisfaction tracking system								indifferent parallel
Complaint management								

Fig. 8.44 Relationship analysis in the context of an integrated quality auditing

indifferent relationships with respect to the external testing instruments due to the different measurement object.

- If the **temporal relationships** are considered, the internal quality measurement can be used in parallel to almost all instruments. Only employee observation has to be carried out beforehand, so that its results can be considered for the assessment as part of the internal quality measurement.
- With regard to the **hierarchical classification** of the internal quality measurement within the totality of the quality control instruments, this instrument has a predominantly strategic character, because its use makes it possible to systematically assess the internal performance of the company.

The following relationships to other instruments of quality control can be identified for the instrument of **complaint analysis** within the context of external quality assurance:

- With regard to the **functional relationships** of complaint measurement, a rather conditional relationship can be established for the content related linking possibilities with thc feature and customer oriented approaches, because with the latter it is possible to analyze to what extent the problem areas observed in complaint analysis are significant for a large number of customers. Complementary relationships usually exist with the other external instruments, because potential problem areas in the context of service provision can also be determined with these (e.g. silent shoppers, incident oriented approaches). In contrast, there are indifferent relationships with the internal instruments of quality control, because the respective objects of quality control differ.
- Due to the requirement of permanently carried out complaint management, **parallel relationships** must exist with all other instruments of quality control with regard to the **temporal relationships** of complaint analysis.
- With regard to the **hierarchical classification** of complaint management within the totality of instruments of quality control, the importance of complaint analysis has rather strategic character.

In order to exemplarily show the use of instruments of quality control, Fig. 8.45 shows various instruments of quality control according to their hierarchical and temporal character in an overview.

Amongst the instruments of quality auditing with a more strategic character are internal quality measurement as well as employee and customer surveys within the framework of multi-attribute approaches, since with these methods the internal and external quality perception is shown in a representative and systematic manner. In order to develop the questionnaires, the criteria whose degree of fulfilment is to be determined in the mentioned surveys are to be agreed on first by means of objective as well as subjective incident and problem oriented methods. The problems uncovered by these

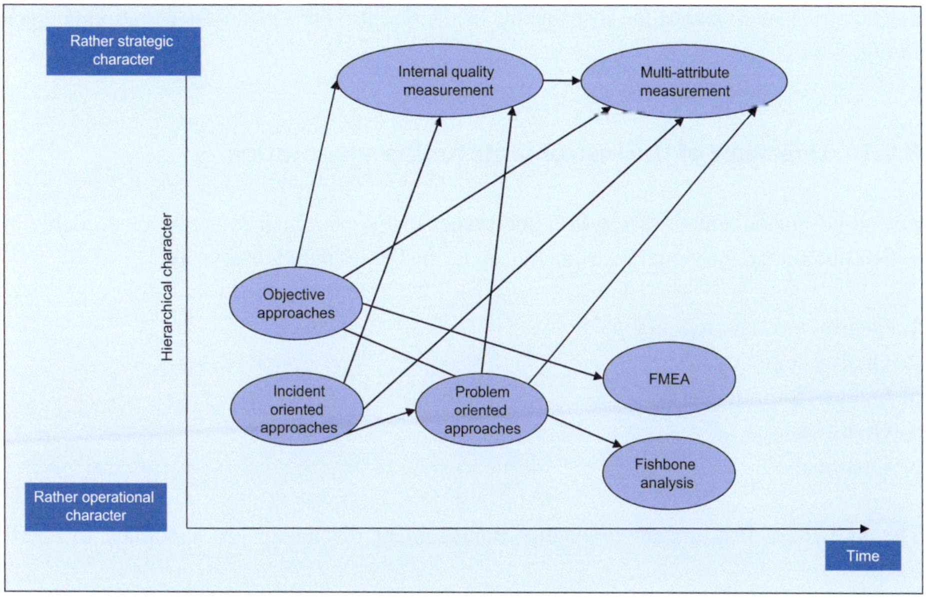

Fig. 8.45 Exemplary use of instruments of quality auditing

measurements in the production of the company's services should finally be intensively investigated by means of methods such as FMEA or Fishbone analysis in order to avoid the occurrence of the respective difficulties in the future.

8.5 Instruments of Quality Management Documentation

Finally, at the "end of the cycle" in the quality management system, the **phase of quality management documentation** is to be considered. This can be defined as follows:

▶ The **quality management documentation** is the part of quality management which documents all planned and systematic activities of the quality management system and provides corresponding evidence of activities carried out and achieved results (DIN EN ISO 9001, 2015, p. 10).

In this case, **internal and external objectives** of the quality management documentation are to be distinguished, as within as well as outside of a company, the management and employees of the service company can create trust in their own quality and thus create motivation. In particular, the quality management documentation may serve profiling purposes towards the outside, create trust and even serve as the basis for the issuance of a certificate. A overview of the instruments of the quality management declaration is given in Fig. 8.46.

8.5.1 Overview of the Instruments for Documentation

For quality management disclosure, comprehensive concepts or systems are available in which numerous individual instruments are used. These include above all:

1. Quality management manuals
2. Quality statistics
3. Quality communication
4. Quality audits
5. Certification

The individual instruments of quality management disclosure are explained in detail below.

(1) Quality management manuals Quality management manuals (e.g. Brüggemann, 2012; Schmitt & Pfeifer, 2015) generally set out the quality policy of the enterprise and also describe the organization's quality management system. These quality manuals either refer to the totality of activities or only to a specific part thereof (Deutsche Gesellschaft für Qualität e. V., 1995b, p. 144; DIN EN ISO 9001, 2015).

In the preparation of such a manual, the entire process of service design and actual service creation is to be set out on the basis of the quality objectives. This includes the documentation of organizational structures of quality management, service elements and quality requirements or responsibilities in the service company, etc. (Churchill, 1993, 313 pp.; Sachs, 1993; Reimann, 2012). Fig. 8.47 shows an exemplary **structure of a quality management manual,** as recommended by the *Deutsche Gesellschaft für Qualität e. V.* Starting from the definition of the purpose of quality implementation, the responsibilities and organizational embedding in the company are clarified. In addition,

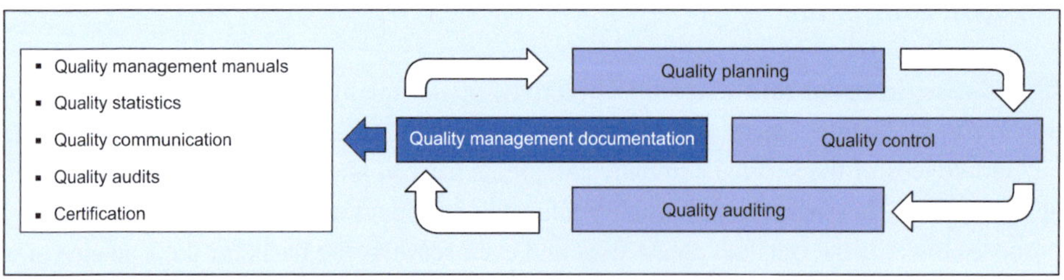

Fig. 8.46 Instruments of quality management documentation of service companies

1. Introduction (Preface, General)

2. Scope

2.1 Purpose and scope
2.2 Structure of the QM manual
2.3 Exclusions of the scope

3. Organization and responsibility

3.1 Company profile
3.2 Organigram
3.3 Corporate philosophy

4. Quality management system

4.1 General requirements (objectives and scope of the QM system)
4.2 Documentation requirements (QM manual, control of documents and quality records)

5. Management responsibility

5.1 Management commitment
5.2 Customer orientation
5.3 Quality policy
5.4 Planning (quality objectives, planning of the QM system)
5.5 Responsibility, authority and communication
5.6 Management review

6. Resource management

6.1 Provide resources
6.2 Human resources (general, skills, awareness, training).
6.3 Infrastructure
6.4 Working environment

7. Product and service realization

7.1 Planning of the realization processes
7.2 Customer related processes (determination and evaluation of requirements)
7.3 Development (development planning, input, output, evaluation, verification, validation, steering of development changes).
7.4 Procurement (procurement process, procurement data, verification of procured products)
7.5 Production and service provision

8. Measurement, analysis and improvement

8.1 General
8.2 Monitoring and measurement (customer satisfaction, internal audits, processes, products)
8.3 Control of defective products/services
8.4 Data analysis
8.5 Improvement (continuous improvement, corrective action, preventive action)

Fig. 8.47 Recommended structure of a quality management manual. (Source: Deutsche Gesellschaft für Qualität e. V., 1995b)

the resources to be provided are documented and finally the measurement and monitoring of quality management is regulated.

The following example shows the structure of the quality management manual of the *OBI Bau- und Heimwerkermärkte GmbH & Co. Franchise Center KG* by way of example.

Structure of the *OBI* quality management manual

OBI is number one in the German construction and DIY market. With a total of 580 *OBI* markets in 13 countries, *OBI* is the fourth largest hardware store operator in the world. In total, the *OBI* markets generated sales of 6.7 billion EUR in the 2015 financial year. The share of sales from abroad is more than 30%. The quality management manual is structured accordingly into the areas of franchise center and markets. At the same time, four application related levels are distinguished, as shown in Fig. 8.48.

(Source: OBI, 2016) ◀

The latest developments in the Internet in connection with web 2.0 offer new technological possibilities in the documentation of quality management. The starting point for the use of web 2.0 in the context of quality management disclosure are so-called **internal knowledge platforms** in companies, through which knowledge available within organizations can be made transparent and available company-wide. Quality management manuals document the quality related knowledge available in a company in relation to the requirements, measures and results of quality management. Wikis are modern forms of internal knowledge platforms that use the technological possibilities of web 2.0 to prepare and provide quality relevant information such as requirements, measures and results of quality management internally via the intranet.

The term wiki comes from the Hawaiian language and means "fast". A wiki is an online page collection that can be read, changed, or extended by the user (Huber, 2008, p. 78). Wikis enable employees to create a quick and user-friendly access to relevant quality management information and also provide the opportunity for employees to publish their own quality relevant information. However, in order to ensure the quality of

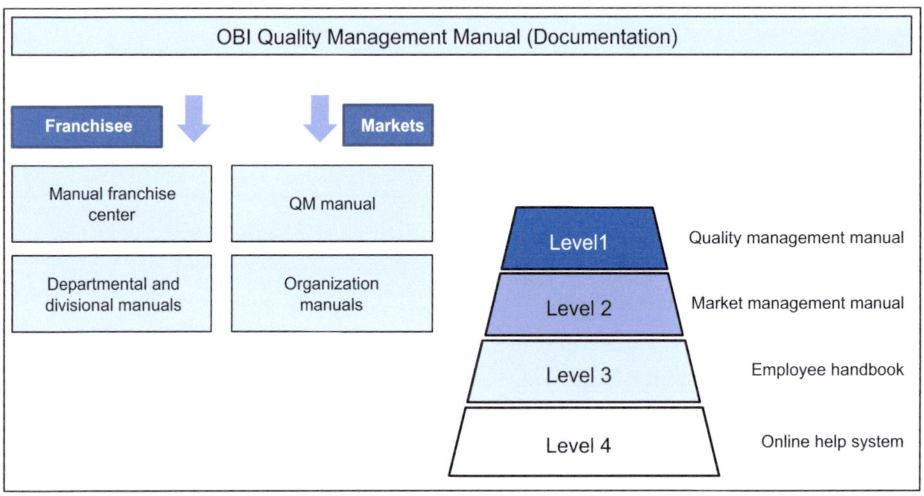

Fig. 8.48 Structure of the *OBI* quality management manual. (Source: Bruhn and Ahlert, 2002, p. 251)

information of the wikis, the use of suitable moderators is required, who regularly check and possibly adjust the content.

Internal knowledge platforms in different companies

Internal knowledge platforms are already in use in many companies of different industries today. A popular example is the platform Bluepedia by *IBM*. This database follows a similar principle to *Wikipedia*. Employees have the opportunity to write articles on certain topics and make them available to all employees. These articles can then be read, commented and supplemented by other employees. Frankfurt Airport *Fraport* maintains the platform *Skypedia*. This platform is also very closely related to *Wikipedia*. In addition to publication on these platforms, the most interesting contributions are regularly published in the intranet and in the employee magazine. ◀

(2) Quality statistics In addition, quality statistics—for example, using achievement rates for ensuring service quality—often provide valuable sources of information for companies. Such statistics often give the results of the feature oriented approaches. Data on customer satisfaction, but also on the cost and earnings situation of the company are included in the statistics. In principle, each company has to decide individually which are the "critical factors" in the service delivery process and how to record them using a quality statistic. For example, in the case of a hotel, these may be the number of incorrect bookings or the frequency of complaints about the unfriendliness of the service staff. It is important that, in addition to the absolute values, the relation to the total number of cases (e.g. the proportion of incorrect bookings to the total number of room bookings) is also shown. This increases the interpretability of the statistics and makes it possible to compare it with other branches or a time comparison.

(3) Quality communication In order to ensure an appropriate external effect of these measures, finally, appropriate implementation of the quality management declaration to the inside and outside is to be ensured by internal and external communication. In this context, targeted measures within integrated communication highlight the seriousness of the quality efforts of the service provider (Bruhn, 1999b, 2002a). This is a concept that is designed to create unity from the differentiated sources of internal and external communication of companies in order to convey a consistent image of the company or an object of reference of communication to the target groups of communication (Bruhn, 2014, p. 38). This applies equally to quality communication. It must be ensured internally and externally that there is the same and uniform understanding of the company's quality perception.

Appropriate measures for **internal communication** are, for example, the publication of particularly positive results of customer satisfaction surveys via employee magazines, notices, newsletters or the intranet. Communication of particularly successful improvement suggestions from employees also contribute to achieving a positive internal representation of quality management. For external communication, various measures of quality communication are to be used. In addition to the classical measures of marketing communication (e.g. advertising, personal communication), these include *Facebook* entries, press reports or PR measures on successful certifications, awards or the successful evaluation in quality tests by external test institutes (e.g. *J.D. Power, Good Housekeeping Institute UK, Choice, Que Choisir* etc.).

Communication of quality awards

On the homepage of the hotel *The Mandala* from Berlin, all prizes won in recent years are listed chronologically under the heading "Awards" (e.g. *Dehoga Environmental Award, Michelin Star* and *Gault Millaud Points*). In addition, the links to the respective homepage of the quality prizes received.

The advertising agencies *TBWA, Serviceplan* and *Jung von Matt* also offer an overview of the awards won on their company homepage. In addition, the press releases justifying the award of the prize are accessible.

Companies such as hairdressers, engineering offices, law firms, etc. communicate their competence and quality standards by making diplomas, certificates, etc. visible to customers at all times in consultation and sales offices (e.g. master craftsman's certificate in hairdressing salon, ISO certificates in engineering office). This creates a feeling of trust in the competence of the employees amongst the customers. ◄

(4) Quality audits In order to uncover weaknesses in the quality management system and to receive suggestions from various employee groups with regard to quality improvements and to check quality control measures carried out, so-called quality audits are increasingly being carried out in companies, which either take place externally, e.g. by a management consultancy, or internally as a quality review (e.g. Gaster 1994; Schmitt & Pfeifer, 2015) and are defined as follows:

▶ A **quality audit** is a systematic, independent and documented process for obtaining audit evidence and for its objective evaluation in order to determine to what extent audit criteria are met (DIN EN ISO 9001, 2015).

Quality audits are generally carried out by people who do not have direct responsibility in the areas to be audited, but who work together with the employees concerned in a meaningful way. The purpose of an audit is primarily to assess the need for improvements and corrective action. Quality audits relate to individual processes, products, services, but also to the entire quality management system (Deutsche Gesellschaft für Qualität e. V., 1995a, p. 141; Gietl & Lobinger, 2009).

Based on the three main quality audit types—system audit, process audit and product audit—designed primarily for manufacturers of consumer and industrial goods, the audit types system audit, process audit and service audit can be distinguished for services.

The main responsibility for quality audits in service companies lies with the management; However, employees of the auditing department and other control or inspection institutions of the service company are preferably involved in the actual implementation. The different quality audits are to be carried out regularly, i.e. semi-annually or at least once a year, in order to continuously record service results in the individual departments or branches of the service company and to carry out deviation analyses. Audits are concluded with an **audit report,** which comments on the individual results and suggests changes for quality improvement. Fig. 8.49 shows an extract of the most frequent questions of quality audits, divided into the most important categories.

A special form of quality audits are so-called **customer audits**. The company is audited by representatives of the customer side in this audit. The auditors prepare a questionnaire and answer catalog tailored to the customer's needs for the assessment of the quality management system. The individual answers are given points, which are added up to a total score at the end. Depending on the total score, point ranks are awarded. This form of auditing is often used by suppliers.

Quality oriented customer audit of *Robert Bosch GmbH*

Bosch obliges its suppliers to sign a quality assurance guideline, according to which suppliers, among other things, must also allow a customer audit. The following excerpt shows how *Bosch* ensures the quality orientation of its suppliers by means of a quality audit:

"The SUPPLIER undertakes to implement and maintain a quality management system at least according to ISO 9001. This includes the commitment to a zero-defect target and the continuous improvement of its services. The SUPPLIER allows *BOSCH* to determine by audits whether its quality assurance measures meet the requirements of *BOSCH*. An audit can be carried out as a system, process or product audit by prior announcement. The SUPPLIER will enable short-term requests for appointments if necessary."

(Source: Bosch, 2011) ◀

A restaurant could set a similar quality assurance guideline including a mandatory customer audit. Here, its suppliers could be checked for their quality management systems. Specifically, this could mean that a butcher shop that supplies meat to the restaurant is checked for food quality. The following questions could be asked as examples during the audit:

- Is the quality awareness of all employees actively promoted?
- Are approved sub-suppliers checked for quality assurance?
- How is the quality management system of the supplier described?

(5) Certification Another instrument of quality management disclosure is certification. Through certification, service companies have the opportunity to undergo an review by

Category	Questions
Organization	▪ Has the organization identified the processes required for the QM system and its applications? (management activities, resource allocation, service realization, measurement of processes). ▪ Has the sequence and interaction of the processes been taken into account? ▪ Does a QM manual exist? ▪ Are statements on quality policy and quality objectives available in the documentation? ▪ Are all necessary documents for the internal organization available (organization chart, job description, internal audit reports, management review)? ▪ Are the frameworks in place to identify customer requirements? ▪ Is there a QM representative? ▪ Has a quality policy been established by top management that meets the following points: Adequacy in form and scope, commitment to fulfill the requirements, commitment to continuous improvement, dissemination and comprehensible publicity of the policy in the company, continuous review of adequacy?
Distribution	▪ Are the following points fulfilled before the conclusion of the contract for deliveries: Determination of the performance requirements, assurance of the fulfillment of requirements by the company? ▪ Are assessment results and derived actions recorded? ▪ Are there uniform rules for communication with the customer? (e.g. type and scope of service information, inquiries, contracts or order confirmations including changes, feedback to the customer including customer complaints and claims).
Development	▪ Are the following points taken into account in development planning: Development phases, assessment, verification and validation measures in the individual phases, clear assignment of responsibility and authority in development? ▪ Is communication and responsibility for interfaces with other departments and areas in development regulated? ▪ Is the appropriateness of development requirements evaluated? ▪ Are the results and measures of the individual development phases adequately documented?
Service provision	▪ Are the following criteria taken into account in the steering conditions for service provision: Service descriptions, availability of adequate work instructions, use of appropriate tools (e.g. machines and tools), use of adequate test equipment, monitoring and measurement of processes, systematic release of performance? ▪ Is the traceability of the individual services during the creation process guaranteed? ▪ Is the handling of customer property regulated in the company? ▪ Are instances of damage, loss, or otherwise improper use of customer property recorded and reported to the customer?
Testing and measuring equipment	▪ Are the necessary test and measurement procedures for monitoring performance planned, developed, and deployed? ▪ Are the planned tests and measurements carried out systematically? ▪ If deviations are detected, are appropriate measures taken for the test equipment used and the performance already measured?
Measurement and improvement	▪ Are audit analysis and improvement processes planned and used that address the following areas: conformance of performance, conformance of QM system, capability of QM system to continuously improve? ▪ Does the analysis of customer satisfaction take customer perception into account as a metric? ▪ Does the company use appropriate methods to measure and monitor the processes of the QM system? ▪ Are the performance characteristics monitored within the service creation process and is the fulfillment of the requirements measured? ▪ Is the data collected in the processes evaluated to demonstrate the suitability and effectiveness of the QM system and to identify opportunities for improvement? ▪ Does the data analysis provide information on the following criteria: Customer satisfaction, compliance with performance requirements, process and result variations, indications for preventive measures, quality of the delivered service?

Fig. 8.49 Most frequent questions of quality audits

an external impartial institution in order to obtain a certificate that documents compliance with certain quality standards (e.g. ISO 9001 or ISO 14001). Due to its growing importance, certification is presented in detail in Chap. 11.

Finally, it should be noted that various instruments can be used in service companies in terms of inward and outward oriented quality management disclosure; However, a combination of individual measures in an overall concept—also from the perspective of efficiency—seems appropriate.

Fig. 8.50 summarizes the individual instruments of quality management disclosure. Here, too, the different description features can be used to check which of the instruments are suitable for a company in the specific situation.

8.5.2 Integration of Instruments for Quality Management Documentation

Similar to quality planning, quality control and quality audit, the relationships between the individual instruments have to be analyzed in the phase of quality management documentation in order to ensure an integrated approach. Fig. 8.51 shows a relationship analysis of the instruments of quality management documentation illustrated.

The networking possibilities with other instruments of this phase are exemplarily shown for the instrument **quality management manual**:

- With regard to the functional relationships, the relationships to quality audits and certification have a conditional character in that a quality management manual forms the basis for an audit or certification.
- If the temporal relationships of quality management manuals are considered within the quality management declaration, the creation of the manual has to take place before the instruments for which the manual defines the strategic framework of quality management are carried out.
- With regard to their hierarchical importance, quality management manuals have a rather strategic character compared to the other instruments.

Fig. 8.52 shows the design of the use of the instruments of quality management documentation according to their hierarchical and temporal character.

Since quality management manuals document the entire quality management of the service provider, they represent the instrument of quality management documentation with the greatest strategic importance. Furthermore, it is advisable to use this instrument continuously over time. Based on these books, quality audits can be prepared, which in turn are a prerequisite for the certification of the service company.

	Quality management manuals	Quality statistics	Quality communication	Quality audits	Certification
Strategic goal	Documentation of the quality policy within the company	Analyses of the degree of target achievement of quality management	Communication of quality competence internally and externally	Detection of weak points in quality management	Documentation of quality competence to the outside
Goals for action (operational goals)	Creation of a quality manual, documentaion of quality related knowledge available within the company	Statistical evaluation of quality management and identification of weaknesses and potential for improvement	Creation of confidence in the company's quality capability among external stakeholders (especially customers); communication to employees to document the importance of quality management.	Systematic assessment of individual measures, know-how of external institutions (e.g. consulting), review of implemented quality measures	Improvement of quality management by obtaining external know-how, obtaining a certificate for the external communication of quality competence.
Subject of investigation	Overall QM system	Quality management measures and instruments	Overall QM system	Overall QM system, partly detailed focus on individual measures	Overall QM system
Individual measures	Mapping of the entire process of service creation, documentation of the structural and procedural organization, quality requirements, responsibilities, accountabilities, etc.	Statistical evaluation of quality management as a whole, of individual measures and instruments. Determination of the contribution to success and derivation of measures for action	Quality oriented behavior among all employees, uniform, competent external appearance, high quality image	Evaluation of individual quality management measures and assessment of the need for improvements and corrective measures.	Review of quality management by external certification body, adjustment, improvement and elimination of weak points
Result/ output	Uniform reference work for all employees for quality questions	Improvement of quality indicators, awareness of quality weaknesses in the company	Building trust with external stakeholders, demonstrating the importance of quality management to employees.	Detection of weak points, suggestions for quality improvements	Efficient and effective quality management, high quality image to the outside world
Main contact in the company	Quality management	Controlling	Communication department, marketing	Quality management	Quality management
Resource allocation	Medium	Medium	Low to medium	Medium to high	High
Frequency of implementation	One-time creation, regular review and adjustment	Regular	Continuous	Regularly, at least annually	Every 3 years
Responsible parties	Management, QM	Controlling	Communication, marketing, management	Management	Management, QM
Cooperation with other agencies	Passing on to all employees, communication by head of department	Passing on the data to the department heads concerned	All departments in the company, externally mainly sales	External entities, internal entities concerned	All departments involved in the certification process

Fig. 8.50 Overview of quality management documentation instruments

Integrated quality management documentation	Quality management manual	Quality statistics	Quality communication	Quality audit	Certification
Quality management manual		Conditional parallel operational	Complementary parallel strategic	Conditional successive strategic	Conditional successive strategic
Quality statistics			Indifferent parallel operative	Conditional successive operative	Conditional successive operative
Quality communication				Indifferent parallel strategic	Indifferent parallel strategic
Quality audit					Conditional successive operative
Certification					

Fig. 8.51 Relationship analysis in the context of an integrated quality management documentation

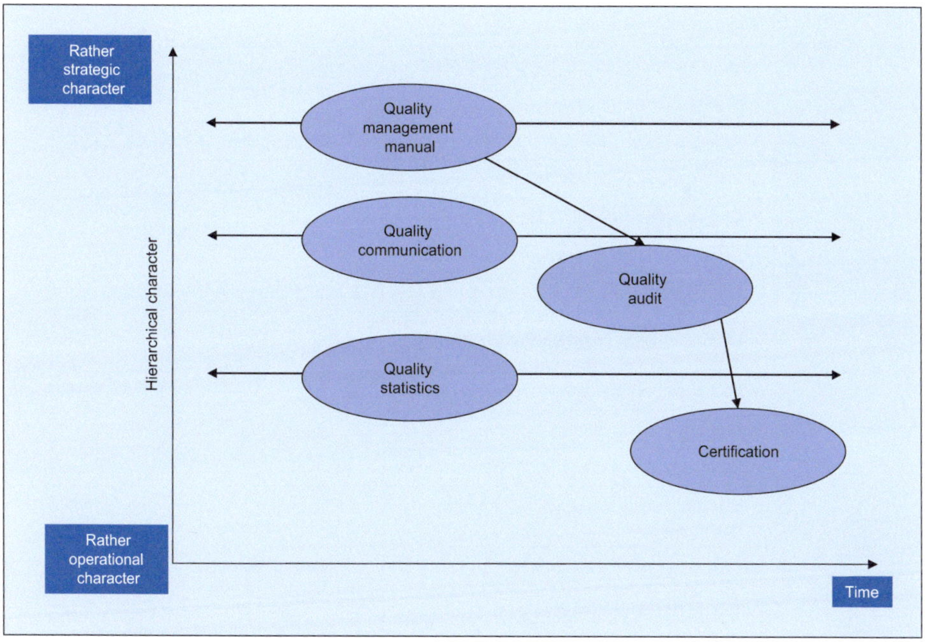

Fig. 8.52 Exemplary use of instruments of quality management documentation

8.6 Integration of the Phases into a Comprehensive Quality Management System

The individual phases as well as the multitude of quality management instruments are not to be seen in isolation from each other. Rather, it is necessary to integrate the partial concepts into a comprehensive quality management system.

First, Fig. 8.53 shows an overview of the essential instruments of the individual phases of the quality management system for service companies, the respective responsible persons or competent employees as well as the respective time horizons in a summary. This overview is general and has to be individually designed for each service company.

In the context of **interphase integration of quality management instruments** the following steps are to be taken:

Phases	Quality tools	Responsible persons/ responsible employees	Period/ time
Quality planning	▪ Customer surveys	Executives (managing directors, branch managers), market research	At least once a year
	▪ Service Blueprints		
	▪ Incident controlled process chain		
	▪ Company suggestion system		
	▪ Benchmarking		
Quality control	▪ Quality oriented personnel management	Personnel/ development	Continuous
	▪ Quality oriented incentive scheme		
	▪ Role model function of managers	Supervisor	
	▪ Vision, mission, principles	Management	
	▪ Quality committee/office for quality		
	▪ Decentralization of quality management	Executives	
	▪ Quality circle	Quality management	
	▪ Information and communication systems	Management/ communication	
Quality auditing	▪ Four-eyes principle/service supervision controls	Executives	At least once a year
	▪ Employee observation		
	▪ Employee assessment/performance review		
	▪ External quality measurement through employee survey	Market research	
	▪ Internal quality measurement		
	▪ Test purchasing methods		
	▪ Customer satisfaction tracking systems		
	▪ Complaint management	Executives	Continuous
Quality management documentation	▪ Quality management manuals	Executives/ management	Continuous
	▪ Quality statistics		
	▪ Quality communication		
	▪ Quality audits		
	▪ Certification		

Fig. 8.53 Instruments of quality management for service companies at a glance

1. Hierarchization of the instruments
2. Interphase linking of the instruments
3. Assessment of the instruments
4. Selection of the instruments

(1) Hierarchization of the instruments In the context of a comprehensive quality management concept, it is always necessary to carry out an **agreement of the individual sub-concepts** in order to optimize the mutually supportive effect. In order to derive rules for the absolutely necessary or merely accompanying use of instruments from the multitude of measures presented in quality management, a distinction between so-called **leading and following instruments of quality management** appears to be expedient. In principle, this is a hierarchization of quality measures, i.e. there are quality measures of first and second priority.

Core quality management instruments represent the central measures of quality management. They have a special strategic importance for the implementation and success of the quality management concept; in addition, they take on the leadership function for other quality measures. After the necessary core quality management instruments have been defined as part of the various phases of quality management, the sufficient **quality follow-up instruments** (also known as quality support instruments) are to be identified. Follow-up instruments include those measures that are strongly influenced by the core quality management instruments, have to be aligned with them accordingly and have a more complementary character. Which quality measures can possibly be used as core instruments and which as follow-up instruments in service companies is shown in an overview in Fig. 8.54.

The hierarchy of quality management instruments established here makes it clear that the core instruments are **must-have instruments** for every service company. They are definitely to be used and represent the starting point and framework for the further quality assurance measures. The follow-up instruments are **should** and **can instruments** and give the degree of professionalization of the service provider with regard to quality management (for the use and degree of professionalization of quality management of different service companies, see Bruhn & Meffert, 2002).

In this context, it should be noted that some quality management instruments can be used in different phases of the quality control cycle. Therefore, the first column in Fig. 8.39 shows the possible phases of use and the importance of the instrument, i.e. its must, should or can character, in the respective phase for each instrument.

(2) Interphase linking of instruments In the second step of the interphase integration of the instruments of quality management, the linking possibilities of each instrument with the instruments of other phases are to be analyzed on the basis of the deployment phase of the respective instrument specified in step 1. It is to be checked for each phase which networking potential exists for the individual instruments in order to achieve synergy effects in the realization of the quality objectives. There are various relationships

	Core quality instsurments	Quality follow-up instruments
Quality planning	▪ Customer surveys ▪ (SERVQUAL approach) ▪ Employee surveys	▪ Fishbone analysis ▪ FRAP analysis ▪ Company suggestion system
Quality control	▪ Customer orientation seminars ▪ Quality circles	▪ Quality oriented education and training ▪ Recruitment of employees ▪ Incentive systems ▪ Information Technology
Quality auditing	▪ Employee appraisals ▪ Tracking systems ▪ Complaint management	▪ Employee observations ▪ Test purchasing methods ▪ Control systems
Quality management documentation	▪ Quality management manuals	▪ Quality techniques ▪ Quality audits ▪ Quality awards ▪ Certification

Fig. 8.54 Core and follow-up instruments of a quality management system of service companies

between the individual instruments of the respective phases. For example, Fig. 8.55 shows the use of quality management instruments within the individual phases over time. Here it can be seen that, for example, instruments of quality planning and control, such as customer surveys and the setting of quality oriented employees, can take place in parallel. However, other instruments are used—depending on the experience—at different times.

Fig. 8.56 shows in the second column for each instrument which phases can be shifted in content and time to the associated instruments of the phase under consideration, while possible upstream phases are shown in the third column.

This interphase linkage is explained below using the instruments identified as must instruments. Fig. 8.57 shows the possible linkages between the instruments:

- Feature and customer oriented approaches, complaint analysis (quality planning and/or control)
- Personnel development, quality responsibility (quality control)
- Employee evaluation (quality auditing)
- Quality management manual (quality management documentation)

The instruments shown in Fig. 8.57 are connected to each other by arrows leading from the instrument that is to be upstream of the second instrument to the latter. The must instruments that can be assigned to several phases are entered in the cutouts of the rectangles representing the respective phases. Thus, the instrument of the feature and customer oriented approaches is entered in the cutout of quality planning and control. As

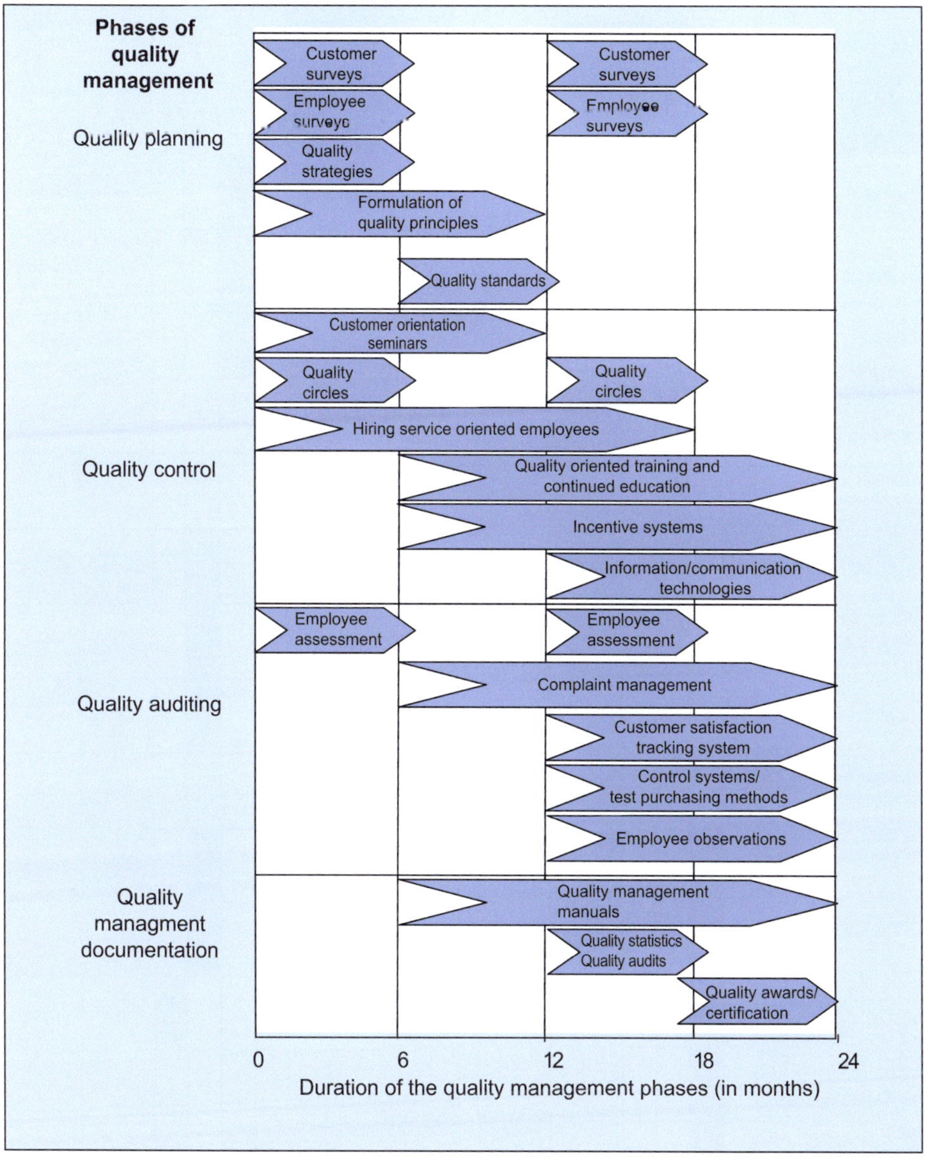

Fig. 8.55 Use of quality management instruments over time

upstream must instruments of these approaches, personnel development and employee evaluations come into question, while the use of quality management manuals can be downstream of the approaches.

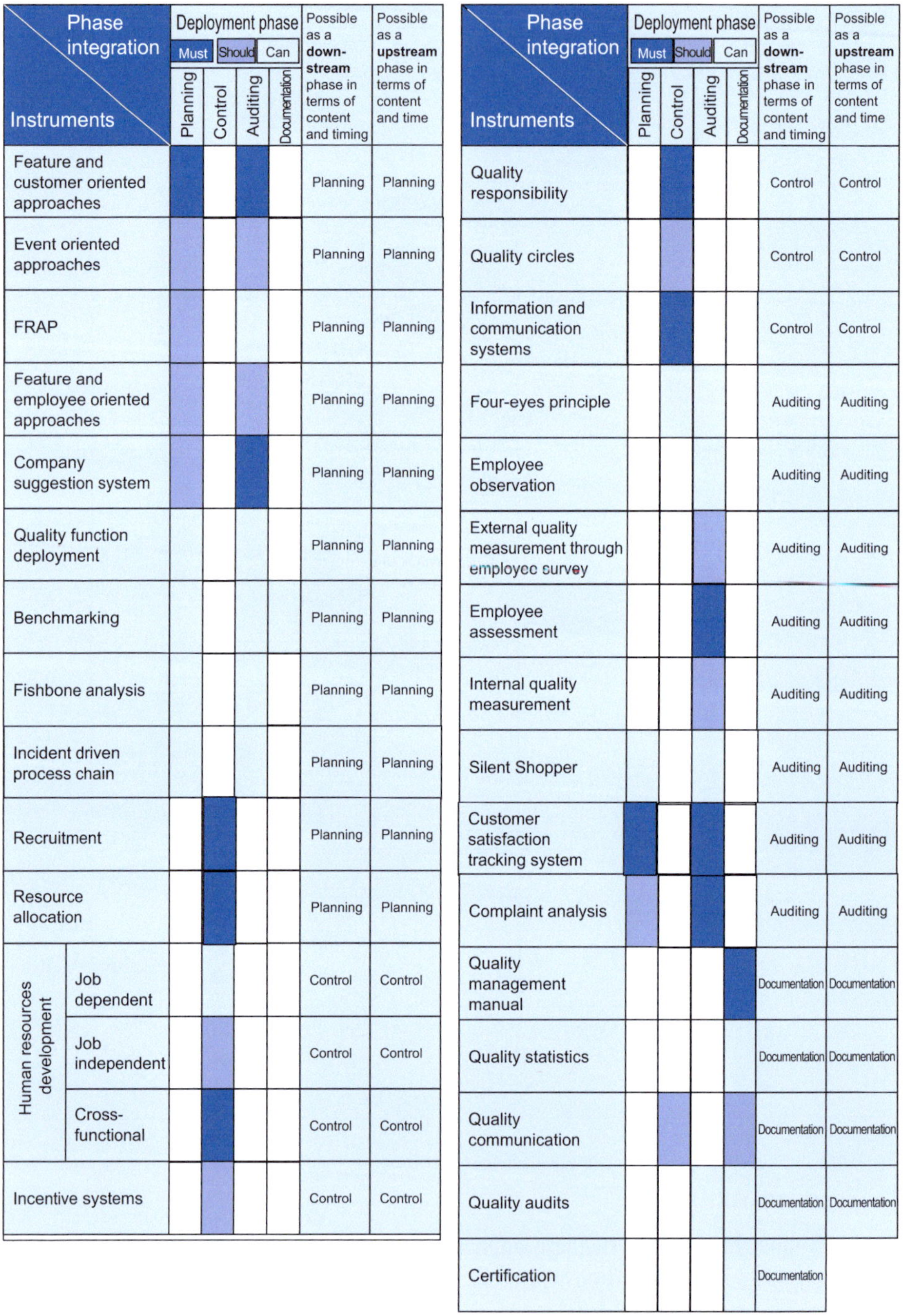

Fig. 8.56 Interphase integration of quality management instruments

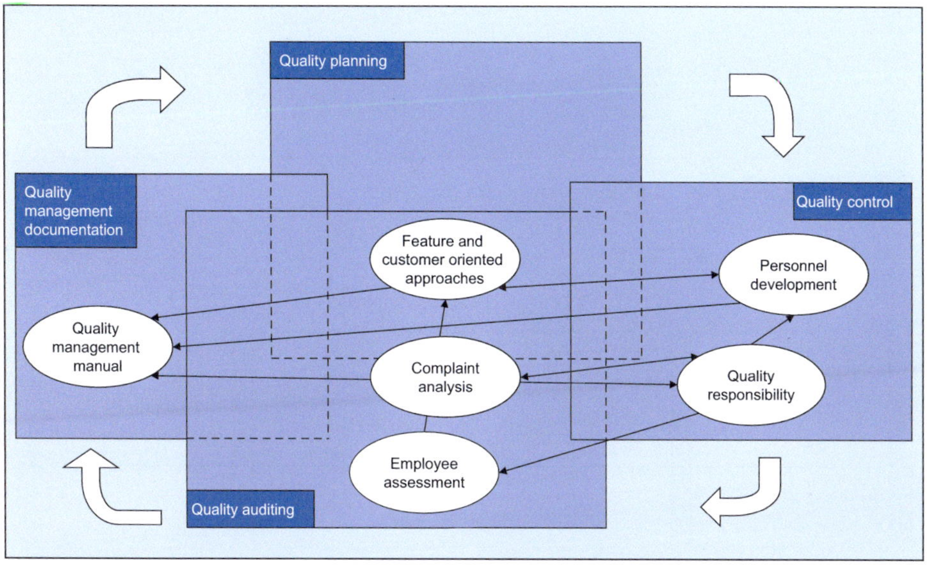

Fig. 8.57 Possible linkages of the must instruments of quality management in the context of interphase integration

After the hierarchy of the instruments and the analysis of the networking potentials, all instruments of quality management are assessed in the following in order to facilitate the selection of individual instruments.

(3) Evaluation of the instruments In a third step, the instruments of all phases are to be evaluated using suitable **evaluation criteria**. Fig. 8.58 gives an overview of the criteria, which can be divided into four criterion groups:

a) Theoretical criteria
b) Preparation related criteria,
c) Implementation related criteria,
d) Result related criteria

(a) Theoretical criteria

- **Quality dimension:** This criterion is not necessarily decisive for the practical implementation of the instruments. It indicates which quality dimensions (potential, process and/or result dimensions) are taken into account by the respective instruments.

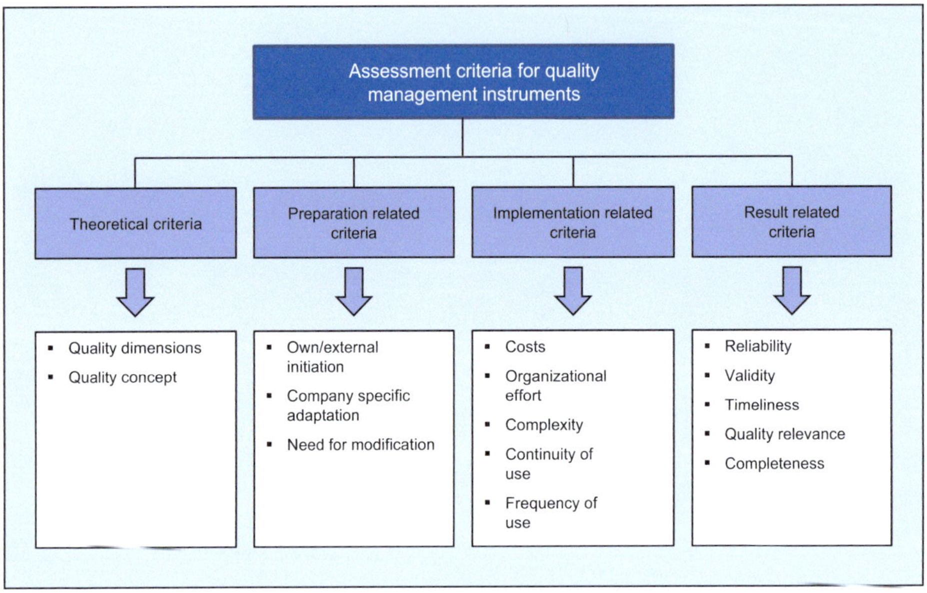

Fig. 8.58 Evaluation criteria for quality management instruments

- **Quality concept:** This criterion also has a more theoretical character. It is to be specified which quality concept—out of the five quality concepts of Garvin (1988), i.e. the objective, product, customer, manufacturing or value oriented concept—is the basis for the respective instrument.

(b) Preparation-related criteria

- **Self-/external initiation:** This criterion expresses whether the company itself can initiate the implementation of an instrument (expression "internal") or the implementation of the instrument is rather stimulated by external groups, e.g. customers (expression "external" or "both"). The more the latter is given, the less the respective instrument is controllable from the company's point of view.
- The need for a **company-specific adaptation:** The methodology of a large part of the procedures is very general. Therefore, the respective methods are often to be adapted to industry or even company specific specifications (expression "required"; e.g. quality criteria in the context of the feature oriented approaches: For hotels, other criteria are important than for car washes). For some instruments, such an adaptation is only partially necessary (expression "partially required").

- **Modification requirement** over time: A large number of procedures require more or less intensive adaptation to changes in the internal and external company environment (e.g. it may be necessary to regularly consider new problem areas in the context of problem oriented approaches, while the feature oriented—and here in particular the multi-attribute—approaches are to be designed in such a way that the queried quality criteria are relevant over a long period of time).

(c) Criteria related to implementation

- **Costs:** The costs incurred for the implementation of each instrument vary to a great extent. Factors influencing the costs are the number of employees involved, the need for external expert knowledge, the duration of the implementation, etc.
- **Organizational effort:** The organizational effort involved in the implementation (and preparation) of each instrument also varies considerably. The organizational effort is closely related to other evaluation criteria mentioned here, such as the need for company specific adaptation, modification requirements, complexity, etc. Other influencing factors of this criterion are—depending on the phase of use—e.g. the number of services offered (in the case of feature and customer oriented approaches in the context of planning), the number of employees (in the case of personnel development measures in the context of control), the need for the representativeness of the test results (in the case of multi-attribute approaches in the context of auditing), the size of the company (in the context of certification in the context of documentation).
- **Complexity:** The methodology of the individual procedures is of different complexity. In this context, it should be noted that in many cases the complexity is correlated with the effectiveness of the instruments (in quality management instruments) or the significance of the results (in quality planning and quality assurance instruments).
- **Continuity of use:** Depending on the objective of an instrument, a more or less sporadic or more continuous use is advisable.
- **Frequency of use:** Depending on the objective and financial as well as organizational effort, a more or less frequent use of an instrument is appropriate.

(d) Criteria related to results The result related criteria are only used in the assessment of the measurement approaches of service quality which are used in the phases of quality planning and control.

- **Reliability:** A measurement is then referred to as reliable when the measuring instrument is consistent, i.e. leads to the same result on repeated measurements for the same value and is thus able to determine the "true" value of the measured variable.
- **Validity:** The validity of a measurement approach indicates whether the respective measuring instrument actually measures what is to be measured.
- **Objectivity:** The objectivity of a measurement approach indicates to what extent the answers or measurement values are independent of the interviewer or auditor.
- **Actuality:** In order to ensure the validity of the results of planning and control instruments, it is necessary to use the most up-to-date assumptions in the procedures.
- **Quality relevance:** The use of measurement approaches in quality management is only meaningful if the respective instruments are actually able to plan or test the quality.
- **Completeness:** The different instruments are different in terms of the services considered and the individual criteria of services. However, it is not necessarily required or desirable to use instruments that are considered to be complete. Depending on the subject of the investigation, it is sometimes certainly sensible to only investigate individual aspects of quality or problem areas in which there is a great need for action.

In view of the specific company situation of the service provider, all quality management instruments are to be assessed by means of these criteria. Fig. 8.59 shows an overview of how such an assessment might look.

(4) Selection of instruments On the basis of the results of the hierarchisation and assessment as well as the analysis of the networking possibilities of the individual instruments, the service company can now select those instruments which it wishes to use within the framework of its quality management system or by means of which it wishes to supplement a possibly already existing quality management system.

After the operational design has been completed on the basis of the control loop, the next step is to implement the defined tasks and measures within the company. The following part D therefore deals with the implementation of quality management within the company. First, the basics of implementing a quality management system are presented in Chap. 9. In Chap. 10 and 11, specific approaches such as quality prices, auditing and certification are presented in detail.

	Instruments of QM / Assessment criteria	Theoretical criteria		Preparation related criteria			Implementation-related criteria					Result-related criteria				
		Quality dimension	Quality term	Initiative	Adaption	Modifications needs	Costs	Organizational effort	Complexity	Continuity	Frequency	Reliability	Validity	Aktuality	Quality relevance	Full constant time
Approaches to the measurement of service quality Deployment phases : Quality planning / or control	Silent Shopper	Process	Product	internal	bed. req.	no	medium	low	low	medium	medium	medium	medium	high	low	low
	Expert Observation	Process	Product	internal	bed. req.	no	medium	low	low	medium	medium	bad	medium	high	low	low
	War Tests	Pr./Erg.	Product	both	required	yes	none	low	high	medium	low	good	good	medium	low	low
	Feature/customer oriented approaches	Pot./Pr./Erg.	Prod./Cust.	internal	required	yes	high	high	high	medium	medium	good	good	low	low	medium
	Story Telling	Process	Customers	internal	bed. req.	no	medium	medium	high	medium	medium	medium	medium	high	high	high
	SEM	Process	Customers	internal	bed. req.	no	medium	medium	medium	medium	medium	medium	medium	high	high	high
	CIT	Process	Customers	internal	bed. req.	no	medium	medium	medium	medium	medium	medium	medium	high	high	high
	Complaint measurement	Pot./Pr./Erg.	Customers	both	bed. req.	no	high	medium	medium	high	high	medium	bad	high	high	low
	FRAP	Pr./Erg.	Prod./Cust.	internal	bed. req.	yes	medium	medium	medium	medium	medium	good	medium	medium	medium	medium
	Problem Detecting	Pr./Erg.	Prod./Cust.	internal	bed. req.	yes	medium	medium	medium	medium	medium	good	medium	medium	medium	low
	Quality Audits	Pot./Pr.	Manufacturer	both	required	yes	high	high	high	low	low	bad	medium	high	low	low
	Cost Analysis	Pot./Pr./Erg.	Manu./Prod	internal	required	yes	medium	high	high	high	high	bad	good	low	low	low
	Statistical Process Control	Process	Product	internal	required	yes	medium	high	high	high	high	medium	good	medium	low	medium
	Quality Function Deployment	Pot./Pr./Erg.	Prod./Cust.	internal	required	yes	high	high	high	medium	medium	good	medium	low	medium	low
	Benchmarking	Process	Product	both	bed. req.	yes	high	high	high	low	low	medium	medium	low	medium	medium
	FMEA	Process	Manu./Prod	internal	bed. req.	yes	medium	medium	medium	medium	medium	medium	medium	low	low	medium
	Fishbone approach	Pot./Pr.	Manu./Prod	internal	required	yes	low	medium	medium	medium	high	medium	medium	low	low	low
	Characteristics/employee oriented approaches	Pot./Pr./Erg.	Customers	internal	required	yes	medium	medium	medium	medium	low	good	medium	medium	low	medium
	Internal Quality Measurement	Pot./Pr./Erg.	Customers	internal	required	yes	medium	medium	medium	medium	medium	good	medium	medium	low	medium
	Four-eyes principle	Process	Product	internal	required	no	low	low	medium	high	medium	medium	medium	high	low	low
	Company suggestion system	Pot./Pr./Erg.	Product	internal	required	no	low	medium	medium	high	high	medium	bad	high	medium	low
	Poka yoke process	Process	Product	internal	required	yes	medium	medium	medium	high	high	medium	medium	medium	medium	medium
Steering instruments	Personnel selection	Potential	Product	both	required	yes	medium	high	high	high	high					
	Personnel development	Pot./Pr.	Product	internal	required	yes	medium	high	high	high	medium					
	Incentive systems	Pr./Erg.	Product	internal	bed. req.	yes	medium	medium	high	high	medium					
	Quality culture	Process	Manufacturer	internal	required	yes	low	medium	high	high	high					
	Quality responsibility	Process	Product	internal	required	yes	low	medium	medium	high	high					
	Quality circles	Pr./Erg.	Product	internal	required	yes	low	medium	medium	high	high					
	Communication/information systems	Pot./Pr.	Product	both	required	yes	medium	medium	medium	high	medium					
Presentation tools	Quality management manuals	Pot./Pr.	Manu./Prod.	both	required	no	medium	high	medium	low	high					
	Quality statistics	Pot./Pr./Erg.	Customers	internal	required	yes	low	high	high	medium	low					
	Quality communication	Pot./Pr.	Product	internal	required	yes	medium	high	medium	high	high					
	Certification	Pot./Pr.	Manu./Prod.	both	bed. req.	no	high	high	medium	low	low					

These criteria are only relevant for the assessment of approaches to measuring service quality.

Fig. 8.59 Evaluation of quality management instruments

References

Agarwala, T. (2003). Innovative human resource practices and organizational commitment: An empirical investigation. *International Journal of Human Resource Management, 14*(2), 175–197.

Ahmed, P. K., & Rafiq, M. (2012). *Internal marketing. Tools and concepts for customer-focused management.* Oxford: Routledge.

Ambros, H. (1991). Entwicklungen, Trends und Gefahren für Sparkassen. *Banking & Finance, 15*(1), 74–87.

Bayer, W. (2009). *Strategie und Planung: 28 Erfolgsbausteine für eine zukunftsorientierte Unternehmensführung.* München: MI-Wirtschaftsbuch.

Bechmann, R. (2010). *Qualitätsmanagement und kontinuierlicher Verbesserungsprozess.* Frankfurt a. M: Bund-Verlag.

Becker, F. G. (1999). Marketingorientierte Ausrichtung der Personalentwicklung in Dienstleistungsunternehmen – Am Beispiel von Finanzdienstleistern. In M. Bruhn (Ed.), *Internes Marketing. Integration der Kunden- und Mitarbeiterorientierung. Grundlagen, Implementierung, Praxisbeispiele* (2nd Edn., pp. 271–292). Wiesbaden.

Becker, W. S., & Wellins, R. S. (1990). Customer-service perceptions and reality. *Training & Development Journal, 44*(3), 49–51.

Benkenstein, M. (1993). Dienstleistungsqualität. Ansätze zur Messung und Implikationen für die Steuerung. *Zeitschrift für Betriebswirtschaft, 63*(11), 1095–1116.

Beriger, P. (1995). *Quality Circles und Kreativität. Das Quality Circle-Konzept im Rahmen der Kreativitätsförderung in der Unternehmung* (3rd Edn.). Stuttgart: Haupt.

Berne, E. (1961). *Transactional analysis in psychotherapy. A systematic individual and social psychiatry.* New York: Martino Fine books.

Berry, L. L. (1984). The employee as customer. In C. H. Lovelock (Ed.), *Services marketing* (pp. 271–178). Englewoods Cliffs: Prentice Hall.

Berry, L. L. (1995). *On great service.* New York: Simon and Schuster.

Berry, L. L., & Parasuraman, A. (1999). Dienstleistungsmarketing fängt beim Mitarbeiter an. In M. Bruhn (Ed.), *Internes Marketing. Integration der Kunden- und Mitarbeiterorientierung. Grundlagen, Implementierung, Praxisbeispiele* (2nd Edn., pp. 69–92). Wiesbaden: Gabler.

Berry, L. L., Zeithaml, V. A., & Parasuraman, A. (1990). Five imperatives for improving service quality. *Sloan Management Review, 31*(4), 29–38.

Bieberstein, I. (2006). *Dienstleistungs-Marketing* (4th Edn.). Ludwigshafen: Kiehl.

Biel, A. (2008). Chancen und Risiken der Bereichskommunikation am Beispiel Controller und Qualitätsmanager. *Controlling & Management, 52*, 18–27.

Bokranz, R. (1989). Arbeitsgestaltung. In H. Strutz (Ed.), *Handbuch Personalmarketing* (pp. 503–518). Wiesbaden: Gabler.

Bolton, R., & Saxena-Iyer, S. (2009). Interactive services: A framework, synthesis and research directions. *Journal of Interactive Marketing, 23*(1), 91–104.

Borg, I. (2003). *Führungsinstrument Mitarbeiterbefragung. Theorien, Tools und Praxiserfahrungen* (3rd Edn.). Göttingen: Hogrefe.

Bosch. (2011). Qualitätssicherungsleitlinie. http://purchasing.bosch.com/download/QSL_Industrie-Betriebsmittel-Dienstleistungen_12-2011-DE.pdf. Accessed: 22. Apr. 2013.

Bowen, D. E., & Lawler, E. E. (1995). Empowering service employees. *Sloan Management Review, 36*(4), 73–84.

Brachetti, H., & Wiegran, A. (2010). Welche Kennzahlen sind im Beschwerdemanagement besonders interessant? In O. Ratajczak (Ed.), *Erfolgreiches Beschwerdemanagement. Wege zu Prozessverbesserungen und Kundenzufriedenheit* (pp. 63–81). Wiesbaden: Gabler.

Braun, G., & Burger, F. (2007). Notdienst. *McKinsey Wissen, 6*(20), 80–85.

Brehm, S. (2001). *Konzepte zur Unternehmensveränderung: Organisationales Lernen in Vorschlagswesen, Qualitätszirkeln und Kaizen-Workshops*. Wiesbaden: Springer.

Breyer, M., & Zinnagl, E. (1995). Quality Banking in österreichischen Kreditinstituten. *Zeitschrift für das gesamte Bank- und Börsenwesen, 6,* 403–412.

Brinkmann, E. P., & Simon, A. (1995). Qualitätssteigerung durch Partizipation der Mitarbeiter. Teil 2: Betriebliches Vorschlagswesen. Grundzüge. In W. Hansen, H. H. Jansen & G. F. Kamiske (Eds.), *Qualitätsmanagement im Unternehmen. Grundlagen, Methoden und Werkzeuge, Praxisbeispiele, Loseblatt-Sammlung, Kapitel 03.08* (pp. 1–27). Berlin: Springer.

Brock, C. (2009). *Beschwerdeverhalten und Kundenbindung. Erfolgswirkungen und Management der Kundenbeschwerde*. Wiesbaden: Gabler.

Bruhn, M. (1982). *Konsumentenzufriedenheit und Beschwerden. Erklärungsansätze und Ergebnisse einer empirischen Untersuchung in ausgewählten Konsumbereichen*. Bern: Lang.

Bruhn, M. (1999a). Internes Marketing als Forschungsgebiet der Marketingwissenschaft. In M. Bruhn (Ed.), *Internes Marketing. Integration der Kunden- und Mitarbeiterorientierung. Grundlagen, Implementierung, Praxisbeispiele* (2nd Edn., pp. 15–44). Wiesbaden: Gabler.

Bruhn, M. (1999b). Sicherstellung der Dienstleistungsqualität durch integrierte Kommunikation. In M. Bruhn & B. Stauss (Eds.), *Dienstleistungsqualität. Konzepte – Methoden – Erfahrungen* (3rd Edn., pp. 405–431). Wiesbaden: Gabler.

Bruhn, M. (2002a). Electronic Services. Eine Einführung in die theoretischen und praktischen Probleme. In M. Bruhn & B. Stauss (Eds.), *Electronic Services. Jahrbuch Dienstleistungsmanagement 2002* (pp. 3–49). Wiesbaden: Gabler.

Bruhn, M. (2002b). *Integrierte Kundenorientierung. Implementierung einer kundenorientierten Unternehmensführung*. Wiesbaden: Gabler.

Bruhn, M. (2004). Interne Servicebarometer als Instrument interner Kundenorientierung. Messung und Steuerung der Qualität und Zufriedenheit interner Dienstleistungen. *Marketing ZFP, 26*(4), 282–294.

Bruhn, M. (2013). *Servicequalität – Konzepte und Instrumente für die perfekte Dienstleistung*. München: Beck.

Bruhn, M. (2014). *Integrierte Unternehmens- und Markenkommunikation: Strategische Planung und operative Umsetzung* (6th Edn.): Schäffer-Poeschel.

Bruhn, M., & Ahlert, M. (2002). Innovatives Qualitätsmanagement bei OBI Bau- und Heimwerkermärkte GmbH & Co. Franchise Center KG. In M. Bruhn & H. Meffert (Eds.), *Exzellenz im Dienstleistungsmarketing* (pp. 231–277): Gabler.

Bruhn, M., & Meffert, H. (2002). *Exzellenz im Dienstleistungsmarketing. Fallstudien zur Kundenorientierung*. Wiesbaden: Gabler.

Bruhn, M., Brunow, B., & Specht, D. (2002). Kundenorientierung durch Mitarbeiterorientierung im Schindlerhof. In M. Bruhn & H. Meffert (Eds.), *Exzellenz im Dienstleistungsmarketing* (pp. 125–176). Wiesbaden: Gabler.

Brüggemann, H. (2012). Qualitätsmanagementsysteme. In H. Brüggemann & P. Bremer (Eds.), *Grundlagen Qualitätsmanagement. Von den Werkzeugen über Methoden zum TQM* (pp. 122–149). Wiesbaden: Vieweg + Teubner.

Brüggemann, H., & Bremer, P. (2012). Total Quality Management (TQM). In H. Brüggemann & P. Bremer (Eds.), *Grundlagen Qualitätsmanagement. Von den Werkzeugen über Methoden zum TQM* (pp. 178–200). Düsseldorf: Vieweg + Teubner.

Bullinger, H. J., & Moll, K. (2003). Qualifizierung der Mitarbeiter für das Total Quality Management (TQM). In W. Hansen & G. F. Kamiske (Eds.), *Qualitätsmanagement. Methoden, Praxisbeispiele, Hintergründe* (pp. 1–19): Symposion.
Chen, W.-H. (1998). Benchmarking quality goals in service systems. *Journal of Services Marketing, 12*(2/3), 113–128.
Chun-Lang, C. (2006). Application of quality function deployment launches to enhancing nursing home service quality. *Total Quality Management & Business Excellence, 17*(3), 287–302.
Churchill, G. F. (1993). Quality management in the national nuclear corporation. In B. G. Dale & J. J. Plunkett (Eds.), *Managing quality* (pp. 307–329). Hertfordshire: Philip Allan.
Deutsche Gesellschaft für Qualität e. V. (1995a). *Begriffe zum Qualitätsmanagement* (6th Edn.). Frankfurt a. M: Beuth.
Deutsche Gesellschaft für Qualität e. V. (1995b). *DQS-Zertifikate. Liste, Zertifizierungsleistungen und Preise*: Hanser.
Deutsches Benchmarking Zentrum. (2016). Benchmarking Service Excellence – Branchenübergreifender Best Practice Austausch. http://benchmarkingforum.de/leistungen/best-practice-benchmarking/projektbeispiel-service-excellence-aerztehaus-hotel/. Accessed: 18. Febr. 2016.
DIN EN ISO 9000:2014. (2014). *Qualitätsmanagementsysteme – Grundlagen und Begriffe*. Berlin: Beuth.
DIN EN ISO 9001:2015. (2015). *Qualitätsmanagementsysteme – Grundlagen und Begriffe*. Berlin: Beuth.
Domsch, M. (1991). Mitarbeiterbefragungen. Ein Instrument zeitgemäßer Personalführung. *IO Management Zeitschrift, 60*(5), 56–58.
Drewes, W., & Schneider, U. (1999). Qualitätsmanagement im Bankgewerbe. In W. Masing (Ed,), *Handbuch Qualitätsmanagement* (4th Edn., pp. 831–853). München: Hanser.
Eder, A., Becker, U., & Nagersheth, A. (2010). Wie können Kundenwünsche erkannt und sichtbar gemacht werden? In O. Ratajczak (Ed.), *Erfolgreiches Beschwerdemanagement. Wege zu Prozessverbesserungen und Kundenzufriedenheit* (pp. 121–137). Wiesbaden: Gabler.
Evans, J. R., & Lindsay, W. M. (2011). *The management and control of quality*. Mason: South-Western Cengage Learning.
Facebook (2016). Telekom-hilft. https://www.facebook.com/telekomhilft/. Accessed: 20. Febr. 2016.
Finsterwalder, J., & Tomczak, T. (2001). EQ und DLQ: Emotionale Intelligenz und Dienstleistungsqualität im Kundenkontakt. In M. Bruhn & B. Stauss (Eds.), *Interaktionen im Dienstleistungsbereich. Dienstleistungsmanagement Jahrbuch 2001* (pp. 375–403). Wiesbaden: Gabler.
Ford, D., & Mouzas, S. (2012). Service and value in the interactive business landscape. *Industrial Marketing Management, 42*(1), 9–17.
Fröhlich, W., & Langecker, F. (1989). Hochschulkontakte. In H. Strutz (Ed.), *Handbuch Personalmarketing* (pp. 152–157). Wiesbaden: Gabler.
Fröhlich, W., & Sitzenstock, K. (1989). Personalimage-Werbung. In H. Strutz (Ed.), *Handbuch Personalmarketing* (pp. 134–142). Wiesbaden: Gabler.
Gabler Wirtschaftslexikon. (2013). Qualitätszirkel. http://wirtschaftslexikon.gabler.de/Definition/qualitaetszirkel.html. Accessed: 13. März 2013.
Garvin, D. A. (1988). *Managing quality. The strategic and competitive edge*. New York: Free Press.
Gaster, D. (1994). Qualitätsaudit. In W. Masing (Ed.), *Handbuch Qualitätsmanagement* (3rd Edn., pp. 927–948). München: Hanser.
George, W. R. (1977). The retailing of services – A challenging future. *Journal of Retailing, 53*(3), 85–98.

George, W. R., & Grönroos, C. (1999). Internes Marketing. Kundenorientierte Mitarbeiter auf allen Unternehmensebenen. In M. Bruhn (Ed.), *Internes Marketing. Integration der Kunden- und Mitarbeiterorientierung. Grundlagen, Implementierung, Praxisbeispiele* (2nd Edn., pp. 45–68): Gabler.

Gietl, G., & Lobinger, W. (2009). *Leitfaden für Qualitätsauditoren: Planung und Durchführung von Audits nach ISO 9001:2008* (3rd Edn.). München: Hanser.

Gleitsmann, B. M. (2007). *Internes Marketing, Unternehmenskultur und marktorientiertes Verhalten – Direkte, indirekte und moderierende Effekte*. Wiesbaden: Gabler.

Gogoll, A. (2003). Service-QFD: Quality Function Deployment für den Dienstleistungsbereich. In W. Hansen & G. F. Kamiske (Eds.), *Qualitätsmanagement. Methoden, Praxisbeispiele, Hintergründe* (pp. 1–19). Düsseldorf: Symposion.

Goleman, D. (2015). *Emotionale Intelligenz* (25th Edn.). München: dtv Verlagsgesellschaft.

Goodman, J., O'Brien, P., & Segal, E. (2000). Turning CFOs into quality champions. *Quality Progress, 33*(3), 47–54.

Grönroos, C. (1985). Internal marketing. Theory and practice. In T. M. Bloch, G. D. Upah, & V. A. Zeithaml (Eds.), *Services marketing in a changing environment* (pp. 41–47). Chicago: American Marketing Associaton.

Grönroos, C. (2007). *Service management and marketing. Customer management in service competition* (3rd Edn.). Chichester: Wiley.

Gruhlich, J. (2016). Transnationale Unternehmensstruktur: Das innerorganisatorische Netzwerk. In J. Gruhlich (Eds.), *Transnationale Unternehmen und Geschlecht* (pp. 167–171). Wiesbaden: Springer.

Gubler, T., Larkin, I., & Pierce, L. (2013). *The dirty laundry of employee award programs: Evidence from the field. Harvard Business School Working Paper*. Boston.

Gummesson, E. W. (1987). Using internal marketing to develop a new culture – The case of ericsson quality. In J. A. Czepiel, C. A. Congram, & J. Shanahan (Eds.), *The service challenge: Integrating for competitive advantage* (pp. 13–15). Chicago: American Marketing Association.

Hadwich, K., & Keller, C. (2015). Interne Servicequalität in Unternehmen: Eine empirische Untersuchung der Einflussfaktoren und Auswirkungen. *Zeitschrift für betriebswirtschaftliche Forschung, 67*(2), 170–205.

Haist, F., & Fromm, H. (1991). *Qualität im Unternehmen. Prinzipien, Methoden, Techniken* (2nd Edn.). München: Hanser.

Haller, S. (1998). *Beurteilung von Dienstleistungsqualität. Dynamische Betrachtung des Qualitätsurteils im Weiterbildungsbereich* (2nd Edn.). Wiesbaden: Gabler.

Haller, S. (2017). *Dienstleistungsmanagement* (7th Edn.). Wiesbaden: Springer Gabler.

Hansen, U., & Jeschke, K. (1999). Beschwerdemanagement für Dienstleistungsunternehmen. Beispiel des Kfz-Handels. In M. Bruhn & B. Stauss (Eds.), *Dienstleistungsqualität. Konzepte – Methoden – Erfahrungen* (3rd Edn., pp. 433–459). Wiesbaden: Gabler.

Hartmann, E. (2012). *Ansätze einer „Dienstleistungskultur"*. Wiesbaden: Gabler.

Hauser, J. R., & Clausing, D. (1988). The house of quality. *Harvard Business Review, 66*(3), 63–73.

Heinen, E., & Dill, P. (1990). Unternehmenskultur aus betriebswirtschaftlicher Sicht. In H. Simon (Ed.), *Herausforderung Unternehmenskultur* (pp. 12–24). Stuttgart: Schäffer-Poeschel.

Hobel, B., & Schütte, S. (2006). Qualitätsmanagement. In *Gabler Business-Wissen A–Z Projektmanagement* (pp. 249–264). Wiesbaden: Gabler.

Hoffmann, F. (1989). Erfassung, Bewertung und Gestaltung der Mitarbeiterqualität. *Zeitschrift für Organisation, 58*(6), 410–414.

Homburg, C., & Fürst, A. (2005). How organizational complaint handling drives customer loyalty: An analysis of the mechanistic and the organic approach. *Journal of Marketing, 69*(3), 95–114.

Homburg, C., & Pflesser, C. (2000). A multiple-layer model of market-oriented organizational culture: Measurement issues and performance outcomes. *Journal of Marketing Research, 37*(4), 449–462.
Homburg, C., & Stock, R. (2001). Der Zusammenhang zwischen Mitarbeiter- und Kundenzufriedenheit. *Die Unternehmung, 55*(6), 377–400.
Homburg, C., & Stock-Homburg, R. (2012). *Der kundenorientierte Mitarbeiter: Bewerten, begeistern, bewegen* (2nd Edn.). Wiesbaden: Gabler.
Homma, N., & Bauschke, R. (2010). *Unternehmenskultur und Führung: Den Wandel gestalten*. Wiesbaden: Springer Gabler.
Horváth, P., & Urban, G. (1990). *Qualitätscontrolling*. Stuttgart: Schäffer-Poeschel.
Huber, R. (1992). Qualitätszirkel. In H. Diller (Ed,), *Vahlens Großes Marketinglexikon* (p. 989). München: Vahlen.
Huber, M. (2008). *Kommunikation im Web 2.0*. Konstanz: UVK.
Kamiske, G. F., & Brauer, J.-P. (2011). *Qualitätsmanagement von A bis Z. Erläuterungen moderner Begriffe des Qualitätsmanagements* (7th Edn.). München: Hanser.
Kamiske, G. F., & Krämer, F. (1994). Der moderne Qualitätsmanager als Leiter des Qualitätswesens. In W. Masing (Ed.), *Handbuch Qualitätsmanagement* (3rd Edn., pp. 991–1000). München: Hanser.
Kövesi Grafinger, E., & Schmoll, A. (1998). Wettbewerbsfaktor interne Servicequalität. *Die Bank, o. V.*(3), 132–136.
Kummer, A., & Genz, H. O. (2004). Betriebliches Vorschlagswesen Als Ideenmanagement. http://hartmut-genz.de/Downloads/Beratung/2004_RGM-Ideenmanagement.pdf. Accessed: 5. Febr. 2016.
Kupitz, G. (2006). Die besten Unternehmens-Webseiten. Financial Times Deutschland. http://www.ftd.de/karriere_management/management/46502.html?eid=24262. Accessed: 15. März 2012.
Ladwig, D. H. (2006). Qualitätsmanagement interner Dienstleistungen – Befragung der internen Kunden. In M. E. Domsch & D. H. Ladwig (Eds.), *Handbuch Mitarbeiterbefragung* (pp. 249–264). Berlin: Springer.
Lee, C. H., & Bruvold, N. T. (2003). Creating value for employees: Investment in employee development. *International Journal of Human Resource Management, 14*(6), 981–1000.
Lehmann, A. P. (1995). *Dienstleistungsmanagement. Strategien und Ansatzpunkte zur Schaffung von Servicequalität* (2nd Edn.). Stuttgart: Schäffer-Poeschel.
Lufthansa. (2012). Aufgabenprofil für Flugbegleiter. http://www.spiegel.de/karriere/berufsleben/eine-stewardess-erzaehlt-von-ihrem-harten-berufsalltag-a-852528.html. Accessed: 31. Aug. 2012.
Madu, C. N., & Kuei, C.-H. (1995). *Strategic total quality management. Corporate performance and product quality*. Westport.
Malhotra, N., & Mukherjee, N. (2004). The relative influence of organisational commitment and job satisfaction on service quality of customer contact employees in banking call centres. *Journal of Services Marketing, 18*(3), 162–174.
McKinsey (2007). House of service quality. *McKinsey Wissen, 6*(20), 58–61.
Meffert, H., Bruhn, M., & Hadwich, K. (2018). *Dienstleistungsmarketing. Grundlagen, Konzepte, Methoden* (9th Edn.). Wiesbaden: Springer.
Mertins, K., & Kohl, A. (2009). *Benchmarking: Leitfaden für den Vergleich mit den Besten* (2nd Edn.). Düsseldorf: Symposion.
Merz, E., & Biehler, B. (1994). *Betriebliches Vorschlagswesen – Professionell und wirksam* (2nd Edn.). Landsberg am Lech: Moderne Industrie.
Methner, H. (1999). Aus- und Weiterbildung. In W. Masing (Ed,), *Handbuch Qualitätsmanagement* (4th Edn., pp. 1093–1105). München: Hanser.

Mills, P. K., & Margulies, N. (1980). Toward a core typology of service organizations. *Academy of Management Review, 5*(2), 255–265.
Mohr, G. (2012). Spiritualität und transaktionsanalytisches Coaching. In M. Hänsel (Ed,), *Die spirituelle Dimension in Coaching und Beratung* (pp. 192–205). Göttingen: Vandenhoeck & Ruprecht.
Nagel, K. (2011). *Employer Branding. Starke Arbeitgebermarken jenseits von Marketingphrasen und Werbetechnik.* Wien: Linde.
Nerdinger, F. W., & Wilke, P. (2009). *Beteiligungsorientierte Unternehmenskultur.* Wiesbaden: Gabler.
Nerdinger, F. W., Blickle, G., & Schaper, N. (2014). *Arbeits- und Organisationspsychologie* (3rd Edn.). Heidelberg: Springer.
Niefind, F., & Ratajczak, O. (2010). Wie sollte der Beschwerdemanagementmitarbeiter mit Beschwerdeführern umgehen? In O. Ratajczak (Ed.), *Erfolgreiches Beschwerdemanagement. Wege zu Prozessverbesserungen und Kundenzufriedenheit* (pp. 99–119). Wiesbaden: Gabler.
Niefind, F., & Wiegran, A. (2010). Wie sollte die Ablauforganisation des Beschwerdemanagements aussehen? In O. Ratajczak (Ed.), *Erfolgreiches Beschwerdemanagement. Wege zu Prozessverbesserungen und Kundenzufriedenheit* (pp. 47–61). Wiesbaden: Springer.
Noé, M. (2016). *Praxisbuch Teamarbeit: Aufgaben, Prozesse, Methoden.* München: Hanser.
OBI. (2016). Unternehmens-Homepage. http://www.obi.com/de/company/de/Unternehmen/index.html. Accessed: 20. März 2016.
Oechsler, W. A. (2011). *Personal und Arbeit. Grundlagen des Human-Resource-Management und der Arbeitgeber-Arbeitnehmer-Beziehungen* (9th Edn.). Oldenburg: Oldenbourg.
Pantouvakis, A. (2012). Internal marketing and the moderating role of employees: An exploratory study. *Total Quality Management & Business Excellence, 23*(2), 177–195.
Parasuraman, A., Zeithaml, V. A., & Berry, L. L. (1985). A conceptual model of service quality and its implications for future research. *Journal of Marketing, 49*(1), 41–50.
Parasuraman, A., Zeithaml, V. A., & Berry, L. L. (1988). SERVQUAL: A multiple-item scale for measuring consumer perceptions of service quality. *Journal of Retailing, 64*(1), 12–40.
Parente, D. H., Pegels, C. C., & Suresh, N. (2002). An exploratory study of the sales-production relationship and customer satisfaction. *International Journal of Operations & Production Management, 22*(9), 997–1013.
Paschen, M. (2004). *Instrumente der Personalentwicklung.* Norderstedt: Books on Demand.
Pepels, W. (2016). *Handbuch des Marketing* (7th Edn., Teilband III). Berlin: Duncker & Humblot.
Rafiq, M., & Ahmed, P. K. (1998). A customer-oriented framework for empowering service employees. *Journal of Services Marketing, 12*(6), 379–396.
Reimann, G. (2012). *Erfolgreiches Qualitätsmanagement nach DIN EN ISO 9001:2008. Lösungen zur praktischen Umsetzung* (2nd Edn.). Berlin: Beuth.
Rothlauf, J. (2014). *Total Quality Management in Theorie und Praxis. Zum ganzheitlichen Unternehmensverständnis* (4th Edn.). München: De Gruyter Oldenbourg.
Saatweber, J. (1994). Quality Function Deployment (QFD). In W. Masing (Ed,), *Handbuch Qualitätsmanagement* (3rd Edn., pp. 445–468). München/Wien: Hanser.
Saatweber, J. (2011). *Kundenorientierung durch Quality Function Deployment. Produkte und Dienstleistungen mit QFD systematisch entwickeln* (3rd Edn.). Düsseldorf: Symposion.
Sachs, M. C. (1993). Praktiziertes Qualitätsmanagement für Dienstleistungen. Tugend oder Vision? *Zeitschrift für industrielle Qualitätssicherung, 38*(4), 217–222.
Salovey, P., & Mayer, J. D. (1990). Emotional intelligence. *Imagination, Cognition and Personality, 9*(3), 185–211.
SAP. (2016). Live-Chat als Beschwerdekanal bei SAP AG. http://help.sap.com/saphelp_crm60/helpdata/de/45/f06323312c3112e10000000a1553f7/content.htm. Accessed: 3. März 2016.

Scheer, A.-W., Grieble, O., & Klein, R. (2006). Modellbasiertes Dienstleistungsmanagement. In H.-J. Bullinger & A.-W. Scheer (Eds.), *Service Engineering. Entwicklung und Gestaltung innovativer Dienstleistungen* (2nd Edn., pp. 19–51). Berlin: Springer.

Schildknecht, R. (1992). *Total Quality Management. Konzeption und State of the Art*. Frankfurt a. M.: Campus.

Schlick, C. M., & Rochhold, S. (2007). Qualitätsmanagement bei Dienstleistungen. In T. Pfeifer & R. Schmitt (Eds.), *Masing – Handbuch Qualitätsmanagement* (5th Edn., pp. 941–954). München: Hanser.

Schmitt, R., & Pfeifer, T. (2015). *Qualitätsmanagement. Strategien, Methoden, Techniken* (5th Edn.). München: Hanser.

Schneider, B., & Bowen, D. E. (1999). The service organization. Human resources management is critical. In J. E. G. Bateson (Ed,), *Managing services marketing. Text and readings* (3rd Edn., pp. 273–283). Texas: The Dryden Press.

Schneider, B., & Bowen, D. E. (2010). Winning the service game: Revisiting the rules by which people co-create value. In P. P. Maglio, C. A. Kieliszewski, & J. C. Spohrer (Eds.), *Handbook of service science* (pp. 31–60). New York: Springer.

Schneider, B., & Schechter, D. (1991). Development of a personnel selection system for service jobs. In S. W. Brown, E. W. Gummesson, & B. Edvardsson (Eds.), *Service quality. Multidisciplinary and multinational perspectives* (pp. 217–235). Lexington: Lexington Books.

Schöler, A. (2009). *Beschwerdeinformation und ihre Nutzung*. Wiesbaden: Gabler.

Schreiber, F. (2007). Intergrierte Managementsysteme QM – UM – SIM. In T. Pfeifer & R. Schmitt (Eds.), *Masing – Handbuch Qualitätsmanagement* (5th Edn., pp. 207–251). München: Hanser.

Schulte, C. (2012). *Personal-Controlling mit Kennzahlen* (3rd Edn.). München: Vahlen.

Schulze, H. S. (1999). Erhöhung der Dienstleistungsqualität durch transaktionsanalytisch orientierte Personalschulungen. In M. Bruhn & B. Stauss (Eds.), *Dienstleistungsqualität. Konzepte – Methoden – Erfahrungen* (3rd Edn., pp. 261–285). Wiesbaden: Gabler.

Seifert, K. (1998). *Prozessmanagement für die öffentliche Verwaltung*. Wiesbaden: Gabler.

Semigator. (2019). www.semigator.de/seminare/beschwerdemanagement.html. Accessed: 26. Apr. 2019.

servmark. (2012). *Interne Bankstudie*. Eching bei München.

Sieber, B. (2008). Ökonomische Betrachtung des Qualitätsmanagements. In P. Kleine & J. Ennker (Eds.), *Qualitätsmanagement in der operativen Medizin. Leitfaden für Ärzte und Pflegekräfte* (pp. 122–131). Darmstadt: Steinkopf.

Sokolakis, C. (2011). *TQM erfolgreich umsetzen. Analyse der Rahmenbedingungen und Ableitung von Maßnahmen* (2nd Edn.). Norderstedt: Books on Demand.

Sommerhoff, B. (2008). Controlling und Qualitätsmanagement. *Controlling & Management, 52*(3), 5–9.

Sondermann, J. P. (1994). Instrumente des Total Quality Managements. Ein Überblick. In B. Stauss (Ed,), *Qualitätsmanagement und Zertifizierung von DIN ISO 9000 zum Total Quality Management* (pp. 223–253). Wiesbaden: Gabler.

Spector, R., & McCarthy, P. D. (1995). *The Nordstrom way: The inside story of America's number 1 customer service company*. Chichester: Wiley.

Stauss, B. (1989). Beschwerdepolitik als Instrument des Dienstleistungsmarketing. *Jahrbuch der Absatz- und Verbrauchsforschung, 35*(1), 41–62.

Stauss, B. (1999). Internes Marketing als personalorientierte Qualitätspolitik. In M. Bruhn & B. Stauss (Eds.), *Dienstleistungsqualität. Konzepte – Methoden – Erfahrungen* (3rd Edn., pp. 204–222). Wiesbaden: Gabler.

Stauss, B., & Hentschel, B. (1991). Dienstleistungsqualität. *Wirtschaftswissenschaftliches Studium, 20*(5), 238–244.

Stauss, B., & Seidel, W. (2014). *Beschwerdemanagement: Unzufriedene Kunden als profitable Zielgruppe* (5th Edn.). München.

Steinle, C., Eggers, B., & Langemann, G. (1995). Möglichkeiten und Grenzen von Qualitätszirkeln. *Management Zeitschrift, 64*(5), 38–41.

Stock-Homburg, R. (2013). *Personalmanagement. Theorien – Konzepte – Instrumente* (3rd Edn.). Wiesbaden: Gabler.

Thom, N. (2003). *Betriebliches Vorschlagswesen: Ein Instrument der Betriebsführung und des Verbesserungsmanagements* (6th Edn.). Bern: Lang.

Thommen, J.-P., & Achleitner, A.-K. (2012). *Allgemeine Betriebswirtschaftslehre. Umfassende Einführung aus managementorientierter Sicht* (7th Edn.). Wiesbaden: Gabler.

Tuzovic, S. (2004). *Kundenorientierte Vergütungssysteme zur Steuerung der Erfolgsgrößen im Relationship Marketing. Anforderungen, Konzeptionalisierung und Institutionalisierung*: Gabler.

von Diemer, R. (1999). Motivation. In W. Masing (Ed.), *Handbuch Qualitätsmanagement* (4th Edn., pp. 1107–1125). München: Hanser.

von Rosenstiel, L. (2010). *Motivation im Betrieb. Mit Fallstudien aus der Praxis* (11th Edn.). Leonberg: Rosenberger Fachverlag.

Weber, W., Mayrhofer, W., & Nienhüser, W. (1993). *Grundbegriffe der Personalwirtschaft*. Stuttgart: Schäffer-Poeschel.

Weiss, H. (2009). *Den Kunden zum König machen: Erkenntnisse aus 10 000 Testkäufen. Ein Praxisbuch über Mystery Shopping im Einzelhandel*. Frankfurt a. M., Books on Demand.

Wesselmann, S., & Hohn, B. (2012). Qualitätsmanagement. In S. Wesselmann & B. Hohn (Eds.), *Public Marketing. Marketing-Management für den öffentlichen Sektor* (3rd Edn., pp. 227–253). Wiesbaden: Gabler.

Wirtz, J., & Lovelock, C. H. (2016). *Services marketing. People, technology, strategy* (8th Edn.). New Jersey: World Scientific Publishing.

Zeithaml, V. A., Bitner, M. J., & Gremler, D. D. (2017). *Services marketing. Integrating customer focus across the firm* (7th Edn.). London: McGraw-Hill.

Zeller, H. (1994). Organisation des Qualitätsmanagements im Unternehmen. In W. Masing (Ed.), *Handbuch Qualitätsmanagement* (3th Edn., pp. 903–926). München/Wien: Hanser.

Part IV

Implementation of Quality Management for Services

Implementation of Quality Management for Services

9

Abstract

A sustainable and systematic implementation of quality management is crucial for the success of the company. As a result, the specifics of the implementation of quality management are examined in detail within this chapter. A central prerequisite for the implementation of quality management is the situational specification of the quality requirements as well as their anchoring in the structures, systems and culture of the company. However, before such a specification of quality requirements can be made, it is necessary to identify resistance and barriers in service companies. Accordingly, this chapter shows necessary measures for the design of the organizational structure, the company systems and the corporate culture for a successful implementation of quality management of services.

9.1 The Need for a Systematic Implementation of Quality Management

Quality management programs often fail not because of conceptual ideas in the company or the commitment of individuals, but because of the weaknesses of an insufficient and unsustainable **implementation of quality management** within the company. This is especially true in customer oriented (and not technically, product oriented) quality management, as is typical for service companies due to the special features of services and, above all, the integration of the external factor. If the goal is to create quality, this requires, for example, from the customer contact employee to behave empathically towards the customer, i.e. to put oneself in the customer's situation and to behave according to his needs. To achieve this behaviour by the contact employee is possibly much

M. Bruhn, *Quality Management for Services*,
https://doi.org/10.1007/978-3-662-67032-3_9

more difficult than to make a machine meet certain specifications. All the more it is surprising that in service industries often a systematic approach to the implementation of quality management and a consistent customer orientation is lacking. Here, significant discrepancies exist between the desire for customer orientation and the actual perception of customer orientation. Figure 9.1 shows the corresponding results of a survey of Austrian companies. Similar results are available from a survey of German companies (Droege & Com., 2000).

▶ The term **implementation** refers to a process that transforms strategies into actionable tasks and thus ensures the implementation and enforcement of the strategy (Bruhn, 2016, p. 82).

In order to successfully implement quality management, acceptance of a quality oriented approach must be created among all employees of the company. The implementation of quality management therefore requires that situational specifications of the quality requirements be made and that company structures, systems and culture be oriented towards quality.

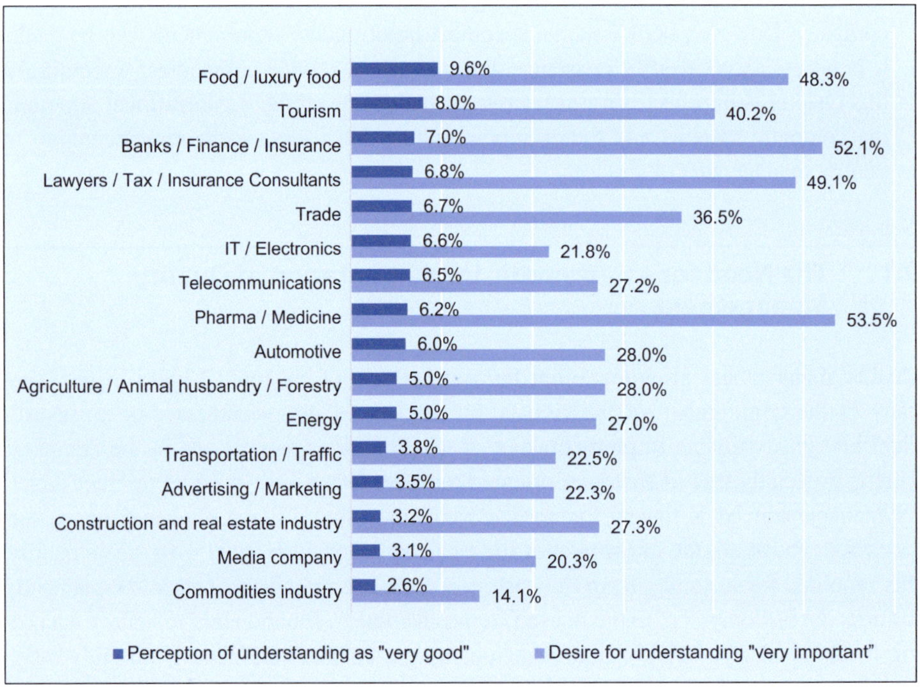

Fig. 9.1 Desire for customer orientation and actual perception of customer orientation by industry. (Source: Statista, 2019)

The difficulties that can arise in the implementation of quality management are based on various factors that show up as "barriers to implementation" of quality management in many companies.

9.2 Barriers to the Implementation of Quality Management

Despite the generally recognized, obvious advantages of a quality orientation of the company (e.g. results of the PIMS database by Buzzell and Gale, 1989) or of a comprehensive quality management, the implementation of this concept often meets with diverse **barriers** (e.g. Oess, 1994, 217 pp.; Bruhn, 2002; Meffert et al., 2018, 458 pp.), which can be divided into content conceptual, organizational structural and personnelcorporate cultural barriers.

Content conceptual barriers is basically directly related to the concept of quality management and can result from the different understanding of the term quality and quality management. Divergent ideas of the employees about the goals and contents of a quality management system appropriate for the company can lead to an inconsistent orientation and lack of coordination of the quality related measures.

Due to the necessary organizational anchoring of quality management in the company, taking into account the existing structures and hierarchies, **organizational structural barriers** can also arise. A lack of organizationally anchored, central quality responsibility as well as the lack of cooperation between departments dealing with service, quality control, etc. can lead to a so-called "departmental thinking" that can prevent the implementation of a comprehensive quality management.

Personnel cultural barriers can arise from employee individual values, ways of thinking and behavior as well as in connection with the existing corporate culture. Departmental thinking, fear of loss of competence or increased control as well as the existence of different subcultures can complicate or impede the development and implementation of quality management in service companies. The so-called "not invented here" syndrome occurs especially when it comes to implementing concepts that have been developed and tested by headquarters or by specialist (staff) departments, now also in other departments or even company-wide, and encountering disinterest or resistance amongst employees.

In Fig. 9.2 central barriers to the implementation of quality management in service companies are summarized.

The basic principles of total quality management described above are to be concretized by means of various planning instruments, leadership techniques and incentive systems and implemented in the service company—taking into account existing barriers. This **step from the philosophy level of TQM to the instrument level** requires a systematic compilation and evaluation of the most important instruments for managing service quality. In this context, quality related activities have two essential objectives:

Barriers of a quality management system	Central barriers of a quality management system in service companies
Barriers in terms of content and concept	▪ Different understanding of the concept of quality (comparison of business units, departments) ▪ Diverging ideas among employees about the goals and contents of a quality management system (comparison of business units, departments, hierarchy levels)
Organizational structural barriers	▪ No institutionalized coordination and decision making rules for quality measures ▪ Lack of organizationally anchored, central quality responsibility in the company ▪ Lack of cooperation between service and quality departments (departmental thinking) ▪ Lack of cooperation between departments or within the overall organization ▪ Independence of individual departments and business units
Personnel cultural barriers	▪ Departmental thinking ▪ Fear of loss of authority or greater control ▪ Existence of different subcultures (departments, divisions, subsidiaries) ▪ "Not invented here" syndrome (resistance of employees to concepts from headquarters or individual departments)

Fig. 9.2 Barriers to the implementation of quality management

On the one hand, an improvement of the level of performance is aimed at in order to increase the quality of service, on the other hand, it is about the target oriented control of customer expectations.

9.3 Approaches to Implementing Quality Management

There are numerous **approaches** for the foundation of a systematic implementation of quality management, which can be divided into three groups (see Fig. 9.3):

- Structure oriented approach
- System oriented approach
- Culture oriented approach.

In this context, it should be noted in the interest of delimitation that there are parallels between the implementation approaches listed here and the instruments of quality control (see Chap. 8), which, for example, also include measures for controlling employee behavior and the organizational design of quality management. However, quality control and the implementation of quality management are located on two different levels. Quality control includes measures that serve to meet customer expectations and thus to ensure quality. The implementation is about how these measures can actually

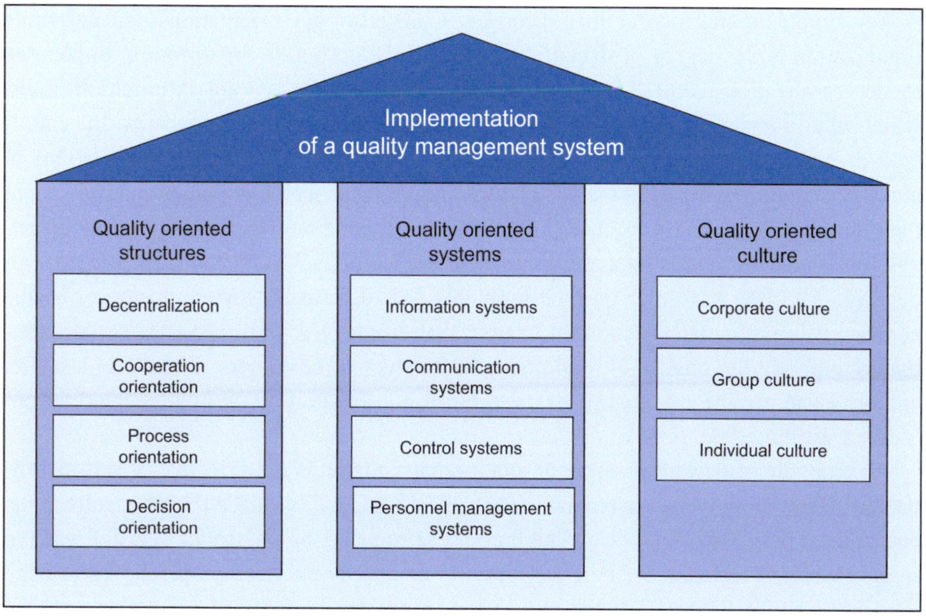

Fig. 9.3 Approaches for an implementation of quality management. (Source: Based on Bruhn, 2002, p. 33)

be implemented in the company. For example, a quality manager is used as part of the organization of quality management in the context of quality control. In the context of implementation, it must then be taken into account how this quality manager and the rest of the company organization interact.

9.3.1 Structure Oriented Approaches

The structural approach focuses on questions of **corporate organization** and in particular the question of which corporate structure favors the implementation of customer oriented quality management. Based on practical experience, various trends can be identified that contribute to a **quality oriented corporate organization**. The following four **trends** should be particularly emphasized (Bruhn, 2016):

- Decentralization
- Cooperation orientation
- Process orientation
- Decision orientation

A key organizational aspect that determines the customer orientation of a corporate organization is the degree of **decentralization** of the corporate organization. In contrast to decentralization, centralization of decision making entails long information paths, which in turn increases the risk of delays in responding to customer concerns. In centralized organizations, for example, the path from knowledge of customer expectations in customer contact to decision makers is very long, so that such companies can only adapt their services very slowly to changing customer expectations. In addition, upper hierarchy levels are overloaded by a multitude of strategically irrelevant problems, while lower departments often have motivational problems due to formally low competence. In this context, it is particularly important to note that employees with customer contact ultimately embody the quality of a company in the eyes of customers. Therefore, a higher authority can prescribe a quality strategy, but it is ultimately to be implemented decentralized (Bruhn, 2002, p. 44).

An example of decentralization of quality management is the principle of **self-examination**. Here, employees are empowered to check the quality of their work results at the end of each partial work process. The purpose of transferring control or decision making rights is mainly to make the best possible use of distributed quality specific knowledge (Brommer, 1999, 85 pp.). This is also evident from the example of complaint management as an instrument of quality management. If incoming complaints have to be forwarded upwards through many hierarchy levels at first and the decision is taken again along this information path, quick reactions to customer complaints are hardly possible. It is not uncommon for there to be a time lag of more than a week between the complaint and the complaint reaction. This means, the chance given by the customer to the company, by filing a complanct with the company is not used. On the contrary, from the customer's point of view, the company delivers poor quality for a second time.

According to ISO 9000:2014, a quality manager is to be established at management level, who is responsible for the coordination and monitoring of quality management in the decentralized units. This gives the other members of the management board the freedom to concentrate on strategic tasks. At the same time, it is ensured that quality management is located at the top level. In the decentralized departments, however, the motivation and the personal commitment of the employees increase through gained competences.

Another important aspect of implementing a customer oriented quality management is ensuring a **cooperation orientation**. This refers to internal cooperation, namely between the functional areas of a company. Due to the company-wide importance of quality, which is attributable to the activities of a large number, if not all, functional areas, a customer oriented quality management is based on cross-departmental cooperation. One component of quality management is, for example, the measurement and analysis of service quality. These quality measurements, e.g. in the form of customer surveys, are often carried out by the market research department—often even without the involvement of other departments in the design of quality measurement. It is aimed that the results of these analyses are used by other departments to ensure a high quality. It is obvious that this only succeeds in a few cases. First of all, it is assumed that the specialist departments have information that can be helpful in designing the customer survey. In addition, the so-called "not invented here syndrome" is a well-known phenomenon in the company, which impedes the implementation of results generated without the involvement of one's own department. An instrument for implementing a cooperation orientation are **"cross functional visits"** (Hilker, 2001; Conti, 2010). Here, the affected employees maintain direct customer contact together. For example, it is conceivable that employees from the back office of a bank are directly confronted with a customer problem.

Cooperation orientation at the *Waldklinik Jesteburg*

The *Waldklinik Jesteburg* near Hamburg holds the following statements on the achievement of a high cooperation orientation within the framework of its target and strategy system for quality improvement:

- Responsibility of all employees for rehabilitation goals through cross-functional, transparent action.
- Further development and implementation of regulated communication structures.
- Implementation of guideline oriented treatment methods.
- Basic orientation to the ICF ideology within therapeutic and medical action.
- Further development of quality management as a systematic basis for the work in the clinic.
- Personnel development with high performance requirements.
- Up to date adaptation of all equipment features.
- Expansion of cooperation.
- Authentic and consistent communication policy of the entire company.
- Optimal utilization of all business areas.
- Economic use of personnel and material resources.

(Source: Waldklinik Jesteburg, 2010) ◄

The **process orientation** of a company is closely related to its cooperation orientation, the realization of which supports the cooperation between different company functions. If a customer oriented view is also taken as a basis, then process orientation means that company processes are initiated by the customer and not by a functional area, such as the product management of a bank. This implies that the conditions must be created so that information about the customer is always available at the appropriate points within the company in order to implement the customer's wishes. While the customer usually only comes into contact with the company at one communication point, his wishes and needs have far reaching effects on the various departments within the company and on the individual participants in the value chain. However, this requires not only a functioning communication between the customer and the company, but also a continuous optimization of all business processes, supported by the necessary technical aids (Bruhn, 2002). The following example illustrates the concrete meaning of a process orientation for corporate organization.

Process orientation for insurance companies

The growing competitive pressure within the insurance industry has led to a reorientation of customer management with regard to the organizational anchoring of the relationship to customers. In the past, the individual organizational units were strictly separated according to products. Each department only considered the customers from the responsibility of their product area, but hardly knew which other transactions the customer maintained with the company. In order to meet the individual needs of customers, a modular product range was put together, which can be combined to form tailor made product bundles—a task that required a complete restructuring of sales. The previously existing product areas, which led an "island existence" in the company, were combined into a meshed product portfolio that ideally leaves no customer wishes unfulfilled. Instead of the internal competition of individual departments, the customers were divided into different segments. Customers now receive a single point of contact for all their interactions with the company, who is responsible for all needs. The customer relationship becomes more transparent. This also significantly increases the quality of service. Based on the 10 Cs of quality management for services (cf. Fig. 6.1), the cost benefit ratio can be optimized by the implementation of a single point of contact, whereby the administrative effort for customers is significantly reduced. In addition, increased consistency is provided and coordination is greatly simplified. Furthermore, the increasing communication offers the contacts the opportunity to develop additional products more easily, which contribute to an increase in customer benefit.

(Source: Bruhn, 2002, 73 pp.) ◄

In quality management, a distinction can be made between core quality processes and support processes within the framework of process orientation. The **quality core processes** represent the central measures of quality management that are directly related to the performance creation. The core quality processes (e.g. implementation of a sustainable quality of performance, ensuring customer benefits, non-imitability of quality, etc.) represent the central measures of quality management that are directly related to the performance creation. The **quality support processes** have the task of supporting the core processes accordingly. These measures of the support processes are to be aligned accordingly with the core processes in order to ensure optimum support. Typical instruments of the quality management support processes include, for example, the operational suggestion system, quality oriented personnel development, employee observations, quality audits, etc. When designing, it is again important to ensure that the individual processes and instruments of quality management are optimally coordinated with each other (see also Chap. 13 on process controlling in more detail). The following example provides an exemplary overview of the division into core quality and support quality processes of a university.

Process map of the *OTH Amberg-Weiden*

The process map provides an overview of all processes at the first level and shows the three essential process areas—management process, core process and support processes—of the *OTH Amberg-Weiden* in an overview. It shows the interaction and interdependence of the university processes (cf. Fig. 9.4).

(Source: OTH, 2016) ◀

Finally, the implementation of a quality management is supported by a **decision making** process This means that the idea of customer and quality orientation can only be implemented if corresponding decisions are made and implemented. This is especially true for decisions in direct customer contact. If customer contact employees have to confirm each of their decisions by one or more superiors, a flexible reaction to customer requirements is not possible. Accordingly, the so-called **empowerment of customers** is considered an important instrument for the implementation of customer oriented quality management. This form of competence expansion has gained particular importance in the service sector, as it is a means of increasing the relationship orientation of customer contact employees. The term empowerment includes all measures that allow the employee to make autonomous decisions in the corresponding customer contact situation (Stewart, 1997; Blanchard et al., 2009; Scholl et al., 2012, 391 pp.). An example is the logistics service provider *TNT Express*, which integrates extensive further training measures for employees into personnel development with the guiding principle "recruiting, equipping, developing and empowering people". The company strives for a long-term customer loyalty through the empowerment of employees in the performance creation process, supported by further training programmes (Logistics Today, 2016). Through empowerment, employees are more motivated, so that they see the customers not only as customers of the company, but also as their own customers (cf. Sect. 8.3.1.2).

However, in order to ensure a high level of service quality, it is not only the empowerment of employees, but also the empowerment of customers that must be considered. The so-called **concept of customer empowerment** offers customers the opportunity to specify services according to their wishes, to choose the preferred method of service creation and to learn from other customers (Wathieu et al., 2002). In the scientific literature, the concept of customer empowerment is defined as a strategy under which companies transfer part of the service creation process to their customers (Fuchs et al., 2010). In this context, customer empowerment is characterized above all by the fact that it is initiated by the company and the way in which customers participate in the service creation process is also determined by this (Wathieu et al., 2002). However, the decision on empowerment lies with the customer and is therefore considered voluntary (Schmitz & Eberhardt, 2009). Customers can therefore choose whether they want to take over parts of the service creation process or prefer the traditional way of service creation without

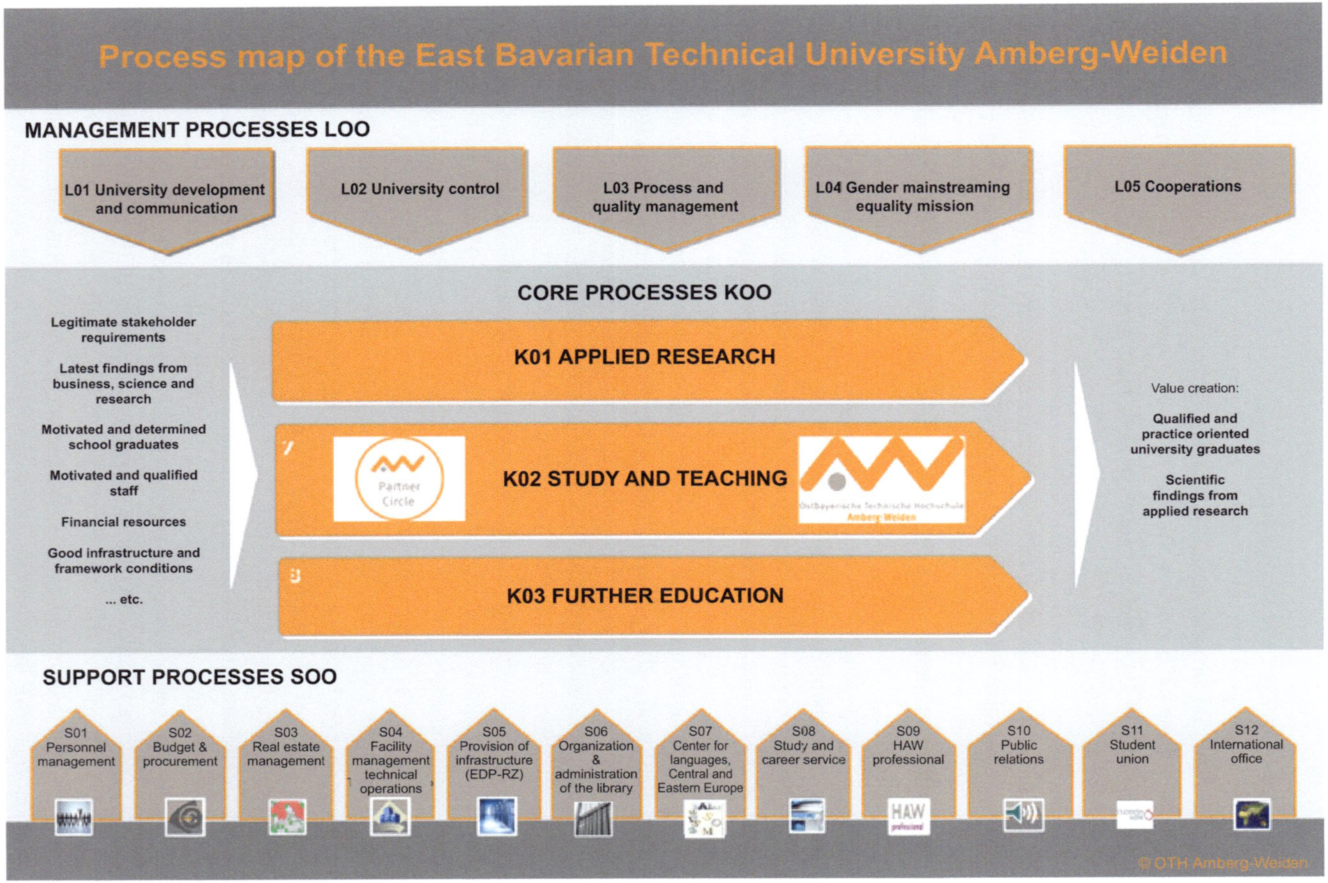

Fig. 9.4 Process map of the *OTH Amberg-Weiden*. (Source: OTH, 2016)

participation. As a result, companies have to offer customers the traditional way of service creation in addition to the possibility of empowerment. In this context, Braun et al. (2016) show in the context of a customer empowerment framework that this is determined by the customer's willingness and competence to take over parts of the service creation process. Accordingly, companies that pursue the strategy of customer empowerment have to support customers with process instructions and advisory staff in order to reduce uncertainties of the customer during the service creation carried out by them (Braun et al., 2016). If these criteria are met, not only financial and economic benefits can be realized for the customer, but also the company benefits from increased efficiency, as processes that were previously carried out by employees are now outsourced to the customer (Fuchs & Schreier, 2011; Braun et al., 2016). As a result, the concept of customer empowerment represents a new emerging strategy of companies to intensify the relationship with the customer, to increase the quality of the services offered and thus finally to increase the company's success (Niininen et al., 2007).

9.3.2 System Oriented Approaches

In addition to the company structure, the implementation of quality management depends on the **company systems**. A key factor in the design of customer oriented quality management is the coordination of different systems so that quality management is kept functional. The following types of systems should be highlighted (Bruhn, 2002):

- Information systems
- Communication systems
- Control systems
- Personnel management systems

With regard to the **information systems** of the service provider, information from various company departments forms the basis for deriving quality measures. For example, the results of customer surveys on customer satisfaction and customer loyalty from the market research department are the basis for improving services with regard to customer expectations. Another source of information for quality management are reports from the complaint management, in which the number and reasons for complaints are analyzed and show points of improvement for the services. Due to the direct customer-employee contact, the customer advisors also have customer related information from the direct customer contact. However, these are rarely recorded systematically, even if they allow for significant conclusions for performance improvements. An example are the reasons for customer churn. In many cases, customers leave because of their dissatisfaction with the quality of the service. Corresponding information about the reasons for customer churn can help to further improve the company's performance and avoid further customer attrition for the same reasons. A key task in terms of information systems is therefore to systematically record the information required for quality management from the various sources within the company and to make it available for quality management.

In addition to information acquisition, information processing in the sense of meaningful data preparation, information provision to the relevant departments, and information storage in a so-called data warehouse are central tasks of information systems. Figure 9.5 shows the individual tasks of an information system.

The demand of companies for integrated quality management systems is increasingly being taken into account by the supplier market of enterprise information systems, as the following example shows.

Quality management function of the *SAP AG*

The software provider *SAP*, for example, offers an integrated quality management function for the entire product life cycle as part of the module *SAP Product Lifecycle Management* (*SAP PLM*). It covers aspects such as quality engineering, quality control, quality reports (processing of all unplanned incidents), management of test equipment and quality certificates. The function enables companies to take quality requirements into account in the program and target management of a supplier as early as the design phase in order to control product quality throughout the value chain. In addition, the module *SAP PLM* can be connected to the quality management function together with other modules such as the customer relationship management module *SAP Customer Relationship Management* (*SAP CRM*), which allows system-supported quality management to be extended to other quality oriented information systems such as problem and complaint management.

(Source: SAP Deutschland AG & Co. KG, 2007) ◄

The **communication systems** of a service provider are to be selected in such a way that an optimal implementation of quality management is ensured by consistent internal and external communication. Communication systems include the involved communication partners, the communication structures and relationships. Its goal is to achieve different communication results (e.g. information, dialogue).

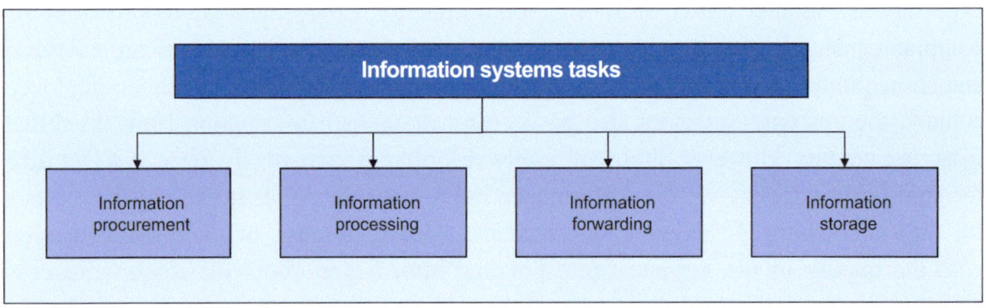

Fig. 9.5 Tasks of information systems

Due to the special importance of employees in service marketing, **internal (employee) communication** plays an important role in the communication system. It is important to ensure that employees are continuously informed about new products, behaviours (e.g. work manuals, guidelines, etc.) and market trends, so that a uniform and competent appearance of employees in customer contact is guaranteed. It is also important to ensure that employees themselves have the opportunity to exchange information with each other (e.g. weekly team meetings) and also receive the opportunity for direct communication with their superiors (feedback meetings). This increases the motivation and satisfaction of employees, which in return directly affects the quality of service provision.

With **external communication** we mean the communication of the company or the employees with the stakeholder groups outside the company, such as customers, intermediaries, suppliers. Here it is necessary that the communication arrives consistently and coordinated across the different instruments (e.g. field sales, customer service, media advertising) as a single message at the receiver.

An **optimal communication system** from the customer's point of view is characterized by the fact that a tailor-made communication offer is provided for the respective customer at the right time. Ideally, the customer can choose himself which form, means and frequency of communication he prefers in order to obtain information about the company or to contact it. In the service sector, companies have now installed communication systems that adapt to the different information and interaction needs of customers. Today, the communication systems of many service providers range from simple newsletters for customers who only want to be informally informed about product news from time to time, to call centres and so-called "customer interaction centres", where customers can interact and consult with company representatives "face to face" (Bruhn, 2002, pp. 130–132).

"Interaction Center" of *SAP CRM* in the personnel department of *EnBW Energie Baden-Württemberg AG*

In connection with the CRM platform *SAP CRM*, which the software provider *SAP* distributes as a module of the enterprise software *SAP Business Suite*, the implementation of a so-called system supported "interaction center" is possible.

The Karlsruhe energy provider *EnBW Energie Baden-Württemberg AG* recently implemented the function of the interaction center in the personnel department in order to increase the quality of personnel services in this way. At the same time, the bundling of all service functions for employees through the interaction center made it possible to implement leaner, more transparent and more cost-effective processes of personnel care.

At the center of the implementation of the interaction center in the personnel department is the implementation of a service center, which serves as a central point of contact for employees and pensioners. This offers various functions that make the interaction center a competent point of contact for employees, through which their concerns can be processed consistently and conclusively as early as the first contact. For example, the interaction center offers the possibility of scanning documents at the workplace and storing them in a digital file. This digital file then bundles all documents, e-mails and word documents that are needed for quick processing of the action.

Through the implementation, the *EnBW AG* was able to benefit from various aspects in addition to the cost savings. For example, the internal service quality towards employees could be significantly increased by better availability. In addition, the different concerns of different groups of employees are taken into account to a greater extent by the introduction of the interaction center, which significantly improves the group specific processing of different groups of employees (e.g. pensioners).

(Source: SAP Deutschland AG & Co. KG, 2007) ◀

With regard to the **control systems**, it must be examined to what extent the existing control systems can be used by quality management or possibly modified. The aim is to use the control systems for the realization of the goals of quality management, i.e. the fulfillment as well as the control of customer expectations or the improvement of service quality. For this purpose, quality management has a wide range of control instruments at its disposal. In the following, above all the complaint and customer loyalty management should be highlighted (cf. also Bruhn, 2002):

The **complaint management system** is used to subsequently meet customer expectations if they are not met during actual service use. In this respect, complaint management must be oriented towards quality in such a way that it is possible to meet customer expectations. This affects the customer expectations of the service itself. A complaint management system is only effective if the customer expectations can actually be met by the complaint handling. In addition, the customer expectations of the complaint behavior of the service provider are a key factor in the customer's quality perception. If the complaint behavior does not meet the customer's requirements (e.g. due to a delayed reaction to the customer complaint or a reaction disproportionate to the customer's point of view), the customer's dissatisfaction is increased.

The **customer loyalty management system** has the goal of increasing the number of customers who choose the company and its services again. Again, an orientation towards customer expectations contributes to the fact that customer loyalty management does not have a conflictual effect on the goals of quality management. For example, it is not useful for customer loyalty management to address mainly customers whose needs are not oriented towards the core service of the company. Vice versa, when designing customer loyalty incentives, it is important to ensure that they are selected according to customer expectations (Wirtz & Lovelock, 2016, p. 489).

With regard to the **personnel management systems**, it must be ensured that the resource personnel is oriented towards a successful quality management. The productivity, motivation and creativity of the staff can be a decisive competitive advantage (Thom,

2003, p. 433; Sia & Appu, 2015). Employees can create an "added value" for customers that is difficult to imitate by the competition (Bruhn, 2002, p. 202). This is especially true for services.

To maintain this "added value" in the long term, the various **fields of activity of human resource management** (recruitment, deployment, development, etc.) must always be processed from a quality perspective. For example, the recruitment process must be structured in such a way that the candidates who are best suited both professionally and personally are recruited and selected for the respective positions to be filled. This poses great challenges for the employer branding (advertising for the company as an employer/attractive job postings) as well as for the candidate assessment and selection processes. Furthermore, it is necessary to ensure that the correct positions are filled with the right people as part of the deployment and development process and that these people are always challenged and promoted in order to make optimal use of their existing potential. This is achieved by regularly carrying out training and seminar measures relevant to the position in order to continuously expand both the professional and, above all, the personal competencies relevant in the service sector with direct customer contact.

Incentive and remuneration systems as part of the human resource management systems can be used to increase the motivation and commitment of employees. It is important to ensure that the behaviour caused by the **incentive system** is compatible with the goals of quality management. After all, these have a significant impact on service quality through their control of employee behaviour. If, for example, sales are the main factor in the composition of the salary of customer contact employees, as is often the case in management consulting, this promotes a sales-oriented approach by employees. If too much emphasis is placed on sales and customer acquisition, the quality of projects with existing customers and thus customer loyalty will inevitably be neglected. In addition, there are a number of non-monetary incentives that counteract this development. These include, for example, opportunities for advancement, status symbols, praise, recognition, etc. In general, the types of incentives can be divided into extrinsic and intrinsic forms of incentive (for the various quality oriented incentive systems, see Sect. 8.3.1.3).

9.3.3 Culture Oriented Approaches

The culture of a service company has a significant influence on the achievability of the goals of quality management.

▶ The **corporate culture** describes a system of long-term stable values and beliefs shared by company members (Kotler & Armstrong, 2018, p. 135).

Because of the latent character of corporate culture, it is often difficult for companies to tell whether their own corporate culture is sufficiently customer oriented or not. The following **indicators of a lack of customer oriented orientation** of the corporate culture can be identified (Homburg & Werner, 1998, 174 pp.):

- Leadership positions are filled with people who represent an exclusively production oriented view.
- From a certain management level, there is no direct customer contact anymore.
- The horizontal and vertical exchange of customer related information does not work.
- There are numerous hierarchies that make coordination within the company more difficult.
- The bureaucracy in the company is very high.
- The production of services does not meet customer needs.

A quality oriented culture is closely related to the **customer oriented behavior** of employees. This is particularly relevant in the service sector, as it documents the corporate culture not only internally, but also to the customer. Customer oriented behavior of employees is only possible if they are satisfied with their job and their work situation. Within a company, the following three **levels of culture** are to be distinguished:

- Corporate level (corporate culture)
- Group level (group culture)
- Individual level (individual culture)

At the **corporate level**, culture is understood to be the self-conception of the company, which is expressed in mission statements, visions, and the philosophy of the company and has the goal of shaping the behavior of all employees in the sense of a quality oriented strategy. The goal of corporate culture is therefore to establish a uniform way of thinking throughout the company and in the minds of all employees. With regard to quality management, culture is to be aligned in such a way that employees develop a self-conception of always delivering and serving customers with the highest quality. This is also illustrated by the following example of the Klinikum Arnsberg.

Klinikum Arnsberg—Quality orientation as the highest priority

The quality oriented provision of services for patients is the highest priority of the *Klinikum Arnsberg,* an academic teaching hospital of the *Westfälische Wilhelms-Universität Münster*. The quality orientation anchored in the corporate culture is noticeable in all actions towards the inside and outside. The employees feel committed to a holistic treatment and service principle. They consistently and regularly analyze their own strengths and weaknesses in all areas of performance and measure themselves against the judgment of their patients, customers and partners. They optimize their performance in a continuous, cross-departmental and cross-professional improvement process.

(Source: Klinikum Arnsberg, 2016) ◀

Especially in larger companies, there are often strong subcultures at the **group level**—for example, in individual departments, at certain hierarchical levels, or in groups of employees of the same ethnic or national origin. The stronger subcultures are pronounced in a company, the weaker is a uniform culture shaping at the corporate level. Often, such different subcultures already arise during the selection of the suitable employees, who often also have a completely different way of thinking. Due to the daily interaction with colleagues, the subculture of employees is experienced much more intensively than the corporate culture. As a result, the difference between the individual subcultures is amplified.

It is therefore the task of management to create a shared way of thinking among employees and to establish a team orientation across departmental boundaries. Increased cross-departmental communication and collaboration helps to break down these different ways of thinking. For example, cross-departmental project teams, such as quality related task forces, represent suitable measures to break down thinking and behaviour within different subgroups and to enable the fusion of culture defining elements. In terms of quality management, cross-departmental events such as quality circles, quality seminars, workshops, etc. are suitable for creating a shared understanding of quality awareness and for defining common quality objectives and measures across the various subcultures.

Quality related task forces in a hotel

If, for example, customer surveys show that the quality criterion "fulfilment of individual customer wishes" has been rated negatively, a cross-departmental task force can help to initiate improvement measures. Here, it is important to involve as many departments as possible, such as the kitchen staff, IT, reception, etc.

The different perspectives and cultures make it possible to develop creative ideas. If the task force were only made up of service employees (e.g. reception, waiters, etc.), the perspective and, as a result, the basis for possible ideas would be more limited. ◀

On the **individual level**, the individual cultural background of each employee is expressed. Due to their geographical, socio-cultural and professional background, each employee brings different ways of thinking and behaving into the group or company. With a view to building a shared corporate culture, it is important to take this individual culture of each employee into account.

Even if the culture of a company has developed historically, the question arises as to whether this culture can be changed systematically in order to adapt it to the needs of quality management. First of all, it is a question of which elements of corporate culture can be controlled. In addition, the time dimension of changeability is relevant. The experiences of companies with a demonstrably quality oriented corporate culture, such as *Ritz-Carlton* or *Hotel Schindlerhof*, show that a systematic approach to **adapting corporate culture** is most likely to be successful. For this purpose, three **steps** are helpful:

1. For the **analysis of the current corporate culture** there are various approaches to choose from. Employee surveys have proven to be effective (Deshpandé et al., 1993; Schein, 2009). The result of the situational analysis is the derivation of characteristic features of the existing corporate culture, on the basis of which it is possible to assign it to a certain culture type. Figure 9.6 shows the characteristics of four culture types and their way of handling relationships and customer orientation. The clan, adhocracy, market and hierarchy cultures are distinguished (Quinn & Rohrbaugh, 1983). Research results in this area show that companies with an adhocracy culture realize the highest relationship and customer orientation.
2. The **cultural change process** set against the results of the first step differentiates between two levels: the individual level (e.g. leaders with a strong role model function) and the collective level. The latter often goes hand in hand with the development of a new corporate identity and the implementation of corporate identity concepts. The design of the corporate culture on these two levels is illustrated by the following example of *Otto's AG*.
3. Finally, within the framework of the **control of cultural change**, it is about checking to what extent the desired culture is actually achieved on the basis of the culture analysis, i.e. to what extent the cultural change process takes effect. Since corporate culture is a dynamic phenomenon, the cultural change process and, consequently, the control of this process must also be dynamic. Continuous controls of the implementation success contribute to the adaptation of the implementation process if necessary.

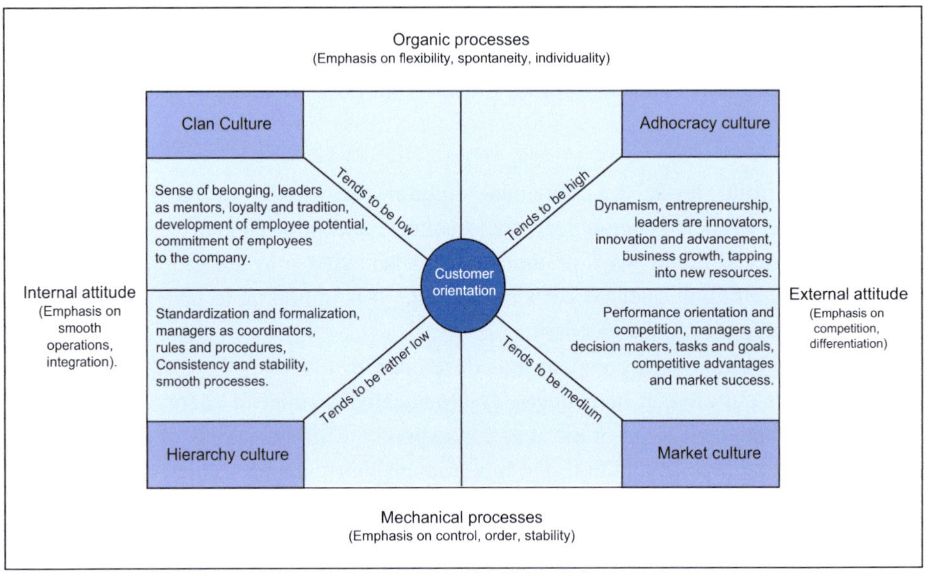

Fig. 9.6 Types of corporate culture. (Source: Based on Deshpandé et al., 1993, p. 27)

Culture of *Otto's AG*

At the Swiss trading company *Otto's AG*, the corporate culture can be described as very customer oriented when positioning itself as a discount provider. This was achieved on the one hand by the clear specification of values by the founder of the company and on the other hand by the corporate culture which is permanently carried forward by the employees. In an analysis of the corporate culture, the following **features**, for example, were identified as particularly important (Kunz, 1998):

- **Performance obligation**: The direction of customer orientation is set by the boss, passed on by the management and lived by all employees in the company.
- **Slim organization**: Aspiration for simple processes and minimal administration.
- **Clear goal agreements**: There are clear goal agreements between the managers and employees of a department.
- **Flexibility**: Everyone can take their breaks at individual times.
- **Loyalty**: The employees show a particularly high identification with the company. ◀

9.4 Design of the Implementation of Quality Management

In view of the complexity of a holistic quality management approach, it becomes immediately clear that the implementation of such a concept in a service company cannot be realized through a short-term transformation process of an office or a branch, but only through a **long-term planned restructuring process** of the entire service company. For this reason, it is expedient to define the development and implementation of the quality management system as a multi-stage phase concept that is divided into different time units. The networking possibilities and the sequence of the individual quality phases and instruments are to be considered here (see the explanations on the coordination of the individual phases of quality management in Sect. 8.6 as well as the corresponding Fig. 8.55 on the use of quality management instruments in the course of time).

Basically, the implementation of the individual phases of the quality management system must take place within the framework of a **detailed project planning** (Döttinger & Klaiber, 1994, p. 263). Various steps that are necessary for the realization of a comprehensive, service specific and effective quality management are taken into account. In Fig. 9.7 an overview of the possible project steps is shown.

In order to implement a quality management system for service companies, it is necessary that a **system of interconnected control loops** is created for all members at all levels of the company, through which the goals, structures, responsibilities, procedures, processes and the measures required for their implementation are specified.

In **quality control loops** specific quality requirements for products and processes are recorded, their compliance is checked and, after analysis and evaluation, they are used

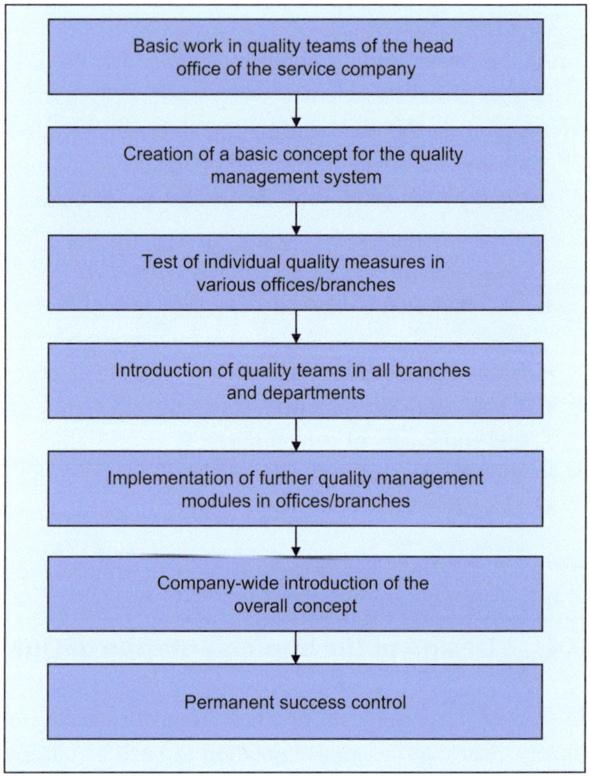

Fig. 9.7 Project steps for the development and implementation of quality management

as a quality data basis for long-term corrections and future quality requirements. Since quality control and quality assurance measures are rarely effective on their own and are often dependent on the results of other methods or themselves provide a basis for further measures, quality control loops should not be considered in isolation in the company, but should "nest" at all hierarchy levels.

In Fig. 9.8 a **quality control loop system for service companies** is shown in a model, in which the different levels of the company and the required cross-connections are shown, which are affected by the implementation of quality management.

Planning considerations must be made for the implementation of a quality management system, which are specifically concerned with the overcoming of the expected implementation barriers. A **basic concept for an implementation program**—based on the decision oriented approach in service marketing (Meffert et al., 2018, 451 pp.)—helps to systematically structure the implementation process.

In this basic concept, the **analysis of the implementation situation** represents the starting point. Here, the implementation need for quality management is specified in the individual company areas—in relation to the quality culture, the strengths and weaknesses of quality work in different functional units—as well as the suitability of the existing structures and processes to ensure internal and external quality.

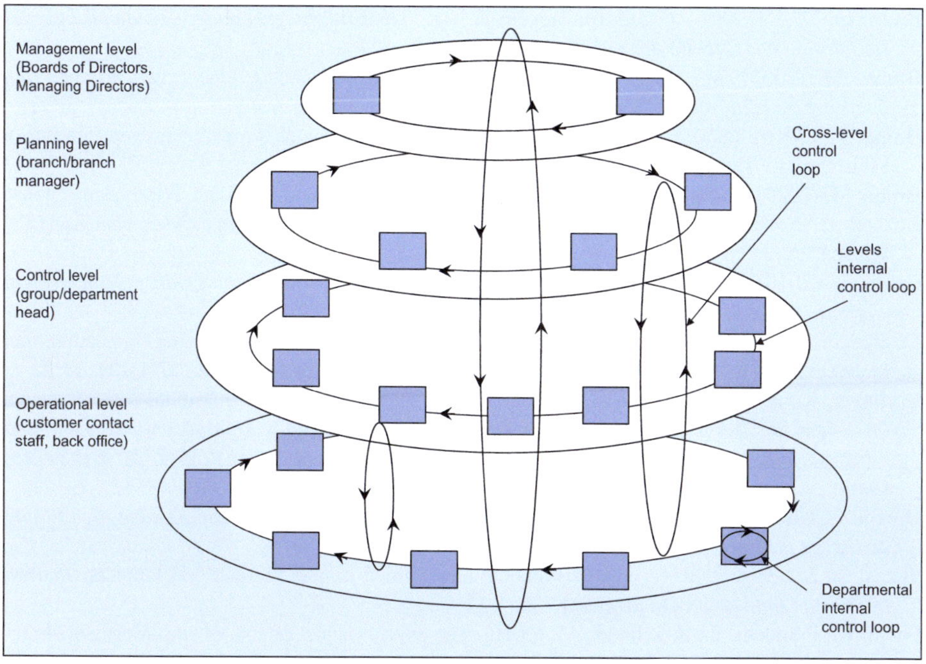

Fig. 9.8 Model of quality control loops for service companies

Then the **goals and target groups of the implementation** are to be defined before specific individual measures can be planned, designed and coordinated with each other. Success control is required in order to continuously measure the degree of achievement of implementation goals and to initiate possible corrective measures. With this basic concept of an implementation program, an iterative approach with constant feedback processes at all levels is recommended. In this case, the management and employees involved are to be actively involved in the analysis and planning of the implementation program.

After this chapter has presented the necessary steps and barriers to the implementation of quality management, the importance of a successful implementation and its review will now be discussed. In the next chapter, the relevance of quality awards for service companies is to be shown before the importance of auditing and certification for quality management of service companies is described in Chap. 11.

References

Blanchard, K. H., Carlos, J. P., & Randolph, W. A. (2009). *Management durch Empowerment. Das neue Führungskonzept: Mitarbeiter bringen mehr, wenn sie mehr dürfen* (3rd Edn.). Reinbek bei Hamburg: Rowohlt.

Braun, C., Popp, M., & Hadwich, K. (2016). *Success Factors of Customer Empowerment from a customer's perspective: a qualitative study and strategic implications*. Proceedings of the International Service Research Conference (AMA SERVSIG). Maastricht.

Brommer, E. (1999). *Qualitätsmanagement im Dienstleistungsbereich: Evaluierung der Einführung von DIN EN ISO-9000*. Berlin: Beuth.

Bruhn, M. (2002). *Integrierte Kundenorientierung. Implementierung einer kundenorientierten Unternehmensführung*. Wiesbaden: Gabler.

Bruhn, M. (2016). *Relationship Marketing. Das Management von Kundenbeziehungen* (5th Edn.). München: dtv Verlagsgesellschaft.

Bruhn, M. (2019). *Marketing. Grundlagen für Studium und Praxis* (14th Edn.). Wiesbaden: Gabler.

Buzzell, R. D., & Gale, B. T. (1989). *Das PIMS-Programm. Strategien und Unternehmenserfolg*. Wiesbaden.

Conti, T. (2010). Systems thinking in quality management. *The Total Quality Management Journal, 22*(4), 352–368.

Deshpandé, R., Farley, J. U., & Webster, F. E. (1993). Corporate culture customer orientation, and innovativeness in Japanese firms. A quadrad analysis. *Journal of Marketing, 57*(1), 23–27.

Döttinger, K., & Klaiber, E. (1994). Realisierung eines wirksamen Qualitätsmanagementsystems im Sinne des Total Quality Managements. In B. Stauss (Ed.), *Qualitätsmanagement und Zertifizierung von DIN ISO 9000 zum Total Quality Management* (pp. 255–273). Wiesbaden: Gabler.

Droege & Comp. (2000). *Triebfeder Kunde IV. Eine Zeitverlaufsstudie zur Kundenorientierung deutscher und internationaler Unternehmen*. Düsseldorf.

Fuchs, C., & Schreier, M. (2011). Customer empowerment in new product development. *Journal of Product Innovation Management, 28*(1), 17–32.

Fuchs, C., Prandelli, E., & Schreier, M. (2010). The psychological effects of empowerment strategies on consumers' product demand. *Journal of Marketing, 74*(1), 65–79.

Hilker, J. (2001). Marketingimplementierung. Grundlagen und Umsetzung für das Dienstleistungsmanagement. In H. Meffert & M. Bruhn (Eds.), *Handbuch Dienstleistungsmanagement. Von der strategischen Konzeption zur praktischen Umsetzung* (2nd Edn., pp. 827–850). Wiesbaden: Gabler.

Homburg, C., & Werner, H. (1998). *Kundenorientierung – Mit System. Mit Customer Orientation Management zu profitablem Wachstum*. Frankfurt a. M.: Campus.

Klinikum Arnsberg. (2016). Werteorientierung Arnsberg. http://www.klinikum-arnsberg.de/klinikum-arnsberg/philosophie-werteorientierung/qualitaetsorientierung/. Accessed: 3. Mar 2016.

Kotler, P., & Armstrong, G. (2018). *Principles of marketing* (17th Edn.). New Jersey: Pearson Education.

Kunz, H. U. (1998). Was bringt Unternehmenskultur? *Der Organisator, o. V.*(1–2), 6–7.

Logistik heute. (2016). TNT Express – Langfristige Bindung durch systematische Weiterbildung. http://www.logistik-heute.de/Logistik-Karriere-Wissen/Blick-in-die-Praxis/6746/TNT-Express-Langfristige-Bindung-durch-systematische-Weiterbildung. Accessed: 25. Febr. 2016.

Meffert, H., Bruhn, M., & Hadwich, K. (2018). *Dienstleistungsmarketing. Grundlagen, Konzepte, Methoden* (9th Edn.). Wiesbaden: Springer.

Niininen, O., Buhalis, D., & March, R. (2007). Customer empowerment in tourism through Consumer Centric Marketing (CCM). *Qualitative Market Research: An International Journal, 10*(3), 265–281.

Oess, A. (1994). Total Quality Management. Eine ganzheitliche Unternehmensphilosophie. In B. Stauss (Ed.), *Qualitätsmanagement und Zertifizierung von DIN ISO 9000 zum Total Quality Management* (pp. 199–222). Gabler.

OTH. (2016). Qualitätsmanagement OTH. http://www.othaw.de/hochschule/qualitaetsmanagement/prozessanalyse/prozesslandkarte/. Accessed: 3. Mar 3016.

Quinn, R. E., & Rohrbaugh, J. (1983). A spatial model of effectiveness criteria. Towards a competing values approach to organizational analysis. *Management Science*, *29*(3), 363–377.

SAP Deutschland AG & Co. KG. (2007). Qualitätsmanagementfunktion der SAP AG. http://www.sap.com/germany/solutions/business-suite/plm/featuresfunctions/index.epx. Accessed: 16. Juli 2012.

Schein, E. H. (2009). *The corporate culture survival guide* (2nd Edn.). San Francisco: Jossey-Bass.

Schmitz, G., & Eberhardt, S. (2009). Empowerment als Ansatzpunkt zur effektiven Steuerung der Kundenintegration? – Eine konzeptionelle Analyse. In M. Bruhn & B. Stauss (Eds.), *Kundenintegration. Forum Dienstleistungsmanagement* (pp. 381–401). Wiesbaden: Gabler.

Scholl, W., Schermuly, C., & Klocke, U. (2012). Wissensgewinnung durch Führung – Die Vermeidung von Informationspathologien durch Kompetenzen für Mitarbeiter (Empowerment). In S. Grote (Ed.), *Die Zukunft der Führung* (pp. 391–413). Gabler.

Sia, S. K., & Appu, A. V. (2015). Work autonomy and workplace creativity: Moderating role of task complexity. *Global Business Review*, *16*(5), 772–784.

Statista (2019). https://de.statista.com/statistik/daten/studie/547256/umfrage/wunsch-nach-und-wahrnehmung-von-kundenorientierung-in-oesterreich-nach-branchen/. Accessed: 26. Apr. 2019.

Stewart, A. M. (1997). *Mitarbeitermotivation durch Empowerment*. Niederhausen: Falken.

Thom, N. (2003). *Betriebliches Vorschlagswesen: Ein Instrument der Betriebsführung und des Verbesserungsmanagements* (6th Edn.). Bern: Lang.

Waldklinik Jesteburg. (2010). Qualitätsmanagement in der Waldklinik Jesteburg. http://www.waldklinik-jesteburg.de/downloads/Q-Bericht-2010-Waldklinik-Jesteburg.pdf. Accessed: 3. Mar 2016.

Wathieu, L., Brenner, L., Carmon, Z., Chattopadhyay, A., Wertenbroch, K., Drolet, A., Gourville, J., Muthukrishnan, A. V., Novemsky, N., Ratner, R. K., & Wu, G. (2002). Consumer control and empowerment: A primer. *Marketing Letters, 13*(3), 297–305.

Wirtz, J., & Lovelock, C.H. (2016). *Services marketing. People, technology, strategy* (8th Edn.). New Jersey: World Scientific Publishing.

Importance of Quality Awards for Service Companies

10

Abstract

To implement quality management, quality models have proven themselves in recent decades that were developed within the framework of so-called quality awards. Originally, these models were used primarily to assess the quality management of service companies and to award quality prizes. Meanwhile, they also serve service companies as a reference framework for implementing and improving their own quality management. To illustrate the background of quality models, their relevance is explained in this chapter and important quality prizes are presented and critically assessed with the *EFQM Model,* the *Ludwig-Erhard-Preis*, the *ESPRIX Award* as well as the *Austrian Quality Award.*

The implementation of quality management for services has to be started at different levels. This concerns the mentioned points of departure for the orientation of structures, systems and culture in order to improve the quality work of the company. On the other hand, it is also important to find out the state of implementation of quality management at certain intervals. One approach is the orientation of quality work to the criteria of quality awards. These criteria for successful quality work are of particular importance to companies because they were developed by different institutions and are the result of intensive discussion. Therefore, a very high relevance results for an assessment of the degree of implementation of quality management of service companies. Before the individual quality awards and models in the European and German speaking area are presented, the basic concepts of these quality awards will be discussed.

M. Bruhn, *Quality Management for Services*,
https://doi.org/10.1007/978-3-662-67032-3_10

10.1 Characteristics and Systematization of Quality Awards

By **quality awards** incentives are set to promote the holistic view of quality management systems and thus to drive their implementation within the company.

▶ **Quality awards** are to be understood as prize awards by special institutions which are awarded for the proof of the promotion of quality, the understanding of quality in the whole company as well as its successful internal and external implementation.

It becomes clear that the **goals of quality awards** are very broadly defined. Quality prizes are initiated and awarded by institutions in order to improve the **level of competitiveness** as a whole.

The focus on improving competitiveness is achieved by awarding companies for outstanding quality management and outstanding performance. The quality awards reflect two significant influencing factors for competitiveness: the provision of constantly improved services for customers and the comprehensive improvement of the overall performance of the company (Reimann & Hertz, 1994, p. 337).

Another important goal of quality awards has didactic reasons: The awards promote the **knowledge transfer** with regard to competitiveness and expand the knowledge base, so that a broad knowledge base is created. The awareness of and understanding of the importance of quality improvements in the company can be promoted and an exchange of information on successful quality strategies can be initiated (Reimann and Hertz, 1994, pp. 339, 362).

The large number of quality awards have, first and foremost, the **comprehensive evaluation criteria** in terms of quality management in common, which take into account the following areas in particular:

- Customer satisfaction
- Business results
- Employees
- Processes
- Leadership
- Resources (finance, information and technology)
- Politics and strategy
- Influence on society

The steps to obtaining a quality award include the **application** by the service company, the formal and practical examination of the management system, and the central award by a significant institution (Peacock, 1992, p. 526; EFQM, 2013). In this context, the *"Malcom-Baldrige Award"*, the *"EFQM Award"* and the *"Deming Prize"* are the three best known quality awards worldwide (EFQM, 2016). Furthermore, the currently best known quality awards are mentioned in the classification given in Fig. 10.1 (see for an overview of national quality awards around the world, e.g. Nist, 2010). The presentation is limited to national prizes as the smallest geographical reference level. Another group are other quality awards that—like the Topit Quality Award—award excellent quality oriented initiatives and projects. However, the classification made in Fig. 10.1 does not make a distinction between the institutions involved in the award, but the respective prize categories, so that a classification based on the criteria "performance independent" and "service specific" quality awards is appropriate.

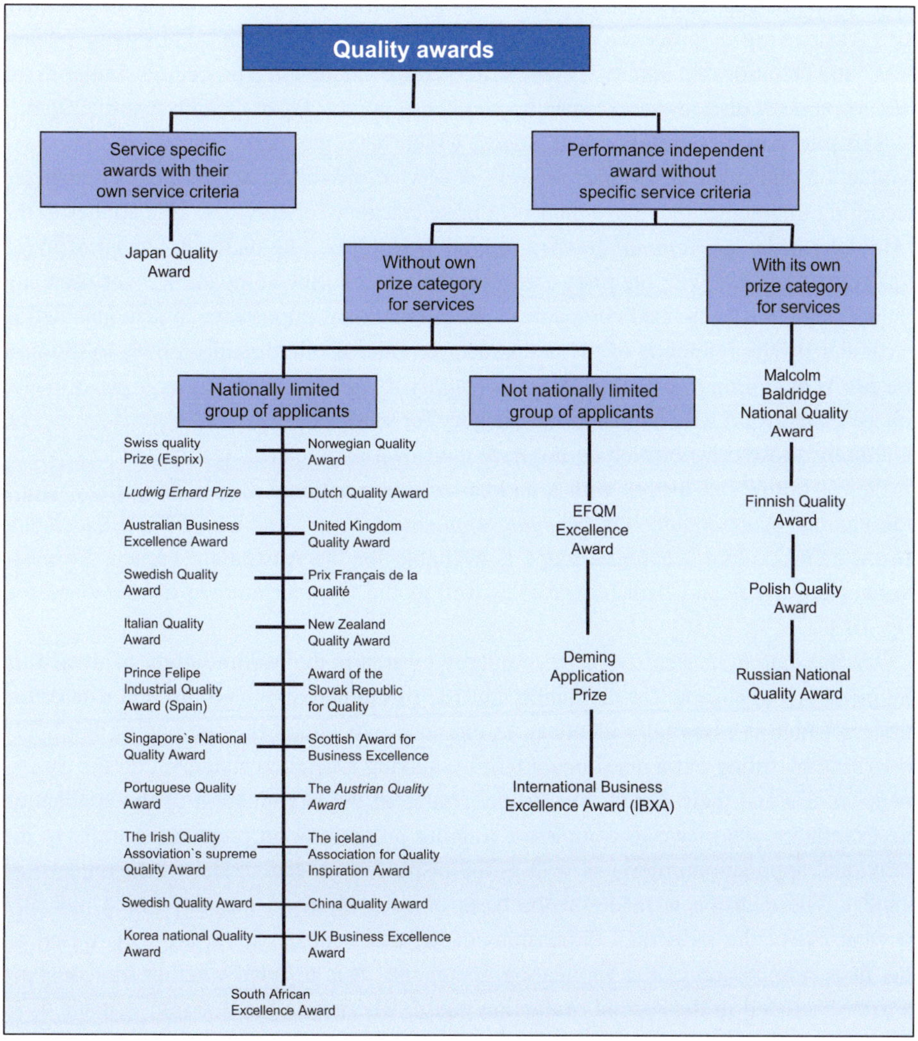

Fig. 10.1 Classification of selected national and international quality awards

In the **service specific awards** only those prizes are recorded, which show specific criteria in relation to the service sector for the award. The specific consideration and assessment of service companies is useful due to the significant differences between goods and services quality, which require other priorities in quality management. So aspects of the customer contact situation play a major role in the service sector, methods of statistical quality control play a less important role than in the manufacturing sector. Through its own quality prize, the service specific requirements for quality management are better taken into account. Only the *"Japan Quality Award"* falls into this category. It has been awarded since 1996 and is conceptually based on the four basic elements of customer and employee orientation, social responsibility and the presence of unique skills ("unique capabilites"). This results in the criteria "customer oriented service quality", "leadership", "process orientation", "knowledge generation", "reactivity", "fairness" and "relationship quality". These criteria are evaluated in a procedure similar to the *EFQM-model* with a respective specific number of points (Japan Quality Award, 2013).

The **performance independent** prizes, which have the same evaluation criteria for production and other companies as well as service providers, are further systematized according to whether they have their own prize category for services. This applies to the *"Malcolm Baldrige National Quality Award" (MBNQA)*. The national, annual *MBNQA* was created in 1987 with the prize categories "manufacturing companies", "service companies" and "medium-sized companies". Its international importance is demonstrated in particular by the reference of the assessment criteria of other quality prizes to those of the *MBNQA*. Furthermore, the *"Finnish Quality Award"* is provided as a performance independent award with its own prize category for services.

Finally, those prizes which do not have their own prize categories for services can be further classified into prizes with a national and non-national scope. The prizes with a non-national scope include the *"Deming Application Prize"* and the *"EFQM Excellence Award"(EEA)*. Only a national scope is available for the Australian, French, Swedish, Norwegian, Swiss and British awards as well as the German *Ludwig-Erhard-Preis* (cf. Fig. 10.1).

Another common element of these quality awards is the **methodology of awarding** the prize. All applicants for the quality awards mentioned above go through a selection process which is essentially identical, in the course of which the companies first apply in writing by filling out a questionnaire or by making a detailed statement on the assessment criteria and their dimensions. In the course of the examination of the application the examiners, who have to complete a training programme in preparation, refer to the individual application criteria as well as the respective criteria model which reflects the context. The decision is made on the basis of an **evaluation matrix,** which indicates to what extent the individual dimensions of the assessment criteria are met. Based on this first examination of the application documents, it is decided whether the company is to be included in the second evaluation phase, whereupon the decision is made as to whether an on-site visit to the company is to be carried out. The final report prepared on the basis of the results of the on-site visit is forwarded to the committee entrusted with the decision, which decides on the winner or winners.

10.2 Overview of Selected Quality Awards and Quality Prize Models

In science and practice, there are some models for the holistic implementation of quality management. However, a globally recognized model has not prevailed. Therefore, many companies use models that are based on quality awards as guidelines for a comprehensive quality orientation (Bou-Llusar et al., 2009, p. 16). In this context, the ***EFQM Excellence Model,*** on the basis of which the *EFQM Excellence Award* is awarded, is a management model developed for European companies for the holistic implementation of quality management (EFQM, 2010). This model is based on the principles of the ***Malcolm Baldrige Quality Awards,*** which is to be seen as the basic model for the following quality awards. However, since this is only intended for American companies, the *EFQM* Excellence Model will be presented in more detail below (Bou-Llusar et al., 2009; EFQM, 2016). This model currently exists in two variants:

- *EFQM Excellence Model 2013*
- *EFQM Model 2020*

These two models will be presented below. The *EFQM-Model* is the basis for the award of quality awards and quality prizes. Therefore, the European *EFQM Global Award,* the German *Ludwig-Erhard-Preis,* the Swiss *ESPRIX Swiss Award for Excellence* as well as the Austrian *StaatspreisUnternehmensqualität(Austria Quality Award)* will also be dealt with.

10.2.1 The EFQM Excellence Model as the Starting Point

The *EFQM Excellence Model* is the starting point for both a holistic management of companies as well as quality awards. It was developed with the aim of creating a European quality award. The *"European Foundation for Quality Management",* the European Commission and the *"European Organisation for Quality" (EOQ)* were involved in the development (Ellis, 1994, p. 280; Malorny & Dicenta, 2014, p. 380). This *EFQM Excellence Model* has always been the subject of intense discussion for the further development and permanent revision of the model by representatives of the European countries. In November 2019, the model was fundamentally redesigned, so that there are currently two models, which are to be subsumed under the terms ***EFQM Excellence Model 2013*** (classic model) and ***EFQM Model 2020*** (newly designed model).

10.2.1.1 The EFQM Excellence Model 2013

The original **basic idea of the *EFQM Excellence Model*** was to improve the quality of entrepreneurial performance. It can therefore be used for the implementation of a quality management system. It continuously integrates checked "best practice" of leading companies, thus allowing companies to carry out a **benchmarking** beyond their own industry. In addition to that, it has developed into a comprehensive **leadership model** that is successfully used by many companies and non-profit organizations.

The **use of the *EFQM Excellence Model*** has continuously and very successfully developed in practice for decades. This applies to every type of company, regardless of industry, company size, development status, degree of internationalization or organizational structure. A survey of industrial companies has shown that especially larger companies use the *EFQM model* (see Fig. 10.2). This is due to the effort that is associated with a systematic use of the model. However, the effort is worth it, because with the use of the model, special achievements have been demonstrated in the areas of quality, flexibility and innovation (Lay et al., 2009). In the meantime, more than 30,000 companies in Europe have used the *EFQM Excellence Model* (EFQM, 2019b, c).

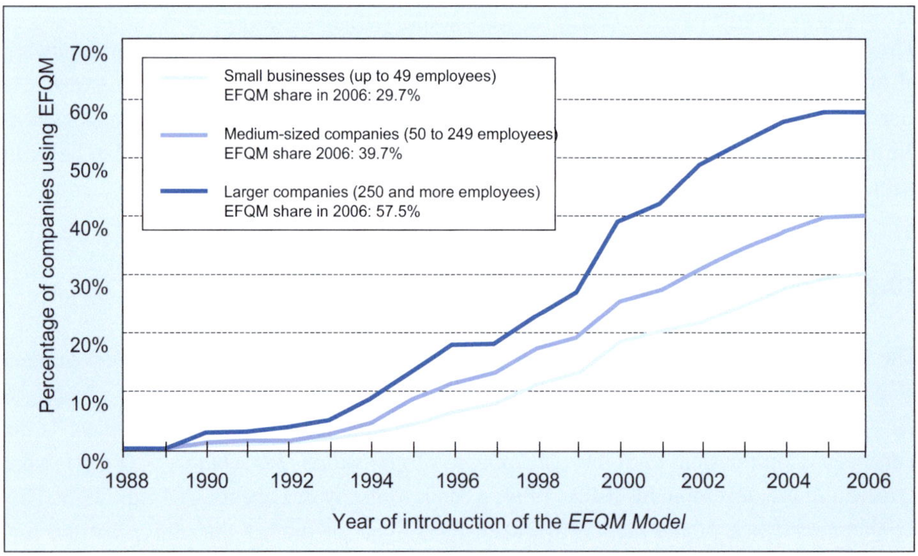

Fig. 10.2 Development of *EFQM* user rates for companies of different sizes. (Source: Lay et al., 2009)

At the center of the *EFQM Excellence Model* are various components that enable service companies to compare themselves with other, sustainable companies, to assess themselves and to derive action implications for quality management independently. These are the following **components of the *EFQM Excellence Model*** (EFQM, 2016):

- The "basic concepts of excellence" as basic principles for sustainable excellence in the company
- The "criteria model" as the basic structure of the *EFQM model*
- The "RADAR logic" as a dynamic evaluation framework for sustainable excellence

The individual components will be explained in more detail below and an overview of the application process for the EFQM Excellence Award will be given.

10.2.1.1.1 Basic Concepts of Excellence in the EFQM Model 2013

The basic concepts of excellence define the basic requirements for a sustainable implementation of excellence. The following **features** characterize an excellent organization (cf. Fig. 10.3):

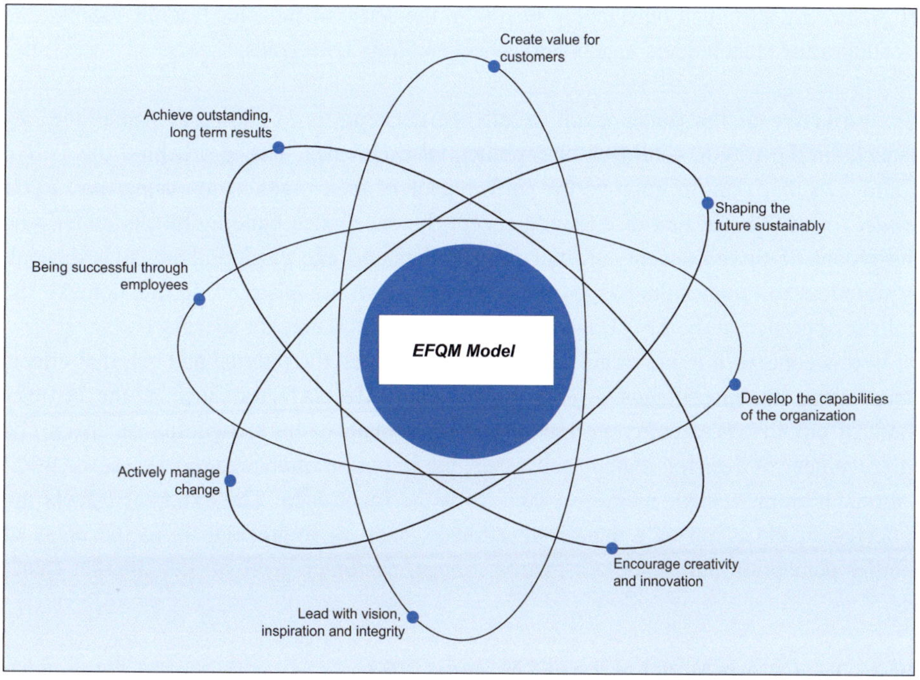

Fig. 10.3 Basic concepts of excellence of the *EFQM Model*. (Source: EFQM, 2016)

- Creating customer value through consistent identification, understanding and meeting of customer needs.
- Sustainable future oriented thinking by simultaneously increasing company performance and improving the economical, ecological and social conditions of its stakeholders.
- Developing organizational skills through effective management of internal and external changes.
- Encouraging creativity and innovation in the sense of value and performance enhancements through continuous and systematic innovation. This includes the continued development of networks and the involvement of all stakeholders in order to expand possible sources of creativity and innovation.
- Leading with vision, inspiration and integrity: The company's continued success should be secured by the ability of leaders to unite the commitment of all stakeholders.
- Active management of changes by early identification and effective and efficient reaction to developments in the company environment.
- Success through employees: appreciation and empowerment of employees. The key to success is the balance between the organization's strategic needs and the personal expectations of each individual employee.
- Achieving sustained outstanding results in the sense of industry leading success and supporting stakeholders in achieving short and long term goals.

The **objective** of the fundamental excellence concepts is to show companies the key aspects on the way to achieving entrepreneurial excellence. The idea behind this is that companies that want to improve the performance of their organization using the *EFQM model* for excellence first discuss and accept the excellence building blocks underlying the model. If these concepts are not fully understood and carried along, it is difficult or pointless to transfer the *EFQM* model to their own organization. In this context, the holistic approach of the *EFQM model* is of central importance for success.

In this context, it is important to distinguish between the internal and external effects when using the *EFQM model.* With the **internal effects,** factors such as the involvement of employees, a common understanding of quality, the recognition of errors, the improvement of decision making processes, the initiation of change processes, the motivation for improvement measures, etc. are to be mentioned. The **external effects** are reflected in the company's market appearance, such as improvements in the areas of quality perception, customer satisfaction, image, customer loyalty, sales, market share, etc.

10.2.1.1.2 Criteria Model of the EFQM Model *2013*

The criteria model represents cause and effect relationships between the activities of a company and its achieved results. It structures the evaluation criteria according to which the *EFQM-Excellence Award* is awarded, according to dimensions and takes into account a weighting of the individual characteristic categories (cf. Fig. 10.4). The criteria are divided into two main groups, enablers and results:

Enablers describe situations that express the efforts of a supplier for high quality, and thus the design of the quality management system. Unlike certification (see Chap. 11) companies are not only judged on whether they meet certain standards, but also on how well they implement quality management. Therefore, the enablers can be used to check which measures a company successfully implements on the way to total quality management.

A basic enabler is the criterion of **leadership** (10%) in the *EFQM model*. The future of an excellent company is determined by leaders who take on a role model function by taking into account ethically justifiable actions and moral values. Through flexibility, forward looking action and timely reaction, they contribute to ensuring sustainable corporate success. For example, the *Bern University of Applied Sciences (BFH)* defines the criterion of leadership in the *EFQM model* as follows: "Our leaders live the quality awareness personally and are committed to a continuous development of the 'excellence' culture" (BFH, 2016). Further requirements for quality management in relation to the leadership of a service company can also be summarized in the following five **sub-criteria**:

1. Development of mission, vision, values and ethics for the company as well as the living of these as role models for a culture of excellence.
2. Ensuring the development, introduction, continuous improvement and leadership of the organization's management system through personal involvement.

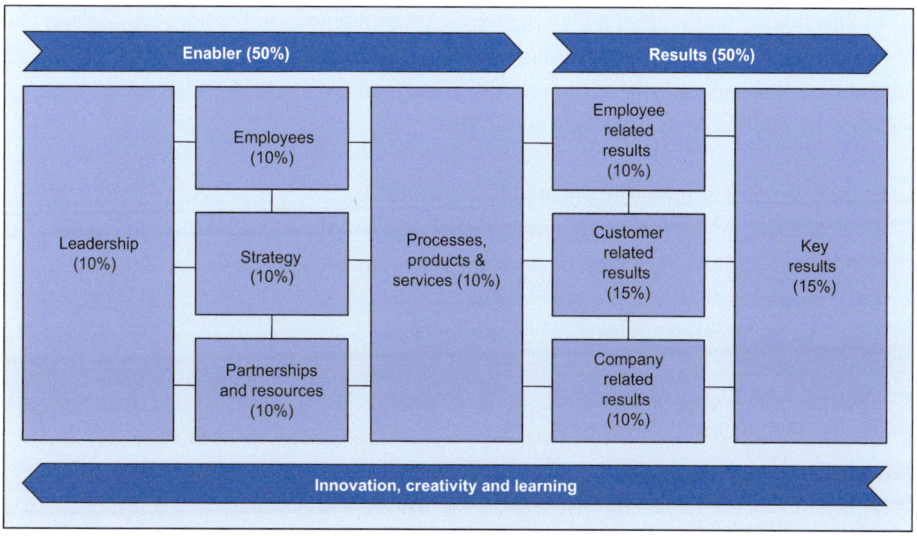

Fig. 10.4 *EFQM Model for excellence*. (Source: EFQM 2016)

3. Dealing with customers, partners and representatives of society.
4. Consolidation of the excellent corporate culture together with the employees.
5. Ensuring the flexibility of the company and effective coping with changes.

The evaluation of **strategy** (10%) deals with the implementation of the company's missions and visions as well as the development and use of methods and processes that are oriented towards the needs of the stakeholder groups. In this context, for example, the *Bern University of Applied Sciences* concretizes strategy as an adaptation of activities and development goals to its corporate identity and its overall strategic orientation (BFH, 2016). The criterion of strategy is divided into four sub-criteria:

1. Strategy based on the current and future needs and expectations of the stakeholder groups and the external environment
2. Strategy based on an understanding of the company's internal performance and capabilities
3. Development, review and update of the strategy and the methods supporting it to ensure economic, social and ecological sustainability
4. Communication of strategy and supporting methods as well as introduction of processes and objectives

In the area of **employees** (10%), it is necessary to create a culture that allows both the achievement of organizational and personal goals. The skills of the employees are developed and at the same time fairness and equality are promoted. Communication and recognition should motivate and create a bond so that employees can use their knowledge and skills in favor of the company. For example, the *Bern University of Applied Sciences* promotes, among employees, a high degree of independence, motivation, satisfaction and good training. Therefore, they are supported in "professional, personal and workplace related" matters and equality of opportunity is ensured (BFH, 2016). Five sub-criteria are used for the requirements regarding employees:

1. Employee plans support the organization's strategy
2. Identification, development and maintenance of employees' knowledge and skills
3. Bonding, involvement of employees and empowerment to act independently
4. Ongoing dialogue between employees and organization
5. Rewards, recognition and support of employees

The **partnerships and resources** criteria category (10%) describes the management, development, and deployment of external partnerships, suppliers, and resources to support strategy, methodology, and process optimization. For the achievement of objectives, high efficiency and effectiveness in the use of resources for the *Bern University of Applied Sciences* are central. In this context, partnerships are also entered into if this appears to be the best alternative (BFH, 2016). The following sub-criteria are distinguished in the area of partnerships and resources:

1. Sustainable management of external partnerships and suppliers
2. Management of finances to secure sustainable success
3. Sustainable management of buildings, facilities, equipment and natural resources
4. Management of technology to support the implementation of the strategy
5. Management of information and knowledge to support effective decision making and the development of organizational skills

In the context of a TQM-led organization, it is necessary to design, implement, and improve **processes, products, and services** (10%) in such a way that value is generated for customers and other stakeholders. For example, the *Bern University of Applied Sciences* designs its work steps to be as uncomplicated and customer and student oriented as possible. For this purpose, processes are designed systematically, continuously adapted and coordinated (BFH, 2016). The requirements for processes, products and services are divided into five sub-criteria:

1. Systematic design and control of processes for the optimization of stakeholder value
2. Development of products and services to create optimal value for the customer
3. Effective marketing of products and services
4. Production, delivery and maintenance of products and services
5. Management and intensification of customer relationships

The **results** relate to the effects of quality management. Of the sizes of the success chain of quality management (cf. Chap. 1 and 2) customer satisfaction and economic success variables are contained in the *EFQM Model.* However, customer loyalty is not taken into account, so that the *EFQM Model* implicitly assumes a direct relationship between customer satisfaction and economic success.

With regard to the results, one can first name **customer related results** (15%). Here, a distinction is made between customer perceptions (e.g. customer satisfaction indices, customer service and customer loyalty) and performance indicators (e.g. evaluation of a specific service or feature or customer integration) (EFQM, 2016). In this context, the *Berner University of Applied Science* specifies the customer related results, for example, as follows: "We use the results of surveys on student and customer satisfaction to continuously develop and enhance the attractiveness of our educational, service and research offerings" (BFH, 2016).

The criterion of **employee related results** (10%) considers direct and indirect measures that serve to determine the satisfaction of employees. Here too, a distinction is made between the effectiveness of the strategy and performance indicators perceived by employees (e.g. evaluation of individual internal services by employees or the aggregated overall satisfaction of employees). For example, the *Bern University of Applied*

Science takes into account, discusses and, if possible, implements comments and feedback from employees (BFH, 2016).

Under the heading of **society related results** (10%), the same pattern is followed as for customer and employee related results, with a consideration of the company's perceptions from the perspective of society and specific performance indicators (e.g. image values or publicly raised opinions on certain services of a service provider). The focus is on better aligning the social and ecological aspects of the strategy and taking into account the influence of society as an interest group in the company's result. For example, the *Bern University of Applied Science* receives suggestions for its activities from various "stakeholder groups from business, society, public sector and culture" (BFH, 2016).

In the context of the consideration of **business results** (15%), the focus is on the specific results of a business strategy, taking into account the importance criterion. The often aggregated data from business reports are distinguished here according to the sub-criteria of important performance related results and performance related indicators (e.g. generated value creation or also specific performance volumes such as number of successful customer contacts). For example, the *Bern University of Applied Sciences* checks whether the set goals have been achieved by evaluating the business results and also obtains indications of how the quality of products and services can be increased (BFH, 2016).

A final overview of the main and sub-criteria of the *EFQM Model* is given in Fig. 10.5.

The ***EFQM Model for excellence*** with its nine evaluation criteria, presented here, was introduced with the aim of providing companies with a structured framework for the assessment and sustainable improvement of their performance. The *EFQM Model* is regularly updated.

In the next step, a connection between the basic concepts and the criteria model is to be shown. The basic concepts of excellence (see Sect. 10.2.1.1) provide the enablers with orientation points for achieving the desired results by means of the nine criteria of the criteria model. An efficient and effective achievement of the target is ensured by a combination. Fig. 10.6 gives an overview of the relationships between the basic concepts of excellence and the sub-criteria of the *EFQM Model*. If deficiencies are identified in an individual basic concept in a company, the overview is suitable to identify which enabler criteria require corresponding measures. The target "Creating value for customers", for example, can only be pursued by sub-criteria of the enabler criterion 5 "Processes, products and services", while, for example, for the target "Creativity and innovation" sub-criteria from almost all areas can be used.

10.2.1.1.3 RADAR logic of the *EFQM Model 2013*

The **RADAR logic** is used to assess the performance of an organization and is the method by which applicants for the *EFQM Excellence Award* are evaluated. It assumes that excellent results are achieved when companies plan, implement, control and later

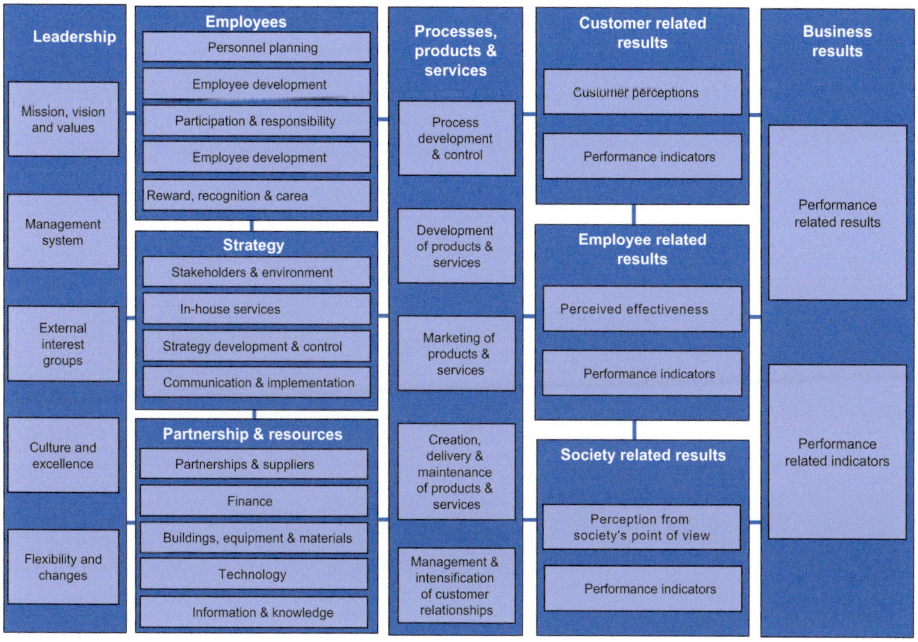

Fig. 10.5 Main and sub-criteria of the *EFQM Model*. (Source: Based on EFQM 2016)

Criteria	1. Leadership					2. Strategy				3. Employees					4. Partnerships and resources					5. Processes, products and services				
Partial criteria (see text)	a	b	c	d	e	a	b	c	d	a	b	c	d	e	a	b	c	d	e	a	b	c	d	e
Create value for customers																					X	X	X	X
Shaping the future sustainably	X		X		X			X								X	X				X			
Develop the capabilities of the organization			X				X		X				X		X									
Encourage creativity and innovation					X				X			X							X					
Lead with vision, inspiration and integrity	X		X	X	X																			
Actively manage change					X	X			X	X								X		X				
Being successful through employees											X	X	X	X										
Achieve outstanding long term results		X	X			X		X	X									X						

Fig. 10.6 Connection between the basic concepts of excellence and the criteria of the *EFQM Model*. (Source: EFQM, 2010, p. 48)

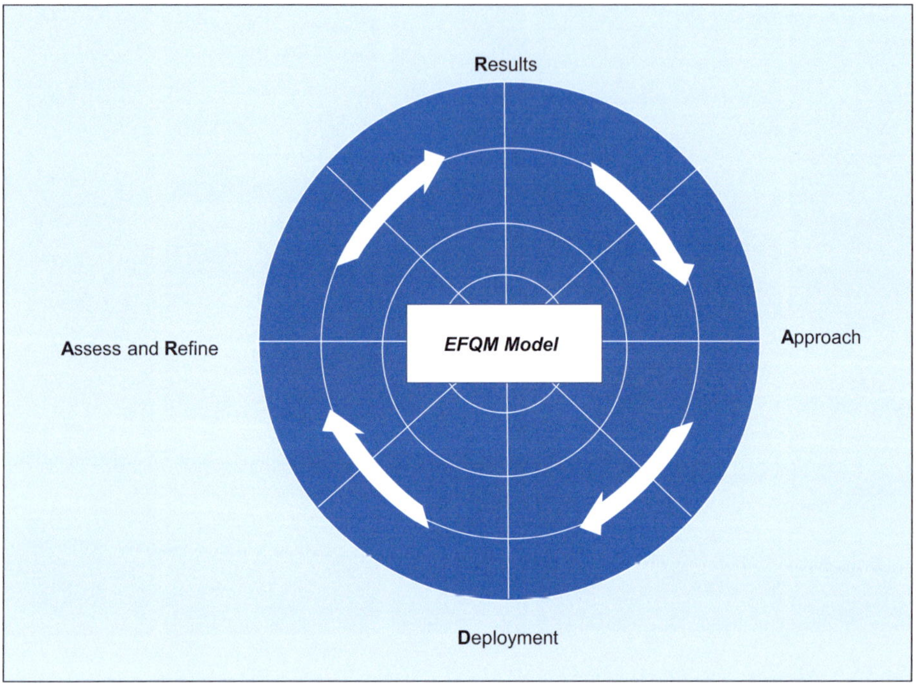

Fig. 10.7 RADAR logic of the *EFQM Excellence Award*. (Source: EFQM 2016)

improve their strategy. The term "RADAR" is made up of the initial letters of the following terms (see Fig. 10.7):

- **R**esults
- **A**pproach
- **D**eployment
- **A**ssessment
- **R**efinement

The RADAR logic generally shows that significant **improvement potential** can be achieved by improvement plans for further actions from already achieved results. If they are implemented, they are to be monitored and evaluated later. This gives companies a point of reference for the strategic implementation of their quality management system and equally enables the implementation of benchmarking and self-evaluations.

The evaluation of applicants for the *EFQM Excellence Award* is carried out within the framework of the *EFQM Model* with the RADAR logic. Here, each point is given attention individually. Thus, in an excellent organization, the results show positive trends and/or good performance over a longer period of time. The results from the criteria model are therefore evaluated, amongst other things, with regard to their **long term relevance.** Excellent organizations should also have sound procedures that integrate well-founded and well-defined processes. These procedures are also to be implemented systematically and as planned. Finally, as part of the evaluation, it is checked which measures the company implements to continuously review and improve its processes and results.

Experience report on the EFQM excellence approach in practice

The *EFQM* (European Foundation for Quality Management) as a European organization works closely with its partner organizations in the individual countries. They have the task of spreading the benefits of the *EFQM* approach to companies and organizations on the basis of the leadership and thinking model in the respective country.

Mr. *Siegfried Schmidt* is head of excellence at the *SAQ Swiss Association for Quality* and has extensive experience in working with companies. His practical experience relates to the following areas:

- Leadership—culture, management system and interest groups

The leadership tools such as vision, mission and corporate identity are to be created and exemplified by all groups in the company. Here, a "values compass" or jointly developed leadership principles can be useful. The management is to define key figures that reflect the core business and are measured continuously. Cross-industry aids are a key figure cockpit, a balanced scorecard or an internal control system. Stakeholder orientation or stakeholder management can be profitably implemented with a partnership concept. The challenge in many service industries is to incorporate political, legal and operational aspects into the strategic process.

- Employees—qualification, participation and recognition

Increasingly, the labour market also needs to be taken into account in the area of personnel policy. The shortage of skilled workers and changing job profiles are in the focus here. "Internally", every entrepreneur needs to know the required competencies, use them for the success of the company, check them and continuously develop them. Practical examples include, in addition to the classic job and position descriptions, above all training and further education concepts, but also a competence and/or skill grade matrix. A central importance is attached to the participatory leadership style, which promotes the leadership philosophy at every level. Supporting tools and methods in practice are KAIZEN, LEANI—but also transparent and systematic information in the form of communication principles.

- Partnerships and resources—successful networking, financial, supplier and knowledge management

  This criterion includes, in addition to sustainability in the form of economic, ecological and social principles (often referred to as a concept for corporate social responsibility), a regulation of the cooperation with suppliers and other market participants.

- Processes, products and services—development, design, steering and improvement of processes
  This area is about creating optimum values for customers with marketable products and services, maintaining customer relationships, but also managing marketing concepts and surveys (including complaint management) in order to ensure the ongoing success of the company. A systematic CRM (customer relationship management) can be helpful here. ◀

10.2.1.2 The EFQM Model 2020

The classic *EFQM Excellence model* has proven itself over decades and has been able to provide many companies and non-profit organizations with concrete orientations for successful management of the company and improvement of quality management. It is also still used today for the award of quality prizes. However, a certain stagnation has set in with the *EFQM Excellence model 2013,* so that, after many years of discussion, a fundamental reorientation and reconceptualization has been carried out.

This stagnation is also due to the fact that, in the meantime, numerous developments have taken place which have led the EFQM to consider a reconceptualization to be necessary. These include, for example, **megatrends** such as the management of demographic diversity, the need for self-organizing and self-regulating systems, increased demand for certain skills and effects of automation, the rise of technologies and digital disruptions, phenomena such as the sharing economy and trust sharing, resource scarcity, environmental crises, etc. (Graf, 2020, 9 pp.).

The *EFQM Model 2020* was published in November 2019 and has been in the introductory phase since then. The following section presents the central components of the new model. Here too, the following **components** can be distinguished:

- Principles of the model
- Segments of the model
- Criteria and sub-criteria of the model
- EFQM diagnosis tool RADAR

These components are explained in the following section (see in particular EFQM, 2019a) and the reference to services is established.

10.2.1.2.1 Principles of the EFQM Model 2020

The **principles of the *EFQM model 2020*** have not changed fundamentally compared to the *EFQM Excellence model 2013.* Even in the new design, the goal is still to have a generic model that can be used for different sectors and industries, company sizes, development stages, etc.. Of particular importance are the three following aspects (EFQM, 2019a):

- **Customer orientation:** The customer's priority in the company's target system.
- **Long term orientation:** The inclusion of different stakeholder groups of the company.
- **Causality:** The consideration of cause and effect relationships in all actions of the company.

These three aspects are of particular importance for service companies. If the management of services is considered as the management of customer relationships under the target of **relationship marketing,** then the customer orientation, sustainability and causality mentioned are central characteristics of the concepts of the customer life cycle and the service success chain (Bruhn, 2016).

In addition to these principles, the *EFQM model 2020* continues to be oriented towards various values of Europe. This **European value canon** is documented by the Charter of Fundamental Rights of the *European Union,* the *European Convention on Human Rights,* the *European Social Charter* and others.

In addition to these European values, the new model explicitly refers to the observance of the **goals of the *United Nations.*** This applies to the *UN Global Compact* from the year 2000, in which ten principles for sustainable and socially responsible

Fig. 10.8 Sustainable development goals of the United Nations. (Source: UN, 2020)

management were formulated (www.unglobalcompact.org) as well as to the goals for sustainable development of the Agenda 2030, known as the Sustainable Development Goals (UN, 2020). These 17 different goals of sustainability are shown in Fig. 10.8.

The principles listed here, the European values and the sustainability goals of the *United Nations* are to be understood in the sense that companies are invited—if relevant for them individually—to orient themselves to these principles, values and goals.

10.2.1.2.2 Segments of the EFQM Model 2020

The *EFQM Excellence model 2013* has made a distinction between the enablers (internal success factors) and the results (external success factors) (see Fig. 10.4), each with a relative importance to the success contribution set at 50:50.

The *EFQM Model 2020* takes a fundamentally different approach here. Based on numerous expert interviews, workshops, research results, etc., the structure of the model has a logic that is aligned with the three questions of the golden circle by Sinek (2014) (see also Fig. 10.9):

- **WHY** does the organization exist, what purpose and what strategy does it pursue?
  - This is the **alignment** in the model.
- **HOW** is the purpose and strategy implemented by the organization?
  - This is the **realization** in the model.
- **WHAT** does the organization want to achieve with the strategy?
  - These are the **results** in the model.

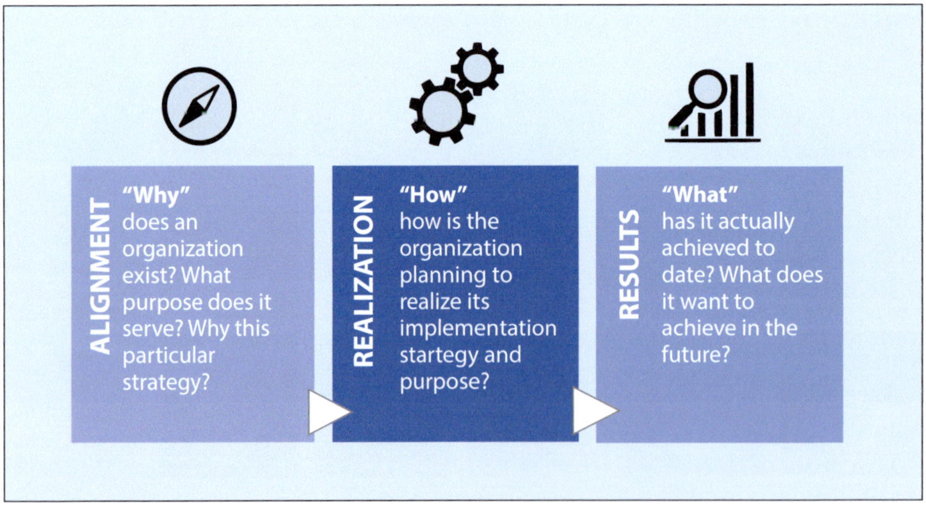

Fig. 10.9 Structure and logic of the *EFQM Model 2020*. (Source: Graf, 2020)

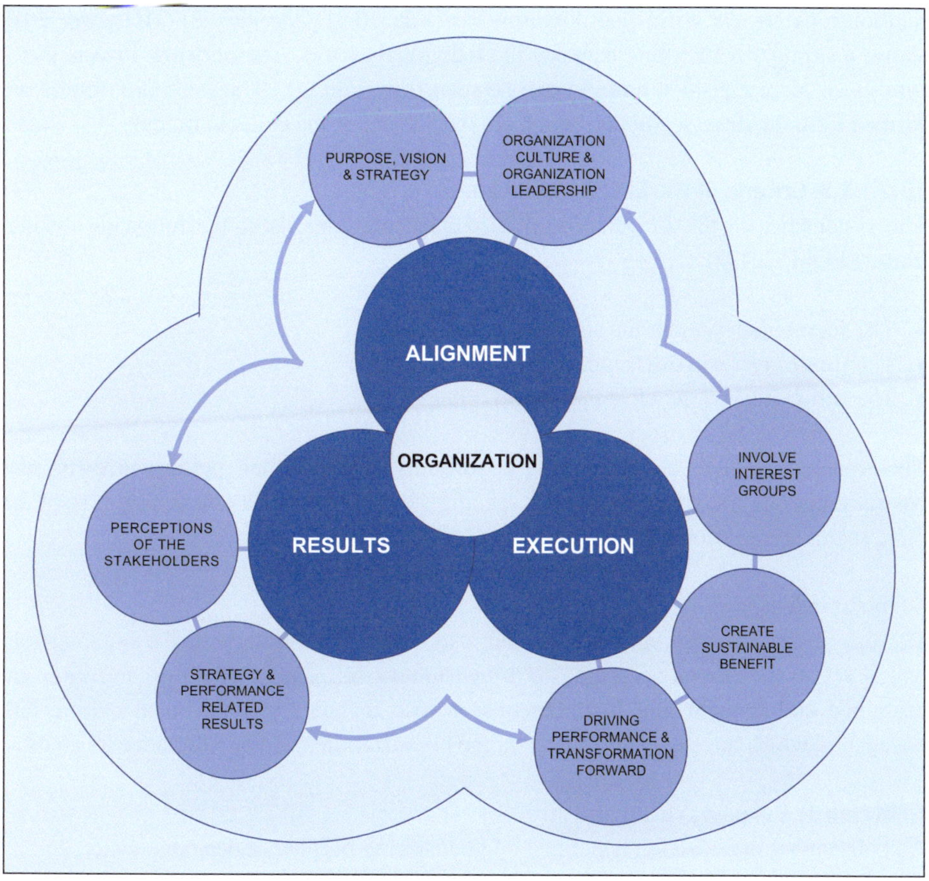

Fig. 10.10 Structure and segments of the *EFQM Model 2020*. (Source: EFQM, 2019a, p. 4)

This distinction sets new focus areas that relate to the increasing demands on corporate governance. This includes a discussion of the **purpose** of companies, which deals with the question of why organizations exist and what their social value is (see, for example, Kirchgeorg et al., 2018). It also picks up the discussion on **ecosystems** by taking into account the need to deal with different stakeholder groups and their relationships and to see the survival and sustainability of an organization in an intensified cooperation to increase the benefits of the participants (Lusch & Nambisan, 2015; Lusch et al., 2016; Nambisan, 2018; Nambisan et al., 2018; Jabobides et al., 2018; Senn & Bruhn, 2019).

The structure of the model with the three segments orientation, realization and results is shown in Fig. 10.10.

If you compare the graphical representation between the 2013 model and the 2020 model, then the 2013 model had a comparatively "deterministic" character in that the relationships always go in one direction (see Fig. 10.4). The new *EFQM model 2020* shows a stronger networking between the individual factors. The **network idea** is therefore given more consideration. As can be seen from Fig. 10.10, seven main criteria are formed from the three segments, which are the subject of the **criteria model.**

10.2.1.2.3 Criteria of the EFQM Model 2020

The systematics of the *EFQM Model 2020* therefore consists of the following distinctions (EFQM, 2019a):

- The **segments** represent the structure of the model.
- The **(main) criteria** differentiate the three segments.
- The **sub-criteria** concretize the (main) criteria.

This systematics gives companies the opportunity to check their performance for successful corporate management in very specific steps. The criteria and sub-criteria of the three segments are shown below (cf. EFQM, 2019a).

Segment orientation

The orientation segment specifies the purpose, vision and strategy of the organization, which act as guidelines according to which almost all aspects of an organization are examined and brought into line. Orientation also includes organizational culture and leadership, which have a major impact on the organization's ability to achieve its goals.

Criterion 1: Purpose, vision and strategy

An outstanding organisation is defined by an inspiring purpose, a desirable vision and an effective strategy. The following five **sub-criteria** are defined:

- Define purpose and vision
- Identify and understand the needs of stakeholders
- Understand ecosystem, own capabilities and key challenges
- Develop strategy
- Develop and implement governance structures and control systems for the organisation's performance

The purpose, vision, strategy and governance structures are present in most service organisations. In recent years, this has also led to the formation of numerous **service ecosystems** (Bruhn et al., 2019, 60 pp.), in which different service providers combine their special skills and competencies to generate additional value for customers (e.g. in the sharing economy).

Criterion 2: Organisational culture and leadership

Organisational culture refers to the values and behavioural norms shared by employees and groups within an organisation that over time shape both their behaviour towards each other and towards the stakeholders outside the organisation who are important for the purpose, vision and strategy. Leadership does not only refer to a person or group that sets the direction from above, but to the organisation as a whole. The criterion is divided into four **sub-criteria:**

- Lead and promote organisational culture
- Shape the conditions for successful change
- Enable creativity and innovation
- Stand together and engage for the purpose, vision and strategy of the organisation

The corporate culture plays a special role for service companies. Due to the need for interaction with customers in the creation of the service, it is essential that customer contact personnel are able to be customer oriented and build a relationship with the customer. A **culture of customer orientation** is therefore a defining feature of service companies.

Segment realization

Within the segment realization, it is checked whether the stakeholder groups and their needs are known and taken into account, the organization generates sustainable benefits and implements improvement processes with regard to the provision of services. These improvements not only relate to the optimization of business processes in order to increase efficiency or productivity, but also to necessary adjustments and changes due to internal and external changes that occur continuously. The following three **criteria** are in the focus (cf. Fig. 10.10).

Criterion 3: Involve stakeholder groups

Although each company has its very specific stakeholder groups and decides on their priority, it is very likely that there are many parallels between the organizations when it comes to application. Accordingly, the following five **sub-criteria** have been formulated in the *EFQM Model 2020* which are to be concretized for service companies:

- Customers—build sustainable relationships
- Employees—win, involve, develop and retain
- Economic and regulatory stakeholder groups—ensure continuous support
- Society—contribute to development, well-being and prosperity
- Partners and suppliers—build relationships and ensure contributions to creating sustainable benefits

The list of interest or stakeholder groups makes it clear that the aspects of **relationship management** play a particularly important role for service companies. It is important to note that the relationship with the customer (relationship marketing) and the relationship with the employees (internal marketing) are a necessary prerequisite for providing good quality services (Bruhn, 2016).

Criterion 4: Creating sustainable value

An outstanding organization understands that creating sustainable value for its long-term success and financial strength is of great importance. For this criterion, four **sub-criteria** have been developed:

- Planning and developing sustainable value
- Communicating and marketing sustainable value
- Delivering sustainable value
- Defining and realizing an overall experience

For service companies, two aspects are particularly emphasized in this criterion with the sub-criteria. This concerns the value that is to be generated for customers and the company. Various value or **value categories** can be distinguished here, such as value in exchange, value in co-creation, value in use or value in context. The possibilities of value creation are particularly diverse for service companies in cooperation with customers. However, the aspect of **experience management** is addressed by the sub-criteria. Services are based on experiences and experiences, which is why experiences are particularly important for the quality of the services (Bruhn et al., 2019, 354 pp.).

Criterion 5: Drive performance and transformation

The conditions are constantly changing and very dynamic for companies. Therefore, the requirements for maintaining and developing performance as well as shaping necessary transformation processes are of particular importance. This requires managing the day to day business while at the same time always recognizing, evaluating and, if necessary, making use of changes that occur in order to remain competitive in the future. Here, five **sub-criteria** are specified:

- Drive performance and manage risks
- Transform the organization for the future
- Promote innovation and use technology
- Effectively use data, information and knowledge
- Manage assets and resources

These sub-criteria primarily address transformation processes that take place in various areas of the service sector and have a significant impact on the quality of services. This affects various aspects of **service transformation** (Bruhn & Hadwich, 2016), such as digitalization, globalization and internationalization. The processes of digitalization in particular have shown that new possibilities for value creation have been generated for services and also new forms of quality assessment have arisen for digital services.

Segment results

Within the segment results, it is checked on the basis of data to what extent the requirements set out in the segments orientation and realization have been met or whether the organization is making promises in this regard. The results are divided into two **criteria:** Perception of stakeholders as well as strategy and performance related results (see Fig. 10.10).

Criterion 6: Stakeholder perceptions

This criterion focuses on those results that are based on feedback from important stakeholders and thus describe their personal perception of the organization. This includes the **perceptions of different groups of stakeholders:**

- Customer perceptions
- Employee perceptions
- Perceptions of economic and regulatory stakeholders
- Perceptions of society
- Perceptions of partners and suppliers

This criterion is closely related to criterion 3, i.e. the involvement of stakeholders. In service companies, the service success chain is a central basic principle for the management of companies and the orientation of quality management (Bruhn, 2016, 71 pp.). The **perception of the performance by the customer** and the other market participants plays an important role, because they are ultimately the drivers of the success of service companies.

Criterion 7: Strategy and performance related results

Results in connection with the performance of the organization focus on the ability of an organization to fulfill its purpose, implement its strategy and create sustainable benefits and be fit for a successful future. These strategic and operational **performance indicators** relate to different areas:

- Indicators related to sustainable benefits in terms of the purpose of the organization
- Indicators for financial results
- Indicators for meeting the expectations of important stakeholders
- Indicators related to the achievement of strategic goals
- Indicators for measuring and improving performance
- Indicators for the progress of transformation
- Indicators for predicting the future

What is particularly noteworthy about these sub-criteria is the diversity of indicators. Here too, the *EFQM Model 2020* has taken significant steps forward compared to the previous model and has not only based financial and other quantifiable figures, but also defined progress as performance related results in the areas of transformation, future orientation, etc. The ability to **further develop the company** plays a special role here and is of outstanding importance in view of the rapid changes in the framework conditions. The list of indicators mentioned here is not exhaustive; in individual cases, other indicators tailored to the situation of a service company can be used.

The **presentation of the *EFQM Model 2020*** (see Fig. 10.11) shows the segments as circles, with the criteria connected to each other by lines and arrows. This shows their mutual influence. This way, the dynamics of the relationships between the segments and criteria in an organization within a constantly changing environment can be reproduced. This is often referred to as an ecosystem. This represents a change from the model of 2013, which gave a more static impression optically.

The relationships between the segments, criteria and sub-criteria are again clearly shown in Fig. 10.11. In particular, the **mutual relationships** between the three segments and their interrelationships are shown in more detail.

In the *EFQM Model 2020*, the **emphasis on the ecosystem approach** is of particular importance, which in the previous model was only marginally effective in the criterion of partnerships and resources. The changes in business models in the service sector caused by digitalization and globalization show the possibilities of cooperation despite existing competition relationships. This aspect in the tension between service cooperation and service coopetition (Bruhn & Hadwich, 2019) is becoming increasingly important for

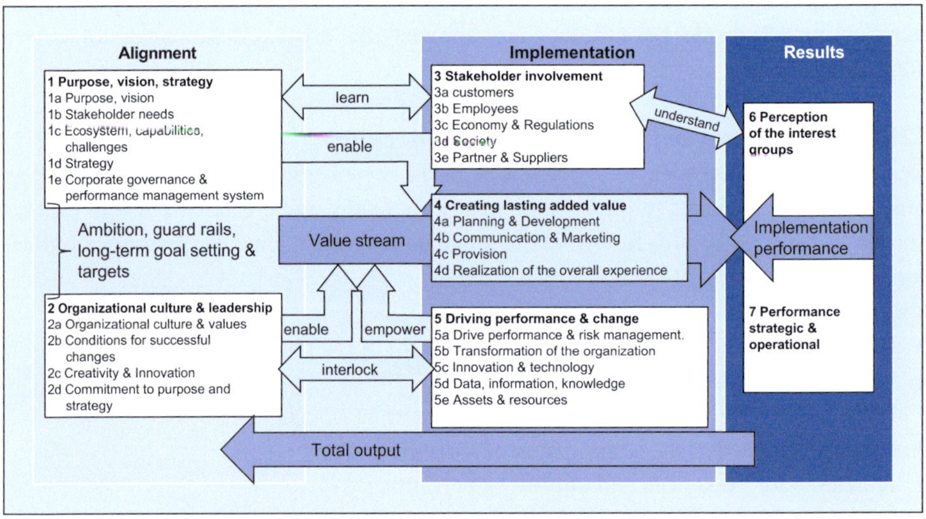

Fig. 10.11 Relationships between the criteria of the *EFQM Model 2020*. (Source: ILEP, 2019, p. 11)

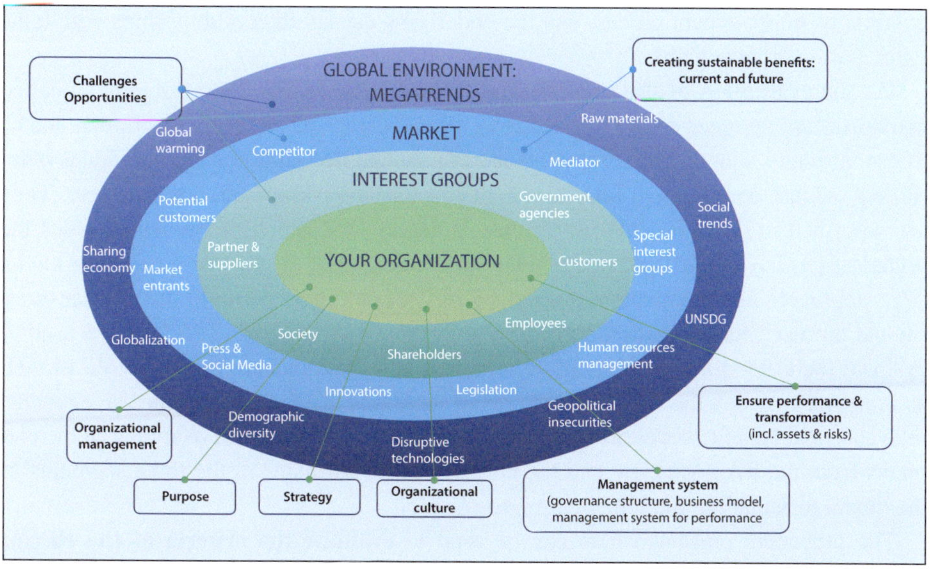

Fig. 10.12 The emphasis on the eco-system in the *EFQM Model 2020* (Source: Graf, 2020)

service offerings and ensuring service quality. The *EFQM Model 2020* emphasizes the importance of an active role in these ecosystems in order to learn and grow together with others (EFQM, 2019a, p. 9). Figure 10.12 illustrates this relationship.

10.2.1.2.4 Function of the EFQM diagnostic tool RADAR in the EFQM Model 2020

As already explained in the presentation of the *EFQM Excellence model 2013* (see Section 10.2.1.2.3), the so-called RADAR logic is used to assess the performance of a company based on the underlying segments as well as the criteria and sub-criteria. According to the acronym **RADAR** (Results, Approach, Deployment, Assessment and Refinement), this is done in different process steps. This guide is to be interpreted as a proposal to determine the status of the company and to identify potential for improvement. This can be done in the form of a self-assessment or when applying for a quality award.

For example, when applying for a quality award, the company has to document the continuous improvement with regard to planning and structured implementation. This enables external assessors to understand to what extent measures for the continuous review and improvement of processes and results have been specifically implemented.

Not only the achievement of targets, but also the way in which they are approached are taken into account in the evaluation. With the demand for intensive dealing and documentation of this aspect, the *EFQM* would like to sharpen the understanding of the continuous improvement process and the underlying cause-effect relationships within the participating organizations.

For the **evaluation of the sub-criteria** of the *EFQM Model 2020* different so-called **attributes** are assigned to the five RADAR process steps or elements mentioned above. These attributes show which properties are considered desirable for the individual criteria, e.g. sound, coordinated, introduced, flexible, analyse, improve, compare, etc. They are described in detail in order to provide indications of which properties or states the organisations have to achieve in this area. In the context of the *EFQM Excellence model 2013* approach, implementation, evaluation and improvement relate to the enabler criteria and are used for evaluation accordingly. In the *EFQM Model 2020* they are used to evaluate the criteria in the segments orientation and implementation. In the old as well as in the new model, the criteria of the result measures are evaluated using the elements relevance and benefit as well as performance. Fig. 10.13 provides an overview of the elements from the RADAR logic and the corresponding attributes of the three segments of the model (orientation, implementation, results).

The procedure presented here can be used to **evaluate the criteria of the *EFQM Model 2020*.** This can be done by *EFQM* as part of an assessment or in the form of a self-assessment. This results in a numerical individual evaluation in the form of points for each of the sub-criteria. Finally, they are consolidated in the **RADAR evaluation matrix** to give an overall result between 0 and 1000 points.

For the **distribution of points** over the seven criteria of the model, the *EFQM* has proposed a numerical evaluation system, which can be seen in Fig. 10.14. This gives up the previous system between enabler (500 points) and results (500 points) and makes a new distribution over the three segments:

- Alignment: 200 points
- Realization: 400 points
- Results: 400 points

In order to receive external recognition of the achievements (e.g. in connection with the application for a quality award), an evaluation by an official assessor team is required. But even through a self-evaluation using the presented evaluation tables, organizations can gain valuable insights into the current development level of their maturity level (EFQM, 2019a, p. 32 f.). Organizations with little experience in dealing with the *EFQM Model* usually achieve results in the range of 200 to 300 points at first. Winners of the *EFQM Award* on the other hand achieve between 700 and 850 points (Herrmann & Fritz, 2016, 28 pp.).

These seven criteria are not independent of each other. Fig. 10.15 shows that they are interrelated. The model tries to look at the different normative, strategic and operational elements in a logical sequence.

The **procedure with the *EFQM*-Diagnostic tool RADAR** is in principle a very simplified system, as is also the case with other systems in various, long-established

Alignment analysis		
Element	**Attribute**	**Description**
Procedure	Sound	The approach is clearly justified and aims to meet the needs of stakeholders important to the purpose, vision and strategy. It is appropriately described and designed for the future.
	Tuned	Attention: Does not apply to alignment.
Implementation	Introduced	The procedure is implemented in the relevant areas in an appropriate time and effective manner.
	Flexibel	Attention: Does not apply in the case of alignment.
Evaluation & improvement	Analysis	Feedback on the efficiency and effectiveness of the approach and implementation is obtained, understood and shared.
	Learn & improve	Insights from trend analysis, measurement, learning and benchmarking are used to stimulate creativity and develop innovative solutions to improve performance in appropriate timeframes.
Realization analysis		
Element	**Attribute**	**Description**
Procedure	Sound	The approach is clearly justified and aims to meet the needs of stakeholders important to the purpose, vision and strategy. It is appropriately described and designed for the future.
	Tuned	The approach supports the alignment of the organization and is linked and aligned with other relevant approaches.
Implementation	Introduced	The approach is implemented in the relevant areas in an appropriate timeframe and effective manner.
	Flexibel	The manner of implementation allows for flexibility and adaptation.
Evaluation & Improvement	Analysis	Feedback on efficiency and effectiveness of approach and improvement of implementation is obtained, understood and shared.
	Learn & Improve	Insights from trend analysis, measurement, learning and benchmarking are used to stimulate creativity and develop innovative solutions to improve performance in appropriate timeframes.
Analysis of the results		
Element	**Attribute**	**Description**
Relevance & Benefit	Scope & Relevance	A set of outcomes that are clearly linked to the organization's purpose, vision, and strategy is identified. It is reviewed and adjusted over time.
	Usable data	Results are collected in a timely manner, are meaningful, data accurate and appropriately segmented. They provide meaningful insights and learnings into performance improvement and transformation.
Power	Trends	There are positive trends or sustained outstanding performance over a strategic cycle.
	Targets	Appropriate targets in line with the strategy are set and consistently achieved.
	Comparisons	Relevant external comparisons are made in order to assess the company's own performance in terms of strategic direction. These turn out to be favorable.
	Focus on the future	Based on current cause-and-effect relationships, as well as analysis of data, performance patterns, and predictions, the organization understands the drivers for outstanding performance in the future.

Fig. 10.13 Overview of the application of the RADAR logic for the segments orientation, implementation and results in the *EFQM Model 2020*. (Source: EFQM, 2019a, 29 pp.)

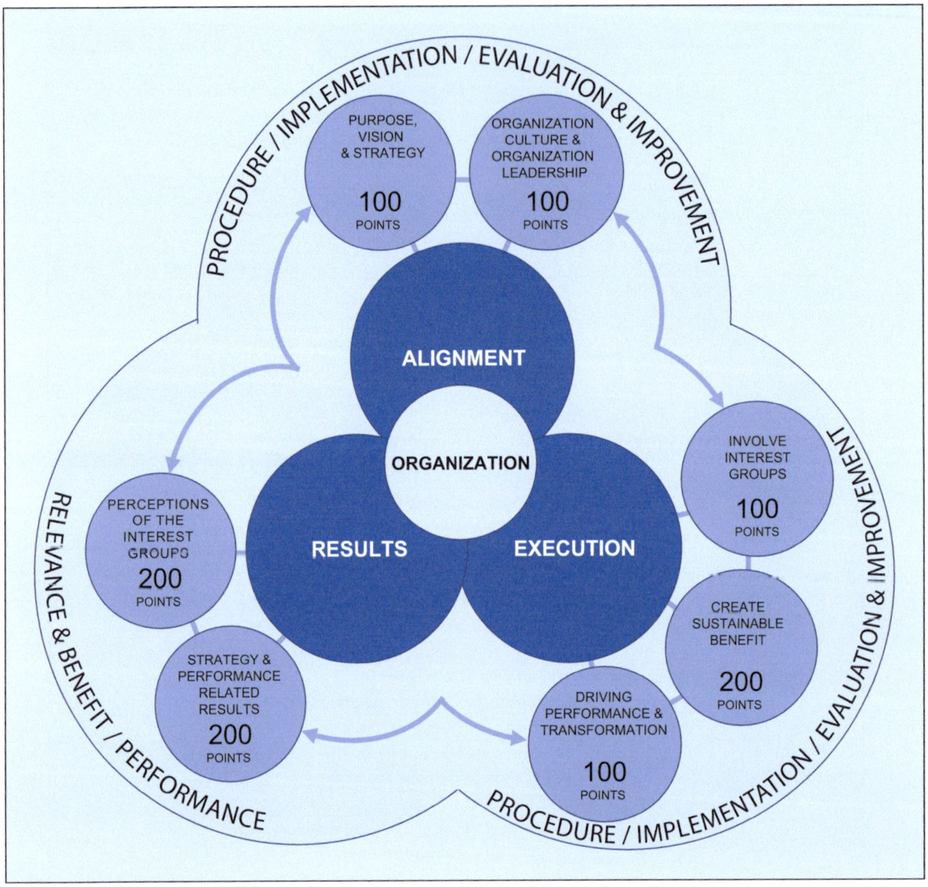

Fig. 10.14 Evaluation of criteria and point distribution according to the *EFQM Model 2020*. (Source: EFQM, 2019a, p. 31)

concepts of Total Quality Management. Here, for example, the process management, the Kaizen concept, the continuous improvement process, the Six Sigma concept, the DMADV method, etc. are mentioned (see, for example, the presentation and discussion of these concepts in Frehr 1994,1999; George & Weimerskirch 1998; Imai 2002; Oakland 2004; Oakland & Porter, 2004; Zink, 2004; Magnusson et al., 2004; Atul et al., 2005; Dahm & Heindl, 2006, 2007; Kosta & Kosta, 2008; Zollondz, 2011; Gimenez-Espin et al., 2013; Schmitt & Pfeiffer, 2015; Horatzek, 2018; Huth, 2018; see also Chap. 3 in this book).

So it is basically a classical method that is propagated as part of quality management. None of the approaches has a theoretical foundation. They are ultimately based on the **decision-oriented management approach,** in which different phases are systematically thought through, although often in a shortened form in quality concepts.

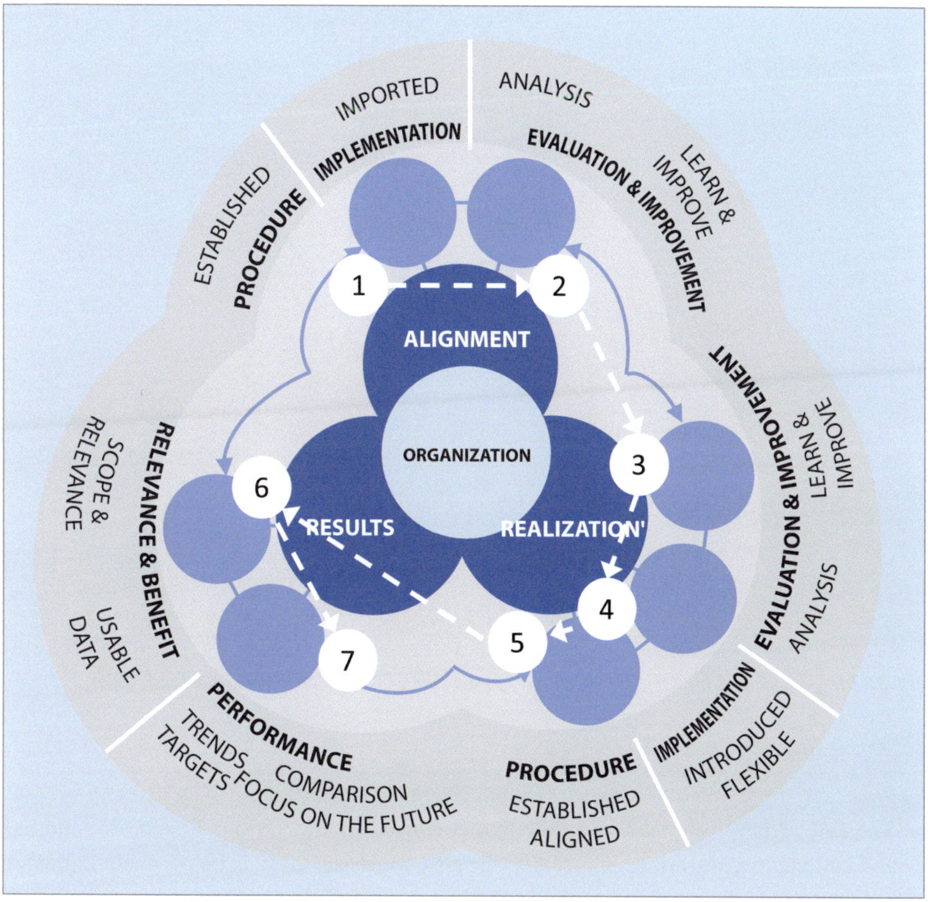

Fig. 10.15 The relationships between the criteria in the *EFQM Model.* (Source: Graf, 2020)

In a **comparison** between the *EFQM Excellence Model 2013* and the *EFQM Model 2020* shows Fig. 10.16, that the new model has been extended. The inclusion of stakeholders, the creation of sustainable benefits and the need to adapt to the various changes through transformation processes are essential elements of the new model.

The *EFQM Model 2020* is relatively new and therefore there is no extensive experience in practice yet. With the model, interesting **further developments** have been made. This includes, for example, the consideration of factors such as ecosystems, sustainability, transformation processes, perception aspects and much more. This makes the model more responsive to the dynamics of the environment and the associated and necessary change processes. It remains to be seen how the model will also establish itself in practice for service companies. Current information about the model is available from the headquarters of the *EFQM* in Brussels (www.efqm.org).

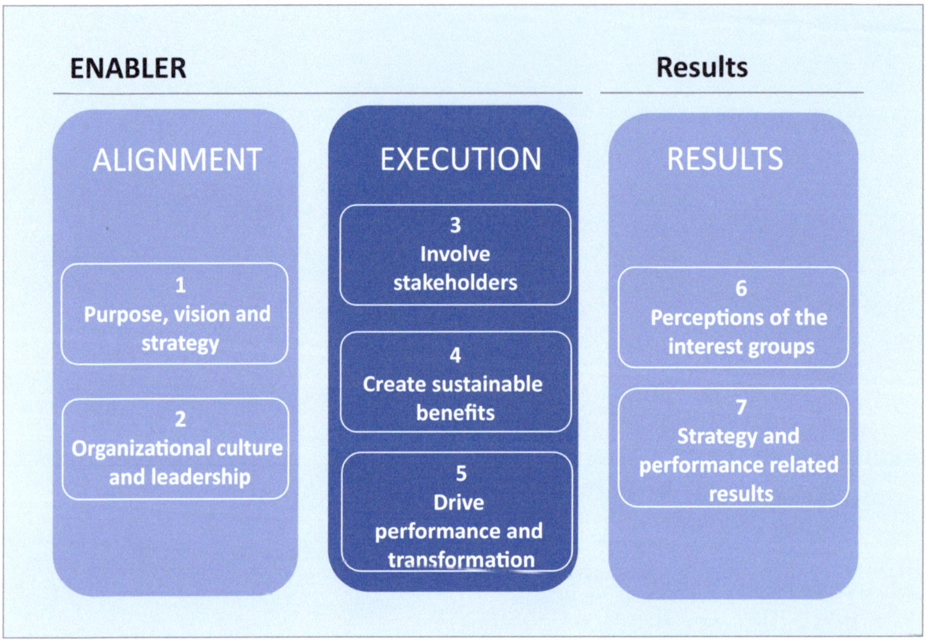

Fig. 10.16 The *EFQM Model 2020* in relation to the *EFQM Model 2013* (Source: Graf, 2020)

Report on the training of assessors for the EFQM Model 2020

The *EFQM* is a European initiative that is carried by the various European countries and national organizations. Currently, the *EFQM Model 2020* is being made known to the interested organizations in the individual countries. *Hugo Graf* is a lecturer at the Lucerne University of Applied Sciences, Business Excellence/*EFQM* mandate leader and *EFQM* certified trainer. He has taken over the task of creating the conditions for the dissemination of the new model in Switzerland as part of the assessor training. His previous experience in assessor training with the new *EFQM Model 2020* is outlined here.

The first transfer courses for assessors who were trained according to the *EFQM Excellence model 2013* and have gained practical experience as *EFQM* assessors have been carried out. Feedback shows that the new *EFQM Model 2020* is rated as a very good revision of the model 2013. Positive aspects are the professional approach to the model renewal or update by the *EFQM*. After an extensive research phase using various quantitative and qualitative methods, a systematic processing of the results followed in different expert groups. The development of the model was only completed after a pilot of the new *EFQM Model 2020* was validated in practice.

The *EFQM Model 2020* is seen by the course participants as a more comprehensive and future oriented management model that is more dynamic, logical and consistent. The new model has been streamlined. So it has fewer criteria, fewer sub-criteria and also fewer orientation points than its predecessor model 2013 and in addition the eight basic concepts have been omitted.

Even if one could suspect this "simplification" of the new model 2020, the activity as an *EFQM* assessor will certainly not be less demanding. For example, it will be central challenges to think in new business models, to identify and assess the relevant megatrends and ecosystems as well as to assess the key competence, how organizations manage fundamental changes and the optimization of the operational business at the same time.

The training of future *EFQM* assessors and the participation in *EFQM* practice assessments is a great pleasure. It can be assumed with certainty that the new *EFQM Model 2020* and the *EFQM Diagnostic tool RADAR* will be of good service to organizations in their positioning and their constant, systematic development if they are applied consistently. ◀

10.2.2 The European EFQM Global Award

Quality prizes and **quality awards** are awarded by specific institutions as evidence of the promotion and successful implementation of quality in organizations. They serve as an incentive for the implementation and sustainable improvement of a quality management at the organizational level. The motivational effect they have on the organization contributes to raising the level of performance and competitiveness. In addition to the globally advertised ***EFQM Global Award,*** the *EFQM Model* still serves as a basis for evaluation for a number of other national quality prizes, such as the German *Ludwig-Erhard-Preis,* the Swiss *ESPRIX Swiss Award for Excellence* and the Austrian *State Prize for Quality in Business.*

With the new 2020 model, the prize will be called the *"EFQM Global Award."* There is no experience yet with the new quality prize, so the procedure for awarding the EFQM excellence award will be described below.

For companies, the following two **requirements** must be met in order to apply for the *EFQM Excellence award* (EFQM 2013).

- Applications from former winners are excluded for the three years following the date of the prize, unless the application relates to a different business unit or brand. If this is the case, the entire company is nevertheless excluded from further applications.
- The maximum number of applications for the *EFQM Excellence award* from the same organization or company is four. This means that up to two business units or brands can apply for each trophy (two for large and two for small business units).

The **assessment** of an application is carried out by an examination team (so-called assessors) of experts from science and practice, from which typical assessment teams of four to six people are formed. In content, the RADAR logic is applied to each of the nine assessment categories of the *EFQM Model* and each of the eight basic concepts of excellence (cf. Fig. 10.3) in the examination. After the completion of this examination, the examination teams make recommendations to the jury. In the end, the jury decides on the choice of finalists and awards the finalist status. Among the finalists, the companies that are considered best by the jury are awarded the status "prize winner". If a company stands out particularly among the prize winners, the highest award status, **"award winner"**, is awarded.

The evaluation of the applicants is based on two dimensions: one is based on the evaluation profile along the nine assessment categories, the other on the question of the type and strength of the role model character of an applicant for other companies within the *EFQM* network. If the basic concepts and the assessment criteria of the *EFQM Model* are integrated into the examination contents, the RADAR logic (cf. Sect. 10.2.1.1.3) is used to evaluate the performance of the organizations. Each of the applicants receives a written examination report after the evaluation process, from which possible quality improvement potentials can be derived. An application for the *EEA* is associated with costs of 8,000 to 15,000 EUR (EFQM, 2013, p. 10).

The prize system of the ***EFQM Excellence awards*** was simplified for the prize round of 2007. Since 2007, the *EFQM* has only awarded a total of four quality prizes for outstanding quality management to organizations. Each applicant is a candidate for one of the four quality prizes. Two awards are reserved for each of private companies and non-profit organizations. Within a category, one award is given to small and medium-sized organizations, the other to large organizations.

Fig. 10.17 provides an overview of the winners of the highest award level of the ***EFQM Excellence award*** in recent years.

The British courier company *TNT Express United Kingdom Ltd.* was the first service company to win the *European Quality Award* in 1998. As part of the introduction of a quality management system, *TNT Express United Kingdom Ltd.* has set itself the goal of winning the *EFQM* excellence award. In doing so, the company consistently orientated itself towards all *EFQM* criteria. For example, in the area of "customer satisfaction", it was analysed which aspects are most important to the customer when choosing an express service. The results of these analyses provide *TNT* with indications of where the company's quality management has to be set from the customer's point of view. It turned out that the aspects mentioned most frequently by far are the punctuality of the delivery, the condition of the delivery and the reliability of the pick-up.

In addition to *TNT Express United Kingdom Ltd.,* service companies such as the information and advertising service provider *YELL* (England), the automobile manufacturer *BMW,* the *Alpenresort Schwarz* (Austria) or the *St. Mary's College Northern Ireland* (Great Britain) are now also winners of the Award.

The former *EFQM Excellence award* has gained wide recognition for decades. In addition, the *EFQM* organizations offer recognition at four different levels (see Fig. 10.18). Depending on the maturity of the previous quality efforts and the willingness

2018	Country of origin	Award
Infineon Technologies Austria AG	Austria	Award Winner
Dubai Police	United Arab Emirates	Price Winner
Sakarya University	Turkey	Price Winner
VAMED-KMB	Austria	Price Winner
Tarsus Municipality	Turkey	Price Winner
2017	**Country of origin**	**Award**
Glasgow Housing Association	United Kingdom	Award Winner
Robert Bosch GmbH, Blaichach Plant	Germany	Award Winner
Dubai Electricity and Water Authority - DEWA	United Arab Emirates	Award Winner
Robert Bosch, Rodez Plant	France	Price Winner
Bosch Car Multimedia Portugal S.A.	Portugal	Price Winner
Sanitas Hospitales	Spain	Price Winner
Geriatric Health Centres of the City of Graz	Austria	Price Winner
Hospital Universitario Infanta Elena	Spain	Price Winner
Tarsus Belediyesi (Municipality)	Turkey	Price Winner
2016	**Country of origin**	**Award**
Ayuntamieto de Alcobendas	Spain	Award Winner
Robert Bosch GmbH, Blaichach Plant	Germany	Price Winner
Bradstow School	England	Price Winner
Aéroports de Lyon	France	Price Winner
City of Glasgow College	United Kingdom	Price Winner
GC Europe N.V.	Belgium	Price Winner
General Directorate of IETT	Turkey	Price Winner
Glasgow Housing Association	United Kingdom	Price Winner
Stavropol State Agrarian University	Russia	Price Winner
2015	**Country of origin**	**Award**
BMW AG, Plant Regensburg	Germany	Award Winner
wdh	United Kingdom	Award Winner
Bosch Car Multimedia	Portugal	Price Winner
Cabinet d'Orthodontie Becker & Associés	Luxemburg	Price Winner
Coca Cola Icecek A.S., Bursa Plant	Turkey	Price Winner
Robert Bosch GmbH, Blaichach Plant	Germany	Price Winner
Robert Bosch GmbH, Rodez Plant	France	Price Winner
Sakarya University	Turkey	Price Winner
Sanítas Hospitales	Spain	Price Winner
Vamed-KMB	Austria	Price Winner
2014	**Country of origin**	**Award**
Bosch Bari Plant	Italy	Award & Price Winner
BMW Plant Regensburg	Germany	Price Winner
One Vision Housing	United Kingdom	Price Winner
Siemens Motion Control, Congleton	United Kingdom	Price Winner
The Cedar Foundation	United Kingdom	Price Winner
2013	**Country of origin**	**Award**
Alpenresort Schwarz	Austria	Award & Price Winner
BMW Plant Regensburg	Germany	Price Winner
Glasgow Housing Association	United Kingdom	Price Winner
Nilufer Municipality	Turkey	Price Winner
St Mary's College Northern Ireland	United Kingdom	Price Winner
Stavropol State Agrarian University	Russia	Price Winner
VAMED-KMB	Austria	Price Winner
Wakefield and Distict Housing	United Kingdom	Price Winner
2012	**Country of origin**	**Award**
Robert Bosch GmbH, Bamberg Plant (BaP)	Germany	Award & Price Winner
BMW AG, Plant Regensburg	Germany	Price Winner
Bosch Technologie, Diesel e Sistemi Frenanti S.p.A.	Italy	Price Winner
Coca Cola Icecek A.S., Ankara Plant	Turkey	Price Winner
JSC Medicina	Russia	Price Winner
pom+Consulting	Germany	Price Winner
Vamed-KMB	Austria	Price Winner
2011	**Country of origin**	**Award**
Bilim Pharmaceuticals	Turkey	Award Winner
Robert Bosch GmbH, Bamberg Plant (BaP)	Germany	Price Winner
Ricoh Deutschland GmbH	Germany	Price Winner
Siemens Healthcare UK	United Kingdom	Price Winner
2010	**Country of origin**	**Award**
VAMED-KMB	Austria	Award Winner
Olabide Ikastola	Spain	Price Winner
Bradstow School	United Kingdom	Price Winner
Stravropol State Agrarian University	Russia	Price Winner

Fig. 10.17 Winners of the *EFQM Excellence Award*. (Source: EFQM, 2018a, b)

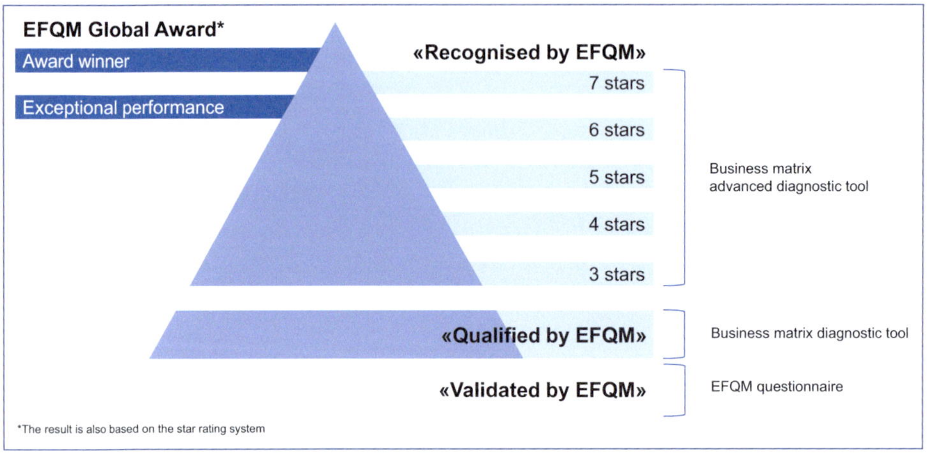

Fig. 10.18 Levels of recognition by *EFQM*. (Source: EFQM Recognition-EFQM, 2020)

to invest in personnel and financial resources, organizations can achieve the following ***EFQM* Recognition levels**:

- Validated by *EFQM* (formerly *EFQM* committed to excellence (C2E) validation),
- Qualified by *EFQM* (formerly *EFQM* committed to excellence (C2E) assessment),
- Recognized by *EFQM* (formerly *EFQM* recognised for excellence (R4E) assessment),
- *EFQM Global award* (formerly *EFQM Excellence award*).

"Validated by EFQM"

In order to validate, organizations that want to initiate improvement processes carry out a self-assessment based on the *EFQM Model 2020,* define improvement projects and implement them. The results of these efforts are summarized in application documents, based on which an external expert, called a validator, plans a one day visit to the organization. During the visit, the organization's competitiveness is evaluated using the RADAR logic. In a subsequent feedback report, applicants receive recommendations for the next steps on their way to becoming an outstanding organization. In addition, the recognition "Validated by *EFQM*" is pronounced and published via the *EFQM* channels.

"Qualified by EFQM"

At this level of recognition by *EFQM,* it is possible to have a comprehensive analysis of the most important management practices of an organization carried out by external assessors of *EFQM* with relatively little time investment. Using the seven criteria of the *EFQM Model 2020,* applicants prepare an online self-assessment. This self-assessment serves directly as an application document. Subsequently, the one to one and a half day visit of two specially trained assessors is planned, prepared and carried out. The resulting report then provides an external view of the organization based on the *EFQM* criteria, information on the maturity level of the organization and extensive feedback to meet their specific challenges and increase value creation. The recognition "Qualified by *EFQM*" is awarded and also published by *EFQM*.

"Recognized by EFQM"

The most comprehensive assessment process of *EFQM* recognition is based on the seven criteria and the associated sub-criteria of the *EFQM Model 2020*. It includes a complete review of all relevant aspects of an organization and, in addition to extensive application documents, depending on the complexity of the organization, requires a two to four day visit on site by a multi-member assessor team. Accordingly, this level of recognition is reserved exclusively for organizations that already have solid experience in quality management and in the application of the *EFQM Model* and want to achieve further progress in this area with the help of a detailed, comprehensive feedback. The process is largely similar to the process for qualification, but is characterized by a much higher level of detail. The goal is the recognition "Recognized by *EFQM*", which is expressed with three to seven stars depending on the score achieved. In addition, organizations receive a detailed report on strategic challenges and goals, a complete assessment profile including strengths and improvement opportunities at the level of the seven criteria of the *EFQM Model 2020* as well as food for thought for the next steps towards an "excellent" organization.

"EFQM Global Award"

The fourth and highest level of recognition is the *EFQM Global award*. It is awarded worldwide to outstanding private, public and non-profit organizations. However, the global award is reserved for organizations that can demonstrate an undisputed track record in the implementation of strategies into practice with continuous improvement of their performance. As a result, award candidates are thoroughly examined and evaluated; this is done by an international team of assessors consisting of three to seven *EFQM* trained and experienced experts. During a week on site, they interview individuals and groups and examine documents to assess the organization's effectiveness in implementing its strategic initiatives and achieving its strategic goals. The result is a detailed feedback report that is submitted to a independent jury for final decision.

Winner of the EFQM Global excellence award

As winner of the *EFQM Global excellence award* 2019, *GC Europe N.V.* from Belgium qualified. It is a subsidiary of *GC Corporation*, a Japanese organization with worldwide activities dedicated to the manufacture and sale of materials and equipment for oral health and dental care. *GC Europe N.V.* employs 430 employees.

On the second place, with seven stars, *PostMail*, a division of the *Swiss Post AG*, was able to position itself. The organization with 15,000 employees is part of a corporation that is owned by the Swiss Confederation. Every day it transports around 16 million shipments within Switzerland and abroad for private and business customers. In addition, it is constantly expanding its range of digital services. Four years earlier, it had already won the Swiss quality award—the *ESPRIX Swiss Award for Excellence.*

Some other companies such as *BBT—BMW Brilliance Shenyang* from China and *Bosch Security Systems, S.A.* from Portugal were also awarded seven stars for outstanding performance in various categories. In the "society" category, for example, the *Wheatley Group* from Scotland was awarded. It is a leading provider of services in the areas of housing, care and property management. A total of 2,500 employees support over 250,000 customers with a wide range of services, including affordable housing and the care and support of needy people. ◀

10.2.3 The German Ludwig-Erhard-Preis

The *EEA* or the *EFQM Model* forms the basis for the development of a number of national quality prizes. Thus, in Germany, since 1997, the *Ludwig-Erhard-Preis* (Initiative *Ludwig-Erhard-Preis* e. V. 2018c) has been awarded, which is closely modeled on the *EFQM* concept. Only different formulations of the main criteria are more precisely defined. The sub-criteria are also largely identical. There is a slight deviation in the criteria "resource use" and "processes", which, however, only refers to the explicitly formulated demand for evidence of the innovation capability of a company (Malorny, 1996, p. 342). The award of the prize is made by the *Ludwig-Erhard-Preis* committee, which consists of representatives of the top associations of the German economy (Kraus 2012).

The ***Ludwig-Erhard-Preis*** is a national partner of the *EFQM.* The evaluation is carried out in the peer assessment by volunteer assessors, who are appropriately trained experts from the economy. The prize is awarded annually in four size categories—large organization, medium-sized company, small organization and smallest organization—since 2018 in gold, silver and bronze. Previously, there were three categories: prize winner, award and finalist. Other applicants are honored in the league principle according to their maturity level. The gold placement automatically qualifies organizations for participation in the *EFQM Global award.* In addition to the various prize categories of the *EFQM,* the initiative *Ludwig-Erhard-Preis* also awards recognition for "digital excellence", "workplace excellence" and "service excellence" (Initiative *Ludwig-Erhard-Preis* 2019a, b; www.ilep.de).

Fig. 10.19 shows an overview of the prize winners and awardees of the *Ludwig-Erhard-Preis* in recent years. Companies and organizations that have achieved a sustainable success in their economic activities are the prize winners of the *Ludwig-Erhard-Preis.* An award is given to the second placed companies or organizations and the finalists receive an honor (Kraus, 2012, pp. 103–104).

Year	Finalists	Award	Award winner
2017	• OBI Baumarkt Franken GmbH & Co. KG • Ingenics AG • Regensburg Tourismus GmbH • cmx Konzepte	• DRK-Krankenhaus Clementinenhaus	• MDK Medizinischer Dienst der Krankenversicherung Berlin-Brandenburg
2016	• VHS Reckenberg-Ems gem. GmbH • FARE gem. GmbH	• Bosch Werk Blaichach • BMW Komponentenfertigung Dingolfing • Gedikom GmbH • Schneider Bräuhaus München	• BMW Werk München
2015	• ABB Business Service GmbH • TESA-Werk Offenburg • Bankwitz Architekten	• BMW Werk München	• Bauunternehmung Jökel
2014	• VHS Reckenberg-Ems gem. GmbH • FARE gGmbH mit Sitz in Rheda-Wiedenbrück	• MDK Berlin-Brandenburg • Sparkasse Ingolstadt • Weisses Bräuhaus München GmbH & Co. KG	• Robert Bosch Fahrzeugelektrik Eisenach GmbH
2013	• DRK-Krankenhaus Clementinenhaus • EBK Krüger GmbH & Co. KG • Daimler AG • Bildungszentrum Haus Lautenbach • Augenfacharztpraxis Helga und Andreas Linke	• Caritasverband Olpe • Schöck AG • Thüringische Weidmüller GmbH • Flow consulting gmbh	• BMW Werk Regensburg • Domino-World™
2012	• Busch-Jaeger Elektro GmbH (GO)	• BMW, Werk Regensburg	• Robert Bosch GmbH, Werk Stuttgart-Feuerbach • Schindlerhof Klaus Kobjoll GmbH • Allresist GmbH
2011	• WISAG Gebäudereinigung Holding GmbH & Co. KG • Neumarkter Lammsbräu, Neumarkt	• Robert Bosch GmbH, Werk Feuerbach • ASSA ABLOY Sicherheitstechnik, Gmbh, Albstadt • Herth+Buss Fahrzeugteile, Heusenstamm • MDK Rheinland-Pfalz • Polizeidirektion Biberach • Schöck AG, Baden-Baden • Thüringische Weidmüller GmbH, Wutha-Famroda	• BMW Motorrad, Werk Berlin • Endress-Hauser Conducta, Gerlingen • WSS AKTIV BERATEN, Rottweil
2010	• Brose Fahrzeugteile GmbH & Co. KG Meerane • Deutsches Zentrum für Luft- und Raumfahrt e.V. • Reha Vita GmbH	• BMW Motorrad, Werk Berlin • TRW Airbag Systems GmbH • TRW Gebäudeservice GmbH • Marc Klejbor Marketing & Merchandising GmbH	• Ricoh Deutschland GmbH • I.K. Hofmann GmbH
2009	• TRW Automotive GmbH • Schöck AG • Praxis für Kinderheilkunde und Jugend-medizin Dr. med. Hans Ibel	• Ricoh Deutschland GmbH • Burton GmbH & Co. KG • Harting Deutschland GmbH & Co. KG • Allresist GmbH	• Robert Bosch GmbH, Werk Homburg • Glow & Tingle Unternehmensberatung
2008	• Burton GmbH & Co. KG • IMD Oberland	• Robert Bosch Fahrzeugtechnik Eisenach GmbH	• Domino-World™ • Voss & Partner AG
2007	• Carl Zeiss SMT AG • Apetito AG	• Daimler AG, Werk Untertürkheim • GRG Services Hamburg GmbH & Co. KG • Glow & Tingle Unternehmensberatung	• Kautex Tetron GmbH & Co. KG • Praxis für Zahn-, Mund- und Kiefer-heilkunde Dirk Schöttelndreier & Dr. Holger Rapsch
2006	• Homag Holzbearbeitungssysteme • GRG Services Group • Seniorenzentrum St. Liborius	• Grundfos Pumpenfabrik GmbH • Domino-World™ • Kautex Tetron GmbH & Co. KG • Praxis für Zahn-, Mund- und Kieferheilkunde Dirk Schöttelndreier & Dr. Holger Rapsch	• Busch-Jaeger Elektro GmbH • T-Systems Multimedia Solutions
2005	• Blaupunkt GmbH • Herth+Bus GmbH & Co. KG	• Busch-Jaeger Elektro GmbH • Emz-Hanauer GmbH & Co. KGaA • Kautex Tetron GmbH & Co. KG • T-Systems Multimedia Solutions • Praxis für Zahn-, Mund- und Kieferheilkunde Dirk Schöttelndreier & Dr. Holger Rapsch	• BMW Group Production Fahrwerk und Antriebskomponenten

Fig. 10.19 Prize winners and awardees of the *Ludwig-Erhard-Preis* (Source: Initiative Ludwig-Erhard-Preis, 2018a)

Year	German Excellence Award in Gold	German Excellence Award in Silver	German Excellence Award in Bronze
2018	• I. K. Hofmann GmbH Nürnberg • VHS Reckenberg-Ems gem. GmbH / FARE gem. Gmbh	• Evenord-Bank eG-KG Nürnberg • carriere & more private Akademie Südwest GmbH	• Deutsche Rentenversicherung Braunschweig-Hannover • thyssenkrupp Hohenlimburg GmbH Hagen • Ehrhorn Vermessung Achim • Spherea GmbH Ulm • Hornhautbank Aachen

Fig. 10.20 Prize winners of the *Ludwig-Erhard-Preis* 2018 (Source: Initiative *Ludwig-Erhard-Preis* 2018b)

Since 2018, a new prize format has been introduced (German excellence award in gold, silver and bronze) (Fig. 10.20).

10.2.4 The Swiss ESPRIX Swiss Award for Excellence

In 1999, the ***ESPRIX Award*** was introduced in Switzerland to promote entrepreneurial competitiveness (www.esprixexcellencesuisse.ch). It is also based on the *EFQM Model* and is awarded annually in two categories: To companies with fewer than 250 employees, so-called small and medium-sized enterprises (category SMEs) and to companies with more than 250 employees (category large companies) (ESPRIX, 2016). The *ESPRIX Award* is also the most important award for sustainable entrepreneurial excellence and competitiveness in Switzerland. For companies and organizations, participation in the award, in terms of comparison with the best and the possibility of economic development, is particularly worthwhile. The award is based on different levels on the way to excellence and is therefore not to be won in contrast to other prizes, but can be worked out by the participants by committing, demonstrating progress and recognizing excellence (2016). In the evaluation of the companies, a distinction is made between finalists, winners and award winners. These are categorized as follows (ESPRIX, 2016):

- **Finalists:** Organizations that have already achieved an above-average level.
- **Winners:** Award for a generally high level and role model function in one of the eight basic concepts of excellence (e.g. achieving balanced results; creating value for customers; leading with vision, inspiration and integrity).
- **Award winners:** Highest award for sustainable entrepreneurial excellence, comprehensive performance and high competitiveness.

Fig. 10.21 shows the finalists, winners and award winners of recent years.

The winner of the *ESPRIX Award 2010* is briefly introduced in the following example.

Year	Finalists	Winner	Award Winner
2017	• Rheinburg-Klinik AG, Walzenhausen	• Noser Engineering AG, Winterthur	• None
2016	• BMW Niederlassung Zürich-Dielsdorf • Höhere Kaderausbildung der Armee • Hochschule Luzern • CKW Conex AG, Luzern • Gewerbliche Berufsschule Chur (GBC), Chur	• Hochschule Luzern (HSLU), Luzern • Höhere Kaderausbildung der Armee (HKA), Luzern	• BMW Niederlassung Zürich-Dielsdorf
2015	• Bundesamt für Bauten und Logistik BBL, Bern (Grossunternehmen) • Post CH AG, Bern, Postmail Supply Chain (Grossunternehmen) • Alterszentrum am Bachgraben, Allschwil/Schönenbuch (Grossunternehmen)	• Alterszentrum am Bachgraben, Allschwil/Schönenbuch (Grossunternehmen)	• Post CH AG, Bern, Postmail Supply Chain (Grossunternehmen)
2014	• Clinique de La Source, Lausanne (Grossunternehmen) • Ospedale Regionale di Locarno La Carità, Locarno (Grossunternehmen) • Raiffeisenbank Mischabell-Matterhorn, St. Niklaus (KMU) • SBB AG, Personenverkehr Operating (P-OP), Bern (Grossunternehmen) • Schurter AG, Luzern (Grossunternehmen) • VEBO Genossenschaften, Oensingen (Grossunternehmen)	• Clinique de La Source, Lausanne (Grossunternehmen) • Ospedale Regionale di Locarno La Carità, Locarno (Grossunternehmen) • Raiffeisenbank Mischabell-Matterhorn, St. Niklaus (KMU) • SBB AG, Personenverkehr Operating (P-OP), Bern (Grossunternehmen)	• Schurter AG, Luzern (Grossunternehmen)
2013	• Forrer Lombriser & Partner, St. Gallen (KMU) • Manor AG, Division Supply Chain Management, Basel (Grossunternehmen)	• Manor AG, Division Supply Chain Management, Basel (Grossunternehmen)	• None
2012	• None	• None	• None
2011	• Grand Casino Baden, Baden (Grossunternehmen) • SBB Personenverkehr Operation (P-OP), Bern (Grossunternehmen) • Zürich Versicherungsgesellschaft, Geschäftsbereich Schaden, Zürich (Grossunternehmen)	• Grand Casino Baden, Baden (Grossunternehmen)	• Zürich Versicherungsgesellschaft, Geschäftsbereich Schaden, Zürich (Grossunternehmen)
2010	• Raiffeisenbank Mischabel-Matterhorn AG, Oberwallis (KMU) • SEEDAMM PLAZA (Hotel Seedamm AG), Pfäffikon SZ (KMU)	• None	• SEEDAMM PLAZA (Hotel Seedamm AG), Pfäffikon SZ (KMU)
2009	• Adullam-Stiftung, Basel (Grossunternehmen) • Biral AG, Münsingen (KMU) • Suva, Luzern, Grossunternehmen	• None	• Biral AG, Münsingen (KMU) • Suva, Luzern, Grossunternehmen
2008	• Amt für Arbeit, St. Gallen (Grossunternehmen) • pom+Consulting AG, Zürich (KMU) • Thurgauer Kantonalbank, Weinfelden, (Grossunternehmen)	• Thurgauer Kantonalbank, Weinfelden Grossunternehmen	• pom+Consulting AG, Zürich (KMU)

Fig. 10.21 Finalists, winners and award winners of the *ESPRIX* Award. (Source: ESPRIX, 2018)

Example of an ESPRIX Award: Seedamm Plaza

The *Seedamm Plaza* is a creative, innovative company that ranks among the top 10 in Europe in the seminar, congress and event sector. The 4-star superior hotel is located in Pfäffikon in Switzerland and employs 220 employees. In total, the *Seedamm Plaza* has 142 guest rooms with 230 beds as well as 42 conference and seminar rooms for up to 500 people.

The customer orientation "everything from one source" is central to the company's guiding principles. Customer needs are specifically determined through conversations, customer surveys and events. At the same time, market research, trend observation and competitor comparisons flow into the development of new, innovative products and services. The award of the *ESPRIX Award* 2010 is the prize for years of consistent quality management with the aim of comprehensive optimization of the entire company and leading to business excellence.

(Source: Seedamm Plaza, 2010) ◄

The *ESPRIX Award winner* was *Campus Sursee* in 2019. The *Tamedia AG* was awarded the *ESPRIX Winner prize* in 2019 in particular for the basic concept "actively manage changes" (Tamedia-Gruppe, 2019). The *Stiftung Rodtegg,* a foundation for people with physical disabilities from Lucerne, was also awarded as an ESPRIX finalist in 2019 (ESPRIX Award, 2019).

ESPRIX Award winner 2019: Campus Sursee

The *CAMPUS SURSEE* is the leading Swiss educational and seminar center for professionals in the construction industry. It was founded in 1972 to promote vocational education in the Swiss construction industry. The training ground covers over 20,000 square meters. Today, the seminar center is also popular with many other industries because of its pleasant learning environment and high customer orientation. The *CAMPUS SURSEE* foundation is the owner and also takes over the central management and joint resource planning of the educational center construction, the seminar center and the foundation. In addition, it pursues the goal of sustainable strengthening of training and further education as well as profiling of the construction industry. Around 270 employees are employed at the campus. After the *CAMPUS SURSEE* was already awarded four stars by *ESPRIX Excellence Suisse* as the best company in Switzerland in 2017, it received the *ESPRIX Swiss Award for Excellence* in 2019 (Campus Sursee, 2020). ◄

10.2.5 The Austrian National Award for Corporate Quality

The Austrian **national award for corporate quality** has been the official national award for top business performance in Austria since 1996 (www.staatspreis.com; www.bmdw.gv.at; www.qualityaustria.com). It rewards the successful implementation of excellence principles, also based on the *EFQM model. Quality Austria* and the *Federal Ministry for Digitalization and Economic Location* award the prize in the categories of large companies with more than 250 employees, medium-sized companies, small companies with up to 50 employees, non-profit organizations and organizations predominantly in public ownership. Outstanding achievements can also be rewarded with special prizes.

An independent jury, made up of representatives from business, science and administration, selects the finalists in each category from all applications. The evaluation of corporate quality is based on the *EFQM Model 2013* and the RADAR logic, evaluated by assessors. The applicants and winners benefit, for example, from the evaluation by assessors, an assessment of the current situation, the strengths and weaknesses of the organization, the advertising opportunities, etc.

The nominees for 2019 in the category of large companies were the *BKS Bank AG,* for medium-sized companies the *B. Braun Austria GmbH,* for small companies the *Kompetenzzentrum Gesundheit Attersee Süd,* for non-profit organizations the *Volkshochschule Götzis,* for organizations predominantly in public ownership the *WIFI Kärnten GmbH* and special prizes were awarded to various service providers such as pension funds and clinics (Quality Austria, 2020).

10.3 Use of Quality Award Models to implement quality management

Independent of the application for a quality award and the subsequent use of the award in corporate communication, the underlying quality model often also serves those companies that do not initially aim for their own application. This way, management can carry out a **self-evaluation** of the quality level by means of an assessment, based on the criteria of the quality award by means of the following **methods** (Stauss & Scheuing, 1994, 314 pp.; Kuhlang, 2011):

1. Self-evaluation by questionnaire
2. Self-evaluation by mock application
3. External audit

The methods will be explained in more detail below.

(1) Self-evaluation by questionnaire A first possibility for self-evaluation offers questionnaires with which the fulfilment of the evaluation criteria of the quality award is recorded in detail. For each individual criterion, various questions or statements are listed, which are to be answered by the company management on a five score scale. In the company's own interest, a realistic assessment is urgently required in this respect. Based on the evaluation of the answers, the strengths and weaknesses of the company are highlighted. This type of self-evaluation makes it possible to assess the company's situation relatively cheaply and quickly. However, the strengths and weaknesses are only roughly identified. The main contribution of this approach is to initiate a discussion process within the company.

(2) Self-evaluation by mock application By means of the method of self-evaluation by mock application, a process is initiated in the service company with the fictitious goal of obtaining a quality award. This requires a more detailed examination and enforcement of a comprehensive quality management system. An evaluation team assembled within the company examines the company on the basis of the application documents for the quality award analogously to the actual procedure. This method often results in a more intensive commitment of the company towards quality management, but still causes less costs and risks than the official application for a quality award.

However, possible deficits, such as the lack of commitment of the management and employees as well as a lack of consequence in the implementation of the identified measures, which are due to the reduced binding character of a mock application, must be pointed out.

(3) External auditing The third way of using the criteria of quality awards for management is external auditing. An external consultant group assesses the internal application documents. The discussion of the results can provide valuable information for the company. This type of self-evaluation offers the most approach scores for an effective company evaluation with regard to comprehensive quality management. However, it is also associated with higher external costs.

Based on a self-evaluation with regard to the quality management of a service company, the criteria catalogue of quality awards serves as a basis for **planning a comprehensive TQM programme.** The strategic corporate planning is aligned with the nine categories and the individual plans are related to the subcategories and the individual criteria. After this target and planning setting, the time dimensioning is carried out in the long and short term with a precise specification of the projects (Stauss, 2001, p. 507).

10.4 Critical Assessment of Quality Awards

In order to answer the question of whether it is expedient for a service company to make use of the possibility of orienting itself to the assessment criteria of quality awards or applying for a quality award, the possible effects and the limits of quality awards are to be discussed (Reimann & Hertz, 1994, pp. 350 et seq.; Madu & Kuei, 1995, pp. 188 et seq.; Hohmann, 2012; EFQM, 2016). The **opportunities** of applying for a quality award are above all to be found in the following aspects:

- Definition of company goals, clarity of goals
- Disclosure and transparency of internal structures
- Promotion of quality awareness
- Improvement and securing of internal and external processes
- Motivation of employees
- Commitment of management
- Creation of team spirit
- Establishment of a pronounced attitude to competition
- Readiness for change

- Focus on customer preferences
- Increase in customer satisfaction
- Reduction of costs
- Shortening of process times
- Systematic design of organization development
- Improvement of image and thus competitive advantages
- Increase in market share

In this context, a recent study compares the development of financial performance, measured by stock price, of *EFQM Excellence award* winners with that of comparable non-winners over a period of ten years (Boulter et al., 2013). The results show a significantly better stock price development of the winners. The authors also emphasize that in their study they use the winning of an *EFQM Excellence award* as an auxiliary variable for the successful implementation of quality management. However, the potential effects mentioned above have to be viewed very differently depending on the industry and company. The opportunities mentioned do not result solely from the company's application for a quality award, but rather from the company's overall and long term efforts to implement consistent total quality management.

Similar results are also achieved by a study of the *University of Leicester.* In a study, the influence of quality award winners on economically relevant key figures was examined. In comparison with companies that do not deal with a consistent total quality management philosophy, the results showed a faster increase in sales, a better development of the stock price and the creation of more jobs with this comparison group (Centre of Quality Excellence, 2005). In the study, above all, the long term effects of key results were emphasized. So apparently quality awards are also worthwhile from an economic point of view.

With regard to the **costs** of an application, the internal costs of establishing a quality assurance system as well as the external costs of the tendering of the application can be listed. A specific quantification depends on numerous influencing factors and has to be calculated in detail.

The main **problem areas** of quality awards are initially in the practical implementation of the application and the **completeness of the evaluation criteria.** In this context, the question of cost-benefit aspects must be raised in the company and the commitment of high personnel capacities must be determined. In the case of an award by the jury, there is the danger that further improvement commitments in the company may be postponed and a continuous process of improvement is not promoted.

Likewise, in the event of an unsuccessful application, a significant loss of motivation of the employees in the company can occur. In addition, the **methodology of measuring customer satisfaction** is to be criticized as insufficiently transparent in formal terms. Central dimensions and peculiarities of service quality are not completely considered. Finally, the lack of objectivity of the measurement and the subjectivity of the assessment of the award winners must be pointed out.

In the context of an assessment of quality awards as an instrument in quality management, it must be noted that the circle of applicants for quality awards is relatively limited. However, above all, the momentum that goes out from the winners to implement a quality management on a national and international level are of particular importance for those service companies that are in strong competition. This creates both an economic and a public economic benefit through quality awards.

After the further development of quality management systems (e.g. participation in a quality award), in addition to a successful implementation, a certification of the quality management is also of importance. Therefore, the following chapter will explain various aspects of certification in detail.

References

Atul, G., Jason, C. M., & Herath, S. K. (2005). Quality management in service firms. Sustaining structures of total quality service. *Managing Service Quality, 15*(4), 389–402.

BFH. (2016). Berner Fachhochschule. http://www.bfh.ch/index.php?id=1289&L=0. Accessed 15. Feb. 2016.

Bou-Llusar, J. C., Escrig-Tena, A. B., Roca-Puig, V., & Beltrán-Martín, I. (2009). An empirical assessment of the EFQM excellence model: Evaluation as a TQM framework relative to the MBNQA model. *Journal of Operations Management, 27*(1), 1–22.

Boulter, L., Bendell, T., & Dahlgaard, J. (2013). Total quality beyond North America: A comparative analysis of the performance of European excellence award winners. *International Journal of Operations & Production Management, 33*(2), 197–215.

Bruhn, M., & Hadwich, K. (Hrsg.). (2016). *Servicetransformation. Entwicklung vom Produktanbieter zum Dienstleistungsunternehmen.* Forum Dienstleistungsmanagement. Wiesbaden: Springer Gabler.

Bruhn, M., & Hadwich, K. (Hrsg.). (2019). *Kooperative Dienstleistungen. Spannungsfelder zwischen Service Cooperation und Service Coopetition.* Forum Dienstleistungsmanagement. Wiesbaden: Springer.

Bruhn, M., Meffert, H., & Hadwich, K. (2019). *Handbuch Dienstleistungsmarketing. Planung—Umsetzung Kontrolle,* 2nd Edn. Wiesbaden: Springer Gabler.

Campus Sursee. (2020). Über uns. Campus Sursee. https://www.campus-sursee.ch/de/Campus-erleben/Ueber-uns.11.html. Accessed 5. Jan. 2020.

Centre of Quality Excellence. (2005). *Report on EFQM and BQF funded study into the impact of the effective implementation of organisational excellence strategies on key performance results.* University of Leicester, Leicester.

Dahm, M., & Heindl, C. (2006). Von Six Sigma zu Lean Sigma. *ZUb Zeitschrift der Unternehmensberatung, 1*(2), 81–88.

Dahm, M., & Heindl, C. (2007). Effizienter werden mit Lean Sigma. *IO New Management, 76*(6), 86–92.

EFQM. (2010). Introducting the EFQM excellence model 2010. www.efqm.org/en/PdfResources/EFQMModel_Presentation.pdf. Accessed 16. July 2012.

EFQM. (2013). EFQM excellence award application. http://www.efqm.org/en/Home/Ourservices/Recognition/EFQMExcellenceAward/tabid/154/Default.aspx. Accessed 21. Feb. 2013.

EFQM. (2016). The EFQM excellence model. http://www.efqm.org/the-efqm-excellence-model. Accessed 10. Feb. 2016.
EFQM. (2018a). EFQM global excellence award 2018 winners overview. http://www.efqm.org/index.php/efqm-recognition/efqm-global-excellence-award/award-history/. Accessed 30. Dec. 2018.
EFQM. (2018b). Previous EFQM award winners. http://www.efqm.org/index.php/efqm-recognition/efqm-global-excellence-award/award-history/previous-efqm-award-winners/. Accessed 30. Dec. 2018.
EFQM. (2019a). Das EFQM Modell 2020. https://www.efqm.org/index.php/efqm-model/download-your-free-short-copy-of-the-efqm-model/. Accessed 1. Jan. 2020.
EFQM. (2019b). EFQM—Our members. EFQM. https://www.efqm.org/index.php/membership/our-members/. Accessed 3. Jan. 2020.
EFQM. (2019c). EFQM Magazine June 2019. https://www.efqm.org/index.php/2019/07/01/efqm-magazine-june-2019/. Accessed 4. Jan. 2020.
EFQM Recognition—EFQM. (2020). https://www.efqm.org/index.php/efqm-recognition/. Accessed 4. Jan. 2020.
Ellis, V. (1994). Der European Quality Award. In B. Stauss (Ed.), *Qualitätsmanagement und Zertifizierung von DIN ISO 9000 zum Total Quality Management* (pp. 277–302). Wiesbaden: Gabler.
ESPRIX. (2016). ESPRIX award. http://www.esprix.ch/stufen_der_excellence/esprix_swiss_award_for_excellence/winners/. Accessed 10. Feb. 2016.
ESPRIX. (2018). ESPRIX winners. http://www.esprix.ch/stufen_der_excellence/esprix_swiss_award_for_excellence/winners/. Accessed 2. Dec. 2018
ESPRIX Award. (2019). ESPRIX Excellence Suisse. https://www.esprixexcellencesuisse.ch/award/. Accessed 4. Jan. 2020.
Frehr, H.-U. (1994). *Total Quality Management. Unternehmensweite Qualitätsverbesserung.* München: Hanser.
Frehr, H.-U. (1999). Total Quality Management. In W. Masing (Ed.), *Handbuch Qualitätsmanagement* (4th Edn., pp. 31–48). München/Wien: Hanser.
George, S., & Weimerskirch, A. (1998). *Total quality management. Strategies and techniques proven at todays´s most successful companies* (2nd Edn.). New York: Wiley
Gimenez-Espin, J. A., Jiménez-Jiménez, D., & Martinez-Costa, M. (2013). Organizational culture for total quality management. *Total Quality Management & Business Excellence, 24*(5–6), 678–692.
Graf, H. (2020). EFQM certified assessor transformation training. Seminarunterlagen. Swiss Association for Quality (SAQ). https://www.hslu.ch/de-ch/wirtschaft/weiterbildung/weiterbildungskurse/ibr/efqm-assessor-training/ Accessed 15. May 2023.
Herrmann, J., & Fritz, H. (2016). *Qualitätsmanagement: Lehrbuch für Studium und Praxis* (2nd Edn.). München: Hanser.
Hohmann, K. (2012). *Unternehmens Excellence Modelle: Das EFQM-Modell.* Hamburg: Diplomica.
Horatzek, S. (2018). *Prozessmanagement.* Weinheim: Hanser.
Huth, M. (2018). *Toolbox Prozessmanagement. Vorgehensmodell und praktische Methoden für Industrie und Dienstleistung.* München: Hanser.
Imai, M. (2002). *Kaizen. Der Schlüssel zum Erfolg im Wettbewerb* (2nd Edn.). München: Econ Ullstein List.
Initiative *Ludwig-Erhard-Preis.* (2018a). *Ludwig-Erhard-Preis.* https://ilep.de/Artikel/cmx4eb00a0ae6125.html. Accessed 28. Dec. 2018.

Initiative *Ludwig-Erhard-Preis*. (2018b). *Ludwig-Erhard-Preis*. https://ilep.de/Artikel/cmx-5899a1ad4936b.html. Accessed 28. Dec. 2018.

Initiative *Ludwig-Erhard-Preis* e. V. (2018c). Excellence für Deutschland. https://ilep.de/. Accessed 28. Dec. 2018.

Initiative *Ludwig-Erhard-Preis* e. V. (2019a). Deutscher Excellence Preis—Ergebnisband 2019. https://ilep.de/Artikel/autowert-cmx5d808982b96a5. Accessed 1. Jan. 2020.

Initiative *Ludwig-Erhard-Preis* e. V. (2019b). Neuerungen des *EFQM-Modells 2020*. https://ilep.de/Artikel/cmx4f13656068d35.html. Accessed 1. Jan. 2020.

Jacobides, M. G., Cennamo, C., & Gawer, A. (2018). Towards a theory of ecosystems. *Strategic Management Journal, 39*(8), 2255–2276.

Japan Quality Award. (2013). Japan quality award. http://www.jqac.com. Accessed 21. Feb. 2013.

Kirchgeorg, M., Meynhardt, T., Pinkwart, A., Suchanek, A., & Zülich, H. (2018). *Das Leipziger Führungsmodell. Führen und beitragen*. (3rd Edn.). Leipzig: HHL Academic Press.

Kosta, C., & Kosta, S. (2008). *Der kontinuierliche Verbesserungsprozess: Methoden des KVP*. München: Hanser.

Kraus, T. (2012). Der *Ludwig-Erhard-Preis*. In H. Schnauber & A. Schuster (Eds.), *Erfolgsfaktor Qualität. Einsatz und Nutzen des EFQM-Excellence-Modells* (pp. 103–113). Düsseldorf: Symposion Publishing.

Kuhlang, P. (2011). Business Excellence, Qualitätspreise und Selbstbewertung. In F. J. Brunnr & K. W. Wagner (Eds.), *Taschenbuch Qualitätsmanagement. Leitfaden für Studium und Praxis* (5th Edn., pp. 337–371). Wien: Hanser.

Lay, G., Schat, H.-D., & Jäger, A. (2009). Mit EFQM zu betrieblicher Exzellenz: Verbreitung, Ausgestaltung und Effekte des Qualitätsmanagementmodells der European Foundation for Quality Management. Fraunhofer ISI. https://hdl.handle.net/10419/29348. Accessed 1. Jan. 2020.

Lusch, R. F., & Nambisan, S. (2015). Service innovation: A service-dominant logic perspective. *MIS Quarterly, 39*(1), 155–175.

Lusch, R. F., Vargo, S. L., & Gustzafsson, A. (2016). Fostering a trans-disciplinary perspectives of service ecosystems. *Journal of Business Research, 69*(8), 2957–2963.

Madu, C. N., & Kuei, C.-H. (1995). *Strategic total quality management. Corporate performance and product quality*. Westport: Praeger.

Magnusson, K., Kroslid, D., & Bergman, B. (2004). *Six Sigma umsetzen. Die neue Qualitätsstrategie für Unternehmen* (2nd Edn.). München: Hanser.

Malorny, C. (1996). *TQM umsetzen: Weltklasse neu definieren, Leistungsoffensive einleiten, Business Excellence erreichen* (2nd Edn.). Stuttgart: Schäffer-Poeschel.

Malorny, C., & Dicenta, M. (2014). Funktion und Nutzen von Qualitätsauszeichnungen (Awards). In T. Pfeifer & R. Schmitt (Eds.), *Masing—Handbuch Qualitätsmanagement* (6th Edn., pp. 362–384). München: Hanser.

Nambisan, S. (Ed.). (2018). *Open innovation, ecosystems and entrepreneurship: Issues and perspectives*. Singapur: World Scientific.

Nambisan, S., Lyytinen, K., & Yoo, Y. (Eds.). (2018). *Handbook of digital Innovation*. Cheltenham.

Nist. (2010). National quality—Business excellence awards in different countries. www.baldrige.nist.gov/Excel_files/National_Quality_Business_Excellence_Awards_in_Different_Countries.xls. Accessed 16. Juli 2012.

Oakland, J. S. (2004). *Oakland on quality management* (3rd Edn.). Oxford: Butterworth-Heinemann.

Oakland, J. S., & Porter, I. J. (2004). *Total quality management* (2nd Edn.). Oxford u. a.: Butterworth-Heinemann.

Peacock, R. D. (1992). Ein Qualitätspreis für Europa. *Qualität und Zuverlässigkeit, 37*(9), 525–528.

Quality Austria. (2020). BKS Bank AG gewinnt den Staatspreis Unternehmensqualität 2019. Quality Austria. https://www.qualityaustria.com/news/bks-bank-ag-gewinnt-den-staatspreis-unternehmensqualitaet-2019/. Accessed 4. Jan. 2020.

Reimann, C. W., & Hertz, H. S. (1994). Malcolm Baldrige National Quality Award und die Zertifizierung gemäß den Normen ISO 9000 bis 9004. Die wichtigsten Unterschiede. In B. Stauss (Ed.), *Qualitätsmanagement und Zertifizierung von DIN ISO 9000 zum Total Quality Management* (pp. 333–364). Wiesbaden: Gabler

Schmitt, R., & Pfeifer, T. (2015). *Qualitätsmanagement. Strategien, Methoden, Techniken* (5th Edn.): Hanser.

Seedamm Plaza. (2010). Medienmitteilung vom 12.03.2010. http://www.seedamm-plaza.ch/download/htm/3540/de/Medienmitteilung-vom-12-Maerz-2010.pdf. Accessed 18. Oct. 2012.

Senn, T., & Bruhn, M. (2019). Digitalisierte Service Ecosysteme—Entscheidungstatbestände und Forschungsbedarf. In M. Bruhn & K. Hadwich (Eds.), *Kooperative Dienstleistungen. Forum Dienstleistungsmanagement* (pp. 201–226). Wiesbaden: Springer.

Sinek, S. (2014). *Frag immer erst: Warum. Wie Führungskräfte zum Erfolg inspirieren.* München: Redline.

Stauss, B. (2001). Die Bedeutung von Qualitätspreisen für Dienstleistungsunternehmen. In M. Bruhn & H. Meffert (Eds.), *Handbuch Dienstleistungsmanagement. Von der strategischen Konzeption zur praktischen Umsetzung* (2nd Edn., pp. 499–524). Wiesbaden: Gabler.

Stauss, B., & Scheuing, E. E. (1994). Der Malcolm Baldrige National Quality Award und seine Bedeutung als Managementkonzept. In B. Stauss (Ed.), *Qualitätsmanagement und Zertifizierung von DIN ISO 9000 zum Total Quality Management* (pp. 303–332). Wiesbaden: Gabler.

Tamedia-Gruppe. (2019, Juni 25). Tamedia Verlagsdienstleistungen Druck & Logistik gewinnen ESPRIX Preis 2019. https://zeitungsdruck.ch/meta/news/news-detail/tx_news/tamedia-verlagsdienstleistungen-druck-logistik-gewinnen-esprix-preis-2019/?tx_news_pi1%5Bcontroller%5D=News&tx_news_pi1%5Baction%5D=detail&cHash=8ace608354a573c0ea6185633bb5b4d1. Accessed 4. Jan. 2020.

UN. (2020). Sustainable development goals. www.un.org/sustainabledevelopment/sustainable-development-goals/. Accessed 16. May 2020.

Vereinte Nationen. (2020). Sustainable development goals: Sustainable development knowledge platform. https://sustainabledevelopment.un.org/?menu=1300. Accessed 1. Jan. 2020.

Watson, G., & Gallagher, K. (2005). *Managing for results.* (2nd Edn.). London: CIPD Kogan Page.

Zink, K. J. (2004). *TQM als integratives Managementkonzept* (2nd Edn.). München: Hanser.

Zollondz, H.-D. (2011). *Grundlagen Qualitätsmanagement. Einführung in Geschichte, Begriffe, Systeme und Konzepte* (3rd Edn.). München: Oldenbourg.

Certification of the Quality Management of Service Companies

11

Abstract

Certification has a high importance for the implementation of quality management and is therefore widespread in many service industries. In order to obtain a quality certificate, a company has to implement the necessary measures for the implementation of quality management in accordance with the certification requirements. This results in an incentive that increases the quality awareness of the employees and thus consistently drives the implementation of a high level of service quality. In order to illustrate the certification of a company, the goals of the certification as well as the ideal type of a certification process in the individual steps with the necessary activities and decisions of the company are set out in this chapter. In addition, the effects and limits of certification are shown.

11.1 Term and Framework Conditions of Certification

Certification has a high importance for the implementation of quality management. For the acquisition of a quality certificate, a company has to implement the necessary measures for the implementation of quality management in accordance with the certification requirements. This results in an incentive that increases the quality awareness of the employees and thus drives the implementation more consistently.

The **origin of certification** of quality management systems goes back to the procurement strategies of the military, which imposed precise production specifications on their civilian suppliers. From such auditable elements of a quality management system, the first international standard ISO 9000 for quality management emerged in 1987. This way, the various trends of total quality management found their way into the field of

M. Bruhn, *Quality Management for Services*,
https://doi.org/10.1007/978-3-662-67032-3_11

international standardization with the result of the definition of clear terminology and understanding in the field of quality management (Zollondz, 2011, pp. 452–453).

The certification of a service company has the goal of checking the implementation of a quality management system in service companies and thus ensuring a certain quality of service. The certification has a high importance for communication with the customer. The use of a service is often associated with uncertainty on the part of the customer due to its constitutive features—such as the immateriality of the service. This creates the need for a service provider to reduce the mentioned uncertainty in the customer by means of trust building measures. Consequently, certification serves primarily the **building of trust** (Zollondz, 2011, p. 452), especially since this communicates the guarantee of a certain quality of service to potential and current customers.

Certification is understood to be a procedure that specifically specifies that a product or service, a process, a quality management system or an environmental management system meets specified requirements. The **certification** is defined in this sense as follows:

▶ The **certification** is an official, written statement by an impartial third party that a certain object meets certain requirements specified by an independent body (conformity in the sense of agreement). The written document that records this fact is called a certificate (based on Zollondz, 2011, p. 452).

The norm DIN EN ISO 9001 has established itself as a **audit basis,** which is internationally recognized and has been institutionalized in over 80 countries (Zollondz, 2011, p. 309). This norm defines the requirements for a quality management system with regard to the ability of a company to provide conformity of the products or constantly error free products (cf. Sect. 11.3.2). It serves the external presentation of a quality management system for the purpose of certification or auditing (Zollondz, 2011, p. 310). Only an organization that has a quality management system that has been documented and implemented according to DIN EN ISO 9001 can be certified (Drechsler, 2007, p. 345).

DIN EN ISO 9001 belongs to the so-called international family of standards ISO 9000ff., which was revised fundamentally in the year 2000. In total, the current standards DIN EN ISO 9001 (edition 2015-09) and DIN EN ISO 9004 (edition 2009–12) belong to the family of standards. Fig. 11.1 gives an overview of the system components of the ISO 9000ff. family of standards in its current version:

The term standard ISO 9000 is also integrated into the series of standards, which contains the **basic terminology on the subject of quality management,** as well as the supplementary standard ISO 19011 (edition 2002), in which it is shown how the auditing or certification takes place (Zollondz, 2011, p. 310). Furthermore, ISO 9004 was implemented as a guideline for the optimization of the overall performance and efficiency of the company in the sense of a continuous improvement process. In order to facilitate its application, it is designed consistently with ISO 9001. Both standards have been developed together in order to complement each other, but can be applied independently of

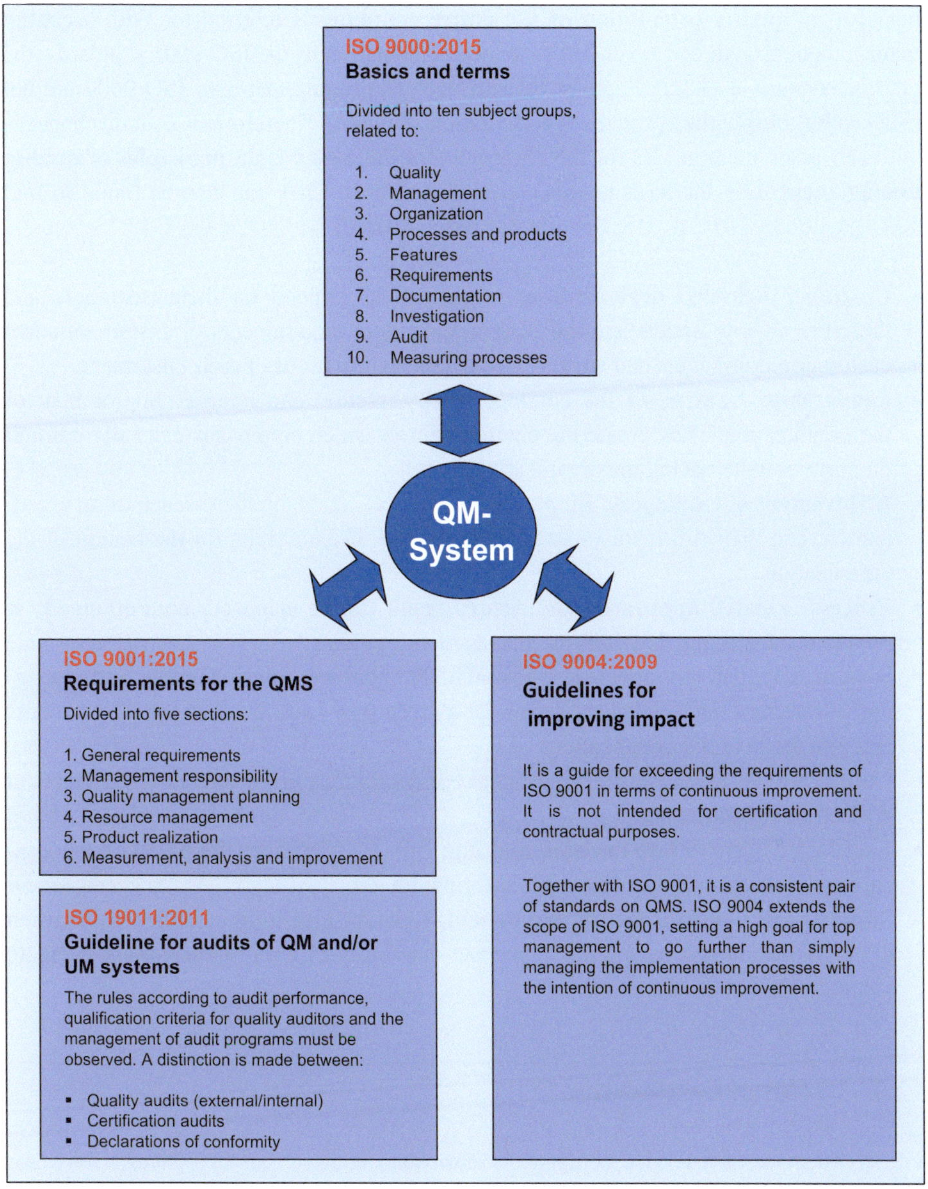

Fig. 11.1 System components of DIN EN ISO 9000ff. (Source: Based on Zollondz, 2011, p. 313)

each other (Brauer, 2009, p. 15). Therefore, it is also referred to as a consistent pair of standards. ISO 9004 helps companies to improve all relevant processes of the entire organization and thus to facilitate the way to business excellence. This shifts the focus of ISO 9004 beyond the requirements of ISO 9001, not only by demanding product quality, but also by **quality orientation of the entire company** is referred to. With a quality management system that is limited to the implementation of the ISO 9001 standards, the goal of business excellence cannot be achieved. The requirements of ISO 9004 are not checked by third parties (it is not a certification standard). Therefore, it contains annexes with self-assessment guides for the company. The following eight **principles of quality management** form the basis of this series of standards. They can also be found in ISO 9004 (Zollondz, 2011, p. 318; Schmitt & Pfeifer, 2015):

- **Customer oriented organization:** Organizations depend on their customers and therefore need to understand the current and future requirements of customers, meet customer requirements and strive to exceed the expectations of their customers.
- **Leadership:** Leaders set the common goal, direction and internal environment of the organization. They create the environment in which employees can fully commit themselves to achieving the organization's goals.
- **Involvement of employees:** Employees at all levels make up the essence of an organization, and their full involvement allows the use of their skills for the benefit of the organization.
- **Process oriented approach:** The desired result can be achieved more efficiently if related resources and activities are managed as a process.
- **System oriented management approach:** Recognizing, understanding and managing a system of interrelated processes for a given goal contributes to the effectiveness and efficiency of the organization.
- **Continuous improvement:** The principle of continuous improvement is a permanent goal of the organization.
- **Fact-based approach to decision making:** Effective decisions are based on the logical and intuitive analysis of data and information.
- **Supplier relationships for mutual benefit:** Relationships for mutual benefit between the organization and its suppliers promote the ability of both organizations to create value.

11.2 Goals of Certification

The certification of a service company is associated with various objectives, which are both internally and externally oriented. These are summarized in Fig. 11.2.

Among the **external objectives,** the first is the company external proof that the service company consistently meets the requirements of a quality management system according to the DIN/ISO standards. This documentation is recognized throughout Europe and therefore the certification of the company offers customers greater transparency and comparability of company performance when choosing suppliers and creates trust in the supplier on the part of customers. In addition, often expands the potential customer base of the company through the documentation of its performance. Due to

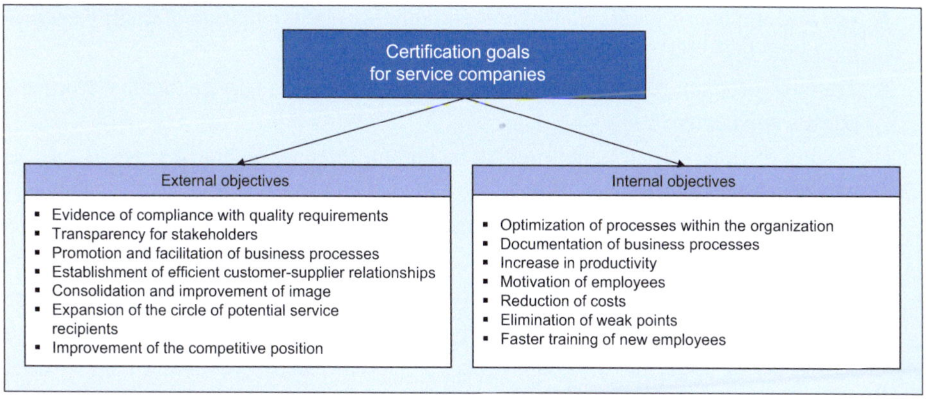

Fig. 11.2 Goals of certification

the easier transparency of company processes and the documentation of its performance, business processes are promoted and simplified. This forms the basis for the establishment of efficient customer-supplier relationships (Döttinger & Klaiber, 1994, pp. 269–270; Zollondz, 2011, p. 455). Since the certificate is used primarily in the company's communication to the outside, certification also serves to expand a positive corporate image. These, externally oriented aspects of certification usually make a significant contribution to the overall competitive position of the service provider.

Goals that a service company wants to achieve externally must be based on the company's internal perspective. **Internal company goals** of certification therefore take on special importance. Certification makes it possible to first openly disclose company processes, which in turn makes it possible to revise and optimize existing company processes. The need to document business processes often supports the improvement of internal processes. As a result of the optimization of company processes, it is possible to reduce costs and thus increase productivity. In particular, a reduction in quality costs is often sought, which often results from the optimization of company processes and potentials. In this context, it is also necessary to promote the elimination of weaknesses in the process of creating value. In addition, the preparation and implementation of a certification often results in a significant motivational impulse for employees, which is transmitted to all employees of the company via the certification team and, in the best case, lasts beyond the duration of the certification. This way, for example, a faster induction of new employees is often achieved by conveying the philosophy of quality management to the company (Saatweber, 2011).

Certification of the *Berlin Plaza Hotel* on Kurfürstendamm

The *Berlin Plaza Hotel* on Kurfürstendamm communicates the goals and motivation of a certification on its homepage with the following points:

- For customers, traceable, transparent and reliable documentation of quality according to the ISO 9001:2015 standards.
- Training of management and employees on the topic "service quality from the guest's perspective".
- Testing of other departments such as reception, kitchen, technical services, etc. as part of internal audits.
- Implementation of more effective processes, increase in product quality and motivation of employees.
- Savings of personnel costs and other resources of up to 1,000 EUR per employee. ◀

It becomes clear that certification is an instrument that supports the **introduction of a long term and profound quality management** in the service company. A comprehensive quality management offers service companies the only way to create long term, comparative competitive advantages.

Essential **conditions** for deciding on a certification project are in particular the time frame of the certification process and the costs of the procedure.

The **certification duration** until receipt of the certificate is dependent on various factors. First and foremost, the size of the company or business area to be certified, the choice of certification body and the determination of the test standards determine the time required until receipt of the certificate. Furthermore, the initial state of the quality management system and the resources provided are decisive for the duration of the certification process. With the certification bodies, there is usually no waiting time for the commencement of the certification process; after the review of the quality manual, an audit date is to be expected within one to two months (Pärsch, 1999, p. 202; Schmitt & Pfeifer, 2015). Experience values show that certification for smaller organizations with simple processes takes at least six months, while the period for larger organizations can also be two years. The actual certification process usually takes two to three months (Drechsler, 2007, p. 350; Holland, 2009).

The **certification costs** are to be differentiated into costs for the contractually specified services of the certification body and costs for activities on the part of the company and possibly costs for the involvement of external consultants. For the latter costs, no guideline value is usually given, as they depend in particular on the company's starting situation and the management and employee potential. The costs for the contractually specified services of the certification body are determined by the scope of the certifier's services and the complexity of the service company. However, internationally binding minimum numbers have been set in relation to the person-days required for an assessment, which is to ensure a sufficient assessment depth and which the accredited certification bodies may not fall below (Drechsler, 2007, p. 15).

The following **influencing factors** are often used to categorize the complexity of companies into “low”, “medium” and “high” in order to then derive guidelines for the costs of certificate creation:

- Variety of services
- Variety of procedures
- Degree of automation
- Organizational structure
- Requirements from legal regulations
- Company size
- Number of employees and business volume
- Distribution of locations

The total costs of the process of certificate issuance with a subsequent period of validity of a total of three years (including the costs of the annual surveillance audits and the re-audit) then reach approximately the following order of magnitude:

- Approximately 4,500 EUR for small companies
- Around 10,000 EUR for medium-sized companies
- From 15,000 EUR for large companies

It should be noted that these are **guidelines** to which the travel expenses of the auditors and value added tax must be added. Ultimately, the actual costs are agreed between the service company and the respective certification body (Pärsch, 1999, p. 198; Schmitt & Pfeifer, 2015).

11.3 Process of Certification

After the decision of the service company to certify, the process of certification essentially comprises the following **steps:**

1. Selection of the certification body
2. Decision on the underlying examination standard
3. Preparation for certification
4. Certificate issuance

11.3.1 Selection of the Certification Body

There is a large number of **independent certification bodies** from which service companies can choose. Since the certification of services is mostly voluntary, the state does not intervene in the market of certifiers. Accordingly, there is competition in this market, with the decisive criterion for the selection of a certification body being the information content of the certificate. As a determinant of the information content of certificates the **quality of the examination** is to be mentioned. The quality of the examination is determined in particular by the training, the independence, the process of monitoring and the underlying examination standards of the examiners.

It also has to be considered that the **information content** of the certificate is significantly influenced by the quality of the certification standards (in particular the quality of the DIN EN ISO standards). In this respect, reference is made to the completeness, relevance, precision and effectiveness of the standards (Bretzke, 1999, 590 pp.; Haller, 2018, 385 pp.).

In order to demonstrate their competence and thus document the quality of the examination, certification bodies often apply for recognition by a umbrella organization—the so-called **accreditation.** For this purpose, the certification institutions carry out an examination in accordance with the international standard ISD/IEC 17011:2005 (e.g. EN 17021, General requirements for bodies operating quality management systems certification), which is accepted by the "accreditation body" within the "German Accreditation Council". The accreditation body then issues the accreditation certificates for the certified certification institutions, which then serve as proof of their competence. In Fig. 11.3 the accreditation certificate of the *German Society for Certification of Quality Management Systems* is shown by way of example. In order for national certificates to be recognized across borders, organizations such as the *"European Organization for Testing and Certification"* strive for harmonization of the country specific standards. These are, for example, in the "European Criteria for Certification Bodies Operating Quality System Certification" (DIN EN ISO/IEC 17021:2001-7) or the "International Guidelines for Auditing Quality Systems and Qualification of Auditors" (ISO 10011-1990) (Deutsche Gesellschaft für Qualität e. V., 1995). Although the certification of management systems, and thus also of quality management systems, falls within the non-regulated area of certification, accreditation of the certification body is important for the credibility and acceptance of a certificate. Therefore, it is recommended for organizations that are aiming for certification of their quality management to pay attention to the fact that the certification body is accredited but also, that the respective industry is covered by the accreditation (Drechsler, 2007, p. 345).

The accredited and also the non-recognized certification institutions differ in addition on the basis of various **criteria.** The degree of national and international recognition is just as important a differentiation criterion as the scope of the services offered and the speed of certification. The decision on the choice of the certification body is therefore dependent on these criteria. If a certificate is needed in particular for the customer-supplier relationship, it is also possible that the customer of the service provider requires certification at an institution specified by him.

TGA - Trägergemeinschaft für Akkreditierung -
German Association for Accreditation GmbH

Mitglied im Multilateralen Abkommen von EA und IAF

vertreten im

Deutschen AkkreditierungsRat

Akkreditierung

Die TGA - Trägergemeinschaft für Akkreditierung - German Association for Accreditation GmbH bestätigt hiermit, dass die Zertifizierungsstelle

DVGW CERT GmbH
Josef-Wirmer-Straße 1-3
53123 BONN

die Kompetenz nach DIN EN ISO/IEC 17021 besitzt, die Zertifizierung von Qualitätsmanagementsystemen nach der Norm DIN EN ISO 9001 in den in der Anlage näher bezeichneten Wirtschaftsbereichen durchzuführen.

Die Akkreditierung ist gültig: **05.08.2009 - 04.08.2014**

Die Anlage ist Bestandteil der Urkunde und besteht aus **1** Seite.

DAR-Registriernummer: **TGA-ZM-11-94-00**

Frankfurt / Main, 13.08.2009

(Dr. Thomas Facklam)
Leiter der Akkreditierungsstelle

Fig. 11.3 Accreditation certificate of the *DVGW CERT GmbH* for certification of quality management systems. (Source: DVGW CERT GmbH, 2012)

11.3.2 Decision on the Certification Standard

Just as the selection of the certification body, the corresponding **certification basis** is to be specified in the form of determining the quality control standards to be met. The most common basis for the certification of quality management is the industry independent, international standard DIN EN ISO 9001:2015 (see Sect. 11.2). In addition, there is a wide range of other industry specific or product specific quality management systems, the certification of which is carried out in a similar way, but partly in compliance with certain procedural requirements (Drechsler, 2007, p. 344).

The 9001 standard describes the core requirements for a quality management system. It thus targets the processes that are needed to meet the given customer requirements. It therefore represents a minimum standard for the implementation of quality management within companies (Wagner, 2007, p. 175; Wagner & Käfer, 2013).

The ISO 9001 standard is characterized by four central focus points:

- Focus "processes"
- Focus "management"
- Focus "customer"
- Focus "continuous improvement"

The **process focus** is reflected in the process orientation underlying the standard, in that the quality management system is consistently aligned with the processes or process steps. At its core, the task for companies is to identify the essential, value adding processes of the company and to control them reliably (Wagner, 2007, p. 179; Wagner & Käfer, 2013, p. 64). The process orientation of quality management according to the DIN EN ISO 9000ff. standards is illustrated graphically in the process model shown in Fig. 11.4. The accompanying text of the standard explains the model as follows: In an

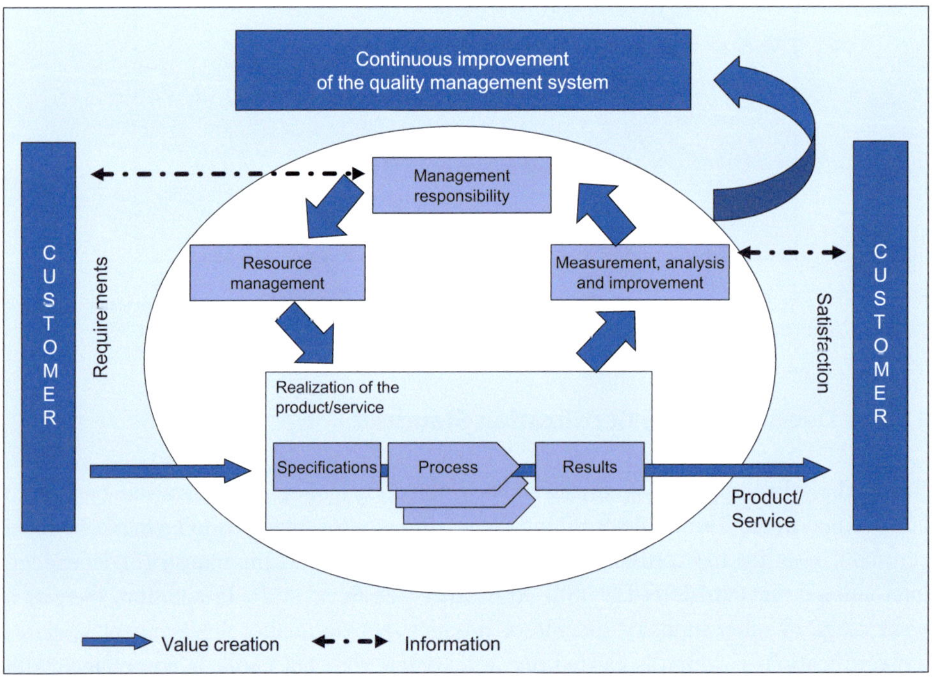

Fig. 11.4 Process model of quality management. (Source: Wagner & Käfer, 2013, p. 127; DIN EN ISO 9001:2015, 2015)

efficient quality management system, both the processes and the responsibilities, procedures and resources associated with the processes are uniformly defined and managed accordingly. In addition, the quality management system requires the coordination and compatibility of associated process components as well as the clear definition of interfaces. The organization is subject to mutual dependencies, which are often complex and result in a network of processes. In this context, the emphasis on the value adding main process chain is significant. In addtion it is necessary to indicate the influence of the individual processes on the ability to meet customer requirements for the service (Zollondz, 2011, p. 233). Another focus of the standard is **"management".** This is particularly evident in the fact that management processes are gaining more and more importance in relation to production processes. The focus on the **"customer"** makes it clear that it is necessary to align the company consistently with customer needs and to bring customer orientation into focus in quality management. The demand for **continuous improvement** is the last essential focus of ISO 9001 (Wagner, 2007, pp. 179–180; Wagner & Käfer, 2013, p. 67).

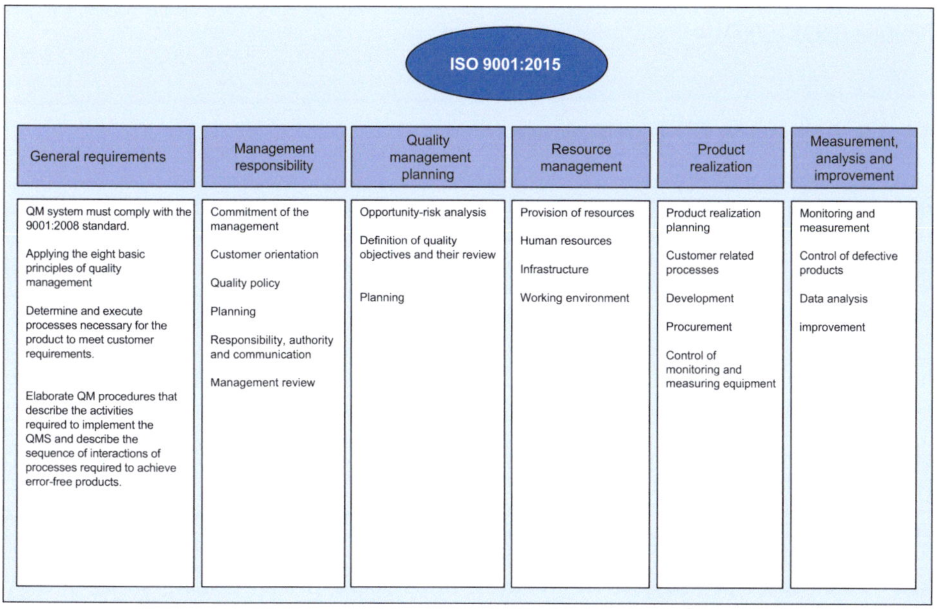

Fig. 11.5 Requirements for a quality management system. (Source: Based on Zollondz, 2011, p. 316; DIN EN ISO 9001:2015)

ISO 9001 sets out the requirements for a quality management system as shown in Fig. 11.5. The standard is applicable if an organization:

- Wants to gain the trust of its customers, suppliers, etc., that its **requirements** for the product will be met and that the organization is able to provide error-free products on a continuous basis.
- Strives to increase **customer satisfaction** through effective application of the system, including processes for continual improvement of the system and assurance of compliance with customer and regulatory requirements.

In order to meet the requirements of the respective organization, the standard leaves a wide **margin of interpretation** for the users. Frequently, the standard is designed to be flexible by terms such as "appropriate", "if necessary", "where applicable", etc. (Schlüter & Dunkhorst, 2003). Those responsible for certification within the organization must independently assess which requirements of the standard are specifically applicable to their organization and to what extent the existing quality management system meets these requirements. If individual requirements cannot be transferred to the organization or its products, they may be excluded if they fall under the "permissible exclusions". For example, if the services do not require development work, these requirements can be omitted (DQS, 2002).

11.3.3 Preparing for Certification

In order to put a service company in a position to meet the certification standards, it is first necessary to initiate or refine quality management within the company. For this purpose, the measures described in Chap. 8 can be used within the scope of the quality management control loop.

When a service company is aiming for a certification of its quality management and has dealt with the measures to be taken to meet the certification standards, it makes sense to record the planned procedure in a schedule within the framework of the certification project in order to be able to work through the steps to be carried out successively or in parallel in a systematic way.

In any case, it is the responsibility of the company to ensure that the quality management system to be certified is already "certification-ready". A failure can quickly lead to loss of confidence and motivation problems; in particular if the certification has already been announced to customers and employees. Essential steps in the preparation are the selection of a competent (ideally accredited) certification body. Furthermore, the preparation phase should be used to familiarize yourself with the profiles and competence of the auditors in question before the actual certification (Drechsler, 2007, p. 346; Holland, 2009).

The following example shows how such a schedule can look up to certification for an institution in the social sector.

Example of a certification project

The typical course of a specific certification project according to DIN EN ISO 9001:2015 for an institution from the social sector is exemplarily shown below (cf. Fig. 11.6): ◀

(1) Decision for a quality management system In a first step, the decision to introduce a quality management system is taken jointly by the management and the quality manager. In a first meeting, the first implementation steps are defined.

(2) Development of documentation techniques At the start of the project, sample procedure manuals for the quality manual are presented and discussed with the quality manager and other involved employees in a first introductory event. In addition, the quality manager takes part in an auditor seminar of the consultant. The aim of this phase is to raise awareness of the need to document the process steps.

(3) Development of the quality management manual In a second information event, the consultant will train the employees in order to promote internalization of the quality policy. In addition, a comparison of the quality management manual, which is largely completed, will be made. The goal of the following training is to deepen and implement

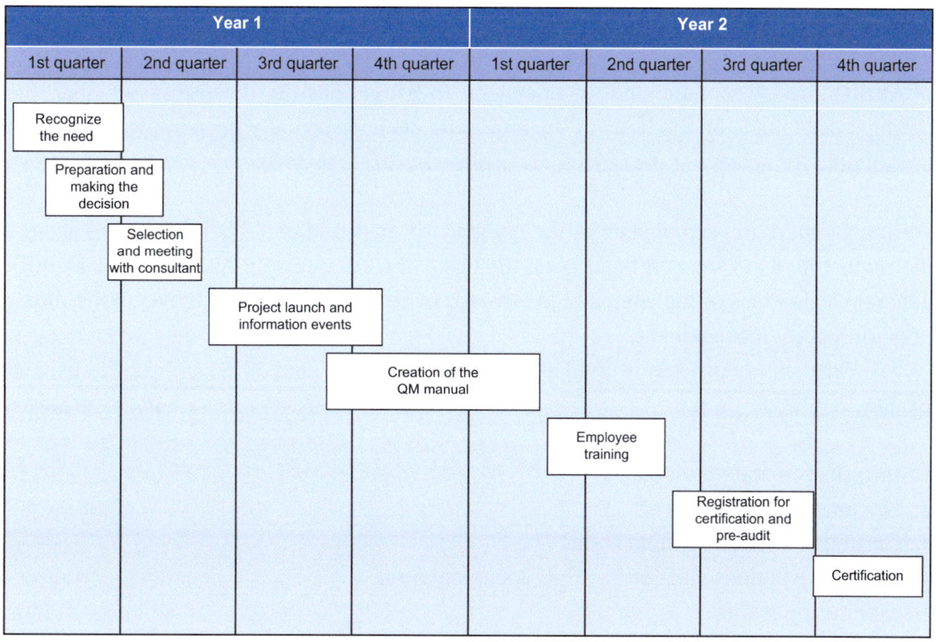

Fig. 11.6 Schedule of a certification project according to DIN EN ISO 9001:2015

the quality policy in daily work. In the last part of this phase, the missing aspects of the quality management manual will be worked out.

(4) Introduction of the quality management manual The quality management manual is finally checked and introduced. In a fourth and final training with the external consultant and the employees, final details will be discussed. After that, eight manual training sessions will be held by the quality manager within the company over the next three months. The goal of these training sessions is the introduction of the quality management system and the concretization of the quality policy in cooperation with all employees.

(5) Registration for certification After the training courses have been completed, the quality management manual is already being used within the company in order to ensure optimum preparation for the upcoming certification. For certification, the certification body is selected, whose auditors check the quality management system and the quality manual for compliance with standards and communicate necessary improvements.

(Source: Based on BBU, 2005 and ISO 9001:2015)

11.3.4 Course of Certification

The certification process is carried out in practice by means of audits, which can be defined as follows:

▶ **Audits** are systematic and independent investigations to determine whether the arrangements made on the basis of the requirements have actually been implemented and are suitable for achieving the objectives (Zollondz, 2011, p. 346).

During an audit, the actual state of the company is analyzed and this is evaluated against the requirements of the certification on the basis of a comparison. Audits aim to identify general weaknesses of the company in advance of certification and to derive action measures for quality management.

The certification process is divided into five **phases** (Brauer, 2009, 41 pp.; Kamiske & Brauer, 2011, 357 pp.):

1. Information and pre-audit
2. Documentation review
3. Certification audit (system audit)
4. Corrective actions and certification documentation
5. Monitoring audits

The phases of certification are presented below.

(1) Information and Pre-audit After the selection of the certification body, the certification standard, the definition of the scope of the company and the preliminary talks with the certification institution, a **contract** is concluded between the service company and the certification body. The scope of the examination, the relevant company areas, the time required and the costs of certification are specified therein. After the conclusion of the contract, the actual certification process begins, which is divided into various sections and is shown in overview form in Fig. 11.7.

After the preliminary information and the conclusion of the contract, the service company receives a **short questionnaire** from the certification body in order to prepare for the **quality audit.**

The company informs the certification body in the form of a formalized self-assessment and answers the questions in writing regarding the implementation of the quality management. Based on the answers to the short questionnaire, the certification body then assesses whether the company can be audited. On request, the results can be communicated to the company in an evaluation meeting, and there is the possibility of a **pre-audit.** The pre-audit is optional; without disadvantages for the actual certification, it is possible to check the implementation of the quality management system. The pre-audit is to be understood as a measure of confidence building while at the same time determining the actual state of the service company. The pre-audit provides indications of application weaknesses, but no specific requirements for changes within the company (Pärsch, 1999, p. 195; Ertl-Wagner et al., 2013; TÜVRheinland, 2013).

(2) Documentation review The formal review of the company begins with the submission of the company documentation on the introduced quality management system to the certification body. This documentation usually consists of a quality management manual and possibly further procedural instructions. These quality management documents are first checked for compliance with the selected standards. The results of the review are communicated to the customer orally and in writing, and if the review is positive, the audit program, the procedure for the certification audit, is created. If weaknesses have been revealed, they must be remedied before the certification audit. If the documentation review has been negative and therefore no certification audit is to be expected, this will be communicated to the company and the further procedure will be coordinated.

(3) Certification audit (system audit) In a next step, a **review** of the implementation of the measures laid out in the quality management documents takes place in the company itself. The auditors of the certification body check that the rules are followed and adhered to in practice by the employees of the company. The standard DIN EN ISO 19011:2002 (guidelines for audits of quality management and/or environmental management systems) can be used for the preparation of the certification audit (Kamiske & Brauer, 2011, 357 pp.). The certification audit is usually divided into the following five steps:

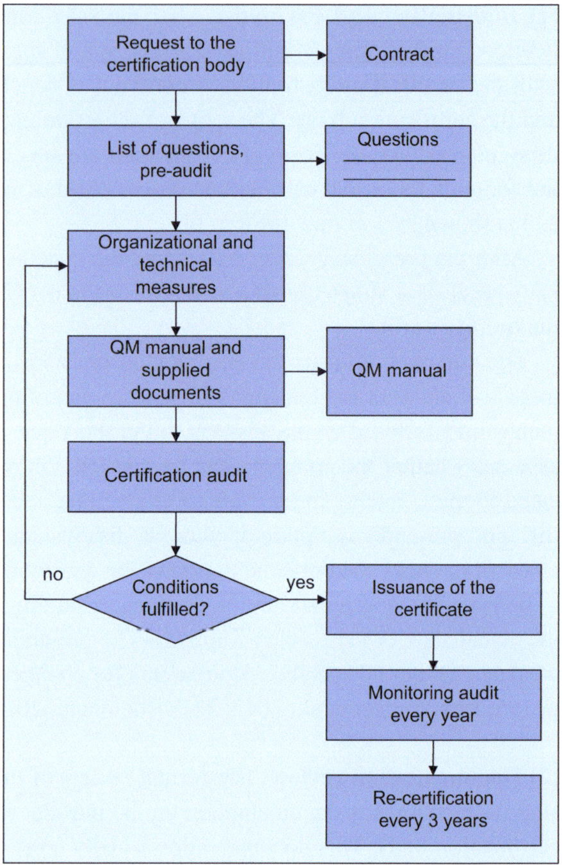

Fig. 11.7 Scheme of the certification of the quality management system. (Source: Brauer, 2009, p. 42)

a) Introductory conversation
b) Review of the implementation of quality management based on a checklist
c) Tour of the company
d) Evaluation of the audit results by the audit team
e) Closing discussion

The examination is carried out on the basis of a **check or audit question list.** Here, significant and rather subordinate questions are distinguished for the company, and the answers are used as an audit protocol. In Fig. 11.8 exemplary questions for a checklist are shown, which are used in the context of the certification audit.

The audit can last several days, the duration depending on the variety of products or services, the size and number of locations of a company (von Schwamen, 1994, p. 194;

Hartz & Meisel, 2011). The assessment of the individual questions, as well as the evaluation of the documentation, results in the following results:

- "fulfilled"
- "partially fulfilled, still acceptable"
- "partially fulfilled, not acceptable"
- "not fulfilled"

On the basis of this formal and practical examination, the certification body draws up the audit report. The audit leader documents the identified compliance and deviations of the company with the certification standard. The findings thus gained are discussed in a final meeting with the management, in which the audit report is explained in more detail. Likewise, the deviations identified in the audit report and the corrective measures intended by the respective company are recorded (von Schwamen, 1994, p. 190; Hartz & Meisel, 2011).

Areas	Questions
Management responsibility	▪ How can the top management demonstrate its commitment regarding the development and improvement of the QM system? ▪ How does the management convey to employees the importance of meeting customer requirements as well as regulatory and legal requirements? ▪ How are the quality policy and quality objectives defined? ▪ How is unintended use of outdated documents prevented?
Resource management	▪ How are the resources needed to improve processes determined and allocated in a timely manner? ▪ Are the personnel to whom responsibilities defined in the QM system are assigned sufficiently competent? ▪ What are the procedures established to identify staff skill needs? ▪ Through which measures do employees become aware of how they contribute to the achievement of quality objectives?
Process management	▪ What are the process needs and which of them still need to be implemented? ▪ What process does the organization use to identify customer requirements? ▪ How is the development of performance planned and managed? ▪ How are the processes of service delivery controlled whose results cannot be verified by subsequent audit?
Measurement and analysis of improvements	▪ What information on customer satisfaction or dissatisfaction is recorded? ▪ Are the specified intervals for conducting audits sufficient to determine whether the QM system is being effectively implemented and maintained? ▪ In what way are the processes required for the continuous improvement of the QM system planned, managed and controlled? ▪ In what way are corrective actions taken to eliminate the causes of errors to prevent recurrence?

Fig. 11.8 Exemplary audit question list. (Source: Based on Meinhold & Matul, 2011)

(4) Corrective measures and certification documentation
If the result of the certification audit is not satisfactory in parts of quality management, **corrective measures** must be carried out within 30 days. The issuance of the certificate by the auditing institution requires the implementation of all necessary corrective measures. Depending on the degree of deviation, either a written statement by the service provider or more extensive individual audits up to a complete follow-up audit is sufficient. If the corrective measures—if necessary—are positively evaluated by the certifier, the audit report is prepared and submitted to an internal, independent review. Less critical deviations are to be remedied within six months, mentioning the corrective measures taken. After that, the certificate is issued, which documents the details of the examination exactly. The certificate is usually valid for three years, subject to positive monitoring audits (Gaster, 1994, p. 927). In the event of changes in the organizational structure or process organization during this time, which have consequences for quality management, the certification body is to be informed so that, if necessary, a review of the further validity of the certificate can be carried out. Examples of quality certificates from *Berger Logistics GmbH,* Radfeld and *Gemeinschaftspraxis für Strahlentherapie,* Singen, are shown in Figs. 11.9 and 11.10.

(5) Surveillance audits In order to ensure that the quality management system is also maintained in full during the validity of a certificate, so-called "surveillance visits" are scheduled at intervals of six months to one year. These surveillance audits are concluded with a short report. In addition, it is possible to extend the validity of the certificate by another three years after the expiry of the three years by means of a new complete evaluation of the quality management system (Bretzke, 1999, p. 586).

11.4 Benefits of Certification

The **benefits of certification** reflect the achievement of internal and external objectives.

In terms of the **external effects of certification**, it should be noted that a certificate does not yet guarantee the actual compliance of a service provider with a specific quality standard for a specific interaction to the customer, but often a benefit of certification is seen in the reduction of **uncertainty on the part of the service seeker.** A certificate provides a whole range of information to customers and simplifies their information search and processing processes, especially against the background of an increasing overload of information. This is all the more true as, in the case of services, the intangible part of the service is of particular importance for the perceived quality of the service. Certification will therefore contribute to increased transparency of service offers for service users.

ZERTIFIKAT ◆ CERTIFICATE ◆ 認証書 ◆ СЕРТИФИКАТ ◆ CERTIFICADO ◆ CERTIFICAT

TÜV SÜD
Management Service

ZERTIFIKAT

Die Zertifizierungsstelle
der TÜV SÜD Landesgesellschaft Österreich GmbH
bescheinigt, dass die Organisation

Berger LOGISTIK

Berger Logistik GmbH
Rettenbach 10a
A-6241 Radfeld

für den Geltungsbereich

Weltweite logistische Dienstleistungen

ein Qualitätsmanagementsystem
eingeführt hat und anwendet.
Durch ein Audit, Bericht-Nr. 153719
wurde der Nachweis erbracht, dass die Forderungen der

ISO 9001 : 2008

erfüllt sind. Dieses Zertifikat ist gültig in Verbindung
mit dem Hauptzertifikat bis Dezember 2013
Zertifikat-Registrier-Nr. Q1530516 / 01

Wien, 2011-02-08

Zertifizierungsstelle
der TÜV SÜD Landesgesellschaft Österreich GmbH
Campus 21, Europaring A04301, A-2345 Businesspark Wien Süd, Austria

TÜV®

Fig. 11.9 Certificate of *Berger Logistics GmbH*. (Source: Berger Logistics GmbH, 2012)

Certification has a positive **advertising and image effect**. Although there are strict limits on the advertising use of certification, as references to certification may not be related to individual services or used in a way that suggests that the certificate relates to a specific service (Rothery, 1991, p. 112; Apel, 2006; Kramer, 2010, p. 279). In terms of the qualitative evaluation of the benefits of certification, there is a **marketing benefit** in the sense of the acquisition of competitive advantages and the avoidance of potential competitive disadvantages. Certification represents at least a competitive advantage for the service provider concerned for as long as the competitors do not have corresponding certificates.

ZERTIFIKAT ◆ CERTIFICATE ◆ СЕРТИФИКАТ ◆ CERTIFICADO ◆ CERTIFICAT

TÜV SÜD Management Service

ZERTIFIKAT

Die Zertifizierungsstelle der TÜV SÜD Management Service GmbH

bescheinigt, dass das Unternehmen

Gemeinschaftspraxis für Strahlentherapie
Virchowstr. 10B, D-78224 Singen
Röntgenstraße 12, D-88048 Friedrichshafen

für den Geltungsbereich

Strahlentherapie

ein Qualitätsmanagementsystem eingeführt hat und anwendet.
Durch ein Audit, Bericht-Nr. **70733783**
wurde der Nachweis erbracht, dass die Forderungen der

ISO 9001:2008

erfüllt sind. Dieses Zertifikat ist gültig bis **2014-01-30**
Zertifikat-Registrier-Nr. **12 100 33142 TMS**

München, 2011-02-08

QMS-TGA-ZM-07-92

TÜV SÜD Management Service GmbH • Zertifizierungsstelle • Ridlerstraße 65 • 80339 München • Germany

Fig. 11.10 Certificate of *Gemeinschaftspraxis für Strahlentherapie*. (Source: Gemeinschaftspraxis für Strahlentherapie, 2012)

Exploitation of cost reduction potentials of *DPD Deutsche Paket Dienst GmbH & Co. KG*

DPD Deutsche Paket Dienst GmbH &Co. KG recorded a total cost saving of 5% for the certification period of one year. The saving is estimated to be 10% in the medium to long term. This results in particular from the

- Reduction of losses
- Reduction of complaints
- Reduction of system errors
- Increase in productivity

(Source: von Schwamen, 1994, p. 194). ◀

In connection with a certification, an increase in **employee motivation** and an **improvement in quality awareness** are also to be mentioned as significant internal benefits. In particular, the task of "obtaining the certificate" has a positive effect on team spirit within the certified areas. As a result of the resulting "sense of change", communication processes on quality issues are often intensified. Often, a quality philosophy is created or improved within the company by means of certification. In companies that implement holistic quality concepts, certification has significantly contributed to the improvement of the anchoring of these concepts. The success of *Deutsche Paket Dienst* lies, in addition to the creation of a sense of change, the improvement of communication and a concentration of know-how, above all in the non-capital-oriented values of the expansion of scope of action for employees, the emergence of a quality awareness and in the motivation and team spirit of the entire company (von Schwamen, 1994, p. 63; Hartz & Meisel, 2011).

In addition, the **results of four empirical studies** are presented with regard to the benefits. Although a large part of the companies investigated are manufacturing companies, the mentioned effects appear to be given for service companies as well.

An empirical study by Specht and Schenk showed that the companies they surveyed saw the following **effects of certification** (Specht & Schenk, 1995, p. 63):

- Possibility of implementing measures that have been under consideration for a long time as part of preparation for certification
- Initial ignition of an increase in quality awareness within the company
- Problem detection and resolution
- Efficiency improvement of the organizational structure and processes through a more specific distribution of responsibility
- Avoidance of duplicate work
- Process optimizations
- Acquisition of new customers through competitive advantages
- Increase in customer loyalty

Similar results are also obtained from the studies by Homburg and Becker (1996) and Vloeberghs and Bellens (1996). For example, the study by Homburg and Becker (1996) showed that, as internal effects, achieving standardized processes (this effect was given by 87.5% of the companies surveyed) and improving the working atmosphere (37.5%) were in the focus, while, as an external effect, an improved quality image (77.5%) was in the focus. Fig. 11.11 provides an overview of the survey results with regard to the positive **effects of certification.** The majority of the companies surveyed by Vloeberghs and Bellens (1996) also stated that, with regard to **external benefit effects** of certification, increasing the trust of customers is in the focus. With regard to **internal benefit effects,** the majority of those surveyed gave particular weight to improving internal quality related performance processes.

Another study by Terlaak and King (2006) also makes it clear in the context of an 11-year panel that certification within quality management serves primarily to reduce

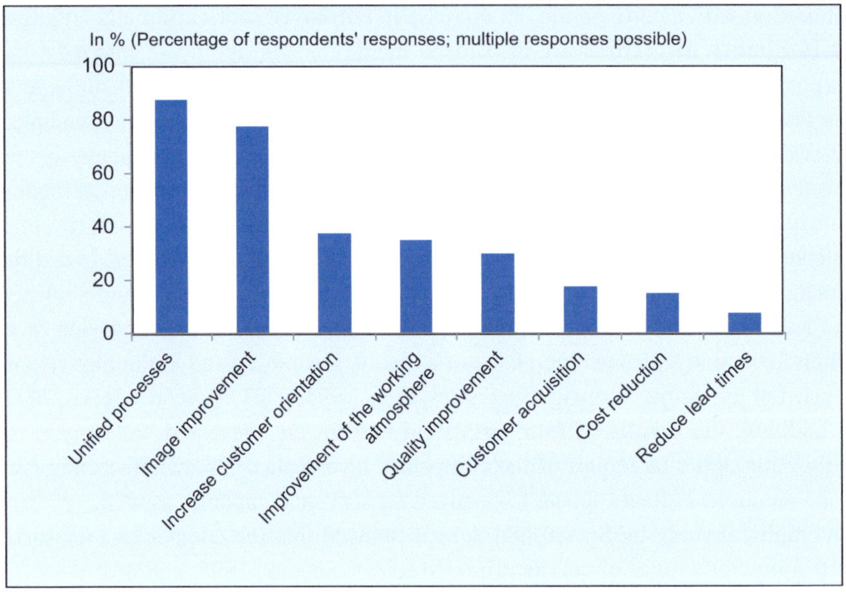

Fig. 11.11 Benefits of certification. (Source: Homburg & Becker, 1996, p. 449)

information asymmetries between providers and customers and to generate a competitive advantage. These results are also confirmed by the study by Terziovski and Guerrero (2014). Here, the authors show that certification can be crucial in increasing the performance of the company. In a study, Starke et al. (2012) finally examine the **effect of certification** on three financial ratios that, according to financial policy considerations, should have a relationship to the implementation of a quality management system: sales revenue, cost of sales vs.sales revenue and turnover. To do this, they measure the changes in the indicators of all Brazilian companies that were certified according to the ISO 9000 standard between 1995 and 2006. The results show a significant increase in sales revenue and turnover as well as a significant decrease in the cost of sales vs.sales revenue ratio.

In summary, it can be said that the **achievement of certification goals** has a generally higher degree of fulfilment in terms of internal effects. In comparison, the fulfilment of requirements with regard to external effects is lower. Only for the respective industry leader does certification offer a chance to achieve a significant external effect. The essential value of an initiated quality process lies above all in the demonstrably increased customer satisfaction and employee motivation, because these are ultimately the significant economic and social success potentials of a service company (Stauss, 2001; Haller, 2018). In this context, the successful certification of the Suva (Swiss Accident Insurance Institution) insurance medical competence center is to be mentioned. The following example shows the initial problem areas as well as the resulting success.

Insurance medical competence center with certification by *Suva*

Suva, the largest provider of (mandatory) accident insurance in Switzerland, operates the first insurance medical competence center in Switzerland with a management system certified according to ISO standard 9001. Insurance medicine plays a decisive role at *Suva.* The doctors affiliated with this area mediate between the insurance and medical service providers, advise employers and employees, and coordinate with other medical experts in the event of damage. In addition, they are in direct contact with sick patients or employees.

The need for certification resulted in particular from the heterogeneous processing of insurancentechnical processes. In addition, customer needs of employers, employees, insurers and medical service providers were often not recorded, and performance goals were not uniformly and customeroriented. As part of the certification process or the change of internal processes, it was often necessary to motivate employees. "Sometimes the sessions were of a technical nature—then there were meetings where it was almost exclusively about motivating people to keep going," describes the head physician of the competence center the situation.

Finally, the expectations of the certification were nevertheless more than fulfilled, especially since the benefit for *Suva* lies in particular in the fact that it now has a system that provides insured persons with services of constant and verifiable quality.

(Source: Piazza, 2003, p. 60). ◀

11.5 Critical Assessment of Certification

In connection with a certification, numerous **problems** can be identified, the consideration of which is often helpful in the systematic planning and successful implementation of a certification. The potential difficulties can be divided into four problem groups, which are summarized in Fig. 11.12:

1. Problems related to certification standards
2. Problems related to the certification process
3. Problems related to the company
4. Problems related to the industry

(1) Problems related to the certification A key criticism is that many companies acquire certification for **self-purpose** and not for a holistic implementation of quality management. Thus, short term, external goals are pursued that do not meet the purpose of quality management (Pfitzinger, 2009; Scholtis, 2013; Brauweiler et al., 2015).

With regard to the ISO 9001 certification standards, it must first be criticized that it will certainly prove to be difficult in practice for companies to identify the essential processes for each organization, as required by the standard. Every company has a **variety**

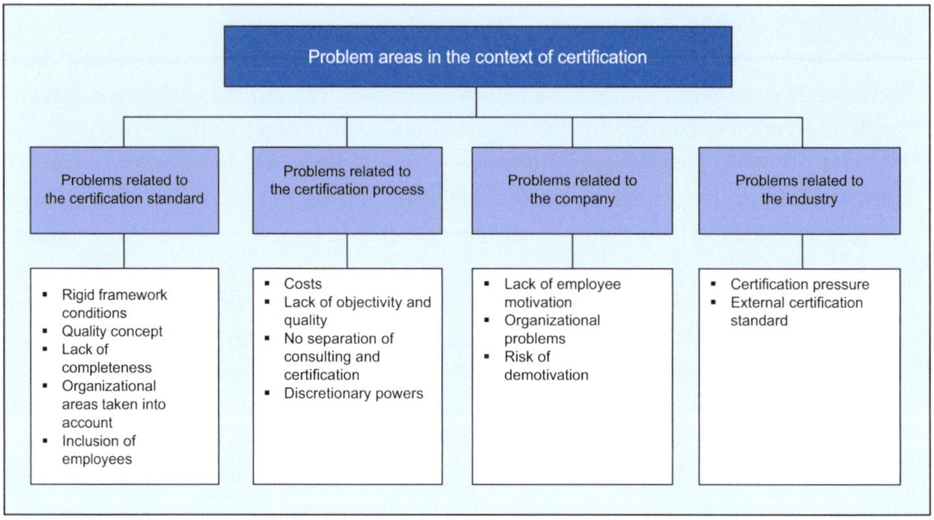

Fig. 11.12 Problem areas in the context of the certification of service companies

of processes that need to be analyzed in terms of their relevance to achieving customer satisfaction and the company's overall goals. The determination of the interaction between the individual processes is particularly problematic in this context.

Further difficulties arise in the **definition of quality objectives.** According to a survey by the *German Society for the Certification of Management Systems,* more than 50% of certified companies had considerable problems in defining measurable quality objectives (DQS, 2002). This also results in difficulties in finding points of improvement for the effectiveness of quality management and guaranteeing continuous improvement.

In addition, no requirements were included in the revised standard with regard to the **employee orientation** (Gietl & Lobinger, 2012; Brauweiler et al., 2015). Without satisfied and motivated employees, however, the explicitly formulated goal of customer satisfaction in the standard will be difficult to achieve.

Furthermore, the implementation of ISO 9001 as the basis for a certification often leads to very high **bureaucratic effort** because processes have to be documented. In this context, critics also point out that the prescribed templates would restrict company activities because they should not violate the standard (Pfitzinger, 2009; Scholtis, 2013; Brauweiler et al., 2015).

(2) Problems related to the certification process Another group of problems related to certification concerns aspects that are related to the **certification process** itself. In this regard, the first thing to mention are the costs of certification, which, together with the other costs of quality management, have to be set against the achievable cost reduction potentials and revenue increases due to certification.

Further weaknesses in this area concern the certification companies. First of all, certification is increasingly becoming a commercial process, so that in some cases **lack of objectivity and quality** in the certification process has to be feared. In addition, there is **no regulation of the separation of consulting and certification services,** so that companies that both advise and certify often run the risk of lack of objectivity in the examination. Finally, there are large discretionary powers for the certification companies in the context of their examinations (Homburg & Becker 1996; Pfitzinger, 2009; Gietl & Lobinger, 2012).

(3) Problems realted to the company The third group of problems in the context of certification are those arguments that are to be found within the company to be certified. The investigation by Homburg and Becker (1996) showed that a large part of the companies interviewed complained about the **lack of motivation of the employees** in the certification preparations. Other problem areas mentioned by the companies concern **organizational problems,** such as unclear task allocation, internal communication difficulties and problems in cooperation with the certification company (Pfitzinger, 2009; Scheibeler, 2014).

Furthermore, from an internal company perspective, it should be noted that the employee motivation induced by a certification project usually relates to the initial phase of the certification process and a flattening or even a **demotivation,** especially in the case of a unsuccessful certification, is to be expected.

(4) Problems related to the industry If you look at the industry of the company to be certified, a weakness can be mentioned that is related to the credibility of the certification. If certification is established within an industry, in some cases this creates a certain **pressure to certify** for the competitors, which is then often interpreted as a standard. In this respect, the expansion of the potential customer base as well as the promotion and facilitation of business processes are to be questioned.

In this context, reference is made to the largely given external motivation for certification. Fig. 11.13 shows the reasons for certification that were collected by Homburg and Becker (1996). Due to the partly existing **pure external certification motivation,** such as the customer's demand for certification or the reaction to the behavior of the competition, it has to be questioned whether the companies concerned are aware of the content of the standards and the actual goals of the certification. Only the documentation of quality management is to be seen as an internal goal of certification.

Finally, it should be noted that there are various benefits as well as various problems in relation to certification. Therefore, the service company concerned must check in each individual case whether certification is an appropriate instrument for implementing quality management.

In the previous chapters, the importance of implementing quality management and its benefits have been highlighted if it is consistently and holistically implemented.

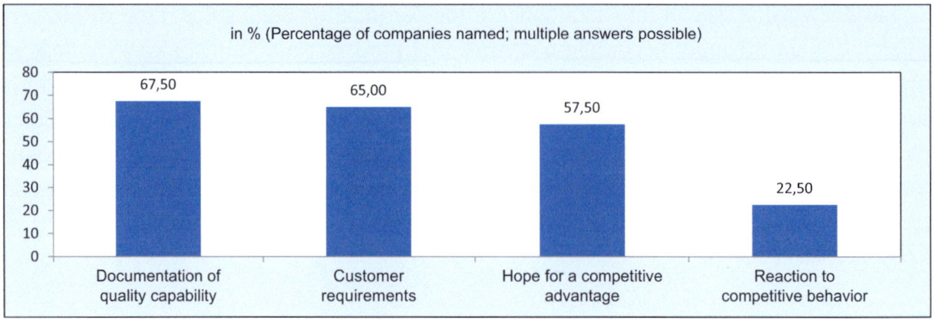

Fig. 11.13 Reasons for certification. (Source: Homburg and Becker, 1996).

However, in order to ensure this, a continuous review is required. Therefore, quality control for services is addressed in the following.

References

Apel, H. (2006). Qualitätssicherung im Kontext einer Bildung für nachhaltige Entwicklung (BNE). In W. Rieß & H. Apel (Eds.), *Bildung für eine nachhaltige Entwicklung* (pp. 129–138). Wiesbaden: VS Verlag für Sozialwissenschaften.

BBU. (2005). Typischer Ablauf eines Zertifizierungsprojektes. http://www.bbu-unternehmensberatung.de/archive/zert.pdf. Accessed 21. Feb. 2013.

Berger Logistik GmbH. (2012). Zertifikat der Berger Logistik GmbH. http://www.berger-logistik.com/berger-logistik/qualitaet/zertifizierte-qualitaet/. Accessed 31. May 2012.

Brauer, J. P. (2009). *DIN EN ISO 9000:2000 ff. umsetzen. Gestaltungshilfe zum Aufbau Ihres Qualitätsmanagementsystems* (5th Edn.). München: Hanser.

Brauweiler, J., Will, M., & Zenker-Hoffmann, A. (2015). *Auditierung und Zertifizierung von Managementsystemen: Grundwissen für Praktiker.* Wiesbaden: Springer Gabler.

Bretzke, W.-R. (1999). Zertifizierung von Qualitätsmanagementsystemen in Dienstleistungsunternehmen. In M. Bruhn & B. Stauss (Eds.), *Dienstleistungsqualität. Konzepte—Methoden—Erfahrungen* (3rd Edn., pp. 573–600). Wiesbaden: Gabler.

Deutsche Gesellschaft für Qualität e. V. (1995). *DQS-Zertifikate. Liste, Zertifizierungsleistungen und Preise.* Frankfurt a. M.: Beuth

DIN EN ISO 9001:2015. (2015). *Qualitätsmanagementsysteme—Grundlagen und Begriffe.* Berlin: Beuth.

Döttinger, K., & Klaiber, E. (1994). Realisierung eines wirksamen Qualitätsmanagementsystems im Sinne des Total Quality Managements. In B. Stauss (Ed.), *Qualitätsmanagement und Zertifizierung von DIN ISO 9000 zum Total Quality Management* (pp. 255–273). Wiesbaden: Gabler.

DQS. (2002). Forderungen an ein Qualitätsmanagementsystem. www.dqs.de. Accessed 16. July 2012.

Drechsler, M. (2007). Zertifizierung von Qualitätsmanagementsystemen. In T. Pfeifer & R. Schmitt (Eds.), *Masing—Handbuch Qualitätsmanagement* (5th Edn., pp. 343–350). München: Hanser.

DVGW CERT GmbH. (2012). Akkreditierungsurkunde der DVGW CERT GmbH zur Zertifizierung von Qualitätsmanagementsystemen. http://mycert.dvgw-cert.com/. Accessed 31. May 2012.

Ertl-Wagner, B., Steinbrucker, S., & Wagner, B. C. (2013). *Qualitätsmanagement und Zertifizierung. Praktische Umsetzung in Krankenhäusern, Reha-Kliniken, stationären Pflegeeinrichtungen* (2nd Edn.). Berlin: Springer.

Gaster, D. (1994). Qualitätsaudit. In W. Masing (Ed.), *Handbuch Qualitätsmanagement* (3rd Edn., pp. 927–948). München: Hanser.

Gemeinschaftspraxis für Strahlentherapie. (2012). Zertifikat der Gemeinschaftspraxis für Strahlentherapie. http://www.strahlentherapie-ostalb.de/. Accessed 31. May 2012

Gietl, G., & Lobinger, W. (2012). *Leitfaden für Qualitätsauditoren: Planung und Durchführung von Audits nach ISO 9001: 2008*. München: Hanser.

Haller, S. (2017). *Dienstleistungsmanagement* (7th Edn.). Wiesbaden: Springer Gabler.

Hartz, S., & Meisel, K. (2011). *Qualitätsmanagement* (3rd Edn.). Bielefeld: Bertelsmann.

Holland, H. (2009). Kundenbindungsmanagement. In H. Holland (Ed.), *Direktmarketing* (pp. 309–322). München: Vahlen.

Homburg, C., & Becker, J. (1996). Zertifizierung von Qualitätssicherungssystemen nach den Qualitätssicherungsnormen DIN ISO 9000 ff. Eine kritische Beurteilung. *Wirtschaftswissenschaftliches Studium, 25*(9), 444–450.

Kamiske, G. F., & Brauer, J.-P. (2011). *Qualitätsmanagement von A bis Z. Erläuterungen moderner Begriffe des Qualitätsmanagements* (7th Edn.). München: Hanser.

Kramer, M. (2010). *Integratives Umweltmanagement. Systemorientierte Zusammenhänge zwischen Politik, Recht, Management und Technik*. Wiesbaden: Gabler.

Meinhold, M., & Matul, C. (2011). *Qualitätsmanagement aus der Sicht von Sozialarbeit und Ökonomie*. Stuttgart: Nomos.

Pärsch, J. (1999). Zertifizierung von Qualitätsmanagementsystemen. In W. Masing (Ed.), *Handbuch Qualitätsmanagement* (4th Edn., pp. 193–202). München: Hanser.

Pfitzinger, E. (2009). *Projekt DIN EN ISO 9001:2008: Vorgehensmodell zur Implementierung eines Qualitätsmanagementsystems*. Berlin: Beuth.

Piazza, K. (2003, May 7). Sprechstunde mit Gütesiegel. *Handelszeitung*, Nr. 19, p. 60.

Rothery, B. (1991). *Quality assurance*. Brookfield.

Saatweber, J. (2011). *Kundenorientierung durch Quality Function Deployment. Produkte und Dienstleistungen mit QFD systematisch entwickeln* (3rd Edn.). Düsseldorf: Symposion.

Scheibeler, A. A. (2014). *Balanced Scorecard für KMU: Kennzahlenermittlung mit ISO 9001: 2000 leicht gemacht* (3rd Edn.). Wiesbaden: Springer.

Schlüter, S., & Dunkhorst, P. (2003). *ISO 9001:2000. Qualitätsmanagement praxisgerecht einführen und weiterentwickeln* (2nd Edn.). Hamburg: B. Behr's.

Schmitt, R., & Pfeifer, T. (2015). *Qualitätsmanagement: Strategien—Methoden—Techniken* (5th Edn.). München: Hanser.

Scholtis, T. (2013). *Vertragsgestaltung bei Informationsasymmetrie: Probleme und Lösungen bei der Zertifizierung von QM-Systemen nach ISO 9000*. Wiesbaden: Springer.

von Schwamen, J. (1994). Effiziente Vorbereitung auf die Zertifizierung sowie deren Durchführung in einem Dienstleistungsunternehmen. Erfahrungsbericht der Deutsche Paket Dienst GmbH. In B. Stauss (Ed.), *Qualitätsmanagement und Zertifizierung von DIN ISO 9000 zum Total Quality Management* (pp. 165–196). Wiesbaden: Gabler.

Specht, G., & Schenk, M. (1995). *Auswirkungen der Zertifizierung nach DIN (EN) ISO 9001 bis 9003. Ein Bericht über eine empirische Studie, Arbeitspapier 7.* Darmstadt.

Starke, F., Eunni, R. V., Manoel Martins Dias Fouto, N., & Felisoni de Angelo, C. (2012). Impact of ISO 9000 certification on firm performance: Evidance from Brazil. *Management Research Review, 35*(10), 947–997.

Stauss, B. (2001). Stichwort: Internes Marketing. In H. Diller (Ed.), *Vahlens Großes Marketinglexikon* (2nd Edn., pp. 477–479). München: dtv Verlagsgesellschaft.

Terlaak, A., & King, A. A. (2006). The effect of certification with the ISO 9000 quality management standard: A signaling approach. *Journal of Economic Behavior & Organization, 60*(4), 579–602.

Terziovski, M., & Guerrero J.-L. (2014). ISO 9000 Quality system certification and its impact on product and process innovation performance. *International Journal of Production Economics, 158,* 197–207.

TÜVRheinland. (2013). Der Zertifizierungsablauf. http://www.tuv.com/de/germanyinfothek/infothek/zertifizierungspartner/zertifizierungsablauf/zertifizierungsablauf.html. Accessed 25. Mar. 2013.

Vloeberghs, D., & Bellens, J. (1996). ISO 9000 in Belgium. Experience of Belgian quality managers and HRM. *European Management Journal, 14*(2), 207–211.

Wagner, K. (2007). Ausgestaltung von QM-Systemen auf Basis der ISO-9000-Reihe. In T. Pfeifer & R. Schmitt (Eds.), *Masing—Handbuch Qualitätsmanagement* (5th Edn., pp. 173–205). München: Hanser.

Wagner, K. W., & Käfer, R. (2013). *PQM-Prozessorientiertes Qualitätsmanagement: Leitfaden zur Umsetzung der ISO 9001* (6th Edn.). München: Hanser.

Zollondz, H.-D. (2011). *Grundlagen Qualitätsmanagement. Einführung in Geschichte, Begriffe, Systeme und Konzepte* (3rd Edn.). München: Oldenbourg.

Part V

Quality Controlling for Services

12 Concept of Quality Controlling for Services

Abstract
Quality controlling, as an important phase of quality management for services, has the purpose of checking the degree to which quality management objectives have been achieved. Accordingly, this is where the economic efficiency of a service company is to be taken into account by means of quality control. In this context, important functions of quality control are explained in this chapter and the individual components of a comprehensive control are presented. A distinction is made between strategic, operational and integrated quality control. On this basis, companies can develop and implement their own concept for quality control. As part of a systematic quality management for services, activities of analysis, planning and implementation in relation to service quality are carried out, as described in the previous parts of this book. In order to take into account the principle of effectiveness and the economic efficiency of a service company, these activities are continuously subjected to quality control.

12.1 Concept and Functions of Quality Control

The term "controlling" is often wrongly equated with the exclusive control of company activities. On the contrary, it characterizes a certain part of corporate management, the task of which is only partially to control company activities. The overriding task of controlling is to coordinate the management system of a company (Horváth et al., 2020, 35 pp.). Thus, the **concept of quality control** is defined as follows (Bruhn, 1998, p. 63):

M. Bruhn, *Quality Management for Services*,
https://doi.org/10.1007/978-3-662-67032-3_12

▶ As **quality control,** the analysis, planning, implementation and control of the support and coordination of quality related activities with a view to aligning quality management with the desired objectives is to be understood.

Based on this definition, the increase in the effectiveness and efficiency of quality management is the **overall goal of quality control** (Horváth et al., 1994, p. 3; Benkenstein & Stenglin, 2006). While **effectiveness** of quality management is to be understood as the production of services in accordance with customer requirements, **efficiency** of quality management refers to the economic implementation of corresponding quality activities (Bruhn, 1998, p. 64). For example, increasing customer satisfaction or customer orientation of employees are to be counted among the goals of quality management (Homburg & Stock, 2005). Quality control serves the effective and efficient realization of these goals.

To ensure the effectiveness and efficiency of quality management, quality control has four **functions** to fulfill (Meffert et al., 2018, pp. 481–482; see Fig. 12.1):

1. Coordination function
2. Information function
3. Planning function
4. Control function

(1) Coordination function The coordination function relates to the central task of quality controlling, coordinating various quality related activities of the company (Horváth & Urban, 1990, p. 12; Tomys, 1995, p. 90; Reinecke & Janz, 2007; Schmitt & Reissiger,

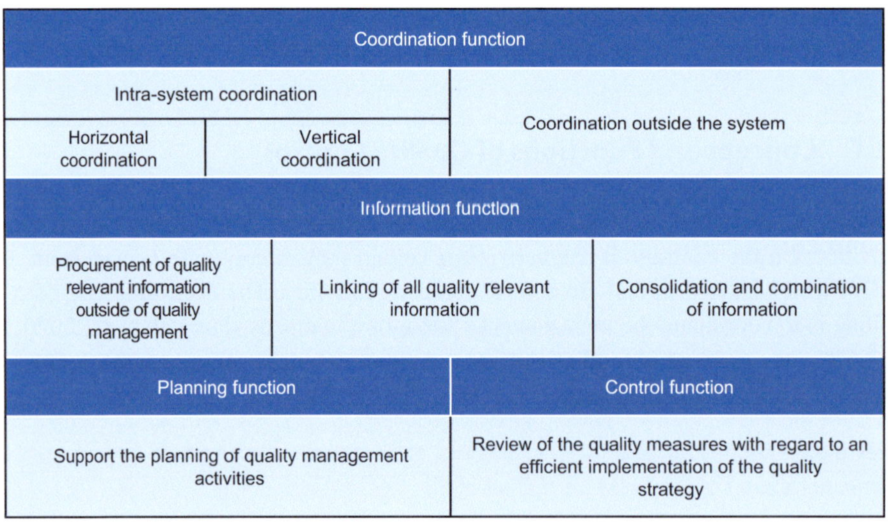

Fig. 12.1 Functions of quality controlling

2010). The requirement for coordination of quality related activities can be traced back in particular to the character of quality management as a cross-sectional function within the company. A need for coordination arises, for example, because employees at different hierarchical levels bear quality responsibility. In addition, the activities of all company departments have an impact on the quality of company performance (Meffert et al., 2018, 481 pp.). Based on these considerations, two **directions of quality related coordination** can be distinguished (Meffert et al., 2018, 481 pp.):

a) The **vertical coordination** serves to coordinate quality related activities at different hierarchical levels.
b) The **horizontal coordination** serves to coordinate quality measures between different company departments.

(2) Information function In the context of quality management of a company, numerous quality relevant information is generated, which are to be made available to the departments involved in quality management at the right place and at the right time. In this context, three **phases of the information process** are to be distinguished (Meffert et al., 2018, p. 482):

- In the phase of the **information demand analysis** it is to be decided which type of information for quality management for which departments or employees is relevant. It is to be defined which information is to be collected in order to ensure a high quality of service.
- In the context of the **information acquisition** it is to be defined which information sources are to be used. In this context, internal (e.g. accounting, customer databases) and external (e.g. market research results) information sources are distinguished.
- The phase of the **information preparation and storage** defines in which form the collected information is stored and made available to the relevant internal departments. An important criterion here is the degree of detail and aggregation. Furthermore, it is to be defined in which time interval (e.g. daily, monthly, quarterly) the information should be available.

(3) Planning function The support of the planning of quality management by quality control is the central task with regard to the planning function. In this context, methods are to be provided with which the planning activities of quality management can take place according to a company-wide system (Reinecke & Janz, 2007; Schmitt & Reissiger, 2010). These include the preparation and interpretation of the relevant quality, time and cost data in so-called quality reports (Zollondz, 2011). Quality control thus takes over a support function for quality management in the planning and evaluation of the quality strategy and the measures to be taken for efficient achievement of the quality objectives.

(4) Control function Finally, quality control also involves the control of quality related activities. Appropriate methods must also be provided for this purpose. In particular, the interactions between planning and control are to be taken into account here, which are not to be interpreted as isolated from each other (Chwolka, 1996; Schmitt & Reissiger, 2010).

The main task of quality control is the coordination of all activities of a service provider that are necessary for the efficient creation of service quality. Its organizational implementation is the quality control in the **quality control system:**

▶ Under a **quality control system** for services, the combination of various bundled measure catalogs is to be understood, which serve the systematic support and coordination of quality related activities of a service provider.

Even though quality control primarily concerns the quality management system (Kamiske & Brauer, 2011), a broader perspective is required to illustrate the content related anchoring of quality control (cf. Fig. 12.2). Since quality control relates to all quality related activities of a service company, two **subsystems of the quality execution system** have to be considered due to the close connection between quality management and performance creation:

1. The **quality management system,** in which all activities for increasing service quality are organized.
2. The **performance creation system,** in which all activities for performance creation are located.

The structure of the quality control system results from the two types of control functions. The subject of system forming coordination is the design of structures and processes with the aim of orienting and adapting the quality related activities of the service provider to expected future developments with quality relevance. With regard to the planning and controlling as well as the information provision function, two **subsystems** are to be formed within the **quality control system:**

1. The **quality related planning and control system** through which the planning and control functions of quality control are carried out.
2. The **quality related information system** through which the information function of quality management is fulfilled.

These two systems cannot be considered independently of each other. For example, the information gained within the framework of the information system flows into the processes controlled by the planning and control system. Therefore, the central task of quality control here is the coordination of the processes of the two subsystems. As illustrated in Fig. 12.2, both the quality related planning and control system and the quality related information system have a close relationship to the quality management system.

The **system coupling functions of quality controlling** relate to the coordination of quality related activities within the system framework formed. Adjustments are to be made in the given system structure, in particular in the case of unforeseeable deviations. In this context, four **elements of quality controlling** are distinguished (Bruhn, 1998, p. 83):

1. Within the framework of the quality related information system, the **quality information center** is in the focus. The aim of this element is to ensure that all external and internal quality related information is collected, analyzed and stored.
2. One element of the quality related planning and control system is the **auditing of quality management**. As part of the auditing, a critical examination of all quality

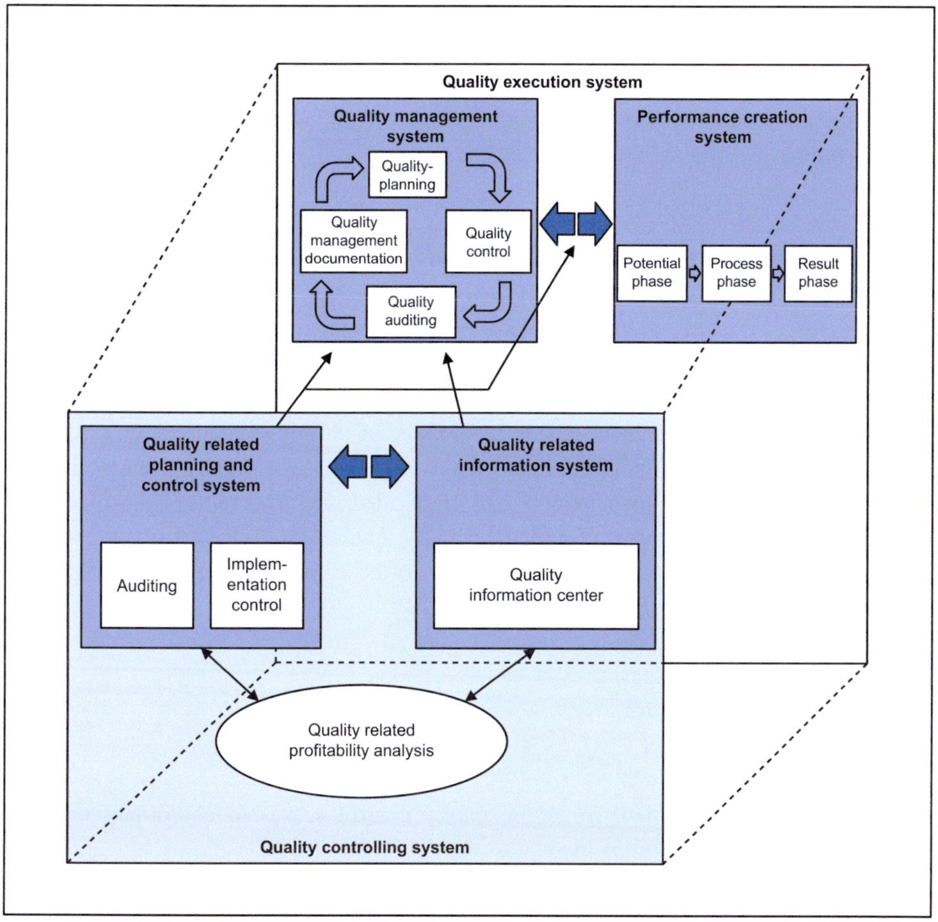

Fig. 12.2 Interrelationships between quality controlling system and quality execution system of a service company. (Source: Bruhn, 1998, p. 84)

related procedures and decision making processes is carried out (Piontek, 2005; Boon et al., 2008; Habib, 2012).

3. Furthermore, the quality related **implementation control** is to be attributed to the quality related planning and control system. Through this element, the quality related objectives and the respective achievement levels are checked.
4. Finally, the quality related **cost-benefit analysis** of quality management is the central element of quality control. Its importance becomes apparent when considering the tasks and functions of quality control. The main task of quality control is to ensure the economic implementation of all quality related activities. Furthermore, the cost-benefit analysis forms the basis for fulfilling both the planning and control as well as the information function. With regard to the information function, the cost-benefit analysis analyses and provides the data required for an assessment of the quality related cost effectiveness. Furthermore, parts of the planning and control functions are carried out by the cost-benefit analysis, in that the results of the cost-benefit analysis form the basis for controlling the quality related processes of the service provider. The results of a quality related cost-benefit analysis therefore form the basis for the activities within the framework of the auditing and the implementation control of quality management.

12.2 Components of Quality Control

Quality control is mainly used for **coordinating the activities of quality management** in service companies. With quality management, activities of strategic quality management and operational quality management can be differentiated. For example, the quality philosophy and strategy of the company are defined as part of strategic quality management. Operational quality management serves to implement the quality strategy in daily work. Operational quality management can be divided into the control loop of quality management—with the phases of quality planning, quality control, quality auditing and quality management documentation—according to the control loop of quality management (see Fig. 12.3; Bruhn & Georgi, 1998, p. 30f.):

1. Strategic quality control
2. Operational quality control
3. Integrated quality control

(1) Strategic quality control Strategic quality control serves to coordinate strategic quality management. The aim is to support the definition of quality philosophy and strategy, and to ensure the quality capability of the company (Horváth & Urban, 1990, p. 14; Simons, 2013). The main goal of strategic control is primarily the **long term security** of the company through the acquisition and retention of potentials and competitive

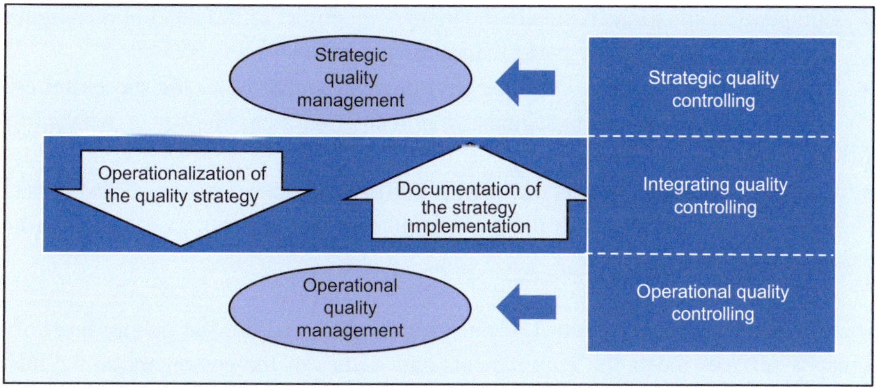

Fig. 12.3 Components of quality controlling. (Source: Bruhn & Georgi, 1999, p. 11)

advantages (Erichsen, 2011, p. 199). The tasks of the control are primarily the detection of opportunities and potentials as well as the identification and avoidance of risks. In this context, it is necessary to take into account both the external and the internal company situation within the framework of strategic quality management and to ensure an adaptation of quality management to changing environmental conditions. A **quality related analysis of the company situation** can, for example, be divided into the following areas:

- Market situation
- Customer situation
- Trading situation
- Supplier situation
- Competition situation
- Environmental situation
- Company situation

Strategic quality control using the example of a hotel

In order to implement a first-class philosophy, it is recommended for a hotel to carry out the following quality related analyses as part of strategic quality controlling:

- Market situation: How is the market for first-class hotels structured in the corresponding region/city? What are the common quality standards for first-class hotels?
- Customer situation: How are the typical customers of a first-class hotel characterized? What expectations do they have of the range of services offered by a hotel in this segment?
- Supplier situation: Are the suppliers able to support the hotel in providing first-class services (e.g. suppliers for food for a planned gourmet kitchen)?

- Competition situation: What are the main competitors of the hotel in this segment? What special promises of performance distinguish these hotels?
- Environmental situation: Does the environment contribute to the successful establishment of a hotel in this segment (e.g. recreation area, number of relevant trade fairs in the city)?
- Company situation: How is the hotel set up? Are the corresponding (also financial) resources and structures available to implement the concept of a first-class hotel internally? ◀

As part of strategic quality control, it must be checked whether the quality level of the company's services meets the requirements and wishes of the customers on the market and allows the company to differentiate itself from the competition. It should be noted that strategic quality control has a time horizon of several years. Therefore, only limited detailed key figure systems are used, instead basic statements and recommendations for further company development are derived from strategic control (Erichsen, 2011, pp. 199–200). These include the definition of the quality position as well as the derivation and implementation of the quality strategy.

The determination of the strategic **quality position** is the essential basis for quality management. Depending on the current and desired positioning in the market and in relation to competitors, different priorities must be set in the design of quality management. In the context of auditing the quality position, the structures and processes of quality management are to be examined with regard to the achievement of the desired positioning. In the context of implementation control, it is checked whether and to what extent the desired positioning has actually been realized. For the control of the quality positioning, instruments such as the quality related SWOT analysis or portfolio analysis are used. The application for a quality award and participation in a certification process are further measures for the control of the quality positioning.

Based on the quality position, the **quality strategy** is to be defined and implemented in a next step. Here too, auditing and implementation control of the quality strategy are used. The audit checks to what extent the structures and processes in the company are available and suitable for implementing the desired quality strategy. Implementation control checks to what extent the quality strategy has actually been implemented. The implementation of the quality strategy proves to be a long term and complex process that lasts for several years. An ex-post control of the quality strategy is therefore only partly suitable, since many deviations are often only visible at a later stage due to the long time horizon (Schreyögg, 2006, p. 102). For this reason, strategic controlling is to be seen as an accompanying process of quality management, which accompanies the activities in all phases of quality management. Fig. 12.4 shows the embedding of strategic quality control in the quality management process.

Strategic quality control thus takes on the function of a monitoring or an **early warning system** which is to be carried out continuously in order to identify impending deviations from the strategic goal at an early stage and to avert them. In principle, the

Fig. 12.4 Embedding of strategic quality control in the quality management process. (Source: Based on Schreyögg, 2006, p. 103)

following types of **strategic quality control** can be distinguished (cf. Schreyögg, 2006, 103 pp.):

- Strategic requirement control
- Strategic implementation control
- Strategic monitoring

A strategic planning process often relies on requirements in order to simplify the complex business environment and to transform it into a decision-making situation that is as specific as possible (Bea & Haas, 2017, p. 242). These requirements relate to external factors influencing the company such as exchange rate fluctuations, technological developments, changes in demand, etc. The task of **strategic requirement control** is to continuously check the validity of these requirements that have been explicitly formulated and to adjust them accordingly in the event of identified deviations. If strategic quality decisions are made on the basis of requirements that are no longer valid, this can lead to misallocations of the resources used by the company and thus have far-reaching financial consequences for the company.

Strategic requirement control using the example of a supermarket

A supermarket has found through market research that customers' preferences tend to be for high-quality organic products and that they value fresh products from the region. Based on this, the owner has decided to pursue a quality leadership strategy and to offer more high-quality products from organic farmers in the immediate vicinity. However, the economic situation in the region has deteriorated in recent months due to factory closures and the resulting increase in unemployment. Customers are therefore forced to switch to cheaper alternatives. The owner of the supermarket now has to check whether his strategy can still be successfully continued in view of the changed requirements. ◄

The **strategic implementation control** is to be carried out during the implementation of quality management and has the task of identifying any deviations from strategic intermediate goals (so-called milestones). This prevents a deviation from the strategic goals at an early stage. The main task of the strategic implementation control is the definition of individual milestones (see also Sect. 13.3.1 for the milestone technique). If it is shown that individual milestones are not achieved at an early stage, it is to be checked whether the measures taken are to be adapted or whether the strategic goal is to be redefined. Fig. 12.5 shows the systematics of the strategic implementation control at a glance.

Strategic implementation control using the example of a hair salon

The owner of a reputable hair salon has the goal of adjusting his quality positioning and offering more cost-effective services. This should result in ten percent new customers by the end of the year. For this purpose, the owner employs two new inexpensive employees with little professional experience. However, the analysis of the first two quarters shows that the salon has lost many regular customers and hardly any new customers due to the decreased quality. Therefore, it is now to be checked at an early stage whether the strategic goal is to be redefined and which measures are to be taken to reverse the negative trend. ◄

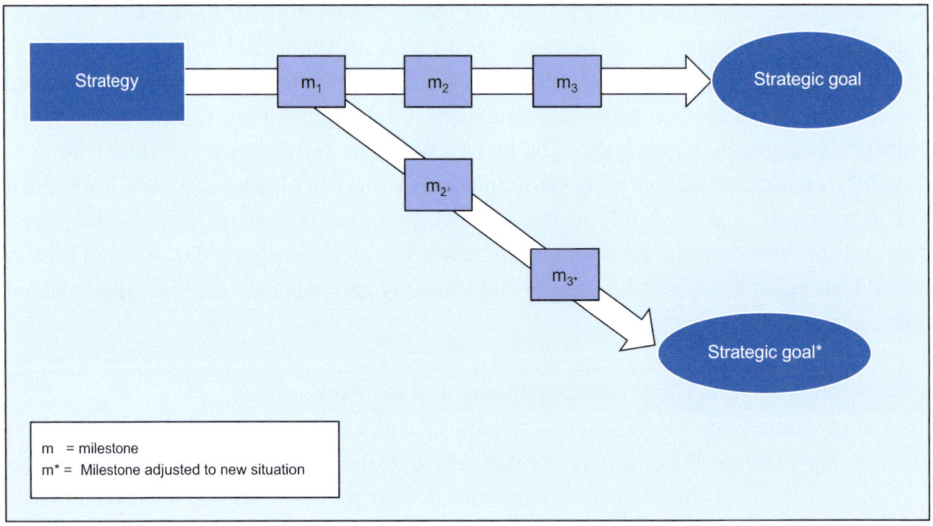

Fig. 12.5 Systematics of the strategic implementation control. (Source: Based on Schreyögg, 2006, p. 106)

Strategic implementation and requirement control are not sufficient for a holistic monitoring of strategic quality management. The task of **strategic monitoring** is to take into account critical incidents and developments that have been neglected or wrongly assessed in the context of requirement control or have not yet or not been fully captured in the implementation control (Schreyögg, 2006, p. 107). The strategic monitoring control therefore takes on a holistic control function over all company activities. Due to the rapid changes in the business environment, strategic monitoring in the sense of an early warning system is of particular importance.

Strategic Monitoring Control at the Example of a Supermarket

In the example of the supermarket, the task of the strategic monitoring control is not to lose the holistic view. The requirement control may focus too much on the design of the product range without taking into account other important quality factors. It is quite conceivable that a decline in sales is not due to the selection of organic products, but to the lack of consulting services or the poor opening times of the supermarket. ◀

The three forms of strategic quality control presented are not to be viewed in isolation, but to be integrated into a strategic control system. Fig. 12.6 shows the temporal use of the different control approaches using the example of a supermarket.

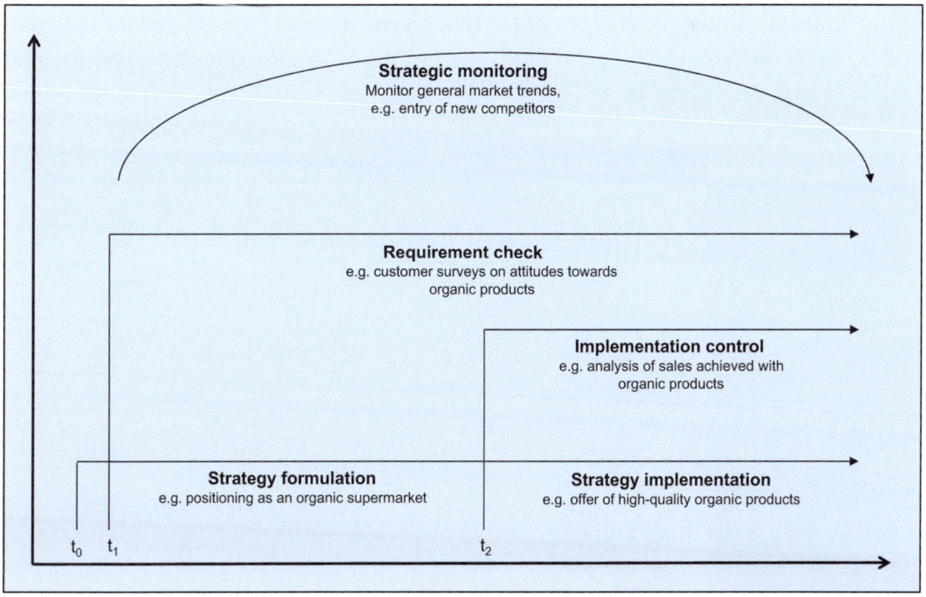

Fig. 12.6 The strategic control process using the example of a supermarket. (Source: Based on Schreyögg, 2006, p. 105)

(2) Operative quality control In the context of operational quality control, it is necessary to support the activities of operational quality management in order to implement the quality strategy. In contrast to strategic, operational control is concerned with a time horizon of only one to two years and is mainly concerned with quantifiable **financial indicators** such as profit, liquidity and profitability of the company (Erichsen, 2011, p. 9). The task of operational quality control is to create quality related indicator systems and reports and to make them available to the relevant departments. This is particularly necessary for quality processes and activities involving several departments within the company. Mainly instruments from accounting are used, such as the contribution margin calculation, investment calculation or the cost-benefit analysis (see Erichsen, 2011 for a detailed description of the individual control instruments). In the context of quality management, the challenge when using operational control instruments lies mainly in the quantifiability of the individual key figures. It is often difficult to assign the individual cost and benefit effects to individual quality management measures. For example, it is hardly possible to map the specific benefits of a complaint management system or training measures in quantifiable or monetary terms. Chap. 15 provides a detailed description of this problem of allocating the costs and benefits of quality management in the context of efficiency control.

Fig. 12.7 shows a comparison of strategic and operational quality control and makes it clear that operational control has a **supportive function** of strategic quality control with the task of ensuring that the strategic goals are implemented in daily work. Strategic control pursues the guiding principle "doing the right thing", while operational control is aimed at "doing the things right". This means that the strategy defines the goal, that is, the general direction, and ensures it. Operational control then has to ensure that

	Strategic quality controlling	Operational quality controlling
Guiding principle	"Doing the right thing"	"Doing the things right"
Time horizon	Long-term (approx. 5 years)	Short-term (1-2 years)
Orientation	Environment and business	Company: Profitability of operational processes
Dimensions	Opportunities/risks Strengths/weaknesses	Expense/income Costs/services
Goal variables	Securing existence, potential for success	Economy, profit, profitability
Goal/measure	Qualitative	Quantitative (indicator-based, monetary)
Instruments (exemplary)	▪ Balanced scorecard ▪ SWOT analysis ▪ Benchmarking ▪ Competitor analysis ▪ Life cycle analysis ▪ Portfolio analysis/potential analysis ▪ Risk controlling ▪ Strategic GAP analysis	▪ ABC analysis ▪ Reporting and key figure systems ▪ Break-even analysis ▪ Cost-benefit analysis ▪ Contribution margin accounting ▪ Investment accounting methods ▪ Liquidity planning ▪ Project controlling

Fig. 12.7 Comparison of strategic and operational quality controlling. (Source: Based on Erichsen, 2011; Horváth, 2011)

these goals are "right", that is, implemented as efficiently as possible. Operational control therefore has a **continuous character** with the goal of ensuring efficient resource use in quality management (Erichsen, 2011, p. 9; see also efficiency control in detail in Chap. 15). It also takes on a more internal perspective and focuses on internal company processes, while strategic control carries out a comprehensive review and also takes into account changes in the company's environment. Fig. 12.7 shows the comparison of strategic and operational quality control and shows a selection of commonly used instruments of the two control areas.

(3) Integrative quality control Finally, the **integrated quality control** is used to coordinate the interaction between strategic and operational quality management. Due to the short term nature of operational quality control, the instruments and measures must be aligned with the long term strategic goals and control instruments (see also Fig. 12.3). The strategic control primarily takes on an information provision and documentation function by providing indicators that give an indication of the extent to which the required quality strategy has been implemented. Therefore, a constant exchange between operational and strategic control must be guaranteed.

The central element of the integrated quality control represents the **economic analysis of quality management.** In a simplification of the relationships between strategic and operational quality control, the economic analysis of quality management has two **tasks:**

- **Operationalization of the quality strategy for operational quality management:** Through the economic analysis of quality management, a framework is set up which shows the "levers" for implementing the quality strategy to operational quality management.
- **Documentation of strategy implementation for strategic quality management:** From the perspective of strategic quality management, the quality strategy is to be translated into economic key figures. By determining the extent of these indicators based on the economic analysis of quality management, it can be shown to what extent the quality strategy is implemented by the activities of operational quality management.

The subject area of the cost-effectiveness of quality management has content related references to different research areas (e.g. quality research, production theory, operations research, marketing research, investment theory and accounting) of business administration. Therefore, there are also different facets in the literature that deal with the cost-effectiveness of quality management. These developments can be divided into four **phases** (cf. Fig. 12.8):

- Activity oriented analysis of quality management
- Effect-oriented analysis of quality management
- Investment analysis of quality management
- Value analysis of quality management

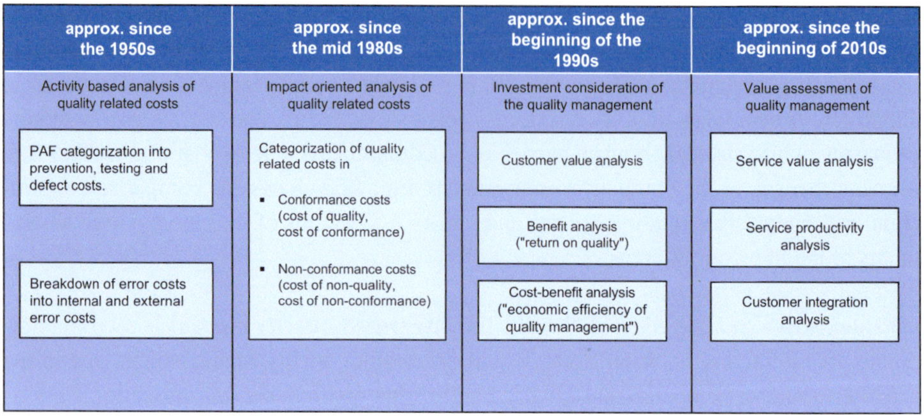

Fig. 12.8 Phases of the cost-effectiveness consideration of quality management

The **quality cost concept** (better: quality-related costs) was founded in the mid 1950s by Feigenbaum (1956). His **PAF categorization** (Prevention—Appraisal—Failure) represents a **task oriented analysis** of quality management, as it captures the costs in the company that arise due to certain quality related activities (cf. for the task oriented classification of quality related costs Sect. 15.1.2). These include activities of prevention (prevention), activities of examination (appraisal) and error induced activities (failure). Thus, the analysis of quality related costs serves to capture all costs that are related to the quality of a company.

The approach was modified in the mid 1980s when the traditional PAF structure was abandoned and a distinction was made between conformance costs (cost of conformance, costs of quality) and non-conformance costs (cost of non-conformance, costs of non-quality, failure costs) in the form of a **performance oriented analysis of quality management** (Crosby & Henniges, 1990; Wildemann, 1992, p. 762; Masing, 1999) (cf. for the performance oriented classification of quality related costs Sect. 15.1.3). This cost categorization differentiates quality related costs more than the task oriented approach with regard to the manifestation of quality. While **conformance costs** arise due to the creation of quality, **non-conformance costs** are due to a non-fulfillment of quality requirements. Although non-conformance costs can be regarded as the (lost) benefits of quality management from an opportunity cost perspective, the term non-conformance costs is still used today. This way, however, a cost-benefit consideration is not possible, since the two cost categories exclude each other by definition. Either conformance costs or non-conformance costs arise from a certain quality related behavior (1998, p. 12).

The **error costs** were later differentiated into internal and external error costs. External error costs are those costs that only arise when the customer notices the error (e.g. through warranty payments) (cf. for error cost accounting Sect. 15.1.4). According to the quality related cost approach, an economic creation of quality leads to a minimization of the total quality related costs (Feigenbaum, 1956, 73 pp.; Dale & Plunkett, 1999, pp. 8-9).

The quality cost concept mainly considers quality related costs that arise within the company. Even in studies based on the performance oriented approach, the external effects of quality management on customer behavior are largely neglected due to measurement problems. It is precisely these customer related quality effects that represent a significant element of the **investment analysis of quality management**.

Studies that deal with the measurement of (primarily external) target variables of quality management are to be counted among these approaches. These include, for example, analyses in the areas of **customer satisfaction research** (Lingenfelder & Schneider, 1991; Meyer & Dornach, 1995; Spreng et al., 1996; Hu et al., 2009; Bitner & Wang, 2014) and the **measurement of perceived service quality** (Parasuraman et al., 1985, 1988; Boulding et al., 1993; Bruhn & Hennig, 1993). In addition to the capture of the characteristics of these target variables, these approaches are also concerned with the effects of customer satisfaction and perceived service quality on behavior related target variables, such as customer loyalty (cf. to the target variables of quality management Sect. 14.1.1).

The value of a customer results from the sum of the coverage contributions that a customer generates over the course of a relationship with a company. Consequently, in the context of **customer value analysis**, the income and expenses that arise over the course of a company-customer relationship are considered. The customer value (e.g. "customer lifetime value") represents the discounted difference between income and expenses in relation to a single customer relationship (see, for example, Krafft, 2007, 43 pp.; Bruhn et al., 2008; Vogel et al., 2008; Wong, 2013). In numerous studies, customer value analysis has been able to demonstrate the relative advantages of a customer loyalty strategy over more traditional marketing strategies (e.g. new customer acquisition through advertising) (Blattberg & Deighton, 1996; Bruhn et al., 2000; Homburg et al. 2008). The goal of customer value management is the analysis and optimization of customer value in every phase of the customer relationship (see Fig. 12.9).

To optimize customer value, it is necessary to increase customer related revenues and reduce customer related costs at the same time. In terms of both determinants of customer value, the measures of a quality management system play an important role.

On the **revenue side,** quality management can contribute to the fact that important customer potentials are maximised through continuous quality improvements or through more effective and efficient customer approach and processing, so that customer specific revenues are increased. The focus is on the potentials of cross-selling, up-selling and the potential for recommendation. By constantly improving the quality of services, the probability increases that customers will not only use the services of another provider (cross-selling), but also take advantage of higher-value services (up-selling) over time. Both have a positive effect on the amount of customer related revenue. Ideally, the quality of services constantly improved by quality management will also have an impact on the recommendation of customers, as the potential for recommendation of a customer is

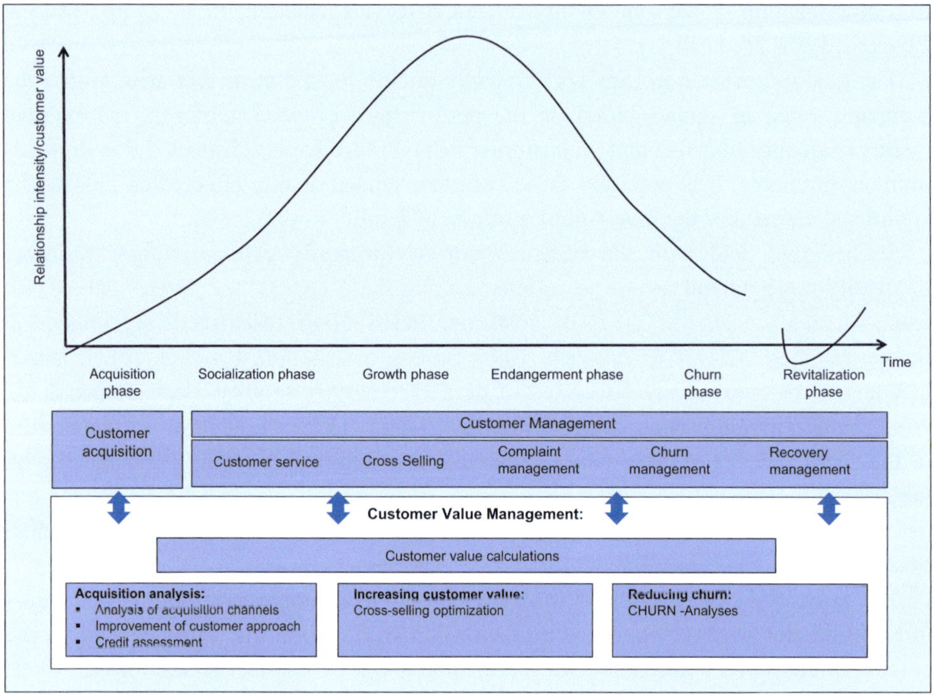

Fig. 12.9 Tasks of customer value management in the life cycle of the customer relationship. (Source: Büschken & Gropp, 2005, p. 5)

also a component of customer value and is to be considered in the context of customer value analysis. The higher the number of people to whom a satisfied customer recommends the services of a service provider, the higher the probability that they will also choose the supplier. This customer acquisition potential is therefore also to be credited to the value of a customer.

On the **cost side,** quality management can also contribute to increasing customer value. Starting points are above all the complaint management regularly used within the framework of quality management as well as the customer churn and customer retention management (cf. Fig. 12.9). For example, customer complaints can be effectively processed and a potential customer churn can be counteracted by complaint management, so that the costs incurred in connection with possible customer retention measures are avoided. In addition, complaints contain important information on possible quality problems, sources of error or improvement potentials. Consequently, the analysis of complaints provides important information for avoiding error costs, thus increasing customer value by reducing costs in connection with inadequate quality.

In analogy to classical investment theory, the investment consideration of quality management is carried out in the—limiting itself to the revenue increasing effects of quality management—benefit analysis and the cost-benefit analysis (see in detail for the cost-benefit analysis Sect. 15.3). Due to the mentioned problems of capturing the benefits of quality management, the positive success effect of quality management is in the focus as part of the **benefit analysis.** In the context of the benefit analysis, the effects of quality management or a high quality on customer behavior and the resulting revenues on the one hand (e.g. customer loyalty, word-of-mouth communication) and in terms of cost savings on the other hand are measured (Rust et al., 1995; Zeithaml et al., 1996). The ratio of these effects and the—assumed given—costs of quality management is discussed in the American language under the term **"Return on Quality"** (Rust et al., 1994). However, this approach has the disadvantage that the costs associated with quality management are not further analyzed (Bruhn, 1998, p. 18).

Since economic variables of quality management are due to both the benefits and the costs associated with it, the concept of **cost-benefit analysis of quality management** goes one step further (Bruhn, 1998; Bruhn & Georgi, 1999; Antony et al., 2007; Goetsch & Davis, 2014). Due to the existence of many "hidden" costs generated by quality measures (e.g. time loss due to training), both the benefits and the costs of quality management have to be considered. In addition to providing methods for capturing and optimizing the benefits of quality management, an instrumentarium for complete capture as well as optimization approaches of the costs of quality management is being developed.

An exemplary cost-benefit analysis with regard to the introduction of a service quality management from the medical sector is shown in Fig. 12.10.

Cost-benefit analysis of the introduction of quality management in the medical sector	
Costs of quality management	**Benefits of quality management**
Failure prevention costs • Personnel costs of mortality and morbidity conferences • Costs for the preparation of SOP ("standard operating procedures") • Costs to take measures to improve the infrastructure.	**Increase in performance quality** • Reduction of reoperations • Lowering of increased medication consumption • Fewer repairs of improperly operated equipment.
Testing costs • Costs due to audits in medical technology • Hygiene monitoring costs • Costs for sterilization of instruments. • Certification costs	**Reputation gain** • Lowering the risk of image loss • Reduction of customer churn
	More effective use of resources • Error prevention
	Increase customer satisfaction

Fig. 12.10 Cost-benefit analysis of service quality management in the medical sector. (Source: Schroeder-Printzen, 2014)

Since the 2010s, an orientation of quality management towards a **value analysis** can increasingly be seen in the scientific literature. Studies are mainly concerned with topics of the service value concept, service productivity and value creation through customer integration (cf., for example, Bruhn & Stauss, 2009; Bruhn & Hadwich, 2011, 2014).

The contributions of Zeithaml (1988), who made a widely cited conceptualization of service value from the customer's perspective, and of Heskett and Schlesinger (1994), who were able to integrate and link the service value perceived by the customer with the supplier side activities and revenues of the service value management, are particularly influential for the current discussion of service value management. Since then, numerous pieces of work have been concerned with the consideration of service value (cf., for example, Bolton & Drew, 1991; Woodall, 2003; Sánchez-Fernández & Iniesta-Bonillo, 2007). The discussion deals with the measurement of service value (Sánchez-Fernández & Iniesta-Bonillo, 2007; Buschak et al., 2014; Heinisch, 2014). In this context, the service value is determined in accordance with the perceived service quality and the procurement costs on the customer's side (Cronin et al., 1997, 375 pp.; Ruiz et al., 2008, 1278 pp.; Park et al., 2013, 622 pp.). In addition to that the scientific literature deals with the question to what extent service value management can be controlled on the supplier side (Sawhney, 2006; Tuli et al., 2007; Benkenstein & Waldschmidt, 2014; Leischnig & Messer, 2014). In this context, in particular, service value strategies and operational service value measures are considered, which can be taken by the supplier (Bruhn & Hadwich, 2014). For example, Roth and Pfisterer (2014) propose to identify and analyze value generating activities for the customer in companies. Such a specification of these activities offers the possibility to differentiate value generating processes according to their structure and their usefulness for the customer. However, the question remains how the value creation is realized by the service value management. When considering the costs and revenues as components of value, the processes for service creation lead to costs first. The use of employees, the design of communication measures and the development of new services are each directly associated with costs (Bruhn & Hadwich, 2014). In addition, the revenue effect of the processes of service value management is to be considered. Ultimately, it is about increasing the revenue through good service and relationship processes. Service value management therefore deals with complex decision problems that have to be dealt with in the context of a systematic planning process (Bruhn & Hadwich, 2014).

In close connection with the topic of service value is the discussion about **service productivity** (Grönroos & Ojasalo, 2004). In general, productivity results from the quotient of a certain product quantity and the production factors consumed for its production in the observation period (Laßmann, 1975). In other words, it is a quantitative relation between output and input variables. Productivity is therefore to be understood as a pure quantity, reflecting the company's performance (Bruhn & Hadwich, 2011). However, with services, input (production factors) and output (production result) cannot be determined as easily as in the case of goods (Parasuraman 2010). Concepts and methods originally developed for the goods sector cannot therefore be transferred to the service sector without further ado (Bruhn & Hadwich, 2011). The essential characteristics of a service, such as immateriality, integration of the external factor and the uno-actu principle, undoubtedly make it more difficult to capture and determine service productivity (Engelhardt et al., 1993, 400 pp.). Specific requirements for the determination of service productivity arise from the essential characteristics of a service (Parasuraman, 2010; Rutkauskas & Paulavičienė, 2015). In the context of the provider's performance, it is therefore the case that all structures and potentials of the service provider can contribute to significantly improving service productivity. In the case of a credit institution, for example, this dimension includes the variable design of the service offering, a high level of competence of the employees and the increased use of innovative information and customer systems. In the context of the process dimension, it becomes clear that the customer shapes the service creation process to a decisive extent and has a significant impact on the productivity of the service (Fliess & Kleinaltenkamp, 2004; Meffert et al., 2018). Accordingly, in this context, the mutual information between supplier and customer as well as customer communication, for example by means of an increased use of information systems, are to be designed in such a way that process and interaction quality are improved (Kleinaltenkamp et al., 2008; Rutkauskas & Paulavičienė, 2015) and thus service productivity is increased (Bruhn & Hadwich 2011, p. 8). The high proportion of intangible components of a service finally means that the result of the service is not tangible and therefore only partially quantifiable (Lasshof, 2006; Meffert et al., 2018). Therefore, the aim of a service provider is always to convince the customer of a good result during the service creation process in order to increase service productivity. As a result, a value oriented service quality management offers the opportunity to form categories of service productivity based on the service marketing value chain. In this context, a distinction is made between customer activity, customer satisfaction and customer loyalty indices (Bruhn & Hadwich, 2011, p. 19). As a result, the measurement and evaluation of services forms the decisive basis for productivity enhancement of the service provider.

Another focal point within the value based perspective of quality management is value creation by customer integration (see also Sect. 3.4.1). Service customers are increasingly being integrated into the creation of services (Bruhn & Stauss, 2009; Edvardsson et al., 2012; Homburg & Kuehnl 2014). They contribute information so that services can be created as needs-based as possible, or they even take care of the configuration of their service themselves (Büttgen, 2009). While value for the customer is to be generated by a needs-based creation of services, the striving for cost reduction on the part of the providers also causes more and more elements of services to be outsourced to the customer (Corsten, 1999; Tat Keh & Wei Teo, 2001; Hsieh et al., 2004; Ramani & Kumar, 2008; Sharma et al., 2014). The higher the share of processes to be carried out by the customer, the more urgent the question of the customer's ability and motivation to provide economic and satisfactory services (Canziani, 1997; Lengnick-Hall et al., 2000; Rodie & Kleine, 2000; Büttgen, 2009; Braun et al., 2016). Accordingly, it is necessary in this context to examine the type and extent of the processes to be externalized in detail and to collect information about the performance relevant skills and abilities of the customer (Enkel et al., 2005; Moeller, 2008; Büttgen, 2009). Accordingly, it is essential for the provision of customer oriented service creation that the integration of the customer is a central concern for the company. It is necessary here to intensively examine the opportunities and risks of customer participation, to understand the decisive mechanisms of a fair participation of customers and to use the resulting possibilities of a positive influence (Bruhn & Stauss, 2009). It is the task of the provider above all to convince and qualify in order to create the necessary conditions or framework conditions so that the customer can effectively participate in the creation of services and generate a satisfactory result for both sides (Büttgen, 2009).

Based on this inventory of quality control for services in the literature, three **types of control** are distinguished, which are presented in detail in the following chapters:

- **Process controls** deal with the control of quality related activities. It must be ensured that the activities are carried out in a process oriented manner over time. This way, progress is controlled and steered and the current state of quality management or quality processes is measured (Chap. 13).
- **Effectiveness controls** examine the quality effects on customers according to different categories of effects. It is checked to what extent the individual quality activities contribute to increasing psychological (e.g. customer satisfaction) and behavioral effects (e.g. customer loyalty, sales growth) (Chap. 14).
- **Efficiency controls** make cost-benefit comparisons for the assessment of quality activities by comparing all costs of these activities with the benefits achieved (Chap. 15).

References

Antony, J., Jiju Antony, F., Kumar, M., & Rae Cho, B. (2007). Six sigma in service organisations: Benefits, challenges and difficulties, common myths, empirical observations and success factors. *International Journal of Quality & Reliability Management, 24*(3), 294–311.

Bea, F. X., & Haas, J. (2017) *Strategisches Management* (9th Edn.). Konstanz: UVK.

Benkenstein, M., & von Stenglin, A. (2006). Prozessorientiertes Qualitätscontrolling von Dienstleistungen. In M. Bruhn & B. Strauss (Eds.), *Dienstleistungscontrolling. Forum Dienstleistungsmanagement* (pp. 56–70). Wiesbaden: Gabler.

Benkenstein, M., & Waldschmidt, V. (2014). Wertkettenanalyse und Service Value—Eine Diskussion im Lichte der Service-Dominant Logic. In M. Bruhn & K. Hadwich (Eds.), *Service Value als Werttreiber. Forum Dienstleistungsmanagement* (pp. 205–221). Wiesbaden: Springer Gabler.

Bitner, M. J., & Wang, H. S. (2014). Service encounters in service marketing research. In R. T. Rust & M. H. Luang (Eds.), *Handbook of service marketing research* (pp. 221–243). Northampton: Edward Elgar Pub.

Blattberg, R. C., & Deighton, J. (1996). Manage marketing by the customer equity test. *Harvard Business Review, 74*(4), 136–144.

Bolton, R. N., & Drew, J. H. (1991). A multistage model of customers' assessments of service quality and value. *Journal of Consumer Research, 17*(4), 375–384.

Boon, K., McKinnon, J., & Ross, P. (2008). Audit service quality in compulsory audit tendering: Preparer perceptions and satisfaction. *Accounting Research Journal, 21*(2), 93–122.

Boulding, W., Kalra, A., Staelin, R., & Zeithaml, V. A. (1993). A dynamic process model of service quality. From expectations to behavioral intentions. *Journal of Marketing Research, 30*(1), 7–27.

Braun, C., Popp, M., & Hadwich, K. (2016). *Success factors of customer empowerment from a customer's perspective: A qualitative study and strategic implications.* Proceedings of the International Service Research Conference (AMA SERVSIG). Maastricht.

Bruhn, M. (1998). *Wirtschaftlichkeit des Qualitätsmanagements. Qualitätscontrolling für Dienstleistungen.* Berlin: Springer.

Bruhn, M., & Georgi, D. (1998). Kundenbezogene Wirtschaftlichkeitsanalyse des Qualitätsmanagements für Dienstleistungen. Konzept, Modellrechnung und Fallbeispiel. *Marketing ZFP, 20*(2), 98–108.

Bruhn, M., & Georgi, D. (1999). *Kosten und Nutzen des Qualitätsmanagements. Grundlagen, Methoden, Fallbeispiele.* München: Hanser.

Bruhn, M., & Hadwich, K. (2011). Dienstleistungsproduktivität—Einführung in die theoretischen und praktischen Problemstellungen. In M. Bruhn & K. Hadwich (Eds.), *Dienstleistungsproduktivität: Forum Dienstleistungsmanagement* (Vol. 1, pp. 3–34). Wiesbaden: Gabler.

Bruhn, M., & Hadwich, K. (2014). Service Value—Eine Einführung in die theoretischen und praktischen Problemstellungen. In M. Bruhn & K. Hadwich (Eds.), *Service Value als Werttreiber. Forum Dienstleistungsmanagement* (pp. 3–31). Wiesbaden: Springer.

Bruhn, M., & Hennig, K. (1993). Selektion und Strukturierung von Qualitätsmerkmalen. Auf dem Weg zu einem umfassenden Qualitätsmanagement für Kreditinstitute. *Jahrbuch der Absatz- und Verbrauchsforschung, 39*(3), 214–238.

Bruhn, M., & Stauss, B. (2009). Kundenintegration im Dienstleistungsmanagement—Eine Einführung in die theoretischen und praktischen Problemstellungen. In M. Bruhn & B. Stauss (Eds.), *Kundenintegration. Forum Dienstleistungsmanagement* (pp. 3–33). Wiesbaden: Gabler.

Bruhn, M., Georgi, D., & Hadwich, K. (2008). Customer equity management as formative second-order construct. *Journal of Business Research, 61*(12), 1292–1301.

Bruhn, M., Georgi, D., Treyer, M., & Leumann, S. (2000). Wertorientiertes Relationship Marketing: Vom Kundenwert zum Customer Lifetime Value. *Die Unternehmung, 54*(3), 167–187.

Buschak, D., Lerch, C., & Gotsch, M. (2014). Messung des Service Value innovativer Dienstleistungen in der Industrie—Eine integrierte Anbieter-Kunden-Perspektive. In M. Bruhn & K. Hadwich (Eds.), *Service Value als Werttreiber. Forum Dienstleistungsmanagement* (pp. 91–113). Wiesbaden: Springer Gabler.

Büschken, J., & Gropp, M. (2005). *Kundenwertmanagement bei Gesetzlichen Krankenversicherungen.* Ingolstadt.

Büttgen, M. (2009). Beteiligung von Konsumenten an der Dienstleistungserstellung: Last oder Lust?—Eine motivations- und dissonanztheoretische Analyse. In M. Bruhn & B. Stauss (Eds.), *Kundenintegration. Forum Dienstleistungsmanagement* (pp. 63–89). Wiesbaden: Gabler.

Chwolka, A. (1996). *Controlling als ökonomische Institution: Eine Agency-theoretische Analyse.* Heidelberg: Physica.

Corsten, H. (1999). Der Integrationsgrad des externen Faktors als Gestaltungsparameter in Dienstleistungsunternehmungen—Voraussetzungen und Möglichkeiten der Externalisierung und Internalisierung. In M. Bruhn & B. Stauss (Eds.), *Dienstleistungsqualität. Konzepte—Methoden—Erfahrungen* (3rd Edn., pp. 145–168). Wiesbaden: Gabler.

Cronin, J. J., Brady, M. K., Brand, R. R., Hightower Jr, R., & Shemwell, D. J. (1997). A cross-sectional test of the effect and conceptualization of service value. *Journal of Services Marketing, 11*(6), 375–391.

Crosby, P. B., & Henniges, B. (1990). *Qualität ist machbar* (2nd Edn.). Hamburg: McGraw-Hill.

Dale, B. G., & Plunkett, J. J. (1999). *Quality costing* (3rd Edn.). London: Routledge.

Edvardsson, B., Kristensson, P., Magnusson, P., & Sundström, E. (2012). Customer integration within service development—A review of methods and an analysis of insitu and exsitu contributions. *Technovation, 32*(7), 419–429.

Engelhardt, W. H., Kleinaltenkamp, M., & Reckenfelderbäumer, M. (1993). Leistungsbündel als Absatzobjekte. Ein Ansatz zur Überwindung der Dichotomie von Sach- und Dienstleistungen. *Zeitschrift für betriebswirtschaftliche Forschung, 45*(5), 395–426.

Enkel, E., Kausch, C., & Gassmann, O. (2005). Managing the risk of customer integration. *European Management Journal, 23*(2), 203–213.

Erichsen, J. (2011). *Controlling-Instrumente von A–Z: Die wichtigsten Werkzeuge zur Unternehmenssteuerung* (8th. Edn.). München: Haufe-Lexware.

Farber Canziani, B. (1997). Leveraging customer competency in service firms. *International Journal of Service Industry Management, 8*(1), 5–25.

Feigenbaum, A. V. (1956). Total quality control. *Harvard Business Review, 34*(6), 93–101.

Fliess, S., & Kleinaltenkamp, M. (2004). Blueprinting the service company: Managing service processes efficiently. *Journal of Business Research, 57*(4), 392–404.

Goetsch, D. L., & Davis, S. B. (2014). *Quality management for organizational excellence.* New York: Pearson.

Grönroos, C., & Ojasalo, K. (2004). Service productivity: Towards a conceptualization of the transformation of inputs into economic results in services. *Journal of Business Research, 57*(4), 414–423.

Habib, A. (2012). Non-audit service fees and financial reporting quality: A meta-analysis. *Abacus, 48*(2), 214–248.

Heinisch, S. (2014). Serviceerfolgsmessung—Ein Modell zur Analyse von Service Value. In M. Bruhn & K. Hadwich (Eds.), *Service Value als Werttreiber. Forum Dienstleistungsmanagement* (pp. 115–136). Wiesbaden: Springer.

Heskett, J. L., & Schlesinger, L. A. (1994). Putting the service-profit chain to work. *Harvard Business Review, 72*(2), 164–174.

Homburg, C., & Kuehnl, C. (2014). Is the more always better? A comparative study of internal and external integration practices in new product and new service development. *Journal of Business Research, 67*(7), 1360–1367.

Homburg, C., & Stock, R. M. (2005). Exploring the conditions under which salesperson work satisfaction can lead to customer satisfaction. *Psychology and Marketing, 22*(5), 393–420.

Homburg, C., Droll, M., & Totzek, D. (2008). Customer prioritization: Does it pay off, and how should it be implemented. *Journal of Marketing,72*(5), 110–130.

Horváth, P., Gleich, R. & Seiter, M. (2020). *Controlling* (14th Edn.). München: Vahlen.

Horváth, P., & Urban, G. (1990). *Qualitätscontrolling.* Stuttgart: Schäffer-Poeschel.

Horváth, P., Gentner, A., & Lingscheid, A. (1994). Qualitätscontrolling. In W. Hansen, H. H. Jansen, & G. F. Kamiske (Eds.), *Qualitätsmanagement im Unternehmen. Grundlagen, Methoden und Werkzeuge, Praxisbeispiele, Loseblatt-Sammlung, Kapitel 08.07* (pp. 1–18). Berlin: Springer.

Hsieh, A. T., Yen, C. H., & Chin, K. C. (2004). Participative customers as partial employees and service provider workload. *International Journal of Service Industry Management, 15*(2), 187–199.

Hu, H. H., Kandampully, J., & Juwaheer, T. D. (2009). Relationships and impacts of service quality, perceived value, customer satisfaction, and image: an empirical study. *The Service Industries Journal, 29*(2), 111–125.

Kamiske, G. F., & Brauer, J.-P. (2011). *Qualitätsmanagement von A bis Z. Erläuterungen moderner Begriffe des Qualitätsmanagements* (7th Edn.). München: Hanser.

Kleinaltenkamp, M., Griese, I., & Klein, M. (2008). Wie Kundenintegration effizient gelingt. *Marketing Review St. Gallen, 25*(2), 40–43.

Krafft, M. (2007). *Kundenbindung und Kundenwert* (2nd Edn.). Heidelberg: Physica.

Lasshof, B. (2006). *Produktivität von Dienstleistungen. Mitwirkung und Einfluss des Kunden.* Wiesbaden: Gabler.

Laßmann, G. (1975). Produktivität. In E. Grochla & W. Wittmann (Eds.), *Handwörterbuch der Betriebswirtschaft* (4th Edn., pp. 3164–3169). Stuttgart: Schäffer-Poeschel.

Leischnig, A., & Messer, U. (2014). Service Value durch Customization—Implikationen für die Gestaltung des Dienstleistungsmanagements. In M. Bruhn & K. Hadwich (Eds.), *Service Value als Werttreiber. Forum Dienstleistungsmanagement* (pp. 245–264). Wiesbaden: Gabler.

Lengnick-Hall, C. A., Claycomb, V., & Inks, L. W. (2000). From recipient to contributor: Examining customer roles and experienced outcomes. *European Journal of Marketing, 34*(3/4), 359–383.

Lingenfelder, M., & Schneider, W. (1991). Die Zufriedenheit von Kunden. Ein Marketingziel?. *Marktforschung & Management, 35*(1), 29–34.

Masing, W. (1999). *Handbuch Qualitätsmanagement* (4th Edn.). München: Hanser.

Meffert, H., Bruhn, M., & Hadwich, K. (2018). *Dienstleistungsmarketing. Grundlagen, Konzepte, Methoden* (9th Edn.). Wiesbaden: Springer.

Meyer, A., & Dornach, F. (1995). Nationale Barometer zur Messung der Qualität und Zufriedenheit bei Dienstleistungen. In M. Bruhn & B. Stauss (Eds.), *Dienstleistungsqualität. Konzepte—Methoden—Erfahrungen* (2nd Edn., pp. 429–453). Wiesbaden: Gabler.

Moeller, S. (2008). Customer integration—A key to an implementation perspective of service provision. *Journal of Service Research,11*(2), 197–210.

Parasuraman, A. (2010). Service productivity, quality and innovation: Implications for service-design practice and research. *International Journal of Quality and Service Sciences, 2*(3), 277–286.

Parasuraman, A., Zeithaml, V. A., & Berry, L. L. (1985). A conceptual model of service quality and its implications for future research. *Journal of Marketing, 49*(1), 41–50.

Parasuraman, A., Zeithaml, V. A., & Berry, L. L. (1988). SERVQUAL: A multiple-item scale for measuring consumer perceptions of service quality. *Journal of Retailing, 64*(1), 12–40.

Park, G. W., Park, K., & Dessouky, M. (2013). Optimization of service value. *Computers & Industrial Engineering, 64*(2), 621–630.

Piontek, J. (2005). *Controlling* (3rd Edn.). München: De Gruyter Oldenburg.

Ramani, G., & Kumar, V. (2008). Interaction orientation and firm performance. *Journal of Marketing, 72*(1), 27–45.

Reinecke, S., & Janz, S. (2007). *Marketingcontrolling: Sicherstellen von Marketingeffektivität und -effizienz.* Stuttgart: Kohlhammer.

Rodie, A. R., & Kleine, S. S. (2000). Customer participation in service production and delivery. In T. A. Swarz & D. Iacobucci (Eds.), *Handbook of services marketing and management* (pp. 111–125). London: Sage Publications.

Roth, S., & Pfisterer, L. (2014). Die Bedeutung von Nutzungsprozessen für den Service Value. In M. Bruhn & K. Hadwich (Eds.), *Service Value als Werttreiber. Forum Dienstleistungsmanagement* (pp. 223–243). Wiesbaden: Springer Gabler.

Ruiz, D. M., Gremler, D. D., Washburn, J. H., & Carrión, G. C. (2008). Service value revisited: Specifying a higher-order, formative measure. *Journal of Business Research, 61*(12), 1278–1291.

Rust, R. T., Zahorik, A. J., & Keiningham, T. L. (1994). *Return on quality. Measuring the financial impact of your company's quest for quality.* Chicago: Probus Professional Pub.

Rust, R. T., Zahorik, A. J., & Keiningham, T. L. (1995). Return on Quality (ROQ): Making service quality financially accountable. *Journal of Marketing, 59*(2), 58–70.

Rutkauskas, J., & Paulavičienė, E. (2015). Concept of productivity in service sector. *Engineering Economics, 43*(3), 35–41.

Sawhney, R. (2006). Interplay between uncertainty and flexibility across the value-chain: Towards a transformation model of manufacturing flexibility. *Journal of Operations Management, 24*(5), 476–493.

Schmitt, R., & Reissiger, W. (2010). Qualitätscontrolling. In G. F. Kamiske (Ed.), *Qualitätsmanagement.* Digitale Fachbibliothek: Symposion.

Schreyögg, G. (2006). Strategische Kontrolle einer marktorientierten Unternehmensführung. In S. Reinecke & T. Tomczak (Eds.), *Handbuch Marketing-Controlling. Effektivität und Effizienz einer marktorientierten Unternehmensführung* (2nd Edn., pp. 99–115). Wiesbaden: Gabler.

Schroeder-Printzen, P. D. I. (2014). Kosten und Nutzen eines Qualitätsmanagements. *Der Urologe, 53*(1), 15–20.

Sharma, S., Conduit, J., & Hill, S. R. (2014). Organisational capabilities for customer participation in health care service innovation. *Australasian Marketing Journal, 22*(3), 179–188.

Simons, R. (2013). *Levers of control: How managers use innovative control systems to drive strategic renewal.* Boston: Harvard Business School Press.

Spreng, R. A., MacKenzie, S. B., & Olshavsky, R. W. (1996). A reexamination of the determinants of consumer satisfaction. *Journal of Marketing, 60*(3), 15–32.

Sánchez-Fernández, R., & Iniesta-Bonillo, M. Á. (2007). The concept of perceived value: a systematic review of the research. *Marketing Theory, 7*(4), 427–451.

Tat Keh, H., & Wei Teo, C. (2001). Retail customers as partial employees in service provision: a conceptual framework. *International Journal of Retail & Distribution Management, 29*(8), 370–378.

Tomys, A.-K. (1995). *Kostenorientiertes Qualitätsmanagement. Qualitätscontrolling zur ständigen Verbesserung der Unternehmensprozesse.* München: Hanser.

Tuli, K. R., Kohli, A. K., & Bharadwaj, S. G. (2007). Rethinking customer solutions: From product bundles to relational processes. *Journal of Marketing, 71*(3), 1–17.

Vogel, V., Evanschitzky, H., & Ramaseshan, B. (2008). Customer equity drivers and future sales. *Journal of Marketing, 72*(6), 98–108.

Wildemann, H. (1992). Kosten- und Leistungsbeurteilung von Qualitätssicherungssystemen. *Zeitschrift für Betriebswirtschaft, 62*(7), 761–780.

Wong, I. A. (2013). Exploring customer equity and the role of service experience in the casino service encounter. *International Journal of Hospitality Management, 32,* 91–101.

Woodall, T. (2003). Conceptualising 'value for the customer': An attributional, structural and dispositional analysis. *Academy of Marketing Science Review, 12*(1), 1–42.

Zeithaml, V. A. (1988). Consumer perceptions of price, quality, and value: A means-end model and synthesis of evidence. *Journal of Marketing, 52*(3), 2–22.

Zeithaml, V. A., Berry, L. L., & Parasuraman, A. (1996). The behavioral consequences of service quality. *Journal of Marketing, 60*(2), 31–46.

Zollondz, H.-D. (2011). *Grundlagen Qualitätsmanagement. Einführung in Geschichte, Begriffe, Systeme und Konzepte* (3rd Edn.). München: Oldenbourg.

Process Controls in Quality Management 13

Abstract

Managing and controlling business processes is a central task of corporate management, because customers are often directly involved in the service delivery process. Accordingly, ensuring a consistently high quality throughout the process is of great importance. Against this background, this chapter first presents different approaches to process control and their objectives. Building on this, individual process types are introduced within the framework of quality management and instruments of process control are explained using examples. Finally, the limits of process control to be considered are shown in a critical assessment.

13.1 Basics of Process Controls in Quality Management

13.1.1 Concept of Process Controls

Managing and controlling business processes is a central task of corporate management. Especially with services, customers are often directly integrated into the entire service creation process (e.g. haircut) and therefore it is of great importance to ensure a consistently high quality throughout the process.

Process controls within the framework of quality management are concerned with the control of the implementation of all quality related activities (e.g. in the areas of potential provision or within service creation) during the entire service creation process. It must be ensured that the individual activities follow a previously defined timetable. A process can be defined as follows:

M. Bruhn, *Quality Management for Services*,
https://doi.org/10.1007/978-3-662-67032-3_13

▶ A **process** is to be understood as the assignment of activities within the framework of a timetable under logical considerations (e.g. with regard to goals, costs, interdependencies).

The partially interdependent activities are always to be aligned with a common process or business goal under consideration of cost aspects. Ultimately, the activities serve the creation of a service that provides the customer with a specific benefit (Ahlrichs & Knuppertz, 2010, p. 10).

If a service company consistently pursues a process oriented approach, then it must integrate both the "customer oriented" and the "process oriented" perspective in the process view (vom Brocke & Sonnenberg, 2011, p. 55). The **customer oriented perspective** implies that a process is basically a sequence of activities with the purpose of generating a specific value for the customer (Hammer & Champy, 2006). This means that even with process orientation, the customer focus is in the center. All processes are therefore to be aligned with customer needs and designed in such a way as to generate the highest possible customer benefit. The **process oriented perspective** considers the temporal sequence of the individual process steps that are to be carried out in order to achieve the goal. The aim is to achieve an optimal process, i.e. an optimal coordination of the sub-processes. The process idea is to be anchored in the service company both horizontally (along the value chain) and vertically (internal customer-supplier relationships) (vom Brocke & Sonnenberg, 2011, p. 55).

The **horizontal processor orientation** is based on the concept of the value chain. The **value chain** represents the company as a collection of activities through which its services are designed, created, distributed, delivered and supported (Porter, 2014, p. 66). Primary and supporting activities can be distinguished. Fig. 13.1 shows an exemplary value chain using the example of a restaurant.

The **primary activities** are directly involved in the creation of the core service and are arranged in a time logical sequence. First, inbound logistics (e.g. purchasing raw materials or ingredients in a restaurant) is required before the further steps such as production operations (preparation of the dishes), outbound logistics (serving the dishes), marketing & sales (marketing activities such as advertisments, flyers, posters) or customer service (e.g. advising guests on certain dishes, ingredients, etc.) begin.

The **supporting activities** are so-called supply functions and have to be carried out continuously throughout the entire process of creating the service. Within the infrastructure, for example, accounting or a functioning IT is to be provided in order to generate and settle customer invoices. Human resources are needed to hire qualified personnel (e.g. service personnel, chefs) and to train them continuously through further training measures. Technology is used to set up customer databases and to provide customer and market research information across departments. Within procurement, it must be ensured that there are always sufficient raw materials and adequate equipment (e.g. kitchen appliances, tables in the dining rooms) available. It is the task of process management to

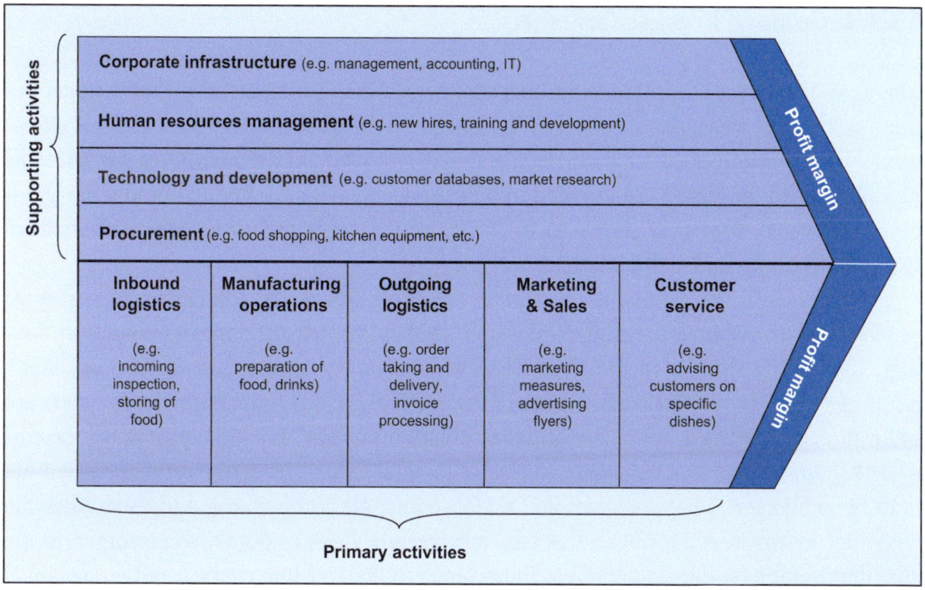

Fig. 13.1 Exemplary value chain using the example of a restaurant. (Source: Based on Porter, 2014, p. 64)

design these supply functions in such a way as to ensure an efficient process flow of the primary activities for service creation.

The **vertical processor orientation** coordinates the cooperation of the individual departments within the company. The entire performance often consists of individual partial performances, which are carried out by different people or departments. The employees act as internal customers or internal suppliers of other departments. In this context, one speaks of **internal customer-supplier relationships**. The chef of a restaurant is, for example, the internal food supplier for the service personnel. The service personnel in turn act as an internal customer by taking the order from the guests and forwarding it to the kitchen. Within the framework of process management, these internal customer-supplier relationships are to be designed in such a way that they run as smoothly and efficiently as possible. Often, these interfaces between the individual departments represent quality critical situations within the performance. Therefore, from a quality point of view, it is important to ensure that the quality standards required by the company are not impaired. In the restaurant example, it is necessary to ensure when coordinating between the kitchen and the service personnel that no long waiting time between "food ready" and "serving the guests" arises. A lack of coordination would result in the dishes being served too cold and the customers being dissatisfied with the performance.

13.1.2 Goals of Process Controls

The goal of process control is to identify and avert faulty processes at an early stage. For this reason, the measures are used in various areas of quality management. The **global goal** is the optimization of processes and the simultaneous reduction of errors. This global goal can be further subdivided into customer related, quality and productivity relevant sub-goals. Fig. 13.2 shows an overview of the goals and tasks of process control using the example of a car repair shop.

First of all, in the context of **customer oriented sub-goals,** the design of corporate processes is to achieve the highest possible **customer satisfaction** and **customer loyalty.** Optimal performance creation processes are to be designed from the customer's point of view. This aspect is particularly relevant for service companies, as customers are often directly involved in the performance creation process. For example, in the case of a workshop service, it is necessary to clarify to what extent the repair service carried out is to be explained to the customer and which points are to be clarified together with the customer in advance. However, this can vary greatly from customer to customer. Some customers, for example, attach great importance to the fact that every detail is explained to them as much as possible and that they are involved in the decision making process as to which individual repair services are to be carried out. Others, on the other hand, prefer that the interaction between the workshop manager and the customer takes place as short as possible and that they receive their vehicle back quickly.

The processes in the company are also suitable for achieving a **stabilization of the customer relationship** or further expanding it. Especially through the personal interaction between the company and the customers, there are many possibilities to establish sustainable customer relationships. Process control has a variety of options for checking

	Goals	Examples with key figures
Global Goal	▪ Process optimization with simultaneous reduction of undesirable developments	▪ Cost savings through process optimization ▪ Number of faulty processes
Customer oriented subgoals	▪ Increase customer satisfaction ▪ Increase customer loyalty ▪ Stabilize customer relationships ▪ Increase in perceived service orientation	▪ Customer surveys on process satisfaction ▪ Customer churn, duration of customer relationships ▪ Loyalty indicators, customer surveys ▪ Service orientation of the company/processes from the customer's point of view
Quality oriented subgoals	▪ Repeatability/reliability ▪ Faultlessness ▪ Traceability/accountability	▪ Time, branch, employee comparisons ▪ Number of complaints/corrections ▪ Error rate of individual sub-processes
Productivity oriented subgoals	▪ Savings in operating costs ▪ Shortening of processing times ▪ Minimization of idle times ▪ Minimization of waiting times ▪ Minimization of interfaces	▪ Energy requirement per repair order ▪ Duration of processing for a repair order ▪ Time span between completion of repairs and invoicing ▪ Waiting time for customers until the repair order is accepted ▪ Number of employees/departments involved per repair order

Fig. 13.2 Goals and tasks of process controls using the example of a car workshop

in customer surveys or collecting image values and loyalty measures (e.g. willingness to recommend, willingness to visit again) whether an orientation of corporate processes towards relationship oriented has the desired effects on customers.

The **quality oriented sub-goals** of process control relate to the assurance of a consistently high level of performance quality. Basically, processes are to be designed so that a constant result is always achieved, regardless of the time, the employees or the branch of the performance creation. McDonald's is an example of this. The company guarantees that the products always have the same quality and taste in all branches. This is achieved by the preparation being divided into very many **standardized partial steps.** Through this standardizability, a **faultless delivery** is being aimed fort. If the individual partial steps are precisely defined and prescribed, the probability of error decreases. At the same time, this makes it possible to guarantee a **traceability** of possible error causes. For example, if an increase in incorrect invoicing is identified, the process control is to check at which point in the process the cause can be found. In the example of the car workshop, for example, the error may lie in the invoicing department when entering incorrect values, but it is also possible that the incorrect invoice was created by an inadequate transfer of the work carried out from the workshop to the invoicing department.

Standardization of processes using the example of *McDonald's Deutschland Inc.*

According to the company, the preparation of a hamburger at *McDonald's* is carried out according to the process shown in Fig. 13.3. It is clear from this that each process step is precisely defined using predetermined key figures. For the toasting process of the hamburger bun, for example, precise specifications for the toasting time and temperature are given. The same applies to the grilling process of the meatball. ◀

The **productivity oriented sub-goals** deal with the efficient design of the processes. It must be checked, for example, whether an efficient use of resources (e.g. energy consumption) or personnel is made per customer order. This can in turn be checked by comparing times and branches. If it becomes apparent that there are significant differences between individual branches in these key figures, the processes must be checked for their efficiency. Furthermore, idle times and interfaces are to be minimized. If different departments are integrated into a service creation process and individual sub-services build on each other, idle times between the sub-services are to be avoided. This applies to employees and customers alike. If a customer is cared for by several departments or employees during a service creation process (e.g. consultation, workshop), it must be ensured that there are as short or no waiting times for the customer between the individual sub-processes. It must be ensured that after the consultation the vehicle is handed over directly to the workshop and after the repair it is handed over directly to the customer. Productivity oriented sub-goals can also be exemplarily represented, for example, in the context of the implementation of a self check-in (see Fig. 13.4).

Partial steps	Description
Wheat roll toasting	▪ Top of wheat roll toast in toaster ▪ 35 seconds ▪ At 215°C
Grill meatball	▪ Cook frozen meatball on the grill ▪ 41 to 47 seconds (depending on core temperature). ▪ At 215°C (upper griddle) or 177°C (lower griddle).
Wheat roll toast	▪ Toast the bottom part of the wheat roll in the toaster ▪ 35 seconds ▪ At 215°C
Garnish bun top	▪ A splash of mustard sauce ▪ A splash of ketchup ▪ Onions one slice of pickled cucumber ▪ For cheesburger: one slice of melted cheese preparation with cheddar cheese
Season meatball	▪ Season cooked beef burger with a seasoning mix of salt and pepper and place on top
Assemble hamburger	▪ Put on the bottom part of the wheat roll
Pack hamburger	▪ Ready prepared hamburger wrap in paper

Fig. 13.3 Process steps in the preparation of a hamburger at *McDonald's*

The representation of the sub-goals makes it clear that **goal divergences** can arise through process control. The productivity and quality oriented sub-goals can mainly be achieved through standardization of the company processes. The higher the standardization, the lower the probability of error and the more efficient the processes can be designed. However, these collide with the customer oriented goals. Through higher standardization, the possibility for companies to satisfy customer specific needs decreases. This has a negative effect on customer satisfaction and can lead to the termination of the customer relationship in the worst case. In addition, standardization often comes at the expense of personal interaction between the company and the customer. As a result, companies lose important contact points for developing a long term sustainable customer relationship. The global goal of process control in the sense of process optimization is therefore only achieved by taking into account both the customer as well as the quality and productivity oriented goals in the long term. The following presents the different types of processes as well as individual instruments of process control that support process management in this respect.

13.2 Types of Processes in the Context of Quality Management

In practice, processes are usually divided into core processes, performance processes, support processes and management and leadership processes (Ahlrichs & Knuppertz, 2010, p. 10; Grabner, 2017). All processes in the company are defined as business

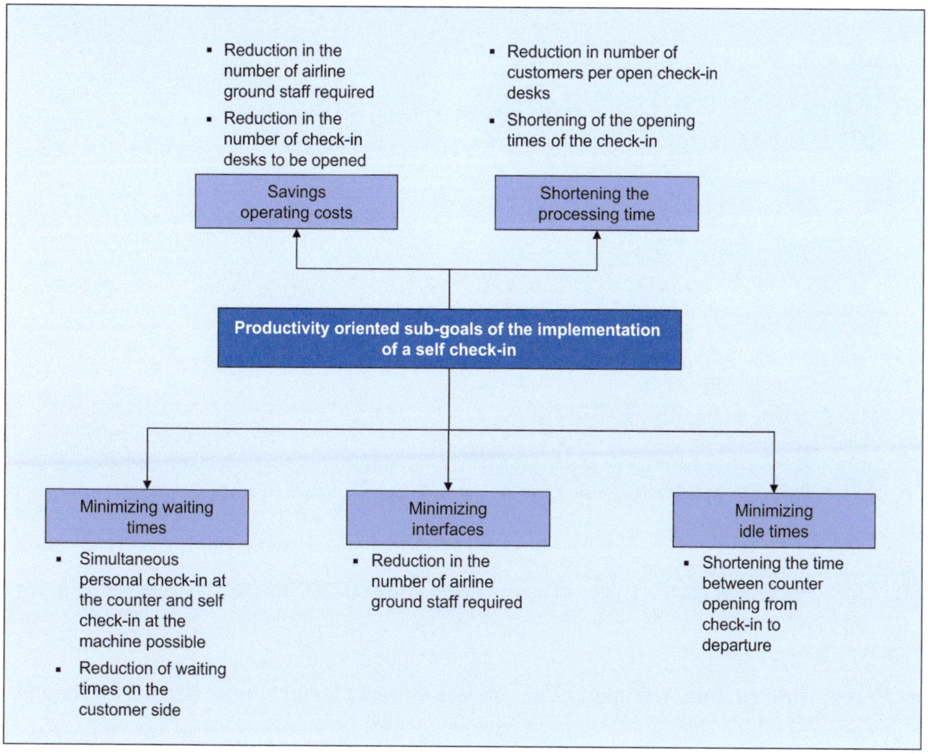

Fig. 13.4 Productivity oriented sub-goals in the context of self check-in

processes that represent a relevant business purpose of the company and are therefore generally referred to as "processes". There are usually higher level relationships and connections between the different types of processes (Ahlrichs & Knuppertz, 2010, p. 11).

In order to characterize their purpose, a distinction is made below between management, performance (core) and support processes (see Fig. 13.5).

13.2.1 Performance Processes (Core Processes)

The **core processes** include all activities, decisions, information and material flows that contribute significantly to the success of the company and are significant for the development of competitive advantages (Kaplan et al., 1991, p. 28; Osterloh & Frost, 2006, p. 36). These are mainly those activities that contribute to a knowledge based and non-imitable core competence of the company. Accordingly, core competencies have to meet the following **requirements** (Osterloh & Frost, 2006, p. 37):

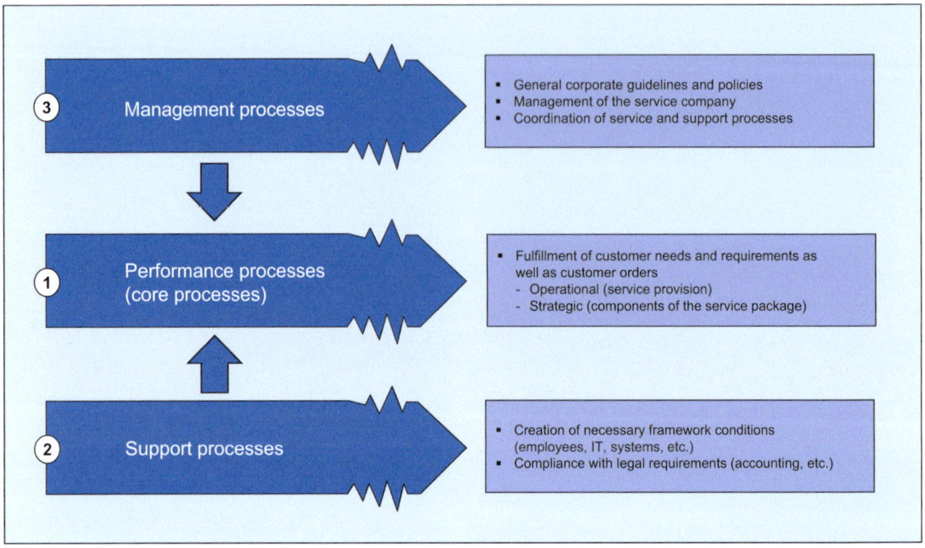

Fig. 13.5 Overview of process types and tasks. (Source: Based on Ahlrichs & Knuppertz, 2010, p. 12)

- **Perceptible customer benefit:** The processes create a perceptible benefit for the customer, for which they are willing to pay.
- **Company specific:** The processes are unique due to a company specific use of resources.
- **Non-imitabel:** The properties of the processes are not easy to imitate.
- **Non-substitutable:** The processes cannot be replaced by other problem solving processes.

In the context of quality management, those processes that contribute to the implementation of a core competence in the sense of a sustainably successful performance quality, generate a decisive benefit for the customer and cannot be copied by other companies are therefore counted as core processes. These core processes are mainly of a strategic nature and are therefore always to be carried out in the company and not outsourced (Osterloh & Frost, 2006, p. 37).

Examples of performance processes in different industries

Depending on the industry, different performance processes have to be designed. In Fig. 13.6 performance processes of different industries are shown.

- For example, the performance process of a pizza service is the quick delivery of a high-quality pizza. Here, the individual sub-services such as (telephone) order taking, preparation of the pizza up to delivery have to be designed in such a way that the pizza arrives in the shortest possible time with the desired quality.

Service company	Performance processes
Pizza service	▪ Fast pizza delivery
Hairdresser	▪ Hair cutting process
Restaurant	▪ Stay in the restaurant from the greeting to the farewell
Theater	▪ Theater performance ▪ Stay before, during and after the event
Car repair shop	▪ Process from making an appointment, consultation to collection
Surgery	▪ Process from making the appointment, counseling, performing the procedure, to leaving/discharging

Fig. 13.6 Performance processes of different service industries

- The performance process or core competence of a hairdresser lies in the creation of the haircut. Here, the individual steps and sub-processes (washing, cutting, blow-drying) have to be designed in such a way that the customer is satisfied with the haircut after it has been carried out.
- In a restaurant, various activities have to be counted as performance processes. If it is an upscale restaurant that is used for celebrations, in addition to the preparation of the food, the care of the guests, the provision of a pleasant ambience, etc. are also to be counted as performance processes.
- The decisive performance at a theater visit is the quality of the performed play. The performance process thus consists of the implementation of the performance. However, a theater visit is often seen as a social event, so that the catering and care of the guests before, during and after the theater performance are also to be counted as performance processes.
- The performance process in a car repair shop is primarily the repair service. Depending on the complexity of the repair needs, however, the competent advice and the quick processing of the repair order are also to be counted as performance processes.
- The performance process in a surgery refers to the surgical procedure to be performed on the patient. This depends very much on the competence of the surgeon. However, here too, the consulting services before the operation are often to be counted as the core services or competencies of the surgeon. ◀

The core processes run through the entire performance creation process. Accordingly, the processes are also present in all phases of quality management. Due to their strategic character, the core processes play a special role in **quality planning.** Here it is necessary to implement corresponding instruments for shaping the core processes. Competitive processes for quality management are to be designed by means of customer and market research analyses (cf., for example, service blueprints, incident driven process chain, etc. in Sect. 8.2.1). In the context of **quality control,** the core processes are used in the specific implementation of quality management. The instruments for this purpose are quality circles for developing solutions to quality problems and the daily implementation of quality requirements by employees (e.g. friendliness of customer contact employees, comprehensive consulting services, etc.). In the phase of **quality auditing,** the complaint management process is to be counted amongst the instruments of quality management in the context of the implementation of the core processes. A successful complaint management has a direct effect on customer satisfaction and, through the successful processing of a complaint (e.g. by apology, compensation), generates direct value for the customer. **Quality management documentation** contains further instruments for the implementation of the core processes with certification and auditing. Through the successful implementation of the certification process, a competitive advantage is generated and the customer is communicated the advantages and quality competence of the company. This gives the customer security when using the company's services. An audit also leads directly to an improvement in quality. Audits identify weaknesses and reveal improvement potentials, the elimination of which contributes to an increase in the quality level.

13.2.2 Support Processes

Support processes perform supporting tasks with the aim of relieving the core processes (Osterloh & Frost, 2006; Ahlrichs & Knuppertz, 2010). These do not generate direct value for the customer and, in contrast to the service processes, do not directly contribute to the implementation of the core competencies. Consequently, support processes serve internal process customers in service creation, set framework conditions (e.g. provision of service employees and customer databases) and fulfil legal requirements (e.g. accounting processes) (Ahlrichs & Knuppertz, 2010, p. 11). These are processes that could be outsourced without significant loss of expertise.

The support processes are also used in all phases of quality management. For example, the process of employee suggestion schemes (examination, award, communication of suggestions) is one of the supporting instruments in the context of **quality planning.** The collection of individual suggestions alone does not lead to an improvement in quality. Only the specific implementation of the improvement suggestions in the context of the core processes has an effect on the performance quality. The suggestion scheme therefore has a supporting character in that it makes a pre selection of the suggestions submitted. The instruments of personnel management (personnel acquisition, training and development, release) are typical supporting measures of **quality control.** Personnel management has the task of building up a staff that is able to meet the strategic requirements of quality management. **Quality auditing** consists to a large extent of support processes. The instruments include the set up and administration of quality statistics, the conduct of test purchases and employee evaluations, etc.. These measures provide indications of quality deficiencies and improvement potentials, but do not directly contribute to an improvement in quality. Support processes in the phase of **quality management**

documentation include, for example, the development and provision of quality statistics as well as the conduct of internal and external quality communication.

Often these support processes are carried out according to standardized routines and are therefore also easily imitable. For this reason, these processes are generally suitable for the use of benchmarking in process control, as no trade secrets regarding the core competencies are revealed and these processes can generally take place similarly in different companies. For example, it can be helpful for a car repair shop to take over a well functioning process for the processing of improvement suggestions from a hotel. In Fig. 13.7 different support processes of the individual service industries are shown.

In comparison to the performance processes, it becomes clear that these support processes do not apply to one industry alone, but are generally applicable in the different industries. With regard to the performance processes, it has been shown, for example, that the performance of a surgical procedure is a core competence of the surgeon and does not provide any added value for the pizza service. However, the support processes are interchangeable. The documentation of the performance creation is not only carried out by a surgeon, but also in a car repair shop or in a restaurant. Even a customer database can be used in a theater to analyze which customer segments are to be addressed in which theater performances.

13.2.3 Leadership Processes

Leadership processes serve the overarching coordination and control of core and support processes. In addition, setting general company policies and strategies is also considered a leadership process. The value substance of the service company is permanently secured through leadership processes.

The task of management is to define the strategic orientation of an organization. The associated leadership processes have the task of shaping the organizational framework, carrying out these core processes economically and successfully (Osterloh & Frost, 2006; Töpfer, 2007, p. 477). The leadership process is also referred to as the

Service company	Support processes and instruments
Pizza service	▪ Database with customer addresses
Hairdresser	▪ Personnel management → Hiring of new hairdressers
Restaurant	▪ Quality inspection → Engaging restaurant testers
Theater	▪ Quality inspection → Collecting and evaluating theater reviews
Car repair shop	▪ Company suggestion system → Improvement of repair processes, resource savings
Surgery	▪ Documentation → Recording of performed process steps for future interventions

Fig. 13.7 Support processes of different service industries

information process for the targeted control and the results oriented use of the employees and resources available in the company (Töpfer, 2007, p. 478).

The main focus of the leadership process is on goal definition, monitoring goal achievement, planning measures and supporting employees in the implementation of individual measures. The leadership process thus affects all phases of the quality management feedback loop, with the focus in the planning phase on goal definition and in the control phase on checking goal achievement. Fig. 13.8 shows different phases and contents of the leadership process based on the quality management feedback loop.

In **quality planning,** it must be specified which goals are to be achieved in general with quality management and which measures are necessary to achieve these goals. In **quality control,** the management processes have a supporting function for the core processes in the context of goal achievement. Appropriate assistance must be given to enable employees to work in a quality oriented manner. A special role is played here by the exemplary and motivational function that the management takes on. The instruments here are the quality vision, mission and quality principles. These are to be established and exemplified at management level in general. **Quality auditing** also contains elements of management processes. While operational quality control is to be counted amongst the supporting processes, the management processes here rather take on monitoring tasks. These must ensure that the relevant quality goals are actually adhered to and that the results of quality auditing are used for quality optimization. The evaluation and interpretation of so-called "customer satisfaction tracking systems" is therefore to be counted amongst the management processes, as these often have a strategic character and can entail far-reaching quality decisions. The decision on which relevant quality information is to be documented and communicated is to be made by the management

Quality management phase	Questions	Contents
Quality planning	▪ What is to be achieved? ▪ Where do we stand? ▪ What are expedient measures? ▪ What are the effects of the planned measures?	▪ Goal planning/objectives ▪ Analysis of the current situation ▪ Search for alternatives ▪ Impact forecasts
Quality control	▪ How are the measures to be implemented? ▪ How is quality management to be organized? ▪ How are the employees to be motivated?	▪ Implementation ▪ Support/behavioral guidelines
Quality auditing	▪ What has been achieved and how well? ▪ How large are the goal deviations?	▪ Implementation control ▪ Goal-performance analyses
Quality management documentation	▪ How are the contents of the quality management to be documented? ▪ How is the level of quality to be communicated credibly?	▪ Quality documentation ▪ Quality communication

Fig. 13.8 Phases and contents of the leadership processes. (Source: Based on Töpfer, 2007, p. 479)

within the context of **quality management documentation.** The design of the individual communication instruments, such as the quality manual, is then the task of the core and support processes. Furthermore, the decision to carry out certification and audits is to be counted amongst the management processes.

In general, it can be said that quality management is designed by the core processes. The management and support processes are supporting in this context. While the support processes are involved in the operational design of the core processes and act as internal suppliers of the core processes, the management processes take on the strategic design and set the framework conditions for the implementation of the core processes. In doing so, they also serve as a framework for the implementation of quality management.

The task of process control is to continuously control and evaluate the company processes. For this purpose, companies have a variety of instruments at their disposal. The instruments are categorized below and selected approaches are presented.

13.3 Methods of Process Control

Different methods are used in the context of process control in quality management. The subject and content of the process controls are initially the analysis of the temporal sequence of the projects (control and schedule planning) as well as the progress within the context of the service creation process (control of activity planning). In addition, the resources and means required are to be supervised in the form of employees and cost budgets made available or required for the projects (control of resource planning).

The methods shown in Fig. 13.9 for controlling the schedule, activity, and resource planning have established themselves as methods in quality management process controls (Bruhn, 2014, p. 371).

The individual methods of process control will be explained in more detail below and possible uses in quality management will be shown.

13.3.1 Methods for Controlling the Schedule

The methods for controlling the schedule have the task of guaranteeing a smooth process flow without idle or waiting times as well as time delays. Of particular importance are the two following procedures:

1. Network planning and bar charts
2. Milestone technique

The two methods are explained below.

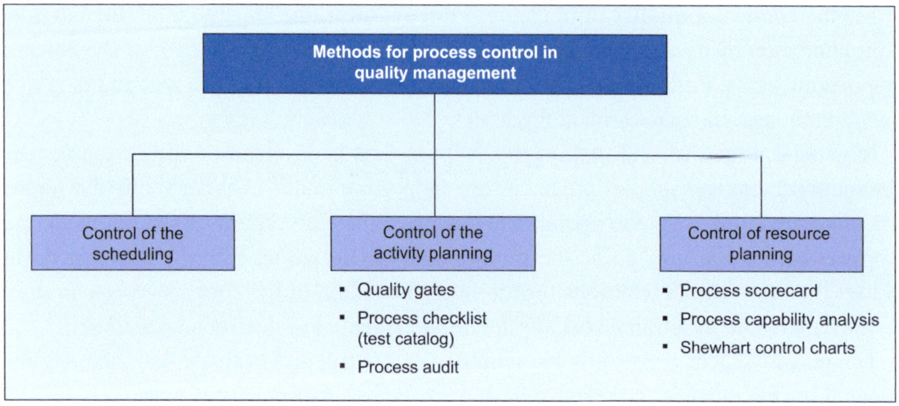

Fig. 13.9 Methods of process control in quality management

(1) Network planning and bar charts

The individual activities of a process are examined and monitored in terms of their temporal and functional dependencies using **bar charts** and **network planning**. A network plan is used to graphically represent complex processes and is often used in project management and in the planning and control of projects in practice (Fliess et al., 2004, p. 27; Holzbaur, 2016). The aim is to obtain an assessment of the expected duration of the overall process and to enable a temporal monitoring of the process flow. First, a graphical representation of the process, consisting of all individual activities, must be created. For this purpose, the functional dependencies, i.e. upstream and downstream processes, are to be identified. If there is a **functional dependency** between process A and process B, this means, for example, that process B can only be started after process A has been completed. If there is no functional dependency, the two processes can be carried out in parallel (see also the following example).

Based on the individual processing times of the individual activities and the functional sequence, start and end times as well as **buffers** (time reserves) can be determined for the individual process steps. Activities with a buffer time of "0" form the so-called **critical path** of the process. Within the scope of process control, a special focus should be placed on the activities of the critical path, as a delay in these activities inevitably leads to a delay in the overall project and a shift in the deadline. The following example illustrates the procedure of network planning using an example.

Network planning technique using the example of a financial consultancy

The following example shows, using financial consultancy, the process of creating an offer for a financing plan for a medium-sized company (Fliess et al., 2004, 42 pp.). The offer process consists of the following sub-activities (including the processing time in brackets):

- Requirement analysis (2 days)
- Protocol creation (1 day)
- Acquisition of brochures (1 day)
- Offer calculation (4 days)
- Compilation of the offer (1 day)
- Contract negotiations (1 day)

First, it must be clarified whether functional relationships between the individual activities exist, or which activities can be processed in parallel. Based on this, the network plan is represented graphically (see Fig. 13.10).

The figure shows that only the activities "acquisition of brochures" and "offer calculation" run in parallel. These require that the activity "protocol creation" and "requirement analysis" have been completed. The buffers are calculated by the difference of the latest and earliest start time or end time (for a detailed description of the calculation, see Schwarze, 2014).

It can be seen that only the activity "acquisition of brochures" has a **buffer** of three days. If there are delays in this activity of one to three days, this will not have a negative effect on the total duration of the process. All other activities have no time buffer and must therefore be processed on time.

The **critical path** is therefore as follows: Analysis → protocol creation → offer calculation → compilation of the offer → contract negotiations. If there are delays in one of these activities, the entire process will be extended and the deadline for the contract conclusion cannot be met.. ◀

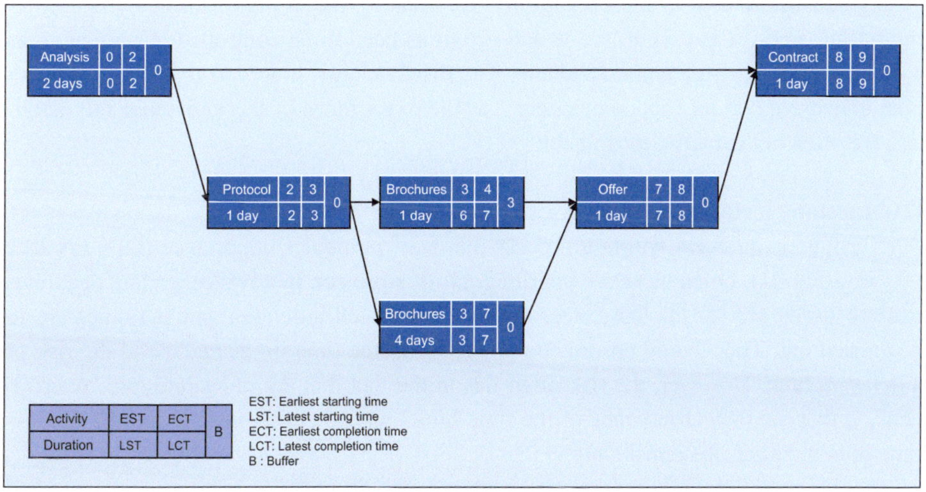

Fig. 13.10 Network plan using the example of a financial consultancy. (Source: Based on Fliess et al., 2004, p. 45)

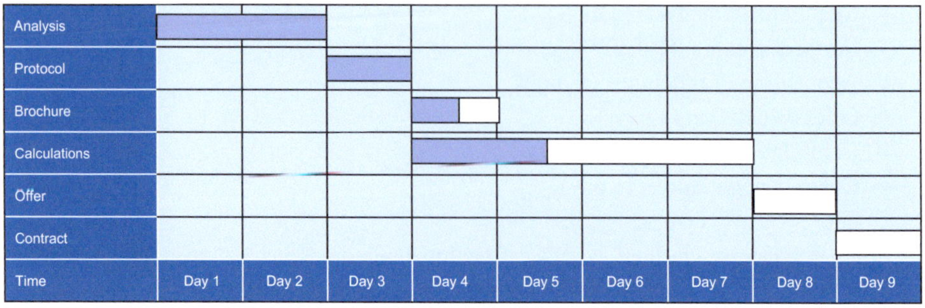

Fig. 13.11 Gantt chart of a financial consulting company (Source: Based on Fliess et al., 2004, p. 35)

A similar procedure is represented by **bar charts**. The best-known procedure from project management is the **Gantt chart** (named after the US-American management consultant Henry L. Gantt). Fig. 13.11 shows the Gantt chart of the above mentioned example of financial consultancy. Here too, a graphical representation of the process flow is made. The individual sub-activities are usually represented by bars. The length of the bars corresponds to the duration of the activities. In addition, the Gantt chart also allows the progress of the overall process and the individual sub-activities to be read off in real time during the process (dark hatching of the bars) (Schwarze, 2014, p. 24). It can be seen that the two activities "analysis" and "protocol" have already been completed, while the activities "brochures" and "calculation" are currently in the creation phase. This timely capture makes it possible for the company to determine the current time buffer of the individual activities at any time in real time.

The advantage of the network planning technique and bar charts is above all in the clarity and in the easy comprehensibility. By creating the critical path or by determining buffer times, it is easy to see which activities need to be controlled in particular, as a delay would shift the entire schedule. The disadvantage, however, is that the network plan conveys a certain "apparent safety", as the exact times of the individual sub-activities are often not exactly determinable.

(2) Milestone technique

A milestone is an event within a process that is of particular importance (DIN EN ISO 69900: 2009-01). Often these are interim goals of a project. In advance, certain deadlines or time limits are set for these interim goals, by which individual sub-activities are to be carried out. This should ensure compliance with the time limits and avoid the risk of a delay in time. However, the risk often lies in the fact that the milestones are "overridden", i. that the non-observance of the time limits has no consequences and the process time thus increases uncontrollably.

The control of the schedule plays an important role within quality management and has the goal of monitoring the time limits of the processes and recognizing and avoiding potential deviations at an early stage. However, the qualitative aspect and the goal orientation remain largely disregarded. For this reason, complementary instruments of activity planning are to be used.

13.3.2 Methods for Controlling Activity Planning

The control of the **activity planning** is used to check predefined quality criteria or sub-activities. In contrast to the schedule, the time component is not in the focus here, but it is checked to what extent specific quality requirements are covered by the process to be examined. The methods of activity planning include, in particular:

1. Quality gates
2. Process checklists (audit catalog)
3. Process audits

The individual instruments and their use in relation to process control are presented below.

(1) Quality gates

Quality gates form a systematically constructed **network of assessment barriers.** These barriers are often characterized by quantitative indicators. If a certain acceptance threshold of the actual values is exceeded in the process flow, the risk is immediately transparent through the assessment barrier. The acceptance threshold can be determined as a size anchored in the plans for the criteria quality, time and cost. At each of the quality gates, an assessment and evaluation of the previous process phase and a look ahead to the activities of the following phase are to be carried out. Often, instruments such as quality checklists or quality audits are used for this purpose (Sondermann, 2007, p. 400; Li, 2014; Wuest et al., 2014).

The concept of **quality gates** is thus an extension or refinement of milestone technique and is often used in software companies (Flohr, 2011, p. 207). In contrast to milestones, quality gates are time flexible, but content strictly formulated (Prefi, 2007, p. 412). Only when a partial goal, or "gate", has been achieved, the next process step is initiated. A quality gate denotes a result oriented point in time, which is defined by process or performance specific contents (Sondermann, 2007, p. 400).

Quality gates using the example of a restaurant

A restaurant plans to expand its offerings to include a vegan menu. First, a business plan with the relevant financial key figures must be developed and, based on this, a feasibility study must be carried out with the aim of developing a first draft of the specific design of the production and marketing of the menu. Fig. 13.12 shows different quality gates in this example.

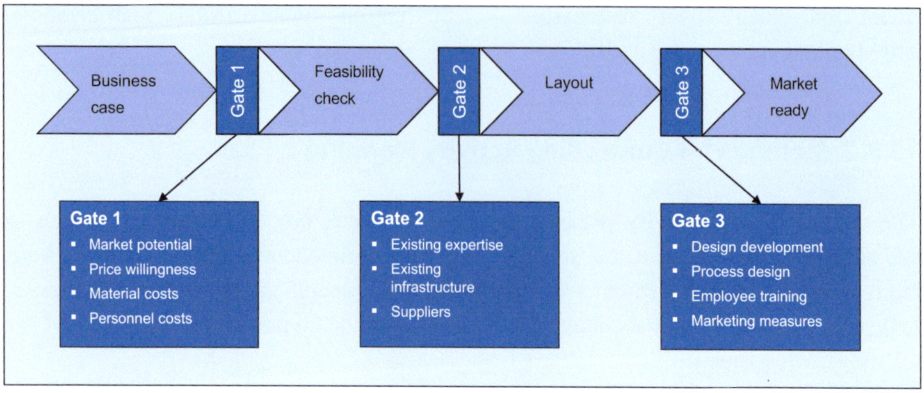

Fig. 13.12 Quality gates in the example of a restaurant

In **gate 1,** it must be checked whether the existing market potential is sufficient for the introduction. In addition, first cost calculations are to be made with regard to the production of the menu. If these requirements are met, gate 1 is passed and, in preparation for **gate 2,** a general feasibility study is to be carried out. Here, information is to be collected on whether the necessary expertise is available, i.e. whether the employees in the kitchen have the necessary knowledge to prepare vegan dishes. It is also to be checked whether special kitchen equipment is to be purchased and whether suppliers are available who offer the special ingredients. If, for example, no suitable suppliers are found in this stage, it is to be decided whether the project is to be pursued further. If the feasibility study is positive, the specific design is to be defined in **gate 3.** Here, first proposals for the menu designs are to be developed as well as concrete marketing measures (e.g. information leaflets, advertisements in the local press) to be derived, and training measures for the staff (e.g. training of the service staff in order to be able to answer guest questions competently on vegan nutrition) to be developed. The key figures for the costs of the planned measures are to be defined and evaluated again in order to avoid a strong burden on the budget. If this gate is also passed successfully, the vegan menu has met all requirements and has the so-called "market maturity", i.e. a market launch can be implemented. ◀

The decision of whether a quality gate is "passed" depends on the degree of fulfillment of the defined key figures and is structured like a traffic light:

- **Green:** Fulfillment of the requirements, i.e. release for the next project phase.
- **Yellow:** No complete fulfillment of the requirements. Only conditional release for the next project phase, development of a concrete action and schedule for ensuring the fulfillment of the requirements.
- **Red:** Requirements not met. The results are not suitable for triggering the next phase. The question here is whether to restart or even cancel the project.

The **advantage** of quality gates is above all to be seen in the fact that several small, better controllable checkpoints are introduced with the possibility of continuously stabilizing the project (Prefi, 2007, p. 418; Wuest et al., 2014). Another advantage is that the assessment of the individual gates is based on quantitative facts and key figures. The **disadvantage** is, however, that the introduction, control and assessment of quality gates is very time consuming and resource intensive. Often so-called **gate committees** are used at the gates, which decide on the continuation of the project on the basis of specially designed documents for the assessment. Therefore, it is recommended to only carry out the concept of quality gates in long term processes that extend over a long period of time and justify a high level of resources. Quality gates are therefore often used in innovation processes, e.g. the introduction of new products and services.

(2) Process checklist (audit catalog) The use of a **process checklist** is intended to ensure that pre-defined requirements are met by the performance creation process. A checklist can be created for different types of processes in general. For example, internal processes (e.g. the introduction of a new menu in a restaurant) and external processes that relate to an interaction with the customer can be controlled using checklists. In a restaurant, for example, a process checklist includes criteria regarding the design of the restaurant, the design of the menu, the availability of information on ingredients (e.g. regional origin of ingredients), and the restaurant's surroundings in terms of parking, cloakroom, terrace, etc.

The process checklist as an instrument of a customer consultation in a credit institution serves to continuously check the strengths and weaknesses of an employee in the sales process. For example, it includes information on the appointment of the customer, further planning information regarding the design of the investment portfolio, such as further customer and product information, the identification of customer needs, information on the offer and cross-selling options, etc..

These checklists not only serve as an instrument of subsequent control, but they must always be present during the performance creation process in order to ensure that all quality relevant points are taken into account within the process. In Fig. 13.13 examples of an internal and external process checklist are shown.

(3) Process audit In order to uncover weaknesses in the process flow, so-called **process audits** are increasingly being carried out in companies, which are mainly conducted by external auditors (e.g. Baumgarth, 2008, p. 359; Bruhn, 2014, p. 371; Schmitt & Pfeifer, 2015, p. 164). The process audit pursues the goal of an evaluation within the framework of seminars and workshops, whereby no closed system is provided, but rather individual important structural features of the audit system are discussed (Baumgarth, 2008, p. 359; Bruhn, 2014, p. 3278). The process audit takes into account the use of various instruments, such as internal standardized surveys on the knowledge of the process flow or content analysis (cf. in detail on audits Sect. 8.5.1). This ensures that the processes

Checklist for internal processes (e.g. introduction of a new menu in a restaurant)		Checklist for external processes (e.g. consultation of a financial service provider)	
▪ Are suitable suppliers available that meet the quality requirements of the ingredients?	☐	**Before the interview:**	
▪ Has an appropriate costing been performed to ensure profitable creation and utilization?	☐	▪ Is all relevant information about the customer available? (income, risk propensity, needs)	☐
▪ Has a market analysis been conducted to assess sales potential?	☐	▪ Is all product and performance information available?	☐
▪ Does the new menu fit into the restaurant's existing service offerings?	☐	**During the interview:**	
▪ Are there employees with appropriate food preparation qualifications?	☐	▪ Is the customer given enough time to express their needs?	☐
▪ Is the kitchen equipment suitable for food preparation?	☐	▪ Are all the customer's open questions clarified?	☐
▪ Is there sufficient information on the court in case of any queries from customers?	☐	**Upon completion of the interview:**	
▪ Is the service staff properly trained to answer questions about the new menu?	☐	▪ Are the next steps/follow-up dates defined?	☐
▪ Were the new dishes added to the menu?	☐	▪ Was a call log created and sent to the customer?	☐

Fig. 13.13 Process checklists for internal and external processes of service creation

Questions
▪ What are the process needs and which of them still need to be implemented?
▪ What process does the organization use to determine customer requirements and expectations regarding quality?
▪ How is the development of performance planned and managed?
▪ How are service delivery processes controlled whose results cannot be verified by subsequent audits?
▪ What information on customer satisfaction or dissatisfaction is recorded?
▪ Are the specified intervals for conducting process audits sufficient to determine whether the QM system is being effectively implemented and maintained?
▪ In what way are the processes required for the continuous improvement of the QM system planned, managed and directed?
▪ In what ways are corrective actions taken to eliminate process variation to prevent recurrence?

Fig. 13.14 Exemplary process audit question list for service companies. (Source: Based on Meinhold & Matul, 2011)

established in the company are aligned with the achievement of the quality management goals. In Fig. 13.14 exemplary questions for a process audit are shown.

As part of process control, process audits are used for both prevention and correction of quality defects. Regular performance of process audits helps **prevent quality defects** by identifying and correcting faulty (sub)processes at an early stage. However, unplanned audits serve the **error correction.** These are then to be carried out if a decline in the quality level or an increase in errors is to be observed.

13.3.3 Methods for Controlling Resource Planning

In the context of resource planning, efficient process flow and the identification of resource savings in terms of cost reduction or reduction of processing time are in the focus. The aim is to identify cause-effect relationships between processes and corporate goals. The methods are therefore used predominantly to achieve the productivity oriented goals of process management. The following methods are used to control resource planning:

1. Process scorecard
2. Process capability analysis
3. Shewhart control charts

The individual methods are explained below and possible uses in the context of process control are shown.

(1) Process scorecard

The process scorecard is to be understood as a further development and modification of the balanced scorecard against the background of an increased process orientation (Wildemann, 2010, p. 23). The basis of the process scorecard is the division of processes into core, support and management processes (cf. process types in Sect. 13.2).

The aim of the process scorecard is to identify **cause-and-effect relationships** between individual company processes and company success on the basis of specified (usually quantitative) key figures (Zangl, 2009, p. 83). The **company success** is divided into the following six **categories:**

- Profitability
- Financial strength
- Innovation strength
- Process strength
- Customer loyalty
- Corporate culture

For the assessment of profitability and financial strength, classical financial key figures such as sales margin, liquidity, cash flow, etc. are used. The other four success factors are the so-called leverage factors. The measure for each of these four success factors is an index value, which is formed from the measurement variables of the company processes for the individual leverage factors. A change in the measurement variables of the company processes affects the index value of the leverage factors (Zangl, 2009, p. 86).

In a next step, relationships between company processes and success categories are identified and displayed in a matrix. It becomes clear which company processes (performance, support and management processes) influence which success categories. The following example explains the procedure for creating a **process scorecard matrix** using the example of a car dealership.

Process scorecard matrix using the example of a car dealership

Building on the success categories and the performance, support and leadership processes identified in the company, the process scorecard matrix of a car dealership shown in Fig. 13.15 is created.

First, the six categories of company success are shown **(step 1).** Based on this, the core processes are defined as well as the objectives to be achieved with the establishment of the core processes **(step 2).** For example, the goal of the customer needs process at the car dealership is to raise the customers' need for additional services (e.g. winter check for the vehicle) in order to generate additional sales. The goal of the marketing process, for example, is to increase the visibility of the car dealership in the

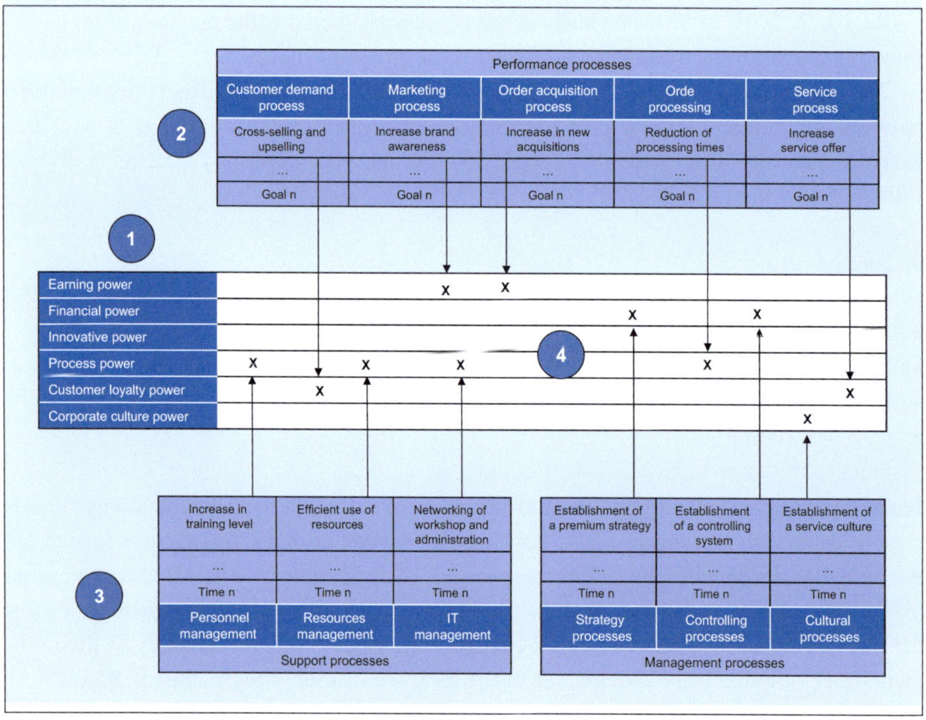

Fig. 13.15 Process scorecard matrix using the example of a car dealership. (Source: Based on Zangl, 2009, p. 87)

region in order to win new customers. The goal of the order processing process could be to simplify the internal process of an order processing and to reduce the interfaces between the employees or departments involved.

Step 3 deals with the definition of the core and leadership processes as well as the derivation of the corresponding objectives. The personnel management as a supporting department has the goal to acquire qualified personnel for the car dealership and to train the existing personnel continuously. For example, the technological developments require new special skills of the mechanics, but also the employees in the administration have to be trained for new systems continuously (e.g. SAP systems). One goal of the IT processes is, for example, the networking of the workshop with the administration. The goal of the strategy processes is to pursue the strategic orientation of the car dealership (e.g. pursuing a premium strategy).

In **step 4,** the company processes are to be linked with the success factors. It is determined which processes and objectives contribute to the achievement of the success factors. This is the decisive step of the process scorecard matrix, as this makes the connection between the company processes and the success factors clear. The relationships are each marked with an "X" in the figure. For example, Fig. 13.14 makes it clear that cross- and up-selling as a goal of the customer needs process contributes to achieving the success factor "customer loyalty" by binding customers to the car dealership through the purchase of additional services. The networking of the workshop with the administration as a goal of the IT management contributes to the increase in process power by simplifying processes, reducing them and making them more efficient overall. ◄

Based on this matrix, the relationships between the company processes and the success categories are identified and a separate **process scorecard** is developed for each of the six success categories. The individual scorecards show which success factors have been achieved and which factors require action. Thus, the process scorecard is suitable as a control instrument, as it can be used to determine which processes contribute to the company's success to what extent or to identify inefficient processes that do not contribute to success.

Process scorecard of earnings power using the example of a car dealership

Fig. 13.16 shows a process scorecard for earnings power as a success factor using the example of a car dealership.

As the example shows, the financial target values (EBIT, equity ratio) are achieved. The target of cost or budget compliance is critical. A closer look at the individual processes shows that the marketing and order processing processes have exceeded the budget. Savings measures are therefore to be planned. In addition, however, it is also clear that the planned order volume has been exceeded and as a result the utilization rate of the workshop is above the target value. On one hand, this is positive, on the other hand, this development is to be observed further in order to avoid possible bottlenecks that would lead to the rejection or delay of customer orders. ◀

The **advantage** of the process scorecard lies - similar to the balanced scorecard - in the clear presentation of extensive target categories as well as in the identification of cause and effect relationships between the processes and the company objectives. In addition, the quantitative approach using key figures and index values is to be emphasized. This makes it possible, for example, to compare processes, companies and branches. The development of individual process key figures over time can also be observed using the process scorecard. The **disadvantage** is again the high effort required for data collection as well as the apparent inaccuracy of the key figures.

(2) Process capability analysis

Process capability analysis has its origins in the automotive industry and is a measure of long term quality variability, which is influenced by different factors such as employees, materials, methods and the working environment (Kamiske & Brauer, 2011, p. 305). The instrument serves as a statistical control instrument, whereby disruptions in the performance creation process are uncovered using identified deviations from pre-defined indicators. This makes it possible to identify critical processes and, based on this, to

Profitability scorecard			Actual/target	Implementation status	Recommendation/ consequence
Key figures/ goals	EBIT		Actual: 2,530,467 EUR Target: 2,500,000 EUR	Goal is achieved	No need for action
	Return on equity		Actual: 18% Target: 15%	Goal is achieved	No need for action
	Total cost/budget		Actual: 1,290,300 EUR Target: 1,000,000 EUR	Goal achievement not possible	Check costs
Performance processes	Customer demand process	Cost/budget	Actual: 42,500 EUR Target: 45,000 EUR	Goal achievement endangered	Check savings
	Marketing process	Cost/budget	Actual: 20,750 EUR Target: 25,000 EUR	Goal achievement not possible	Reduce marketing costs
	Car acquisition process	Order volume	Actual: 530,000 EUR Target: 400,000 EUR	Goal is achieved	No need for action
		Cost/budget	Actual: 32,800 EUR Target: 25,000 EUR	Goal achievement not possible	Reduce costs
Support processes	Human resources management process		Actual: 17,000 EUR Target: 20,000 EUR	Goal is achieved	No need for action
	Resource management process	Utilization rate workshop	Actual: 100% Target: 95%	Goal achievement endangered	Observe; avoid bottlenecks
		Stocks material	Actual: 340,500 EUR Target: 300,000 EUR	Goal achievement endangered	Observe; reduce stocks
	IT management process	Cost/budget	Actual: 74,300 EUR Target: 90,000 EUR	Goal is achieved	No need for action
Management processes	Strategy process	Growth path	Actual: +25% Target: +20%	Goal is achieved	No need for action

Legend: Goal is achieved; Goal achievement endangered; Goal achievement not possible

Fig. 13.16 Example of an earnings power scorecard using the example of a car dealership. (Source: Based on Zangl, 2009, p. 88)

eliminate the underlying weaknesses. For process capability analysis, target values and tolerance ranges are to be determined for individual process steps (see also the example in Sect. 5.2.1). If the tolerance ranges are exceeded or fallen short of, appropriate countermeasures and process adjustments are to be initiated. If it turns out that a permanent exceedance of the tolerance limit exists, it is to be checked whether the process behind it still meets the current requirements.

Process capability analysis of a call center

If you look at process capability analysis using the example of a call center (see also Sect. 5.2.1), the target value for the duration until the call is received is three rings. If it is now identified through the control that the waiting time for the duration until the call is accepted has increased significantly, it is to be clarified which reasons are mainly responsible for this and how these deviations are to be eliminated. One conceivable cause could be the increased need for information on the part of the customers. As a possible consequence, the number of employees is to be increased. It is also possible to relieve the call center by expanding alternative communication channels (e.g. homepage, e-mail requests, chats, etc.). ◀

Goal of process capability analysis is to establish stable, robust and reproducible processes that contribute to high performance quality (Syska, 2006, p. 110). This is often only possible through standardization of the processes carried out. Therefore, this instrument is to be adapted accordingly for service companies. Due to the integration of the external factor and the high personnel intensity, standardization in services is very difficult. Nevertheless, the instrument is a tried and tested instrument that is often used in practice due to its easy implementability and quantifiability using indicators in standardized service processes. For services, the instrument is therefore particularly suitable for the control of standardized processes that are subject to few fluctuations and external influences.

(3) Shewhart control charts

The Shewhart control chart is an extension of the process capability analysis and is named after the American physicist and engineer Walter A. Shewhart. For the individual processes, target key figures are also developed and a tolerance zone is defined around this target value (cf. also Sect. 5.2.2). The deviation of the actual performance from the specified target values is graphically represented, so that a development of the processes over time is visible. The goal is to recognize deviations early and to take countermeasures in good time to prevent a negative development. The following example shows the use of the Shewhart control chart in a restaurant.

Shewhart control chart for determining the waiting time in a restaurant

In a restaurant, a Shewhart control chart can be used to control the waiting times for an order. The waiting time is continuously measured and compared with a norm that has been set in advance. The measured time should never exceed or fall below a certain deviation from this norm. The Shewhart control chart shows the course of the waiting times over a certain period of time. If the norm is, for example, 20 minutes waiting time and the tolerated deviation 10 minutes, a guest should never have to wait more than 30 and never less than 10 minutes for his food. If it takes too long for the food to come, the guest will be very angry. If the food is already on the table shortly after the order, the guest might assume that the dishes were not prepared fresh. If a deviation of more than 10 minutes is measured, it is to be checked how this error came about. A fictitious example of a Shewhart control chart is shown in Fig. 13.17.

The Shewhart control chart shows when and how often a non-tolerable deviation occurred. From the example it is clear that the waiting times are always at the upper tolerance limit in the evening. On weekends, i.e. Friday and Saturday evenings, the tolerance limit is then clearly exceeded. Measures must therefore be taken to reduce these tolerance violations. However, it must first be checked which causes the long waiting times. These can be caused by the service personnel, but also by a long preparation time in the kitchen. Accordingly, the personnel in the service or in the kitchen is to be increased. ◀

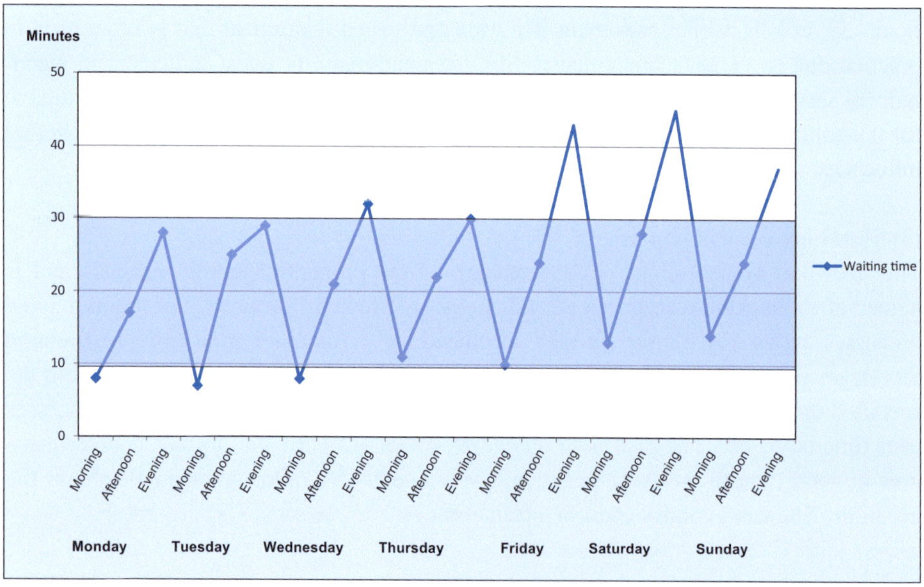

Fig. 13.17 Shewhart control chart for controlling the waiting time in a restaurant

The process control instruments presented here are characterized above all by the fact that the company processes are considered and evaluated as holistic processes. The process idea plays a particularly important role in the creation of services, since the customer often becomes part of the service creation and thus perceives the entire creation process. With the instruments presented, companies have a comprehensive tool available to achieve different customer, quality and productivity oriented objectives. Finally, individual weaknesses of process control will be discussed on the basis of a critical assessment.

13.4 Critical Assessment of Process Controls in Quality Management

There is still a great lack of established structures and processes in service companies in general and quality management within these companies in particular. Therefore, an **advantage** of process controls is that a process oriented thinking is generally promoted in these companies. Through process control it is achieved that company processes are perceived and evaluated as a whole. With the help of process controls it is possible to detect and correct faulty processes at an early stage. This prevents deficiencies from becoming visible only at the end of the service creation process, which have already occurred at an early stage.

The central **disadvantage** of process controls is the transferability to certain types of services. The origin of process orientation and process control lies in manufacturing. The aim is to establish as standardized, reproducible and efficient processes as possible and to control their implementation. However, these standardization efforts are often not implemented to this extent in service companies. Through the integration of the customer into the service creation process and the tendency for a high degree of individualization of services, the extent of process standardization is often very limited. This is especially true for services that are characterized by a high degree of individuality (e.g. medical, social and cultural services).

Nevertheless, it has to be stated that process controls are of outstanding importance for service companies. This is especially true for those services that are characterized by a low degree of integration, immateriality, interaction and individualization. So in principle all services that have a high degree of standardization of their processes.

In addition, it should be mentioned that the identified **cause and effect relationships** between the company processes and the success factors can not be clearly demonstrated. Here, an apparent inaccuracy arises, which does not exist to this extent in reality. Many process control procedures are very complex and can only be mastered with great effort. Therefore, it must be checked in each case for which processes it is reasonable to carry out a process control. In principle, companies are obliged to develop their own control instruments, which appear adequate in the individual company situation and the processes established in the company.

Process controls alone are not sufficient to master the complexity of the quality management system comprehensively. In addition, it is recommended to carry out effectiveness and efficiency controls. The special features and fields of application of these forms of control are explained in more detail in the following chapters.

References

Ahlrichs, F., & Knuppertz, T. (2010). *Controlling von Geschäftsprozessen* (2nd Edn.). Stuttgart: Schäffer-Poeschel.

Baumgarth, C. (2008). Möglichkeiten und Grenzen der Kontrolle einer integrierten Markenkommunikation. In A. Hermanns, T. Ringle, & P. C. van Overloop (Eds.), *Handbuch Markenkommunikation. Grundlagen, Konzepte, Fallbeispiele* (pp. 351–365). München: Vahlen.

Bruhn, M. (2014). *Integrierte Unternehmens- und Markenkommunikation: Strategische Planung und operative Umsetzung* (6th Edn.). Stuttgart: Schäffer-Poeschel.

Fliess, S., Lasshof, B., & Meckel, M. (2004). Möglichkeiten der Integration eines Zeitmanagements in das Blueprinting von Dienstleistungsprozessen. http://www.fernunihagen.de/www2bonsai/FBWIWI/forschung/beitraege/pdf/dp362.pdf. Accessed 22 Apr. 2013.

Flohr, T. (2011). Assessing the quality of quality gate reference processes. In D. Hutchison, T. Kanade, J. Kittler, J. M. Kleinberg, F. Mattern, & J. C. Mitchell (Eds.), *Software engineering techniques – Lecture notes in computer science* (pp. 207–217). Berlin.

Grabner, T. (2017). *Operations Management – Auftragserfüllung bei Sach- und Dienstleistungen* (3rd Edn.). Wiesbaden: Springer Gabler.

Hammer, M., & Champy, J. (2006). *Reengineering the corporation. A manifesto for business revolution*. New York: Harper Business.

Holzbaur, U. (2016). *Events nachhaltig gestalten Grundlagen und Leitfaden für die Konzeption und Umsetzung von Nachhaltigen Events*. Wiesbaden: Springer Gabler.

Kamiske, G. F., & Brauer, J.-P. (2011). *Qualitätsmanagement von A bis Z. Erläuterungen moderner Begriffe des Qualitätsmanagements* (7th Edn.). München: Hanser.

Kaplan, R. B., Murdock, L., & Ostroff, F. (1991). Core process redesign. *The McKinsey Quarterly, o. V.*(2), 27–44.

Li, Z. Q. (2014). A research on quality risk controlling principles and methods in automotive products' design based on the quality gates. *Advanced Materials Research, 909*, 141–147.

Meinhold, M., & Matul, C. (2011). *Qualitätsmanagement aus der Sicht von Sozialarbeit und Ökonomie* (2nd Edn.). Stuttgart: UTB .

Osterloh, M., & Frost, J. (2006). *Prozessmanagement als Kernkompetenz* (5th Edn.). Wiesbaden: Springer Gabler.

Porter, M. E. (2014). *Wettbewerbsvorteile. Spitzenleistungen erreichen und behaupten* (8th Edn.). Frankfurt a. M.: Campus

Prefi, T. (2007). Qualitätsmanagement in der Produktentwicklung. In T. Pfeifer & R. Schmitt (Eds.), *Masing – Handbuch Qualitätsmanagement* (5th Edn., pp. 405–440). München: Hanser.

Schmitt, R., & Pfeifer, T. (2015). *Qualitätsmanagement. Strategien, Methoden, Techniken* (5th Edn.). München: Hanser.

Schwarze, J. (2014). *Projektmanagement mit Netzplantechnik* (11th Edn.). Berlin: NWB.

Sondermann, J. P. (2007). Interne Qualitätsanforderungen und Anforderungsbewertung. In T. Pfeifer & R. Schmitt (Eds.), *Masing – Handbuch Qualitätsmanagement* (5th Edn., pp. 387–404). München: Hanser.

Syska, A. (2006). *Produktionsmanagement*. Wiesbaden: Gabler.

Töpfer, A. (2007). *Betriebswirtschaftslehre. Anwendungs- und prozessorientierte Grundlagen* (3rd Edn.). Wiesbaden: Springer.

vom Brocke, J., & Sonnenberg, C. (2011). Prozesstransparenz als Grundlage für das Management und Controlling von Geschäftsprozessen. *Controlling & Management, 55*(2), 55–68.

Wildemann, H. (2010). Controlling im Total Quality Management. In G. F. Kamiske (Ed.), *Digitale Fachbibliothek Qualitätsmanagement* (pp. 1–41). Symposion.

Wuest, T., Liu, A., Lu, S. C. Y., & Thoben, K. D. (2014). Application of the stage gate model in production supporting quality management. *Procedia CIRP, 17,* 32–37.

Zangl, H. (2009). Process Scorecard. Unternehmen ganzheitlich über Prozesse steuern. *Controlling & Management, 5,* 83–93.

Effectiveness Controls in Quality Management

14

Abstract

Effectiveness control plays an important role in quality management. The goal of effectiveness control is to check whether the desired effects have been achieved through quality management. An effective implementation of quality management is the central prerequisite for consistent high quality. Accordingly, this chapter deals primarily with the degree of achievement, that is, the effects of the measures taken to achieve a high level of service quality. In doing so, relevant effect components are derived as target variables of effectiveness control and instruments for checking the achievement of targets are shown using specific examples. A final critical assessment shows the advantages and disadvantages of effectiveness control within quality management.

14.1 Basics of Effectiveness Control

14.1.1 Effect Components of Effectiveness Control

In addition to process control, **effectiveness control** plays an important role within quality management. The goal of effectiveness control is to check whether the desired effects have been achieved through quality management. Therefore, the term **effect control** is also used in this context. The individual effect components result from the goals defined in the planning phase.

Depending on the company's objectives, different effects are to be achieved with the help of quality management. Monetary and non-monetary quality objectives are used as

M. Bruhn, *Quality Management for Services*,
https://doi.org/10.1007/978-3-662-67032-3_14

control variables. In principle, different **effect categories** can be distinguished, which are listed in Fig. 14.1.

On the first level, the effect components can be divided into psychological and behavioral effects. The **psychological effects** can be further differentiated into cognitive and conative effects. **Cognitive effects** include, for example, the company's or service's awareness or the general knowledge of consumers' performance. A possible cognitive goal of quality management could be to achieve a high level of awareness or to communicate a high level of quality to the outside. A high level of service quality will eventually pay off for the company only if it is perceived by consumers, i.e. consumers have the appropriate knowledge. **Affective effects** include, for example, consumers' attitude towards the company, the company's image, customers' satisfaction with the company's services or complaints and / or recommendations expressed accordingly. Effective quality management has a significant impact on these variables. Quality management measures can directly increase the company's image and contribute to high customer satisfaction.

In addition to the psychological effects, it is necessary to influence the **behavioral effects** through quality management. While the psychological effects are of a non-monetary nature, the conative behavioral effects can be further divided into monetary and non-monetary effects. **Non-monetary behavioral effects** include, for example, the intention to use, customer loyalty or the intention to inform, such as positive word-of-mouth communication. **Monetary behavioral effects** of a high level of service quality are sales, profit and market share. As a rule, the behavioral effects are downstream of

Control object			Control instruments
Psychological effects	Cognitive effects	Non-monetary effects	▪ Awareness ▪ Information level ▪ Performance skills ▪ etc.
	Affective effects	Non-monetary effects	▪ Attitudes ▪ Image ▪ Satisfaction ▪ Preferences ▪ Complaints and recommendations ▪ etc.
Behavioral effects	Conative effects	Non-monetary effects	▪ Customer loyalty ▪ User intension ▪ Recommendation intension ▪ Usage intensification ▪ etc.
		Monetary effects	▪ Sales volume ▪ Profit ▪ Share of wallet ▪ Market share ▪ etc.

Fig. 14.1 Systematics of effect components of effectiveness control

the psychological effects. For example, it is assumed that a long term relationship with a customer can only be achieved if the customer is satisfied with the quality and performance of the company.

14.1.2 Instruments of Effectiveness Control

Companies have a variety of instruments for effectiveness control available. Basically, in the effectiveness control **goal vs. actual comparisons** are made which determine the **degree of goal achievement** of quality management. It is checked whether and to what extent the goals defined in the planning phase of quality management were actually achieved. The individual key figures are then each to be subjected to a detailed **deviation analysis** and, based on the results of the deviation in the event of deviations, to identify the causes of the deviations. Fig. 14.2 shows possible control instruments for effectiveness control for the respective categories of effects and control variables.

14.1.2.1 Instruments for the collection of the psychological effects of quality management

A variety of market research methods are available to assess the cognitive effects on customers, such as awareness, level of information or knowledge about the quality level of the service company. For example, the general knowledge of a service can be determined with a **recognition test** with test subjects. They are presented with studied material (e.g. a specific selection of insurance companies) together with other material (e.g.

Control object			Control parameters	Control instruments
Psychological effects	Cognitive effects	Non-monetary effects	▪ Awareness ▪ Information level ▪ Performance skills knowledge, etc. ▪ etc.	▪ Recognition test ▪ Recall test ▪ Evoked set analysis ▪ Projective techniques ▪ etc.
	Affective effects	Non-monetary effects	▪ Attitudes ▪ Image ▪ Satisfaction ▪ Preferences ▪ Complaints and recommendations ▪ etc.	▪ Attitude measurement ▪ Image profiles ▪ Satisfaction measurement ▪ Preference studies ▪ Opinion portals / web 2.0 ▪ etc.
Behavioral effects	Conative effects	Non-monetary effects	▪ Customer loyalty ▪ User intension ▪ Recommendation intension ▪ Usage intensification ▪ etc.	▪ Customer data bases ▪ Customer surveys (e.g. NPS) ▪ Complaints department data
		Monetary effects	▪ Sales volume ▪ Profit ▪ Share of wallet ▪ Market share ▪ etc.	▪ Accounting data ▪ Sales data ▪ Industry data ▪ etc.

Fig. 14.2 Instruments of effectiveness controlling in quality management (Source: Based on Bruhn & Hadwich, 2011, p. 17)

other insurance companies) and asked to remember which material or insurance companies they remember. This test is used very often in the evaluation of the effectiveness of certain advertisements or campaigns (Fantapié-Altobelli, 2017). The test can be extended with regard to quality management in that the test subjects are asked to assign the mentioned insurance companies to certain quality categories (e.g. high, medium, low) or to identify service companies with a particularly high quality standard. Since it is particularly difficult to make an objective assessment of quality in the service sector and since the subjective perception of quality by the consumer is therefore to be seen as a purchase decision, the recognition test is particularly suitable for checking whether the quality measures used have achieved the desired effect on the customer (recognizing one's own company as being of high quality).

In a **recall test**, on the other hand, the test subjects are asked to name companies of a performance category, e.g. different insurance companies, freely. With regard to quality, the procedure can be adapted in that the test subjects are asked, for example, to name high-quality insurance companies. In addition, the procedure can be further specified by asking for companies that are considered to be particularly successful in specific quality features (e.g. consulting quality, friendliness, etc.) (cf. Fig. 14.3).

The **evoked set analysis** asks customers about all alternatives they would consider when buying a certain service. Often, a consumer's evoked set consists of a limited number of services with which positive experiences have already been made or which are known through advertising (Homburg, 2017, pp. 107-108). In addition, it is assumed that actively known service providers (e.g. insurance companies) are selected with a higher probability than passively known providers and that these are tendentially rated as higher quality. Recall values (unsupported awareness levels) are therefore much more significant than recognition values (supported awareness levels).

Quality feature	Recognition test	Recall test
Global quality judgment	Which of the following insurance companies would you consider to be of high quality?	Name insurance companies that you consider to be of high quality.
High advisory quality	In your opinion, which of the following insurance companies provide high quality advice?	Name insurance companies that you believe provide high quality advice.
High quality of care	Which of the following insurance companies do you think have a high quality of care?	Name insurance companies that you believe have a high quality of care.
Friendliness of the staff	Which of the following insurance companies do you think stand out as having particularly friendly staff?	Name insurance companies that stand out for having particularly friendly staff.
Attractive condition system	In your opinion, which of the following insurance companies are characterized by an attractive system of conditions?	Name insurance companies that stand out for their attractive conditions system.
Uncomplicated claims handling	Which of the following insurance companies do you think have a straightforward claims process?	Name insurance companies that you feel have a straightforward claims process.

Fig. 14.3 Examples of recall and recognition tests with quality reference

Projective techniques are used to uncover the "true" opinions and feelings of a test person towards a company with regard to quality aspects, even if the test persons are not initially willing or able to make reliable statements (Fantapié-Altobelli, 2017). The aim of these techniques is to distract the test persons from the actual question in order to obtain more open and informative answers (Kepper, 2014, p. 184). The stimuli are often characterized by a certain uncertainty, for example by presenting not clear and unambiguous situations. The respondents then have the task of interpreting these situations according to their own experiences and values (Fantapié-Altobelli, 2017). This way, the level of knowledge with regard to the quality level or the customer's knowledge about the company can be better identified by obtaining more information than the test person would originally be willing to communicate (for a detailed description of the individual methods see Fantapié-Altobelli, 2017).

The most widespread projective methods include the **completion techniques.** In the context of completion techniques, the person being interviewed is presented with partial sentences that are content related to the object to be examined, which they are to complete spontaneously (Fantapié-Altobelli, 2017). This indirect questioning yields statements about ideas, opinions or attitudes regarding the quality of the company to be examined, which would not be articulated to this extent in a direct interview, for example due to response tendencies in the direction of social desirability. With regard to quality assessment, the partial sentences can be formulated accordingly so that the respondents take specific reference to the quality level or interesting sub-aspects of quality in their answers. Possible examples of a sentence completion test with a quality reference are shown in Fig. 14.4.

Control instruments such as attitude measurement, quality profile, satisfaction measurements and (quality) preference studies are used to check the **affective effects.** Especially in service companies, these affective effects such as quality image, attitudes, etc. are of particular interest, since the quality of the services can usually not be estimated before the service is used and therefore these variables play a special role in the purchase decision compared to the assessment of consumer goods.

For the execution of the **attitude measurement** different methods are available, which are all carried out by means of customer surveys. In the simplest case, a measurement of opposite quality related statements about the examined company is made on a rating scale (e.g. "The trading company stands for high / low service quality"). In addition, a differentiated analysis can be carried out via so-called multi-item measurements, i.e. a measurement over several quality properties, which are relevant for the overall assessment of the performance. The summary of the relevant property assignment then provides an **image profile** of the assessed performance or the assessed company. The image of a company consists of more or less evaluative impressions of a company or a performance, which are combined to form a holistic image (Trommsdorff & Teichert, 2011, p. 133). In the context of quality management, the goal is often to build up a high quality image or to achieve the image of a high-quality supplier. The regular creation of an image profile provides important indications as to whether the various goals of quality

Fig. 14.4 Examples of a sentence completion test. (Source: Based on Fantapié-Altobelli, 2017)

management in terms of the company's external appearance have been achieved over time.

Image profile of the textile retail group *H&M – Hennes und Mauritz*

Fig. 14.5 shows an exemplary image profile of the textile retail chain *H&M*. The image was determined by querying various performance components such as, for example, the friendliness of the staff, the quality of the goods, the quality of the advice, etc.

For example, it can be concluded from the image profile that there is potential for improvement, in particular with regard to the quality of advice and process quality during shopping (no queue at the checkout). What is decisive here is also to observe the development over time. For example, if, as part of quality management, the sales staff were trained in terms of advice quality, the comparison of image profiles at different times makes it possible to determine whether this measure has led to the desired quality improvement from the customer's point of view. ◀

Customer satisfaction is closely linked to attitude and image. Satisfaction is the result of a comparison between quality related requirements and the quality of service perceived by the customer. In the context of **satisfaction measurements,** either an overall satisfaction rating is requested (e.g. "I am very satisfied/not satisfied at all with the quality

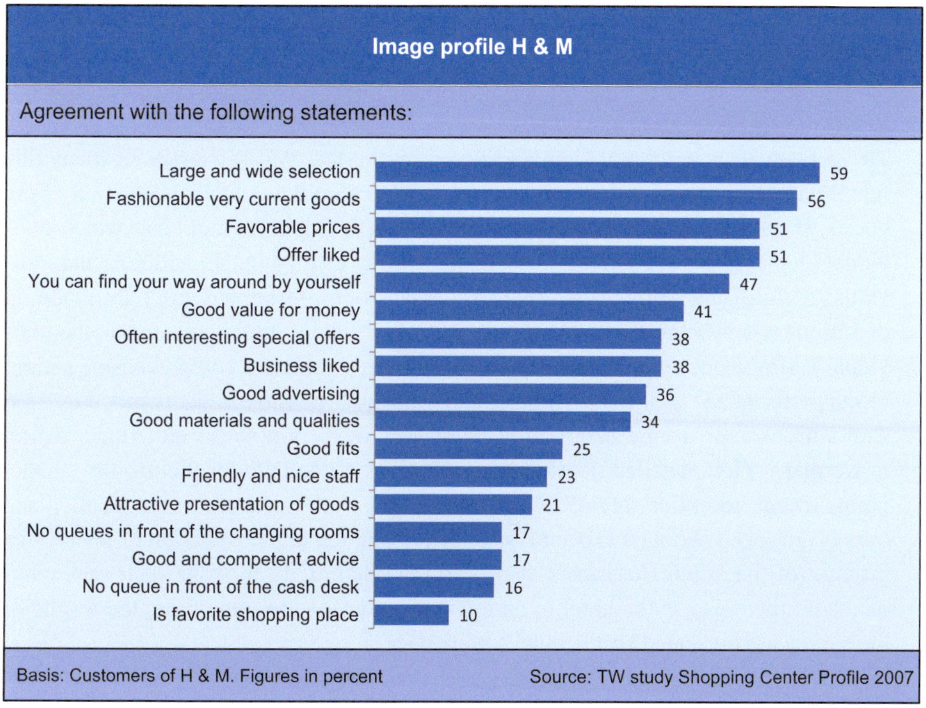

Fig. 14.5 Image profile of the textile retail group *H&M*

of company XY"), or various partial satisfactions are requested with regard to individual quality aspects (see Sect. 5.1.2.1 for the different types of satisfaction measurement). In contrast to attitude, satisfaction measurement is only possible with customers who have already used a company's service and are able to make statements about the quality of service they have experienced. Satisfaction measurement therefore allows conclusions to be drawn about the effectiveness of quality management, as ultimately it is the quality of service in the minds of customers that is judged. A high level of customer satisfaction therefore indicates a high level of service quality. In **preference studies,** it is further determined whether and why a certain service company is preferred over comparable suppliers.

Finally, an analysis of opinion portals and social media platforms in the context of web 2.0 can contribute to using freely expressed customer opinions for the purpose of improving and developing services (Decker & Gnibba, 2009). This way, informally written customer reviews can be converted into structured data using natural language processing techniques and made available for evaluation using relevant data analysis methods (Decker & Gnibba, 2009). For example, the relative importance of product attributes can be determined using a Poisson regression, while a KeyGraph approach provides indications for the design of new services by identifying potentially relevant attributes (Decker & Gnibba, 2009).

Satisfaction measurement of a hotel

The performance of a hotel is characterized by the fact that it consists of many different partial performances which have an influence on the overall satisfaction of the guests. If only the global satisfaction is determined, it is often not clear which individual elements are specifically responsible for the evaluation. In addition, the individual performance components (e.g. reception, room, restaurant, etc.) are made up of various other components. Therefore, it is of particular importance to ask the individual performance components separately in order to obtain a comprehensive picture of the performance quality of the hotel and to identify possible weak points. Fig. 14.6 shows an excerpt from a satisfaction questionnaire for a hotel in the Allgäu region in Germany. First, detailed questions are asked about the individual customer contact points (room, reception, breakfast, restaurant, bar, wellness, etc.). Subsequently, customers are asked about their overall satisfaction and the implementation of the service promise of the hotel. Customers also have the opportunity to make their own comments on their stay at the hotel in an open question. At the same time, the results of the survey are presented on the hotel's homepage

(Source: Hotel Mohren, 2016). ◀

14.1.2.2 Instruments for measuring the behavioral effects of quality management

However, the mere determination of customer satisfaction is not enough to conclude on the success of quality management. Because it is not automatically assumed that a high level of customer satisfaction also leads to the desired behavioral effects such as customer loyalty, re-purchase, intensification of use (e.g. cross buying), recommendation, etc..

Therefore, corresponding control instruments have to be implemented which capture these behavioral effects. A distinction is made here between instruments for determining the non-economic and economic effects. The non-economic effects are primarily concerned with the performance use by the customers, i.e. how often they use the service. In addition, the intensification of use (e.g. cross buying), information and use intention are in the focus. In the first instance, **customer surveys** are suitable here - similar to the satisfaction measurement - to determine the non-monetary effects. The satisfaction or an assessment of the service quality is not queried here, but specific statements about the future behavior of the consumers are determined. Thus, it can be stated to what extent the measures of quality management lead to the desired behavioral effects with the customers. In addition, weaknesses can be identified which, if necessary, are to be improved in the future by means of quality management. The following example shows how the different non-monetary behavioral effects can be determined by customer surveys with an airline.

Your rating

Please rate the following criteria according to the School grading system (grades: 1=very good, 2=good, 3=satisfactory, 4=sufficient, 5=deficient, 6=insufficient

Reservation	1	2	3	4	5	6
Procedure/handling of the reservation	O	O	O	O	O	O
Advice and information about the hotel	O	O	O	O	O	O
Competence of the staff	O	O	O	O	O	O
Friendliness of employees	O	O	O	O	O	O
Consideration of special requests	O	O	O	O	O	O

Room	1	2	3	4	5	6
Equipment & atmosphere	O	O	O	O	O	O
Room size	O	O	O	O	O	O
Room cleanliness	O	O	O	O	O	O
Friendliness and attentiveness employees	O	O	O	O	O	O

Reception	1	2	3	4	5	6
Check-in, e.g. reception, escort to the room ...	O	O	O	O	O	O
Check-out, e.g. farewells	O	O	O	O	O	O
Friendliness and attentiveness of the employee	O	O	O	O	O	O
Competence of employees	O	O	O	O	O	O

Breakfast	1	2	3	4	5	6
Furnishings and ambience of the breakfast room	O	O	O	O	O	O
Variety and diversity of the food offer	O	O	O	O	O	O
Food quality	O	O	O	O	O	O

Restaurant/half board	1	2	3	4	5	6
Furnishings & ambience of the restaurants	O	O	O	O	O	O
Variety and diversity of the food offer	O	O	O	O	O	O
Quality of the food	O	O	O	O	O	O
Friendliness and attention of the employee	O	O	O	O	O	O
Speed of service	O	O	O	O	O	O

Bar	1	2	3	4	5	6
Range of drinks	O	O	O	O	O	O
Friendliness and attentiveness of the employees	O	O	O	O	O	O
Speed of service	O	O	O	O	O	O

Wellness	1	2	3	4	5	6
Sauna, steam bath	O	O	O	O	O	O
Rest area	O	O	O	O	O	O
Cleanliness of the wellness area	O	O	O	O	O	O
Atmosphere of the wellness area	O	O	O	O	O	O

Family	1	2	3	4	5	6
Family friendly	O	O	O	O	O	O

Guest program

How satisfied were you with the guest program?'
(Ap?ro, hut evening, hikes, golf ...)

O Very satisfied
O Satisfied
O Less satisfied
O Not satisfied
O I cannot judge

Overall satisfaction

Overall, how satisfied were you with your stay at the hotel Mohren?

O Very satisfied
O Satisfied
O Less satisfied
O Not satisfied

Our promise

Have you found at the hotel Mohren what was promised to you in our advertisement?
we promised in our advertising materials?

O Yes
O No
O Why not?

Fig. 14.6 Example of satisfaction measurement in a hotel. (Source: Hotel Mohren, 2016)

Non-monetary effects with customers of an airline

Using the example of a passenger who, for example, regularly books the Munich-Hamburg connection with an airline for professional reasons, the non-monetary effects of quality management become clear. Through successful quality management, it is possible that the effects shown in Fig. 14.7 are achieved for the customer. ◄

The intention to recommend represents another non-monetary customer behavior effect (as it is also represented in the questionnaire in Fig. 14.6). For the purpose of measuring the intention to recommend, the so-called **Net Promoter Score** has gained specific importance in practice. In the Net Promoter Score (NPS), customers are asked to answer the following question on a 10-point scale:

"On a scale of 0-10, how likely is it that you would recommend (product/service/company) to a friend or colleague?"

Impact category	Expression
Intensification of use	▪ More frequent use of the Munich-Hamburg connection ▪ Use of further connections for business purposes, which were previously covered by train, for example
Information intent	▪ Willingness to participate in customer surveys ▪ Willingness to pass on information regarding special service preferences and willingness to pay ▪ Willingness to share information about planned air travel
Intended use	▪ Willingness e.g. also to plan vacation trips with the airline

Fig. 14.7 Non-monetary effects using the example of an airline

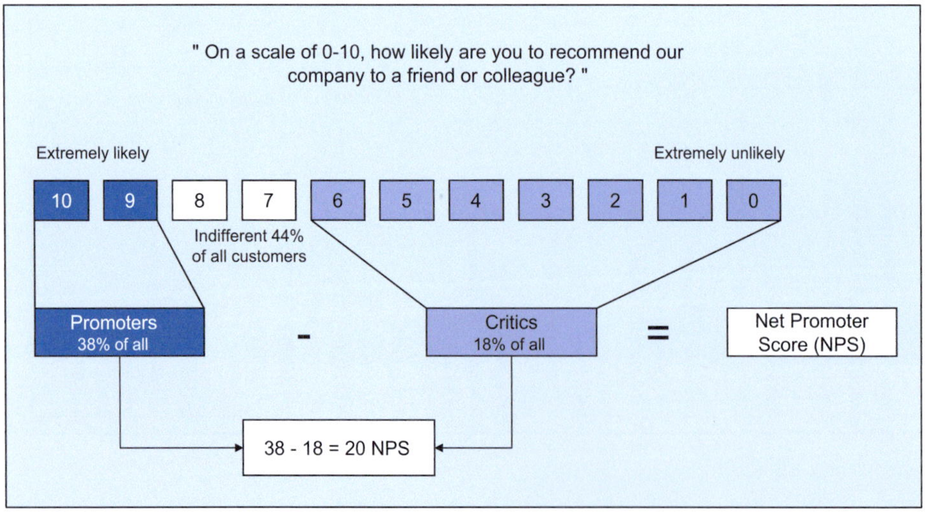

Fig. 14.8 Calculation of the Net Promoter Score

In the calculation of the NPS, the responses on the scale at 9 and 10 are rated as "promoters" (those who would definitely recommend the company), the responses at 7 and 8 as "indifferent" (those who would neither positive nor negative recommend it) and the responses 6 to 0 as "critics" (those who would not make a positive recommendation). The NPS results from the difference between the promoters and the critics.

Calculation of the Net Promoter Score

A survey in a hotel on the intention to recommend on a 10-point scale resulted in a total of 38% promoters (values 9 and 10), 44% indifferent (values 8 and 7) and 18% critics (values 6 to 0).

The NPS results from the difference between the critics and the promoters (38-18), i.e. an NPS of 20 (see Fig. 14.8). This NPS can then be compared with past values, other branches, other customer segments, other services, other regions, etc.. ◀

The **advantage** of the NPS lies primarily in its simplicity, i.e. only a single question is required for the calculation. The **disadvantages** are methodological weaknesses (e.g. the arbitrary division of the groups) as well as the lack of convincing evidence of the behavioral relevance of recommendations. In this respect, the NPS will not replace customer satisfaction measurement, but rather supplement it. Of decisive importance for companies will be the analysis of the reasons for the lack of recommendation and the NPS itself can not provide any information about this. For this additional analyses are necessary.

Net Promoter Score Benchmark

An advantage of the Net Promoter Score is its comparability. The figure is calculated from the answers to only one question. This is identical in almost all NPS surveys. However, it is important to note the selection of the test subjects. As a result, a benchmark is possible, but only subject to reservations (see Fig. 14.9).

(Source: Net Promoter Score, 2016). ◀

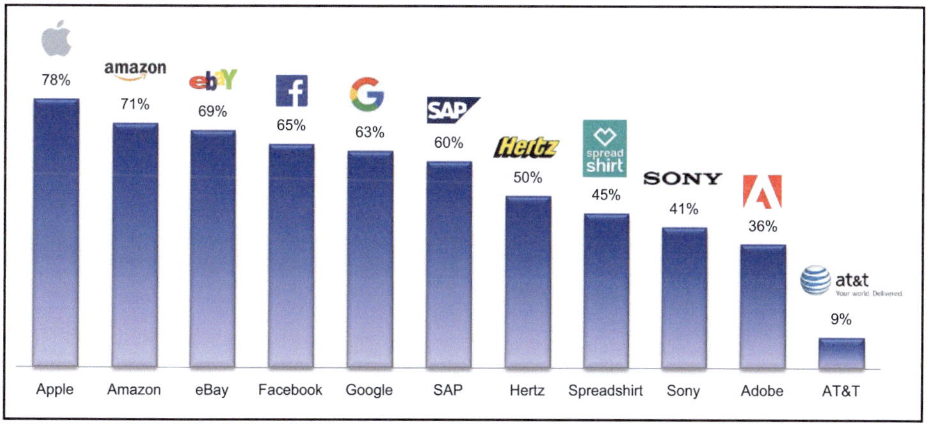

Fig. 14.9 Net Promoter Score Benchmark. (Quelle: Net Promoter Score, 2016)

Further information on customer behavior is provided by **customer databases.** The key here is that a functioning customer relationship management is implemented in the company. The goal of these databases is to gain insight into future customer behavior and customer potential from past behavior (Bruhn, 2013, p. 8). Consequently, methods of **data mining** are used here, which make it possible to classify customer types, for example according to their preferences or customer value, or to identify customers at risk of churning (Töpfer, 2008, p. 633). In this context, the term **analytical CRM** is often used (Wehrmeister, 2001, p. 246). Fig. 14.10 shows the central components of CRM in relation to quality management.

The successful use of analytical CRM requires a comprehensive data base. The data pool has to meet the following requirements (Töpfer, 2008, p. 634):

- Error-free
- Relevance
- Actuality
- Completeness

With the help of data mining, quality relevant contexts and **implications for quality management** can be identified (Strawe et al., 2002, p. 34; Baumann et al., 2012; Demirkan & Delen, 2013). For example, if stagnant or declining sales figures are seen in a certain customer segment, it is necessary to check whether systematic quality defects exist in this segment, which have led to this decline and whether appropriate measures

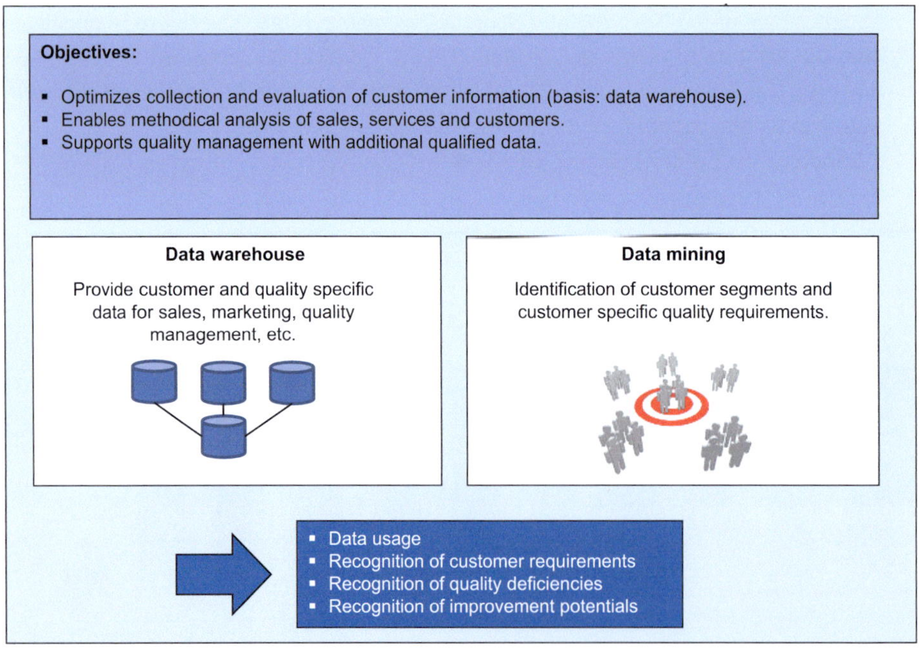

Fig. 14.10 Basic structure of an analytical CRM. (Source: Based on Töpfer, 2008, p. 633)

Task	Quality features of complaint management
Complaint evaluation	▪ User oriented implementation of quantitative evaluations with quality reference ▪ Precision of root cause analysis
Complaint reporting	▪ User friendly provision of information ▪ Timely provision of information ▪ Adherence to defined reporting deadlines
Complaint information use	▪ Use of complaint information for quality improvements ▪ Use of complaint information for the development of new services

Fig. 14.11 Tasks and quality characteristics of complaint management with regard to quality controlling. (Source: Based on Stauss & Seidel, 2014, p. 307)

need to be taken to keep these customers at risk of churning. By storing customer data centrally with access for all employees involved in quality management, it is ensured that the relevant customer information is always available in order to ensure an effective and customer oriented implementation and optimization of quality management.

The **complaints department** is another important instrument of effectiveness control. Thus, not only customer specific, but also on a global level, regular evaluations of complaint behavior are to be carried out across the entire customer base (cf. for complaint management Sect. 8.4.2). Analogous to the problem oriented determination of service quality, a frequency relevance analysis of complaints can be carried out on the basis of the collected complaint data (cf. for the frequency relevance analysis of problems Sect. 5.1.2.3). In accordance with the procedure of the FRAP, the quality deficiencies articulated in the complaints are prioritized and action implications for quality management are derived therefrom (Stauss & Seidel, 2014). The consideration is that the quality deficiencies are to be eliminated all the more urgently, the more often they are articulated via complaints and the more they lead to customer annoyance. Analogous to the customer databases, the information from the complaints department is to be collected centrally and made available to the relevant persons and departments of quality management. With regard to service controlling, particular attention is to be paid to the quality characteristics shown in Fig. 14.11 of complaint management.

However, the behavioral effects must ultimately lead to financial success for the company in order for the company to be successful in the long term. Therefore, the monetary effects on corporate success are also to be examined as part of efficiency control. Here, mainly internal data sources from **accounting** are used. These are primarily **sales statistics,** but also profit figures, etc.. The decisive factor here is that these are to be prepared as meaningfully as possible, so that implications for quality management can be derived from them. It is therefore not enough to simply look at the company's total sales at regular intervals. While the development over time is clearly visible, no information is available for quality management as to where quality deficiencies may exist. Therefore, target oriented evaluations are to be carried out which document the degree of target achievement. If, for example, it was decided in the planning phase of quality management of a restaurant that the quality of the dessert offer should be improved, only the sales generated by dessert products are to be taken into account from accounting. Something similar applies if a quality target was to increase customer loyalty of regular customers. It is to be examined to what extent the proportion of regular customers to total sales has increased. However, information on sales to new customers does not help to assess this target achievement.

In addition to the internal monetary key figures, other information plays a role, which is to be evaluated rather by using external company information sources at the same time. These include, for example, the **market share, share of wallet, market penetration** etc.. In contrast to the figures of accounting, not only the success of the company is considered in isolation here, but this is interpreted in comparison to the total market or to the competition. A goal could be, for example, to increase the market share by 2 percentage points in a performance category. Pure sales figures are not enough for this. It is quite possible that the company's sales increase by 2%, but the total market reports an increase of 5%. This would mean that the company had lost market share despite an increase in sales. For this reason, it is crucial to also consider the economic key figures in the context of the total market in order to avoid potential misinterpretations.

In terms of quality management, it is always important to ensure that a connection is made between the conative non-economic and the economic effects and the measures of quality management. It must therefore be determined whether a certain intensification of use and the increase in sales and profit figures are actually due to an increased quality level of the services offered or whether other factors may influence the conative variables and economic effects. In Fig. 14.12 different **key figures** of the psychological and behavioral effects of quality management are shown.

Control variables		Operationalization
Market shares	Quantity	Share of own sales volume in the total sales volume of all suppliers in the relevant market.
	Value	Share of own sales in total sales of all suppliers in the relevant market.
	Field Share	Proportion of the number of own customers to the total number of demand (or targeted) customers.
Pricing	Relative price or price premium achieved	Ratio of market share by value to market share by volume.
	Price band compliance (in terms of volume)	Proportion of own sales achieved within the target price band.
	Price band compliance (in value)	Proportion of own sales generated within the target price band.
Awareness	Unaided awareness (recall)	Proportion of target customers who spontaneously recognize the company.
	Supported awareness (recognition)	Proportion of target customers who recognize the company.
Image position	(Brand) sympathy	Percentage of customers in the relevant market who rate the company or brand as likeable
	(Brand) status	Relationship between awareness, (brand) liking and (brand) use.
	(Brand) image	The nature and expression of the quality characteristics and competencies associated with the company, brand or services.
Customer satisfaction	Customer satisfaction index	Percentage of customers who are (very) satisfied with the company or its quality.
	Relative customer satisfaction	Own customer satisfaction index in relation to the customer satisfaction index of the main competitor.

Fig. 14.12 Selection of central key figures of quality management. (Source: Based on Reinecke & Tomczak, 2006, p. 901)

The essential **advantage** of the effectiveness controls by means of the presented control instruments is that the control variables provide information about the effect of quality activities. The focus is not only on the actual behavior of the customers, but also on the rational and emotional effects of quality management. However, in this context, it must be mentioned as a **disadvantage** that the assumed cause and effect relationships can be influenced by factors not taken into account, such as competition activities.

14.2 Effectiveness Controls by Success Chain Controlling

A systematic and efficiency oriented quality management is oriented towards the success chain of quality management, in which the effects of quality management on customer behavior and thus on corporate success are structured (cf. Chap. 1). Thus, these effects also play an important role in the context of quality management controlling. The goal of **success chain controlling** is to check to what extent the effects desired by quality management, i.e. customer satisfaction and customer loyalty in particular, have been achieved by a company.

In this respect, it is not a matter of an isolated measurement of effects, but of an **integrated effect measurement,** by considering the relationships between different effect sizes. Its conceptual basis and origin are the so-called national customer barometers.

14.2.1 National Customer Barometers as a Conceptual Basis

National customer barometers were first introduced at the end of the 1980s. Since then, they have been a key element in the discussions of research and practice on the measurement of the effects of quality management. A key aspect of the concept of national customer barometers is that it builds on the success chain of quality management and thus represents a predecessor of a company specific success chain controlling.

14.2.1.1 Concept of national customer barometers

National customer barometers are taking on an increasingly important role as an economically relevant factor. The beginning of this development can be seen in the early 1980s. In 1989, the Swedish Customer Satisfaction Barometer (SCSB) started with the investigation of the satisfaction with the largest companies in Sweden in different industries (Meyer & Dornach, 1994, 444 pp.; Turkyilmaz et al., 2013). Then, in 1994, the American Customer Satisfaction Index (ACSI) was determined for the first time, which is closely based on the concept of the Swedish model. For the **terminological formulation of national customer barometers** the following definition can be used as a basis:

▶ A **national customer barometer** is an industry wide study by a neutral institution that, by means of periodic surveys, measures various perceived quality characteristics in terms of psychological and behavioral effects of customers in numerous sectors, industries and companies of a nation or an economic area.

In Germany, national satisfaction data have been collected since 1992 as part of the customer barometer Germany (Meyer & Dornach, 1995a). Fig. 14.13 shows the results of the customer barometer Germany from 2015.

Goals of the national customer barometer often include economic goals in addition to business goals. In this context, social, corporate and customer related goals can be distinguished. Fig. 14.14 shows possible goals of national customer barometers in an overview.

Company related goals can be illustrated by the following example. The German drugstore chain *dm* uses the positive results of the customer barometer Germany aggressively in its public relations work in order to draw attention to its good performance in terms of customer orientation in the industry comparison and to differentiate itself from the competition.

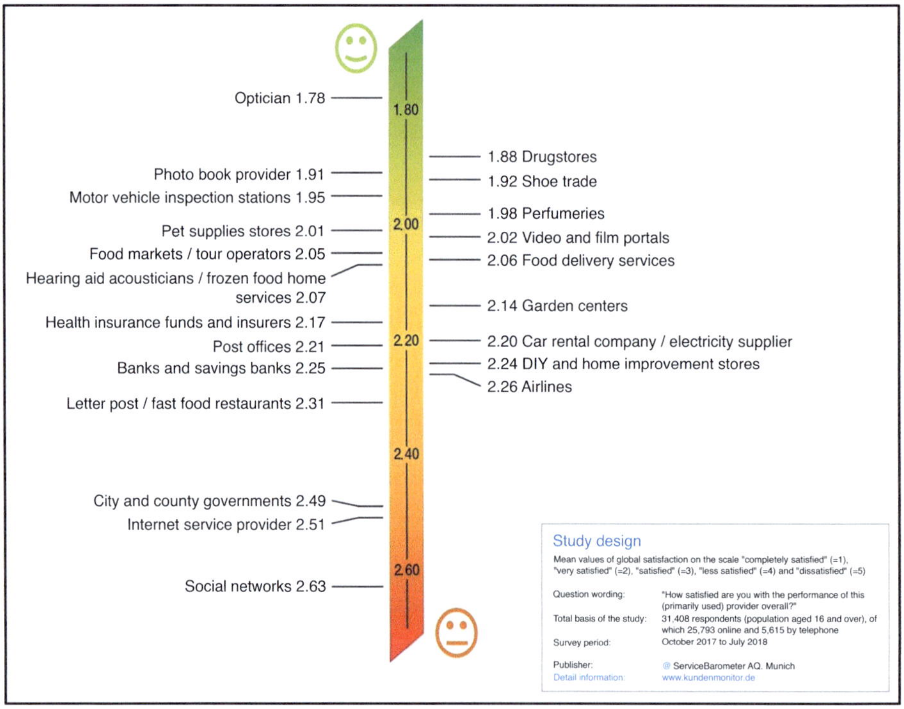

Fig. 14.13 Results of the customer barometer Germany in 2018. (Source: Service Barometer AG, 2018)

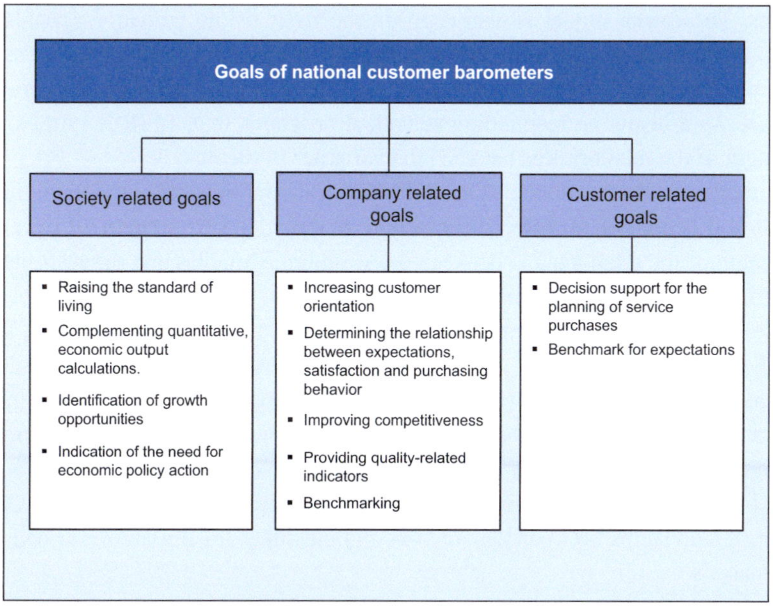

Fig. 14.14 Goals of national customer barometers

Press release by *dm-drogerie markt GmbH + Co. KG* on the results of the customer barometer Germany 2012

The German drugstore chain *dm* has communicated a detailed report on its positive performance in the customer barometer Germany on its homepage. This explains in detail in which categories the company takes a leading position in comparison to the competition. In addition, links to detailed test reports and to graphics are available. The continuous improvement of customer satisfaction with *dm* over time is particularly emphasized. Quotes from the management about the importance of this quality award underline the importance and relevance of customer satisfaction within the company.

(Source: dm drogerie-markt, 2012) ◀

In order to determine satisfaction, individual questions (so-called indicator variables) about a company, a product or the performance of an institution are presented to the respondents as part of a national customer monitor, which are answered with whole number satisfaction values between 1 (very satisfied) and 5 (very dissatisfied). The **overall satisfaction** is determined using a causal model based on the results of the individual questions. The causal model is estimated on the basis of the indicator variable values obtained from the customer using **causal analysis**, which in numerous iterations identifies the best possible representation of the "real" relationships between the variables by the model. As a result, corresponding statistical programs (e.g. LISREL, Amos) deliver the strength of the relationships between the variables of the model.

The strength of the relationships between the structural variables (e.g. the influence of the perceived benefit on customer satisfaction) is quantified by **structural parameters.** The strength of the relationships between the structural variables and the respective indicator variables is expressed by measurement parameters.

These measurement parameters can be interpreted as **importance weights** of the individual indicator variables. They are used to form indices (e.g. quality index, satisfaction index, customer loyalty index) over the structural variables by aggregating the mean values of a structural variable using weights with the help of measurement parameters. Through the **index formation** a comparability, e.g. the customer satisfaction, both over time and between different providers, is made possible. In addition to the overall satisfaction, relationships between satisfaction and customer loyalty are examined using a causal analysis.

The general problems of measuring service quality and customer satisfaction as well as the experiences of the already institutionalized barometers result in different **requirements for national customer barometers** (cf. e.g. Meyer & Dornach, 1995b; extensively Bruhn & Murmann, 1998; Hansen & Hennig-Thurau, 1999; Ertl, 2013):

- The first requirement for national customer barometers relates to the **validity of the measurement** of customer satisfaction. This criterion specifies whether the individual measurements actually reflect the underlying construct of "customer satisfaction". A significant determinant of the validity of a satisfaction barometer is the number of categories of the scale used, which is positively correlated with the validity of the measurement (Andrews, 1984).
- In addition, the **reliability** of the satisfaction measurements must be taken into account. A measurement is all the more reliable the narrower the range of measurements is due to changes in customer satisfaction and not to other factors. A very reliable measurement is reflected in stable results over time or the same results when using the same measuring instrument several times.
- Furthermore, when assessing satisfaction indices, the **predictive power** of the results is to be considered. This is about the effects of changed satisfaction values on customer loyalty in the form of re-purchase intention, recommendation behavior, cross-selling, etc..
- Furthermore, the **representativeness** of the results must be taken into account. This requires that the companies considered make up as large a proportion of economic variables as possible, e.g. gross domestic product, with respect to an entire economy or an industry.

- In order to ensure an appropriate approach when carrying out the satisfaction measurements, it is also necessary to ensure a certain **simplicity of the survey and evaluation**.
- To increase the significance of the results obtained, the data related to the measurement should be kept as **up-to-date** as possible.
- Another criterion is the **simplicity of interpretation** of the results, i.e. it is necessary to ensure that the indicator values can be read in a as simple as possible way.
- Finally, it is necessary that the measuring instrument provides a **comparability** of the results within an economy or industry.

14.2.1.2 Use of national customer barometers for effectiveness control

The goal of effectiveness control is to determine the degree of goal achievement. As explained in Sect. 14.1, it is very costly and time consuming to determine psychological effects of quality management (e.g. awareness, image values, customer preferences, etc.). National customer barometers offer the opportunity to determine various qualitative and quantitative target variables in a **cost-effective** way and to evaluate them with regard to their effective target achievement by means of a **wide data base.**

In addition, the indirect measurement of the constructs examined by means of indicator variables contributes to the identification of specific statements about the **importance of individual performance features** for the overall quality in a specific industry or for a single company within the framework of effectiveness control. This way it can be checked whether individual quality measures are effectively used. For example, if it is shown that the quality of the service hotline plays an important role in the overall satisfaction of mobile phone providers, it is to be evaluated within the framework of effectiveness control whether corresponding quality measures contribute to an increase in the quality of performance of the service hotline and can be effectively implemented.

In addition, national customer barometers are suitable for effectiveness control due to the ability to make comparisons between one's own company and the largest competitors. One's own market research efforts usually relate to one's own company, and comparable benchmarking data from competitors is hardly available. This isolated view therefore significantly reduces the interpretability of the indicators. National customer barometers enable **industry and company comparisons.** By comparing with competing companies within the same industry, one can identify deficiencies of one's own company as well as competitive advantages over the competition and derive recommendations for quality management. The objectives of effectiveness control can therefore be extended beyond absolutely formulated goals (e.g. increase customer satisfaction by 2 percentage points) to **relative goals** (e.g. increase customer satisfaction compared to the largest competitor by 2 percentage points).

In addition, as part of effectiveness control, the effects of individual quality measures can be viewed in terms of **time course.** A comparison of satisfaction values over different years makes it possible to conclude whether certain quality initiatives had the desired effect in the long term.

Comparisons with companies from unrelated industries through national customer barometers also enable, in connection with effectiveness control, the possibility to identify potential benchmarking partners. As Fig. 14.13 shows, companies in the book order business have particularly successful satisfaction values. Companies from other industries, such as electricity or gas suppliers, thus have the opportunity to identify improvement potentials with regard to an efficient target achievement through comparisons with particularly successful companies.

14.2.1.3 Representation of the European Performance Satisfaction Index (EPSI)

A difficulty with national customer barometers has been so far that an international comparison of the data of different national customer barometers was almost impossible, as different approaches to measuring customer satisfaction are pursued in most countries. This is particularly unsatisfactory for companies that are internationally active.

Against this background, in 1999 - initiated by the *European Commission* and organized by the *European Organisation for Quality (EOQ)* and the *European Foundation for Quality Management (EFQM)* - a pilot measurement was carried out as part of the ***European Performance Satisfaction Index (EPSI).*** In addition to measuring satisfaction across industries, this European barometer also enables international comparability of the data.

In order to establish the ***European Performance Satisfaction Index (EPSI*-rating),** customer satisfaction and customer loyalty as well as their influencing factors were investigated in parallel in 12 European countries in spring 1999 - Germany did not participate in the pilot study. Since then, the surveys have been carried out at regular intervals. In the survey in 2015, more than two million telephone interviews were conducted with private and business customers and the following ten nations participated in the survey (*EPSI* Rating, 2015):

- Denmark
- Estonia
- Finland
- Latvia
- Lithuania
- Norway
- Russia
- Sweden
- Czech Republic

The *EPSI* rating is based on the ***EFQM Model* for excellence,** which highlights the evaluation category "customer related results" with a weighting of 15% as an important driver for the performance capabilities of organizations (cf. Sect. 10.2.1.1). Based on this thought, the model of the *EPSI* rating is consistently oriented towards a customer perspective (*EPSI* Rating, 2015).

Fig. 14.15 shows the ***EPSI*** **Model,** in which the image of the company, customer expectations, perceived product and service quality as well as the customer perceived benefits are included as interdependent influencing factors of customer satisfaction and loyalty (Sütőová & Šolc, 2009). Compared to the pilot study of 1999, the model has undergone some changes compared to the original model. The models differ essentially in that now multiple cause and effect relationships between the enabler and the outcome dimension are assumed. In addition, complaints from some countries are now partly explicitly modeled as an effect of satisfaction and as an influencing factor of customer loyalty. Also in newer models of the *EPSI* the modeling of the financial success of companies as an outcome variable is dispensed with (*EPSI* Rating, 2015).

Within the causal model, the **structural variables** are understood as latent variables, that is, as not directly measurable, which are multi-dimensionally measured by so-called **indicator variables**; The values of the latent variables being determined via the customer's perceptions of the indicator variables. For the structural variable customer satisfaction, the indicators "product features", "accessibility of the sales channels", "image of the supplier", "price", "fulfillment of customer expectations" and "performance" are used in this model (Sütőová & Šolc, 2009). For the operationalization of the expectations and the quality, performance features of the supplier are used, similar to the feature oriented methods (e.g. SERVQUAL). Typical indicator variables of customer loyalty are the willingness to re-elect and the willingness to recommend.

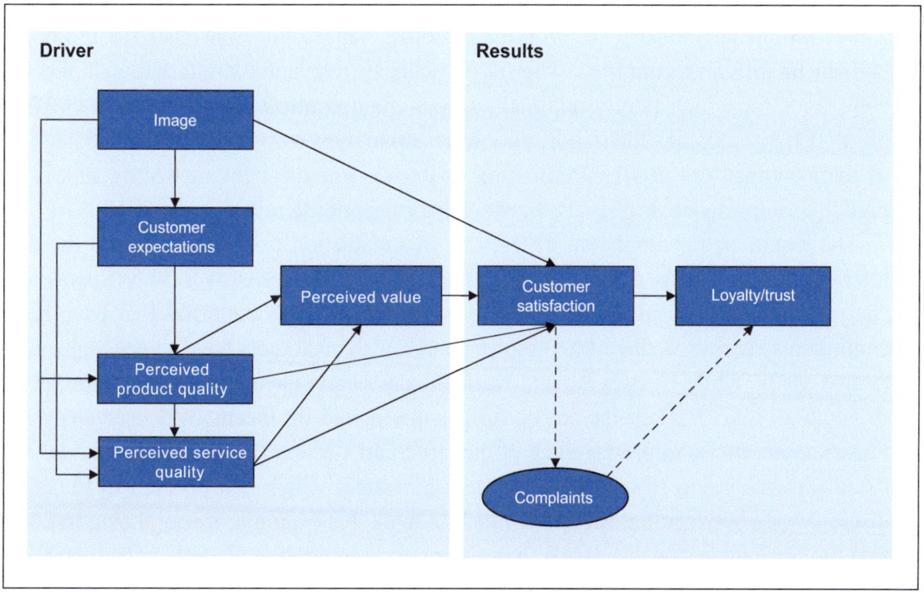

Fig. 14.15 Structure model of the European Performance Satisfaction Index (EPSI). (Source: EPSI Rating, 2015)

The primary goal of the European Customer Barometer *EPSI* is the **development of a system of performance oriented indicators** in order to improve the competitiveness of European companies. The *European Performance Satisfaction Index (EPSI)* thus aims to generate benefits for the individual company, its industry and society as a whole.

In order to make this possible, the *EPSI* collects data on customer satisfaction and customer loyalty as well as their influencing factors on the basis of a cross-border identical methodology, uniform questionnaires, time parallel surveys and uniform evaluation procedures, thus enabling meaningful international comparisons of customer satisfaction. The data collection itself takes place via telephone interviews with customers. A customer is defined here as a person who is randomly selected from the total pool of potential buyers and has recently made experiences with specific providers of products and services in the respective country.

Specific **benefits** arise from the *European Performance Satisfaction Index* for both companies and consumers. Thus, for a country's and an industry's supplier, the comparison of the satisfaction of the national customers with the customer satisfaction in this industry in the other participating countries of the study is possible. The following example shows the international comparison with regard to customer satisfaction for the banking industry. The supplier not only receives information on its own position in the international satisfaction comparison, but can also identify starting points for improving its range of services in the direct company comparison.

Customer satisfaction with banks in Europe

The European Customer Barometer *EPSI* not only allows for the comparison of companies within an industry in terms of customer satisfaction, but also for the comparison of different countries. This is particularly relevant for international service companies for which purely national success chain controlling is of little use. *EPSI* offers these companies the opportunity to examine their own, but also the satisfaction of their competitors in all participating countries. For the banking sector, this is of particular importance. Fig. 14.16 shows the corresponding results from 2018.

The results show significant differences in satisfaction over time between the different countries. The best value is 83.0 (Georgia) and the weakest is 54.3 (Spain). An analysis of customer satisfaction values over time for those countries that have been continuously recorded since 1999 shows that Finnish customers have a very high satisfaction index, while Norwegian and Icelandic customers have a low satisfaction level. In addition, the *EPSI* results can be further interpreted by means of deeper analyses. For example, index values for each of the structural variables studied in the model are listed separately and so-called profiles are generated, which can be used to illustrate discrepancies between the variables. For Lithuania, for example, it was shown in 2006 that the banking sector as a whole has a very good reputation (image) with customers with a index value of 80.5, but customer loyalty is relatively low with a value of 78.

(Source: EPSI Rating, 2012). ◀

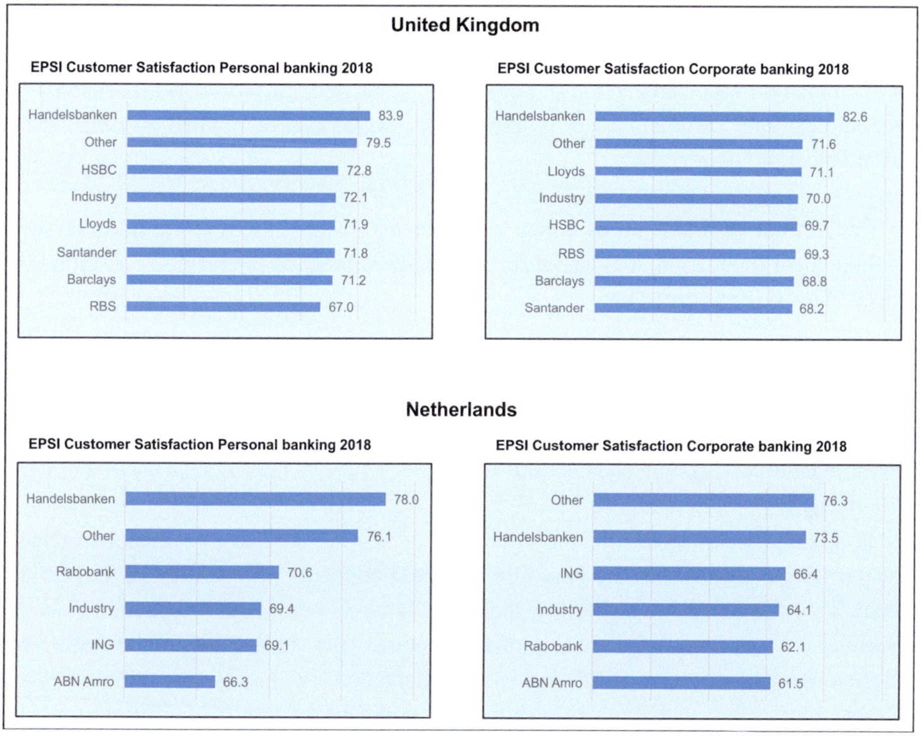

Fig. 14.16 EPSI customer satisfaction personal and corporate banking. (Source: EPSI Rating, 2018a,b)

This creates **advantages for companies and consumers.** One benefit is, that companies receive information on which areas they can improve their range of services to consumers or where they absolutely have to make improvements. This can increase customer orientation. Consumers on the other hand can actively choose those providers who obviously have a particularly good range of services, for example on the basis of the published results.

Against the background of the implementation of the *EPSI*, comparisons between industries and benchmarking studies (cf. Sect. 14.2.1.2) will no longer be limited to the national level in future. Rather, the international comparison of the results for customer satisfaction, customer loyalty and other success factors will experience a strong increase in importance in the coming years.

In particular, for internationally operating companies is an interesting option for comparing their own performance with that of competitors in other countries and thus relating their own performance and identifying potential for improvement. International benchmarking studies, which are of great importance for many industries, will no longer focus on individual, but on a multitude of success factors.

14.2.1.4 Critical assessment of national customer barometers

While the main benefit of national customer barometers is to be seen in a cross-industry comparison of the satisfaction of consumers, some general problems associated with national customer barometers are outlined below, which can be divided into theoretical and practical problem areas.

(1) Theoretical problem areas Since customer barometers are very complex surveys, the theoretical foundation is essential for the interpretability of the results. The following aspects must be considered here:

- Concept of satisfaction
- Interpersonal satisfaction comparisons
- Inter-service satisfaction comparisons
- Aggregation of different satisfaction values

It has already become clear in the course of presenting various customer barometers that the various approaches are based on **different concepts of satisfaction** to a certain extent. This is mainly due to the fact that there is no agreement in satisfaction research regarding a generally valid definition of satisfaction. This creates the problem of a very different operationalization of the concept of satisfaction and thus of very different measurement approaches, which makes it difficult to compare the results of different satisfaction surveys.

In addition, the **interpersonal satisfaction comparisons** necessary for the creation of customer barometers are problematic in two respects. Different survey respondents understand different things under the concept of satisfaction. Also, satisfaction judgments made by test persons always contain measurement errors, which can, however, vary individually to a very different extent.

Furthermore, the feasibility of **inter-service satisfaction comparisons** is to be illuminated. Individuals take their satisfaction formation with regard to certain services on the basis of self-chosen criteria (Johnson & Fornell, 1991). In this context, the problem arises that different criteria are set for different services. For this reason, a comparability of different services or industries is not always guaranteed.

Since the **aggregation of individual satisfaction values** with regard to different services leads to a multiplication of the problems described before, it is evident that the statement of customer barometers is further restricted by the step of aggregation.

(2) Practical problem areas The practical problem areas of national customer barometers mainly concern the usability of the results for different target groups. In this context, social, corporate and consumer related weaknesses are in the focus.

With regard to the **society related problems** it should be noted that the customer barometers presented do not show completeness with regard to the totality of economic sectors, industries and companies, so that a comprehensive supplement to national accounts is questionable. Furthermore, the identification of growth opportunities and the determination of economic policy measures on the basis of customer barometers does not take into account all industries. Finally, the relationship between aggregated customer satisfaction and a high standard of living has not been proven, so that it is currently questionable whether the results of customer barometers allow conclusions to be drawn about the prosperity of an economy.

When considering the **corporate related problems** of national customer barometers, it is about the usability of the results for individual companies. Although it is possible to read the position of a company within an industry from the barometer reports; However, this does not necessarily mean that the companies leading in the barometer have greater economic success. In this context, inter-industry comparability is particularly problematic.

Finally, it is necessary to examine the **consumer related problems** of satisfaction barometers. When describing the benefits of national customer barometers for the individual consumer, it is often mentioned that the satisfaction values determined provide a basis for decision-making when planning future purchases. Since the industry reports of particular interest in this context are only available at very high costs, it is doubtful whether this benefit actually arises for the individual consumer.

In summary, it can be said that national customer barometers simplify satisfaction related comparisons between the services of different industries and companies for different target groups, but this is also associated with some problems. Part of the discussed weaknesses is circumvented when service companies develop individual customer barometers in which an orientation is made to the basic concept of national customer barometers, as well as company specific characteristics are included.

14.2.2 Effectiveness Controls Through Company Specific Customer Barometers

The methodology of the national customer barometer (e.g. *EPSI* or *ACSI*) can be transferred to an **individual company success chain controlling** within the framework of effectiveness control. Success chain controlling is understood to be an integrated and simultaneous success control of quality management that extends over all links of the success chain (cf. Sect. 1.2). This simultaneous control of the effectiveness of quality management with regard to the fulfilment of the pre-economic, psychological goals of quality management (effects on service quality and customer satisfaction), the economic, behavioural goals (effects on customer loyalty and customer value) and the causal relationships between the individual links of the success chain takes place. Success chain controlling therefore takes into account the need for **interdependent effectiveness analyses**, since individual variables (e.g. customer satisfaction) are of no importance to the company when considered on their own, but only when their effects on financial company success can be made measurable.

The procedure of **individual company customer barometers** corresponds to that of the national customer barometers. The focus is on so-called **index systems,** which are also suitable for individual company controlled effectiveness control of quality management within the framework of a success chain controlling.

However, there are differences between the two methods with regard to the consideration of company specifics and the use of results. Since the index system is specifically developed for a single company, the performance features to be included are more precisely aligned with the characteristics of the company and its customers. With regard to the use of results within the framework of effectiveness control, no direct comparison with competitors is possible - unless the company under consideration also collects the corresponding values from customers of the competitors. The specific design of the model on the other hand allows for the derivation of more concrete action requirements as well as the (internal) comparison of different company areas.

Company specific index systems also follow the methodology of the national Ccustomer barometer with **causal or structural equation models.** These allow the relationships between affective-cognitive variables (e.g. service quality, customer satisfaction) and behaviour related variables (e.g. customer loyalty) to be represented as a whole. Within the causal models, these structural variables are understood as latent variables or constructs that are multi-dimensional and measured using so-called indicator variables (e.g. service features as indicators of perceived service quality).

For the development and use of a company specific index system as part of an effectiveness control of quality management for services, the following **procedure** is recommended (cf. Georgi, 2000; Rossiter, 2002; Hadwich, 2003; Siems, 2003; Hsu, 2008; Kaveh et al., 2012):

1. Definition of services and customer segments
2. Collection of potential measurement indicators
3. Selection of relevant measurement indicators
4. Carrying out a causal analysis
5. Creation of activity portfolios (importance-performance portfolios)

The individual sub-steps are explained in more detail below.

(1) Definition of services and customer segments The starting point for the development of a company specific scale for measuring service quality is the determination of the services whose quality is to be measured and the definition of the customers or customer segments whose quality requirements and judgments as well as customer satisfaction and loyalty are to be taken into account.

(2) Collection of potential measurement indicators In the next step, all measurement indicators that are generally relevant for the company specific service quality are to be determined in order to assess the effectiveness of quality management. An important goal of this phase is the completeness of the aspects to be captured, i.e. that many different characteristics are collected that are relevant to the effectiveness of service quality to the extent possible. Various sources can be used to identify the individual characteristics for service quality. First of all, studies from the literature are often used, in the context of which the effectiveness of service quality has already been measured, for example, in the same or a similar industry (secondary research). In addition, the use of oral customer surveys is useful, such as focus group interviews or in-depth interviews with individuals (qualitative market research). As a result of this exploratory approach, an item battery with a number of individual characteristics results, which are used to a large extent for the measurement of service quality in a further step. A confirmatory approach and theoretical definition of measurement indicators is possible for constructs for which validated measurement scales (e.g. SERVQUAL) are available in the literature.

(3) Selection of relevant measurement indicators Then, based on a quantitative study, the relevant measurement indicators for service quality as well as the underlying factor structures and dimensions are to be identified. The reliability and validity of the measurement models are to be checked within the framework of exploratory and confirmatory analyses. To check the conceptualization of the selected construct indicators and to investigate the factor structure of the constructs and the underlying dimensions, an iterative process is carried out using exploratory and confirmatory factor analysis. Various quality measures with fixed minimum values are used as selection criteria. At the end of this process is a set of indicators that is suitable for measuring the perceived (internal) service quality reliably and validly.

(4) Conducting a causal analysis The causal model underlying the customer barometer is estimated on the basis of the indicator variables recorded at the customer (cf. Sect. 14.2.1.1 for the procedure of causal analysis). In one step, the strength of the relationships between the structural variables (e.g. influence of customer satisfaction on customer loyalty) is quantified by so-called structural parameters. In a second step, the strength of the relationships between the structural variables (e.g. customer satisfaction) and the respective indicator variables is expressed by measurement parameters (cf. Fig. 14.17). The strength of the relationships provides helpful information in the context of effectiveness control on how to achieve an effective increase, for example, in customer satisfaction. The higher the measurement parameters of the individual indicator variables, the higher the influence on satisfaction. In Fig. 14.17, for example, it is clear that the indicators "competence" and "reactivity" have a very strong influence on quality (0.9), while "flexibility" has a lower influence (0.5).

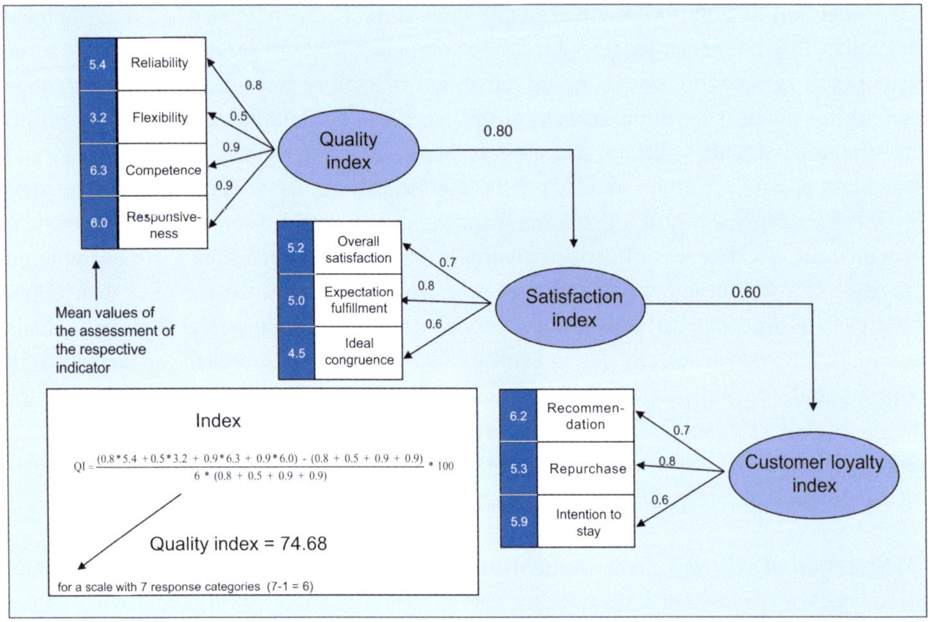

Fig. 14.17 Example of an index system for a company specific customer barometer. (Source: Bruhn, 2016, p. 266)

Therefore, special attention must be paid to these indicators as part of the effectiveness control. The **use of the results of the causal analysis** as part of effectiveness control can be done in two ways:

a. First, the formation of indices (e.g. quality index, satisfaction index, customer loyalty index) can be done by aggregating the mean values of a structural variable using the measurement parameters. Indexing makes it possible to compare the effectiveness of achieving goals between companies, but also over time.
b. In addition, a correlation analysis can be carried out between the sizes of the success chain. This way, for example, it becomes apparent to what extent the affective cognitive target variables (e.g. customer satisfaction) contribute to the realization of the behavioral target variables (e.g. customer loyalty) and therefore take on a special importance with regard to an effective achievement of the goals.

Customer relationship duration and customer satisfaction

The regular conduct (e.g. annually) of company specific customer barometers opens up further analysis options such as, for example, the investigation of the development of service quality and its effects (e.g. relationship quality, customer satisfaction,

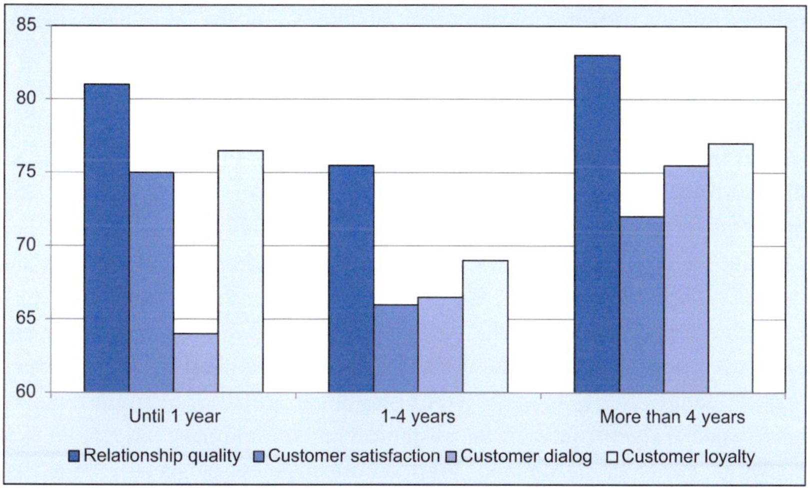

Fig. 14.18 Effects of service quality on key figures of efficiency control in dependence on the duration of the customer relationship. (Source: Bruhn & Frommeyer, 2004, p. 70)

customer loyalty) over the course of a customer relationship. Fig. 14.18 shows, by way of example, the development of the corresponding values in dependence on the length of customer relationships of customers.

In this example, the phenomenon which is frequently observed in practice can be recognised, namely that the customers' statements do not develop linearly, i.e. the customers' perceptions do not improve linearly with increasing duration of the relationship. In the specific example, the following "relationship pattern" can be identified. At the beginning of the customer relationship ("up to 1 year"), the values usually turn out to be relatively good ("honeymoon"), while they fall sharply in the following years ("first crisis"). In long term customer relationships ("more than 4 years"), the values are often highest in relation to most of the variables ("old love"). ◄

(5) Creation of activity portfolios (importance-performance portfolios) As part of the success chain controlling, the effects of quality management can be determined and analyzed by means of index systems within the framework of customer barometers. Based on the evaluation of the indicator variables, measures can be initiated that focus on the central deficits. Activity portfolios or importance-performance portfolios are used as instruments for determining the need for action, which makes it possible to identify first strategic directions of future instrumental strategies.

Within the activity portfolio, the customer perceptions establised in relation to the quality dimensions and their importance from the customer's point of view (importance) are compared. For this purpose, the indicator variables (e.g. reliability) requested are related to customer satisfaction. If there is a high correlation between indicator variables and customer satisfaction, it is an important feature. If, on the other hand, the correlation is relatively low, the feature is to be classified as less important. Depending on the position of individual quality indicators in the importance-performance matrix, corresponding action implications are to be derived (see in detail the importance-performance analysis and the action recommendations in Sect. 5.1.2.1). Fig. 14.19 shows an activity portfolio of a bank branch.

In terms of effectiveness control, it is shown that activities in quadrant I have the greatest influence on the effectiveness of quality management. These activities are classified as particularly important and at the same time the customers are particularly dissatisfied with the performance of the company. Therefore, in order to achieve effective goals, it is necessary to improve these activities with the highest priority. However, the performance features in quadrant IV are of only minor importance because they do not take on any special significance for the customers and the company has already achieved a high level of customer satisfaction. From an effectiveness perspective, it is therefore not recommended to focus on these features.

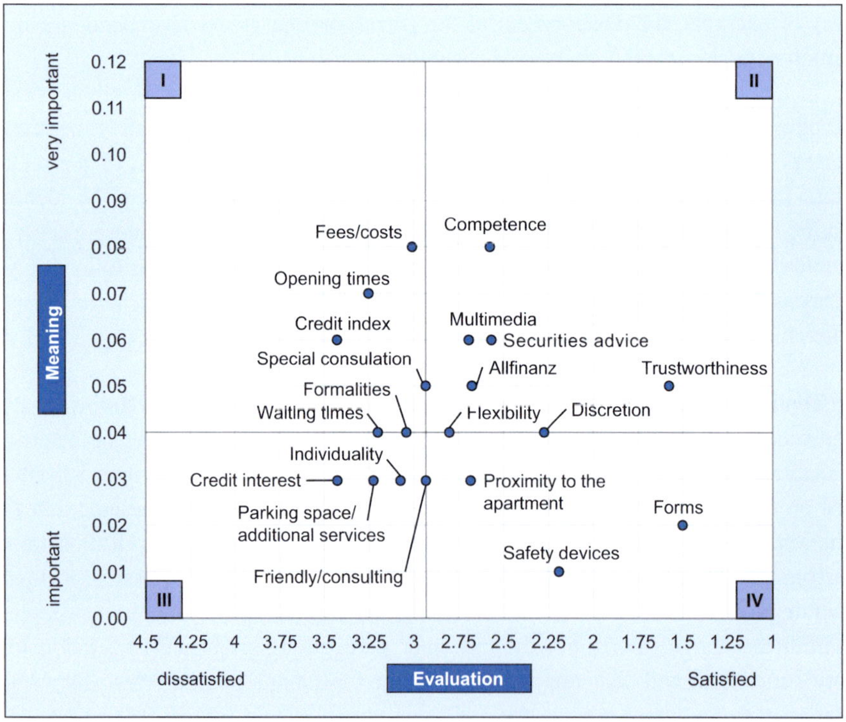

Fig. 14.19 Importance performance analysis using the example of a bank branch. (Source: Bauer et al., 2002)

Use of a company specific customer barometer using the example of the DIY store chain *OBI GmbH und Co. Deutschland KG*

The DIY store chain *OBI* has been conducting more than 65,000 customer surveys annually in all hardware stores for several years as part of the *OBI* customer barometer. This provides concrete statements on overall customer satisfaction, but also on individual aspects of service delivery, such as finding a contact person, product range, friendliness of the staff, quality of advice, etc.. The questionnaires are issued by the cashiers at the end of the shopping trip. In addition, *OBI* has introduced an Internet survey program (Winkelmann, 2012, p. 145). This makes it possible to continuously check the achievement of goals in relation to individual sub-qualities and satisfaction indicators as part of the effectiveness control. ◀

The success chain of quality management for services structures the effects of quality management on customer behavior and thus on corporate success. In addition, the effectiveness and success of quality management can also be explained by the effect of internal service quality or the success chain can be transferred to an internal company perspective. In this context, the use of so-called **"internal service barometers"“** is to be seen (Bruhn & Siems, 2004; Bruhn, 2010) (cf. for internal service barometers also Sects. 5.2.2 and 8.4.1). These are customer barometers specific to the company in the sense of internal marketing, that is, a marketing orientation focused on the internal company perspective (cf., for example, Bruhn, 2004; Hadwich & Keller, 2015).

In the context of effectiveness control, special attention must be paid to the effective operation of internal processes, after all, the effectiveness of internal processes and services largely determines the service quality perceived by the customer (cf. process control Chap. 13). In addition, in the sense of an employee orientation, the extension of quality management to internal services is also required, which initially implies a measurement of internal service quality (Frost & Kumar, 2000; Bruhn & Siems, 2004; Hadwich & Keller, 2015).

▶ **Internal service barometers** are measurements of customer satisfaction, their determinants and their effects on internal departments that are regularly carried out by a neutral institution in a company (Bruhn, 2004, p. 285).

Methodologically, internal service barometers correspond to externally oriented customer barometers, whereby the dimensions of the structural equation models used can be additionally adapted to the specificities of internal services, that is, those dimensions of internal service provision are to be identified which are particularly important for an effective service provision.

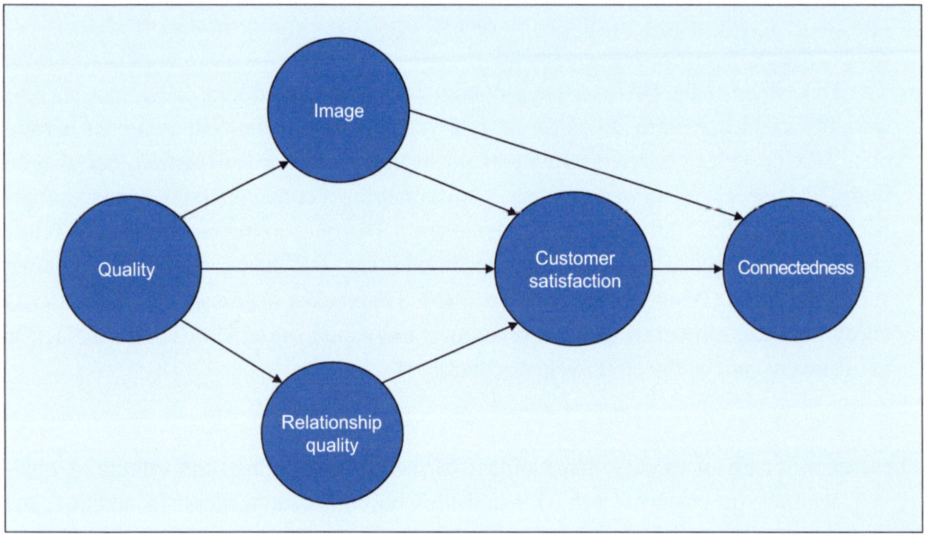

Fig. 14.20 Causal model of the internal service barometer for measuring customer satisfaction with internal IT services. (Source: Bruhn, 2010, p. 313)

Internal service barometer using the example of an internal IT service provider

Since 2003, an internal service barometer has been used at an energy supply company to measure internal satisfaction with the internal service offerings of the conglomerate's own IT company. The internal service barometer used for this purpose is based on the causal model shown in Fig. 14.20 (see also Bruhn, 2010, pp. 312-313 below).

Since the implementation of the internal service barometer in 2003, an annual survey has been carried out. Fig. 14.21 shows the development of the quality assessment of various internal services, with the assessment of information services only being asked for the first time in 2006. With regard to the effectiveness control of quality management, conclusions can be drawn from the results as to whether effective use has been made of various quality management measures. For example, if quality initiatives have been established with regard to improving hardware since 2006, the curve in Fig. 14.21 shows that this has not led to an improvement in the quality perception from the customer's point of view. Rather, a declining trend can be seen since 2007. The measures have therefore not led to the desired results, i.e. an effective use of resources is not apparent here.

Fig. 14.22 shows an importance-performance portfolio for IT support as an example, in which the assessments of the individual quality indicators (performance) are set against the respective weighting factors (importance). In addition, the figure shows the development of the individual assessment criteria, the average values and weights over time. For example, the assessment of the availability of the support hotline has only changed slightly in terms of assessment and weighting between 2006 and 2007,

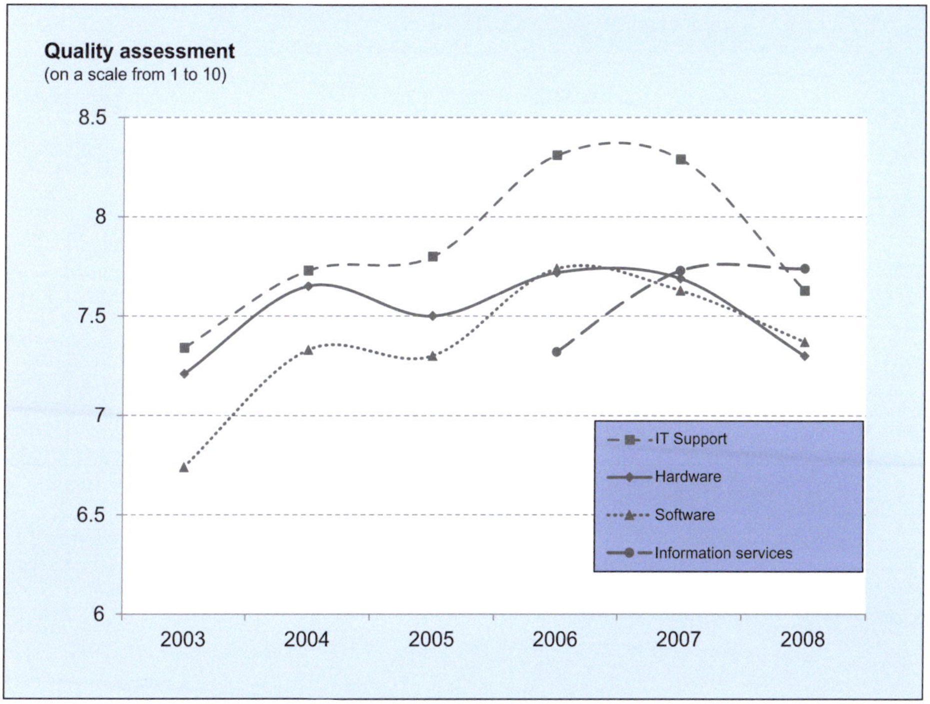

Fig. 14.21 Development of quality assessment of various internal services over time. (Source: Bruhn, 2010, p. 315)

whereas in 2008 the importance has increased sharply. In terms of efficiency control, the results give indications of how the weighting and preferences of internal customers have changed over time. These changes must be taken into account accordingly in terms of efficiency control and the control variables may need to be adjusted. ◀

Internal service barometers can be used in various ways as part of **efficiency control**. For example, the satisfaction of customers and, as a result, the loyalty of employees to services provided internally can be determined. This is partly problematic because internal service providers usually have a monopoly position. Nevertheless, it can be determined whether internal customers would choose the services of the internal provider again in the event of a choice (Bruhn, 2004, pp. 287-288). Thus, low customer satisfaction with internal services can indicate weaknesses in the internal service delivery process, which often have a negative effect on the external service result. In addition, internal service barometers can be used to determine at which point in the service delivery process quality management measures are primarily suitable for increasing customer satisfaction as effectively as possible.

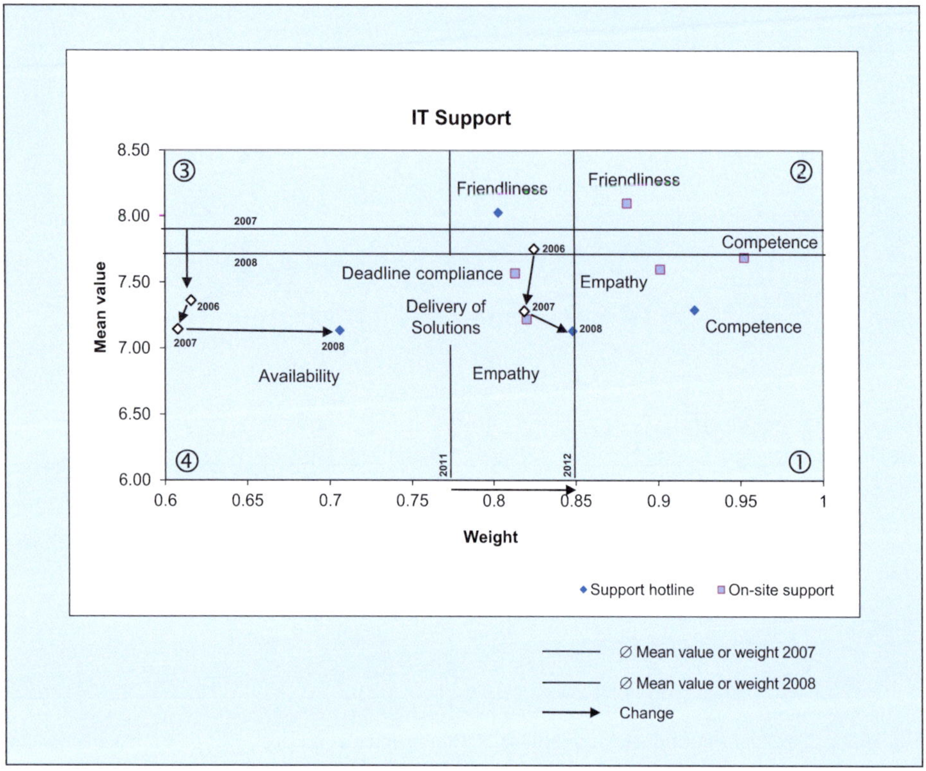

Fig. 14.22 Derivation of action requirements from the comparison of assessment and importance of individual indicators. (Source: Bruhn, 2010, p. 315)

Assuming that only satisfied employees - especially employees in customer contact - can ensure high customer satisfaction, the **measurement of employee satisfaction** by internal service barometers is also of high importance. Against this background, it is necessary to regularly determine employee satisfaction and its influence on employee retention as part of efficiency control.

Since ultimately the external and financial control variables such as market share, sales/turnover, profit, etc. (cf. Fig. 14.1) are decisive for the success of the company, the results are to be checked in the context of efficiency control to see if and to what extent the individual internal quality indicators have an impact on these external success variables. In the context of a **holistic efficiency control,** the parallel and coordinated use of external and internal company specific customer barometers offers the advantage of being able to capture the success or benefit effects of quality management from an internal and external perspective as completely as possible.

14.3 Critical Assessment of Efficiency Controls

Efficiency controls have been established to check the degree of goal achievement. As part of quality management, the influence of quality measures on the cognitive, affective and conative target variables of the company is determined using efficiency controls. Companies have various instruments available, which are associated with different resource requirements. The monetary effects such as sales figures, market share values, etc. are generally available in the company and can therefore be easily collected. The cognitive and affective effects (popularity, image, satisfaction, etc.) are in the focus of the efficiency control of quality management. These are usually to be determined by customer surveys and therefore require more resources. For example, customer surveys can be used to determine customer satisfaction with the quality of the service. In addition, cause-and-effect relationships can be identified between the instruments of quality management and the relevant output variables. The use of customer barometers also offers an added value through the consideration of relationships within the value chain. National customer barometers support efficiency control with regard to a comparison with competing companies or different industries. In addition, time comparisons can be made using the barometers.

However, one of the **criticisms** is that the identified causal relationships are often not clear. For example, it is often not possible to prove that, for example, the increase in customer satisfaction is due solely to activities of quality management. Here, the different confounding variables have to be considered, which can influence the idealized relationships of the success chain (service quality → psychological effects → behavioral effects → economic success) (for the confounding variables of the success chain, see also Sect. 1.2). In the context of the observed behavioral effects, it should also be noted that the behavioral intentions articulated by the customers do not always represent the actual purchase behavior.

Therefore, it is recommended for each company to develop its own system for effectiveness control, which meets the company's own needs. The goal should be to develop a standardized procedure that is used in the long term to enable **continuous tracking** of quality management.

However, the main criticism is that effectiveness controls only consider the output component. While the degree of goal achievement is determined, the amount of resources used is not considered. Thus, by effectiveness controls it can not be determined whether an optimal resource allocation has been achieved, or whether resources have been wasted, i.e. the goal achievement would have been possible with less effort. Therefore, in addition to effectiveness, quality management is also to be controlled under **efficiency aspects,** by comparing the costs with the benefits. In the following chapter, therefore, the individual cost items will be presented in detail before a cost-benefit analysis is used to compare the cost and benefit components.

References

Andrews, F. M. (1984). Construct validity and error components of survey measures: A structural modeling approach. *Public Opinion Quarterly, 48*(2), 409–442.

Bauer, H. H., Grether, M., König, T., & Schlieder, B. (2002). Die Entwicklung eines Instruments zur Messung der Servicequalität von Bankfilialen. *Jahrbuch der Absatz- und Verbrauchsforschung, 48*(4), 345–365.

Baumann, C., Elliott, G., & Burton, S. (2012). Modeling customer satisfaction and loyalty: survey data versus data mining. *Journal of Services Marketing, 26*(3), 148–157.

Bruhn, M. (2004). Interne Servicebarometer als Instrument interner Kundenorientierung. Messung und Steuerung der Qualität und Zufriedenheit interner Dienstleistungen. *Marketing ZFP, 26*(4), 282–294.

Bruhn, M. (2010). Das Konzept des Internen Servicebarometers – Bestandsaufnahme, Methodik und empirische Befunde. In M. Bruhn & B. Stauss (Eds.), *Serviceorientierung im Unternehmen. Forum Dienstleistungsmanagement* (pp. 297–322). Wiesbaden: Gabler.

Bruhn, M. (2013). *Servicequalität – Konzepte und Instrumente für die perfekte Dienstleistung*. München: Beck.

Bruhn, M. (2016). *Relationship Marketing. Das Management von Kundenbeziehungen* (5th Edn.). München: dtv Verlagsgesellschaft.

Bruhn, M., & Frommeyer, A. (2004). Development of relationship marketing constructs over time: Antecedents and consequences of customer satisfaction in a business-to-business environment. *Journal of Relationship Marketing, 3*(4), 61–76.

Bruhn, M., & Hadwich, K. (2011). Dienstleistungsproduktivität – Einführung in die theoretischen und praktischen Problemstellungen. In M. Bruhn & K. Hadwich (Eds.), *Dienstleistungsproduktivität. Forum Dienstleistungsmanagement* (Vol. 1, pp. 3–34). Wiesbaden: Gabler.

Bruhn, M., & Murmann, B. (1998). *Nationale Kundenbarometer. Messung von Qualität und Zufriedenheit. Methodenvergleich und Entwurf eines Schweizer Kundenbarometers*. Wiesbaden: Springer.

Bruhn, M., & Siems, F. (2004). Interne Servicebarometer zur Messung und Verbesserung von internen Dienstleistungen – Ein Erfahrungsbericht aus der Pharma-Branche. In H. Hippner & K. D. Wilde (Eds.), *Management von CRM-Projekten. Handlungsempfehlungen und Branchenkonzepte* (pp. 559–583). Wiesbaden: Gabler.

Decker, R., & Gnibba, K. (2009). Konsumentenforschung im Web 2.0 – Analyse von Online-Rezensionen zur kundenorientierten Produktgestaltung. *Marketing ZFP, 31*(2), 117–136.

Demirkan, H., & Delen, D. (2013). Leveraging the capabilities of service-oriented decision support systems: Putting analytics and big data in cloud. *Decision Support Systems, 55*(1), 412–421.

dm drogerie-markt. (2012). Pressemitteilung vom 26.09.2012. http://www.dm.de/cms/servlet/linkableblob/de_homepage/214602/data/pressemitteilung-data.pdf?v=1348674157000. Accessed 22 Apr. 2013.

EPSI Rating. (2012). Customer satisfaction 2012 – Pan European Banking benchmark. http://www.epsi-rating.com/images/stories/results/Banking/epsi%20pan-european%20banking_12_press-final. Accessed 18 Oct. 2012.

EPSI Rating. (2015). Customer satisfaction 2015. http://www.epsi-rating.com/. Accessed 29 Jan. 2016.

EPSI Rating. (2018a). Smaller and niche banks have the most satisfied customers in the UK banking market. http://www.epsi-rating.com/wp-content/uploads/2016/07/EPSI-Rating-UK-Banking-2018.pdf. Accessed 02 Dec. 2018.

EPSI Rating (2018b). Smaller and niche banks have the most satisfied customers in the Dutch banking market. http://www.epsi-rating.com/wp-content/uploads/2016/07/EPSI-Rating-Dutch-Banking-2018.pdf. Accessed 02 Dec. 2018.

Ertl, R. (2013). *Nationale Kundenbarometer im Vergleich: Theorie – Empirie – Gestaltungsvorschläge*. Wiesbaden: Springer.

Fantapié-Altobelli, C. (2017). *Marktforschung. Methoden – Anwendungen – Praxisbeispiele* (3rd Edn.). Stuttgart: UVK.

Frost, F. A., & Kumar, M. (2000). INTSERVQUAL – An internal adaptation of the GAP model in a large service organisation. *Journal of Services Marketing, 14*(4/5), 358–377.

Georgi, D. (2000). *Entwicklung von Kundenbeziehungen. Theoretische und empirische Analysen unter dynamischen Aspekten*. Wiesbaden: Gabler.

Hadwich, K. (2003). *Beziehungsqualität im Relationship Marketing. Konzeption und empirische Analyse eines Wirkungsmodells*. Wiesbaden: Gabler.

Hadwich, K., & Keller, C. (2015). Interne Servicequalität in Unternehmen: eine empirische Untersuchung der Einflussfaktoren und Auswirkungen. *Zeitschrift für betriebswirtschaftliche Forschung, 67*(2), 170–205.

Hansen, U., & Hennig-Thurau, T. (1999). National customer satisfaction indices: A critical investigation from an application perspective. In P. Kunst, J. Lemmink, & B. Stauss (Eds.), *Service Quality and Management* (pp. 25–53). Wiesbaden: Gabler.

Homburg, C. (2017). *Marketingmanagement. Strategie – Instrumente – Umsetzung – Unternehmensführung* (6th Edn.). Wiesbaden: Springer.

Hotel Mohren. (2016). Zufriedenheitsbefragung in einem Hotel. http://www.hotel-mohren.de/hotel/service/gaestebefragung.html. Accessed 29 Jan. 2016.

Hsu, S. H. (2008). Developing an index for online customer satisfaction: Adaptation of American customer satisfaction index. *Expert Systems with Applications, 34*(4), 3033–3042.

Johnson, M. D., & Fornell, C. (1991). A framework for comparing customer satisfaction across individuals and product categories. *Journal of Economic Psychology, 12*(2), 267–286.

Kaveh, M., Mosavi, S. A., & Ghaedi, M. (2012). The application of European Customer Satisfactiol index (ECSI) model in determining the antecedents of satisfaction, trust and repurchase intention in five-star hotels in Shiraz. *African Journal of Business Management, 6*(1), 6103–6113.

Kepper, G. (2014). Methoden der qualitativen Marktforschung. In A. Herrmann, C. Homburg, & M. Klarmann (Eds.), *Handbuch Marktforschung. Methoden, Anwendungen, Praxisbeispiele* (3rd Edn., pp. 175–212). Wiesbaden: Gabler.

Meyer, A., & Dornach, F. (1994). *Das Deutsche Kundenbarometer 1994 – Qualität und Zufriedenheit. Eine Studie zur Zufriedenheit in der Bundesrepublik Deutschland*. Düsseldorf.

Meyer, A., & Dornach, F. (1995a). *Das Deutsche Kundenbarometer 1995 – Qualität und Zufriedenheit. Eine Studie zur Zufriedenheit in der Bundesrepublik Deutschland*. Düsseldorf.

Meyer, A., & Dornach, F. (1995b). Nationale Barometer zur Messung der Qualität und Zufriedenheit bei Dienstleistungen. In M. Bruhn & B. Stauss (Eds.), *Dienstleistungsqualität. Konzepte – Methoden – Erfahrungen* (2nd Edn., pp. 429–453). Wiesbaden: Gabler.

Net Promotor Score. (2016). Net promotor score benchmark. http://images.google.de/imgres?imgurl=https%3A%2F%2Fnetpromoterscore.files.wordpress.com%2F2010%2F04%2Fnps-benchmarks-blog.jpg&imgrefurl=https%3A%2F%2Fnet-promoterscore.wordpress.com%2Fpage%2F2%2F&h=260&w=454&tbnid=B8UD4_KllBpjmM%3A&docid=_dABKkX3OSeBKM&ei=hSmrVvHkM4qTsAHn3I-wDA&tb-m=isch&iact=rc&uact=3&dur=335&page=1&start=0&ndsp=18&ved=0ahUKEwix-u9r-S087KAhWKCSwKHWfuA8YQrQMITjAP. Accessed 29 Jan. 2016.

Reinecke, S., & Tomczak, T. (2006). Marketingkennzahlensysteme. In S. Reinecke, T. Tomczak, & W. Geis (Eds.), *Handbuch Marketingcontrolling: Marketing als Motor von Wachstum und Erfolg* (pp. 891–913). Wiesbaden: Gabler.

Rossiter, J. R. (2002). The C-OAR-SE procedure for scale development in marketing. *International Journal of Research in Marketing, 19*(4), 305–335.

Service Barometer AG. (2018). Kundenmonitor Deutschland 2018. https://www.servicebarometer.net/kundenmonitor/de/. Accessed 2. Dec. 2018.

Siems, F. (2003). *Preiswahrnehmung von Dienstleistungen. Konzeptualisierung eines theoretischen Konstrukts und Integration in das Relationship Marketing*. Wiesbaden: Gabler.

Stauss, B., & Seidel, W. (2014). *Beschwerdemanagement: unzufriedene Kunden als profitable Zielgruppe* (5th Edn.). München: Hanser.

Strawe, O. V., Saddei, D., Knapp, P., Plehwe, K., & Scheib, M. (2002). CRM: Allheilmittel oder teurer Trend. In Deutscher Direktmarketing Verband (Ed.), *Jahrbuch Dialogmarketing 2002* (pp. 32–34). Düsseldorf: Varus .

Sütőová, A., & Šolc, M. (2009). Utilization of statistical methods within the European Performance Satisfaction Index (EPSI) rating methodology. *Quality Innovation Prosperity,8*(1), 77–82.

Töpfer, A. (2008). Erfolgsfaktoren, Stolpersteine und Entwicklungsstufen des CRM. In A. Töpfer (Ed.), *Handbuch Kundenmanagement* (3rd Edn., pp. 627–650). Berlin: Springer.

Trommsdorff, V., & Teichert, T. (2011). *Konsumentenverhalten* (8th Edn.). Stuttgart: Kohlhammer.

Turkyilmaz, A., Oztekin, A., Zaim, S., & Fahrettin Demirel, O. (2013). Universal structure modeling approach to customer satisfaction index. *Industrial Management & Data Systems, 113*(7), 932–949.

Wehrmeister, D. (2001). *Customer-Relationship-Management. Kunden gewinnen und an das Unternehmen binden*. Köln: Deutscher Wirtschaftsdienst Verlag.

Winkelmann, P. (2012). *Vertriebskonzeption und Vertriebssteuerung – Die Instrumente des integrierten Kundenmanagements (CRM)* (5th Edn.). München: Vahlen.

Efficiency Controls of Quality Management

15

Abstract

Companies are required to ensure an efficient use of the resources available. In the context of so-called efficiency controls, the achievement of selected performance categories is taken into account, but in particular the efficiency of the quality measures carried out. Here, it is important to consider the economic efficiency of the use of quality measures. In the context of this chapter, the cost and benefit components of quality management of services will be presented and discussed in detail. Based on this, the focus is on a cost-benefit analysis as a proven instrument of efficiency control. Finally, the use of efficiency control in the context of quality management is critically assessed.

As part of efficiency checks, not only the achievement of goal categories of effect, but in particular the efficiency of quality measures carried out is taken into account. This efficiency control is carried out by means of **cost-benefit analysis,** by comparing the costs incurred by quality management with the benefits generated. Cost determination often takes place using existing key figures, such as the budget used for a customer survey, the costs of certification, etc.. The quantification of the benefit components turns out to be significantly more complex and multi-layered. First, the cost and benefit aspects of quality management will be considered in detail below, before a comprehensive cost-benefit comparison for quality management will be made on this basis.

M. Bruhn, *Quality Management for Services*,
https://doi.org/10.1007/978-3-662-67032-3_15

15.1 Costs of Quality Management

15.1.1 Concept of Quality Costs

In the industrial sector, economic considerations were already being made in the 1950s as to what financial consequences activities of a company's quality management would have. The so-called **quality cost concept** is still often in the focus in literature and practice today (Bruhn, 1998, p. 762; Mitra, 2008, 23 pp.; Wendehals, 2013). Within the framework of this concept (Dale & Plunkett, 1999; Campanella, 2000; Bröckelmann, 2013), the term quality costs subsumes all costs that are "mainly caused by quality requirements, i.e. (…) by activities to prevent failures, by planned quality checks as well as by internal or external failures and by external quality management" (Dale & Plunkett, 1999; Mitra, 2008, 23 pp.).

A somewhat broader understanding of quality costs subsumes the term quality costs all costs incurred by companies due to the pursuit of their quality objectives (Männel, 2000, p. 6). The avoidance of failures or deviations is the essential quality goal in the sense of a total quality management (Sasse, 2001, p. 76; Mitra, 2008).

The term **"quality costs"** is, however, vigorously disputed in the literature. Therefore, the term **"quality related costs"** is used below, i.e. without the term "quality costs". When considering the definition of the term given above, it becomes clear that this concept covers all costs associated with the quality of a company and its services. A terminological consideration of the term "quality costs" would only make sense if it were exclusively about costs that arise from the quality of the company's services.

This approach differentiates on an abstract level so-called **categories of quality related costs,** which indicate the general character of the cost categories assigned to them. On a more specific level, so-called **elements of quality related costs** are considered, which represent activities, tasks or expenses that are the basis for the cost incurred (Campanella, 2000; Wood, 2013).

There are various **classifications of cost elements** in the literature (see, for example, Dale & Plunkett, 1999; Campanella, 2000; Wood, 2013). It should be noted that the lists contain numerous elements that are not relevant in all industries. In addition, there are specific quality related cost elements that are not listed for certain industries, companies or company divisions. When individually determining quality related cost elements, the following specific characteristics of a company must be considered, amongst others:

- Company organization
- Type of services offered
- Width of the service program
- Degree of controllable processes within service provision
- Automation level of service provision

In terms of the categorization of quality related costs, three **approaches** can be distinguished:

1. Activity oriented classification of quality related costs
2. Effect oriented classification of quality related costs
3. Failure cost accounting

These approaches are explained in more detail below.

15.1.2 Activity Oriented Classification of Quality Related Costs

The **activity oriented classification** of quality related costs is the classic classification and is based on the so-called PAF scheme (Feigenbaum, 1951, 1956), which differentiates quality related costs into prevention costs, appraisal costs and failure costs.

Prevention or failure prevention costs are generated by activities that are specifically undertaken to avoid non-compliance with quality requirements (Campanella, 2000; Hellmich, 2010, p. 14). Examples include costs incurred through quality planning measures, supplier studies, process capability tests, quality improvement team meetings, quality improvement measures or quality training (cf. Fig. 15.1).

In terms of identifying the costs of prevention, two **misinterpretations** can arise (Campanella, 2000):

1. Certain parts of the appraisal or failure costs generated by activities for time related avoidance of higher failure costs could be counted as part of the costs of prevention. However, only those costs that serve the purpose of preventing certain failures from

Marketing/customer/user ▪ Market research ▪ Customer surveys ▪ Review of contracts	**Service provision** ▪ Review of creation processes ▪ Creation related quality planning ▪ Quality training
Performance development ▪ Review of quality improvements ▪ Design activities ▪ Design tests ▪ Field tests	**Quality management** ▪ Salaries ▪ Administrative expenses ▪ Quality program planning ▪ Quality statistics ▪ Quality training ▪ Quality improvement ▪ Quality audits
Procurement ▪ Review of supplier ▪ Supplier ratings ▪ Review of purchase order technologies ▪ Quality planning	**Other prevention costs**

Fig. 15.1 Prevention costs elements. (Source: Campanella, 2000, p. 112; Hellmich, 2010, p. 140)

occurring in the future and thus reducing failure costs over time are to be attributed to the costs of prevention.
2. Certain parts of the "normal" production costs (e.g. monitoring of measuring equipment of production machines, proofreading of a document by a bank employee) could also be attributed to the costs of prevention by their nature. However, since these activities are an integral part of the service creation, these cost items are to be attributed to the production costs and not to the costs of prevention.

Appraisal costs arise from measurement, evaluation or audit activities that serve to ensure the performance conforms to certain norms, standards and/or requirements (Campanella, 2000; Hellmich, 2010, p. 140). Examples of such activities are incoming tests, source checks, process and result controls, auditing of processes and results, definition of measuring and appraisal instruments or customer satisfaction surveys (cf. Fig. 15.2).

Depending on the reason for the appraisals carried out, two different **types of appraisal processes** can be distinguished:

1. **Planned appraisals** are used to examine quality characteristics whose target achievement cannot be predicted with certainty through preventive measures.
2. **Unplanned appraisals** are used when certain processes are not under control.

In connection with the appraisal costs, it must be taken into account, in the case of services, that it is sometimes difficult to check a service for conformity with a standard due

Fig. 15.2 Appraisal cost elements. (Source: Campanella, 2000, p. 112)

1. Procurement appraisals
- Purchasing controls
- Measurement instruments
- Delivery audits
- Origin controls

2. Service provision appraisals
- Planned tests
- Plant tests
- Special tests
- Process controls
- Laboratory tests
- Measuring instruments
- Certificates

3. External appraisals
- Field experiments
- Special product tests
- Testing of competing products

4. Test data appraisals

5. Other quality appraisals

to their specific characteristics. So in the case of services, production and consumption usually take place at the same time ("Uno-Actu-Principle"), a service tends to have an intangible character and, due to the involvement of the external factor in the service creation process, the quality of the service can vary greatly. These **service specific features** in turn have an effect on the recording of appraisal costs and require a differentiated view according to the potential, process and result dimension of services.

In terms of the **potential dimension** of services, the material or tangible elements used in the context of service creation can be relatively easily controlled for their conformity using standardized quality appraisal methods from the area of consumer goods. Appraisals with regard to the potential of customer contact employees is somewhat more difficult. Appraisal costs arise in this context, for example, through measures to check relevant employee qualifications.

The actual service creation process **(process dimension)** can vary greatly depending on the degree of involvement of the external factor, making it more difficult to define and check standards and compliance. Here, the appraisal costs for compliance with service standards with regard to the service creation process, which are defined in accordance with customer expectations (e.g. costs for the implementation of a mystery shopping procedure to check the friendliness of the customer contact person or the duration of the service creation process), are in the focus.

With regard to the **result dimension** of services, it must be taken into account that the quality of the service result can fluctuate due to customer involvement and is perceived very individually. Here, the definition of appraisal standards or the checking of the service result for conformity is particularly difficult. Nevertheless, in this context, for example, the costs for the determination of customer requirements with regard to service quality or for the regular determination of customer satisfaction can be assigned to the appraisal costs.

Failure costs arise from activities that are caused by the lack of correspondence of a performance with certain norms, standards and/or requirements (Campanella, 2000). Depending on the place of origin of the failure costs, two **types of failure costs** are distinguished (Hellmich, 2010, p. 140) (see Fig. 15.3):

1. **Internal failure costs** arise from the consequences of failurs that occur before the customer comes into contact with the supplier and his performance. Examples of the causes of internal failure costs are scrap, rework, re-examination, re-control of used materials or the downgrading of faulty performances.
2. **External failure costs** are generated by the consequences of failures that only occur after contact between the customer and the supplier as well as his performance. Exemplary causes of external failure costs are complaint handling, warranty payments, product liability payments, product recalls or customer churn. With regard to services, it should be noted that a subsequent failure correction is only difficult because in most cases the service consumption takes place at the same time as the service creation (e.g. a failed haircut at the hairdresser). As a result, the focus of

Cost type	Example
1. Construction errors ▪ Construction corrections ▪ Rework ▪ Rejects **2. Purchasing errors** ▪ Material rejection ▪ Material replacement ▪ Delivery corrections ▪ Rework due to delivery rejections ▪ Uncontrolled material losses **3. Performance manufacturing failures** ▪ Material testing ▪ Rework/repairs ▪ Repeat testing ▪ Additional processes ▪ Committee ▪ Downgrading the service **4. Other internal error costs**	1. Complaint management 2. Returns of services 3. Product rejects 4. Guarantee payments 5. Product liability 6. Contractual penalties 7. Loss of goodwill 8. Lost revenues 9. Other external failure costs

Fig. 15.3 Failure costs elements. (Source: Campanella, 2000, p. 112)

Cost type	Example
Prevention costs	▪ Establish a hygiene plan to avoid quality deficiencies due to contamination. ▪ Train employees for kitchen equipment (e.g. fryers, grills) to ensure consistent high quality, regardless of the employee.
Appraisal costs	▪ Incoming inspection of food to ensure the use of impeccable quality in the preparation of meals. ▪ Regular customer surveys in the restaurant to identify potential areas for improvement.
Failure costs	▪ Internal defect costs: costs of destroying incorrectly prepared intermediate products that cannot be used to create the final products. ▪ External error costs: costs for processing customer complaints and replacements for unsatisfactory products.

Fig. 15.4 Activity oriented quality costs using the example of a fast food restaurant

capturing external failure costs is on those activities and measures that are taken to restore customer satisfaction.

Activity oriented quality costs using the example of a fast food restaurant

In large fast food chains, many different quality measures are used in parallel to ensure a consistently high quality across all branches. The chain *McDonald's*, for example, guarantees that the BigMac tastes exactly the same in every *McDonald's* restaurant in the world, whether ordered in Germany, Spain or Australia. Fig. 15.4 shows different activity oriented quality costs in this area as an example. ◀

An extension of the activity oriented division of quality related costs is due to Giakatis et al. Their view is that while the PAF scheme provides a systematic approach to structuring quality related costs, it is not complete (Giakatis et al., 2001, pp. 189–190). Rather, a distinction is needed between **quality related costs** and **quality related losses** or quality losses. This distinction is ultimately necessary to reduce quality related costs at all. The idea behind this is that quality related costs generate value, while quality losses often have a value destroying effect. In contrast to quality related costs, no future benefit is to be expected from quality related losses (Giakatis et al., 2001, p. 182).

In addition to the basic losses from failures, quality losses can also be distinguished in **prevention and inspection losses** (e.g. costs for inspections, quality audits or quality acceptance tests). Losses from prevention and inspection measures arise, for example, when they are ineffective and the capital invested in them is lost (e.g. costs for quality training of employees, costs for quality planning). In addition, there are further hidden quality losses: **production and development related quality losses.** The former occur when production is reduced in order to reduce failures (e.g. costs incurred in the course of production, such as repairs, replacements or rework), the latter are the result of investments to achieve a quality level that unnecessarily exceeds the quality expectations from the customer's point of view (e.g. costs to avoid complaints, recall actions and sales losses) (Giakatis et al., 2001, p. 183).

In the context of a **critical assessment** of the activity oriented classification of quality related costs, it should be noted that the costs of deviations **(deviation costs)** are also taken into account, although strictly speaking they are not quality costs. After all, they do not arise as a result of pursuing quality objectives, but as a result of not achieving them (Männel, 2000; Sasse, 2001, p. 78; Gabler, 2009). As a consequence of these considerations, in the context of an alternative activity oriented classification **quality related costs in the narrower sense** only include those costs that arise as a result of **measures to prevent deviations.** Of the measures to prevent deviations, the **measures of quality control or deviation identification** are to be distinguished and assigned to a separate quality cost category. Consequently, the costs of planned quality control measures can be interpreted as **quality costs in a wider sense.** Deviation costs are not included in the alternative understanding of quality costs (Sasse, 2013, 112 pp.).

15.1.3 Effect Oriented Classification of Quality Related Costs

In response to criticism of the activity oriented division of quality related costs, the effect oriented cost allocation was developed. In the context of the effect oriented cost allocation, two **cost categories** are differentiated: conformance and non-conformance costs (Crosby & Henniges, 1990, p. 102; Brunner, 1991, p. 35; Wildemann, 1992, p. 762, 1995b, pp. 268–269; Hellmich 2010, p. 140).

The **conformance costs** contribute to the success of the company by serving to meet customer expectations. Conformance costs include those costs that arise from measures to permanently eliminate failures and avoid failure risks. Thus, the prevention costs according to the classical PAF cost allocation represent an important part of the conformance costs. Furthermore, the conformance costs include those appraisal costs that are necessarily incurred due to customer expectations (e.g. certification). Three **areas of conformance costs** can be distinguished (Wildemann, 1995b, p. 269; Hoeck, 2007):

1. **Training oriented conformance costs** are predominantly of a preventive nature, in that they are generated by measures to train company members to avoid failures.
2. **Method oriented conformance costs** can have both preventive and testing character, as methods for ensuring compliance with requirements are used for prevention and appraisal.
3. **Control oriented conformance costs** are predominantly of a testing nature, as quality controls serve to check the quality requirements.

The **cost of non-conformance** (cost of deviation, cost of inefficiency) represents a waste of resources (Wildemann, 1995b, pp. 268–269; Brüggemann & Bremer, 2012). They arise primarily from additional effort that is required due to inadequate performance, i.e. when process and result of performance do not meet the quality requirements placed on them. Most often it is assumed in the literature that costs of non-conformance always arise when failures occur during performance, so that the failure costs make up a large part of the costs of non-conformance according to the classical PAF structure.

The relationships between the two categorizations of quality related costs are shown in Fig. 15.5.

Similar to failure costs, internal and external costs of non-conformance can be distinguished (Wildemann, 1995b, p. 269; Brüggemann & Bremer, 2012). **Internal costs of non-conformance** represent the respective additional effort that is incurred before the customer is confronted with the company and its services. In contrast, **external costs of non-conformance** are generated by measures that serve to satisfy a customer afterwards, i.e. after the customer has been confronted with the failures in performance. The following example shows which categories of costs are used by *Xerox* to analyze quality related costs.

Quality costs of the *Xerox corporation*

The copier manufacturer *Xerox* evaluates its quality costs using the classification into conformance and non-conformance costs (cf. Fig. 15.6).

Prevention costs and (planned) appraisal costs are included in conformance costs. Examples of prevention costs are costs for training or internal communication. Examples of appraisal costs are incoming material controls and internal audits.

Non-conformance costs are first divided into underperformance and overperformance of customer requirements. The costs of underperformance are classical

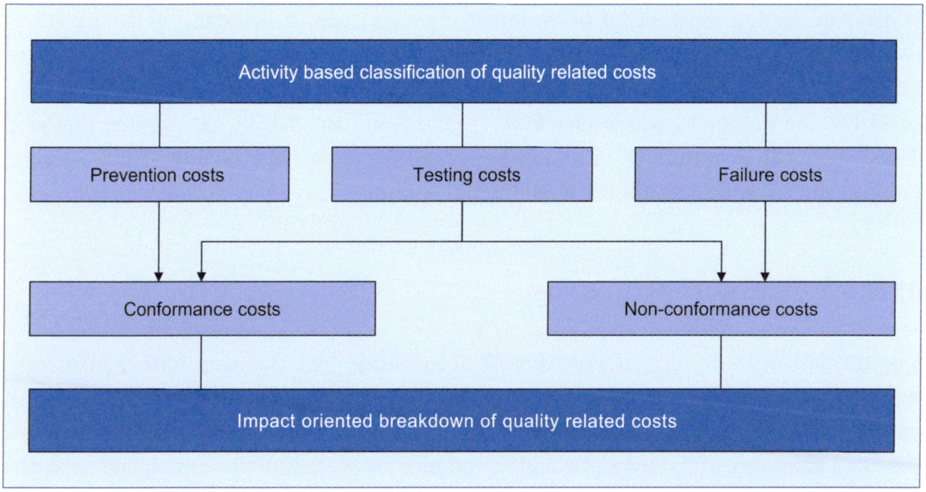

Fig. 15.5 Activity oriented versus output oriented cost structure. (Source: Based on Wildemann, 1995a, p. 409)

Category	Definition	Example
Conformance costs	Expenses necessary to meet customer requirements	
Prevention costs	Error prevention expenses	Training, communication
Audit costs	Work control expenses	Incoming inspections, audits
Non-conformance	Expenses incurred due to a failure to meet customer requirements that are not necessary	
Cost of under-fulfillment of customer requirements	Rework and waste expenses	Warranty payments, replacement costs
Cost of overfulfillment of customer requirements	Expenses for non-essential extras	Exaggerated presentations
Opportunity costs (lost opportunities)	Lost profit due to customer churn caused by a failure to meet customer requirements	Customer churn due to poor service

Fig. 15.6 Categories of quality related costs at *Xerox*. (Source: Carr, 1992, p. 73)

and correspond to expenses for rework, etc., for example, replacement costs. The approach of also calculating the costs of overfulfilling the quality requirements (for example, due to exaggerated presentations) is interesting. In addition to these two groups, opportunity costs, such as lost profit due to customer churn, are also included.

This categorization of costs not only clearly separates the costs that arise from creating quality (conformance costs) from the costs that arise from poor quality. In addition, the customer's behavior due to these quality defects is explicitly included in the calculation. ◀

When transferring the concept of customer oriented division of quality related costs to the service sector, the need for a precise definition of customer requirements for service quality becomes apparent once again (cf. Chap. 4 and 5). In this context, conformity describes the degree to which customer requirements are met by the service provider. Thus, the clear definition of customer requirements for service quality is the essential prerequisite for an exact capture of quality costs within a quality cost controlling.

15.1.4 Failure Cost Accounting

The theoretical basic idea of **failure cost accounting** goes one step further in its criticism of the classical activity oriented division of quality related costs than the one of the customer oriented division. Failure cost accounting is based on the consideration that neither a summary of failure and quality assurance costs (failure prevention and appraisal costs) nor a summary of conformity and non-conformity costs is meaningful (cf. also Seghezzi, 1997, 185 pp. in the following).

Starting from the principle that quality is part of the product or service, quality related cost accounting is replaced by failure cost accounting. In the context of failure cost accounting, an allocation of failure costs is made using two **dimensions** (see Fig. 15.7):

- According to the location of the failure, **internal and external failure costs** are differentiated.
- According to the consequences of the failure, **direct and indirect failure costs** as well as **failure related opportunity costs** are distinguished.

While the location of the failure is also used in the activity oriented division of quality related costs to differentiate failure costs, the consideration of the consequences of the failure leads to a further specification of the failure costs.

The **direct failure costs** include those costs whose occurrence can be clearly attributed to an error and which can usually be clearly assigned to cost centers and cost carriers. **Indirect failure costs** are generated by the analysis and correction of failures. Finally, the **failure related opportunity costs** do not represent actual costs incurred, but rather opportunity costs or profits lost due to poor quality.

Based on the cost allocations according to the concept of quality related costs, different theoretical relationships between the categories and the quality related total costs are hypothesized.

Cost of failure analysis of the *Ritz-Carlton Hotel Company*

The luxury hotel group *Ritz-Carlton* relies on so-called cost of failure analyses to illustrate to employees the costs that arise when a process does not run failure-free. The aim of this measure is to create awareness amongst employees of the cost sensitivity of failures and to understand the countermeasures taken.

Origin of failure \ Failure consequence	Direct failure costs	Indirect failure costs	Failure related opportunity costs
Internal failure costs	▪ Rework costs ▪ Scrap costs ▪ Costs to correct a service	▪ Effort for failure handling ▪ Additional effort in correcting failure consequence (e.g. weekend work to complete rework)	▪ Loss of upset employees ▪ Costs for increased warehousing for products that are susceptible to defects
External failure costs	▪ Impairment due to defects ▪ Costs due to contractual liability ▪ Costs due to legal liability	▪ Costs due to delayed delivery caused by failure (e.g. delayed payment) ▪ Service and replacement within the scope of goodwill	▪ Loss of future orders due to failures in previous deliveries ▪ Loss of upset customers ▪ Additional expenses for advertising, etc. due to quality losses

Fig. 15.7 Extended allocation of failure costs according to failure cost accounting (Source: Based on Seghezzi, 1997, p. 181)

Fig. 15.8 shows an exemplary cost of failure calculation for the case that a room is not ready on time for check-in. Failure costs and corresponding probability of occurrence are assigned to the consequences caused by the failure. In the worst case, the failure leads to the non-arrival of the overnight stay, resulting in costs of 337.50 USD. With an assumed ten percent probability of occurrence of the failure, a total of 33.75 USD in opportunity costs would remain. Internal billing rates are used to quantify the failure costs due to rectification, with the price of an overnight stay representing the lost revenue. In this context, a distinction is made between "soft costs" and "hard costs". "Soft costs", for example, arise from inefficient workflows and the measures taken to rectify them. "Hard costs" can be directly assigned. An example of this is a complimentary cocktail offered to a guest as compensation. ◀

After the various types of quality management costs have been presented, the next step is to go into how the individual quality management costs are determined.

15.1.5 Determining the Costs of Quality Management

The **costs of quality management** correspond to the assessed consumption of goods that arise from activities to ensure the provision of services in accordance with customer requirements (Stauss, 1992, p. 112; Zollondz, 2011, 123 pp.). A task related **cost categorization** according to the phases of the quality control loop (Bruhn, 1998, 155 pp.) leads to the distinction between costs of quality planning (e.g. for the implementation of quality related SWOT analyses or customer surveys to collect customer requirements), costs of quality control (e.g. for a quality oriented personnel selection and development), costs of quality assurance (e.g. for the implementation of customer surveys to measure perceived service quality and customer satisfaction) and costs of quality management disclosure (e.g. for the implementation of a certification or the creation of a quality management manual).

With regard to the **cost capture**, two types of quality related activities are generally distinguished with isolated and integrated activities (Bruhn & Georgi, 1999, 99 pp.). If there are positions in the company that are solely responsible for quality management tasks (e.g. head of quality management or a quality department), so-called **isolated quality activities** are carried out by these, the costs of which can be recorded in cost accounting.

In contrast, **integrated quality activities** are carried out by employees whose area of responsibility implicitly affects quality management (e.g. carrying out a sample test of raw materials by the goods receiving department). The costs attributable to integrated quality activities are often divided into direct costs and overhead costs (Bruhn & Georgi, 1999, p. 100). **Quality related direct costs** (e.g. costs due to the time required for employee training) are generally–like the costs for isolated quality activities–directly taken from cost accounting. In contrast, cost accounting cannot be directly used to determine **quality related overhead costs** (e.g. costs incurred by employees for quality control).

From the special features of services arise problems in terms of cost accounting, which must be taken into account in the context of a quality related cost accounting system. From the immateriality property of services result in a lower variable cost per unit of service in relation to goods and an associated high overhead cost. This in turn leads to the difficulty of allocating overheads to individual cost centers or cost bearers in a cause related manner (Fischer, 2000, p. 3; Benkenstein & von Stenglin, 2006, p. 62; Hellmich, 2010, p. 142). From the limitations of the applicability of the classical cost accounting system for service providers in general and the difficulties in determining quality related costs of service providers in particular, the use of a process cost accounting is recommended.

The **process cost accounting** aims to consider, control and assign the overhead costs of companies to the services rendered in a cause and effect manner as far as possible. Consequently, those processes are examined which cause the overhead costs. Accordingly, process cost accounting serves to map the consumption of resources in the indirect areas and its influencing factors. The information gained this way can contribute to an improvement in corporate governance in two ways. They can be used to specifically design the overhead costs of the company. By identifying factors that influence these processes and the costs incurred through them, the indirect areas are to be analysed and structured and subjected to targeted control. In addition to that, this information is used for the calculation of services. By using quantity variables as reference variables and taking into account information on the number of service variants, order sizes and the complexity of the services, process cost accounting makes it possible to calculate the costs relatively in line with the cause (Götze, 2009, 217 pp.). An example of process cost accounting for the cost centre service production planning can be found in Fig. 15.9.

1	Failure description
	Hotel room not ready in time for check-in

	2 Consequences	3 Costs	4 Probability	5 Calculated costs: Material	Personell	Opportunity cost
A	Guest will not come again	337.50	0.1			33.75
B	Head of house data department loses 15 min.	1.75	0.9		1.58	
C	Hotel manager loses 15 min.	2.00	0.9		1.80	
D	Receptionist loses 15 min.	2.00	0.5		1.00	
E	Free cocktail for the guest	3.00	0.9	2.70		
F	Small gift for the guest	12.00	0.5	6.00		
G	Hotel management writes apology letter	10.00	0.25	2.50		

7 Possible Failure/year	8 Failure rate
50.000	0.02

		Material	Personell	Opportunity cost
6	Total	11.20	4.38	33.75
9	Probable failure	1.000		
10	Failure costs	11.200	4.380	33.700
11	Loss	49.330		

Fig. 15.8 Exemplary cost of failure analysis in the hotel *Ritz-Carlton*. (Source: Beckett, 2008, p. 960)

Cost center production planning									
Sub-processes		Reference values		Cost allocation	Legal costs			Process cost rate	
No.	Title	Type (number of...)	Quantity [process quantity]	Base [man-years]	Service quantity induced [process quantity]	Service quantity neutral [process quantity]	Total	Service quantity induced (total costs per process quantity)	Total (total costs per process quantity)
1	Routings change	Product change	200	4	400,000.-	30,769.-	430,769.-	2,000.-	2,153.85
2	Manufacturing support	Versions	100	6	600,000.-	46,154.-	646,154.-	6,000.-	6,461.54
3	Production orders control	Manufacturing orders	1,000	3	300,000.-	23,077.-	323,077.-	300.-	323.08
4	Department direct			1		100,000.-			
				14			1,400,000.-		

Fig. 15.9 Process cost accounting for the cost centre service production planning. (Source: Based on Götze, 2009, p. 224)

In order to capture the share of overhead costs attributable to integrated activities, the **quality related process cost accounting** is used. Based on the traditional process cost accounting (Schuh et al., 1995; Horváth & Mayer, 2011; Haller, 2018, 205 pp.), so-called **cost drivers** are first to be determined in quality related process cost accounting, which determine the amount of quality management costs incurred by integrated activities to a significant extent (Bruhn, 1998, 168 pp.; Töpfer, 2011; Grabner, 2017). Examples of cost drivers are the number of customer classes (differentiation of customer requirements), the number of failure possibilities (failure proneness of services) or the number of service variants (complexity of the service program).

In the next step, the **quality related core processes** are to be determined on the basis of the applied quality management concept. According to the quality management cycle, the core processes quality planning, quality control, quality auditing and quality management documentation are identified (Bruhn, 1998, p. 169).

Based on the quality activities in the individual business areas, quality related **sub-processes** are determined by activity analyses. The analysis and identification of quality related activities can, for example, be carried out using a so-called service blueprint–that is, a graphical representation of the service process and its sub-activities from the perspective of the customer (see detailed description of the service blueprinting method in Sect. 5.1.2.2). Starting from the service blueprint, it is necessary to take into account all activities above the "line of perception"–that is, the process steps of service creation that are perceptible to the customer–in the context of quality inspections and, in addition, to include all other processes of the blueprint in the determination of quality costs (Benkenstein & von Stenglin, 2006, pp. 58-59). The quality activities thus identified are then assigned to the quality related core processes, thus making it possible to represent a process hierarchy of quality management (see Fig. 15.10).

When considering personnel costs as the main cost factor in the context of quality management (Shenkar et al., 1992, p. 36), the sub-processes for determining the **quality related process costs** are valued with the corresponding personnel expenditure. Finally,

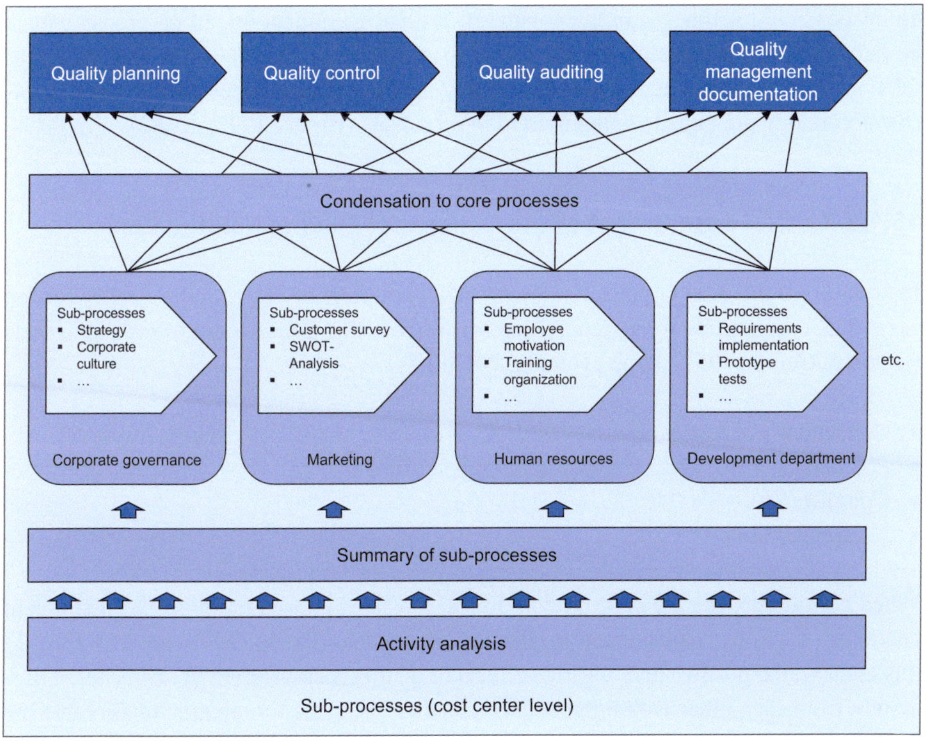

Fig. 15.10 Process hierarchy of quality management. (Source: Bruhn, 1998, p. 171)

the **process cost rate** results from the division of the process costs by the intensity of the corresponding cost driver.

The **main process costs** for detailed model calculations to determine the process costs of quality management in the consumer goods, investment goods and services sector (Bruhn & Georgi, 1999, 163 pp.) result from the addition of the costs incurred in the various company areas as part of a main process.

In the context of a comprehensive cost and benefit analysis of quality management, process cost accounting for quality has the central **advantage** that it contributes to an improvement in the allocability of–difficult to allocate with traditional cost accounting methods–costs to the costs of quality management. In addition to enabling cost capture, it facilitates long term planning (Reckenfelderbäumer, 1998, 125 pp.), so that process cost accounting takes into account the investment character of quality measures (Bruhn, 1998, p. 181).

Process cost accounting for quality is also associated with some **problems** for which the high organizational and financial effort required for its implementation is in the focus. The prerequisites for the implementation of process cost accounting for quality are responsible for this. So in the context of process identification, all participants must be familiar with both quality thinking and process thinking. For these reasons, the introduction of process cost accounting for quality is easier if traditional methods of process cost accounting are already being used in the company (Bruhn, 1998, p. 182).

15.1.6 Critical Assessment of the Concept of Quality Related Costs

In addition to the high organizational and financial effort of determining quality related costs, this concept has some general **problem areas** that concern the following aspects (Bruhn & Georgi, 1999, 64 pp.; Hennig, 2013):

- Conceptual coherence
- Practicality
- Completeness
- Interpretability

When considering the **conceptual coherence**, a lack of consideration of the investment character of quality management measures is noticeable (Bruhn, 1998, pp. 135-136). In this context, the positive and negative effects of quality measures are not considered separately from each other within the concept. Rather, the costs for creating quality and the costs that are due to poor quality are added together. The corresponding sum is referred to as quality related total costs. This way, a consistent comparison of positive and negative financial consequences of quality management is not possible. Furthermore, it is not consistently clear that the costs incurred due to quality management serve a certain purpose, which is represented by the benefit of quality management. Thus, the cause and effect relationship between the negative financial consequences that arise from the implementation of the measures and the positive consequences is not apparent. This argument is reinforced when considering the often used term "quality costs" (Bruhn, 1998, p. 136), which gives the impression that the financial consequences of quality management are only in the form of costs. In particular, in view of the opportunity costs due to lower customer loyalty, etc., the term "quality costs" is–as already mentioned–misleading.

The lack of conceptual coherence means that no statements can be made about the profitability of quality measures on the basis of the concept of quality related costs with regard to the **interpretability** of the analysis results. This is because the costs associated with quality are considered, but no consistent cost and benefit analysis is carried out. In particular, an evaluation of measures is not possible. For example, at a certain point in time, either conformance or non-conformance costs arise in relation to a specific matter. This way, a consistent comparison of the positive and negative financial consequences of quality management is not possible.

With regard to the **practicability** of the approach, it is first of all positive to mention that the respective divisions are comprehensible in their presentation and have a high level of clarity. These arguments also explain the high level of dissemination of the concept in practice (see, for example, Wildemann, 1995b; Hellmich, 2010). It must be noted however, that the categorisation of quality related costs is mostly carried out after their actual recording. Thus, above all, data presentation is the goal of quality related cost consideration. As a result, quality related costs are not directly related to the company's business activities and are not structured according to their origin (Dale & Plunkett, 1999).

With regard to the **completeness** of the categorisations of quality related costs, it should first be noted that, due to the past oriented nature of the concept, a proactive economic analysis is not possible. Even if, according to the results oriented classification of costs, expenditure on quality improvements (conformance costs) can be more clearly separated from benefit-relevant variables (non-conformance costs as opportunity costs) (Wildemann, 1995b, p. 269; Schmitt & Reissiger, 2010), this cost classification does not allow for a proactive economic analysis. Therefore, the quantification of quality related costs can only be used to identify and illustrate quality related improvement potentials, but less for the purpose oriented planning and control of corresponding improvement measures. Furthermore, even with the results oriented categorisation, the non-conformance costs only represent part of the benefits of quality management (in the form of opportunity costs), since customer behaviour related effects, such as increased purchase frequency or word-of-mouth communication, are completely neglected. The criticisms of the concept of quality related costs result in the need to carry out a consistent cost and benefit analysis of the costs and benefits of quality management as part of efficiency controls of quality management. Therefore, the benefits of quality management will be explained in more detail in the next step before a comprehensive cost andbenefit analysis is carried out.

15.2 Benefits of Quality Management

The **benefit of quality management** is the measure of the achievement of objectives through quality related activities. Thus, the benefit results from the degree of achievement of objectives through quality activities (see also Chap. 14 on effectiveness control). Based on the objectives of quality management, an internal and an external benefit can be distinguished (Bruhn & Georgi, 1999; Haller, 2018).

15.2.1 Internal Benefit of Quality Management

The **internal benefit** of quality management has the improvement of performance as its object and is based on the internal effects of quality management, which primarily have a cost reducing character. Internal benefit categories include process improvements (e.g. reduction of idle times) and failure avoidance (e.g. reduction of scrap).

With regard to the **recording of internal benefits,** process improvements can be determined by a process cost analysis. The objects of analysis are those processes that have potential for improvement. By recording and comparing the process costs at two different points in time, between which quality measures are implemented, the internal benefit that affects the corresponding processes can be determined. In order to record failure avoidance, failure analyses are used. As part of the failure analysis, for example, the instruments of the Pareto diagram and the fishbone analysis are used. Information on the costs incurred internally as a result of the corresponding failures can be obtained from controlling. By comparing these costs before and after a quality investment is made, the corresponding internal benefit can be determined (Bruhn & Georgi, 1999; Töpfer, 2011).

15.2.2 External Benefit of Quality Management

An **external benefit** of quality management is achieved by realizing external goals that affect customer behavior. The central–by quality management influenced–determinant of customer behavior is, according to the success chain of quality management, customer satisfaction (Boulding et al., 1993, p. 23; Liljander & Strandvik, 1995, 51 pp.; Oliver, 2010) (cf. to the categories of benefits of quality management Sect. 14.1). By controlling customer satisfaction, an (indirect) influence on customer loyalty or word-of-mouth communication and thus an increase in sales can be achieved (Boulding et al., 1993, p. 12), so that the external benefit of quality management can be divided into a customer loyalty benefit and a (word-of-mouth) communication benefit. When analyzing the benefits of quality management, the results of effectiveness control can therefore be used (cf. Chap. 14).

The **customer loyalty benefit** results from sales increases attributable to an increase in customer satisfaction, which–through the re-election of the company's services, an increase in the purchase frequency or a cross buying (use of services of a supplier that have not been used so far) of satisfied customers–can be concretized in an increase in sales or (through the increase in the willingness to pay of satisfied customers) in a price increase. In connection with the measurement, the construct of customer loyalty is usually operationalized on the basis of two dimensions. The first reflects the **previous behavior** of a customer in terms of buying behavior and recommendation, the second the **future behavior intentions** in terms of re-purchase or additional purchase and the intention to recommend (Krafft, 2007, pp. 33 and 67 for an overview and an intensive discussion of methods of measuring customer loyalty). The positive relationship between customer satisfaction and the individual customer loyalty variables is confirmed in numerous empirical studies (Bruhn, 1998). The positive relationship between customer loyalty and economic corporate success has also been demonstrated in various studies, but there is a risk of overestimating this effect as long as a causal attribution, as is the case with most companies, of cost and revenue effects of customer related measures is omitted (Krafft, 2007, pp. 42-43).

The **communication benefits** (Stauss & Seidel, 2014, p. 353) arise from the expression of positive word-of-mouth communication by satisfied customers–i.e. the recommendation of services to friends and acquaintances–and the reduction of negative word-of-mouth communication by dissatisfied customers–i.e. advising not to use the services to friends and acquaintances. With regard to the relative importance of the two categories of communication benefits, negative word-of-mouth communication is assigned a higher relevance, since negative experiences appear to customers in general to be more worth communicating than positive experiences.

15.2.3 Determining the Benefits of Quality Management

Within the framework of **external benefit capture,** it is possible to measure both actual and intentional customer behavior. With actual benefit capture, actual customer reactions are measured, while with intentional benefit capture, customers' intentions are surveyed.

In a critical **comparison** of the two capture methods, intentional benefit capture is associated with the disadvantage that intentions do not necessarily lead to actual behavior and thus are not directly related to the benefit relevant financial consequences. In contrast, there are survey technical problems with actual benefit capture. For example, in order to determine the customer loyalty benefit, information about the actual behavior of the individual customer must be available. However, in many industries this is either not possible for data protection reasons or for reasons of data collection effort. Even if actual benefit capture is possible–for example, through the use of customer databases–this only allows the customer behavior to be measured retrospectively.

For the above mentioned reasons, the **elicitation of behavioral intentions** represents a helpful and realistically executable alternative (Bruhn, 1998, p. 253). This is further confirmed by the fact that a strong relationship has been found empirically between purchase intentions and behavior (Morrison, 1979, p. 71; Ajzen, 2011). Surveys of new, current and potential customers are used in particular to intentionally capture the benefit related variables, which often contain possible information about customers' intentions with regard to the respective benefit categories.

The determination of both the external and the internal benefits of quality management is subject to a twofold **attribution problem** (Bruhn, 1998). First, the corresponding success effects are not solely attributable to activities of quality management. If, for example, the re-election behavior is considered, then in the absence of quality management not all customers of a company only use its services once. Secondly, it is often not possible to clearly attribute the success effects to the measures and thus to the costs of quality management, because additional expenses arise in other areas of the company for their implementation (e.g. and above all production costs).

To deal with this attribution problem, an **operationalization of the effects of quality management** must be carried out, which is often realized by a comparison of times (Bruhn, 1998; Stauss & Seidel, 2014). From the investment time t_0 the phases of **activity absence and activity presence** are differentiated (cf. Fig. 15.11). The benefits observable in the activity absence are the result of all activities before the quality investment is carried out. The considerations are based on the assumption that the quality measures only become effective at a later time t_1. The period between t_1 and t_2 represents the activity presence, in which the revenue increasing and cost reducing effects of quality management are realized. To isolate the consequences resulting from the quality investment, the activity absence and presence are to be compared in terms of benefits.

To monetize customer behavior as part of external user determination, the customer numbers resulting from a comparison of activity presence and absence are to be valued with sizes of a **customer success calculation**. In this case, the profit per customer is to be taken into account, which can be determined as average total profit, average profit with customers of a certain customer group or on an individual customer basis. Fig. 15.12 shows an example calculation for determining the customer loyalty benefit of quality management (for detailed model calculations for the consumer goods, industrial

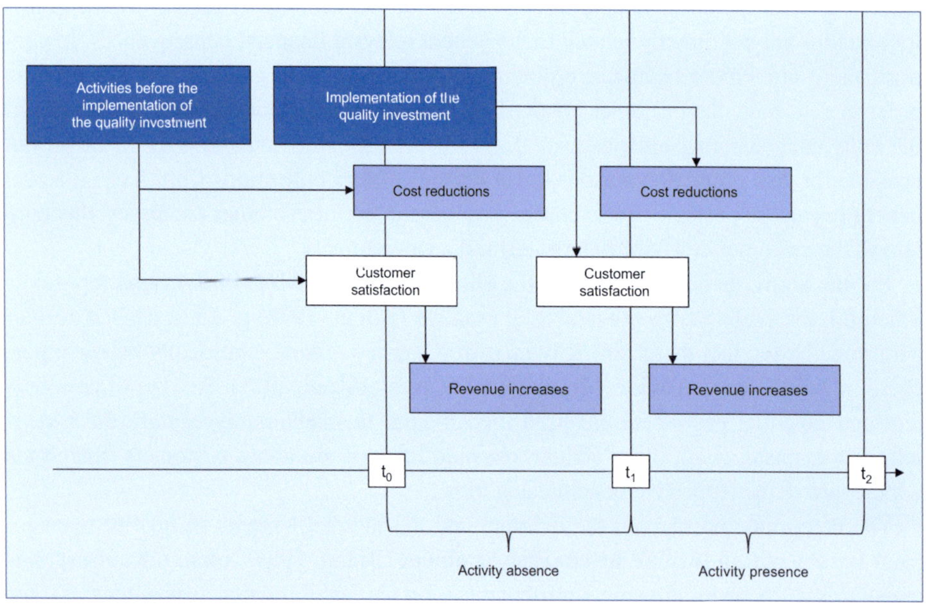

Fig. 15.11 Activity absence and presence of quality management. (Source: Bruhn and Georgi 1998, p. 124)

Calculation steps	Activity absence	Activity presence
Total process count	2,000	2,000
x Error rate	0.20	0.15
x Rework rate	0.80	0.60
Number of reworks	320	180
x Cost per rework	1,000 EUR	1,000 EUR
Absolute expression	320,000 EUR	180,000 EUR
Internal benefit regarding reworks	320,000 EUR - 180,000 EUR = 140,000 EUR	

Fig. 15.12 Exemplary determination of the customer loyalty benefit. (Source: Bruhn & Georgi 1998, p. 128)

goods and services sectors, see Bruhn & Georgi, 1999, 163 pp.). The following example shows an exemplary representation in the banking sector.

Determining the benefits of quality management in a bank

The benefits of quality management in a bank can be determined in a model based manner on the basis of customer satisfaction surveys (Bruhn & Georgi, 1998). A time comparison was carried out, for example in terms of customer loyalty benefits, a comparison of the turnover with the loyal customers at time t_0 with the turnover of the loyal customers at time t_1. Fig. 15.13 shows the corresponding calculation. The procedure can be illustrated by way of example with the re-election benefit: In t_0 the bank had 216,810 "re-election customers", i.e. customers who can be referred to as regular customers and who would definitely choose the bank again. This number increases to 229,259 by t_2. If one values the number of customers with a constant (and fictitiously assumed) profit per customer of 150 EUR, this results in a re-election benefit of approximately 1.9 million EUR.

When interpreting these results, it must be taken into account that the customer's binding and their turnover are attributed entirely to quality management. This is a simplification that is often circumvented by more complex approaches. ◀

15.3 Cost-Benefit Analysis of Quality Management

In the previous explanations, the costs and benefits of quality management were each considered in isolation. However, for a holistic efficiency consideration, an integrated consideration of costs and benefits is necessary. Therefore, a cost-benefit comparison

Baseline data	t0	t0 --> t1	t1
Conviction rate ($ü^{HB}$)	0,73	0,05	0,78
Doubt rate (z^{HB})	0,27	–0,05	0,22
Re-election rate (w^{HB})	0,88	0.00	0,88
Dropout rate (a^{HB})	0,49	–0,03	0,46
Neg. customer status change rate (sw^{HB})	0,1	–0,05	0,05
Positive communication rate (pk^{HB})	0,89	0,00	0,89
Pos. communication partner (KP^{HB})	10	0	10
Purchase probability (kw^{HB})	0,02	0,00	0,02
Negative communication rate (nk^{HB})	0,18	0,00	0,18
Neg. communication partner (KP^{HB})	20	0	0
Avoidance probability (vw^{HB})	0,04	0,00	0,04
Customer movements1	**t0**	**t1**	**t2**
Number of customers (KA^{HB})[2]	337.500	334.003	
Repeat customers (WK^{HB})		216.81	229.259
Exit customers (AK^{HB})		44.651	33.801
Expansion customers (SW^{HB})		6.412	9.283
Degrading customers (SW^{HB})		9.112	3.674
New customers (NK^{HB})		43.854	46.372
Customers not won (NGK^{HB})		13.122	10.581
Rating	**t1**	**t2**	**Nutzenwert**
Re-election benefits			
Re-election customers	216.810	229.259	
Profit per customer	150.00	150.00	
Absolute value	32.521.500	34.388.850	1.867.350
Termination avoidance benefit			
Exit customers	–44.651	–33.801	
Profit per customer	150.000	150.000	
Absolute value	–6.697.650	–5.070.150	1.627.500
Mining avoidance benefits			
Degradation customers	–9.112	–3.674	
Profit per customer	100.00	100.00	
Absolute value	–911.200	–367.400	543.800
Recommendation benefit			
New customers	43.854	46.372	
Profit per customer	120	120	
Absolute value	5.262.480	5.564.640	302.160
Warning avoidance benefit			
Non-acquired customers	–13.122	–10.584	
Profit per customer	120	120	
Absolute value	–1.574.640	–1.269.720	304.920
Total benefits			**4.645.730**

Fig. 15.13 Benefits of quality management in a bank (model calculation). (Source: Bruhn & Georgi, 1998, p. 105)

will be elaborated below. For the evaluation of the efficiency of quality management, economic efficiency key figures are often used.

Quality related efficiency key figures, like other business efficiency key figures, provide concentrated and precise information about important quantifiable facts within the company (Groll, 1991, p. 11; Bayraktar et al., 2013). With regard to the economic analysis, quality related economic efficiency key figures can be used to represent quality related relationships in the company. These key figures can be divided into static and dynamic key figures.

15.3.1 Static Efficiency Key Figures

In the formation of static cost-benefit key figuress, the change in the respective efficiency expressions within a certain period of time (e.g. a quarter, a year or the investment period) is not taken into account. Rather, these expressions flow into the key figuress as an average value for the entire selected observation period. In quality related key figures systems, quality related key figuress are represented in such a way that they are in a meaningful relationship to each other, complement each other and capture the economic object in an overview.

Similar to traditional performance measurement systems, key performance figures and subsidiary performance figuress can be differentiated on the basis of a cost-benefit analysis of quality management. **Quality related key performance figures** provide aggregated statements on the efficiency of quality management. In contrast, **quality related subsidiary performance figures** serve a more intensive analysis of the efficiency of quality management and the derivation of efficiency improvement measures. Fig. 15.14 shows the basic structure of a quality related performance measurement system with quality return on investment and quality gain as key performance figures, which can be derived from the total costs and benefits of quality management.

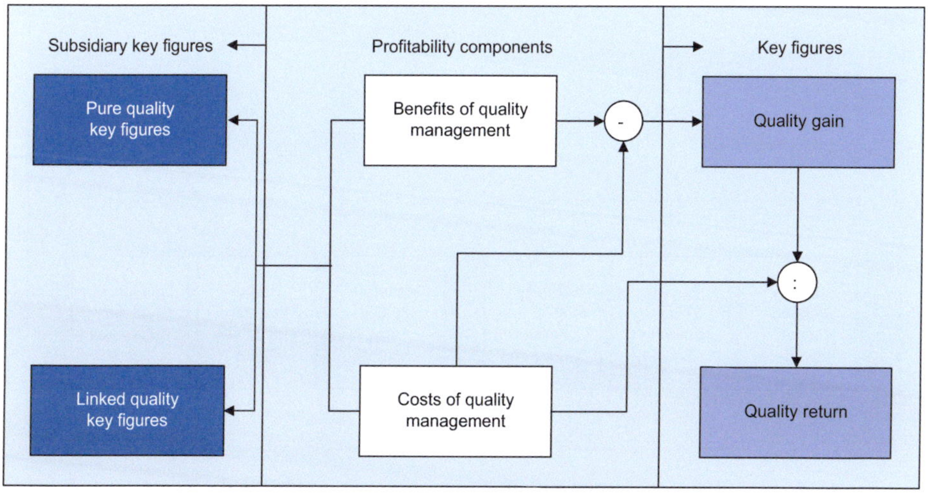

Fig. 15.14 Basic structure of a quality related performance measurement system

The **quality return on investment** results from the quotient of quality gain and the costs of quality management. The higher the quality return on investment, the more efficient the quality investment under consideration is classified. The **quality gain** results from the difference between the benefits of quality management and its costs. With increasing quality gain, an increasing efficiency of quality management can be assumed. Both the quality return on investment and the quality gain depend on the level of cost and benefit expression of quality management. Higher costs lead ceteris paribus to efficiency losses, while a higher benefit of quality management ceteris paribus leads to an increase in efficiency (Bruhn, 1998, p. 271).

While the main key figures serve to identify efficiency trends of quality management, the central **task of the subsidiary key figures** is the identification of efficiency related improvement potentials. Different types of subsidiary key figures can be distinguished, which are divided into pure and linked quality figures (cf. Fig. 15.15).

Pure quality key figures are characterized by the exclusive consideration of quality related elements, i.e. in both the numerator and the denominator of each key figure, the total cost or benefit expressions or expressions of cost or benefit categories are set. This way, quality related cost, benefit and cost-benefit key figures can be differentiated.

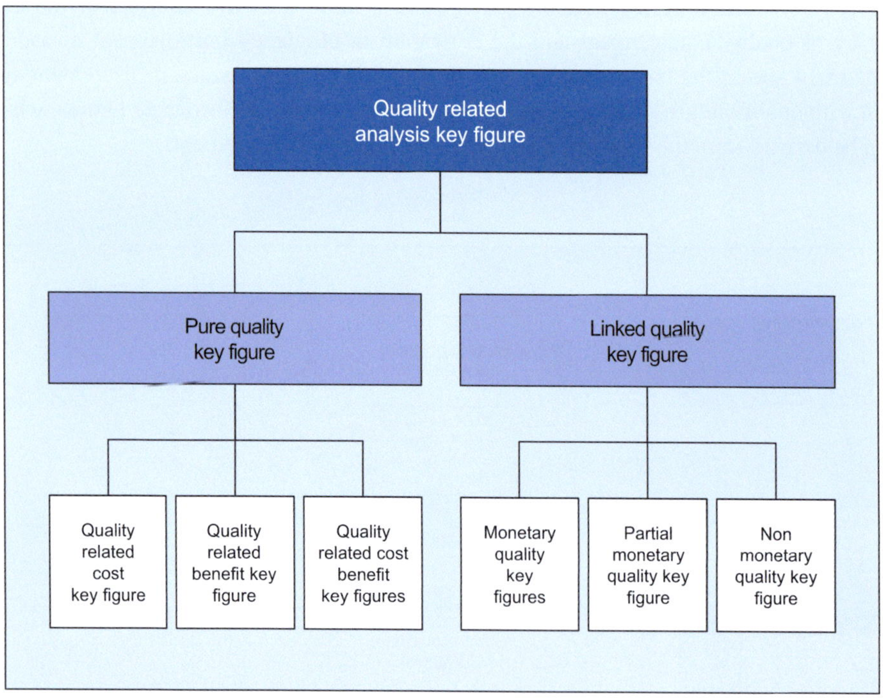

Fig. 15.15 Overview of quality related analysis key figures

$\frac{\text{Costs of quality planning}}{\text{Costs of quality management}}$	$\frac{\text{Costs of quality control}}{\text{Costs of quality management}}$
$\frac{\text{Costs of quality auditing}}{\text{Costs of quality management}}$	$\frac{\text{Costs of quality management documentation}}{\text{Costs of quality management}}$
$\frac{\text{Quality related employee turnover}}{\text{Total employee turnover}}$	$\frac{\text{Costs for customer surveys}}{\text{Costs for quality circles}}$
$\frac{\text{Satisfaction survey costs}}{\text{Total quality audit costs}}$	$\frac{\text{Costs for quality-oriented training}}{\text{Total costs of quality control}}$
$\frac{\text{Time spent on quality circles per employee}}{\text{Costs for quality circles}}$	$\frac{\text{Costs of quality auditing}}{\text{Costs of quality control}}$

Fig. 15.16 Quality related cost key figures

In **quality related cost key figures** cost elements of quality management are set against each other. These key figures therefore relate to the expressions of the cost categories that arise directly from quality investments made (e.g. costs for the external moderator of a quality circle system). Fig. 15.16 shows some such relations by way of example.

When forming **quality related benefit key figures,** the manifestations of the quality management benefit categories are set in relation to each other. In Fig. 15.17 possible benefit related efficiency key figures of quality management are shown. A study on the collection of benefit key figures shows (Kanadaouroff, 1994), that few key figures are specified and that this is only done in relatively few companies (see Fig. 15.18).

Finally, **quality related cost-benefit key figuress** serve a direct cost-benefit analysis of quality management. In this case, characteristics of both cost and benefit categories are related to each other, so that conclusions can be drawn on certain cost-benefit relationships. Fig. 15.19 shows examples of cost-benefit key figures.

Unlike pure quality key figures, **linked quality key figures** relate cost or benefit characteristics to other business entities. Here, monetary, combined and non-monetary quality key figures are distinguished.

The **monetary key figures** compare the cost or benefit characteristics and a monetary amount (e.g. sales). The semi-monetary key figures compare the monetary cost or benefit characteristics of a non-monetary amount (e.g. number of employees). Finally, the

$\frac{\text{Number of retained customers}}{\text{Number of churned customers}}$	$\frac{\text{Number of recommendations}}{\text{Number of purchase warnings}}$
$\frac{\text{Time for error correction}}{\text{Number of failures}}$	$\frac{\text{Number of loyal customers}}{\text{Number of satisfied customers}}$
$\frac{\text{Failure rate in } t_n}{\text{Failure rate in } t_o}$	$\frac{\text{Number of cross-buying customers}}{\text{Number of loyal customers}}$
$\frac{\text{Number of repeated processes}}{\text{Total number of processes}}$	$\frac{\text{Working time for failure corrections}}{\text{Number of failure}}$

Fig. 15.17 Quality related benefit key figures

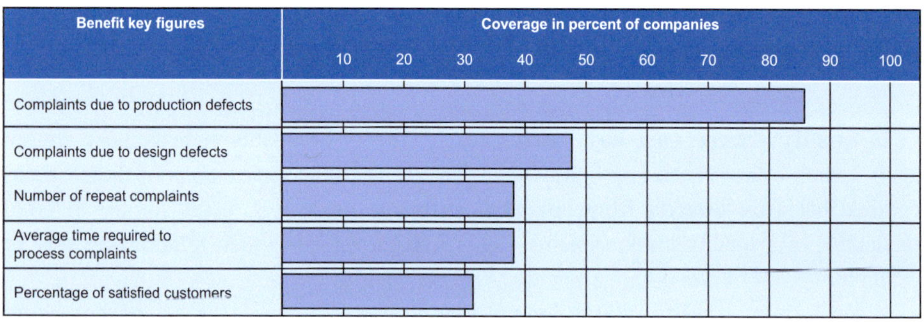

Fig. 15.18 Collection of quality related benefit key figures in practice (Source: Kanadaouroff, 1994, p. 781)

non-monetary key figures represent the so-called quality capability key figures, where only non-monetary amounts (e.g. number of quality training days as a quality related amount and one month as a non-quality related amount) are considered. Fig. 15.20, 15.21 and 15.22 show examples of these types of quality related key figures. With regard to quality capability key figures, the above mentioned study found a higher number of key figures collected, but they were also designed less specifically (see Fig. 15.23).

The following example from *Deutsche Post AG* shows the assessment of the company's suggestion process by key figures in practice.

$\frac{\text{Number of loyal customers}}{\text{Costs for customer orientation seminars}}$	$\frac{\text{Time savings}}{\text{Costs for company suggestion system}}$
$\frac{\text{Number of satisfied customers}}{\text{Costs for satisfaction surveys}}$	$\frac{\text{Cost of quality control}}{\text{Number of defects}}$
$\frac{\text{Quality planning costs}}{\text{Number of satisfied customers}}$	$\frac{\text{Number of recommendations}}{\text{Costs of complaint management}}$
$\frac{\text{Number of controlled processes}}{\text{Costs of quality control}}$	$\frac{\text{Number of failures avoided}}{\text{Costs for fishbone analysis}}$

Fig. 15.19 Quality related cost-benefit key figures

The benefits of an employee suggestion system at Deutsche Post AG

A study by the *Deutsches Institut für Betriebswirtschaft(dib),* which examined the employee suggestion system of 246 companies from various industries, found that *Deutsche Post AG* operates the most successful ideas management through its suggestion system. According to the study, in 2009 the company achieved a benefit of 256 million EUR, which was attributable to the suggestion system. This represents an additional benefit of 1,434 EUR per Post employee. To calculate the benefits of the suggestion system, the following factors were taken into account, among others:

- Prevention of damage
- Avoidance of accidents
- Minimization of risks
- Increasing operational safety
- Improvement of the safety of people and goods
- Activation and improvement of environmental protection
- Rationalization of administrative work

(Source: dib, 2009) ◀

To comprehensively assess the economic viability of quality management, the **quality gain** and the **quality return** are particularly important. These key figures provide a basis for determining the profitability of quality measures and are used in both absolute and relative economic viability analyses. Quality management is considered economically viable in an absolute economic viability analysis if the quality gain or quality return is positive. In a relative economic viability analysis, the economic viability over time can

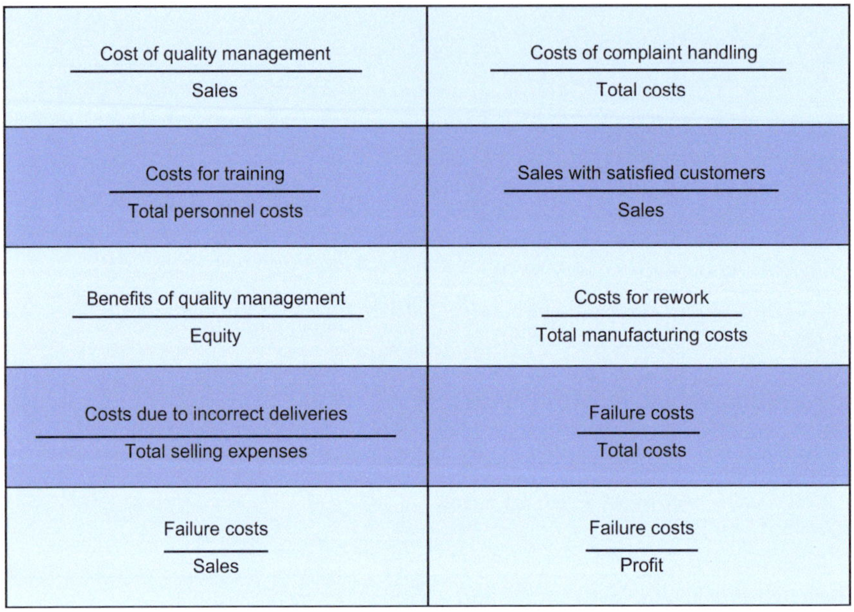

Fig. 15.20 Monetary quality key figures

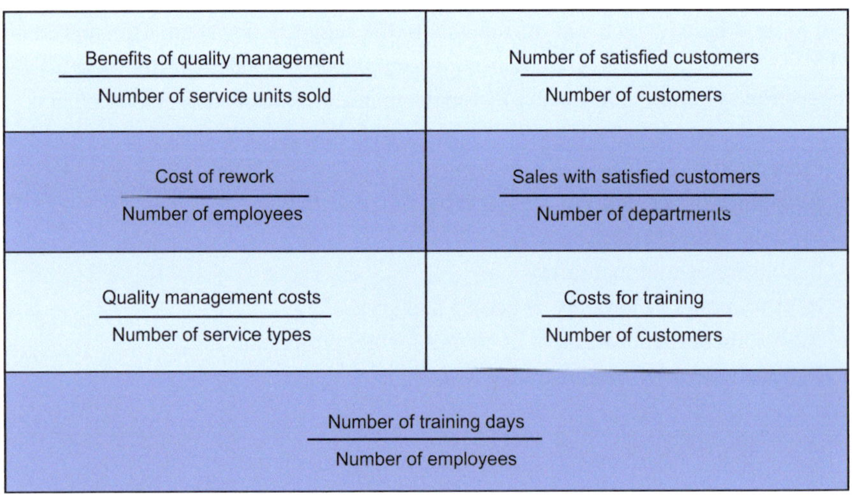

Fig. 15.21 Partial monetary quality key figures

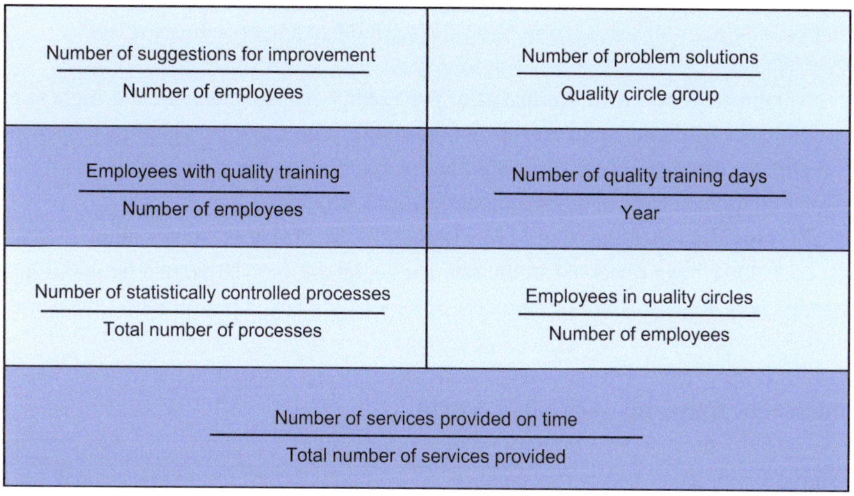

Fig. 15.22 Non-monetary quality key figures

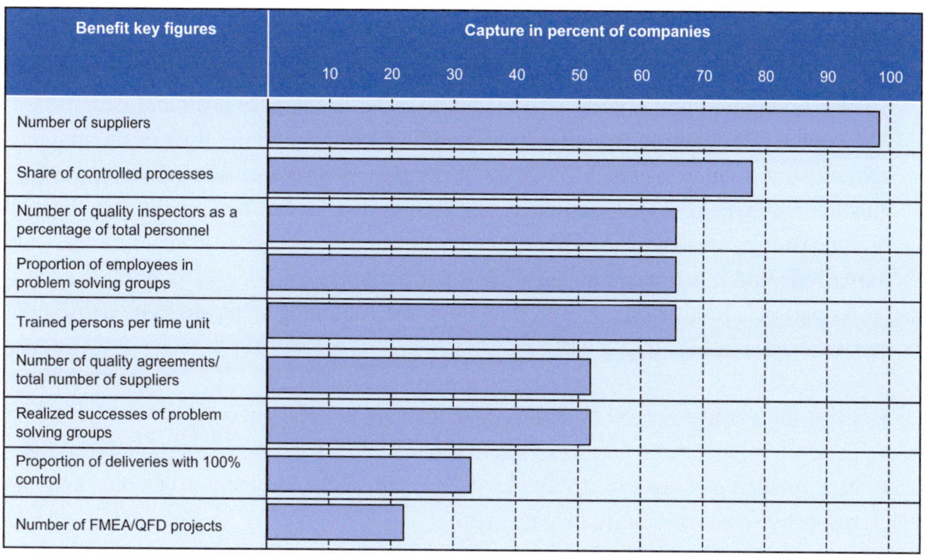

Fig. 15.23 Capturing quality metrics in practice (Source: Kanadaouroff, 1994, p. 781)

be examined in the form of a so-called economic viability tracking or compared with other investments in the form of a profit or return comparison.

To carry out a quality related profit comparison, the profits resulting from two quality measures are compared with each other. The following example illustrates this approach.

Quality related profit comparison using the example of a telecommunications provider

In a quality related profit comparison, the quality investment with the highest quality gain is considered more economically viable. The steps in a quality related profit comparison calculation are illustrated below using a comparison between two quality management measures of a telecommunications provider.

The provider has found that in recent months, declining customer numbers and rising costs have been observed in the company's operations. Now, two potential quality management measures are to be examined for efficiency. The following measures are available:

Measure 1: Company suggestion system

The introduction of an company suggestion system is to improve the company's internal processes. The management of the company hopes to gain new ideas for efficiency improvements in the internal processes through the involvement of employees. It is assumed that this will result in a benefit of around 8,400 EUR per period through cost savings. The costs of the measure are estimated at around 5,000 EUR per period.

Measure 2: Employee training in customer contact

The aim of the measure is to train the employees in customer contact with regard to service orientation and product knowledge in order to increase customer satisfaction. The costs of the training (training fees including loss of working time of the employees) amount to approximately 5,600 EUR per period. The aim is to prevent customer attrition through increased customer satisfaction and to win new customers through high consulting quality. The expected benefit from new customer acquisition and customer retention is estimated at 9,000 EUR per period.

As part of a profit comparison calculation, the following steps are to be carried out and are shown in Fig. 15.24:

- First, the average benefit of quality management is to be given.
- In addition, the average costs of quality management are to be taken into account.
- The quality gain results from the difference of the respective benefits and the respective costs of quality management.

A comparison of the two alternatives in terms of their quality gain shows that, according to the results of the quality related profit comparison, both quality investments have the same efficiency. A profit of 3,400 EUR per period is expected for both the proposal system and the employee training. ◀

The **quality related return on investment** represents an extension of the profit comparison calculation and calculates the funds tied up in the form of quality management costs

Profit comparison	Company suggestion scheme	Employee training
Benefits of quality management (per period, in EUR)	8,400	9,000
Quality management costs (per period, in EUR)	5,000	5,600
Quality gain (Per period, in EUR)	3,400	3,400

Fig. 15.24 Exemplary profit comparison between the company's suggestion system and measures of employee training

over the course of the investment period in the efficiency assessment of a quality investment. When comparing alternative quality investments, the investment is more efficient if the quality return is higher. The example given above is therefore extended below.

Quality related return on investment using the example of a telephone service provider

The procedure for quality related return on investment is again shown in Fig. 15.25 using the comparison of the company's suggestion system and employee training (Bruhn, 1998, pp. 288-289):

- First, the quality gain of the two alternatives is to be considered.
- In addition, the average costs of quality management are to be given.
- The quality return results as the quotient of the quality gain and the average costs of quality management.

By comparing the two alternatives in terms of their respective quality return, it becomes apparent that the company's suggestion system is the more efficient alternative due to the lower level of resources required. ◀

15.3.2 Dynamic Efficiency Key Figures

In contrast to the static efficiency key figures, the importance of time for the value of the relevant variables is taken into account in the determination of dynamic key figures (Lücke, 1991, p. 57).

As part of the dynamic analysis, the economic characteristics are not considered as a whole for a certain period of time (e.g. investment period), but for different **sub-periods within the entire period** (e.g. years within the investment period). Thus, temporal differences in the occurrence of costs and benefits of quality investment are not neglected as in the static key figures or leveled by the use of average values, but are explicitly taken into account in the result of the quality related economic feasibility analysis.

Yield comparison	Company suggestion scheme	Employee training
Quality gain (Per period, in EUR)	3,400	3,400
Quality management costs (per period, in EUR)	5,000	5,600
Quality return (Per period, in EUR)	68%	61%

Fig. 15.25 Example return comparison between the company's suggestion system and measures of employee training

For this purpose, the corresponding quality related cost and benefit characteristics are evaluated by means of **compound interest** (Blohm et al., 2012) with the help of compound interest (Blohm et al., 2012), taking into account temporal aspects. This makes it possible to use effective cost and benefit manifestations in their different temporal occurrence for the feasibility consideration of quality management instead of average values (Bruhn, 1998, p. 293).

In the case of dynamic methods, the financial consequences of quality management are taken into account in the invoice at the **time of their occurrence.** This means that the implementation costs are included in full at the time of the quality investment. For each year (generally: each period), the running costs and the assessed benefits are summarized to form the annual quality gain (-deficit), which is then compared to the implementation costs. Often, the costs and benefits arising at different times are presented in a clear manner in the form of a so-called quality related "complete financial plan" (Bruhn, 1998, p. 294; see Fig. 15.26).

The (net) **quality value** is considered as the central dynamic cost-benefit key figure of a quality investment (Bruhn, 1998, p. 295). It results from the difference between the sum of the discounted (annual) quality surpluses (deficits) as well as the implementation costs. If constant cost or benefit levels, so-called annuities, are assumed, the calculation of the quality value is simplified (Altrogge, 1996, pp. 80-81). The higher the quality value, the more economical the corresponding quality investment. The formulas for calculating the quality value are shown in Fig. 15.27.

Example of a dynamic cost-benefit analysis

For the **illustration of the dynamic cost-benefit analysis** the corporate suggestion system and employee training are compared with each other in terms of efficiency (Bruhn, 1998, 295 pp.; May & Oberdörster, 2012, 59 pp.; see Fig. 15.28). Some assumptions have to be made for this. It is assumed that the respective current cost and benefit expressions remain constant in their absolute amount. The implementation costs of the two quality measures are entered in period zero, which arise from the implementation of the corporate suggestion system or the design of the employee training. The current costs (administrative costs of the proposal processing or costs

Period	0	1	2	...	n
Implementation costs (IK_{QM})	IK_{QM}			...	
Running costs (LK_{QM})		LK_{QM1}	LK_{QM2}	...	LK_{QMn}
Customer loyalty benefits (KBN_{QM})		KBN_{QM1}	KBN_{QM2}	...	KBN_{QMn}
Communication benefit (KN_{QM})		KN_{QM1}	KN_{QM2}	...	KN_{QMn}
Internal benefit (IN_{QM})		IN_{QM1}	IN_{QM2}	...	IN_{QMn}
Quality gain (QG)	QG_0	QG_1	QG_2	...	QG_n
Quality gain discounted (QG_d)	QG_{d0}	QG_{d1}	QG_{d2}	...	QG_{dn}
Cumulative quality gain (QW_k)	QW_{k0}	QW_{k1}	QW_{k2}	...	QW_{kn}

Fig. 15.26 Basic structure of a quality related "complete financial plan". (Source: Bruhn, 1998, p. 294)

$$QW_0 = \sum_{t=0}^{n} (N_{QM} - K_{QM}) \cdot \frac{1}{(1+i)^t}$$

$$\text{or} \quad QW_0 = -IK_{QM} + \sum_{t=1}^{n} (N_{QM} - LK_{QM}) \cdot \frac{1}{(1+i)^t}$$

$$\text{For annuities:} \quad QW_0 = (N_{QM} - K_{QM}) \cdot \frac{(1+i)^n - 1}{i\,(1+i)^n}$$

Legend:
QW_0 = Quality value
N_{QM} = Benefits of quality management (per period)
K_{QM} = Quality management costs (per period)
IK_{QM} = Quality management implementation costs
LK_{QM} = Current costs of quality management (per period)
$1/(1+i)^t$ = Discount factor of period t
n = Planning horizon

Fig. 15.27 Formulas for calculating the quality value. (Source: Bruhn, 1998, p. 296)

of the regular training) are incurred from the first period. According to the respective time lag duration, the effect on the external benefit of the suggestion system (customer loyalty and communication benefits through quality improvements) only starts in the third period, while the internal benefit (cost savings through process optimization) starts in the second period. With employee training, however, both the external (customer loyalty and communication benefits through increased customer orientation) and the internal (cost savings through more competent and faster advice) benefits take effect in the second period.

Company suggestion system						
Period	0	1	2	3	4	5
Implementation costs	–10,000					
Current costs		–3,000	-3,000	–3,000	–3,000	–3,000
Customer loyalty benefits				2,000	2,000	1,000
Communication benefits				2,500	2,500	2,500
Internal benefit			7,500	7,500	7,500	6,000
Quality gain	–10,000	–3,000	4,500	9,000	9,000	7,500
Quality gain (discounted)	–10,000	–2,857	4,082	7,775	7,404	5,876
Quality value (cumulative)	–10,000	–12,857	-8,776	–1,001	6,403	**12,280**
Quality management costs (discounted and accumulated)	10,000	12,857	15,578	18,170	20,638	**22,988**
Quality return	–100%	–100%	-56%	–6%	31%	**53%**
Employee training						
Period	0	1	2	3	4	5
Implementation costs	–8,000					
Current costs		–4,000	–4,000	–4,000	–4,000	–4,000
Customer loyalty benefits			7,000	7,000	7,000	7,000
Communication benefits			4,000	4,000	4,000	4,000
Internal benefit			1,000	1,000	1,000	2,000
Quality gain	–8,000	–4,000	8,000	8,000	8,000	7,500
Quality gain (discounted)	–8,000	–3,810	7,256	6,911	6,582	5,876
Quality value (cumulative)	–8,000	–11,810	–4,553	2,357	8,939	**14,815**
Quality management costs (discounted and accumulated)	8,000	11,810	15,438	18,893	22,184	**25,318**
Quality return	–100%	–100%	–29%	12%	40%	**59%**

Fig. 15.28 Dynamic cost-benefit analysis of quality management using an example. (Source: Bruhn, 1998, p. 297)

In the context of a **dynamic cost-benefit analysis,** the respective quality gain per period is initially considered, which results from the subtraction of the costs per period from the benefits in the respective period. In order to take into account the importance of time for the value of the economic components, the respective quality gains are discounted at the calculation interest rate (in the example 5%). The quality value results for each period from the cumulation of the discounted quality gains. After five periods, the quality value of the employee training is 14,815 EUR and is higher than that of the internal suggestion system (12,280 EUR). Thus, the employee training is more economical taking into account the assumptions made (Bruhn, 1998, 296 pp.).

In addition, by determining quality returns, the funds committed can be taken into account by including the discounted and cumulative costs of quality management in the calculation. The division of the respective quality value by the quality related costs results in the **quality return**. Although employee training has higher quality related costs (25,318 EUR compared to 22,988 EUR), the quality return of employee training is higher than that of the internal suggestion system due to the higher quality value (59% compared to 54%). With regard to the quality return, employee training is the more efficient quality investment. ◀

In order to illustrate the procedure, some simplifications have been made in the example calculation. These are often lifted by means of two central **variation options** (Bruhn, 1998, 298 pp.):

- In the example calculation, it was assumed that both procedures had the same investment period. This simplification can be lifted in the context of the dynamic quality related cost-benefit analysis by taking into account the remaining period for the investment with the shorter duration.
- In reality it is not to be expected that the manifestations of the cost and benefit components will be constant over the entire investment period. In the context of the dynamic cost-benefit analysis, it is possible to take into account cost and benefit values whose absolute amount varies over the course of the investment period.

The following **example** shows how to determine the quality value of a quality circle program.

Quality value of a quality circle program

An industrial goods manufacturer had introduced a quality circle program and wanted to assess five years later to what extent the introduction of the program had paid off economically. Therefore, a quality value calculation was carried out, the results of which are shown in Fig. 15.29.

The costs of the program increased significantly over time as more and more employees were involved in the program. The main effect of the program was

Year	0	1	2	3	4	5
Costs of the QZ program	3	40	162	389	577	802
Time savings	–	–	122	751	485	858
Quality gain	–3	–40	–41	362	–92	56
Quality gain (discounted)	–3	–37	–37	312	76	44
Quality value (cumulative)	–3	–40	–77	235	159	203

Fig. 15.29 Calculation of the quality value of a quality circle program. (Source: Shenkar et al., 1992, p. 39)

identified as time savings, which also increased over time. Since the time savings also increased more than the costs, the cumulative quality value is clearly positive from year 3 onwards.

(Source: Shenkar et al., 1992, p. 39) ◀

The essential **advantages** of dynamic efficiency key figures over static key figures are as follows:

- The consideration of the quality value over the entire investment period rather than an (average) quality gain is more in line with the investment character of a quality measure.
- The dynamic key figures allow a more meaningful operationalization of the entire investment period.
- Finally, the dynamic key figures facilitate the consideration of quality investments with cost and benefit expressions that vary over time.

In contrast, two central **problem areas** can be identified in the dynamic economic key figures of quality management:

- The determination of the dynamic efficiency key figures is associated with a high amount of time and financial expenditure. Therefore, it must be assessed in each individual case whether their use is preferable to the static quality related economic calculation (e.g. return comparison), which already allows valuable conclusions to be drawn about the economic efficiency of a quality investment.
- In order to benefit from the advantages of dynamic efficiency key figures, it is necessary to ensure the availability of the data required for this purpose. This implies the collection of various cost and benefit values over a longer period of time in the case of

an ex-post analysis. The use of suitable forecasting models is required in the case of a forward looking analysis.

15.4 Critical Assessment of Efficiency Control

In view of a **critical assessment** of the approach of cost-benefit analysis of quality management, it must first be noted that the intentions collected do not always represent the actual purchasing behavior. Furthermore, when selecting a measurement approach for the profit per customer, the heterogeneity of the customer base of the respective company must be taken into account. Finally, according to the phase model presented, effects of non-quality related measures occur, which must be isolated by appropriate operationalization (Bruhn & Georgi, 1998).

These potential limitations lead to the definition of various **assumptions** that must be taken into account when interpreting the results of the cost-benefit analysis. However, by means of a company specific adaptation of the approach, an assessment of quality measures according to economic calculations is possible. Thus, comparability can be achieved both between quality measures and between quality related and other activities of a company. The quality related cost-benefit analysis represents a central component of a consistent economic orientation of quality management, which must be optimized continuously in the future after initial results are available.

Even if **practicability and accuracy arguments** have to be brought into line in a cost-benefit analysis of quality management, the quantification of the economic consequences is a basic prerequisite for the systematic design of quality management against the background of corporate goals. Therefore, in further research, additional knowledge about the relationships within the success chain of quality management must be gained, taking into account moderating factors. Furthermore, quality management activities must be differentiated to a greater extent with regard to economic calculations in view of a relationship orientation on the basis of certain customer groups.

In order to implement the cost-benefit analysis of quality management in the company, a more intensive **cooperation between marketing and controlling** is required. For controllers, it is necessary to better understand and follow the upstream processes of customer satisfaction and customer loyalty, marketing has to define appropriate success variables in individual phases. Finally, it is necessary for both areas to develop a system that serves the control of quality measures.

With the explanations of quality controlling, the last phase of the quality management process, consisting of the phases of analysis, planning, implementation and control, was presented. All phases and methods of quality management presented in this book serve the quality optimization with the aim of generating value for the customer and the company. The final chapter summarises ten steps to successful quality management for services.

References

Ajzen, I. (2011). The theory of planned behaviour: Reactions and reflections. *Psychology & Health, 26*(9), 1113–1127.

Altrogge, G. (1996). *Investition* (4th Edn.). München: Oldenbourg.

Bayraktar, E., Tatoglu, E., & Zaim, S. (2013). Measuring the relative efficiency of quality management practices in Turkish public and private universities. *Journal of the Operational Research Society, 64*(12), 1810–1830.

Beckett, N. P. (2008). Qualitätsbewusstsein und Kundenorientierung der Mitarbeiter als Schlüssel zum Erfolg: Qualitätsmanagement bei der Ritz-Carlton Hotel Company. In A. Töpfer (Ed.), *Handbuch Kundenmanagement. Anforderungen, Prozesse, Zufriedenheit, Bindung und Wert von Kunden* (3rd Edn., pp. 948–973). Heidelberg: Springer.

Benkenstein, M., & von Stenglin, A. (2006). Prozessorientiertes Qualitätscontrolling von Dienstleistungen. In M. Bruhn & B. Strauss (Eds.), *Dienstleistungscontrolling. Forum Dienstleistungsmanagement* (pp. 56–70). Wiesbaden: Gabler.

BGW. (2008). Berufsgenossenschaft für Gesundheitsdienst und Wohlfahrtspflege: Betriebliches Vorschlagswesen als Ideenmanagement. https://www.bgw-online.de/SharedDocs/Downloads/DE/Medientypen/BGW%20Broschueren/BGW04-07-005_Betriebliches-Vorschlagswesen-als-Ideenmanagement_Download.pdf?__blob=publicationFile. Accessed 2 Dec. 2018.

Blohm, H., Lüder, K., & Schaefer, C. (2012). *Investition* (10th. Edn.). München: Vahlen.

Boulding, W., Kalra, A., Staelin, R., & Zeithaml, V. A. (1993). A dynamic process model of service quality. From expectations to behavioral intentions. *Journal of Marketing Research, 30*(1), 7–27.

Bröckelmann, J. (2013). *Entscheidungsorientiertes Qualitätscontrolling: Ein ganzheitliches Instrument der Qualitätssicherung*. Wiesbaden: Gabler .

Brüggemann, H., & Bremer, P. (2012). Total Quality Management (TQM). In H. Brüggemann & P. Bremer (Eds.), *Grundlagen Qualitätsmanagement. Von den Werkzeugen über Methoden zum TQM* (pp. 178–200). Wiesbaden: Vieweg+Teubner.

Bruhn, M. (1998). *Wirtschaftlichkeit des Qualitätsmanagements. Qualitätscontrolling für Dienstleistungen*. Berlin: Springer.

Bruhn, M., & Georgi, D. (1998). Kundenbezogene Wirtschaftlichkeitsanalyse des Qualitätsmanagements für Dienstleistungen. Konzept, Modellrechnung und Fallbeispiel. *Marketing ZFP, 20*(2), 98–108.

Bruhn, M., & Georgi, D. (1999). *Kosten und Nutzen des Qualitätsmanagements. Grundlagen, Methoden, Fallbeispiele*. München: Hanser.

Brunner, F. (1991). Steigerung der Effizienz durch Qualitätskostenanalyse. *IO Management Zeitschrift, 60*(7/8), 35–38.

Campanella, J. (Ed.). (2000). *Principles of quality costs. Principles, implementation and use* (3rd Edn.). Milwaukee: McGraw-Hill.

Carr, L. P. (1992). Applying cost of quality to a service business. *Sloan Management Review, 33*(4), 72–77.

Crosby, P. B., & Henniges, B. (1990). *Qualität ist machbar* (2nd Edn.). Hamburg: McGraw-Hill.

Dale, B. G., & Plunkett, J. J. (1999). *Quality costing* (3rd Edn.). London: Routledge.

dib. (2009). dib-Report 2009. www.dib.de. Accessed 15 Apr. 2013.

Feigenbaum, A. V. (1951). *Quality control*. New York: McGraw-Hill.

Feigenbaum, A. V. (1956). Total quality control. *Harvard Business Review, 34*(6), 93–101.

Fischer, R. (2000). *Dienstleistungs-Controlling. Grundlagen und Anwendungen*. Wiesbaden: Gabler.

Gabler, R. (2009). *Qualitätsmanagement bei Sozialdienstleistungsunternehmen. Ein Unternehmensführungsmodell für soziale Berufe*. Norderstedt: Books on Demand.

Giakatis, G., Enkawa, T., & Washitani, K. (2001). Hidden quality costs and the distinction between quality cost and quality loss. *Total Quality Management, 12*(2), 179–190.

Götze, U. (2009). *Kostenrechnung und Kostenmanagement* (5th Edn.). Heidelberg: Springer.

Grabner, T. (2017). *Operations Management – Auftragserfüllung bei Sach- und Dienstleistungen* (3rd Edn.). Wiesbaden: Springer Gabler.

Groll, K.-H. (1991). *Erfolgssicherung durch Kennzahlensysteme* (4th Edn.). Freiburg: Haufe.

Haller, S. (2017). *Dienstleistungsmanagement* (7th Edn.). Wiesbaden: Springer Gabler.

Hellmich, C. (2010). *Qualitätsmanagement und Zertifizierung im Rettungsdienst. Grundlagen – Techniken – Modelle – Umsetzung*. Springer.

Hennig, B. (2013). *Prozessorientiertes Qualitätsmanagement von Dienstleistungen: ein informationswirtschaftlicher Ansatz*. Heidelberg: Springer.

Hoeck, M. (2007). Analyse der Konformitätskosten – dargestellt am Beispiel des Kundenservice eines Versandhandelsunternehmens. *Zeitschrift für Planung & Unternehmenssteuerung, 18*(1), 61–82.

Horváth, P., & Mayer, R. (2011). Was ist aus der Prozesskostenrechnung geworden? *Controlling & Management, 55*(2), 5–10.

Kanadaouroff, A. (1994). Qualitätskosten. Eine theoretisch-empirische Analyse. *Zeitschrift für Betriebswirtschaft, 64*(6), 765–786.

Krafft, M. (2007). *Kundenbindung und Kundenwert* (2nd Edn.). Heidelberg: Physica.

Liljander, V., & Strandvik, T. (1995). The nature of customer relationships in services. In T. A. Swartz, D. E. Bowen, & S. W. Brown (Eds.), *Advances in services marketing and management research and practice* (4th Edn., pp. 141–167). Greenwich: JAI Press.

Lücke, W. (1991). *Investitionslexikon* (2nd Edn.). München: Vahlen.

Männel, W. (2000). Gedanken zum Qualitätskostenbegriff. *Kostenrechnungspraxis, 44*(1), 5–9.

May, M., & Oberdörster, T. (2012). Der Geschäftsbericht – alles andere als die Visitenkarte des Unternehmens. In O. Everling, P. Schaub, & R. Stephan (Eds.), *Transparenzrating. Wege zur effizienten Analyse und Bewertung der Rechnungslegung von Unternehmen* (pp. 59–73). Wiesbaden: Gabler.

Mitra, A. (2008). *Fundamentals of quality control and improvement* (3rd Edn.). New Jersey: Wiley.

Morrison, D. G. (1979). Purchase intentions and purchasing behavior. *Journal of Marketing, 43*(2), 65–74.

Oliver, R. L. (2010). *Satisfaction. A behavioral perspective on the consumer* (2nd Edn.). New York: Routledge.

Reckenfelderbäumer, M. (1998). *Entwicklungsstand und Perspektiven der Prozesskostenrechnung* (2nd Edn.). Wiesbaden: Gabler.

Sasse, A. (2001). Kosten- und nutzenorientiertes Qualitätscontrolling. *Kostenrechnungspraxis, 45*(3), 76–79.

Sasse, A. (2013). *Ganzheitliches Qualitätskostenmanagement: Ein Konzept zur wirtschaftlichen Planung, Steuerung und Umsetzung*. Wiesbaden: Gabler.

Schmitt, R., & Reissiger, W. (2010). Qualitätscontrolling. In G. F. Kamiske (Ed.), *Qualitätsmanagement*. Digitale Fachbibliothek: Symposion .

Schuh, G., Hermann, U., & Martini, C. (1995). Auftragskalkulation mit der ressourcenorientierten Prozesskostenrechnung. *Fördertechnik, 64*(1), 3–8.

Seghezzi, H. D. (1997). Fehlerkosten und Qualitätskostenrechnung. In H. Siegwart (Ed.), *Jahrbuch zum Finanz- und Rechnungswesen* (pp. 167–198). Zürich: WEKA media.

Shenkar, O., Hattem, E., & Globerson, S. (1992). Cost-benefit analysis of quality circles. A case study. *Human Systems Management, 11*(11), 35–40.

Stauss, B. (1992). Dienstleistungsqualität contra Kostensenkung? *Betriebswirtschaftliche Blätter, 41*(2), 111–115.

Stauss, B., & Seidel, W. (2014). *Beschwerdemanagement: unzufriedene Kunden als profitable Zielgruppe* (5th. Edn.). München: Hanser.

Töpfer, A. (2011). Prozessoptimierung: Von der Theorie zur konkreten Umsetzung. *Zeitschrift für Evidenz, Fortbildung und Qualität im Gesundheitswesen, 104*(6), 436–446.

Wendehals, M. (2013). *Kostenorientiertes Qualitätscontrolling: Planung – Steuerung – Beurteilung* (2nd Edn.). Wiesbaden: Gabler.

Wildemann, H. (1992). Kosten- und Leistungsbeurteilung von Qualitätssicherungssystemen. *Zeitschrift für Betriebswirtschaft, 62*(7), 761–780.

Wildemann, H. (1995a). Der Bayerische Qualitätspreis. *Qualität und Zuverlässigkeit, 40*(4), 408–412.

Wildemann, H. (1995b). Qualitätskosten- und Leistungsmanagement. *Controlling, 7*(5), 268–276.

Wood, D. C. (2013). *Principles of quality costs: Financial measures for strategic implementation quality management* (4th Edn.). Milwaukee: Quality Press.

Zollondz, H.-D. (2011). *Grundlagen Qualitätsmanagement. Einführung in Geschichte, Begriffe, Systeme und Konzepte* (3rd Edn.). München: Oldenbourg.

Summary: Ten Steps to Successful Quality Management for Services

16

Abstract

The quality management of services requires extensive strategic and operational considerations. This applies to the analysis, planning, implementation and control of quality management alike. In this final chapter, the central aspects of successful quality management of services are summarized in ten steps. They represent the most important findings and recommendations from the perspective of science and practice. Based on these ten steps, the logical relationships are illustrated and the central success factors for a consistent and professional quality policy of companies are worked out.

The previous chapters presented approaches, systems, concepts and methods of a quality management for services in a comprehensive way. It becomes clear that a systematic quality management for services is structured in the phases of analysis, planning, implementation and control of quality management. This division forms the logical structure of the present book. The **knowledge of the logical relationships** of the different activities of quality management is essential for service companies, as only this way can the instruments be selected sensibly in the different phases. However, the specific implementation of a quality management, a chronological "processing" of the aspects of this book is not useful. Rather, the experience of practice shows that a **pragmatic approach** is successful in introducing quality management. A proposal for such a pragmatic approach includes the following **ten steps:**

1. Definition of service quality
2. Measurement of service quality
3. Involvement of top management,
4. Internal communication of the quality initiative
5. Determination of customer expectations

M. Bruhn, *Quality Management for Services*,
https://doi.org/10.1007/978-3-662-67032-3_16

6. Determination of quality characteristics
7. Internal communication of quality goals
8. Initiation of quality measures
9. Communication of the quality strategy to the outside
10. Assessment of economic viability

The ten steps are explained in detail.

(1) Definition of service quality The starting point of any quality initiative in a service company is to find out what quality means for the company. This is not about designing a theoretical definition of service quality, but a very specific definition that is suitable for the specific company case. Only when the responsible people realize at the beginning how quality is to be understood in the specific case, will the subsequent steps contribute to successful quality management. Based on the general definition of quality, which distinguishes between performance and expectation as components of the perceived service quality, it is necessary to define company specific quality relevant aspects. First, the question must be asked which customer expectations are relevant for the company. For example, it is necessary to specify which customer expectations of which customer groups are decisive for the company. On the performance side, first **knowledge about relevant performance and quality characteristics** is to be collected that are important in the industry or for the specific company. This step is usually carried out by a small group of people who are intensively involved in quality issues in their work, as a conceptual know-how in terms of quality management is required in this phase, bearing in mind that the involvement of too many people quickly leads to the fact that the quality initiative is not sufficiently focused defined in this early phase.

(2) Measurment of service quality After the specific definition of service quality for the relevant company, the creation of quality relevant facts by first quality measurements makes sense. The goal here is to make a so-called "zero measurement" that serves various purposes in the further course of the quality initiative. For example, it forms the basis for subsequent controls of the quality initiative (see also Step 10). It also forms the starting point for the following steps, as it makes it clear from a practical point of view which the quality relevant aspects of the company are. When carrying out the quality measurement, a variety of methods are available from which the service provider selects those that appear to be particularly suitable for his specifc case according to various criteria. Ideally, qualitative methods are used first, such as the sequential event method, which takes into account the process character of a service. The qualitative methods bring to light very specific service experiences from the customer's point of view. These are not only important in analytical terms, but also in communicative terms for the further process, as they provide starting points for the internal indication of specific quality issues of the service company. This provides important content for the internal communication in connection with quality management. The qualitative methods contribute further to the concretization of the dimensions of service quality from the customer's point of view. They thus also form the basis for the quantitative measurement of service quality, in that they capture the assessment of customers with regard to various individual characteristics of the quality dimensions. With the quantitative methods, objectively measurable quality values can be generated, which, for example, serve the emphasis of individual quality characteristics or act as output values for subsequent controls or a benchmark for customer oriented compensation.

(3) Involvement of top management Often, in the first two phases, top management will not be involved, because the first two steps are too far from a specific strategic or operational implementation. This changes after the systematic preparation of the measurement results in terms of service quality. At this point, it is important to expand the quality initiative to a broader foundation. If this does not happen, there is a danger that the few initiators will work more intensively on the topic without taking into account the ideas of other members of the company, in particular the top management. In addition, the measurement results for service quality from the second step are important for winning over the top management. With appropriate preparation of the results, they contribute significantly to convincing the top management of a need for action, as they show the need for quality improvement quantitatively. However, this is not only about the "involvement" of the top management. Much more important is the awareness that the initiation and control of a quality management is ultimately a central task of the management of a service company.

(4) Internal communication of the quality initiative By supporting top management, the basis is created to launch the quality initiative company wide. Since quality is a company wide task and in most service companies all employees contribute more or less directly to the quality of the service, information and involvement of all company areas, functions and employees is an important prerequisite for the later success of the quality initiative. At this point, other departments within the company contribute to the fact that the communication campaign is carried out as effectively as possible. Above all, the human resources department is relevant, as most companies attribute competence in questions of internal communication to it. But also the communication departments of the service provider provide assistance in planning the internal communication campaign - after all, it is about convincing the employees of the company of the usefulness of the quality initiative and to inspire them with it. The role of the top management is not to be neglected in this phase. If the employees feel the commitment and commitment of the top management for the quality initiative, they will convince themselves - consciously or unconsciously - much more of it. Therefore, one of the most important points in this phase is that the top management communicates as much as possible directly with the employees and tries to win them over for the quality initiative.

(5) Determination of customer expectations Even if quality measurement in the second step already provides first indications with regard to customer expectations, it is essential to carry out a systematic capture and analysis of customer expectations at this point. The knowledge of customer expectations is the starting point for planning quality measures. Only when it is known which customer groups have which expectations can priorities be set in quality management. For most industries, a differentiated view of customer expectations is recommended. Not only a differentiation according to expectation types, but above all a differentiation according to customer groups contributes significantly to achieving the best possible and at the same time efficient realization of customer requirements in the sum, i.e. over all customer groups. In many cases, the expectations of different customer groups of the same company differ significantly. For example, completely different aspects are important for private banking customers than for retail or M&A customers of a bank. The identification of these differences forms the basis for a differentiated quality management as well as marketing and customer management in general.

(6) Determination of quality characteristics The analysis of customer expectations serves, amongst others, the prioritization of quality characteristics. Even if quality measurement already provides first information about which quality characteristics are in need of improvement. In principle, in particular, weakly pronounced quality characteristics are starting points for quality improvements. But in many cases, the intensity is not decisive alone. Characteristics that are weakly perceived by the customers but are also not considered important by them - in the sense of customer expectations - are therefore also less relevant for the company. When prioritizing quality characteristics, those characteristics that the customer considers important and in which the service provider has so far had problems are clearly in the focus. During this step, quality management becomes specific. Here it is determined what the future areas of activity of quality management for services are. If the wrong prioritization is made at this point, future budgets will be used incorrectly.

(7) Internal communication of quality goals The prioritized quality features form the content for specific quality goals. The definition of specific quality goals is important because company practice - and this is not only true for quality management, but for the implementation of management concepts in general - shows that only the specific fixation of goals provides an incentive to consistently implement quality management. In addition to the definition of quality goals, their internal communication is the main success factor in this phase. It is particularly important that the company's employees know, understand and accept the quality goals. A corresponding justification of the respective goal is essential for understanding the quality goals. The acceptance is determined to a large extent by the credibility of the goals and the initiators. Here too, the involvement of top management is a key success factor.

(8) Initiation of quality measures In parallel to the communication of quality goals, initial quality measures are used. The earliest possible implementation of initial measures contributes to documenting the relevance and credibility of the quality initiative. If, after the presentation of the quality initiative and goals, nothing happens at first, the message of the internal communication is forgotten. When initiating quality measures, it is important to involve the relevant people. Since quality management is a company wide task, a correspondingly large number of departments and employees are involved. For the success of quality management, it is not helpful if relevant contact persons - for example, due to area egoism - are not involved in the implementation of quality measures, i.e. if, for example, customer satisfaction surveys are carried out without involving the market research department or the call center employees of a bank are evaluated without involving the call center manager. Furthermore, when defining the measures, a balance must be struck with regard to the quality goals. Often it happens that - for good reasons - different quality goals are defined, but when defining the measures later, one quality goal is unconsciously focused on. If then the other quality goals contribute to the achievement of this goal, these interdependence effects are neglected.

(9) Communication of the quality strategy to the outside In the end, the measures of quality management are directed at the customer. By taking into account the customer's expectations and implementing the customer's requirements through a company and its services, the customer's perception of quality and the customer's image of quality are secured in the long term. A quality management is more effective if the quality measures are not only implemented "in secret", in the hope that the customers will notice the improved quality of performance, but are also actively communicated to them. This creates a kind of "self-fulfilling prophecy": If the customers are already prepared for quality improvements, they will notice them more quickly - provided, of course, that the improvements are implemented to a certain extent.

(10) Assessment of economic viability Quality management for services has a special position. Unlike quality management in the area of goods, where often only technical and statistical methods of quality management are used, quality management of a service company cannot rely on objective facts. It must not be labelled as a "soft fact" in the sense of "L'art pour l'art". This would be the farewell to systematic quality management. Rather, an integration of "soft facts" and "hard facts" is to be strived for. Of course, quality management requires a certain amount of creativity. Numerous measures (e.g. quality circles, internal communication, but even qualitative quality measurement methods) cannot be used profitably without creativity. However, this cannot be used as an alibi to base the quality initiative on numbers and not to control it. What already starts with the measurement of service quality in the second of the steps shown, continues in this tenth step on the way to comprehensive quality management. Service companies achieve financial success through quality management if its effects are continuously subjected to an efficiency test. If the company succeeds in making efficiency considerations in time, problems can be avoided, as numerous examples from practice show, in which economic considerations were only made when it was already too late.

These ten steps to successful quality management are not to be misinterpreted in their chronological order as a "law". However, all aspects are addressed in the ten steps, which must not be neglected in the construction of a systematic quality management. In particular, the cooperation of the different company levels and departments is made clear: Due to the company wide responsibility for the service quality of a company, quality management for services will only be successful if all company members contribute to high service quality.